HANDBOOK OF BATTERIES AND FUEL CELLS

Other McGraw-Hill Handbooks of Interest

HANDBOOK OF BATTERIES AND FUEL CELLS

David Linden
Editor in Chief

McGraw-Hill Book Company

New York St. Louis San Francisco Auckland Bogotá Hamburg
Johannesburg London Madrid Mexico Montreal New Delhi
Panama Paris São Paulo Singapore Sydney Tokyo Toronto

Library of Congress Cataloging in Publication Data
Main entry under title:

Handbook of batteries and fuel cells.

Bibliography: p.
Includes index.
1. Electric batteries—Handbooks, manuals, etc.
2. Fuel cells—Handbooks, manuals, etc.   I. Linden,
David.
TK2901.H36 1983   621.31′242   82-23999
ISBN 0-07-037874-6

Parts of Chapter 18 were taken from various publications of Gould, Inc.,
Portable Battery Division, by permission.

1234567890   KGP/KGP   8987654

ISBN 0-07-037874-6

The editors for this book were Harold B. Crawford and Geraldine Fahey,
the designer was Mark E. Safran, and the production supervisor was
Teresa F. Leaden. It was set in Times Roman by Haddon Craftsmen.

Printed and bound by The Kingsport Press.

Contents

PART 2   PRIMARY BATTERIES   4-1

PART 3   SECONDARY BATTERIES

HYDROGEN SYSTEMS

OTHER RECHARGEABLE SYSTEMS—PORTABLE TYPES

PART 4   ADVANCED SECONDARY BATTERIES   26-1

PART 5 RESERVE AND SPECIAL BATTERIES — 33-1

PART 7   APPENDICES

Index follows Appendices

List of Contributors

Austin Attewell • *Royal Aircraft Establishment*
Steven F. Bender • *Gould, Inc.*
A.A. Benderly • *Consultant (formerly with Harry Diamond Laboratories, U.S. Army)*
Navnit Bharucha • *Bright Star Industries*
John Broadhead • *Bell Laboratories*
Ralph J. Brodd • *Consultant (formerly with Exide Corp.)*
Arthur J. Catotti • *General Electric Co.*
Allen Charkey • *Energy Research Corp.*
D.L. Chua • *Honeywell, Inc.*
John F. Cooper • *Lawrence Livermore National Laboratory*
John W. Cretzmeyer • *Medtronic, Inc. (formerly with Gould, Inc.)*
Jack Davis • *Bright Star Industries (deceased)*
James M. Dines • *Eagle-Picher Industries, Inc.*
James D. Dunlop • *Comsat Laboratories*
W.J. Eppley • *Honeywell, Inc.*
John M. Evjen • *General Electric Co.*
S. Uno Falk • *SAB NIFE AB, Sweden*
G.H. Farbman • *Westinghouse Electric Corp.*
Arnold P. Fickett • *Electric Power Research Institute*
Kenneth Franzese • *Bright Star Industries*
Norman H. Hagedorn • *NASA Lewis Research Center*
Robert P. Hamlen • *Exxon R & E Co.*
Ronald O. Hammel • *Gates Energy Products*
R.J. Horning • *Honeywell, Inc.*
Martin Klein • *Energy Research Corp.*
Ralph F. Koontz • *Magnavox Co.*
Karl V. Kordesch • *Institut fur Chemische Technologie, Austria*
Peter Lensi • *Honeywell, Inc.*
David Linden • *Consultant (formerly with U.S. Army Electronics Command)*
Ernest L. Littauer • *Lockheed Missiles and Space Co, Inc.*
Paul Malachesky • *Exxon Enterprises*
Michael Manolios • *Union Carbide Corp.*
George E. Mayer • *Carnegie-Mellon Institute (formerly with St. Joe Minerals Corp.)*
Elliott M. Morse • *Eagle-Picher Industries, Inc.*

Marjorie L. McClanahan • *Ford Aerospace & Communications Corp.*
Dennis Naylor • *Duracell International, Inc.*
Paul Nelson • *Argonne National Laboratory*
Boone B. Owens • *Medtronic, Inc.*
Rod K. Quinn • *Sandia National Laboratory*
Baruch Ravid • *Tadiran Israel Electronic Industries Ltd.*
Ronald A. Rizzo • *Gates Energy Products (formerly with Johnson Controls, Inc.)*
Alvin J. Salkind • *Rutgers University*
Stephen F. Schiffer • *RCA Astro-Electronics (formerly with Yardney Electric Corp.)*
Akio Shimizu • *Hitachi-Maxell, Ltd.*
Paul M. Skarstad • *Medtronic, Inc.*
Douglas H. Spencer • *Berec Group Ltd.*
Robert K. Steunenberg • *Argonne National Laboratory*
Frederick Tepper • *Catalyst Research Corp.*
Warren L. Towle • *Consultant (formerly with Globe Battery Div., Johnson Controls, Inc.)*
Yoshio Uetani • *Hitachi-Maxell, Ltd.*
Darrel F. Untereker • *Medtronic, Inc.*
Charles J. Warde • *Gulf & Western*
J.A. Wiseman • *Gould, Inc.*
David Yalom • *Catalyst Research Corp.*

Preface

During the 40 years since World War II, significant advances have been made in battery and electrochemical devices technology. At the start of World War II, the Leclanche zinc-carbon primary cell and the lead-acid and the Edison nickel-iron secondary batteries were the predominant types, with few other battery systems available. Applications were limited mainly to automotive, signalling, and radio. While these original battery systems are still important and have greatly improved performance characteristics, many other battery systems and fuel cells are now in use, with a wide range of size and configuration. The number and variety of battery applications have increased substantially, as evidenced by the remarkable growth of the battery market, from under half a billion (at the manufacturer's level) in 1947 to almost $10 billion annually in the early 1980s.

This interest in batteries and electrochemical technology, heightened by the recent emphasis on fuel cell and battery development for portable electronics, energy storage and electric vehicles, has resulted in substantial literature in electrochemical research and development, including the publication of several texts covering different phases of battery technology. This literature is supplemented by the publications and catalogs of the battery manufacturers which give the characteristics and performance data of their battery products.

No Handbook has been published, however, which thoroughly covers all batteries and battery technology, with the emphasis on providing detailed up-to-date information on the advantages/disadvantages, characteristics, properties, and performance of all types of electric batteries and fuel cells. This Handbook is intended to fill that need. It is directed toward the battery developer and technologist and especially toward the designer/engineer of battery-operated equipment and the battery user to enable them to: determine the performance characteristics of batteries under all conditions of use, establish the conditions and proper operating procedures to achieve optimum use of each battery system, and select the most suitable battery for a given battery application. The emphasis is on providing the reader with specific and practical working information and data, rather than theory; but sufficient descriptive and background material is included for a complete presentation of each battery system. To achieve this objective, maximum use is made of tables, charts, graphs, drawings, illustrations, and other graphics to support the descriptive text. The comprehensive volume has 1088 pages, over 300,000 words of text, figures, tables, and bibliographic references.

An outstanding group of over 50 scientists and engineers has contributed to the Handbook, each a leading and experienced expert in the battery and fuel cell field. The Handbook, however, is not merely a compilation of these individual contributions;

rather, a special effort was made, by each contributor and the editors, to use a similar format and style of presentation to facilitate the use of the Handbook and to enable the reader to readily locate information and data and to compare the characteristics of the different electrochemical systems.

The Handbook is organized into six major divisions which cover the general principles of electrochemical and battery technology and the performance characteristics of the different types of battery and fuel cell systems:

Part 1, *Principles of Operation,* covers the basic concepts of electrochemical devices, electrochemical principles and reactions, comparative performance, applications, and a detailed discussion of the factors affecting battery performance and considerations for selecting a battery.

Part 2, *Primary Batteries,* covers the general characteristics and applications of primary batteries, including a comparison of the performance characteristics and a discussion of the advantages/disadvantages of each system. Major chapters are devoted to the features and performance of the zinc-carbon, alkaline, magnesium, mercury, silver and air cells, and to a comprehensive treatment of the new lithium primary batteries.

Part 3, *Secondary Batteries,* similarly covers the secondary battery systems with major chapters on the lead-acid, nickel-cadmium, nickel-iron, nickel-zinc, silver-zinc, and the newer hydrogen, air, and lithium batteries.

Part 4, *Advanced Secondary Batteries,* is devoted to the new high-performance aqueous, high-temperature, and other special rechargeable battery systems being developed for electric vehicles and for energy storage for utility and electrical power systems.

Part 5, *Reserve and Special Batteries,* covers the various types of reserve batteries developed for special, usually high-rate, applications. Major chapters cover the water-activated, automatically activated, nonaqueous, and thermal battery systems.

Part 6, *Fuel Cells,* covers the general features and performance characteristics of fuel cell systems with major chapters devoted to the low power systems (for portable applications, lightweight electrical power sources, spacecraft, and electric vehicles) and the larger power plants being developed for utility applications.

Several appendices have been included: a summary list of the various battery systems, the important electrochemical and conversion constants, definitions, a bibliography, and the major worldwide battery manufacturers. All data are presented in the metric system in accordance with the International System of Units (abbreviated SI). These standard symbols and constants are listed in Appendix D; conversion factors to other forms are listed in Appendix E.

Electrochemical couples or cell systems are identified in the Handbook by first listing the anode active material, then the electrolyte, and finally the active cathode material—each separated by a slash (/). The electrolyte may not be included in all cases. The cell system may also be identified by its common name. For example, the conventional alkaline dry cell may be identified as the alkaline-manganese dioxide cell and the couple shown as zinc/potassium hydroxide/manganese dioxide or zinc/manganese dioxide if the electrolyte is omitted. In addition, in the presentation of the cell reaction mechanisms or other references, 'e' has been used to represent the singly negatively charged electron, rather than the form 'e'.

In a new work of this scope and size, particularly with a large group of contributors and in a rapidly advancing technology, it is difficult to achieve the optimum balance in providing the technical information and data on each battery system and being completely up to date in all instances. In addition, space allocations limited the amount of material that could be included. This situation will be rectified in expanded future editions of this Handbook. Further, while great care was taken by the contributors and editors during the preparation and proofreading, it is inevitable that some errors may remain. The editor solicits your comments and suggestions for future editions and would appreciate being advised of any errors or serious omissions and areas where major updating or revision is warranted.

The substantial contribution of each of the authors to the Handbook is gratefully acknowledged, as well as their cooperation in preparing their manuscripts in the style and format selected. I also wish to express my appreciation to the companies, associations, and government agencies who supported the contributing authors and willingly provided their technical information and data and permitted its use in the Handbook. These contributions are identified in the text.

The important contribution of Dr. Jack Davis, who served as co-editor of the Handbook until his untimely death in 1980, must be singled out. His encouragement and participation were instrumental in the initiation and planning stages of the Handbook, the selection of the expert contributors, and in the early editing of the text.

I also wish to thank the staff at McGraw-Hill for providing the needed guidance and assistance, particularly to Harold B. Crawford, Editor in Chief, Technical Books, and Geraldine Fahey, Senior Editing Supervisor, for her aid in the editing and production of the Handbook.

And finally, my deepest thanks and appreciation to my wife, Rose, for her patience and encouragement which were invaluable for the completion of this volume.

David Linden

HANDBOOK OF BATTERIES AND FUEL CELLS

PART 1 PRINCIPLES OF OPERATION

1

Basic Concepts

by
David Linden

1.1 COMPONENTS OF CELLS AND BATTERIES

A battery is a device that converts the chemical energy contained in its active materials directly into electrical energy by means of an electrochemical oxidation-reduction (redox) reaction. This type of reaction involves the transfer of electrons from one material to another through an electrical circuit. In a nonelectrochemical redox reaction, the transfer of electrons occurs directly and only heat is involved.

While the term "battery" is often used, the basic electrochemical unit being referred to is the "cell." A battery consists of one or more of these cells, connected in series or parallel, or both, depending on the desired output voltage and capacity.

The cell consists of three major components as shown schematically in Fig. 1.1:

1. The anode or negative electrode—the reducing or fuel electrode—which gives up electrons to the external circuit and is oxidized during the electrochemical reaction.

2. The cathode or positive electrode—the oxidizing electrode—which accepts electrons from the external circuit and is reduced during the electrochemical reaction.

3. The electrolyte—the ionic conductor which provides the medium for transfer of electrons, as ions, inside the cell between the anode and cathode. The electrolyte is typically a liquid, such as water or other solvents, with dissolved salts, acids, or alkalis to impart ionic conductivity. Some batteries use solid electrolytes which are ionic conductors at the operating temperature of the cell.

The most advantageous combinations of anode and cathode materials are those that will be lightest and give a high cell voltage and capacity (see Sec. 1.3). Such combinations may not always be practical, however, due to reactivity with other cell components, polarization, difficulty in handling, high cost, and other such deficiencies.

In a practical system, the anode is selected with the following properties in mind: efficiency as a reducing agent, good conductivity, stability, ease of fabrication, and low cost. Metals are mainly used as the anode material. Zinc has been a predominant anode because it has these favorable properties. Lithium, the lightest metal, is now becoming a very attractive anode as suitable and compatible electrolytes and cell designs have been developed to control its activity.

The cathode must be an efficient oxidizing agent, stable when in contact with the electrolyte, and have a useful working voltage. Most of the cathode materials are metallic oxides, but recently other cathodes have been used for advanced battery systems giving high voltages and capacity.

The electrolyte must have good ionic conductivity but not be electrically conductive as this would cause internal short-circuiting. Other important characteristics are non-reactivity with the electrode materials, little change in properties with change in temperature, safeness in handling, and low cost. Most electrolytes are aqueous solutions, but there are important exceptions, as, for example, in thermal and lithium anode batteries where molten salt and other nonaqueous electrolytes are used to avoid the reaction of the anode with the electrolyte.

Physically, the anode and cathode electrodes are electronically isolated in the cell to prevent internal short-circuiting but are surrounded by the electrolyte. In practical cell designs, a separator material is used to mechanically separate the anode and cathode electrodes; the separator, however, is permeable to the electrolyte in order to maintain the desired ionic conductivity. In some cases, the electrolyte is immobilized for a nonspill design. Electrically conducting grid structures or materials may also be added to the electrodes to reduce internal resistance.

The cell itself can be built in many shapes and configurations, cylindrical, button, flat, and prismatic, and the cell components are designed to accommodate the particular cell shape. The cells are sealed in a variety of ways to prevent leakage and dry-out. Some cells are provided with venting devices or other means to allow accumulated gases to escape. Suitable cell cases or containers and means for terminal connection are added to complete the cell.

1.2 OPERATION OF A CELL

1.2.1 Discharge

The operation of a cell during discharge is also shown schematically in Fig. 1.1. When the cell is connected to an external load, electrons flow from the anode, which is oxidized, through the external load to the cathode, where the electrons are accepted and the cathode material reduced. The electrical circuit is completed in the electrolyte by the flow of anions (negative ions) and cations (positive ions) to the anode and cathode, respectively.

The discharge reaction can be written, assuming a metal as the anode material and a cathode material such as chlorine (Cl_2), as follows:

Negative electrode: anodic reaction (oxidation, loss of electrons)
$$Zn \rightarrow Zn^{2+} + 2e$$
Positive electrode: cathodic reaction (reduction, gain of electrons)
$$Cl_2 + 2e \rightarrow 2Cl^-$$
Overall reaction (discharge):
$$Zn + Cl_2 \rightarrow Zn^{2+} + 2Cl^- \ (ZnCl_2)$$

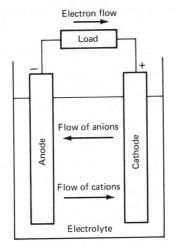

FIG. 1.1   Electrochemical operation of a cell (discharge).

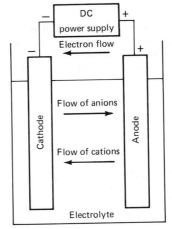

FIG. 1.2   Electrochemical operation of a cell (charge).

1.2.2 Charge

During the recharge of a rechargeable or storage battery, the current flow is reversed and oxidation takes place at the positive electrode and reduction at the negative electrode, as shown in Fig. 1.2. As the anode is, by definition, the electrode at which oxidation occurs and the cathode the one where reduction takes place, the positive electrode is now the anode and the negative the cathode.

In the example of the Zn/Cl_2 cell, the reaction on charge can be written:

Negative electrode: cathodic reaction (reduction, gain of electrons)

$$Zn^{2+} + 2e \rightarrow Zn$$

Positive electrode: anodic reaction (oxidation, loss of electrons)

$$2Cl^- \rightarrow Cl_2 + 2e$$

Overall reaction (charge):

$$Zn^{2+} + 2Cl^- \rightarrow Zn + Cl_2$$

1.2.3 Specific Example: Nickel-Cadmium Cell

The processes that produce electricity in a battery are chemical reactions which either release or consume electrons as the electrode reaction proceeds to completion. This can be illustrated with the specific example of the reactions of the nickel-cadmium battery. At the anode (negative electrode), the discharge reaction is the oxidation of cadmium metal to cadmium hydroxide with the release of two electrons:

$$Cd + 2OH^- \rightarrow Cd(OH)_2 + 2e$$

At the cathode, nickel oxide (or more accurately nickel oxyhydroxide) is reduced to nickel hydroxide with the acceptance of an electron:

$$2NiOOH + 2H_2O + 2e \rightarrow 2OH^- + 2Ni(OH)_2$$

When these two "half-cell" reactions occur (by connection of the electrodes to an external discharge circuit), the overall cell reaction converts cadmium to cadmium hydroxide at the anode and nickel oxyhydroxide to nickel hydroxide at the cathode.

$$Cd + 2NiOOH + 2H_2O \rightarrow Cd(OH)_2 + 2N_i(OH)_2$$

This is the discharge process. If this were a primary battery, at the end of discharge the device would be exhausted and discarded. The nickel-cadmium battery system is, however, a secondary (rechargeable) system, and on recharge (when one returns electrical energy to the device) the reaction at the negative electrode is:

$$Cd(OH)_2 + 2e \rightarrow Cd + 2OH^-$$

At the positive electrode the corresponding reaction is:

$$2Ni(OH)_2 + 2OH^- \rightarrow NiOOH + 2H_2O + 2e$$

After recharge, the secondary battery reverts to its original chemical state and is ready for further discharge. These are the fundamental principles involved in the charge-discharge mechanisms of a typical secondary battery.

1.2.4 Fuel Cell

The operation of the fuel cell is similar to that of a battery except that one or both of the reactants are not permanently contained in the electrochemical cell but are fed into it from an external source when power is desired. The fuels are usually gaseous or liquid (compared with the metal anodes generally used in batteries), and oxygen or air is the oxidant. The electrode materials of the fuel cells are inert in that they are not consumed during the cell reaction, but they have catalytic properties which enhance the electroreduction or electrooxidation of the fuel cells' active materials. A typical fuel cell reaction is illustrated by the hydrogen/oxygen fuel cell. In this device, hydrogen is oxidized at the anode, electrocatalyzed by platinum or platinum alloys, while at the cathode oxygen is reduced, again with platinum or platinum alloys as an electrocatalyst. The simplified anodic reaction is:

$$2H_2 \rightarrow 4H^+ + 4e$$

while the cathodic reaction is:

$$O_2 + 4H^+ + 4e \rightarrow 2H_2O$$

The overall reaction is the oxidation of hydrogen by oxygen to water:

$$2H_2 + O_2 \rightarrow 2H_2O$$

1.3 THEORETICAL CELL VOLTAGE AND CAPACITY

The theoretical voltage and capacity of a cell are a function of the anode and cathode materials (see Chap. 2 for detailed electrochemical theory).

1.3.1 Free Energy

Whenever a reaction occurs, there is a decrease in the free energy of the system, which is expressed as

$$\triangle G^0 = -nFE^0$$

where F = a constant known as the Faraday ($\approx$ 96,500 C or 26.8 Ah)
  n = number of electrons involved in stoichiometric reaction
  E^0 = standard potential, V

1.3.2 Theoretical Voltage

The standard potential of the cell is determined by its active materials and can be calculated from free-energy data or obtained experimentally. A listing of electrode potentials (reduction potentials) under standard conditions is given in Table 1.1. A more complete list is given in Appendix B.

The standard potential of a cell can be calculated from the standard electrode potentials as follows (the oxidation potential is the negative value of the reduction potential):

Anode (oxidation potential) + cathode (reduction potential)
= standard cell potential

For example, in the reaction

$$Zn + Cl_2 \rightarrow ZnCl_2$$

$$\begin{aligned} Zn &\rightarrow Zn^{2+} + 2e & -(-0.76 \text{ V}) \\ Cl_2 &\rightarrow 2Cl^- - 2e & \underline{1.36 \text{ V}} \\ & & 2.12 \text{ V} \end{aligned}$$

TABLE 1.1 Characteristics of Electrode Materials

Material	Atomic or molecular weight, g	Standard potential, 25°C, V	Valence change	Melting point, °C	Density, g/cm³	Electrochemical equivalents		
						Ah/g	g/Ah	Ah/cm³*
Anode materials								
H_2	2.01	0	2	. . .	. . .	26.59	0.037	
Li	6.94	−3.01	1	180	0.54	3.86	0.259	2.06
Na	23.0	−2.71	1	98	0.97	1.16	0.858	1.14
Mg	24.3	−2.38	2	650	1.74	2.20	0.454	3.8
Al	26.9	−1.66	3	659	2.69	2.98	0.335	8.1
Ca	40.1	−2.84	2	851	1.54	1.34	0.748	2.06
Fe	55.8	−0.44	2	1528	7.85	0.96	1.04	7.5
Zn	65.4	−0.76 −1.25†	2	419	7.14	0.82	1.22	5.8
Cd	112.4	−0.40	2	321	8.65	0.48	2.10	4.1
Pb	207.2	−0.13	2	327	11.34	0.26	3.87	2.9
Cathode materials								
O_2	32.0	1.23	4	. . .	. . .	3.35	0.30	
Cl_2	71.0	1.36	2	. . .	. . .	0.755	1.32	
SO_2	64.0	. . .	1	. . .	. . .	0.419	2.38	
MnO_2	86.9	1.23	1	. . .	5.0	0.308	3.24	1.54
NiOOH	91.7	0.49*	1	. . .	7.4	0.292	3.42	2.16
CuCl	99.0	0.14	1	. . .	3.5	0.270	3.69	0.95
AgO	123.8	0.57*	2	. . .	7.4	0.432	2.31	3.20
HgO	216.6	0.10*	2	. . .	11.1	0.247	4.05	2.74
Ag_2O	231.7	0.35*	2	. . .	7.1	0.231	4.33	1.64
PbO_2	239.2	1.69	2	. . .	9.4	0.224	4.45	2.11

†Basic electrolyte; all others, aqueous acid electrolyte.
*Based on density values shown.
NOTE: Also see Appendices B and C.

The cell voltage also is dependent on other factors, including concentration and temperature as expressed by the Nernst equation (covered in detail in Chap. 2).

1.3.3 Theoretical Capacity

The capacity of a cell is expressed as the total quantity of electricity involved in the electrochemical reaction and is defined in terms of coulombs or ampere-hours.

The "ampere-hour capacity" of a battery is directly associated with the quantity of electricity obtained from the active materials. Theoretically 1 gram-equivalent weight of material will deliver 96,487 C or 26.8 Ah (a gram-equivalent weight is the atomic or molecular weight of the active material in grams divided by the number of electrons involved in the reaction). The electrochemical equivalence of typical materials is listed in Table 1.1 and Appendix C.

The theoretical capacity of a battery system, based only on the active materials participating in the electrochemical reaction, is calculated from the equivalent weight of the reactants; hence, the theoretical capacity of the Zn/Cl_2 system is 0.394 Ah/g:

$$\text{Zn} \quad + \quad \text{Cl}_2 \quad \rightarrow \quad \text{ZnCl}_2$$

0.82 Ah/g	0.76 Ah/g	
1.22 g/Ah	1.32 g/Ah	= 2.54 g/Ah or
		0.394 Ah/g

The capacity of batteries is also considered on an energy (watthour) basis by taking the voltage as well as the quantity of electricity into consideration:

$$\text{Watthour (Wh)} = \text{voltage (V)} \times \text{ampere-hour (Ah)}$$

In the Zn/Cl_2 cell example, if the standard potential is taken as 2.12 V, theoretical watthour capacity per gram of active material (theoretical gravimetric energy density) is:

$$\text{Watthour/gram capacity} = 2.12 \text{ V} \times 0.395 \text{ Ah/g}$$
$$= 0.838 \text{ Wh/g}$$

Similarly, the ampere-hour or watthour capacity on a volume basis can be calculated by using the appropriate data for ampere-hours per cubic centimeter.

The theoretical voltages and capacities of a number of the major electrochemical systems are given in Table 1.2. The values given are based on the active anode and cathode material only; water, electrolyte, or any other material that may be involved in the cell reaction are not included in the calculation.

1.4 CLASSIFICATION OF CELLS AND BATTERIES

Electrochemical cells and batteries are identified as primary (nonrechargeable) or secondary (rechargeable), depending on their capability of being electrically recharged. Within this classification, other classifications are used to identify particular structures or designs. The classification used in this handbook for the different types of electrochemical cells and batteries is described below.

1.4.1 Primary Cells or Batteries

These are not capable of being easily recharged electrically and, hence, are discharged once and discarded. Many primary cells, in which the electrolyte is contained by an absorbent or separator material (there is no free or liquid electrolyte), are termed "dry cells."

The primary battery is a convenient, usually relatively inexpensive, lightweight source of packaged power for portable electronic and electrical devices, lighting, photographic equipment, toys, memory backup, and a host of other applications giving freedom from utility power. The general advantages of primary batteries are good shelf life, high energy density at low to moderate discharge rates, little, if any, maintenance, and ease of use. Although large, high-capacity primary batteries are used in military applications, signaling, standby power, etc., the vast majority of primary batteries are the familiar cylindrical and flat cells.

1.4.2 Secondary or Rechargeable Cells or Batteries

These can be recharged electrically, after discharge, to their original condition by

TABLE 1.2 Theoretical Voltage and Capacity of Major Battery Systems*

Battery system	Anode	Cathode	Reaction mechanism	Theoretical		
				V	Capacity†	
					g/Ah	Ah/kg
Primary						
Leclanche	Zn	MnO_2	$Zn + 2MnO_2 \rightarrow ZnO \cdot Mn_2O_3$	1.6	4.46	224
Magnesium	Mg	MnO_2	$Mg + 2MnO_2 + H_2O \rightarrow Mn_2O_3 + Mg(OH)_2$	2.8	3.69	271
Alkaline MnO_2	Zn	MnO_2	$Zn + 2MnO_2 \rightarrow ZnO + Mn_2O_3$	1.5	4.46	224
Mercury	Zn	HgO	$Zn + HgO \rightarrow ZnO + Hg$	1.34	5.27	190
Mercad	Cd	HgO	$Cd + HgO + H_2O \rightarrow Cd(OH)_2 + Hg$	0.91	6.15	163
Silver oxide	Zn	Ag_2O	$Zn + Ag_2O + H_2O \rightarrow Zn(OH)_2 + 2Ag$	1.6	5.55	180
Zinc/air	Zn	O_2 (air)	$Zn + \frac{1}{2}O_2 \rightarrow ZnO$	1.65	1.55	800
Li/SO_2	Li	SO_2	$2Li + 2SO_2 \rightarrow Li_2S_2O_4$	3.1	2.64	379
Li/MnO_2	Li	MnO_2	$Li + Mn^{IV}O_2 \rightarrow Mn^{III}O_2(Li^+)$	3.5	3.50	286
Reserve						
Cuprous chloride	Mg	CuCl	$Mg + Cu_2Cl_2 \rightarrow MgCl_2 + 2Cu$	1.6	4.14	241
Zinc/silver oxide	Zn	AgO	$Zn + AgO + H_2O \rightarrow Zn(OH)_2 + Ag$	1.81	3.53	283
Secondary						
Lead-acid	Pb	PbO_2	$Pb + PbO_2 + 2H_2SO_4 \rightarrow 2PbSO_4 + 2H_2O$	2.1	8.32	120
Edison	Fe	Ni oxide	$Fe + 2NiOOH + 2H_2O \rightarrow 2Ni(OH)_2 + Fe(OH)_2$	1.4	4.46	224
Nickel-cadmium	Cd	Ni oxide	$Cd + 2NiOOH + 2H_2O \rightarrow 2Ni(OH)_2 + Cd(OH)_2$	1.35	5.52	181
Silver-zinc	Zn	AgO	$Zn + AgO + H_2O \rightarrow Zn(OH)_2 + Ag$	1.85	3.53	283
Nickel-zinc	Zn	Ni oxide	$Zn + 2NiOOH + 2H_2O \rightarrow 2Ni(OH)_2 + Zn(OH)_2$	1.73	4.64	215
Nickel-hydrogen	H_2	Ni oxide	$H_2 + 2NiOOH \rightarrow 2Ni(OH)_2$	1.5	3.46	289
Silver-cadmium	Cd	AgO	$Cd + O + H_2O \rightarrow Cd(OH)_2 + Ag$	1.4	4.41	227
Zinc/chlorine	Zn	Cl_2	$Zn + Cl_2 \rightarrow ZnCl_2$	2.12	2.54	394
High temperature	Li(Al)	FeS	$2Li(Al) + FeS \rightarrow Li_2S + Fe + 2Al$	1.33	2.99	345
High temperature	Na	S	$2Na + 3S \rightarrow Na_2S_3$	2.1	2.65	377
Fuel cell						
H_2/O_2	H_2	O_2 (or air)	$H_2 + \frac{1}{2}O_2 \rightarrow H_2O$	1.23	0.336	2975

*Table 3.3 presents data on the practical operating characteristics of these battery systems.
†Based on active anode and cathode materials only.

passing current through them in the opposite direction to that of the discharge current. They are storage devices for electrical energy and are known also as "storage batteries" or "accumulators."

The applications of secondary batteries fall into two main categories:

1. Those applications in which the secondary battery is used as an energy-storage device, generally being electrically connected to and charged by a prime energy source and delivering its energy to the load on demand. Examples are automotive and aircraft systems, emergency no-fail and standby power sources, and stationary energy storage (SES) systems for electric utility load leveling.

2. Those applications in which the secondary battery is used or discharged essentially as a primary battery, but recharged after use rather than being discarded. Secondary batteries are used in this manner in electric vehicles, for cost savings (as they can be recharged rather than replaced), and in applications requiring power drains beyond the capability of primary batteries.

Secondary batteries are characterized (in addition to their ability to be recharged) by high power density, high discharge rate, flat discharge curves, and good low-temperature performance. Their energy densities are generally lower than those of primary batteries. Their charge retention also is poorer than most primary batteries, although the capacity of the secondary battery that is lost on standing can be restored by recharging.

Some batteries, known as "mechanically rechargeable types," are "recharged" by replacement of the discharged or depleted electrode, usually the metal anode, with a fresh one. The metal-air batteries, proposed for electric vehicle use (Part 4, Chap. 30), are representative of this type of battery.

1.4.3 Reserve Batteries

In these primary types a key component is separated from the rest of the battery prior to activation. In this condition, chemical deterioration or self-discharge is essentially eliminated, and the battery is capable of long-term storage. Usually, the electrolyte is the component that is isolated; in other systems, such as the thermal battery, the battery is inactive until it is heated, melting a solid electrolyte, which then becomes conductive.

The reserve battery design is used to meet extremely long or environmentally severe storage requirements that cannot be met with an "active" battery designed for the same performance characteristics. These batteries are used primarily to deliver high power for relatively short periods of time needed as in missiles, torpedoes, and other weapon systems.

1.4.4 Fuel Cells

The fuel cell can be considered a primary-type battery but one in which the reactants are fed into the cell from an external source when power is desired. Thus, the cell can operate continuously as long as reactants are supplied and the internal cell electrodes and components remain unchanged. The anode material, or fuel, is usually gaseous or liquid (compared with metal anodes generally used in batteries), and oxygen or air is the oxidant.

The fuel cell provides a means of electrochemically converting hydrogen and/or other types of conventional fuels without the Carnot cycle limitation of heat engines. Major uses have been in applications requiring electrical energy for long periods of time (such as in a space flight), as an alternate for moderate-power engine generators, and for utility load leveling.

BIBLIOGRAPHY

See Bibliography in Appendix H.

Electrochemical Principles and Reactions

by
John Broadhead

2.1 HISTORY

The basic principles of primary and secondary batteries and fuel cells can be traced back to the work of Volta[1] in 1800. This early work established the nucleus of what is now known as the electromotive series of the elements. A few years later, Gautherot[2] and Ritter[3] discovered and demonstrated the polarization and reversibility of platinum, gold, and silver in several systems, and Ritter constructed what was probably the first battery. In 1839 and 1842, Grove[4] discovered and described the "gas battery" which some might consider the forerunner of the fuel cell. In 1859, Planté[5] began his studies which led to the development of the first practical secondary battery, the lead-acid battery, which to this day is the most widely used and cost-effective secondary battery system. Since this early work began, many new battery systems have been discovered and developed.[6] It is the intention of this chapter to outline the basic principles and electrochemical processes common to primary and secondary batteries and fuel cells as a prelude to the following chapters which cover, in detail, many of the different battery and fuel-cell systems.

2.2 THERMODYNAMIC BACKGROUND

In a cell, reactions essentially take place at two areas or sites in the device. These reaction sites are the electrodes. In generalized terms, the reaction at one electrode (reduction in forward direction) can be represented by

$$aA + ne \rightleftarrows cC \qquad (1)$$

where a molecules of A take up n electrons, e, to form c molecules of C. At the other electrode, the reaction (oxidation in forward direction) can be represented by

$$bB - ne \rightleftarrows dD \tag{2}$$

The overall reaction in the cell is given by addition of these two half-cell reactions:

$$aA + bB \rightleftarrows cC + dD \tag{3}$$

The change in the standard free energy, $\triangle G^0$, of this reaction is expressed as follows:

$$\triangle G^0 = - nFE^0$$

where F is a constant known as the Faraday (96,487 C) and E^0 is the standard electromotive force. When conditions are other than in the standard state, the voltage E of a cell is given by the Nernst equation,

$$E = E^0 - \frac{RT}{nF} \ln \frac{\mu C^c \mu D^d}{\mu A^a \mu B^b} \tag{4}$$

where μ = activity of relevant species
R = gas constant
T = absolute temperature

The change in standard free energy, $\triangle G^0$, of a cell reaction is the driving force which enables a battery to deliver electrical energy to an external circuit. The measurement of electromotive force, incidentally, also makes available data on changes in free energy, entropies, and enthalpies together with activity coefficients, equilibrium constants, and solubility products.

Direct measurement of single (absolute) electrode potentials is considered practically impossible.[7] To establish a scale of half-cell or standard potentials, a reference potential "zero" must be established against which single electrode potentials can be measured. By convention, the standard potential of the H_2/H^+(aq) reaction is taken as zero and all standard potentials are referred to this potential. Table 2.1 and Appendix B list the standard potentials of a number of anode and cathode materials.

2.3 ELECTRODE PROCESSES

Reactions at an electrode are characterized by both chemical and electrical changes and are heterogeneous in type. Electrode reactions may be as simple as the reduction of a metal ion and incorporation of the resultant atom into the electrode structure. Despite the apparent simplicity of the reaction, the mechanism of the overall process is relatively complex and often involves several steps. Electroactive species must be transported to the electrode surface by migration and/or diffusion prior to the electron transfer step. Adsorption of electroactive material may be involved both prior to and after the electron transfer step. Chemical reactions may also be involved in the overall electrode reaction. As in any reaction, the overall rate of the electrochemical process is determined by the rate of the slowest step in the whole sequence of reactions.

The thermodynamic treatment of electrochemical processes given in the previous section describes the equilibrium condition of a system but does not present information on nonequilibrium conditions such as current flow resulting from electrode polarization (overvoltage) imposed to effect electrochemical reactions. Experimental determination

TABLE 2.1 Standard Potentials of Electrode Reactions at 25°C

Electrode reaction	E^0, V	Electrode reaction	E^0, V
$Li^+ + e \rightleftarrows Li$	-3.01	$Tl^+ + e \rightleftarrows Tl$	-0.34
$Rb^+ + e \rightleftarrows Rb$	-2.98	$Co^{2+} + 2e \rightleftarrows Co$	-0.27
$Cs^+ + e \rightleftarrows Cs$	-2.92	$Ni^{2+} + 2e \rightleftarrows Ni$	-0.23
$K^+ + e \rightleftarrows K$	-2.92	$Sn^{2+} + 2e \rightleftarrows Sn$	-0.14
$Ba^{2+} + 2e \rightleftarrows Ba$	-2.92	$Pb^{2+} + 2e \rightleftarrows Pb$	-0.13
$Sr^{2+} + 2e \rightleftarrows Sr$	-2.89	$D^+ + e \rightleftarrows \frac{1}{2}D_2$	-0.003
$Ca^{2+} + 2e \rightleftarrows Ca$	-2.84	$H^+ + e \rightleftarrows \frac{1}{2}H_2$	0.000
$Na^+ + e \rightleftarrows Na$	-2.71	$Cu^{2+} + 2e \rightleftarrows Cu$	0.34
$Mg^{2+} + 2e \rightleftarrows Mg$	-2.38	$\frac{1}{2}O_2 + H_2O + 2e \rightleftarrows 2OH^-$	0.40
$Ti^{2+} + 2e \rightleftarrows Ti$	-1.75	$Cu^+ + e \rightleftarrows Cu$	0.52
$Be^{2+} + 2e \rightleftarrows Be$	-1.70	$Hg^{2+} + 2e \rightleftarrows 2Hg$	0.80
$Al^{3+} + 3e \rightleftarrows Al$	-1.66	$Ag^+ + e \rightleftarrows Ag$	0.80
$Mn^{2+} + 2e \rightleftarrows Mn$	-1.05	$Pd^{2+} + 2e \rightleftarrows Pd$	0.83
$Zn^{2+} + 2e \rightleftarrows Zn$	-0.76	$Ir^{3+} + 3e \rightleftarrows Ir$	1.00
$Ga^{3+} + 3e \rightleftarrows Ga$	-0.52	$Br_2 + 2e \rightleftarrows 2Br^-$	1.07
$Fe^{2+} + 2e \rightleftarrows Fe$	-0.44	$O_2 + 4H^+ + 4e \rightleftarrows 2H_2O$	1.23
$Cd^{2+} 2e \rightleftarrows Cd$	-0.40	$Cl_2 + 2e \rightleftarrows 2Cl^-$	1.36
$In^{3+} + 3e \rightleftarrows In$	-0.34	$F_2 + 2e \rightleftarrows 2F^-$	2.87

of the current voltage characteristics of many electrochemical systems has shown that there is an exponential relation between current and applied voltage. The generalized expression describing this relationship is called the Tafel equation,

$$\eta = a \pm b \log i \qquad (5)$$

where η = overvoltage
i = current
a and b = constants
Typically the constant b is referred to as the Tafel slope.

The Tafel relationship holds for a large number of electrochemical systems over a wide range of overpotential. At low values of overvoltage, however, the relationship breaks down and results in curvature in plots of η vs. log i. Figure 2.1 represents a schematic presentation of a Tafel plot showing curvature at low values of overvoltage.

Success of the Tafel equation's fit to many experimental systems encouraged the quest for a kinetic theory of electrode processes. Since the range of validity of the Tafel relationship applies to high overvoltages, it is reasonable to assume that the expression does not apply to equilibrium situations but represents the current-voltage relationship of a unidirectional process. In an oxidation process this would mean that there is a negligible contribution from re-reduction processes. Rearranging Eq. (5) into exponential form, we have

$$i = \exp - \frac{a}{b} \exp \frac{\eta}{b} \qquad (6)$$

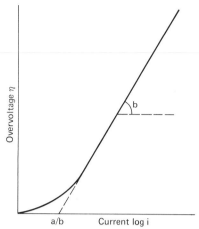

FIG. 2.1 Schematic representation of a Tafel plot showing curvature at low overvoltage and indicating the significance of the parameters a and b.

To consider a general theory, one must consider both forward and backward reactions of the electroreduction process shown, in simplified form, in Fig. 2.2. The reaction is represented by the following equation:

$$O + ne \rightleftarrows R \qquad (7)$$

where O = oxidized species
    R = reduced species
    n = number of electrons involved in electrode process
The forward and backward reactions can be described by heterogeneous rate constants, k_f and k_b, respectively. The rates of the forward and backward reactions are then given by the products of these rate constants and the relevant concentrations which typically are those at the electrode surface. As will be shown later, concentrations of electroactive species at the electrode surface often are dissimilar from the bulk concentration in solution. The rate of the forward reaction is $k_f C_O$ and that for the backward reaction is $k_b C_R$. For convenience, these rates are usually expressed in terms of currents for the forward and backward reactions, i_f and i_b, respectively:

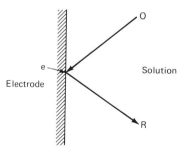

FIG. 2.2 Simplified representation of electroreduction process at an electrode.

$$i_f = nFAk_f C_O \tag{8}$$

$$i_b = nFAk_b C_R \tag{9}$$

where A is the area of the electrode and F the Faraday.

Establishing these expressions is merely the result of applying the law of mass action to the forward and backward electrode processes. The role of electrons in the process is established by assuming that the magnitudes of the rate constants depend on the electrode potential. The dependence is usually described by assuming that a fraction αE of the electrode potential is involved in driving the reduction process, while the fraction $(1-\alpha)E$ is effective in making the reoxidation process more difficult. Mathematically, these potential dependent rate constants are expressed:

$$k_f = k_f^0 \exp \frac{-\alpha nFE}{RT} \tag{10}$$

$$k_b = k_b^0 \exp \frac{(1-\alpha)nFE}{RT} \tag{11}$$

where α is called the transfer coefficient and E is the electrode potential relative to a suitable reference potential.

A little more explanation regarding what the transfer coefficient α (or the symmetry factor β, as it is referred to in some texts) means in mechanistic terms is appropriate since the term is not implicit in the kinetic derivation.[8] The transfer coefficient determines what fraction of the electrical energy resulting from the displacement of the potential from the equilibrium value affects the rate of electrochemical transformation. To understand the function of the transfer coefficient α, it is necessary to describe an energy diagram for the reduction and oxidation process. Figure 2.3 shows an approximate potential energy curve (Morse curve) for an oxidized species approaching an electrode surface together with the potential energy curve for the resultant reduced species. For convenience, consider the hydrogen ion reduction at a solid electrode as the model for a typical electroreduction. According to Horiuti and Polanyi,[9] the potential energy diagram for reduction of the hydrogen ion can be represented by Fig. 2.4 where the oxidized species O is the hydrated hydrogen ion and the reduced species R is a hydrogen atom bonded to the metal (electrode) surface. The effect of changing the electrode potential by a value E is to raise the potential energy of the Morse curve

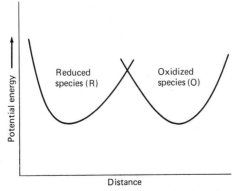

FIG. 2.3   Potential energy diagram for the reduction-oxidation process taking place at an electrode.

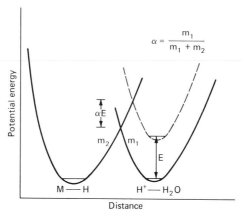

FIG. 2.4 Potential energy diagram for the reduction of a hydrated hydrogen ion at an electrode.

of the hydrogen ion. The intersection of the two Morse curves forms an energy barrier, the height of which is αE. If the slope of the two Morse curves is approximately constant at the point of intersection, then α is defined by the ratio of the slope of the Morse curves at the point of intersection:

$$\alpha = \frac{m_1}{m_1 + m_2} \tag{12}$$

where m_1 and m_2 are the slopes of the potential curves of the hydrated hydrogen ion and hydrogen atom, respectively.

There are inadequacies to the theory of transfer coefficients. It assumes that α is constant and independent of E. At present there are no data to prove or disprove this assumption. The other main weakness is that the concept is used to describe processes involving a variety of different species such as (1) redox changes at an inert electrode (Fe^{2+}/Fe^{3+} at Hg); (2) reactant and product soluble in different phases [Cd^{2+}/Cd(Hg)]; and (3) electrodeposition (Cu^{2+}/Cu). Despite these inadequacies, the concept and application of the theory are appropriate in many cases and represent the best understanding and description of electrode processes at the present time. Examples of a few values of α are shown in Table 2.2.[10]

TABLE 2.2 Values of the Transfer Coefficient α at 25°C

Metal	System	α
Platinum	$Fe^{3+} + e \rightarrow Fe^{2+}$	0.58
Platinum	$Ce^{4+} + e \rightarrow Ce^{3+}$	0.75
Mercury	$Ti^{4+} + e \rightarrow Ti^{3+}$	0.42
Mercury	$2H^+ + 2e \rightarrow H_2$	0.50
Nickel	$2H^+ + 2e \rightarrow H_2$	0.58
Silver	$Ag^+ + e \rightarrow Ag$	0.55

From Eqs. (10) and (11), we can derive parameters useful for evaluating and describing an electrochemical system. Equations (10) and (11) are compatible both with the Nernst equation [Eq. (4)] for equilibrium conditions and with the Tafel relationship [Eq. (5)] for unidirectional processes.

Under equilibrium conditions, no net current flows and

$$i_f = i_b = i_0 \tag{13}$$

where i_0 is the exchange current. From Eqs. (8) to (11) together with (13), the following relationship is established:

$$C_O k_f^0 \exp\frac{-\alpha nFE}{RT} = C_R k_b^0 \exp\frac{(1-\alpha)nFE}{RT} \tag{14}$$

or rearranging,

$$E = \frac{RT}{nF} \ln\frac{k_f^0}{k_b^0} + \frac{RT}{nF} \ln\frac{C_O}{C_R} \tag{15}$$

From this equation we can establish the definition of formal standard potential E_C^0, where concentrations are used rather than activities:

$$E_C^0 = \frac{RT}{nF} \ln\frac{k_f^0}{k_b^0} \tag{16}$$

For convenience, the formal standard potential is often taken as the reference point of the potential scale in reversible systems. This then leads to the definition of the standard heterogeneous rate constant k:

$$k_f^0 = k_b^0 = k \tag{17}$$

This is usually referred to as the rate constant* of the charge transfer step.

Combining Eqs. (15) and (16), we can show consistency with the Nernst equation:

$$E = E_C^0 + \frac{RT}{nF} \ln\frac{C_O}{C_R} \tag{18}$$

except that this expression is written in terms of concentrations rather than activities.

The exchange current as defined in Eq. (13) is a parameter of interest to researchers in the battery field. This parameter may be conveniently expressed in terms of the rate constant k by combining Eqs. (8), (10), (15), and (17):

$$i_0 = nFAkC_O^{(1-\alpha)}C_R^{\alpha} \tag{19}$$

The exchange current i_0 is a measure of the rate of exchange of charge between the oxidized and reduced species at any equilibrium potential without net overall change. The rate constant k, however, has been defined for a particular potential, the formal standard potential of the system. It is not in itself sufficient to characterize the system unless the transfer coefficient is also known. A schematic representation of the forward and backward currents, as a function of overvoltage, is shown in Fig. 2.5.

For situations where the net current is not zero, i.e., where the potential is sufficiently different from the equilibrium potential, the net current approaches that of the forward current (or for anodic overvoltages, the backward current). One can then write

---

*The rate constant is variously referred to as k_s, k_h, or k_{sh} in standard texts and publications on electrode processes.

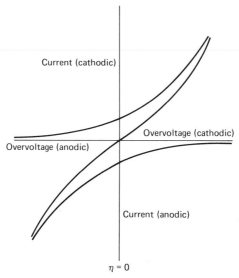

FIG. 2.5 Schematic representation of the relationship between overvoltage and current.

$$i = nFAkC_O\exp\frac{-\alpha nF\eta}{RT} \tag{20}$$

Now when $\eta = 0$, $i = i_0$, then

$$i = i_0\exp\frac{-\alpha nF\eta}{RT} \tag{21}$$

and

$$\eta = \frac{RT}{\alpha nF}\ln i_0 - \frac{RT}{\alpha nF}\ln i \tag{22}$$

which is the Tafel equation introduced earlier in a generalized form as Eq. (5).

It can now be seen that the kinetic treatment here is self-consistent with both the Nernst equation (for equilibrium conditions) and the Tafel relationship (for unidirectional processes).

To present the kinetic treatment in its most useful form, a transformation into a net current flow form is appropriate. Using

$$i = i_f - i_b \tag{23}$$

substitute Eqs. (8), (11), and (17):

$$i = nFAk[C_O\exp\frac{-\alpha nFE}{RT} - C_R\exp\frac{(1-\alpha)nFE}{RT}] \tag{24}$$

When this equation is applied in practice, it is very important to remember that C_O and C_R are concentrations at the surface of the electrode, or are the effective concentrations. These are not necessarily the same as the bulk concentrations. Concentrations at the interface are often (almost always) modified by differences in electrical potential between the surface and the bulk solution. The effect of potential differences that are manifest at the electrode-electrolyte interface will be given in the following section.

2.4 ELECTRICAL DOUBLE-LAYER CAPACITY AND IONIC ADSORPTION

When an electrode (metal surface) is immersed in an electrolyte, the electronic charge on the metal attracts ions of opposite charge and orients the solvent dipoles. There exist a layer of charge in the metal and a layer of charge in the electrolyte. This charge separation establishes what is commonly known as the "electrical double layer."[11]

Experimentally, the electrical double-layer effect is manifest in the phenomenon named "electrocapillarity." The phenomenon has been studied for many years, and there exist thermodynamic relationships that relate interfacial surface tension between the electrode and electrolyte solution to the structure of the double layer. Typically, the metal used for these measurements is mercury since it is the only conveniently available metal that is liquid at room temperature (although some work has been carried out with gallium, Wood's metal, and lead at elevated temperature).

Determinations of the interfacial surface tension between mercury and electrolyte solution can be made with a relatively simple apparatus. All that are needed are (1) a mercury-solution interface which is polarizable, (2) a nonpolarizable interface as reference potential, (3) an external source of variable potential, and (4) an arrangement to measure the surface tension of the mercury-electrolyte interface. An experimental system which will fulfill the above requirements is shown in Fig. 2.6. The interfacial surface tension is measured by applying pressure to the mercury electrolyte interface by raising the mercury "head." At the interface, the forces are balanced as shown in Fig. 2.7. If the angle of contact at the capillary wall is zero (typically the case for clean surfaces and clean electrolyte), then it is a relatively simple arithmetical exercise to show that the interfacial surface tension is given by

$$\gamma = \frac{h\rho g r}{2} \qquad (25)$$

where γ = interfacial surface tension
ρ = density of mercury
g = force of gravity
r = radius of capillary

The characteristic electrocapillary curve that one would obtain from a typical electrolyte solution is shown in Fig. 2.8. From such measurements and more accurately by ac impedance bridge measurements, the background of the structure of the electrical double layer has been built.[11]

Consider a negatively charged electrode in an aqueous solution of electrolyte. Assume that at this potential no electrochemical charge transfer takes place. For simplicity and clarity, the different features of the electrical double layer will be described individually.

Orientation of solvent molecules, water for the sake of this discussion, is shown in Fig. 2.9. The water dipoles are oriented, as shown in the figure, so that the majority of the dipoles are oriented with their positive ends (heads of the arrows) toward the surface of the electrode. This represents a "snapshot" of the structure of the layer of water molecules since the electrical double layer is a dynamic system which is in equilibrium with water in the bulk solution. Since the representation is statistical, not all dipoles are oriented the same way. Some dipoles are more influenced by dipole-dipole interactions than dipole-electrode interactions.

Next, consider the approach of a cation to the vicinity of the electrical double layer.

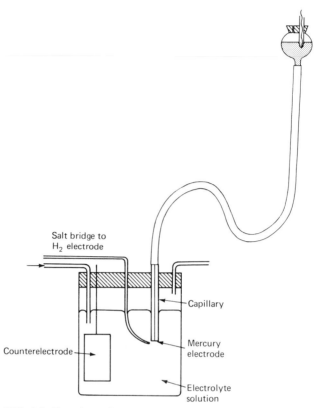

FIG. 2.6 Experimental arrangement to measure interfacial surface tension at the mercury-electrolyte interface.

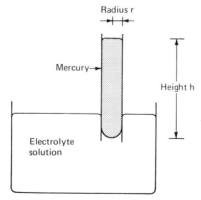

FIG. 2.7 Close-up of the mercury-electrolyte interface in a capillary immersed in an electrolyte solution.

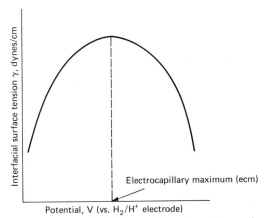

FIG. 2.8 Generalized representation of an electrocapillary curve.

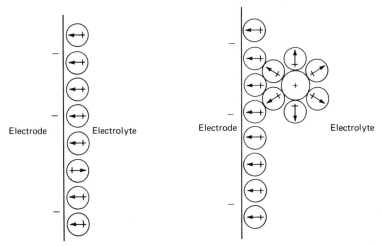

FIG. 2.9 Orientation of water molecules in the electrical double layer at a negatively charged electrode.

FIG. 2.10 Typical cation situated in the electrical double layer.

The majority of cations are strongly solvated by water dipoles and maintain a sheath of water dipoles around them despite the orienting effect of the double layer. With a few exceptions, cations do not approach right up to the electrode surface but remain outside the primary layer of solvent molecules and usually retain their solvation sheaths. Figure 2.10 shows a typical example of a cation in the electrical double layer. The establishment that this is the closest likely approach of a typical cation comes partly from experimental ac impedance measurements of mixed electrolytes and mainly from calculations of the free energy of approach of an ion to the electrode surface. In considering water-electrode, ion-electrode, and ion-water interactions, the free energy of approach of a cation to an electrode surface is strongly influenced by the hydration of the cation. The general result is that cations of very large radius (and thus of low hydration) such as Cs^+ can contact-adsorb on the electrode surface, but for the majority of cations the change in free energy on contact adsorption is positive and thus is against the mechanism of contact adsorption.[12] Figure 2.11 shows the example of the ion, Cs^+, contact-adsorbed on the surface of an electrode.

It would be expected that because anions have a negative charge, contact adsorption of anions would not occur. In analyzing the free-energy balance of the anion system, it is found that anion-electrode contact is favored because the net free-energy balance is negative. Both from these calculations and from experimental measurements, anion contact adsorption is found to be relatively common. Figure 2.12 shows the generalized case of anion adsorption on an electrode. There are exceptions to this type of adsorption. Calculation of the free energy of contact adsorption of the fluoride ion is positive and unlikely to occur. This is supported by experimental measurement. This property is utilized, as NaF, as a supporting electrolyte to evaluate adsorption properties of surface-active species devoid of the influence of adsorbed supporting electrolyte.

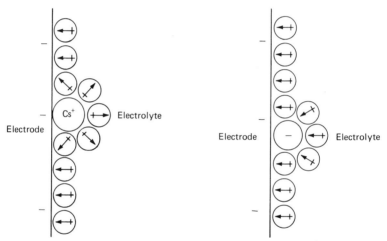

FIG. 2.11 Contact adsorption of the Cs^+ on an electrode surface.

FIG. 2.12 Contact adsorption of an anion on an electrode surface.

Extending out into solution from the electrical double layer (or the compact double layer, as it is sometimes known) is a continuous repetition of the layering effect but with diminishing magnitude. This "extension" of the compact double layer toward the bulk solution is known as the Gouy-Chapman diffuse double layer.[11] Its effect on electrode kinetics and the concentration of electroactive species at the electrode surface is manifest when supporting electrolyte concentrations are low or zero.

The end result of the establishment of the electrical double-layer effect, and the various types of ion contact adsorption, is to directly influence the real (actual) concentration of electroactive species at an electrode surface and indirectly to modify the potential gradient at the site of electron transfer. In this respect it is important to understand the influence of the electrical double layer and allow for it where and when appropriate.

The potential distribution near an electrode is shown schematically in Fig. 2.13. The inner Helmholtz plane corresponds to the plane which contains the contact-adsorbed ions and the innermost layer of water molecules. Its potential is defined as ϕ^i with the zero of potential being taken as the potential of the bulk solution. The outer Helmholtz plane is the plane of closest approach of those ions which do not contact-adsorb but approach the electrode with a sheath of solvated water molecules surrounding them. The potential at the outer Helmholtz plane is defined as ϕ^0 and is again referred to the potential of the bulk solution. In some texts ϕ^i is defined as ϕ^1 and ϕ^0 is called ϕ^2.

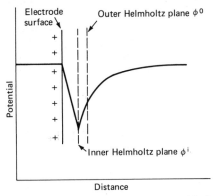

FIG. 2.13   Potential distribution at a positively charged electrode.

As mentioned previously, the bulk concentration of an electroactive species is often not the value to be used in kinetic equations. Species which are in the electrical double layer are in a different energy state from those in bulk solution. At equilibrium, the concentration C^e of an ion or species that is about to take part in the charge transfer process at the electrode is related to the bulk concentration by

$$C^e = C^B \exp\frac{-zF\phi^e}{RT} \tag{26}$$

where z is the charge on the ion and ϕ^e the potential of *closest approach* of the species to the electrode. It will be remembered that the plane of closest approach of many species is the outer Helmholtz plane, and so the value of ϕ^e can often be equated to ϕ^0. However, as noted in a few special cases, the plane of closest approach can be the inner Helmholtz plane, and so the value of ϕ^e in these cases would be the same as ϕ^i. A judgment has to be made as to what value of ϕ^e should be used.

The potential which is effective in driving the electrode reaction is that between the species at its closest approach and the potential of the electrode. If E is the potential of the electrode, then the driving force is $E - \phi^e$. Using this relationship together with Eqs. (24) and (26), we have

$$\frac{i}{nFAk} = C_O \exp\frac{-z_O F\phi^e}{RT} \exp\frac{-\alpha nF(E-\phi^e)}{RT}$$
$$- C_R \exp\frac{-z_R F\phi^e}{RT} \exp\frac{(1-\alpha)nF(E-\phi^e)}{RT} \tag{27}$$

where z_O and z_R are the charges (with sign) of the oxidized and reduced species. Rearranging Eq. (27) and using

$$z_O - n = z_R \qquad (28)$$

yields the following:

$$\frac{i}{nFAk} = \exp\frac{(\alpha n - z_O)F\phi^e}{RT}\left[C_O\exp\frac{-\alpha nFE}{RT} - C_R\exp\frac{(1-\alpha)nFE}{RT}\right] \qquad (29)$$

In experimental determinations, the use of Eq. (24) will provide an apparent rate constant k_{app}, which does not take into account the effects of the electrical double layer. Taking into account the effects appropriate to the approach of a species to the plane of nearest approach,

$$k_{app} = k\,\exp\frac{(\alpha n - z_O)F\phi^e}{RT} \qquad (30)$$

For the exchange current the same applies:

$$(i_0)_{app} = i_0\exp\frac{(\alpha n - z_O)F\phi^e}{RT} \qquad (31)$$

Corrections to the rate constant and to the exchange current are not insignificant. Several calculated examples are shown in the text by Bauer.[13] The differences between apparent and true rate constants can be as great as two orders of magnitude. The magnitude of the correction also is related to the magnitude of the difference in potential between the electrocapillary maximum for the species and the potential at which the electrode reaction occurs; the greater the potential difference, the greater the correction to the exchange current or rate constant.

2.5 MASS TRANSPORT TO THE ELECTRODE SURFACE

We have considered the thermodynamics of electrochemical processes, studied the kinetics of electrode processes, and investigated the effects of the electrical double layer on kinetic parameters. An understanding of these relationships is an important ingredient in the repertoire of the researcher of battery technology. Another very important area of study which has major impact on battery research is the evaluation of mass transport processes to and from electrode surfaces.

Mass transport to or from an electrode can occur by essentially three processes. These are (1) convection and stirring, (2) electrical migration in an electric potential gradient, and (3) diffusion in a concentration gradient. The first of these processes can be handled relatively easily both mathematically and experimentally. If stirring is required, flow systems can be established, while if complete stagnation is an experimental necessity, this can also be imposed by careful design. In most cases, if stirring and convection are present or imposed, they can be handled mathematically.

The migration component of mass transport can also be handled experimentally (reduced to close to zero or occasionally increased in special cases) and described mathematically, provided certain parameters such as transport number or migration current are known. Migration of electroactive species in an electric potential gradient can be reduced to zero by addition of an excess of inert "supporting electrolyte" which effectively reduces the potential gradient to zero and thus eliminates the electric field which produces migration. Enhancement of migration is more difficult. This requires that the electric field be increased so that movement of charged species is increased. Electrode geometry design can increase migration slightly by altering electrode curvature. Fields at convex surfaces are greater than those at flat or concave surfaces, and thus migration is enhanced at convex curved surfaces.

The third process, diffusion in a concentration gradient, is the most important of the three processes and is the one which typically is dominant in mass transport in batteries. The analysis of diffusion uses the basic equation due to Fick[14] which defines the flux of material crossing a plane at distance x and time t. The flux is proportional to the concentration gradient and is represented by the following expression:

$$q = D\frac{\delta C}{\delta x} \tag{32}$$

where q = flux
  D = diffusion coefficient
  C = concentration
The rate of change of concentration with time is defined by

$$\frac{\delta C}{\delta t} = D\frac{\delta^2 C}{\delta x^2} \tag{33}$$

This expression is referred to as Fick's second law of diffusion. Solution of Eqs. (32) and (33) requires that boundary conditions be imposed. These are chosen according to the electrode's expected "discharge" regime dictated by battery performance or boundary conditions imposed by the electroanalytical technique.[15] Several of the electroanalytical techniques are discussed in Sec. 2.6.

For application directly to battery technology, the three modes of mass transport have meaningful significance. Convective and stirring processes can be employed to provide a flow of electroactive species to reaction sites. Examples of the utilization of stirring and flow processes in batteries are the circulating zinc-air system (see Part 4, Chap. 30), the vibrating zinc electrode (Part 3, Chap. 19), and the zinc chlorine hydrate battery (Part 4, Chap. 28). In some types of advanced lead-acid batteries, circulation of acid is provided to improve utilization of the active materials in the battery plates.

Migration effects are in some cases detrimental to battery performance, in particular those caused by enhanced electric fields (potential gradients) around sites of convex curvature. Increased migration at these sites tends to produce dendrite formations which eventually lead to shorts and battery failure.

Diffusion processes are typically the mass transfer processes operative in the majority of battery systems where the transport of species to and from reaction sites is required for maintenance of current flow. Enhancement and improvement of diffusion processes are an appropriate direction of research to follow to improve battery performance parameters. Equation (32) may be written in an approximate, yet more practical, form, remembering that $i = nFq$, where q is the flux through a plane of unit area. Thus

$$i_l = nF\frac{DA(C_B - C_E)}{\delta} \tag{34}$$

where C_B is the bulk concentration of electroactive species, C_E the concentration at the electrode, A the electrode area, and δ is the boundary layer thickness, i.e., the layer at the electrode surface in which the majority of the concentration gradient is concentrated (see Fig. 2.14). This expression defines the maximum diffusion current that can be sustained in solution under a given set of conditions. It tells us that, to increase i_l, one needs to increase the bulk concentration, increase the electrode area, or increase the diffusion coefficient. In the design of a battery, an understanding of the implication of this expression is important. Specific cases can be analyzed quickly by applying Eq. (34), and parameters such as discharge rate and likely power densities of new systems may be estimated.

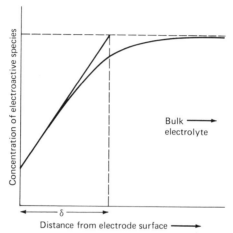

FIG. 2.14 Boundary layer thickness at an electrode surface.

2.6 ELECTROANALYTICAL TECHNIQUES

Many steady-state and impulse electroanalytical techniques are available to the experimentalist to determine electrochemical parameters and assist in both improving existing battery systems and evaluating couples as candidates for new batteries.[16] A few of these techniques are described below.

2.6.1 Cyclic Voltammetry

Of the electroanalytical techniques, cyclic voltammetry (or linear sweep voltammetry as it is sometimes known) is probably one of the more versatile techniques available to the electrochemist. The derivation of the various forms of cyclic voltammetry can be traced to the initial studies of Matheson and Nichols[17] and Randles.[18] Essentially, the technique applies a linearly changing voltage (ramp voltage) to an electrode. The scan of voltage might be ± 2 V from an appropriate rest potential such that most electrode reactions would be encompassed. Commercially available instrumentation provides voltage scans as wide as ± 5 V.

To describe the principles behind cyclic voltammetry, for convenience let us restate Eq. (7) which describes the reversible reduction of an oxidized species O:

$$O + ne \rightleftarrows R \tag{7}$$

In cyclic voltammetry the initial potential sweep is represented by

$$E = E_i - vt \tag{35}$$

where E_i = initial potential
$\quad\; t$ = time
$\quad\; v$ = potential change, V/s
The reverse sweep of the cycle is defined by

$$E = E_i + v't \tag{36}$$

where v' is often the same value as v. By combining Eq. (36) with the appropriate form of the Nernst equation [Eq. (4)] and with Fick's laws of diffusion [Eqs. (32) and (33)], an expression can be derived which describes the flux of species to the electrode surface. This expression is a complex differential equation and can be solved by the summation of an integral in small successive increments.[19-21]

As the applied voltage approaches that of the reversible potential for the electrode process, a small current flows, the magnitude of which increases rapidly but later becomes limited at a potential slightly beyond the standard potential by the subsequent depletion of reactants. This depletion of reactant establishes concentration profiles which spread out into the solution as shown in Fig. 2.15. As the concentration profiles extend into solution, the rate of diffusive transport at the electrode surface decreases and with it the observed current. The current is thus seen to pass through a well-defined maximum as shown in Fig. 2.16. The peak current of the reversible reduction, Eq. (7), is defined by

$$i_p = \frac{0.447 F^{3/2} A n^{3/2} D^{1/2} C_O v^{1/2}}{R^{1/2} T^{1/2}} \tag{37}$$

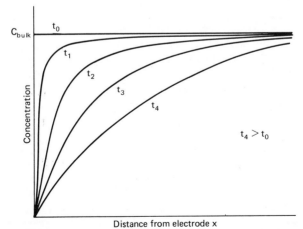

FIG. 2.15   Concentration profiles for the reduction of a species in cyclic voltammetry.

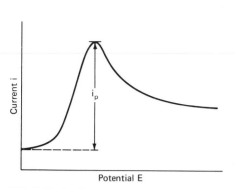

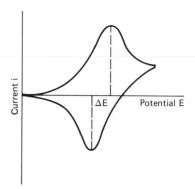

FIG. 2.16 Cyclic voltammetry peak current for the reversible reduction of an electroactive species.

FIG. 2.17 Cyclic voltammogram of a reversible, diffusion-controlled process.

The symbols have the same identity as before while i_p is the peak current and A is the electrode area. It may be noted that the value of the constant varies slightly from one text or publication to another. This is because, as previously mentioned, the derivation of peak current height is performed numerically.

A word of caution is due regarding the interpretation of the value of the peak current. It will be remembered from the discussion on the effects of the electrical double layer on electrode kinetics that there is a capacitance effect at an electrode-electrolyte interface. Consequently, the "true" electrode potential is modified by the capacitance effect as it is also by the ohmic resistance of the solution. Equation (35) should really be written in a form which describes these two components. Equation (38) shows such a modification:

$$E = E_i - vt + r(i_f + i_c) \tag{38}$$

where r = cell resistance
   i_f = faradic current
   i_c = capacity current

At small values of voltage sweep rate, typically below 1 V/s, the capacity effects are small and can in most cases be ignored. At greater values of sweep rate, a correction needs to be applied to interpretations of i_p as described by Nicholson and Shain.[22] With regard to the correction for ohmic drop in solution, typically this can be handled adequately by careful cell design and positive feedback compensation circuitry in the electronic instrumentation.

Cyclic voltammetry provides both qualitative and quantitative information on electrode processes. A reversible, diffusion-controlled reaction such as presented by Eq. (7) exhibits an approximately symmetrical pair of current peaks as shown in Fig. 2.17. The voltage separation $\triangle E$ of these peaks is

$$\triangle E = \frac{2.3RT}{nF} \tag{39}$$

and the value is independent of voltage sweep rate. In the case of the electrodeposition of an insoluble film which can be, subsequently, reversibly reoxidized and which is not governed by diffusion to and from the electrode surface, the value of $\triangle E$ is considerably less than that given by Eq. (39) as shown in Fig. 2.18. In the ideal case, the value of $\triangle E$ for this system is close to zero. For quasi-reversible processes, the current peaks are separated more, and the shape of the peak is less sharp at its summit and is generally more rounded as shown in Fig. 2.19. The voltage of the current peak is dependent on the voltage sweep rate, and the voltage separation is much greater than that given by Eq. (39). A completely irreversible electrode process produces a single peak as shown in Fig. 2.20. Again the voltage of the peak current is sweep-rate dependent, and, in the case of an irreversible charge-transfer process for which the back reaction is negligible, the rate constant and transfer coefficient can be determined. With negligible back reaction, the expression for peak current as a function of peak potential is[22]

$$i_p = 0.22nFC_o k_{app} \exp[-\alpha \frac{nF}{RT} (E_m - E^0)] \tag{40}$$

where the symbols are as before and E_m is the potential of the current peak. A plot of E_m vs. $\ln i_p$, for different values of concentration, gives a slope which yields the transfer coefficient α and an intercept which yields the apparent rate constant k_{app}. Though both α and k_{app} can be obtained by analyzing E_m as a function of voltage sweep

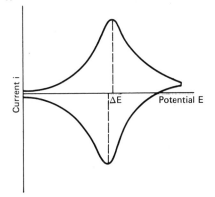

FIG. 2.18   Cyclic voltammogram of the electroreduction and reoxidation of a deposited, insoluble film.

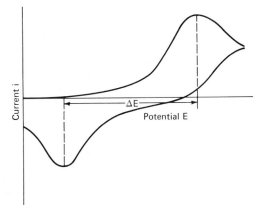

FIG. 2.19   Cyclic voltammogram of a quasi-reversible process.

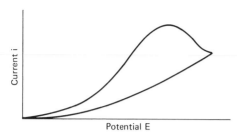

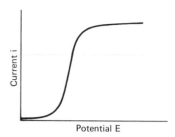

FIG. 2.20   Cyclic voltammogram of an irreversible process.

FIG. 2.21   Cyclic voltammogram of the electroreduction of a species which is controlled by a preceding chemical reaction.

rate v by a reiterative calculation, analysis by Eq. (40) (which is independent of v) is much more convenient.

For more complex electrode processes, cyclic voltammetric traces become more complicated to analyze. An example of one such case is the electroreduction of a species controlled by a preceding chemical reaction. The shape of the trace for this process is shown in Fig. 2.21. The species is formed at a constant rate at the electrode surface, and, provided the diffusion of the inactive component is more rapid than its transformation to the active form, it cannot be depleted from the electrode surface. The "peak" current is thus independent of potential and resembles a plateau.

Cyclic voltammograms of electrochemical systems can often be much more complicated than the traces presented here. It often takes some ingenuity and persistence to determine which peaks belong to which species or processes. Despite these minor drawbacks, the cyclic voltammetric technique is a versatile, and relatively sensitive, electroanalytical method appropriate to the analysis of systems of interest to battery development. The technique will identify reversible couples (needed for secondary batteries), it provides a method for measuring the rate constant and transfer coefficient of an electrode process (a fast rate constant indicates a process of possible interest for battery development), and it can provide a tool to help unravel complex electrochemical systems.

2.6.2 Chronopotentiometry

Chronopotentiometry involves the study of voltage transients at an electrode upon which is imposed a constant current. It is sometimes alternately known as galvanostatic voltammetry. In this technique, a constant current is applied to an electrode, and its voltage response indicates the changes in electrode processes occurring at its interface. Consider, for example, the reduction of a species O as expressed by Eq. (7). As the constant current is passed through the system, the concentration of O in the vicinity of the electrode surface begins to decrease. As a result of this depletion, O diffuses from the bulk solution into the depleted layer, and a concentration gradient grows out from the electrode surface into the solution. As the electrode process continues, the concentration profile extends further into the bulk solution as shown in Fig. 2.22. When the surface concentration of O falls to zero (at time t_6 in Fig. 2.22), the electrode process can no longer be supported by electroreduction of O. An additional cathodic reaction must be brought into play and an abrupt change in potential occurs. The period of time between the commencement of electroreduction and the sudden change in potential is called the transition time τ. The transition time for electroreduction of a species in the

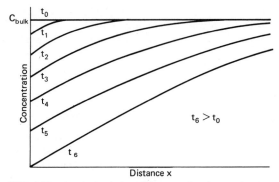

FIG. 2.22 Concentration profiles extending into bulk solution during constant current depletion of species at the electrode surface.

presence of excess supporting electrolyte was first quantified by Sand[23] who showed that the transition time τ was related to the diffusion coefficient of the electroactive species:

$$\tau^{1/2} = \frac{\pi^{1/2} n F C_O D^{1/2}}{2i} \tag{41}$$

where D is the diffusion coefficient of species O and the other symbols have their usual meanings.

Unlike cyclic voltammetry, the solution of Fick's diffusion equations [Eqs. (32) and (33)] for chronopotentiometry can be obtained as an exact expression by applying appropriate boundary conditions. For a reversible reduction of an electroactive species [Eq. (7)], the potential-time relationship has been derived by Delahay[24] for the case where O and R are free to diffuse to and from the electrode surface, including the case where R diffuses into a mercury electrode:

$$E = E_{\tau/4} + \frac{RT}{nF} \ln \frac{\tau^{1/2} - t^{1/2}}{t^{1/2}} \tag{42}$$

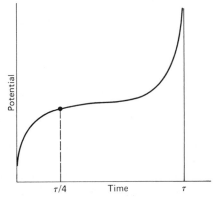

FIG. 2.23 Potential curve at constant current for reversible reduction of an electroactive species.

In this equation $E_{\tau/4}$ is the potential at the quarter-transition time (the same as the polarographic half-wave potential in the case of a mercury electrode) and t is any time from zero to the transition time. The trace represented by this expression is shown in Fig. 2.23.

The corresponding expression for an irreversible process[25] with one rate-determining step is

$$E = \frac{RT}{\alpha n_a F} \ln \frac{nFC_0 k_{app}}{i} + \frac{RT}{\alpha n_a F} \ln \left[1 - \left(\frac{t}{\tau} \right)^{1/2} \right] \tag{43}$$

where k_{app} is the apparent rate constant, n_a is the number of electrons involved in the rate-determining step (often the same as n, the overall number of electrons involved in the total reaction), and the other symbols have their usual meanings. A plot of the logarithmic term vs. potential yields both the transfer coefficient and the apparent rate constant.

In a practical system, the chronopotentiogram is often less than ideal in the shape of the potential trace. To accommodate variations in chronopotentiometric traces, measurement of the transition time can be assisted by use of a construction technique shown in Fig. 2.24. The transition time is measured at the potential of $E_{\tau/4}$.

To analyze two or more independent reactions separated by a potential sufficient to define individual transition times, the situation is slightly more complicated than with cyclic voltammetry. Analysis of the transition time of the reduction of the nth species has been derived[26,27] and is reproduced below:

$$(\tau_1 + \tau_2 + \ldots + \tau_n)^{1/2} - (\tau_1 + \tau_2 + \ldots + \tau_{n-1})^{1/2} = \frac{\pi^{1/2} n_n FD_n^{1/2} C_n}{2i} \tag{44}$$

As can be seen, this expression is somewhat cumbersome.

An advantage of the technique is that it can conveniently be used to evaluate systems with high resistance. The trace conveniently displays segments due to the IR component, the charging of the double layer, and the onset of the faradic process. Figure 2.25 shows these different features of the chronopotentiogram of solutions with significant

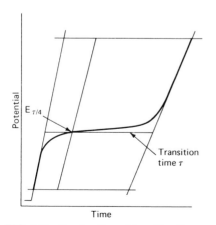

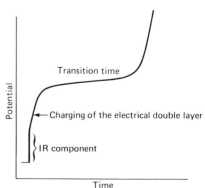

FIG. 2.24 Construction of transition time τ for a chronopotentiogram.

FIG. 2.25 Chronopotentiogram of a system with significant solution resistance.

resistance. If the solution is also one which does not contain excess supporting electrolyte to suppress the migration current, it is possible to describe the transition time of an electroreduction process in terms of the transport number of the electroactive species[28,29]:

$$\tau^{1/2} = \frac{\pi^{1/2} n F C_O D_s^{1/2}}{2i(1 - t_O)} \tag{45}$$

where D_s is the diffusion coefficient of the salt (not the ion) and t_O is the transport number of the electroactive species. This expression can be useful in battery research since many battery systems do not have supporting electrolyte.

2.6.3 AC Techniques (Impedance Methods)

The two electroanalytical techniques, one in which the measured value was current during imposition of a potential scan and the other a potential response under an imposed constant current, owe their electrical response to the change in impedance at the electrode-electrolyte interface. A more direct technique for studying electrode processes is to measure the change in electrical impedance of an electrode by an ac impedance bridge method. To relate the impedance of the electrode-electrolyte interface to electrochemical parameters, it is necessary to establish an equivalent circuit to represent the dynamic characteristics of the interface.

Establishment of equivalent circuits attempting to describe electrode processes dates to the turn of the century with the derivation by Warburg[30] of an equation for the faradic impedance of diffusion-controlled processes at a planar electrode. In fact, the impedance of the diffusion process is often referred to as the "Warburg impedance." In 1903, Krüger[31] realized that the double-layer capacity had influence on the impedance of the electrode interface and derived an expression for its effect. Much later, the technique was developed and adapted to the study of electrode kinetics,[32-36] with emphasis on the charge-transfer process. The technique has been found to be extremely useful in evaluating several different electrode processes at a planar electrode and has been analyzed by several authors. The charge-transfer step has been considered and analyzed by Randles,[33] Grahame,[37] Delahay,[38] Baticle and Perdu,[39,40] and Sluyters-Rehbach and Sluyters,[41] while Gerischer[42,43] and Barker[44] considered coupled homogeneous and heterogeneous chemical reactions. Adsorption processes can be studied by this technique, and the description of the method used to measure adsorption has been given by Laitinen and Randles,[45] Llopis et al.,[46] Senda and Delahay,[47] Sluyters-Rehbach et al.,[48] Timmer et al.,[49] Barker,[50] and Holub et al.[51]

The expressions defining the various electrochemical parameters are relatively straightforward to derive but are complex in format, in particular when capacity effects are considered in the presence of strongly adsorbed electroactive species. Derivation of expressions relevant to the previously mentioned processes will not be given here; only one of the more applicable analyses, and the one which is the most straightforward to handle, will be given. The problem is that the majority of the processes require a transmission-line type of analysis to give a closed-system solution for a nonplanar electrode. We shall consider the system without adsorption and without complications of homogeneous series reactions where the impedance can be represented by a circuit diagram shown in Fig. 2.26. In this analysis due to Sluyters and his coworkers, the electrode process is evaluated by the analytical technique of complex plane analysis.

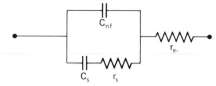

FIG. 2.26 Equivalent circuit for a cell where the cell impedance is kinetically controlled and is localized at the working electrode by using a large, unpolarized counter electrode. C_{nf} is the nonfaradic capacitance, C_s and r_s are the faradic components of the impedance, and r_e is the electrolyte resistance.

In this technique, the capacitive component $1/\omega C$ is plotted vs. the resistive component of the cell. Figure 2.27 shows a typical plot which displays kinetic control only. The interdependence of the capacitive and resistive components yields a semicircle with the top of the semicircle yielding the charge-transfer resistance r_{ct}:

$$\omega_m = \frac{1}{r_{ct}C_{nf}} \tag{46}$$

where C_{nf} is the nonfaradic capacitance and $r_{ct} = RT/nF(i_O)_{app}$. The intercept of the semicircle and the abscissa gives the electrolyte resistance r_e and again r_{ct}. If the electrode process is governed by both kinetic and diffusion control, a somewhat different plot is observed as shown in Fig. 2.28. In this plot, the linear portion of the curve corresponds to the process where diffusion control is predominant. From this plot, in addition to the previously mentioned measurement, the extrapolated linear portion gives a somewhat complex expression involving the diffusion coefficients of the oxidized and reduced species:

$$\text{Intercept} = r_e + 2S^2C_{nf} \tag{47}$$

where
$$S = \frac{RTL}{n^2F^2\sqrt{2}} \tag{48}$$

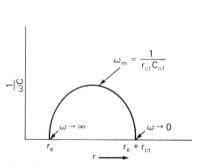

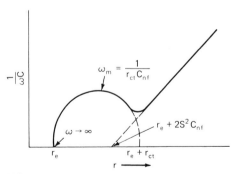

FIG. 2.27 Complex plane analysis of cell impedance for a charge-transfer process with kinetic control at a planar electrode.

FIG. 2.28 Complex plane analysis of cell impedance for a charge-transfer process with both kinetic and diffusion control at a planar electrode.

and
$$L = \frac{1}{C_O\sqrt{D_O}} + \frac{1}{C_R\sqrt{D_R}} \tag{49}$$

where D_O is the diffusion coefficient of the oxidized species and D_R is the diffusion coefficient of the reduced species. Treatment of this system assumes that we can write the equivalent circuit of kinetic and diffusion control as shown in Fig. 2.29 where the diffusion component of the impedance is given by the Warburg impedance W. It should also be noted that the derivation applies to a planar electrode only. Electrodes with more complex geometries such as porous electrodes require a transmission-line analysis.

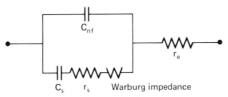

FIG. 2.29   Equivalent circuit for an electrode process which is limited by both charge-transfer kinetics and diffusion processes. The diffusion portion of the impedance is represented by W, the Warburg impedance, and the other circuit components are the same as in Fig. 2.26.

The ac impedance technique coupled to the complex plane method of analysis is a powerful electroanalytical tool to determine a variety of electrochemical parameters. To make the measurements, instrumentation is somewhat more complex than with other techniques. It requires a Wheatstone bridge arrangement with series capacitance and resistance in the comparison arm, a tuned amplifier/detector, and an oscillator with an isolation transformer. A Wagner* ground is required to maintain bridge sensitivity, and a suitably large inductance should be incorporated in the electrode polarization circuit to prevent interference from the low impedance of this ancillary circuitry.

This technique has found considerable acceptance in measuring fast rate constants, and it has been extended to include faradic rectification and second-harmonic generation and detection. These advanced techniques extend the range of ac impedance measurements to the evaluation of charge-transfer rate constants with values of 10 cm/s or greater. Unfortunately the description of these techniques is beyond the scope of this chapter but can be found in advanced texts on electroanalytical methods.[16]

2.6.4 Polarography

The technique of polarographic analysis is probably one of the most used and the electroanalytical technique with the longest history of routine use. Polarography [54–57] utilizes a dropping mercury electrode (DME) as the sensing electrode to which is

---

*A Wagner ground maintains a corner of a bridge at ground potential without actually connecting it directly to ground. This helps to eliminate stray capacitances to ground and maintains bridge sensitivity over a wide range of frequency.

applied a slowly changing potential, usually increasing in the negative direction, such that for each mercury drop, the potential remains essentially constant. The output current from the DME is typically displayed on a chart recorder, and the current magnitude is a measure of the concentration of electroactive species in solution. Because of the periodic nature of the DME, the current-measuring circuit includes a damping capacitor with a time constant such that the oscillations in current due to the charging current of the DME are minimized. Figure 2.30 shows a typical polarogram of an electroactive species where i is the polarographic mean diffusion current. The expression relating concentration of electroactive species to the mean diffusion current is[52,53]

$$i = 607nD^{1/2}Cm^{2/3}t_d^{1/6} \tag{50}$$

where i = mean diffusion current, μA
  n = number of electrons involved in overall electrode process
  D = diffusion coefficient of electroactive species
  C = concentration of electroactive species, mmol/L
  m = mercury flow rate, mg/s
  t_d = mercury drop time, s

Analysis of solutions usually is performed by calibration with known standards rather than by using Eq. (50). Occasionally the polarogram of a solution may show a current spike at the beginning of the current plateau. This effect is due to a streaming phenomenon around the DME and can be suppressed by the addition of a *small* quantity of surface-active component such as gelatin or Triton X-100.

Solutions containing several electroactive species can be conveniently analyzed from one polarogram, provided the potential separation, the half-wave potential, is sufficient to distinguish the current plateaus. Figure 2.31 shows a polarogram of an aqueous solution containing several species in a potassium chloride supporting electrolyte. The polarographic method is often found as a standard instrumental technique in electrochemical laboratories.

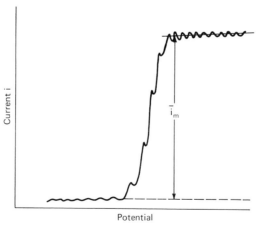

Potential

FIG. 2.30 Polarogram of an electroactive species where i is the polarographic mean diffusion current.

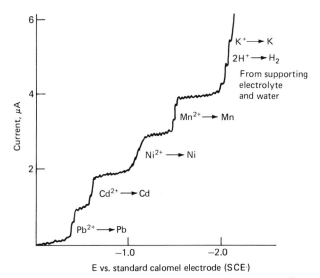

FIG. 2.31   Polarogram of the reduction of several electroactive species in a solution of $0.1M$ potassium chloride as supporting electrolyte.

2.6.5 Electrodes

Several electroanalytical techniques have been discussed in the preceding sections, but little has been mentioned about electrodes or electrode geometries used in the various measurements. This section deals with electrodes and electrode systems.

A typical electroanalytical cell has an indicator electrode (sometimes called a working electrode), a counter electrode, and, in most cases, a reference electrode. The counter electrode is a large, inert, relatively unpolarized electrode situated a suitable distance from the indicator electrode and sometimes separated from it by a sintered glass disk or some other porous medium which will allow ionic conduction but prevent gross mixing of solutions surrounding the counter and working electrodes. The reference electrode provides an unpolarized reference potential against which the potentials of the indicator electrode can be measured. Reference electrodes are usually constructed in a separate vessel and are connected to the cell via a salt bridge composed of an electrolyte with a common anion and/or cation with the supporting electrolyte solution in the cell. The concentration of the salt bridge electrolyte is also chosen to be approximately the same as the supporting electrolyte. This minimizes liquid junction potentials which are established between solutions of different composition and/or concentration. Contamination of the cell solution with material from the salt bridge is minimized by restricting the orifice of the salt bridge by a Luggin capillary. Typical reference electrode systems are $Ag/AgCl$, Hg/Hg_2Cl_2, and Hg/Hg_2SO_4. For a comprehensive treatise on the subject of reference electrodes, the reader is referred to an excellent text by Ives and Janz.[58]

Indicator electrodes have been designed and fabricated in many different geometrical shapes to provide various performance characteristics and to accommodate the special requirements of some of the electroanalytical techniques. The simplest of the indicator electrode types is the planar electrode. This can be simply a "flag" elec-

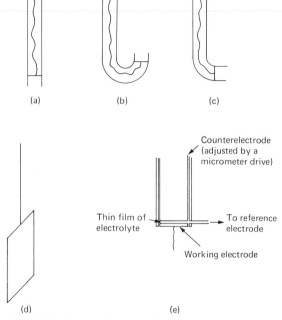

(a)   (b)   (c)

(d)   (e)

FIG. 2.32 Various planar working electrodes, some with shielding to maintain laminar diffusion of electroactive species to the surface.

trode, or it can be shielded to provide for linear diffusion during long electrolysis times. Figure 2.32 shows a selection of planar types of indicator electrode. Included in this selection is a thin-film cell (Fig. 2.32e) in which two planar electrodes confine a small quantity of solution within a small gap between the working and counter electrodes. The counter electrode is adjusted by a micrometer drive, and the reference salt bridge is located, as shown, as a radial extension of the cell. This cell has the advantage that only a small quantity of solution is required, and in many cases it can be used as a coulometer to determine n, the overall number of electrons involved in an electrode reaction.

Indicator electrodes not requiring a planar surface can be fabricated in many different geometries. For simplicity and convenience of construction, a wire electrode can be sealed into an appropriate glass support and used with little more than a thorough cleaning. For more symmetrical geometry, a spherical bead electrode can often be formed by fusing a thin wire of metal and shriveling it into a bead. In some cases it is possible to form a single crystal of the metal. Figure 2.33 shows various electrodes with nonplanar electrode-electrolyte interfaces.

Several electroanalytical methods require special electrodes or electrode systems. A classic example of this is the DME, first used extensively in polarography (Sec. 2.6.4). The DME is established by flowing ultrapure mercury through a precision capillary

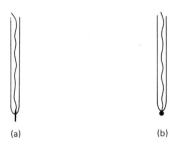

(a)   (b)

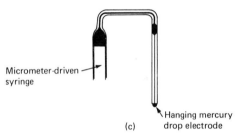

Micrometer-driven
syringe

Hanging mercury
(c)   drop electrode

FIG. 2.33   Working electrodes with nonplanar surfaces.

from a mercury reservoir connected to the capillary by a flexible tube. Mercury flow rates and drop times are adjusted by altering the height of the mercury head so that mercury drops detach every 2 to 7 s with typical polarographic values being in the range 3 to 5 s. Figure 2.34 shows a schematic representation of a DME system. In addition to use of the DME in polarography, the electrode finds application when a fresh, clean mercury surface is necessary for reproducible electroanalytical determinations. In this respect, the DME can be used as a "stationary" spherical mercury electrode suitably "frozen" in time by a timing circuit initiated by the voltage pulse at the birth of the drop and triggered at a time short of the natural drop time of the DME. With a suitably short transient impulse (voltage, current, etc.), an output response can be obtained from what is essentially a hanging mercury drop electrode.

Another special type of electrode is the rotating-disk electrode or rotating-ring-disk electrode. Figure 2.35 shows both rotating-disk and ring-disk electrodes. In both cases, rotation of the electrode establishes a flow pattern which maintains a relatively constant diffusion layer thickness during reduction or oxidation of an electroactive species and provides a means of hydrodynamically varying the rate at which electroactive species are brought to the outer surface of the diffusion layer. In the ring-disk electrode, the ring can be used either as a "guard ring" to ensure laminar diffusion to the disk or as an independent working electrode to monitor species generated from the disk electrode and/or the generation of transient intermediate species. Details of the derivation of current-voltage curves as a function of rotation speed have been given by Riddiford[59] for the disk electrode, while more detail is given by Yeager and Kuta[60] and Pleskov and

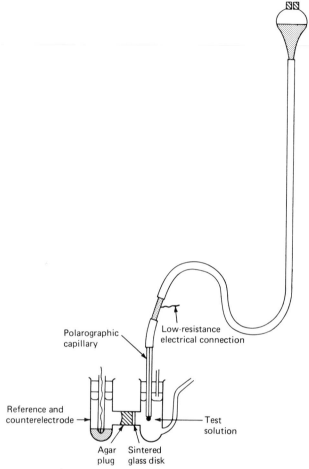

FIG. 2.34 Dropping mercury electrode complete with polarographic H cell.

Filinovskii[61] on the ring-disk electrode, with emphasis being given to detection of transient species by the latter authors. Details of the current-voltage characteristics will not be given here; it is sufficient to say that limiting currents for a fixed potential can be obtained by varying the rotation speed of the electrode from which potential dependent rate constants can be determined. Ring currents may be determined either at a fixed potential such as to reverse the process leading to the formation of the species of interest or scanning through a range of potentials to detect a variety of species electroactive over a range of potentials. Quantitative measurements with the ring of the ring-disk electrode are difficult to interpret because the capture fraction N of the ring is difficult to evaluate.[62]

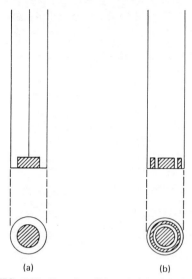

(a)   (b)

FIG. 2.35   Rotating disk and ring-disk electrodes. (a) Disk electrode; (b) ring-disk electrode.

REFERENCES

1. A. Volta, *Phil. Trans. R. Soc.* **90**:403 (1800).
2. N. Gautherot, *J. Phys.* **56**:429 (1802).
3. J. W. Ritter, *J. Phys.* **57**:345 (1803).
4. W. R. Grove, *Phil. Mag.* **14**:447 (1839); **21**:417 (1842).
5. G. Planté, *Compte Rendu* **50**:640 (1860).
6. L. R. McCoy, "Advanced Electrochemical Energy Systems," in P.A. Rock (ed.), *Special Topics in Electrochemistry,* Elsevier, Amsterdam, 1977, chap. 1.
7. J. O'M. Bockris and A. K. N. Reddy, *Modern Electrochemistry,* Plenum, New York, 1970, vol. 2, p. 644.
8. H. H. Bauer, *J. Electroanal. Chem.* **16**:419 (1968).
9. J. Horiuti and M. Polanyi, *Acta Physicochim. U.S.S.R.* **2**:505 (1935).
10. J. O'M. Bockris and A. K. N. Reddy, op. cit., p. 918.
11. P. Delahay, *Double Layer and Electrode Kinetics,* Interscience, New York, 1965.
12. J. O'M. Bockris and A. K. N. Reddy, op. cit., p. 742.
13. H. H. Bauer, *Electrodics,* Wiley, New York, 1972, p. 54, table 3.2.
14. A. Fick, *Ann. Phys.* **94**:59 (1855).
15. P. Delahay, *New Instrumental Methods in Electrochemistry,* Interscience, New York, 1954.
16. E. B. Yeager and J. Kuta, "Techniques for the Study of Electrode Processes," in *Physical Chemistry,* vol. IXA: *Electrochemistry,* Academic, New York, 1970, p. 346.
17. L. A. Matheson and N. Nichols, *J. Electrochem. Soc.* **73**:193 (1938).

18. J. E. B. Randles, *Trans. Faraday Soc.* **44**:327 (1948).

19. A. Sevcik, *Coll. Czech. Chem. Comm.* **13**:349 (1948).

20. T. Berzins and P. Delahay, *J. Am. Chem. Soc.* **75**:555 (1953).

21. P. Delahay, *J. Am. Chem. Soc.* **75**:1190 (1953).

22. R. S. Nicholson and I. Shain, *Anal. Chem.* **36**:706 (1964).

23. H. J. S. Sand, *Phil. Mag.* **1**:45 (1901).

24. Delahay, *New Instrumental Methods in Electrochemistry,* p. 180.

25. P. Delahay and T. Berzins, *J. Am. Chem. Soc.* **75**:2486 (1953).

26. C. N. Reilley, G. W. Everett, and R. H. Johns, *Anal. Chem.* **27**:483 (1955).

27. T. Kambara and I. Tachi, *J. Phys. Chem.* **61**:1405 (1957).

28. M. D. Morris and J. J. Lingane, *J. Electroanal. Chem.* **6**:300 (1963).

29. J. Broadhead and G. J. Hills, *J. Electroanal. Chem.* **13**:354 (1967).

30. E. Warburg, *Ann. Physik* **67**:493 (1899).

31. F. Krüger, *Z. Phys. Chem.* **45**:1 (1903).

32. P. Dolin and B. V. Ershler, *Acta Physicochim. U.S.S.R.* **13**:747 (1940).

33. J. E. B. Randles, *Disc. Faraday Soc.* **1**:11 (1947).

34. B. V. Ershler, *Disc. Faraday Soc.* **1**:269 (1947).

35. B. V. Ershler, *Zh. Fiz. Khim.* **22**:683 (1948).

36. K. Rozental and B. V. Ershler, *Zh. Fiz. Khim.* **22**:1344 (1948).

37. D. C. Grahame, *J. Electrochem. Soc.* **99**:370C.

38. Delahay, *New Instrumental Methods in Electrochemistry,* p. 146.

39. A. M. Baticle and F. Perdu, *J. Electroanal. Chem.* **12**:15 (1966).

40. A. M. Baticle and F. Perdu, op. cit., **13**:364 (1967).

41. M. Sluyters-Rehbach and J. H. Sluyters, *Rec. Trav. Chim.* **82**:525, 535 (1963).

42. H. Gerischer, *Z. Phys. Chem.* **198**:286 (1951).

43. H. Gerischer, op. cit., **201**:55 (1952).

44. G. C. Barker, in E.B. Yeager (ed.), *Trans. Symp. Electrode Processes,* Philadelphia, 1959, Wiley, New York, p. 325.

45. H. A. Laitinen and J. E. B. Randles, *Trans. Faraday Soc.* **51**:54 (1955).

46. J. Llopis, J. Fernandez-Biarge, and M. Perez-Fernandez, *Electrochim. Acta* **1**:130 (1959).

47. M. Senda and P. Delahay, *J. Phys. Chem.* **65**:1580 (1961).

48. M. Sluyters-Rehbach, B. Timmer, and J. H. Sluyters, *J. Electroanal. Chem.* **15**:151 (1967).

49. B. Timmer, M. Sluyters-Rehbach, and J. H. Sluyters, *J. Electroanal. Chem.* **15**:343 (1967).

50. G. C. Barker, op. cit., p. 325.

51. K. Holub, G. Tessari, and P. Delahay, *J. Phys. Chem.* **71**:2612 (1967).

52. D. Ilkovic, *Coll. Czech. Chem. Comm.* **6**:498 (1934).

53. D. Ilkovic, *J. Chim. Phys.* **35**:129 (1938).

54. I. M. Kolthoff and J. J. Lingane, *Polarography* 2d ed., Interscience, New York, 1952.

55. J. Heyrovsky and J. Kuta, *Principles of Polarography,* Czechoslovak Academy of Sciences, Prague, 1965.

56. L. Meites, *Polarographic Techniques,* 2d ed., Interscience, New York, 1965.

57. G. W. C. Milner, *The Principles and Applications of Polarography and Other Electroanalytical Process,* Longmans, London, 1957.

58. D. J. G. Ives and G. J. Janz, *Reference Electrodes, Theory and Practice,* Academic, New York, 1961.

59. A. C. Riddiford, "The Rotating Disc System," in P. Delahay and C. W. Tobias (eds.), *Advances in Electrochemistry and Electrochemical Engineering,* Interscience, New York, 1966, p. 47.

60. E. B. Yeager and J. Kuta, op. cit., p. 367.

61. Y. V. Pleskov and V. Y. Filinovskii, *The Rotating Disc Electrode,* H. S. Wroblowa and B. E. Conway (eds.), Consultants Bureau, New York, 1976, chap. 8.

62. W. J. Albery and S. Bruckenstein, *Trans. Faraday Soc.* **62**:1920, 1946 (1966).

BIBLIOGRAPHY

General

Bauer, H. H.: *Electrodics,* Wiley, New York, 1972.

Bockris, J. O'M., and A. K. N. Reddy: *Modern Electrochemistry,* Plenum, New York, 1970, vols. 1 and 2.

Conway, B. E.: *Theory and Principles of Electrode Processes,* Ronald Press, New York, 1965.

Sawyer, D. T., and J. L. Roberts: *Experimental Electrochemistry for Chemists,* Wiley, New York, 1974.

Transfer Coefficient

Bauer, H. H.: *J. Electroanal. Chem.,* **16**:419 (1968).

Electrical Double Layer

Delahay, P.: *Double Layer and Electrode Kinetics,* Interscience, New York, 1965.

Electroanalytical Techniques

Delahay, P.: *New Instrumental Methods in Electrochemistry,* Interscience, New York, 1954.

Yeager, E. B., and J. Kuta: "Techniques for the Study of Electrode Processes," in *Physical Chemistry,* vol. IXA: *Electrochemistry,* Academic, New York, 1970.

Polarography

Heyrovsky, J., and J. Kuta: *Principles of Polarography,* Czechoslovak Academy of Sciences, Prague, 1965.

Kolthoff, I. M., and J. J. Lingane: *Polarography,* 2d ed., Interscience, New York, 1952.

Meites, L.: *Polarographic Techniques,* 2d ed., Interscience, New York, 1965.

Milner, G. W. C.: *The Principles and Applications of Polarography and Other Electroanalytical Processes,* Longmans, London, 1957.

Reference Electrodes

Ives, D. J. G., and G. J. Janz: *Reference Electrodes, Theory and Practice,* Academic, New York, 1961.

Electrochemistry of the Elements

Bard, A. J. (ed.): *Encyclopedia of Electrochemistry of the Elements,* vols. I–XIII, Dekker, New York, 1979.

Organic Electrode Reactions

Meites, L., and P. Zuman: *Electrochemical Data,* Wiley, New York, 1974.

Considerations for Selection and Application of Batteries

by
David Linden

3.1 GENERAL CHARACTERISTICS

The many and varied requirements for battery power and the different environmental and electrical conditions under which they must operate necessitate the use of a number of different types of batteries and designs, each having superior performance under certain operational conditions. Although many advances have been made in battery technology in recent years, there is still no one "ideal" battery that gives optimum performance under all operating conditions. As a result, many electrochemical systems and battery types have been and are being investigated and promoted. However, a relatively small number have achieved wide popularity and large production and sales. The less conventional systems are typically used in military and industrial applications requiring the specific capabilities offered by the special batteries.

The "ideal" electrochemical cell or battery is obviously one that is inexpensive, has infinite energy, can handle all power levels, can operate over the full range of temperature and environmental conditions, has unlimited shelf life, and is completely safe and "consumer-proof." In practice, energy limitations do exist as materials are consumed during the discharge of the battery, shelf life is limited due to chemical reactions and physical changes that occur, albeit slowly in some cases, during storage, and temperature and discharge rate affect performance. The use of energetic component materials and special designs to achieve high energy and power densities increases costs and requires precautions during use to avoid electrical and physical abuse.

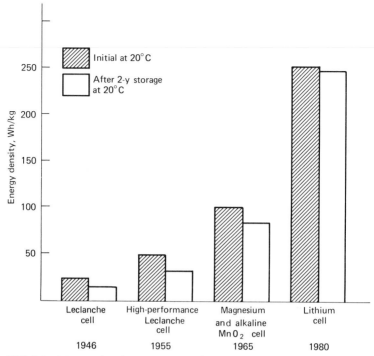

FIG. 3.1   Advances in primary battery performance.

Significant improvements and advances in performance have been made, as illustrated in Fig. 3.1, in the capacity and shelf life of batteries. The selection of the most effective battery and the proper use of this battery are important in an application to achieve optimum performance.

3.2 FACTORS AFFECTING BATTERY CAPACITY

In practice, only a fraction of the theoretical capacity of a battery is realized. This is due not only to the presence of nonreactive components (containers, separators, electrolyte) that add to the weight and volume of the battery, but also to many other factors that prevent the battery from performing at its theoretical level. Figure 3.2 shows the theoretical capacity of several major electrochemical systems (based on the active anode and cathode materials only), the theoretical capacity of a practical cell, and the actual capacity of these batteries when discharged at 20°C under typical discharge conditions.

Many factors influence the operational characteristics, capacity, and performance of a battery. The effect of these factors on battery performance is discussed in this section. It should be noted that these effects can be presented only as generalizations because of the many possible interactions and that the influence of each factor is usually greater under the more stringent operating conditions. For example, the effect of storage is more pronounced not only with high storage temperatures and long storage

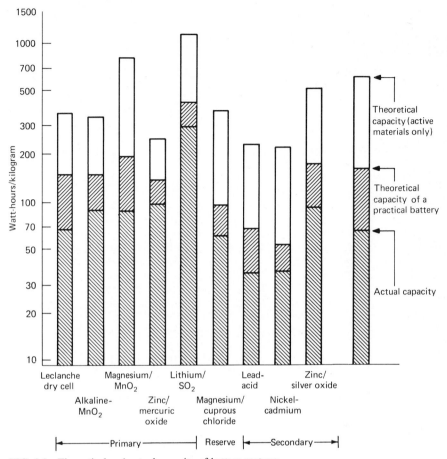

FIG. 3.2   Theoretical and actual capacity of battery systems.

periods, but also under more severe conditions of discharge following storage. After a given storage period, the observed loss of capacity (compared with a fresh battery) will be greater under heavy discharge loads than under light discharge loads. Similarly, the observed loss of capacity at low temperatures (compared with normal temperature discharges) will be greater at heavy than at light or moderate discharge loads. Specifications and standards for batteries list the specific test or operational conditions on which the standards are based because of their influence on battery performance.

Further, it should be noted that even within a given cell or battery design, there will be performance differences from manufacturer to manufacturer and between different versions of the same battery (e.g., standard, heavy-duty, premium). There are also performance variabilities within a production lot, and from production lot to production lot, that are inherent in any manufacturing process. The extent of the variability is dependent on the process controls as well as on the application and use of the battery. Performance variability is usually greater under the more stringent conditions of use. Manufacturer's data should be consulted to obtain specific performance characteristics.

3.2.1 Voltage Level

When a cell or battery is discharged, its voltage is lower than the theoretical voltage. The difference is caused by IR losses due to cell (and battery) resistance and polarization of the active materials during discharge. This is illustrated in Fig. 3.3. In the idealized case (curve 1), the discharge of the battery proceeds at the theoretical voltage until the active materials are consumed and the capacity fully utilized. The voltage then drops to zero. Under actual conditions, the discharge curve is similar to the other curves. The initial voltage is lower than theoretical and the voltage drops off as the discharge progresses. Curves 2 to 5 represent discharges at increasing discharge rates. As the discharge current is increased, the discharge voltage decreases and the discharge shows a more sloping profile.

Different references are made to the voltage of a cell or battery. The theoretical voltage is a function of the anode and the cathode materials, the composition of the electrolyte, and the temperature. The open-circuit voltage is the cell voltage under a no-load condition and is usually a close approximation of the theoretical voltage. The nominal voltage is one that is generally accepted as typical of the operating voltage of the cell, as, for example, 1.5 V for a zinc-carbon cell. The working voltage is more representative of the actual operating voltage of the cell under load and will be lower than the open-circuit voltage. The midpoint voltage is the average or central voltage during a discharge of the cell. Finally, the end voltage is designated as the end of the discharge and usually is the voltage above which most of the capacity of the cell has been delivered. The end voltage may also be dependent on the application requirements. With the lead-acid battery used as an example, the theoretical and open-circuit cell voltages are 2.1 V, the nominal voltage is 2.0 V, the working voltage is between 1.8 and 2.0 V, and the end voltage is typically 1.75 V on moderate and low drain discharges and 1.5 V for engine-cranking loads. On charge, the cell potential may range from 2.3 to 2.8 V.

The shape of the discharge curve can vary depending on the electrochemical system, constructional features, and other discharge conditions. Typical discharge curves are shown in Fig. 3.4. The flat discharge (curve 1) is representative of a discharge where the effect of change in reactants and reaction products is minimal until the active

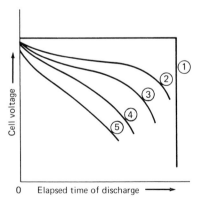

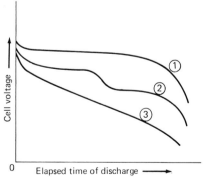

0   Elapsed time of discharge ⟶

0   Elapsed time of discharge ⟶

FIG. 3.3 Cell discharge characteristics—voltage levels.

FIG. 3.4 Cell discharge characteristics—voltage profiles.

materials are nearly exhausted; the plateau profile (curve 2) is representative of two-step discharge indicating a change in the reaction mechanism and potential of the active material(s); the sloping discharge (curve 3) is typical of a discharge where the composition of the active materials, reactants, internal resistance, etc., changes during the discharge to similarly affect the shape of the discharge curve.

3.2.2 Current Drain of Discharge

As the current drain of the battery is increased, the IR loss increases, the discharge is at a lower voltage, and the service life of the battery is reduced. Figure 3.3 also shows typical discharge curves as the current drain is changed. At extremely low current drains (curve 2) the discharge can approach the theoretical voltage and theoretical capacity, although on a very long discharge period, chemical deterioration during the discharge can become a factor and cause a reduction of capacity (Sec. 3.2.10). With increasing current drain (curves 3–5) the discharge voltage decreases, the slope of the discharge curve becomes more pronounced, and the service life is reduced.

A common method for indicating the discharge current for a battery is the C rate, expressed as

$$I = \frac{C_n}{N}$$

where I = discharge current, amperes
$\quad C_n$ = capacity rating of cell, Ah, at n-hour rate
$\quad N$ = hours of discharge
For example, the 0.1 C or $C/10$ discharge rate for a battery rated at 2 Ah at the 20-h rate is

$$I = \frac{C_{20}}{N} = \frac{2}{10} = 0.2 \text{ A}$$

It is to be noted that as the capacity of a battery generally decreases with increasing discharge current, this battery will operate for less than 10 h when discharged at the $C/10$ rate of 0.2 A.

Another method for indicating the discharge current is the "hourly" rate. This is the current at which the battery will discharge for the specified number of hours.

3.2.3 Type of Discharge (Continuous, Intermittent, etc.)

When a battery stands idle after a discharge, certain chemical and physical changes take place which can result in voltage recovery. Thus the voltage of a battery which has dropped during a heavy discharge will rise after a rest period, giving a sawtooth-shaped discharge as shown in Fig. 3.5. This improvement resulting from the intermittent discharge is generally greater at the higher current drains (as the battery has the opportunity to recover from polarization effects that are more pronounced at the heavier loads). In addition to current drain, the extent of recovery is dependent on many other factors such as the particular battery system and constructional features, discharge temperature, end voltage, and length of recovery period.

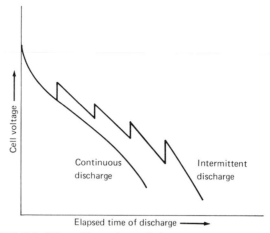

FIG. 3.5   Effect of intermittent discharge on battery capacity.

3.2.4 Type of Discharge (Constant Current, Constant Load, Constant Power)

A battery can be discharged under different modes depending on the equipment load. Three typical modes are constant resistance (the equipment resistance remains constant during discharge), constant current, and constant power (the current load on the battery increases as the voltage drops to maintain a constant power output, $I \times V$). Assuming that the discharge current is the same at the start of the discharge, the current will be different during the discharge under the different discharge modes as shown in Fig. 3.6a. The constant-resistance curve reflects the drop in the battery voltage. Figure 3.6b shows the voltage vs. time discharge curves for the three modes of discharge. Under the conditions shown, the service time is longest in the constant-resistance mode because the average discharge current is lowest under this mode. Figure 3.6c and 3.6d show the same relationships assuming the same average current during the discharge. Under these conditions, the service time is about the same, but the voltage regulation for the constant-resistance mode is best. The constant-power mode has the advantage, however, of providing the most uniform equipment performance throughout the life of the battery and, hence, makes most effective use of the battery's energy.

3.2.5 Temperature of Battery during Discharge

The temperature at which the battery is discharged has a pronounced effect on its service life (capacity) and voltage characteristics. This is due to the reduction in chemical activity and the increase in battery internal resistance at lower temperatures. This is illustrated in Fig. 3.7 which shows discharges at the same current drain but at progressively reduced temperatures (T_4 to T_1). Lowering of the discharge temperature will result in a reduction of capacity as well as an increase in the slope of the discharge curve. The specific characteristics as well as the discharge profile vary for each battery system, design, and discharge rate, but generally best performance is obtained between

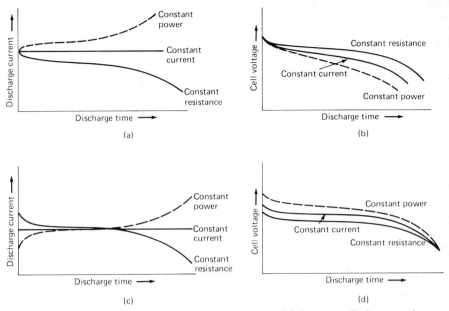

FIG. 3.6 Discharge profiles under different discharge modes. (*a*) Current profile (same starting current); (*b*) voltage profile (same starting current); (*c*) current profile (same average current); (*d*) voltage profile (same average current).

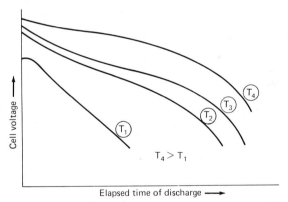

FIG. 3.7 Effect of temperature on battery capacity. (T_1 to T_4 represent increasing temperature.)

20 and 40°C. At higher temperatures, chemical deterioration may be rapid enough during the discharge to cause a loss of capacity, the extent again being dependent on the battery system and temperature.

Figures 3.8 and 3.9 summarize the effect of temperature and discharge rate on the cell's discharge voltage and capacity. As the discharge rate is increased, the cell voltage (for example, midpoint voltage) decreases; the rate of decrease is usually more rapid at the lower temperatures. Similarly, the cell's capacity falls off most rapidly with

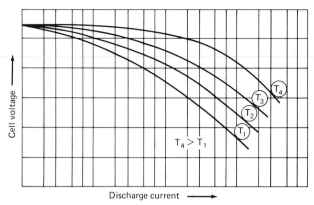

FIG. 3.8 Effect of discharge load on cell midpoint voltage at various temperatures (T_1 to T_4 represent increasing temperatures).

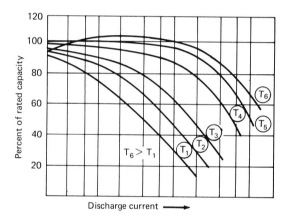

FIG. 3.9 Effect of discharge load on battery capacity at various temperatures (T_1 to T_5 represent increasing temperatures).

increasing discharge load and decreasing temperature. Curve T_6 of Fig. 3.9 shows the loss of capacity at high temperatures and low rate or long discharges due to self-discharge or chemical deterioration.

3.2.6 Service Life

A useful graph employed in this handbook, summarizing the performance of each battery system, presents the service life at various discharge loads and temperatures, normalized for unit weight (amperes per kilogram) and unit volume (amperes per liter). Typical curves are shown in Fig. 3.10. In this type of presentation of data, curves with the sharpest slope represent a better response to increasing discharge load than those which are flatter or flatten out at the high current drain discharges.

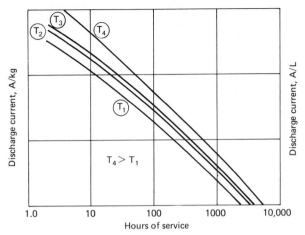

FIG. 3.10   Battery service life (at various discharge loads and temperatures).

Data of this type can be used to approximate the service life of a given cell or battery under a particular discharge condition or to estimate the weight or size of a cell required to meet a given service requirement. In view of the linearity of these curves on a log-log plot, mathematical relationships have been developed to estimate the performance of batteries under conditions that are not specifically stated. Peukert's equation,[1]

$$I^n t = C$$

where I is the discharge rate and t the corresponding discharge time, has been used in this manner to describe the performance of a battery. The value n is the slope of the straight line. Peukert's equation is reasonably accurate for many batteries over a wide range as seen in Fig. 3.10 (and in similar presentations of data in the handbook). The curves are linear on a log-log plot of discharge load vs. discharge time but taper off at both ends because of the cell's inability to handle very high rates and the effect of self-discharge at the lower discharge rates. Other mathematical relationships have been developed to describe battery performance.[2]

3.2.7 Voltage Regulation

The voltage regulation required by the equipment is most important. As is apparent from the various discharge curves, design of equipment to operate to the lowest possible end voltage results in the highest capacity and longest service life. Similarly, the upper voltage limit of the equipment should be established to take full advantage of the battery characteristics. In some applications, where only a narrow voltage range can be tolerated, the selection of the battery may be limited to those systems having a flat discharge profile—or voltage regulators may have to be employed to take full advantage of the battery's capacity.

The advantage of a low end or cutoff voltage is shown graphically in Fig. 3.11. The

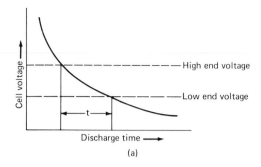

(a)

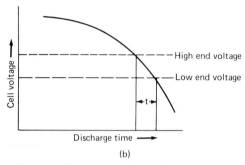

(b)

FIG. 3.11 Effect of end voltage on battery capacity.
(a) Sloping discharge curve; (b) flat discharge curve.

advantage is more pronounced with a sloping discharge curve than with a flat discharge curve which ends with a sharp voltage drop at the end of battery life. Battery performance data (capacity, service life) are usually reported to a specific end voltage.

Another consideration is the response of the cell of battery voltage when the discharge current is changed during the discharge. Examples of this type of discharge are a pulsing requirement against a background current, backlighting for an LCD watch application, or changing loads from receive to transmit in the operation of a radio transceiver. A typical pulse discharge is plotted in Fig. 3.12. The drop in voltage of a battery with lower internal resistance and better response to changes in load current will be less than one with higher internal resistance. It is also noted that the voltage spread widens as the cell is discharged.

3.2.8 Charging Voltage

If a storage battery is used (for example, as a standby power source) in conjunction with another energy source which is permanently connected in the operating circuit, allowances must be made for the voltage required to charge the battery, as illustrated in Fig. 3.13. The applied charging voltage must exceed the battery voltage on charge.

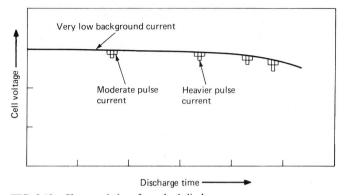

FIG. 3.12   Characteristics of a pulsed discharge.

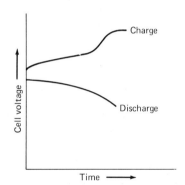

FIG. 3.13   Voltage profile on charge and discharge.

3.2.9 Effect of Cell and Battery Design

The constructional features of the cell strongly influence its performance characteristics.

Electrode Design   Cells that are designed, for example, for optimum service life or capacity at relatively low or moderate discharge loads contain maximum quantities of active material. On the other extreme, cells capable of high-rate performance are designed with large electrode or reaction surfaces and features to minimize internal resistance and current density (amperes per area of electrode surface), often at the expense of capacity or service life.

In Fig. 3.14, the performance of a cell designed for high-rate performance is compared with one using the same electrochemical system but optimized for capacity. The high-rate cells have a lower capacity but deliver a more constant performance as the discharge rate increases.

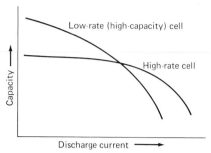

FIG. 3.14 Comparison of cells designed for high- and low-rate service.

Shape and Configuration The shape or configuration of the cell will also influence battery capacity as it affects such factors as internal resistance and heat dissipation. For example, a tall, narrow-shaped cylindrical cell will generally have a lower internal resistance than a wide, squat-shaped cell of the same design and may outperform it, in proportion to its volume, particularly at the higher discharge rates. Similarly, heat dissipation will be better from cells with a high surface-to-volume ratio or with internal components that can conduct heat to the outside of the cell.

Volumetric Efficiency vs. Energy Density The size and shape of the cell and the ability to effectively use the internal volume of the cell influence the energy output of the cell. The volumetric energy density (e.g., watthours per liter) decreases with decreasing cell volume as the percentage of "dead volume" for containers, seals, etc. increases for the smaller cells. This relation is illustrated for several "button" type cells in Fig. 3.15. The shape of the cell (e.g., wide or narrow diameter) may also influence the volumetric efficiency as it relates to the amount of space lost for the seal and other cell constructional materials.

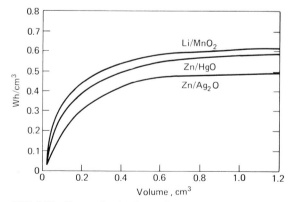

FIG. 3.15 Energy density, in watthours per cubic centimeter, of button cells as a function of cell volume. (*From Paul Ruetschi, "Alkaline Electrolyte-Lithium Miniature Primary Batteries," J. Power Sources, Vol. 7, 1982.*)

Effect of Size on Capacity   Cell size influences the voltage characteristics by its effect on current density. A given current drain may be a severe load on a small battery giving a discharge profile similar to curve 4 or 5 of Fig. 3.3, but it may be a mild load on a larger battery with a discharge curve similar to curve 2 or 3. Often it is possible to obtain more than a proportional increase in the service life by increasing the size of the cell, as the current density is lowered. The absolute value of the discharge current, therefore, is not the key influence, although its relation to the size of the cell, i.e., the current density, is significant.

In this connection, the advantage of the use of a multicell battery vs. a larger cell(s) with a voltage converter to obtain the required high voltage can be compared. An important consideration is the relative advantage of the more efficient larger cell vs. the energy losses of the voltage converter.

Battery Design   The design of a multicell battery (packaging techniques, container material, etc.) will influence the performance as it affects the environment and temperature of the individual cells. Battery designs that retain the heat dissipated by the cells can improve performance at low temperatures; on the other hand, excessive buildup of heat can be injurious to the battery's performance, life, and safety.

3.2.10 Age and Storage Condition of the Battery

Batteries are a perishable product and deteriorate as a result of chemical action that proceeds during storage. The type of cell design, electrochemical system, temperature, and length of the storage period are factors which affect the shelf life of the battery. As self-discharge proceeds at a lower rate at reduced temperatures, refrigerated or low-temperature storage extends the shelf life and is recommended for some battery systems. Refrigerated batteries should be warmed before discharge to obtain maximum capacity.

This self-discharge can also become a factor on long-term discharges and cause a reduction in capacity. This effect is illustrated in Fig. 3.16. A light load will deliver higher capacity than a heavy load; however, on an extremely light load over a long discharge period, capacity may be reduced due to self-discharge.

Some battery systems develop protective films or coatings on the active material during storage. These films can substantially improve the shelf life of the battery. However, when the battery is placed on discharge after storage, the initial voltage may

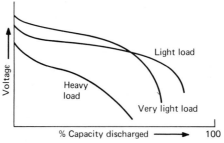

FIG. 3.16   Effect of self-discharge on battery capacity.

be low until the coating is broken or penetrated as a result of the electrochemical reaction. This phenomenon is known as "voltage delay" and is illustrated in Fig. 3.17. The extent of the voltage delay is dependent on and increases with increasing storage time and storage temperature. The delay also increases with increasing discharge current and decreasing discharge temperature. In practice, voltage delay is usually significant only when the battery is first discharged after being stored for a long period at elevated temperatures, but the specific performance is difficult to predict.

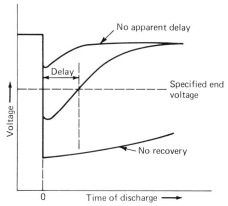

FIG. 3.17   Voltage delay.

3.3 MAJOR CONSIDERATIONS IN SELECTING A BATTERY

A number of factors must be considered in selecting the best battery for a particular application. There is still no one battery that combines optimum performance under all operating conditions with light weight, safety, reliability, low cost, and the other required features. The characteristics of each available battery must be weighed against the equipment requirements, and one selected that best fulfills the requirements.

The factors that affect battery performance should be considered in making a battery selection, and requirements should be established whenever possible to ensure optimum battery performance. It is important that the selection of the battery be considered at the beginning of equipment development rather than at the end, when the hardware is fixed. In this way, the most effective compromises can be made between battery capabilities and equipment requirements.

The considerations that are important and influence the selection of the battery include:

1. *Type of battery:*   Primary, secondary, or reserve system.

2. *Electrochemical system:*   Matching of advantages/disadvantages and battery characteristics with major equipment requirements.

3. *Voltage:*   Nominal or operating voltage, maximum and minimum permissible voltages, voltage regulation, profile of discharge curve, start-up time, voltage delay.

4. *Load current and profile:*   Constant current, constant resistance, or constant power; value of load current, single-valued or variable load, pulse load.

5. *Duty cycle:*   Continuous or intermittent, cycling schedule if intermittent.

6. *Temperature requirements:*   Temperature range over which operation is required.

7. *Service life:*   Length of time operation is required.

8. *Physical requirements:*   Size, shape, weight; terminals.

9. *Shelf life:*   Active/reserve battery system; charged/discharged state; storage time, temperature, and other conditions.

10. *Charge-discharge cycle (if rechargeable):*   Float or cycling service; life or cycle requirement; availability and characteristics of charging source; charging efficiency.

11. *Environmental conditions:*   Vibration, shock, spin, acceleration, etc; atmospheric conditions (pressure, humidity, etc.).

12. *Safety and reliability:*   Permissible variability, failure rates; freedom from outgassing or leakage; use of potentially hazardous or toxic components; type of effluent or signature gases, high temperature, etc.; operation under severe or potentially hazardous conditions.

13. *Unusual or stringent operating conditions:*   Very long term or extreme-temperature storage, standby, or operation; high reliability for special applications; rapid activation for reserve batteries, no voltage delay; special packaging for batteries (pressure vessels, etc.); unusual mechanical requirements, e.g., high shock or acceleration, nonmagnetic.

14. *Maintenance and resupply:*   Ease of battery acquisition, accessible distribution; ease of battery replacement; available charging facilities; special transportation, recovery, or disposal procedures required.

15. *Cost:*   Initial cost; operating or life cycle cost; use of critical or exotic (costly) materials.

3.4 BATTERY APPLICATIONS

Electrochemical batteries are an important power source and are used in a wide variety of consumer, industrial, and military applications. Annual worldwide sales approach $10 billion.

The use of batteries is increasing at a rapid rate, much of which can be attributed to advancing electronics technology, lower power requirements, and the development of portable equipment which can best be powered by batteries. Other contributing factors are the increased demand for battery-operated equipment, the opening of new areas for battery application, e.g., electric vehicles and utility power load leveling, and the significant improvement of battery performance.

Batteries have many advantages over other power sources, as outlined in Table 3.1. They are efficient, convenient, and reliable, need little maintenance, and can be easily configured to user requirements. As a result, batteries are used in an extremely wide range and variety of sizes and applications—from as small as 3 mAh for watches and memory backup to as large as 20,000 Ah for submarine and standby power supplies. A number of important battery applications are listed in Table 3.2.

TABLE 3.1   Application of Batteries

Advantages	Limitations
Self-contained power source	High cost (compared with utility power)
Adaptable to user configuration:	
Small size and weight— portability	Use of critical materials
Variety of voltages, sizes, and configurations	Low energy density
Compatible with user requirements	Limited shelf life
Ready availability	
Reliable, low maintenance, safe, minimum, if any, moving parts	
Efficient conversion over a wide range of power demands	
Good power density (with some types)	
Efficient energy-storage device	

TABLE 3.2   Typical Battery Applications

Memory backup	Vehicle (SLI)*
Watches, clocks	Engine-starting*
Calculators	Aircraft*
Implants (heart pacemakers)	Industrial trucks*
Medical appliances	Auxiliary and emergency (standby)
Meters, test equipment, and	power*
instrumentation	Uninterruptible power systems*
Photographic equipment	Load leveling*
Audio and communication equipment	Electrical energy storage*
Appliances and convenience equipment	Electric vehicles*
Toys	Submarine and underwater propulsion*
Signals and alarms	Satellite and spacecraft
Lighting	Military applications
Tools	Communications
Navigation aids	Surveillance and detection
Oceanographic equipment	Munitions and missiles
Meteorological equipment	
Emergency transmitters	
Railway signaling	

*A secondary battery application. Other applications use primary or secondary batteries, with the primary batteries predominant in the smaller-size, lower-capacity applications.

A generalized summary of battery applications, listing the various battery types and identifying the power level and operational time in which each finds its predominant use, is shown in Fig. 3.18. As with any generalization, there are many instances in which the application of a particular battery will fall outside the limits shown.

Primary batteries are used typically from low to moderately high power drain

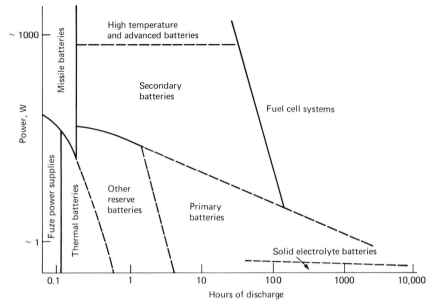

FIG. 3.18 Predominant application field for various types of batteries.

applications, to a large extent with the familiar flat, button, or cylindrical cell configurations. They are a convenient, usually relatively inexpensive, lightweight source of packaged power and, as a result, are used in a variety of portable electric and electronic equipment. The cylindrical "dry" cell is widely used in lighting devices, toys, radios, and other such consumer products; flat or button cells are popular in watches, calculators, photographic equipment, etc., and, more recently, as a battery backup for complementary metal-oxide semiconductor (CMOS) memory preservation. Similarly, primary batteries are used extensively in military applications to power portable communication, radar, night vision, surveillance, and other such equipment. Larger-sized primary batteries are also produced, mainly for special applications such as navigation aids, standby power, remote-area uses, etc., where their high capacity and energy density, long shelf life, and freedom from recharging and maintenance are important requisites.

Secondary batteries are used as an energy storage device, generally connected to and charged by a prime energy source and delivering their energy to the load on demand. Examples of this type of service are the lead-acid automotive starting, lighting, ignition (SLI) battery, which is, by far, the major secondary battery application, standby electrical systems, and load leveling. Secondary batteries are also used in applications where they are discharged and recharged subsequently from a separate power source. Examples of this type of service are electric vehicles and many applications where the secondary battery is used in place of primary batteries, either for cost saving or to handle power levels beyond the capability of conventional primary batteries.

The special and reserve primary batteries are used in selected applications, such as thermal and missile batteries which are capable of high-rate discharges for short periods of time after storage in an inactivated or "reserve" condition. The high rate capability of the reserve thermal and zinc/silver oxide batteries is illustrated in Fig. 3.19.[3] The

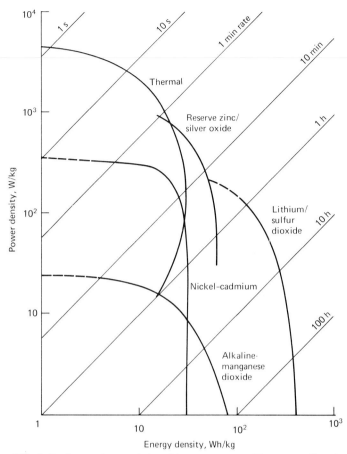

FIG. 3.19 Power characteristics of battery systems. (*From Attewell and Clark.*)

solid electrolyte cells are low-rate cells, operating in the microampere range but with extremely long operational or shelf life. They are used in computer memory backup, cardiac pacemakers, and other applications requiring high reliability and extremely long life.

The fuel cell is used in those applications requiring long-term continuous operation. Its major application has been as the power source in space flights; larger sizes are now in development as an alternative to moderate-power engine generators and utility load leveling and possibly for electric vehicle propulsion.

A comparison of the advantages of each type of battery system is given in Fig. 3.20, which examines one performance parameter—the total weight of a power source to deliver 100 W for different times of operation. For very short missions, the weights of the secondary battery systems, because of their high rate capability, are lightest. At moderate loads, the primary batteries become the lightest system due to their higher

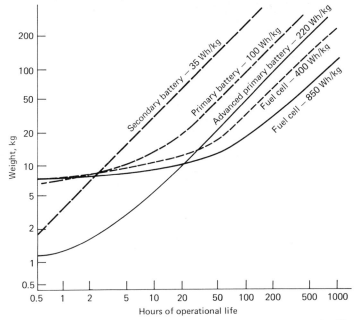

FIG. 3.20 Comparison of electrochemical systems—weight vs. service life (based on 100-W output and stated energy-weight ratio).

energy densities at these loads. On extended missions, the fuel cell systems are lightest once the fixed weight of the fuel cell converter becomes less significant and the weight of the fuel becomes predominant. (Note: The slope of each curve represents the optimum energy density of the battery or the fuel consumption for the fuel cell; the curves flatten out for the short operational periods because of the inefficiency of battery systems under high discharge rates and the weight of the fixed fuel cell converter for the fuel cell systems.)

The specific types of batteries in each classification and the major applications and characteristics of the batteries are covered in the appropriate chapters of this handbook.

3.5 COMPARATIVE FEATURES AND PERFORMANCE CHARACTERISTICS

3.5.1 General Characteristics

The performance of the conventional and advanced primary and secondary batteries, on a theoretical and practical basis, is summarized in Tables 1.2 and 3.3, respectively. Comparison of the theoretical and practical capacities for several of these battery systems is plotted in Fig. 3.2.

As discussed in Sec. 3.2, the actual capacity available from a battery is significantly

TABLE 3.3 Characteristics of Major Battery Systems

Battery system	Anode	Cathode	Typical working voltage, V	Capacity Wh/kg	Capacity Wh/L
			Practical battery		
Primary					
Leclanche	Zn	MnO_2	1.2	65	140
Magnesium	Mg	MnO_2	1.7	100	195
Alkaline-MnO_2	Zn	MnO_2	1.15	95	210
Mercury	Zn	HgO	1.2	105	325
Mercad	Cd	HgO	0.85	50	180
Silver oxide	Zn	Ag_2O	1.5	130	515
Zinc/air	Zn	$O_2(air)$	1.2	290	905
Li/SO_2	Li	SO_2	2.8	280	440
Li/MnO_2	Li	MnO_2	2.7	200	400
Solid electrolyte	Li	$I_2(P2VP)$	2.8	150	400
Reserve					
Cuprous chloride	Mg	CuCl	1.3	60	80*
Zinc/silver oxide	Zn	AgO	1.5	30	75†
Thermal	Ca	$CaCrO_4$	2.4	5	15†
Secondary					
Lead-acid	Pb	PbO_2	2.0	35	80
Edison	Fe	Ni oxide	1.2	30	60
Nickel-cadmium	Cd	Ni oxide	1.2	35	80
Silver-zinc	Zn	HgO	1.5	90	180
Nickel-zinc	Zn	Ni oxide	1.6	60	120
Nickel-hydrogen	H_2	Ni oxide	1.2	55	60
Silver-cadmium	Cd	AgO	1.1	60	120
Zinc/chlorine	Zn	Cl_2	1.9	100	130‡
High temperature	Li(Al)	FeS	1.2	60	100‡
High temperature	Na	S	1.7	100	150‡

*Water-activated.
†Automatically activated 2- to 10-min rate.
‡Estimated.
NOTE: See Table 1.2 for the theoretical data on these battery systems.

less than the theoretical capacity of the active materials and less than even the theoretical capacity of a practical cell which includes the weight of the non-energy-producing materials of construction as well as the active materials.

In general, the capacity of the conventional secondary batteries is lower than the capacity of the primary batteries, but they are capable of performance on discharges at higher current drains and lower temperatures and have a flatter discharge voltage profile (Figs. 3.23 to 3.25). The advanced battery systems, using lighter active materials with higher operating voltages, deliver much higher energy and power densities than the conventional systems. This is illustrated in Fig. 3.21, which shows the performance capabilities of several primary, secondary, and advanced battery systems.

The primary batteries have a distinct advantage over the secondary batteries in charge retention or shelf life. Most primary batteries can be stored for several years or more even at high temperatures and still retain a substantial portion of their capacities;

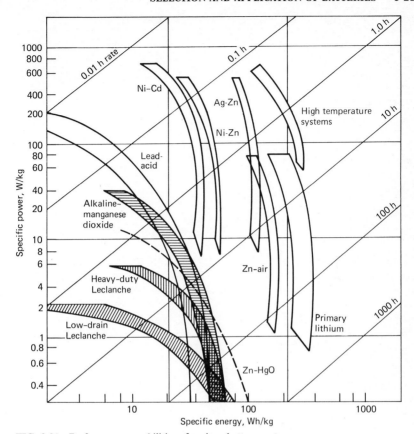

FIG. 3.21 Performance capabilities of various battery systems.

secondary batteries lose their charge more rapidly and must be kept charged during storage or charged just prior to use.

The costs of the secondary batteries are generally higher than those of the primary cells—on a cost per watthour basis—although the size of the cell has an important influence on its cost. Specific data on the cost of the cells and batteries are given in the appropriate sections of the handbook.

The general characteristics of the primary and secondary batteries also are compared in Table 3.4 and in the next section. More detailed characteristics of the different types of batteries are presented in the appropriate sections of this handbook. These data, rather than the more generalized data presented in this section, should be used to evaluate the specific performance of each battery.

3.5.2 Characteristics of Batteries for Portable Equipment

Portable, battery-operated, electrical and electronic equipment are typically powered by primary batteries. In the last decade, however, the development of small, rechargea-

TABLE 3.4 Comparison of Primary and Secondary Batteries*

	Primary batteries	Secondary batteries	Advanced secondary batteries
Energy density	2	3	1
Power density	3	2	1
Shelf life (charge retention)	1	2	2
Flat discharge profile	3	1	1
Operating temperature	2	1	. . .
Maintenance and system complexity	1	2	3
Cost**	1	2	3

*Rated on a scale of 1, best, to 3, poorest.
**Rating based on initial battery cost, not life-cycle cost.

ble, maintenance-free sealed nickel-cadmium and lead-acid secondary batteries made it possible for secondary batteries to be used in applications which had been almost exclusively the domain of the primary battery, the advantage being a lower life cycle cost because the secondary battery can be recharged and reused. High-power applications, such as tools and special types of military uses, were drawn first to secondary batteries because of their power advantage over conventional types of primary batteries. However, recently developed primary batteries particularly those using the lithium anode, which have a high energy density and good power density, can be used in some of the high-power applications. The advantage of the primary batteries is that the need to charge them to maintain them in a state of readiness is eliminated. Table 3.5 compares the performance of the major primary and secondary batteries used for portable applications (approximately 1 to 20 Ah in size).

These comparisons show that:

1. The primary cells, particularly the newer lithium types, deliver up to 8 times the capacity on a weight basis and 4 times on a volume basis of the secondary batteries on a moderate discharge load. This can be translated to much longer service life for the primary battery if similarly sized batteries are compared—or much smaller size and weight for a primary battery giving the same performance as the secondary battery.

2. The charge retention of the primary cell is much superior to that of the secondary battery. Hence, the secondary battery has to be maintained in a charged condition, or periodically recharged to maintain it in a state of readiness, with no such maintenance required for the primary battery. Figure 3.22 compares the charge retention (shelf life) of different battery types under different storage conditions.

3. The secondary batteries generally have better high-rate performance than primary batteries. The newer lithium batteries, however, provide output superior to the secondary batteries even at fairly high rates because of their usually high normal-temperature performance and generally good performance at high discharge rates. This is illustrated in Fig. 3.23, which compares the performance of the different

TABLE 3.5 Characteristics of Batteries for Portable Equipment

	Primary batteries			Secondary batteries	
	Zn/alkaline/MnO$_2$	Li/MnO$_2$	Li/SO$_2$	Nickel-Cadmium	Lead-acid
Nominal cell voltage, V	1.5	3.0	3.0	1.2	2.0
Energy density					
Wh/kg	95	200	260	35	35
Wh/L	210	400	420	80	80
Charge retention at 20°C	2–3 y	5–8 y	5–10 y	3–6 months	6–9 months
(shelf life)					
Calendar life, y	. . .	. . .	. . .	3–5	3–8
Cycle life, cycles	. . .	. . .	. . .	300–500	200–250
Operating temperature, °C	−20 to 45	−20 to 70	−40 to 70	−20 to 45	−40 to 60
Relative cost per kilowatthour (initial unit cost)	1	4	3.5	20	10

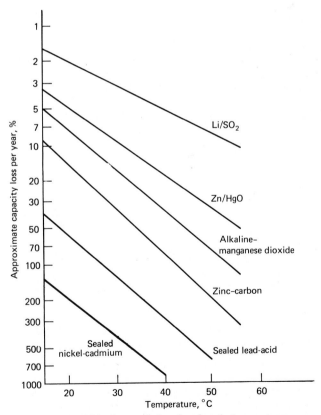

FIG. 3.22 Shelf life characteristics of various battery systems.

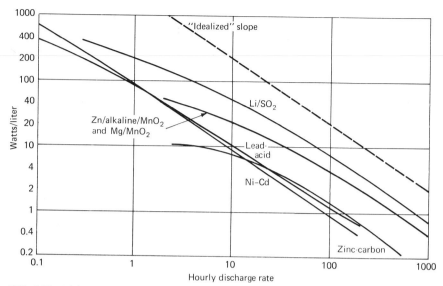

FIG. 3.23 High rate capability of primary and secondary cells, 20°C, unit volume basis.

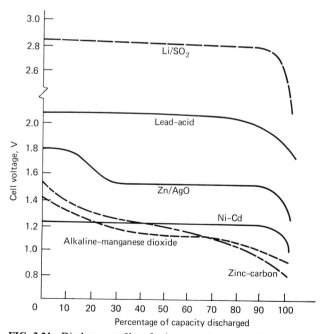

FIG. 3.24 Discharge profiles of primary and secondary batteries.

battery types at different discharge rates. In this figure, the service that each of the battery types will deliver at various power levels (watts per liter) is plotted. A slope parallel to the idealized line indicates that the capacity of the cell, in watt hours, is invariant regardless of the discharge load. A flatter slope, or one that levels off as the load is increased, indicates a loss of capacity as the discharge rate is increased. This figure shows the advantage of the Li/SO$_2$ primary and the secondary batteries at high rates over the conventional primary batteries, whose capacity begins to drop off significantly at the 20- to 50-h rate. Figure 3.23 is a volumetric comparison; a similar plot on a gravimetric basis will show a greater advantage for the primary systems because of their lighter weight (see Fig. 3.25).

4. The secondary batteries generally have a flatter discharge profile than the conventional primary batteries. Figure 3.24 compares the discharge curves of several primary and secondary batteries and illustrates this characteristic.

5. The secondary batteries generally have better low-temperature performance than the conventional aqueous primary batteries. The comparative performances on a gravimetric basis (watthour per kilogram) and a volumetric basis (D size cylindrical cells) of the various types of cells at a moderate 20-h rate are shown in Figs. 3.25 and 3.26, respectively. While the capacity of the primary cells is higher at room temperature, their performance drops off more significantly as the temperature is reduced, again with the exception of some of the lithium systems.

6. The secondary batteries are capable of being recharged to their original condition after discharge and reused rather than being discarded. For those applications where recharging facilities are available, convenient, and inexpensive, a lower life cycle cost

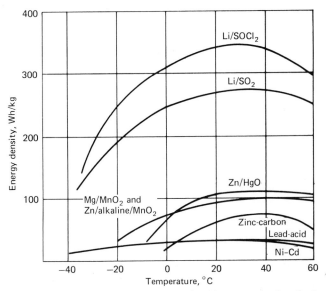

FIG. 3.25 Effect of temperature on gravimetric energy density of primary and secondary cells. Based on D size cells.

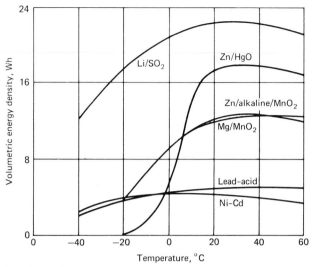

FIG. 3.26   Effect of temperature on volumetric energy density of primary and secondary cells (D size cells).

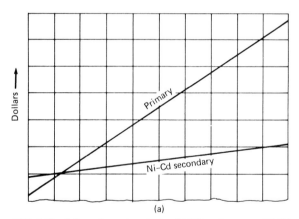

(a)

FIG. 3.27   Life cycle cost analysis—Li/SO$_2$ primary vs. Ni-Cd secondary cells. (a) Training situation.

can be realized with secondary batteries, even with their higher initial cost, if their full cycle life is utilized.

Specific cost effectiveness–life cycle analyses can be used to evaluate the best choice —primary or secondary battery—for a particular application or deployment. Figure 3.27 summarizes such an analysis comparing lithium/sulfur dioxide primary and nickel-cadmium secondary batteries.[4] Figure 3.27a depicts a training situation in which charging facilities are readily available and inexpensive, recharging is convenient, and the batteries are used regularly (thus employing the full cycle life of the secondary battery). In this situation, the initial higher cost of the secondary battery is easily

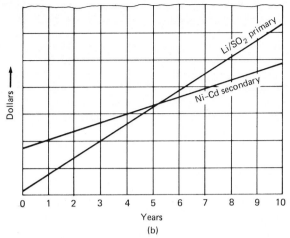

FIG. 3.27 (Continued) Life cycle cost analysis—Li/SO₂ primary vs. Ni-Cd secondary cells. (*b*) Field situation.

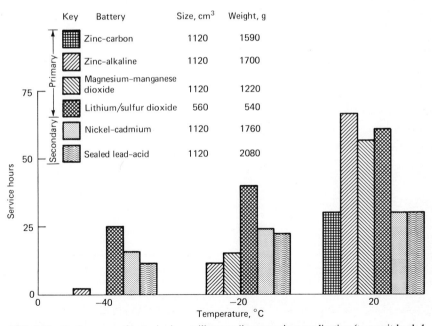

FIG. 3.28 Performance of batteries in a military radio transceiver application (transmit load, 1 A; receive load, 50 mA; duty cycle, 9:1 receive-to-transmit).

recovered, and the payback or break-even point occurs early. Figure 3.27*b* presents a field situation in which batteries are not used regularly, special charging facilities are required, and recharging is not convenient. In this case, the payback time occurs much later, and the use of secondary batteries may not be cost-effective, as the break-even time is close to or beyond the calendar life of the battery.

Figure 3.28 provides another illustration comparing the performance of several types of primary and secondary batteries in a typical portable radio transceiver application at several temperatures. The primary batteries give the longest service at 20°C, but only the secondary batteries and the Li/SO_2 primary battery are capable of performance at the very low temperatures.

REFERENCES

1. G. W. Vinal, *Storage Batteries,* 4th ed., Wiley, New York, 1955, p. 216.

2. R. Selim and P. Bro, "Performance Domain Analysis of Primary Batteries," *Electrochemical Technology, J. Electrochem. Soc.* **118** (5):829 (1971).

3. A. Attewell and A. J. Clark, "A Review of Recent Developments in Thermal Batteries," *Power Sources,* vol. 8, Academic, London, 1980.

4. "Cost Effectiveness Comparison of Rechargeable and Throw-away Batteries for the Small Unit Transceiver," U.S. Army Electronics Command, Ft. Monmouth, N.J., January 1977.

PART 2 PRIMARY BATTERIES

Introduction

by
David Linden

4.1 GENERAL CHARACTERISTICS AND APPLICATIONS OF PRIMARY BATTERIES

The primary battery is a convenient source of power for portable electrical and electronic devices, lighting, photographic equipment, watches and calculators, memory backup, and a wide variety of other applications, giving freedom from utility power. Major advantages of the primary battery are that it is convenient, simple, and easy to use, requires little, if any, maintenance, and can be sized and shaped to fit the application. Other general advantages are good shelf life, reasonable energy and power density, reliability, and acceptable cost.

Primary batteries have existed for over 100 y, but up to 1940, the zinc-carbon cell was the only one in wide use. During World War II and the post-war period, significant advances were made, not only with the zinc-carbon system, but with new and superior types of batteries. Capacity was improved from less than 50 Wh/kg with the early zinc-carbon cells to more than 500 Wh/kg now obtained with the lithium cells. The shelf life of cells at the time of World War II was limited to about 1 y when stored at moderate temperatures; the shelf life of present-day conventional cells is from 2 to 4 y. The shelf life of the new lithium cells is predicted to be 10 y, with a capability of storage at temperatures as high as 70°C. Low-temperature operation has been extended from 0 to −40 to −55°C, and the power density of the new primary cells has been improved manyfold from about 10 to over 250 W/kg. Some of the advances in primary battery performance are shown graphically in Fig. 3.1 in Part 1.

These improved characteristics have opened up many new opportunities for the use of primary batteries. The higher energy density has resulted in a substantial reduction in the size and weight of batteries. This reduction, taken with the advances in electronic technology, has made many new portable radio, communication, and electronic devices practical. The higher power density has made possible the design of portable devices, such as range finders, radar and surveillance systems, transmitters, and other high-power applications, that heretofore had to be powered by secondary batteries or utility power which do not have the convenience and freedom from maintenance of primary batteries. The long shelf life that is now characteristic of many primary batteries has similarly resulted in new uses in medical electronics, memory backup, and other long-term applications as well as an improvement in the lifetime and reliability of battery-operated equipment.

The worldwide primary battery market has now reached more than $2 billion annually, with a growth rate exceeding 10% annually. The vast majority of these primary batteries are the familiar cylindrical and flat or button cells with capacities below 20 Ah. A small number of larger primary batteries, ranging in size up to several thousand ampere-hours, are used in signaling applications, for standby power where independence from utility power is mandatory, and other military and special applications.

Typical characteristics and applications of the different types of primary batteries are summarized in Table 4.1. The trends in the development of primary cells and the advances achieved in energy density are shown graphically in Fig. 4.1.

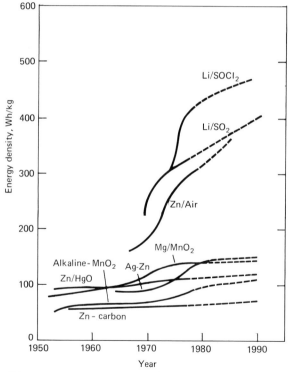

FIG. 4.1 Advances in the development of primary cells, continuous discharge at 20°C, 40- to 60-h rate, D or similar size cell (solid line, historical data; broken line, forecast). (*From Ref. 1.*)

4.2 TYPES AND CHARACTERISTICS OF PRIMARY BATTERIES

Although a number of anode/cathode combinations can be used as primary battery systems (see Part 1), only a relatively few have achieved practical success. Zinc has been, by far, the most popular anode material for primary cells because of its good electrochemical behavior, high electrochemical equivalence, compatability with aqueous electrolytes, reasonably good shelf life, low cost, and availability. Aluminum is attractive because of its high electrochemical potential and electrochemical equivalence and availability, but, due to passivation and generally poor electrochemical performance, it has not been developed successfully into a practical primary cell system. It is now being considered in a mechanically rechargeable or refuelable aluminum/air cell and in reserve battery systems. Magnesium also has attractive electrical properties and low cost and has been used successfully in an active primary cell, particularly for military applications, having a high energy density and good shelf life. Commercial interest has been limited. Magnesium also is popular as the anode in reserve batteries. More recently, the focus has been on lithium, which has the highest gravimetric energy density and standard potential of all the metals. The lithium anode battery systems, using a number of different nonaqueous electrolytes in which lithium is stable and different cathode materials, offer the opportunity for significant advances in the performance characteristics of primary systems.

The Leclanche or carbon-zinc dry cell has existed for over 100 y and is still the most widely used of all the dry cell batteries because of its low cost, reliable performance, and ready availability. Cells and batteries of many sizes and characteristics have been manufactured to meet the requirements of a wide variety of applications. Significant improvements in capacity and shelf life were made with this cell system in the period between 1945 and 1965 through the use of new materials (such as beneficiated manganese dioxide and zinc chloride electrolyte) and cell designs (such as the paper-lined cell). The low cost of the Leclanche cell is a major attraction, but recently it has been losing its market share because of the development of new primary batteries with superior performance characteristics.

TABLE 4.1   Major Characteristics and Applications of Primary Batteries

System	Characteristics	Applications
Zinc/carbon (Leclanche) (Zn/MnO_2)	Popular, common, low-cost, primary battery; available in a variety of sizes	Flashlight, portable radios and electronics, toys, novelties, instruments, etc.
Magnesium (Mg/MnO_2)	High-capacity primary battery; long shelf life	Military receiver-transmitters, aircraft emergency transmitters.
Mercury (Zn/HgO)	Highest capacity (by volume) of conventional types; flat discharge; good shelf life	Hearing aids, medical (pacemakers), photography, detectors, receiver-transmitters, military sensor and detection equipment.
Mercad (Cd/HgO)	Long shelf life; good low- and high-temperature performance; low energy density	Special applications requiring operation under extreme temperature conditions and long life.

TABLE 4.1 *(Continued)*

Alkaline (Zn/alkaline electrolyte/MnO_2)	Popular general-purpose premium battery; good low-temperature and high-rate performance; moderate cost	Cassettes and tape recorders, calculators, radio and TV, etc.; popular for high-drain primary-battery application.
Silver/zinc (Zn/Ag_2O)	Highest capacity (by weight) of conventional types; flat discharge; good shelf life	Hearing aids, photography, electric watches, missiles and space application (larger sizes).
Lithium (Li/SO_2)	High-performance primary battery; excellent low-temperature and high rate performance; long shelf life	Will have wider general-purpose application. First uses are military and special industrial applications needing high-capacity and low-temperature performance.
Lithium (Li/$SOCl_2$)	Highest energy density; long shelf life; good performance over wide temperature range, mainly at low to moderate rates (present state-of-the-art)	Wide range of applications (capacity from 1 to 10,000 Ah) requiring high energy density, long shelf life, e.g., from memory keep alive to standby electrical power applications.
Lithium/solid cathode	High energy density; good rate capability and low-temperature performance; long shelf life; competitive cost	Cost-effective replacement for conventional button and small cylindrical cell applications.
Lithium/solid electrolyte	Extremely long shelf life; low-power battery	Medical electronics, memory circuits, fusing.

In the past 10 to 15 y, an increasing portion of the primary battery market has been shifting to the Zn/alkaline/MnO_2 cell because of its superior performance at high current drains, continuous discharge, and low temperatures and its better shelf life. While more expensive than the Leclanche cell on a unit basis, it is more cost-effective for those applications requiring the high-rate, continuous-drain, or low-temperature capability, where the alkaline cell can outperform the Leclanche cell by a factor of 2 to 10 times (see Table 4.8). In addition, because of the advantageous shelf life of the alkaline cell, it is often selected for applications in which the battery is used intermittently and exposed to uncontrolled storage conditions (e.g., consumer flashlight, smoke alarm, etc.) but must perform dependably when required. The alkaline-MnO_2 cell will be a dominant primary battery system in the 1980s.

Another important zinc anode primary system is the zinc/mercuric oxide cell. This cell was developed during World War II for military communication applications because of its good shelf life and high volumetric energy density. In the postwar period it has been used in small button, flat cell, or cylindrical configurations as the power source in electronic watches, calculators, hearing aids, photographic and similar applications requiring a reliable, long-life miniature power source. Recently, however, with the development of the lithium cells and their superior performance and competitive cost, a trend away from the mercuric oxide cell to the newer technologies has begun.

The substitution of cadmium for the zinc anode (the cadmium/mercuric oxide cell) results in a lower-voltage but very stable system, with a shelf life of up to 10 y as well as performance at high and low temperatures. The watthour capacity of this cell, because of the lower voltage, is about 60% of the zinc/mercuric oxide cell capacity.

The primary zinc/silver oxide cell is similar in design to the small zinc/mercuric oxide button cell but has a higher energy density (on a weight basis) and performs better at low temperatures. These characteristics also make this battery system desirable for use in hearing aids, photographic applications, and electronic watches. However, because of its high cost and the development of replacement cells using the lithium technology, the growth of this cell will be limited.

The zinc/air battery system is noted for its high energy density, but until recently it was used only in large, low-power cells for signaling and navigational-aid applications. With the development of improved cathode structures the high rate capability of the system was improved, and small button-type cells have been introduced for hearing aids, electronic watches, and similar applications. These cells have a very high volumetric energy density for no active cathode material is needed. Larger primary cells, using the new technology, have not become practical because the cells' limitations (limited performance at extreme temperatures and humidity, low shelf life, and low power density) are critical in the applications where the larger cells would be used.

While magnesium has attractive electrochemical properties, there has been relatively little commercial interest in magnesium primary batteries because of the generation of hydrogen gas during discharge and the poor storageability of a partially discharged cell, although recently advances have been made towards eliminating or minimizing these deficiencies. Magnesium dry cells have been used successfully in military communications equipment, taking advantage of the long shelf life of a cell in an undischarged condition, even at high temperatures, and its higher energy density. Magnesium is one of the more popular anode materials for reserve cells (see Part 5).

The lithium anode cells are a relatively recent development (since 1970) and have the advantage of the highest energy density, operation over a very wide temperature range, and long shelf life. The lithium cells are gradually replacing the conventional battery systems, and this trend will accelerate as the lithium technology matures, costs are reduced, and greater confidence develops in safety and reliability of the various types of lithium primary cells.

As with the zinc systems, there are a number of lithium cells under development, ranging in capacity from less than 5 mAh to 20,000 Ah, using various designs and chemistries, but having in common the use of lithium metal as the anode.

The lithium primary cells can be classified into three categories (see Table 11.1 in Part 2). The smallest are the low-power solid-state cells with excellent shelf life, used in applications such as cardiac pacemakers, watches, and battery backup for volatile computer memory where reliability and long shelf life are paramount requirements. In the second category are the solid cathode cells which are designed in flat or small cylindrical configurations. These cells have begun to replace the conventional primary cells in watches, calculators, memory circuits, photographic equipment, communication devices, etc. and should account for a major share of the market in the period between 1985 and 1990. The soluble-cathode cells (using gaseous- or liquid-cathode materials) constitute the third category. These cells are typically constructed in a cylindrical configuration, as flat disks or in prismatic containers, using flat plates. These cells, up to about 30 Ah in size, are used in military and industrial applications, lighting products, and other devices where small size and low weight and operation over a wide

TABLE 4.2 Characteristics of Primary Batteries

System	Carbon-zinc (Leclanche)	Carbon-zinc (zinc chloride)	Mg/MnO$_2$	Zn/Alk./MnO$_2$	Zn/HgO	Cd/HgO
Chemistry:						
Anode	Zn	Zn	Mg	Zn	Zn	Zn
Cathode	MnO$_2$	MnO$_2$	MnO$_2$	MnO$_2$	HgO	HgO
Electrolyte	NH$_4$Cl and ZnCl$_2$ (aqueous solution)	ZnCl$_2$ (aqueous solution)	MgBr$_2$ or Mg(ClO$_4$) (aqueous solution)	KOH (aqueous solution)	KOH or NaOH (aqueous solution)	KOH (aqueous solution)
Cell voltage, V/:						
Nominal	1.5	1.5	1.6	1.5	1.35	0.9
Open-circuit	1.5–1.75	1.6	1.9–2.0	1.5	1.35	0.9
Operating	1.25–1.15	1.25–1.15	1.8–1.6	1.25–1.15	1.3–1.2	0.85–0.75
End	0.9	0.9	1.2	0.9	0.9	0.6
Operating temperature, °C	−5 to 45	−10 to 50	−20 to 60	−20 to 55	0 to 55	−55 to 80
Energy density/ at 20°C:						
Button size:						
Wh/kg				38	100	45
Wh/L				135	470	200
Cylindrical size:						
Wh/kg	65	75	105	95	105	
Wh/L	100	135	195	220	325	
Discharge profile (relative)	Sloping	Sloping	Moderate slope	Moderate slope	Flat	Flat
Power density	Low	Low to moderate	Moderate	Moderate	Moderate	Moderate
Storage:						
Recommended storage temperature, °C^g	−40 to 20	−40 to 20	−40 to 50	−40 to 25	−20 to 25	−20 to 30
Permissible storage temperature, °C	−40 to 45	−40 to 55	−40 to 70	−40 to 50	−20 to 55	−55 to 80
Self-discharge rate at 20°C, % loss per year	15	15	3	7	4	3
Leakage	Moderate	Low	Medium	Very low	Some salting	Negligible salting
Gassing	Moderate	High	High (on discharge)	Low	Very low	None
Shock resistance	Fair to good	Good	Fair to good	Fair to good	Good	Good
Advantages	Lowest cost; good for noncritical use under moderate conditions; variety of shapes and sizes; availability	Low cost; better performance than regular carbon-zinc	High capacity compared with zinc-carbon; good shelf life (undischarged)	High capacity compared with zinc-carbon; good low-temperature, high-rate performance	High volumetric energy density; flat discharge; stable voltage	Good performance at high and low temperatures; long shelf life
Limitations	Low energy density; poor low-temperature, high-rate performance	High gassing rate; performance lower than premium alkaline batteries	High gassing (H$_2$) on discharge; delayed voltage	Moderate cost	Expensive, moderate gravimetric energy density	Expensive, low-energy density
Status	High production	High production	Moderate production, mainly industrial and military	High production	In production	In limited production
Major cell types available	Cylindrical bobbin cells to 30 Ah; flat cells	Cylindrical bobbin cells to 6 Ah	Cylindrical bobbin cells to 60 Ah	Button and cylindrical cells to 20 Ah	Button and cylindrical cells to 20 Ah	Button cells to 3 Ah
Approximate cost, $/kWh^h:						
Cylindrical cells	70	70	100	100	450	
Button cells				400	2500	5000
Representative manufacturers	Union Carbide Corp.; Ray-O-Vac; Bright Star; Panasonic; Tadiran	Union Carbide Corp.; Ray-O-Vac; Bright Star; Panasonic	Marathon; Ray-O-Vac; ACR Electronics; Tadiran	Duracell; Union Carbide Corp.; Ray-O-Vac; Panasonic; Varta	Duracell; Ray-O-Vac; Union Carbide Corp.; Panasonic; Varta	Duracell; ELCA

^{a}Data on Zn/Ag$_2$O system, button cells.
^{b}Zinc/air data for button cell.
^{c}Wide range of size and design; see Part 2, Sec. 11.6.
^{d}See Table 11.6 in Part 2 for other lithium systems.
^{e}Solid-state batteries are made in several chemistries. See Part 2, Chap. 12.
^{f}Data presented are for 20°C, under favorable discharge condition. See details in appropriate chapter.
^{g}Best to store refrigerated, or not above 20°C.
^{h}Costs shown in 1982 dollars are necessarily approximate due to the wide range of sizes.

Zn/Ag$_2$O[a]	Zinc/air[b]	Li/SOCl$_2$	Li/SO$_2$	Lithium/solid cathode[d]		Solid state[c]
				Li/MnO$_2$	Li/(CF)$_n$	
Zn	Zn	Li	Li	Li	Li	Li
Ag$_2$O or AgO	O$_2$ (air)	SOCl$_2$	SO$_2$	MnO$_2$	(CF)$_n$	P2VP PbI$_2$
KOH or NaOH (aqueous solution)	KOH (aqueous solution)	SOCl$_2$ salt solution	Organic solvent, salt solution	Organic solvent, salt solution	Organic solvent, salt solution	Solid LiI
1.5	1.5	3.6	3.0	3.0	2.6	2.8, 1.9
1.6	1.45	3.6	3.1	3.5	3.1	—
1.6–1.5	1.4–1.2	3.5–3.3	2.9–2.7	2.8–2.7	2.7–2.6	
1.0	0.9	2.0	2.0	2.0	2.0	
0 to 55	0 to 50	−40 to >70	−55 to 70	−20 to 55	−20 to 55	0 to 200
130	290			200	200	100–250
515	905			400	400	300–600
		300	280		240	
		650	440		450	
Flat	Flat	Flat	Very flat	Flat	Flat	Moderately flat (at low discharge rates)
Moderate	Low	Low to moderately high (depending on construction)	High	Moderate	Moderate	Very low
0 to 30	−20 to 25	−20 to 40	−20 to 40	−20 to 40	−20 to 40	0 to 40
−40 to 60	−20 to 40 (sealed)	−55 to 70	−55 to 70	−40 to 60	−40 to 60	−20 to 80
6	3 (if sealed)	2	2	3	3	< 1
Some salting	"Water transfer"	None	None (hermetically sealed)	None	None	None
Very low	Low	None	None	None	None	None
Good	Good	Good	Good	Good	Good	Good
High energy density; good high-rate performance	High volumetric energy density; long shelf life (sealed)	Highest energy density; long shelf life	High energy density; best low-temperature, high-rate performance; long shelf life	High energy density; good low-temperature, high- rate performance; cost-effective replacement for small conventional type cells		Excellent shelf life (10–20 y); wide operating temperature range (to 200°C)
Expensive, but cost-effective on button cell applications	Not independent of environment— flooding, drying out; limited power output	Proven in low to moderate rate designs	High-cost pressurized system	Available in small sizes or for low-drain applications		For very low discharge rates; poor low temperature performance
In production	Moderate production	Increasing production	Moderate, but increasing, production, mainly industrial and military	Increasing consumer production		In production for special applications
Button cells to 200 mAh	Button cells to 400 mAh	Button, flat, cylindrical, and prismatic cells, 500 mAh to 15 kAh	Cylindrical cells to 30 Ah	Button small cylindrical cells to 1.1 Ah	Button cylindrical cells to 5 Ah	Button cells to 350 mAh; D-shaped cells to 3.8 Ah
		250	250	600	600	10,000
4000	1500					
Duracell; Ray-O-Vac; Union Carbide Corp.; Panasonic; Varta	Gould, SAFT	Union Carbide Corp.; Tadiran; GTE Sylvania; SAFT; Altus; W. Greatbach; Honeywell, Inc.	Duracell; Power Conversion; Honeywell, Inc.; Silberkraft; Crompton-Parkinson	Duracell; Sanyo; General Electric; Varta; SAFT; Ray-O-Vac	Panasonic; Eagle-Picher	W. Greatbach; Catalyst; Duracell

temperature range are important. The larger cells are being developed for special military applications or as standby emergency electrical power sources.

The solid-electrolyte cells are different from other battery systems in that they depend on the ionic conductivity, in the solid state, of an electronically nonconductive salt rather than the ionic conductivity of a liquid electrolyte. Cells using these solid electrolytes are low-power (microwatt) devices but have extremely long shelf life and the capability of operating over a wide temperature range, particularly at high temperatures. These batteries are used in medical electronics, for memory circuits, and for other such applications requiring a long-life, low-power battery. The first solid-electrolyte batteries used a silver anode and silver iodide for the electrolyte. Recently, lithium has been used as the anode and lithium iodide for the electrolyte, giving cells with a higher voltage and higher energy density.

4.3 COMPARISON OF THE PERFORMANCE CHARACTERISTICS OF THE PRIMARY BATTERY SYSTEMS

4.3.1 General

The characteristics of the major primary batteries are summarized in Table 4.2. This table is supplemented by Table 1.2 in Part 1, which lists the theoretical electrical characteristics of these primary battery systems. The practical electrical performance characteristics for these batteries are also given in Table 3.3 in Part 1. A graphical comparison of the theoretical and practical performance of various battery systems is given in Fig. 3.1 in Part 1 and shows that only about 25% of the theoretical capacity is attained under practical conditions as a result of design and discharge requirements.

It should be noted, as discussed in detail in Part 1, Chap. 3, that these types of data and comparisons (as well as the performance characteristics shown in this section) are necessarily approximations, with each system presented under favorable discharge conditions. The specific performance of a battery system is very dependent on the cell design and all the detailed and specific conditions of the use and discharge of the battery.

A qualitative comparison of the various primary battery systems is given in Table 4.3. This listing illustrates the performance advantages of the newer lithium anode cells and the reason for their selection in critical military and industrial applications and in consumer applications where they are cost-effective. Nevertheless, the conventional primary batteries, because of their low cost, availability, and generally acceptable performance in many consumer applications, will maintain their significant share of the market.

4.3.2 Voltage and Discharge Profile

A comparison of the discharge curves of the major primary batteries is given in Fig. 4.2. The zinc anode batteries generally have a discharge voltage between about 1.5 and 0.9 V. The lithium anode cells have a higher voltage, many in the order of 3 V, with an end or cutoff voltage of about 2.0 V. The cadmium/mercuric oxide cell operates at a lower voltage level of 0.9 to 0.6 V. The discharge profiles of these batteries also show different characteristics. The conventional zinc-carbon and zinc/alkaline/MnO_2 cells have sloping profiles; the magnesium/manganese dioxide and lithium/manganese dioxide cells have less of a slope (although at lower discharge rates the lithium/manganese dioxide cell shows a flatter profile); most of the other battery types have a relatively flat discharge profile.

4.3.3 Energy Density

Figure 4.3 presents a comparison of the gravimetric energy density of the different primary battery systems at various discharge rates at 20°C. This figure shows the hours of service each battery type (unitized to 1 kg battery weight) will deliver at various power (discharge current × midpoint voltage) levels. The energy density can then be determined by

$$\text{Energy density} = \text{power density} \times \text{hours of service}$$

or
$$\text{Wh/kg} = \text{W/kg} \times \text{h} = \frac{A \times V \times h}{kg}$$

The conventional zinc-carbon cell has the lowest energy density of the primary batteries shown with the exception, at low discharge rates, of the cadmium/mercuric oxide cell due to the low voltage of the latter electrochemical couple. The zinc-carbon cell performs best at light discharge loads. Intermittent discharges, providing a rest or recovery period at intervals during the discharge, significantly improve its service life compared with a continuous discharge, particularly at high discharge rates.

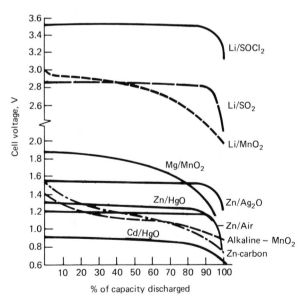

FIG. 4.2 Discharge profiles of primary battery systems (30- to 100-h rate).

TABLE 4.3 Comparison of Primary Batteries*

System	Voltage	Energy density (gravimetric)	Power density	Flat discharge profile	Low-temperature operation	High-temperature operation	Shelf life	Cost
Zinc/carbon	5	4	4	4	5	6	8	1
Zinc/alkaline/manganese dioxide	5	3	2	3	4	4	7	2
Magnesium/manganese dioxide	3	3	2	2	4	3	4	3
Zinc/mercuric oxide	5	3	2	2	5	3	5	5
Cadmium/mercuric oxide	6	5	2	2	3	2	3	6
Zinc/silver oxide	4	3	3	2	4	3	6	6
Zinc/air	5	2	3	2	5	5	..	3
Lithium/soluble cathode	1	1	1	1	1	2	2	6
Lithium/solid cathode	1	1	2	2	2	3	2	4
Lithium/solid electrolyte	2	1	5	2	6	1	1	7

*Note: 1 to 8—best to poorest.

The performance of a zinc-carbon cell falls off sharply with increasing discharge rates. This is shown graphically in Fig. 4.3 by the drop in the slope at the higher discharge rates. (Note: The 1000-Wh/kg line indicates the slope at which the capacity or energy density remains constant at all discharge rates.)

The heavy-duty zinc chloride version of the zinc-carbon cell (not plotted but covered in Part 2, Chap. 5) gives better performance under the more stringent discharge conditions. The zinc/alkaline/manganese dioxide cell, the zinc/mercuric oxide cell, the zinc/silver oxide cell, and the magnesium/manganese dioxide cell all have about the same energy density and performance at 20°C. The zinc/air system has a higher energy density at the low discharge rates, but it falls off sharply at moderately high loads. The lithium cells are characterized by their high energy density, due in part to the higher cell voltage; and the lithium/sulfur dioxide cell and other soluble cathode cells are distinguished by their ability to deliver this high capacity at the high discharge rates.

Volumetric energy density is, at times, a more useful parameter than gravimetric energy density, particularly for button and small cells, where the weight is insignificant. The denser batteries, such as the zinc/mercuric oxide cell, improve their relative position when compared on a volumetric basis, as shown in Figs. 4.7 and 4.8. The chapters on the individual battery systems include a family of curves giving the hours of service each battery system will deliver at various discharge rates and temperatures.

4.3.4 Comparison of Performance of Representative Primary Cells

Table 4.4 gives a comparison of the performance of a number of primary battery systems in a typical button cell, size 44 IEC standard. The data are based on the rated capacity at 20°C at about the C/500 rate. The performance of each system can be compared, but one should recognize that cells with different capacities may be fabricated with a given system (e.g., the Zn/HgO system), depending on the application requirements and the particular market segment the manufacturer is addressing. The discharge curves for these cells are given in Fig. 4.4.

Table 4.5 lists the performance obtained with the different primary battery systems for several cylindrical cells. The discharge curves for the N and D size cells are shown in Fig. 11.1. The discharge curves for some of these systems in the NEDA 1604 9-V battery are compared in Fig. 4.5.

4.3.5 Effect of Discharge Load and Duty Cycle

Another comparison, showing the effect of discharge load on the cell's capacity and how this influences the selection of a battery for an application, is illustrated in Fig. 4.6. (Table 4.6 lists the current drains for various battery-operated portable devices.) In this figure the energy density of several primary battery systems is shown under a continuous current discharge at 20°C. The zinc-carbon cell performs best under light discharge loads, but its performance falls off sharply with increasing discharge rates. The zinc/alkaline/manganese dioxide system has a higher energy density at light loads which does not drop off as rapidly with increasing discharge loads. The lithium cell has the highest energy density with reasonable retention of

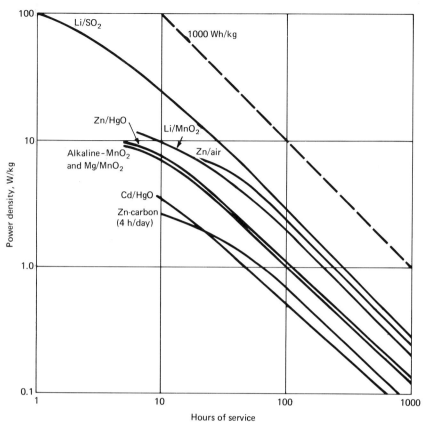

FIG. 4.3 Comparison of performance of primary battery systems—power density vs. hours of service.

TABLE 4.4 Comparison of Primary Batteries (Button Cells)

Size: "44" IEC, 1154; 11.6 mm diam. × 5.4 mm high; volume: 0.55 cm³.

| System | Voltage, V | | Capacity* | | Weight, | Energy density* | |
	Nominal	Working	mAh	mWh	g	mWh/g	Wh/L
Zn/alk/MnO₂	1.5	1.25	60	75	2.0	37.5	135
Zn/HgO	1.35	1.3	180–230	260	2.6	100	470
Zn/Ag₂O	1.5	1.55	175–200	285	2.2	130	515
Zn/AgO	1.5	1.55	245	380	2.2	170	690
Zn/air	1.25	1.25	400	500	1.7	290	905
Li/FeS†	1.5	1.4	100	140	1.1	125	370
Li/FeS₂	1.5	1.4	160	220	1.7	130	400
Li/CuO	1.5	1.4	225	315	1.7	135	570
Li/Bi₂Pb₂O₅	1.5	1.5	185	275	1.85	145	500
Li/MnO₂‡	3.0	2.85	160	450	3.0	150	410
Li/Ag₂CrO₄	3.0	3/2.7	130	370	1.7	215	670

*At approximately C/500 rate.
†Experimental 11.6 mm diam. × 3.6 mm high.
‡⅓M, equivalent to two each "44" cells, 11.6 mm diam. × 10.8 mm high.

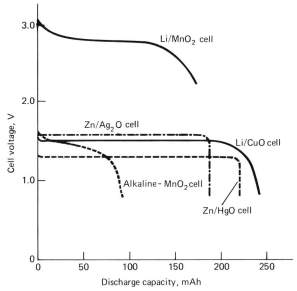

FIG. 4.4 Typical discharge curves for primary battery systems, button cell: 11.6 mm diameter, 5.4 mm high.

this performance at the higher discharge rates. For low-power applications, the service ratio of the lithium:zinc (alkaline):zinc-carbon is in the order of 4:2:1. At the heavier loads, however, such as those required for toys, motor-driven applications, and pulse discharges, the ratio can widen to 24:8:1 or greater. At these heavy loads, the selection of the premium batteries is desirable on both a performance and cost basis. Table 4.7 presents additional comparative data between the zinc-carbon and zinc-alkaline cells.

4.3.6 Effect of Temperature

The performance of the various primary batteries over a wide temperature range is shown in Fig. 4.7 (on a gravimetric basis) and in Fig. 4.8 (on a volumetric basis). The lithium/soluble cathode systems ($Li/SOCl_2$ and Li/SO_2) show the best performance throughout the entire temperature range, with the higher-rate Li/SO_2 system having the best capacity retention at the very low temperatures. The zinc/air system has a high energy density at normal temperatures, but only at light discharge loads. The lithium/solid-cathode systems, represented by the Li/MnO_2 system, show high performance over a wide temperature range, superior to the conventional zinc anode systems. Figure 4.8 shows the improvement in the relative position of the denser, heavier battery systems when compared on a volumetric basis. Figures 3.23 and 3.24 in Part 1 show similar comparative information for several secondary batteries. (Note: As stated in Part 1, Sec. 3.2, these data are necessarily generalized and present each battery system under favorable discharge conditions. With the variability in performance due to manu-

TABLE 4.5 Comparison of Primary Batteries (Cylindrical Cells)

	Zinc-carbon (standard)	Zinc-carbon (heavy-duty ZnCl₂)	Zn/MnO₂ (alkaline)	Zn/HgO	Mg/MnO₂	Li/SO₂	Li/SOCl₂ (bobbin type)	Li/MnO₂	Li/CuO	Secondary cells Sealed Pb-acid	Sealed Ni-Cd
Working voltage, V	1.2	1.2	1.2	1.25	1.75	2.8	3.3	2.8	1.5	2.0	1.2
						D-size cells (54 cm³)					
Ah	4.5	6	10	14	7	8	10.2			2.7	3.5
Wh	5.4	7.2	12	17.5	12.2	22.4	34			5.4	4.3
Weight, g	85	93	125	165	105	85	100			180	140
Wh/kg	65	75	95	105	115	260	340			30	31
Wh/L	100	135	220	325	225	415	675			100	80
						N-size cells (3.0 cm³)					
Ah	0.42	0.65			0.5			1.0*			
Wh	0.5	0.78			0.87			2.8			
Weight, g	6.3	9.5			5.0			13			
Wh/kg	75	80			170			215			
Wh/L	160	260			290			410			
						AA-size cells (7.7 cm³)					
Ah	1.0	1.7				1.0	1.6		3.4		0.5
Wh	1.2	2.0				2.8	5.2		5.0		0.6
Weight, g	14.7	23				14	19		17.4		28
Wh/kg	80	86				200	275		275		21
Wh/L	170	250				360	670		650		200

*2N size.

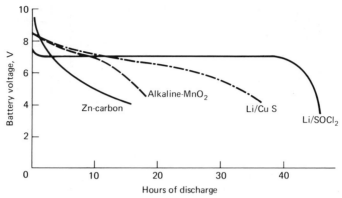

FIG. 4.5 Typical discharge curves for primary battery systems, NEDA 1604 cell, 9-V, 250 Ω discharge load.

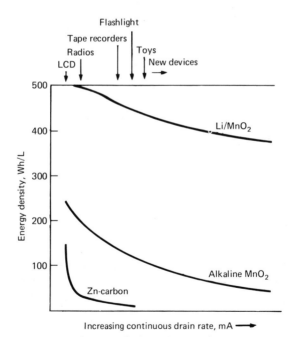

FIG. 4.6 Comparison of primary battery systems under various applications and continuous discharge loads at 20°C. (*Courtesy of Duracell, Inc.*)

TABLE 4.6 Current Drain in Battery-Operated Portable Devices

Device	Current drain, mA
Radios:	
With 9-V battery	8–12 (low volume)
	10–15 (medium)
	15–45 (high)
With cylindrical cells	10–20 (low volume)
	20–30 (medium)
	30–100 (high)
Cassette recorders	70–130 (low)
	90–150 (medium)
	100–200 (high)
Calculators:	
LED (9-V battery)	20–30
LED (AA batteries)	40–100
LCD	Under 1
Fluorescent lamp (6-W bulb)	500–1000
Flashlights	250–1000
Toys:	
Motorized type	400–2000
Electronic games	20–200
Video games	20–200
Cameras:	
Photoflash	1000–2000
Autowind	200–300
Watches:	
LCD	10–25 (backlighted)
LED	10–40
Portable TV	500–1500

TABLE 4.7 Impedance and Efficiency Comparison of Zinc/Alkaline/Manganese Dioxide vs. Zinc-Carbon Cells

	D size cells, continuous discharge at 20°C	
	Zinc-carbon*	Alkaline
Theoretical capacity, Ah	6.0	10.9
Initial impedance, Ω at 1000 Hz	0.3	0.04
Final impedance, Ω at 1000 Hz	2.5	0.045
Hours to 0.65 V at 0.5 Ω	0.5	3.0
1.0 Ω	1.0	6.5
2.25 Ω	3.5	19.0
4.0 Ω	9.0	36.0
10.0 Ω	47.0	85.0
% efficiency to 0.65 V at 0.5 Ω	12	47
1.0 Ω	14	53
2.25 Ω	23	73
4.0 Ω	35	81
10.0 Ω	70	85

*General-purpose flashlight cell.

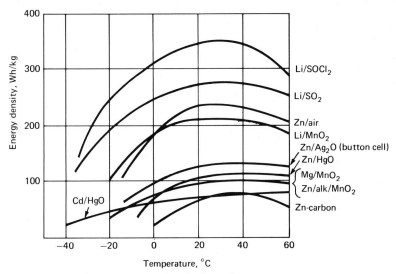

FIG. 4.7 Gravimetric energy density of primary battery systems.

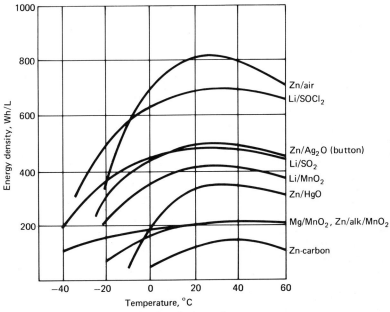

FIG. 4.8 Volumetric energy density of primary battery systems.

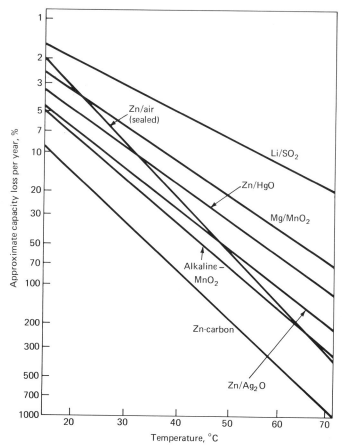

FIG. 4.9 Shelf life characteristics of primary battery systems.

facturer, design, size, discharge conditions, and other factors, they may not apply under specific conditions of use. For these details refer to the appropriate chapter for each battery system.)

4.3.7 Shelf Life of Primary Batteries

The shelf life characteristics of the major primary battery systems are plotted in Fig. 4.9 and show the rate of loss (in terms of percentage capacity loss per year) from 20 to 70°C. The relationship is approximately linear when the log of capacity loss is plotted against the log 1/Temperature (Kelvin). The data assume that the rate of capacity loss remains constant throughout the storage period, which is not necessarily the case with most battery systems. For example, as shown in Part 2 in Figs. 11.16 and 11.47 for several lithium batteries, the rate of loss tapers off as the storage period is extended. The data are also a generalization of the capability of each battery system because of the many variations in battery design and formulation. The discharge conditions and size also have an influence.

The storageability of batteries improves as the storage temperature is lowered. Cold storage of batteries is used to extend their shelf life. Moderately cold temperatures, e.g., 0°C, are usually used as freezing could be harmful for some battery systems and/or designs.

4.3.8 Comparison of Service Life

A comparison of the performance of several types of primary batteries (as well as the sealed lead-acid and nickel-cadmium batteries) in a radio transceiver at several temperatures is given in Part 1, Fig. 3.26. This shows the advantage of the newer battery systems in a typical and key application.

4.3.9 Cost

The approximate initial cost of the different primary battery systems in terms of dollars per kilowatthour and based on the rated capacity is shown in Table 4.2. In practice, there is a considerable range in cost, depending on the size or capacity of the cell because of the higher cost per kilowatthour of the smaller cells.

In the selection of the most cost-effective battery for a given application, other factors in addition to cost should also be considered. These include the battery's performance under the specfic conditions of use, operation under other temperature and environmental conditions (if applicable), shelf life and other such parameters which could effect the battery's capabilities. The impact of the discharge rate and duty cycle on the cost of battery operation (cost per hour of service) is shown in Table 4.8, which compares the service life and cost per hour of service of the general-purpose and premium (zinc chloride) zinc-carbon cells with the zinc/alkaline/manganese dioxide cell under various regimes. Based on the unit cell cost shown in the table, the general-purpose zinc-carbon cell has the lowest hourly service cost only on the low-drain intermittent radio application, the conditions most to its liking. However, even with its significantly higher unit cost, the alkaline cell is, by far, the more economical battery to use under the stringent continuous-drain applications, such as in toys. The benefit of the intermittent duty cycle for the zinc-carbon system is evidenced by its relatively better performance on the intermittent flashlight test when compared with the alkaline cell. Also evident is the cost advantage of using the larger cell (D size compared with C size) because the unit cell cost is not proportional to capacity. (Note: In the case shown, in fact, the same unit cost was used for both cell sizes.)

4.4 RECHARGING PRIMARY BATTERIES

Recharging primary batteries is a practice that should be avoided because the cells are not designed for that type of use. In most instances it is impractical, and it could be dangerous with cells that are tightly sealed and not provided with any mechanism to permit the release of gases that form during charging. Such gassing could cause a cell to leak, rupture, or explode, resulting in personal injury, damage to equipment, and other such hazards. Most primary cells and batteries are labeled with a cautionary notice advising that they should not be recharged.

Some Leclanche zinc-carbon cells can be recharged for several cycles under care-

TABLE 4.8 Comparison of Battery Costs—Zinc-Carbon vs. Zinc/Alkaline/Manganese Dioxide Cells

Type of test	Performance			Cost per hour of service, $		
	General-purpose zinc-carbon	Premium zinc-carbon	Alkaline-manganese dioxide	General-purpose zinc-carbon	Premium zinc-carbon	Alkaline-manganese dioxide
			C size cell			
4-Ω flashlight[a]	250 min	550 min	880 min	0.096	0.055	0.068
4-Ω toy[b]	95 min	280 min	700 min	0.253	0.107	0.086
40-Ω radio[c]	75 h	100 h	165 h	0.0053	0.0050	0.0061
8-Ω cassette[d]	6 h	17 h	32 h	0.0666	0.0294	0.0313
0.5-mA clock	250 days	375 days	425 days	0.00160[e]	0.00133[e]	0.00235[e]
Approximate unit cell cost,$	0.40	0.50	1.00			
			D size cell			
2.25-Ω flashlight	270 min	510 min	835 min	0.089	0.059	0.072
2.25-Ω toy	105 min	320 min	780 min	0.228	0.093	0.077
25-Ω radio	110 h	130 h	180 h	0.00364	0.00385	0.00555
8-Ω cassette	22 h	37 h	60 h	0.0182	0.0135	0.0166
1.5-A electronic flash	N.A.	N.A.	1260 flashes			
Approximate unit cell cost,$	0.40	0.50	1.00			

[a]Flashlight test (light): 4 min/h, 8 h/day to 0.9 V.
[b]Toy test: Continuous discharge to 0.8 V.
[c]Radio test: 4 h/day to 0.9 V.
[d]Cassette test: 2 h/day.
[e]Cost per day.

fully controlled conditions. For successful recharging, the battery voltage on discharge should not be below 1.0 V per cell when it is removed from service for charging. The battery should be put on charge very soon after removal from service and charged at a low rate (the charging rate should be lower than the C/15 rate). The battery should be returned to service soon after recharging for the shelf life after recharging is poor. Charging should not be attempted with any other primary battery system.

Several of the typical primary battery systems, such as the zinc/alkaline/manganese dioxide and zinc/silver oxide systems, also have been designed as rechargeable batteries. These battery systems are covered in Part 3, Chap. 25.

4.5 STANDARDIZATION OF PRIMARY BATTERIES

The standardization of primary batteries was started about 1920 in order to ensure uniformity in the manufacture of batteries and proper fit between the battery and the battery-operated equipment. Several standards now exist, including:

1. "Specifications for Dry Cells and Batteries," American National Standards Institute.[3]
2. "Primary Batteries—Part 1: General," 1976 and "Primary Batteries—Part 2: Specification Sheets," 1977, with amendments, International Electrotechnical Commission.[2]
3. "NEDA Battery Cross-Reference Guide," National Eletronic Distributors Association.[4]
4. U. S. Military Standards, Department of Defense, Washington, D.C.
5. Military and commercial standards of various countries, for example, Japanese Industrial Standards (JIS).
6. Manufacturers' type designations.

Unfortunately, these systems do not use the same nomenclature, a situation exacerbated by the independent nomenclature system of each battery manufacturer. Cross references are generally available, however, and several are included in Tables 4.9 to 4.13.

The International Electrotechnical Commission (IEC)[2] system is based on three unit cell types with a characteristic shape:

Cylindrically shaped cells, designated by letter R (round)

Prismatic-shaped cells, designated by letter S (square)

Flat, prismatic-shaped cells, designated by letter F (flat)

Each series consists of unit cells of different sizes; the size is defined by nominal dimensions to which the letter R, S, or F and a following number specifically relate. Table 4.9 lists the round cells in the standard. The IEC standard covers six electrochemical systems; the electrochemical system in a cell or battery is denoted by a letter before the size designation. Table 4.10 lists the letters assigned to the electrochemical system. Multicell batteries are identified by preceding the cell designation with a

TABLE 4.9   IEC Designation and Dimensions of Round Cells and Batteries*

Designation		Nominal cell dimensions, mm		Maximum battery, dimensions, mm		Approx. weight,
IEC	ANSI	Diameter	Height	Diameter	Height	kg
R 08		. . .	. . .	11.6	3.5	
R 06		10	22	. . .	. . .	
R 03	AAA	. . .	. . .	10.5	44.5	8.2
R 01		. . .	. . .	12.0	14.7	
R 0		11	19	. . .	. . .	
R 1	N	. . .	. . .	12.0	30.2	
R 3		13.5	25	. . .	. . .	
R 4	R	13.5	38	. . .	. . .	10.4
R 6	AA	. . .	. . .	14.5	50.5	15
R 8	A	16	47.8	. . .	. . .	21
R 9		. . .	. . .	16.0	6.2	
R 10		. . .	. . .	21.8	37.3	
R 12	B	. . .	. . .	21.5	60.0	35
R 14	C	. . .	. . .	26.2	50.0	45
R 15		24	70	. . .	. . .	
R 17		25.5	17	. . .	. . .	
R 18		25.5	83	. . .	. . .	
R 19		32	17	. . .	. . .	
R 20	D	. . .	. . .	34.2	61.5	100
R 22	E	32	75	. . .	. . .	132
R 25	F	32	91	. . .	. . .	159
R 26	G	32	105	. . .	. . .	181
R 27	J	32	150	. . .	. . .	272
R 40	6	. . .	. . .	67.0	172.0	998
R 41		. . .	. . .	7.9	3.6	
R 42		. . .	. . .	11.6	3.6	
R 43		. . .	. . .	11.6	4.2	
R 44		. . .	. . .	11.6	5.4	
R 45		. . .	. . .	9.5	3.6	
R 48		. . .	. . .	7.9	5.4	
R 50		. . .	. . .	16.4	16.8	
R 51		. . .	. . .	16.5	50.0	
R 52		. . .	. . .	16.4	11.4	
R 53		. . .	. . .	23.2	6.1	

*Where a single-cell battery is specified, the maximum dimensions are given instead of nominal dimensions; weight is approximate, not specified in standards.

SOURCE: International Electrotechnical Commission (IEC Standard)[2] and American National Standards Institute C 18.1 (1965).

number denoting the number of cells in series. If cells are connected in parallel, a number denoting the quantity of parallel groups follows the cell designation and is connected to it by a hyphen. For example, a battery designated 3R20-2 consists of two parallel groups of cells, each group comprising three cylindrical cells, 20-size, of the zinc-carbon type, connected in series.

The American National Standard Specification[3] uses the same designation for the electrochemical system and for the flat cells. Multicell batteries are listed by their NEDA codes. The National Electronic Distributors Association[4] uses a different code for designating the electrochemical system. The NEDA code is also listed in Table 4.10.

The predominant specification for United States military primary batteries is Specification MIL-B-18 entitled "Batteries, Dry," although a number of other specifications have been published covering the special requirements of the military services. Military dry batteries are type-designated as Battery BA-0000; Leclanche or zinc-carbon batteries use a number from 1-999 following the hyphen; a four-digit number is used for the other electrochemical systems as listed in Table 4.10.

TABLE 4.10   Designations Assigned to Electrochemical Systems

System	Letter designation*			U.S. DOD
	IEC	ANSI	NEDA	
Zinc-carbon	No identifying letter			0-999
Zinc-oxygen ($NH_4Cl,ZnCl_2$ electrolyte)	A	——	——	——
Mg/MnO_2	——	——	——	4000 series
Alkaline-MnO_2	L	L	A	3000 series
Zn/HgO	M	M	M	1000 series
Zn/HgO with MnO_2	N	N	M	1000 series
Zn/Ag_2O	S	S	SO	——
Industrial	——	——	C	——
Heavy duty	——	——	D	——
Economy	——	——	D	——
Flashlight	——	——	F	——
Fahnestock	——	——	FC	——
Zinc-carbon (low temperature)	——	——	——	2000 series
Lithium	——	——	L	5000 series
Mercury premium (Zn/HgO)	——	——	MD	——
Mercury photographic (Zn/HgO)	——	——	MP	——
Photographic	——	——	P	——
Silver oxide photographic (Zn/Ag_2O)	——	——	SOP	——

*Note: —— means not specified.
SOURCE: IEC,[2] ANSI,[3] NEDA,[4] and U.S. DOD, U.S. Dept. of Defense nomenclature.

The Japanese and other standards may also use a four-digit number to describe flat or coin cells. In this system, the first two digits represent the diameter of the cell, in millimeters, and the last two digits represent the height or thickness of the cell, in tenths of millimeters. For example, the BR 2016 lithium cell (Table 4.13) is 20 mm in diameter and 1.6 mm high.

A cross reference of the unit cell batteries, taken from ANSI C18.1, 1979, and listing the cell designations and dimensions is given in Table 4.11. Table 4.12 lists similar information for multicell batteries.[3] The appropriate standards should be consulted for further details and test specifications. Table 4.11 (Part 2, Chap. 5) is an excerpt from the ANSI C18.1, 1979 standard,[3] giving the performance requirements for zinc-carbon unit cells and batteries. The NEDA cross reference, listing the manufacturers' type designations, is reproduced as Table 4.13.[4]

TABLE 4.11 Unit Cell Batteries—Cell Designations, Voltages, and Dimensions

Cell designation*				Diameter, mm		Height overall, mm	
ANSI	IEC	NEDA	Voltage,* V	Max	Min	Max	Min
			Carbon-Zinc				
N	R-1	910	1.5	12.0	10.7	30.2	28.2
AAA	R-03	24	1.5	10.5	9.5	44.5	42.5
AA	R-6	15	1.5	14.5	13.5	50.5	48.4
C	R-14	14	1.5	26.2	24.7	50.0	47.6
D	R-20	13	1.5	34.1	32.2	61.5	58.7
6	R-40	905,906,911	1.5	67.0	63.0	172	162.0
			Alkaline Manganese Dioxide				
L15	LR-53	1129A	1.5	23.2	22.6	6.1	5.4
L20	LR-1	910A	1.5	12.0	10.7	30.2	28.2
L30	LR-03	24A	1.5	10.5	9.5	44.5	42.5
L40	LR-6	15A	1.5	14.5	13.5	50.5	48.4
L70	LR-14	14A	1.5	26.2	24.6	50.0	47.6
L90	LR-20	13A	1.5	34.1	32.2	61.5	58.8
			Silver Oxide				
S4	. . .	. . .	1.5	5.6	5.3	3.4	3.0
S5	SR-41	. . .	1.5	7.9	7.5	3.6	3.3
S6	SR-48	. . .	1.5	7.9	7.5	5.3	4.8
S11	SR-43	. . .	1.5	11.6	11.2	4.2	3.8
S15	SR-44	1107SO	1.5	11.6	11.2	5.3	4.8
			Mercuric Oxide*				
N4	. . .	. . .	1.4	5.6	5.3	3.4	3.0
N5	NR-41	. . .	1.4	7.9	7.6	3.6	3.3
N6	NR-48	. . .	1.4	7.9	7.6	5.3	4.8
N10	NR-42	1106M	1.4	11.6	11.2	3.6	3.0
N11	. . .	. . .	1.4	11.6	11.2	4.2	3.8
N12	. . .	. . .	1.4	12.6	–	7.3	–
N15	NR-44	. . .	1.4	11.6	11.2	5.3	5.1
N20	NR-9	1104M	1.4	15.6	15.2	6.1	5.6
N25	NR-01	. . .	1.4	11.6	-	14.5	–
N30	NR-52	1105M	1.4	15.9	15.5	11.2	10.9
N35	NR-1	910M	1.4	11.9	11.2	29.0	28.2
N36	. . .	. . .	1.4	10.5	9.5	44.5	42.5
N40	NR-50	1100M	1.4	15.9	15.5	16.8	16.0
N55	NR-6	15M	1.4	14.2	13.5	50.0	49.0
N60	. . .	. . .	1.4	25.0	24.6	16.8	16.3
N70	. . .	. . .	1.4	16.3	15.5	50.0	49.3
N100	. . .	. . .	1.4	32.0	31.2	61.1	58.7
			Watch—Silver Oxide (WS Type)				
WS3	. . .	. . .	1.5	15.6	15.2	5.0	4.4
WS4	SR-41	. . .	1.5	7.9	–	3.7	–
WS6	SR-48	. . .	1.5	7.9	7.6	5.3	4.8
WS10	SR-42	. . .	1.5	11.6	11.2	3.6	3.0
WS11	SR-43	. . .	1.5	11.6	11.2	4.2	3.8
WS16	. . .	. . .	1.5	11.6	11.2	5.6	5.0
			Watch—Mercuric Oxide (WN and WM Types)				
WN5	NR-14	. . .	1.4	7.9	7.5	3.7	3.3
WN8	. . .	. . .	1.4	11.6	–	3.7	–
WN10	NR-42	. . .	1.4	11.6	11.2	3.6	3.0
WN15	NR-44	. . .	1.4	11.6	11.2	5.3	5.1

*All mercuric oxide cells are listed as N (1.4 V) type only; however, they may also be supplied by manufacturers as M (1.35 V) type. Test table data are for both M- and N-type cells.
SOURCE: Extracted from ANSI C18.1 1979.[3]

TABLE 4.12 Multicell Batteries—Battery Designations, Voltages, and Dimensions

| Battery designation | | Voltage, V | Diameter, mm | | Height overall, mm | | Length, mm | | Width, mm | |
NEDA	IEC		Max	Min	Max	Min	Max	Min	Max	Min
900	...	1.5	...	...	111.1	107.1	68.2	65.1	68.2	65.1
6	4R25	6.0	...	...	103.2	96.0	68.2	65.1	68.2	65.1
920	4R25	6.0	...	...	103.2	96.0	68.2	65.1	68.2	65.1
908,924,915D	4R25	6.0	...	...	101.6*	96.0	68.2	65.1	68.2	65.1
915,925,908D	4R25	6.0	...	...	111.1	107.1	68.2	65.1	68.2	65.1
918	4R25-2	6.0	...	...	125.4	122.2	136.5	132.5	73.0	69.0
907	4R40Y	6.0	...	...	190.5	181.8	269.9	263.5	71.4	67.5
903	...	7.5	...	...	163.5	160.3	185.7	181.0	103.2	100.0
1611	...	9.0	19.1	...	50.8	...	...	...	...	...
1604	6F22	9.0	...	...	49.2	...	26.5	24.5	17.5	15.5
1600	6F24	9.0	25.4	...	50.0	...	...	...	...	...
1602	...	9.0	...	...	69.9	67.5	35.7	34.1	34.9	32.5
1605	...	9.0	...	...	61.9	60.3	46.0	44.5	46.0	44.5
1603	6F100	9.0	...	...	80.2	78.6	65.1	63.5	51.6	50.0
904	...	9.0	...	...	163.5	...	217.9	...	103.2	...
922	...	12.0	...	...	183.4	...	265.1	...	69.1	...
926	...	12.0	...	...	125.4	122.2	136.5	132.5	73.0	69.0
923	...	12.0	...	...	106.4	102.4	136.5	132.5	73.0	69.0
220	10F15	15.0	...	...	34.9	...	15.9	...	15.1	...
221	15F15	22.5	...	...	50.8	...	15.9	...	15.1	...
215	15F20	22.5	...	...	50.8	...	27.0	...	15.9	...
225	...	22.5	...	...	56.4	...	34.9	...	27.0	...
710	...	22.5	...	...	77.8	...	89.7	...	55.5	...
210	20F20	30.0	...	...	65.1	...	27.0	...	15.9	...
201	30F40	45.0	...	...	95.3	...	63.9	...	24.5	...
205	...	45.0	...	...	106.4	...	79.4	...	60.3	...
200	45F40	67.5	...	...	96.8	...	71.4	...	34.9	...
204	60F40	90.0	...	...	95.3	...	94.6	...	34.9	...
1306A	...	4.5	16.8	...	50.8	...	...	...	...	...
1412A	...	6.0	...	...	48.2	47.6	35.6	...	9.0	...
1604A	...	9.0	...	...	49.2	...	26.2	24.6	17.5	15.9
1200M	...	2.8	16.8	...	33.4	...	...	...	...	...
1306M	...	4.2	16.8	...	49.9	...	...	...	...	...
1300M	...	4.2	16.8	...	49.9	...	...	...	...	...
1404M	...	5.6	16.7	16.5	45.2	43.3	...	...	...	...
1611M	...	8.4	19.1	...	50.8	...	...	...	...	...
1604M	...	8.4	...	...	49.2	...	26.5	24.5	17.5	15.5
1600M	...	8.4	...	...	49.2	...	26.5	24.5	17.5	15.5
1810M	...	12.6	25.4	24.6	61.1	59.9	...	...	...	...
1801M	...	11.2	25.4	...	61.1	...	...	...	...	...

*Height over body exclusive of terminal.
SOURCE: Extracted from ANSI C18.1, 1979.³

TABLE 4.13 Primary Battery Cross-Refence Guide

Voltage	NEDA Number	Berec	Bright Star	Burgess	Duracell	Eveready	Maxell	Panasonic	Ray-O-Vac	Other
6	2	...	...	...	...	724	...	...	...	VS068*
4.5	3	...	...	...	...	736	...	...	...	VS067* VS004*
1.5	4	...	462	...	...	742	...	...	...	BA-65* VS345*
6	6	PJ944*	646	F4PI	...	744	...	...	...	BA-210/U VS065*
7.5	9	...	591	...	...	717	...	...	...	BA-207/U
1.5	11	...	...	...	...	W353	...	...	...	VS141*
1.5	13	R20	...	...	...	1050*	...	...	13*	BA-802/U VS336*
1.5	13A	LR20	7520	...	MN1300	E95	AM-1	AM1	813	VS1336* BA3202/U 1200
1.5	13C	R20PP	10MC	...	...	1150	...	...	3D	HS150 WM30
1.5	13D	R20PP	110C	200	M13HD	1250	SUM-1(SG)	UMID	6D	VS343* 75 800 4650
1.5	13F	R20	10M	205	M13F	950	SUM-1(G)	UM1	2D	VS030* VS2050* HS50 BA-30 WM-10
1.5	14	R14	...	...	...	1035*	...	...	14*	VS335*
1.5	14A	LR14	7522	...	MN1400	E93	AM-2	AM2	814	VS1335*
1.5	14D	R14PP	...	100	M14HD	1235	SUM-2(SG)	UM2D	4C	BA3042/U

	14F	R14	11-M	105	M14F	935	SUM-2(G)	UM2	1C	700
1.5	15	R6		930					15*	VS035A* HS35 BA-42 WM-15
1.5	15A	LR6	7524		MN1500	E91	AM-3	AM3	815	VS334* VS1334*
1.5	15D	R6PP			M15HD	1215	SUM-3(SG)	UM3D	5AA	BA3058/U
1.5	15F	R6	59	910	M15F	915* 1015	SUM-3(G)	UM3	7AA	VS034A* HS15 WM40 BA58
1.4	15M	ZM9	7526		ZM9	E9				VS9*
1.5	24A	LR03			MN2400	E92	AM-4	AM4	824	VS1074*
1.5	24F	HP16		7	M24F	912	UM-4(SG)		400	VS074*
67.5	200			XX45		467				VS016* BA-51
45	201					455				VS055* BA-56
45	202					482				VS013* BA-59
67.5	203					457				VS082* BA-314/U
90	204			V60		490				VS090 BA-805/U
22.44.45	205					738				VS015* BA-63
45	207					484				VS344* BA223/U
15	208			U10		411				VS083* BA331/U
30	210					413				VS085* BA305
63	211					477				VS218*

TABLE 4.13 Primary Battery Cross-Refence Guide (Continued)

Voltage	NEDA Number	Berec	Bright Star	Burgess	Duracell	Eveready	Maxell	Panasonic	Ray-O-Vac	Other
45	213	...	...	U30	...	415	...	...	...	VS086*
90	214	...	...	...	...	479	...	...	...	VS219*
22.5	215	15F20	...	U15	M215	412	...	015	215	VS084* / BA-261/U
67.5	217	...	...	...	...	416	...	...	...	VS318*
15	220	10F15	...	Y15	M504	504	...	W10E	220	VS704* / BA332/U
22.5	221	15F15	...	...	M505	505	...	MV15E	221	VS705*
30	223	...	...	Y20	...	417	...	...	...	20F15R
15	224	...	...	K10	...	420	...	...	...	
22.5	225	...	...	K15	...	...	...	...	...	
30	226	...	...	K20	...	...	...	...	...	
1.5	700	...	...	2FH	...	711	...	...	...	VS101* / BA-15A
3	701	...	...	F2BP	...	...	...	...	...	VS100* / BA-205*
3	703	...	...	...	...	W356	...	...	...	VS136
3	704	...	...	...	...	750	...	...	...	VS134* / BA-208
4.5	706	3R12	...	532	...	...	...	...	...	VS346* / BA-9
3-22.5	708	...	...	5156SC	...	...	...	...	...	VS131* / BA-230*
22.5,45	709	...	...	...	...	762S	...	...	...	VS112* / BA-36
22.5	710	...	...	4156	...	763	...	...	...	VS102* / BA-2
22.5,45	711	...	...	Z30NX	...	...	...	...	...	VS114* / BA-53

1.5 to 7.5	713	...	...	...	5540	...	...	...	VS029*
4.5	714	...	...	...	5360	...	...	...	VS028*
22.5,45	716	...	...	...	10308SC	...	...	...	VS127W* / VS093*
300	722	...	...	...	...	493	...	...	BA-291/U
225	728	...	...	...	PF489	489	...	N150	VS789*
240	729	...	...	...	PF491	491	...	1010	VS791*
450	740	...	...	...	...	496	...	...	
510	741	...	...	...	PF497	497	...	1012	VS797*
1.5	900	993PP		410	M900	735	...	900	VS106* / BA35
3	901		863*	4F2H	...	W357	...	...	BA225/U / VS103*
6	902		...	4F4H	...	706	...	...	BA-222U / VS139*
7.5	903		155	4F5H	...	715	...	903	BA804/U
9	904		164	4F6H	...	716	...	904	VS140* / BA-207/U
1.5	905	6	6Ign	...	M905S	IS6	...	6Ign.S	VS006S* / HS6 / WM50*
1.5	905FC		6Ign	...	...	IF6	...	6IgnC	VS006S*
1.5	906	6	6Tel	...	...	EA6	...	6TelS	
1.5	906FC		6 Tel	...	...	EA6F	...	6TelC	VS006C* / GLF6*
6	907		146	S461	...	1461	...	641	VS039* / BA-249/U
6	908	996	460	F4M	M908	509	...	941	VS040C* / BA200/U
6	908C		...	F4H Ind.	...	...	...	WM-20	600 / 4072 / HS90
6	908D	996PP		F4M-X	...	1209	...	944	
1.5	910A	LR 1		...	MN9100	E90	AM5	810	VS1073*

TABLE 4.13 Primary Battery Cross-Refence Guide (Continued)

Voltage	NEDA Number	Berec	Bright Star	Burgess	Duracell	Eveready	Maxell	Panasonic	Ray-O-Vac	Other
1.5	910F	D23	...	...	M910	904	UM-5 (SG)	...	716	
1.4	910M	RM401	...	...	RM401H	E401E	...	...	R401	VS401*
7.5	912	...	...	...	...	1562				
69	914D	...	...	Z46HD	...	646				
6	915	PJ992	460S	F4BP	M915	510S	...	...	942	VS040S* HS10S BA803/U
6	915D	...	...	...	...	...	...	...	945	
6	918	991	158	TW1	M918	731	...	...	918	VS317* HS31
6	918D	...	...	TW1X						
6	920	...	646	F4BW	...	2744N	...	...	920	
6	921	...	...	2F4BW						
12	922	...	187	2G8H	...	1463	...	...	922	VS340*
12	923	...	...	...		2780N				
6	924	...	...	...		2745N				
12	926	...	125	TW2	M926	732	...	...	926	VS342*
12	926D	...	...	TW2X						
6	930A	...	...	...	...	520				
				All batteries in the following NEDA 1100 series are button types						
1.4	1100M	...	...	...	RM1	E1	...	HM-P	...	VS1* VS12*
1.4	1101M	...	...	...	RM12	E12	...	...	...	BA1328/U
1.4	1104M	...	...	...	RM630H	E630	...	...	...	VS630*
1.4	1105M	RM400	...	...	RM640	E640	...	HM-N	T640	VS640*
1.4	1106M	...	...	...	...	...	...	...	...	VS400*
1.5	1107SO	...	...	...	MS76	EPX76	SR44 (P)	G-13	RS76	VS1176*
1.5	1108A	...	...	...	...	E89*	...	...	...	Z89*
1.35	1109M	...	...	...	RM1R	E1N	...	...	...	BA1312/U
1.35	1110MP	RM1N	...	...	PX1	EPX1	...	H-P	RPX1	

Voltage	No.										
1.35	1111M	...	...	...	...	...	...	...	H-G	...	
1.35	1112M	...	...	RM4R	...	E4	...	...	H-L	...	BA1128/U
1.35	1113M	...	...	RM12R	...	E12N	...	...	H-U	RPX13	
1.35	1114MP	...	...	PX13	...	EPX13	...	...	HS-D	...	
1.35	1115M	...	...	RM42R	...	E42N	...	...	...	...	BA-1030/U
1.35	1116M	...	...	RM400R	...	E400N	...	...	H-B	T400N	
1.35	1117M	...	...	RM401R	...	E401N	...	...	H-R	...	
1.4	1118M	...	...	MP401H	...	EP401E	...	...	...	RP401	BA-1425/U
1.35	1119M	...	...	...	...	...	...	...	H-T	...	BA-1058/U
1.35	1121M	...	...	RM601R	...	E601	...	...	...	...	
1.4	1122M	...	...	RM625H	...	...	...	...	...	R625	
1.35	1123M	RM625N	...	RM625R	...	E625	...	...	...	...	BA-106/U
1.35	1124MP	PX625	...	PX625	...	EPX625	...	...	H-D	RPX625	
1.35	1125M	RM640	...	RM640R*	...	...	...	...	...	...	
1.35	1126MP	...	...	PX640	...	EPX640	...	...	H-N	...	
1.4	1127M	RM675H	...	RM675H	...	E675E	...	...	HM-C	R675	
1.4	1127MD	MP675H	...	MP675H	...	EP675E	...	...	H-CP	RP675	
1.35	1128MP	PX675	...	PX675	...	EPX675	...	...	H-C	RPX675	
1.5	1129A	PX825	...	PX825	...	EPX825	...	...	PX825	RPX825	
1.5	1130SO	BSR44L	...	D303	...	303	...	SR44SW	WS-14	RW42	Type A / GS14
1.5	1131SO	BSR44H	...	D357	...	357	...	SR44W	WL-14	RW42	Type J / GS13 / 228
1.5	1132SO	BSR43L	...	D301	...	301	...	SR43SW	WS-11	RW44	Type D / GS12 / 120TC / 226
1.5	1133SO	BSR43H	...	D386	...	386	...	SR43W	G-12	RW44	Type H / 260
1.5	1134SO	BSR41L	...	D384	...	384	...	SR41SW	WS-1	RW37	GS3
1.5	1135SO	BSR41H	...	D392	...	392	...	SR41W	WL-1	RW47	Type K / G3 / 247

TABLE 4.13 Primary Battery Cross-Refence Guide (Continued)

Voltage	NEDA Number	Berec	Bright Star	Burgess	Duracell	Eveready	Maxell	Panasonic	Ray-O-Vac	Other
1.5	1136SO	BSR48L	...	...	...	309	...	WS-6	RW48	GS5
1.5	1137SO	BSR48H	...	...	D393	393	SR754W	WL-6	RW48	Type F, G5
1.5	1138SO	BSR54H	...	...	D389	389	SR1130W	G-10	RW49	Type M
1.5	1139SO	BSR42L	...	...	D344	344	...	...	RW36	242
1.5	1140SO	...	...	...	D355	355	...	...	RW25S	
1.35	1151M	...	...	...	D387	387	...	...	RW51	214
1.35	1152M	...	...	...	D313	313	...	WH-3	RW52	HS-C
1.35	1153M	...	...	...	...	354	...	...	RW54	
1.35	1154M	...	...	...	D343	343	...	WH-12NM	RW56	Type B, 12UECD, HS-B, 218
1.35	1155M	...	...	...	D325	325	...	WH-1	RW57	6UDC, Type C, HS-t-5
1.35	1156M	...	...	...	D323	323	...	...	RW58	221
1.35	1157M	...	...	...	D388	388	...	...	...	
1.5	1158SO	...	...	...	D362	362	SR721SW	...	RW310	
1.5	1159SO	BSR54L	...	...	D390	390	SR1130SW	...	RW49	
1.5	1160SO	...	...	...	D391	391	SR1120W	G-8	RW40	317
1.5	1161SO	...	...	...	D394	394	SR936SW	...	RW33	
1.5	1162SO	...	...	...	D395	395	SR926SW	...	RW313	
1.5	1163SO	...	...	...	D396	396	SR726W	...	RW411	
1.5	1164SO	...	...	...	D397	397	SR726SW	...	RW311	
1.5	1165SO	...	...	...	...	399	SR926W	...	RW413	
1.5	1166A	BLR44	...	...	...	A76	LR44	LR44	RW82	
1.5	1167A	BLR43	...	...	...	186	LR43	LR43	RW84	
1.5	1168A	BLR54	...	...	...	189	LR1130	LR1130	RW89	
1.5	1169A	...	...	...	...	191	LR1120	LR1120	...	
1.5	1170SO	...	...	...	...	381	SR1120SW	...	RW30	317
1.5	1171SO	...	...	...	...	371	SR920SW	...	RW315	605
1.5	1172SO	...	...	...	...	373	SR916SW	...	RW317	

1.5	1173SO	...	...	...	361	SR721W	...	RW410	
1.5	1174SO	...	...	...	321	SR616SW	...	WIV	
1.5	1175SO	...	...	...	364	SR621SW	...	RW320	T
1.5	1176SO	...	...	...	377	SR626SW	...		
1.5	1177SO	...	...	...	366	SR1116SW	...		

All batteries in the preceding NEDA 1100 series are button types

2.8	1200M	...	...	TR132H	E132	...	...	R132	VS132*
2.7	1201MP	PX14	...	PX14	EPX14	...	H-2D	RPX14	
3.0	1202AP	...	...	PX30	EPX30	...	PX30	RPX30	
2.7	1203M	...	...	TR132R	E132N	...	H-2P		
3	1205A	...	...	9K62					
4.5	1306AP	...	...	PX21	523	...	PX21	RPX21	VS1149*
4.2	1306M	...	...	TR133	E133	...	...		VS133*
4.5	1307AP	...	...	PX19	531	...	PX19	RPX19	VS1337*
3	1308AP	...	...	PX24	532	...	PX24	RPX24	VS1339*
4.05	1311MP	...	...	PX25	EPX25	...	H-3D	RPX25	
4.05	1312MP	...	...	...	...	...	...	RPX29	
Size K	1313AP	...	...	7R31	538	...	7R31	RPX31	
4.05	1314M	...	...	TR133R	E133N	...	H-3P	...	BA1098/U
5.6	1404M	...	...	TR164	E164	...	HM-4N	T164*	VS164*
6.0	1406SOP	PX28	...	PX28	544	4SR44P	4G13	RPX28	
5.6	1407MP	PX23	...	PX23	EPX23	...	PX23	RPX23	
5.6	1408M	...	...	TR134	E134	...	...	T134*	
5.4	1409M	...	...	TR134R	E134N	...	...	T134N*	
5.4	1410M	...	...	TR164R	E164N	...	...		
Size J	1412AP	PX27	...	7K67	539	...	...	867	
5.6	1413M	...	...	PX27	EPX27	...	...	RPX27	
6.0	1414A	4LR44	...	...	A544	4LR44	4LR44		
7.0	1500M	...	...	TR165	E165	...	...	T165*	VS165*
7.0	1501M	...	...	TR175	E175	...	HM-5C	T175	
7.0	1501MD	...	...	...	EP175				
6.75	1505M	...	...	TR135R	E135N	...	...		
9	1600	PP4	...	...	226				
8.4	1600M	...	...	TR286	E286	...	...	...	VS300A*
9	1602	PP6	2N6	...	246	...	...	...	VS305*

TABLE 4.13 Primary Battery Cross-Refence Guide (Continued)

Voltage	NEDA Number	Berec	Bright Star	Burgess	Duracell	Eveready	Maxell	Panasonic	Ray-O-Vac	Other
9	1603	PP9	0920	D6	...	276	...	...	...	VS306*
9	1604	BF22(PP3)		2 U6	M1604	216		006P	1604	BA90/U
9	1604A	...	7590	...	MN1604	522	...	6AM6	A1604	VS1323*
9	1604D	6F22PP	...	2MN6	M1604HD	1222	S-006P(SG)	006PD	D1604	
8.4	1604M	...	...	...	TR146X	E146X	...	TR146	...	VS164X*
9	1605	PP7	...	B266/M6	...	266	...	...	...	BA1090/U
9	1606	...	...	Y-6	...	...	...	...	...	VS322*
9.8	1606M	...	...	...	TR177	E177	...	...	...	VS177*
9	1611	...	...	B206/L6	...	206	...	...	...	VS327*
8.4	1611M	...	...	...	TR128	E126	...	...	...	VS126*
9	1612	...	...	C6X	...	2356N	...	...	...	VS330*
9	1613	...	...	...	...	2709N	...	...	...	VS339*
8.4	1615M	...	...	...	TR136	E136	...	...	...	
9.0	1617M	...	...	...	5K65	...	...	...	...	
8.4	1619M	...	...	...	303996	E303996	...	...	...	
11.2	1801M	...	...	...	TR431	E431	...	...	T431	VS329*
12	1810	...	...	...	...	228	...	...	...	
12.6	1810M	...	...	...	TR289	E289	...	...	...	
12.6	1902M	...	...	...	304116	E304116	...	...	T304116	
3.0	5000L	LiM2016	DL2016	...	...	CR2016	BR2016	BR2016	...	
3.0	5001L	LiM2320	...	...	...	...	BR2320	BR2320	...	
3.0	5002L	...	...	...	...	...	BR2325	BR2325	...	
3.0	5003L	...	DL2025	...	...	CR2025	CR2025	...	...	
3.0	5004L	LiM2032	DL2032	...	...	CR2032	...	...	...	
6.0	5005L	...	PX28L	...	PX28L	...	...	...	...	
1.5	5006L									
1.5	5007L	LiM110	...	803	...					
3.0	5008L	...	DL1/3N	...	GL2-76					
3.0	5009L	...	DL1620							
3.0	5010L	...	DL2420							
3.0	5011L	...	DL2430							

*Discontinued number shown for reference purposes only.   SOURCE: National Electronic Distributors Association, Park Ridge, Il.[4] (Reprinted with permission.)

REFERENCES

1. H. Hazkany, E. Peled, and B. Raz, "Primary Batteries—A Forecast of Performance," *Proc. 29th Power Sources Conf.,* Electrochemical Society, Pennington, N.J., 1980.

2. International Electrotechnical Commission (IEC Standard), "Primary Batteries—Part 1: General," Publication 86-1, 1976, "Primary Batteries—Part 2: Specification Sheets," Publication 86-2, 1977, and amendments, Bureau Central de la Commission Electrotechnique Internationale, Geneva.

3. "Specifications for Dry Cells and Batteries," ANSI C18.1, American National Standards Institute, Inc., New York, NY, 10018, 1979.

4. *NEDA Battery Cross Reference Guide,* published annually by National Electronic Distributors Association, Park Ridge, IL 10068.

5

The Zinc-Carbon (Leclanche) Cell

by
Kenneth Franzese and Navnit Bharucha

5.1 GENERAL CHARACTERISTICS

The Leclanche or zinc-carbon (or carbon-zinc) cell, known for over a hundred years, is still the most widely used of all the primary batteries because of its low cost, acceptable performance, and ready availability. The zinc-carbon battery industry has been growing at a rate of about 5 to 10%/y, even with the introduction of the zinc-alkaline and other new types of primary batteries. Flashlights, using C and D size cells, were the major application of the zinc-carbon cell; now, mainly due to the many new portable electrical and electronic devices, the smaller sizes are becoming more popular.

The basic cell, which bears his name, was produced by Georges Leclanché in 1866. This cell was unique in that it was the first practical cell that used only one low-corrosive fluid as the electrolyte and a cathode in solid form. This made it relatively inactive until the external circuit was connected, resulting in good shelf or storage life, a distinct advantage over cell systems of prior origin. This cell consisted of an amalgamated zinc bar serving as the negative electrode or cell anode, a solution of ammonium chloride as the electrolyte, and a one-to-one mixture of manganese dioxide and powdered carbon packed around a carbon rod as the positive electrode or cell cathode. The positive electrode was placed in a porous pot which was, in turn, placed in a square glass jar along with the electrolyte and zinc bar. Later, in 1876, Leclanché removed the need for a porous pot by adding a resin (gum) binder to the manganese dioxide–carbon mix and forming this composition into a compressed block by the use of hydraulic pressure and a temperature of 100°C.

In 1888 Dr. Carl Gassner constructed the first "dry" cell. It was similar to the Leclanche system except that manganese dioxide was not used as the cathode. Gassner's cell also differed in physical construction in four important areas. First, the zinc element was formed into a cup and therefore took on the additional duty of containing all the other ingredients. Second, the electrolyte was immobilized by a paste (plaster of paris and ammonium chloride). The cylindrical block of cathode mix (called a bobbin) was wrapped in cloth which was saturated with the electrolyte. Third, zinc chloride was added to the ammonium chloride electrolyte solution, reducing the local chemical action and improving the shelf life. Fourth, this construction was capable of commercial development, which started soon after.

It is of interest to note that except for a few changes the cells just described are closely related to modern zinc-carbon batteries. However, the advances that have been made with respect to the cathode material, the electrolyte, the seal, and the overall manufacturing process have extended the discharge life and storage life of the Leclanche-type battery over 400% since 1910.[1,2]

The advantages and disadvantages of the Leclanche cell, compared with other primary cell systems, are summarized in Table 5.1. A comparison of the more popular primary battery systems is given in Part 2, Chap. 4. The capacity figures are taken for discharge conditions favorable to each system. The cost figures are based on 1980 data.

5.2 CHEMISTRY

The zinc-carbon cell uses a zinc anode, a manganese dioxide cathode, and an electrolyte of ammonium chloride and/or zinc chloride dissolved in water. Powdered carbon (acetylene black) is mixed with the manganese dioxide to improve conductivity and

TABLE 5.1 Major Advantages and Disadvantages of the Leclanche Cell and Zinc Chloride Cell

Advantages	Disadvantages	General comments
	Standard Leclanche cell	
Low cell cost	Low energy density	Good shelf life if
Low cost per watthour	Poor low-temperature	refrigerated
Large variety of shapes,	service	For best capacity the
sizes, voltages, and	Poor leakage resistance	discharge should be
capacities	under abusive conditions	intermittent
Various formulations	Low efficiency under	Capacity decreases as the
Wide distribution and	high current drains	discharge drain increases
availability	Comparatively poor shelf	Steadily falling voltage is
Long tradition of	life	useful if early warning
reliability	Voltage falls steadily with	of end of cell life is
	discharge	important
	Zinc chloride cell	
Higher energy density	High gassing rate	Steadily falling voltage
Better low-temperature		with discharge
service		Good shock resistance
Good leak resistance		Low to medium initial
High efficiency under		cost
heavy discharge loads		

retain moisture. As the cell is discharged, the zinc is oxidized and the manganese dioxide is reduced. A simplified overall cell reaction is

$$Zn + 2MnO_2 \rightarrow ZnO \cdot Mn_2O_3$$

The chemical processes which occur in the Leclanche cell are more complicated, and, despite the 100 y of its existence, there is still controversy over the details of the electrode reactions.[3]

The chemical reaction is dependent on such things as electrolyte concentration, cell geometry, discharge rate, discharge temperature, depth of discharge, diffusion rates, and type of MnO_2 used. Furthermore, the chemistry is complex because MnO_2 is a nonstoichiometric oxide and is more accurately represented as $MnO_{1.9}$. A more comprehensive description of the cell reaction is as follows:[2]

1. For cells with ammonium chloride as the primary electrolyte:
   Light discharge: $Zn + 2MnO_2 + 2NH_4Cl \rightarrow 2MnOOH + Zn(NH_3)_2Cl_2$
   Heavy discharge: $Zn + 2MnO_2 + NH_4Cl + H_2O \rightarrow 2MnOOH + NH_3 + Zn(OH)Cl$
   Prolonged discharge: $Zn + 6MnOOH \rightarrow 2Mn_3O_4 + ZnO + 3H_2O$

2. For cells with zinc chloride as the primary electrolyte:
   Light or heavy discharge: $Zn + 2MnO_2 + 2H_2O + ZnCl_2 \rightarrow 2MnOOH + 2Zn(OH)Cl$
   Or: $4 Zn + 8MnO_2 + 9H_2O + ZnCl_2 \rightarrow 8MnOOH + ZnCl_2 \cdot 4ZnO \cdot 5H_2O$
   Prolonged discharge: $Zn + 6MnOOH + 2Zn(OH)Cl \rightarrow 2Mn_3O_4 + ZnCl_2 \cdot 2ZnO \cdot 4H_2O$

(Note: $2MnOOH$ is sometimes written as $Mn_2O_3 \cdot H_2O$ and Mn_3O_4 as $MnO \cdot Mn_2O_3$.)

In the theoretical case, as discussed in Part 1, Chap. 1, the specific capacity calculates to 224 Ah/kg, based on Zn and MnO_2 and the simplified cell reaction. On a more practical basis, the electrolyte, carbon black, and water are ingredients which cannot be omitted from the system. If typical quantities of these materials are added to the "theoretical" cell, a specific capacity of 96 Ah/kg is calculated. This is the highest specific capacity a general-purpose cell can have and is, in fact, approached by some of the larger Leclanche cells under certain discharge conditions. The actual specific capacity of a practical cell, considering all the cell components and the efficiency of discharge, can range from 75 Ah/kg on very light loads to 35 Ah/kg on heavy-duty intermittent discharge conditions.

5.3 TYPES OF CELLS

During its history, the zinc-carbon cell has undergone a gradual change in formulation, design, and performance. Recently, with the tendency toward specialization, or maximizing performance for particular uses, a number of different types have been introduced into the market. Although there are many variations, the cylindrical zinc-carbon cell can be classified into three different types:

1. The traditional, regular cell, which is not too different from the one introduced in the late nineteenth century, uses zinc as the anode, ammonium chloride (NH_4Cl)

as the main electrolyte component, a starch paste separator, and natural manganese dioxide (MnO_2) ore as the cathode. Batteries of this formulation and design are the least expensive and are recommended for general-purpose use. Hence, they are called "general-purpose" cells. They are best when used intermittently, for low-rate discharge, or when the cost is more important than superior service or performance.

2. The industrial "heavy-duty" zinc-carbon cell characteristically utilizes ammonium chloride and/or zinc chloride ($ZnCl_2$) as the electrolyte and synthetic electrolytic manganese dioxide (EMD) alone or with natural ore as the cathode. Its separator may be of the starch paste or paper liner type. This grade is good for heavy intermittent service, industrial applications, or medium-rate continuous discharge.

3. Known in the trade as the "zinc-chloride" cell, this type is also referred to as the "heavy-duty" battery. (For the purpose of clarity it will be identified as the "extra-heavy-duty" carbon-zinc cell.) This cell is composed mainly of an electrolyte of zinc chloride with perhaps a small amount of ammonium chloride. The ore used for the cathode is exclusively electrolytic (EMD). In addition, the cell is almost always constructed with a special separator (paper liner) in place of the paste type, although many manufacturers today use the special separator in almost all their Leclanche-type batteries. This cell performs well on heavy intermittent service and is particularly good for medium and heavy continuous discharge. It also has improved low-temperature characteristics and reduced electrolyte leakage.

Generally, the higher the grade or class of carbon-zinc cell, the lower the cost per minute of service. The price difference between classes is about 10 to 25%, but the performance difference can be from 30 to 100% in favor of the higher class.

5.4 CONSTRUCTION

The zinc-carbon cell is made in many sizes and a number of designs but in two basic constructions: cylindrical and flat. Similar chemical ingredients are used in both constructions.

5.4.1 Cylindrical Cell

In the common cylindrical cell (Figs. 5.1 and 5.2), the zinc can serves as the cell container and anode. The manganese dioxide is mixed with acetylene black, wet with electrolyte, and shaped into the form of a bobbin. A carbon rod is inserted into the bobbin. The rod serves as the current collector for the positive electrode. It also provides structural strength and is porous enough to permit the escape of gases, which accumulate in the cell, without allowing leakage of electrolyte. The separator, which physically separates the two electrodes and provides the means for ion transfer through the electrolyte, can be a cereal paste wet with electrolyte (Fig. 5.1) or a treated absorbent kraft paper in the newer "paper-lined" cell (Fig. 5.2). This provides a thinner separator spacing and lower internal resistance. Single cells are covered with metal, cardboard, or plasticized paper jackets for aesthetic purposes and to minimize the effect of electrolyte leakage.

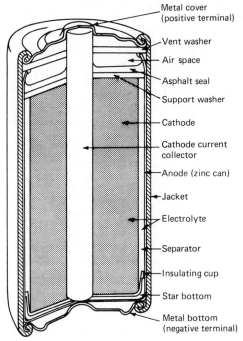

Metal cover
(positive terminal)

Vent washer

Air space

Asphalt seal

Support washer

Cathode

Cathode current
collector

Anode (zinc can)

Jacket

Electrolyte

Separator

Insulating cup

Star bottom

Metal bottom
(negative terminal)

FIG. 5.1   Cross section of a Leclanche cylindrical cell
(paste separator).

5.4.2 Inside-Out Cylindrical Cell

Another cylindrical cell is the "inside-out" construction shown in Fig. 5.3. In this cell, an injection-molded impervious inert carbon wall serves as the container of the cell and as the current collector. The zinc anode, in the shape of vanes, is located inside the cell and is surrounded by the cathode mix. This structure results in more efficient zinc consumption and, because zinc is not used as the container, a high degree of leakage resistance.

5.4.3 Flat Cell

The flat cell is illustrated in Fig. 5.4. In this construction, a duplex electrode is made by coating a zinc plate with carbon so that the zinc of one cell is in intimate contact with the carbon of the next cell. No expansion chamber or carbon rod is used as in the cylindrical cell. This construction increases the available space for the cathode mix and therefore increases the energy density. In addition, a rectangular construction reduces wasted space in multicell assemblies (where, in fact, the flat cell is only used). The volumetric energy density of an assembled battery using flat cells is nearly twice that of assemblies of cylindrical cells.

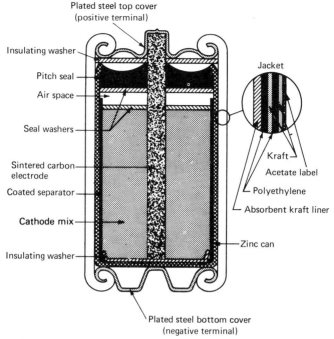

FIG. 5.2 Cross section of a Leclanche cylindrical cell (paper-liner separator). (*Courtesy of Bright Star Industries.*)

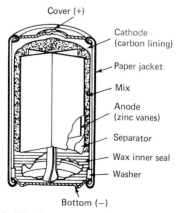

FIG. 5.3 Cross section of a Leclanche inside-out cell.

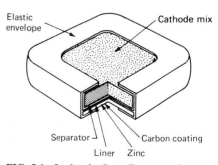

FIG. 5.4 Leclanche flat-cell construction.

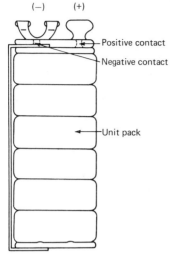

FIG. 5.5   Leclanche flat-cell battery assembly.
(*From Eveready Battery Engineering Data.*)

Metal straps are used to attach the ends of the assembled battery to the battery terminals (e.g., 9-V transistor battery). The entire cell assembly is usually encapsulated in wax or plastic. Figure 5.5 illustrates this construction.[2]

5.4.4 Special Designs

Several new designs for special applications have recently been introduced. These are covered in Sec. 5.7.

5.5 CELL COMPONENTS

5.5.1 Zinc

Battery grade zinc is 99.99% pure. However, for zinc cans an alloy of zinc containing up to about 0.3% of cadmium and somewhat more of lead is used. Lead contributes to the forming qualities of the can, although too much lead softens the zinc. Cadmium makes the zinc corrosion-resistant to ordinary dry cell electrolytes and adds strength to the zinc. For drawing, only 0.1% or less of cadmium is used because more would make the zinc difficult to draw. Zinc cans are commonly made by three different processes:

1. Zinc is rolled into a sheet, formed into a cylinder, and, with the use of a punched-out zinc disk for the bottom, soldered together.

2. Zinc is deep-drawn into a can shape.

3. Zinc is subjected to impact extrusion from a thick, flat calot.

It should also be noted that metallic impurities such as copper, nickel, iron, and cobalt cause corrosive reactions with the zinc and must be avoided. In addition, iron makes zinc harder and less workable. Tin, arsenic, antimony, magnesium, etc., make the zinc brittle.[2,4]

5.5.2 Bobbin

The bobbin is the positive electrode and is also called the black mix, depolarizer, or cathode. It is a wet powder mixture of MnO_2, powdered carbon black, and electrolyte (NH_4Cl, $ZnCl_2$, and water). The powdered carbon serves the dual purpose of adding electrical conductivity to the MnO_2 which has high electrical resistance and of holding the electrolyte.

The bobbin usually contains ratios of MnO_2 to powdered carbon from 3:1 to as much as 10:1 by weight. Also, 1:1 ratios are used in batteries for photo-flash service where high bursts of current are more important than capacity.

5.5.3 Manganese Dioxide (MnO₂)

The types of MnO_2 used in dry cells are generally categorized as natural manganese dioxide, activated manganese dioxide, chemically synthetic manganese dioxide, and electrolytic manganese dioxide (EMD). The latter, more expensive, material gives a higher cell capacity with improved rate capability and is used in heavy or industrial applications. As shown in Fig. 5.6, polarization is much less with the electrolytic material than with the chemical or natural ores.[5] Naturally occurring ores (in Gabon,

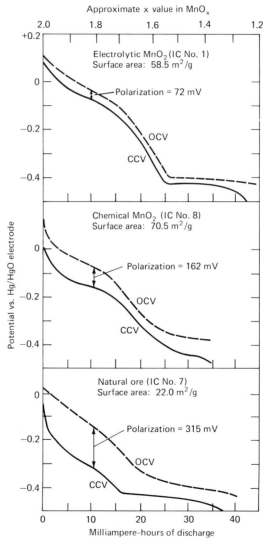

FIG. 5.6 Open- and closed-circuit voltage of three types of MnO_2 (in $9M$ KOH). (*From Kozawa and Powers.*)

Greece, and Mexico), high in battery-grade material (70 to 85% MnO_2), and synthetic forms (90 to 95% MnO_2) generally give capacities proportional to their MnO_2 content. However, service depends also on the crystalline state, the state of hydration, and the activity of the MnO_2, the electrolyte, and the separator as well as the overall construction of the cell.[2,6]

5.5.4 Carbon Black

Carbon black or powdered carbon, although chemically inert, is added to the MnO_2 to improve its conductivity because MnO_2 is a poor electrical conductor. It also serves the important functions of holding the electrolyte and providing compressibility and elasticity to the black mix.

Graphite was once used as the principal carbon black and is still used to some extent. Acetylene black, by virtue of its properties, has displaced graphite in this role. One advantage of acetylene black is its ability to hold more electrolyte in the black mix. Another is that cells containing acetylene black usually give superior intermittent service, which is the way most dry cells are used. Graphite, on the other hand, serves better for high flash currents or for continuous drains.[2,7]

5.5.5 Electrolyte

The ordinary Leclanche cell uses a mixture of ammonium chloride and zinc chloride, with the former predominating. The $ZnCl_2$ Leclanche cell uses only $ZnCl_2$ with perhaps a small amount of NH_4Cl.

Examples of electrolyte formulation for Leclanche cells are listed in Table 5.2.

TABLE 5.2   Electrolyte Formulations*

Constituent	Weight, %
Electrolyte I	
NH_4Cl	26.0
$ZnCl_2$	8.8
H_2O	65.2
Zinc-corrosion inhibitor	0.25 to 1.0
Electrolyte II	
$ZnCl_2$	15–40
H_2O	60–85
Zinc-corrosion inhibitor	0.02–1.0

*Electrolyte I based on Ref. 3; electrolyte II based on Ref. 6.

5.5.6 Corrosion Inhibitor

The zinc-corrosion inhibitor may be mercuric or mercurous chloride, which forms an amalgam with the zinc; potassium chromate or dichromate, which forms an oxide film on the zinc, or surface-active organic compounds, which coat the zinc. The inhibitor is usually part of the electrolyte or part of the coating on the paper separator.[2,6]

5.5.7 Carbon Rod

The carbon rod used in round cells is inserted into the bobbin and acts as the current collector. It also serves the function of venting hydrogen gas which might build up on the carbon rod during heavy discharge.

Raw carbon rods are porous and therefore are treated with enough oils or waxes to prevent water loss (very harmful to cell shelf life) and electrolyte leakage, but they are kept porous enough to continue to pass the hydrogen. Ideally, the treated carbon should pass hydrogen but not pass oxygen which could add to zinc corrosion during storage. Normally a positive internal pressure exists in a cell, and oxygen ingress through the carbon rod is a minor consideration.[2,4,8]

5.5.8 Separator

The separator physically separates and electrically insulates the zinc (negative) from the bobbin (positive) but permits electrolytic or ionic conduction to occur via the electrolyte. The two major types are the gelled paste and the paper coated with cereal (or other gelling agent).

The paste type is flowed into the zinc can, and the preformed bobbin (with the carbon rod) is inserted, pushing the paste up the can walls between the zinc and the bobbin. After a short time the paste sets or gelatinizes. Some paste formulations need to be stored at low temperatures in two parts. The parts are then mixed; they must be used immediately, as they can gel at room temperature. Other paste formulations need elevated temperatures (60 to 96°C) to gel. The gelling time and the temperature depend upon the concentration of the electrolyte constituents. A typical paste electrolyte uses zinc chloride, ammonium chloride, water, and starch and/or flour as the gelling agent.

The coated-paper type uses a special paper coated with cereal or other gelling agent on one or both sides. The paper cut to the proper length and width is shaped into a cylinder and, with the addition of a bottom paper, is inserted into the cell up against the zinc wall. The black mix is then metered into the can forming the bobbin, or, if the bobbin is preformed in a die, it is pushed into the can. At this time the carbon rod is inserted into the center of the bobbin and the bobbin is tamped or compressed, pushing against the paper liner and carbon rod. The compression releases some electrolyte from the black mix which soaks the paper liner, completing the operation.

By virtue of the fact that a paste separator is relatively thick compared with the paper liner, about 10% or more MnO_2 can be accommodated in a paper-lined cell, resulting in a proportional increase in capacity.[2,4,9]

5.5.9 Seal

The seal used to enclose the active ingredients can be asphalt pitch, wax and resin, or plastic (polyethylene or polypropylene). Normally, an air space is left between the seal and the top of the bobbin to allow for expansion. The seal is important to make the cell portable, prevent the evaporation of moisture, and prevent the phenomenon of "air line" corrosion from oxygen ingress.[2]

5.5.10 Jacket

The cell jacket can be made of various components: metal, paper, plastic, Mylar, plain or asphalt-lined cardboard, or foil in combination or alone. The jacket provides strength, protection, leakage prevention, electrical isolation, decoration, and site for manufacturer's label.

5.5.11 Electrical Contacts

The top and bottom of most cells are capped with tin-plated steel terminals (or brass) to aid conductivity and to finish the closure of the cell, preventing exposure of any zinc.

5.6 PERFORMANCE CHARACTERISTICS

5.6.1 Voltage

Open-Circuit Voltage   The open-circuit voltage (OCV) of the Leclanche cell is nominally given as 1.5 V; in reality it ranges from 1.5 to 1.75 V (and sometimes to 1.8 V) for fresh, unused cells.

Closed-Circuit Voltage   The closed-circuit voltage (CCV), or working voltage, of the Leclanche cell is a function of the load or current drain the cell is required to deliver. The heavier the load or the smaller the circuit resistance, the lower the closed-circuit voltage. Table 5.3 illustrates this for the D size cell.

TABLE 5.3   Initial Closed-Circuit Voltage of a Typical D size Zinc-Carbon Cell as a Function of Load Resistance at 20°C

Voltage	Load resistance, Ω	Initial current, mA
1.60	∞	0
1.60	100	16
1.59	50	32
1.58	25	63
1.56	10	156
1.50	4	375
1.41	2	705

The exact value of the CCV is determined mainly by the internal resistance of the cell as compared with the circuit or load resistance. It is, in fact, proportional to $R_l/(R_l + R_{in})$ where R_l is the load resistance and R_{in} is the cell's internal resistance. Temperature, age, and depth of discharge greatly affect the CCV and internal resistance as well.

As a Leclanche cell is discharged, the CCV, and to a lesser extent the OCV, drop in magnitude. The discharge curve is a graphical representation of the CCV as a function of time and is neither flat nor linearly decreasing but, as seen in Fig 5.7, has the character of a single- or double-S curve.

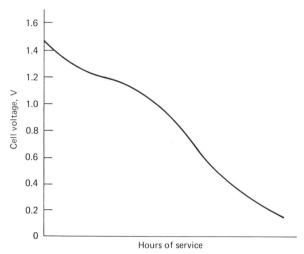

FIG. 5.7   Typical discharge curve of a Leclanche zinc-carbon cell.

End Voltage   The end voltage, or cutoff voltage (COV), is defined as a point along the discharge curve below which no usable energy can be drawn for the specified application.

Typically 0.9 V has been found to be the COV for a 1.5-V cell when used in a flashlight. Some radio applications can utilize the cell down to 0.75 V or lower, while other electronic devices may tolerate a drop to only 1.2 V. Obviously, the lower the voltage is able to drop and still operate the instrument, the more total energy the battery will deliver. Devices that can operate only within a narrow voltage range would do better with a battery system noted for its flat discharge curve.

Although a CCV which steadily decreases may present a disadvantage in some applications, it is quite popular where sufficient warning of the end of cell life is required —as in a flashlight.

5.6.2 Discharge Characteristics

The many factors influencing the performance of a battery (see Part 1, Chap. 3, Sec. 3.2), the sensitivity of the Leclanche cell to these factors, and the variety of grades of cells that are available make it necessary to be specific about cell design, discharge conditions, etc., in determining the performance of a particular battery.

Typical discharge curves for a general-purpose D size cell, discharged 2 h per day at 20°C, are shown in Fig. 5.8. These curves are characterized by a sloping curve and a substantial reduction in voltage with increasing discharge current.

5.6.3 Effect of Intermittent Discharge

The performance of the Leclanche cell is particularly sensitive to the type of discharge; performance is usually significantly better under intermittent discharge as the cell has an opportunity to recover during the rest period. This effect is particularly noticeable

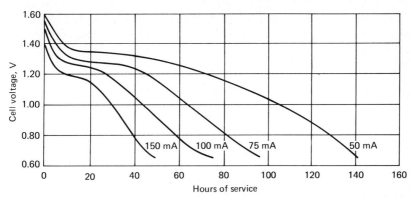

FIG. 5.8 Typical discharge curves for Leclanche zinc-carbon cell (D size), discharged 2h/day at 20°C. (*From Eveready Battery Engineering Data.*)

at the heavier discharge loads at lower temperatures. For these reasons, most of the performance characteristics of the Leclanche cell are based on a specific intermittent discharge test. Table 5.4 lists several of the typical tests.

TABLE 5.4 Standard Performance Tests: American National Standards Institute (ANSI) C 18*

Test	Discharge schedule
General purpose intermittent	One 5-min continuous discharge at 24-h intervals
Heavy intermittent	Two 1-h discharges, 6 h intervening, 16-h rest period, at 24-h intervals
Heavy industrial flashlight (HIF)	One 4-min discharge at 15-min intervals, for 8 consecutive hours each day, with 16-h rest period (128 discharge minutes daily)
Light industrial flashlight (LIF)	One 4-min discharge at 1-h intervals for 8 consecutive hours each day, with 16-h rest period (32 discharge minutes daily)
Transistor radio and electronic equipment battery	One 4-h continuous period of discharge daily
Electronic photoflash	Discharge 15 s each minute for 1-h at 24-h intervals, 5 consecutive days per week
Toy battery	Continuous discharge
Cassette	One 2-h continuous period of discharge daily

*Based on Ref. 10.

The three-dimensional graph shown in Fig. 5.9 illustrates the general effects of intermittency and discharge rate on the capacity of a general-purpose D size cell.

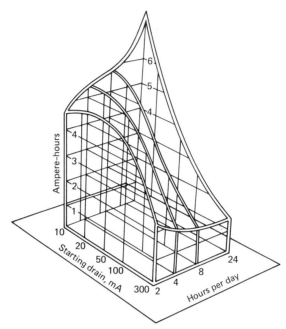

FIG. 5.9   Battery service life as a function of initial current drain and duty cycle for a general-purpose D size zinc-carbon cell at 20°C. (*From Eveready Battery Engineering Data.*)

On low-current discharges, the benefit of intermittent rest and discharge is minimal; in fact, if the discharge is very low, other factors, such as age, will reduce the total delivered capacity. Most discharges, however, fall in the moderate- (radio) to high-rate (flashlight) categories, and for these the energy delivered can more than triple when the cell is used intermittently as compared with continuously.

The standard flashlight current drains are 300 mA (4 Ω per cell) and 500 mA (2¼ Ω per cell), which correspond to two-cell flashlights using PR2 and PR6 lamps, respectively, or three-cell flashlights using PR3 and PR7 lamps, respectively. The beneficial effects of intermittent discharge are clearly shown in Figs. 5.10 and 5.11, which compare general-purpose D cells on four different discharge regimens: continuous, light industrial flashlight, heavy industrial flashlight, and general-purpose intermittent tests.

5.6.4 Comparative Discharge Curves— General-Purpose Cells

A comparison of D, C, AA, and AAA cells (see Table 5.10 for a listing of cell sizes) discharged through a relatively high resistance of 150 Ω (about 10 mA) continuously

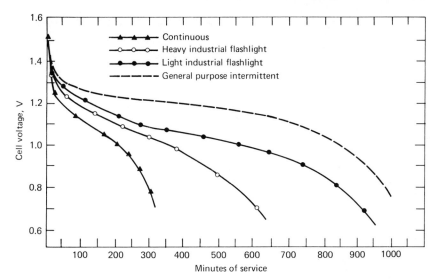

FIG. 5.10   General-purpose D size zinc-carbon cell discharged through 4.0 Ω at 20°C.

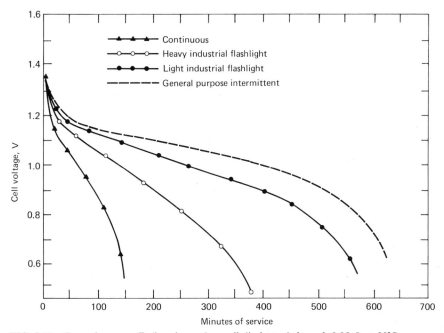

FIG. 5.11   General-purpose D size zinc-carbon cell discharged through 2.25 Ω at 20°C.

at 20°C is shown in Fig. 5.12. The performance of the D and C size cells on this continuous discharge would not be too different from that of an intermittent discharge, because, for cells of these sizes, the current drain is low. For the smaller AA and AAA

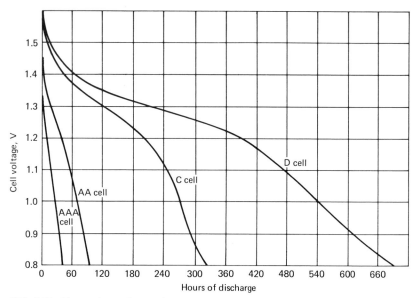

FIG. 5.12   Zinc-carbon cells, continuous discharge through 150 Ω at 20°C.

sizes, however, discharge through a 150-Ω load is a heavier burden, and an intermittent discharge would increase their service life.

Figure 5.13 shows the same four cell sizes discharged through a relatively low resistance of 10 Ω (about 150 mA) continuously. At this heavier discharge load, all cell

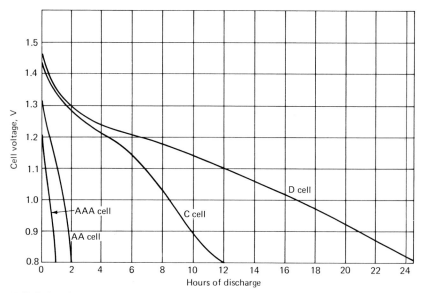

FIG. 5.13   Zinc-carbon cell, continuous discharge through 10-Ω load.

types would deliver at least twice the service on an intermittent discharge with adequate rest periods between discharge, with the smaller cells having a proportionally greater increase in service.

It is to be noted that the relative performances of the AAA, AA, C, and D sizes roughly follow a 1:2:8:16 proportion to the 0.9-V cutoff for the low rate and a 1:2:12:24 proportion for the high-rate drain, illustrating the advantage of the lower current density discharge of the larger cells.

The high discharge rate for the general-purpose C and D size cells at 375 mA (4 Ω) is shown in Fig. 5.14 compared with the performance of the general purpose No. 6 cell for which this discharge rate is not particularly high.

5.6.5 Comparative Discharge Curves—Different Cell Types

Figure 5.15 shows a comparison of the general-purpose (GP), the heavy-duty (HD), and the extra-heavy-duty (EHD) zinc chloride D size cell formulations as defined in Sec. 5.3, discharged continuously through a 2¼-Ω load. A performance ratio of 1:2:3 is obtained to the 0.9-V cutoff.

Figure 5.16 shows a comparison of the same cell formulations discharged intermittently through a 2¼-Ω load on the American National Standards Institute (ANSI) light industrial flashlight (LIF) test. On this regimen, the performance ratio is 1:1.8:2.4 to the 0.9-V cutoff. The intermittent discharge, allowing a rest period for recovery, has decreased the difference in performance between the cell formulations.

Figure 5.17 shows a comparison of the same cell formulations discharged continuously through a 4-Ω load. A ratio of 1:1.6:2.2 is obtained to the 0.9-V cutoff, less of a difference than that obtained at the heavier 2¼-Ω discharge rate. A comparison of

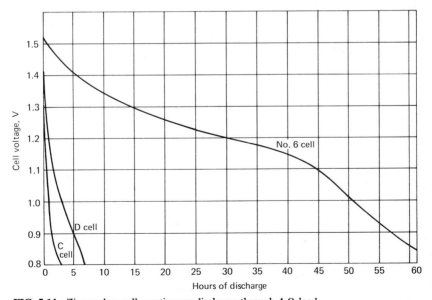

FIG. 5.14   Zinc-carbon cell, continuous discharge through 4-Ω load.

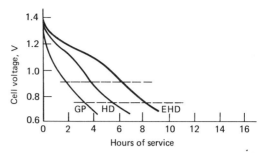

FIG. 5.15 Discharge curve of D size zinc-carbon cells continuously discharged through 2.25 Ω at 20°C. [GP: general purpose; HD: heavy duty; EHD: extra-heavy duty (zinc chloride cell).]

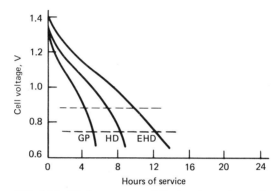

FIG. 5.16 Discharge curve of D size zinc-carbon cells discharged on the ANSI LIF test (4 min/h, 8 h/day) through 2.25 Ω at 20°C.

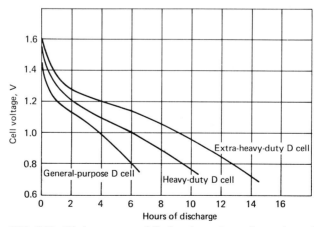

FIG. 5.17 Discharge curve of D size zinc-carbon cells continuously discharged through 4 Ω at 20°C.

an intermittent discharge at 4 Ω on the ANSI light industrial flashlight (LIF) test is shown in Fig. 5.18. On this regimen, the performance ratio drops to 1:1.5:1.8.

These cell grades are compared once again in Fig. 5.19 on a moderate discharge through a 25-Ω resistor for 4 h continuously with 20 h of rest on the ANSI transistor radio and electronic equipment battery test. At this more moderate discharge load, the performance ratio is 1:1.3:1.5 to 0.9-V cutoff.

Continuous discharge tends to increase the difference in performance between the different types of cells of the same size, while intermittent discharges tend to reduce the difference. Similarly, higher discharge currents tend to increase the performance difference.

Figure 5.20 summarizes the performance of a general-purpose D size cell discharged continuously to different end voltages. The performance of the same cell, but discharged intermittently (4 h each day), is shown in Fig. 5.21. Figures 5.22 and 5.23 present the same performance relationships but for a heavy-duty D size cell.

Performance differences between the batteries of the same grade but offered by

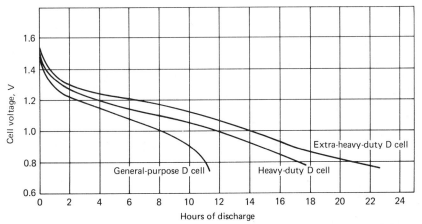

FIG. 5.18 Discharge curve of D size zinc-carbon cells discharged on the ANSI LIF test through 4 Ω at 20°C.

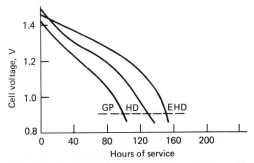

FIG. 5.19 Discharge curve of D size zinc-carbon cells discharged through 25 Ω for 4 h/day at 20°C.

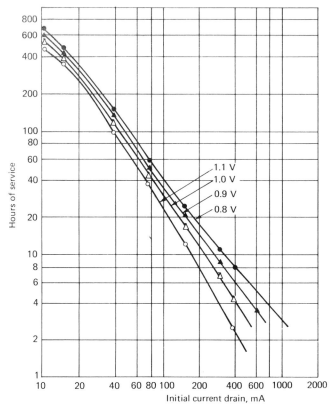

FIG. 5.20 Initial current drain vs. hours of discharge to different cutoff (end) voltages at 20°C, general-purpose D size zinc-carbon cell, continuous discharge.

different manufacturers are shown in Fig. 5.24. There is a difference of about 30% between the best and poorest cell to the 0.9-V cutoff.

5.6.6 Internal Resistance and Flash Current

The internal resistance R_{in} depends on the size and construction as well as all the other variables affecting the electrical qualities of the cell (temperature, age, depth of discharge, etc.). It can be calculated from a measurement of the OCV with the use of a high-resistance voltmeter, and a measurement of the flash current (FC) using an ammeter whose total resistance including leads does not exceed 0.01 Ω and is no more than 10% of the cell's internal resistance, as follows:

$$R_{in} = \frac{OCV}{FC}$$

where R_{in} is expressed in ohms.

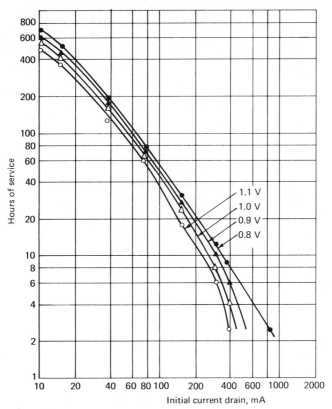

FIG. 5.21   Initial current drain vs. hours of discharge to different cutoff (end) voltages at 20°C, general-purpose D size zinc-carbon cell, discharged 4 h/day.

Another and probably more accurate method of calculation is to take the OCV as before and then shunt the battery with an external resistance R (e.g., 1 Ω) and note the instantaneous voltage reading V. The internal resistance is now calculated by

$$R_{in} = \frac{(OCV - V) \times R}{V}$$

where R_{in} is expressed in ohms. Note that if the resistance of the circuit for which the battery is to be used is known, it would be advisable to make R the value of that resistance.

Table 5.5 gives a comparison of the flash current and internal resistance of the cells most common in the industry. The internal resistance of a cell increases and flash current decreases with storage time, use, and decreasing temperature. Figure 5.25 shows the decline of the flash current and working voltage (voltage under load) of the cell as the discharge progresses. The load resistance is 300 Ω which is very high for the No. 6 cell compared to its internal resistance. Therefore, even when the cell cannot support an FC of 2% of its initial value, the working voltage (with 300 Ω) is still high, and much additional service is still available.

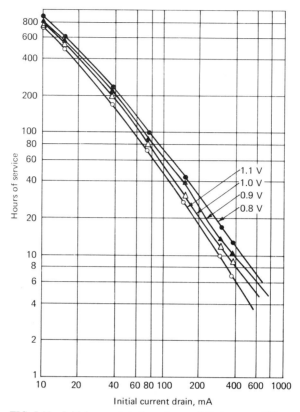

FIG. 5.22 Initial current drain vs. hours of discharge to different cutoff (end) voltages at 20°C, heavy-duty D size zinc-carbon cell, continuous discharge.

5.6.7 Effect of Temperature

Leclanche cells operate best at normal temperatures of about 20 to 30°C. The energy output of the cell increases with higher operating temperatures, but prolonged exposure to high temperatures (50°C) will cause rapid deterioration. The capacity of the Leclanche cell falls off rapidly with decreasing temperatures, and this cell is essentially inoperative below −20°C. The effects are more pronounced at heavier current drains; a low current drain would tend to result in a higher capacity at the lower temperatures than a higher current drain (except for a beneficial heating effect that may occur at the higher current drains).

The effect of temperature on the performance of the Leclanche cell is shown graphically in Fig. 5.26 for both the general-purpose (ammonium chloride electrolyte) cell and the extra-heavy-duty (zinc chloride electrolyte) cell. The data represent performance at flashlight-type current drains (300 to 600 mA for a D size cell). A lower current drain would result in a higher capacity than shown. Typical discharge curves for a general-purpose D size cell at various temperatures are given in Fig. 5.27. Additional characteristics of this D size cell at various temperatures are listed in Table 5.6.

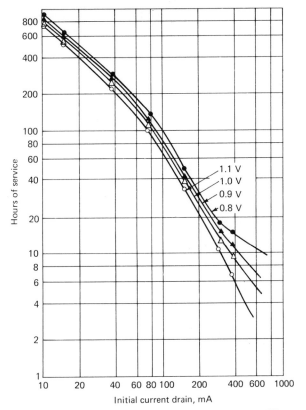

FIG. 5.23   Initial current drain vs. hours of discharge to different cutoff (end) voltages at 20°C, heavy-duty D size zinc-carbon cell, discharged 4 h/day.

These general statements may be qualified because of the influence on performance of specific formulations as shown in Figs. 5.28 to 5.30. The cell formulations shown are as follows:

A. General purpose type I is made of only natural ore and NH_4Cl as the electrolyte.

B. General purpose type II is made of two parts natural ore to one part electrolytic MnO_2 with $ZnCl_2$ as the electrolyte.

C. Heavy duty is made of one part natural ore and two parts electrolytic MnO_2 with NH_4Cl as the principal electrolyte.

D. Extra heavy duty type I is made of only electrolytic MnO_2 with $ZnCl_2$ as the electrolyte. (This is the so-called zinc chloride cell.)

E. Extra heavy duty type II is made of only electrolytic MnO_2 with mainly NH_4Cl as the electrolyte.

F. Extra heavy duty type III is made identical to formulation E except that the electrolytic MnO_2 is from a different manufacturer.

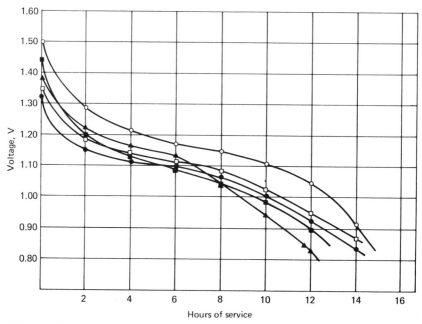

FIG. 5.24   Comparison of five general-purpose D size zinc-carbon cells manufactured by different companies (discharged on the ANSI LIF test at 20°C, through 4-Ω load).

TABLE 5.5   Cell Size vs. Flash Current vs. Internal Resistance*

| | Average flash current A | | Approximate internal resistance, Ω | |
| | Standard | Zinc-chloride | Standard | Zinc-chloride |
ANS cell size	cell	cell	cell	cell
N	2.5	. . .	0.6	
AAA	3.8	. . .	0.4	
AA	5.3	4.5	0.28	0.33
C	3.9	6.5	0.39	0.23
D	5.6	8.5	0.27	0.18
F	9.0	11.3	0.17	0.13
G	12.0	. . .	0.13	
No. 6	30.0	. . .	0.05	

*Based on Ref. 11.

Figure 5.28 shows the initial-load voltage on a 2¼-Ω load at various temperatures. It is significant that the cells with zinc-chloride electrolyte have the higher voltage than those fabricated with ammonium chloride electrolyte. Cells made with electrolytic MnO_2, however, show different behavior at low temperatures, obviously influenced by the other cell components.

Figure 5.29 shows the discharge curves for the six cell formulations on discharge at 20°C on the 4-Ω HIF test, and Fig. 5.30 shows the same for a discharge at −20° C. These discharge tests confirm the better performance of the zinc-chloride cells.

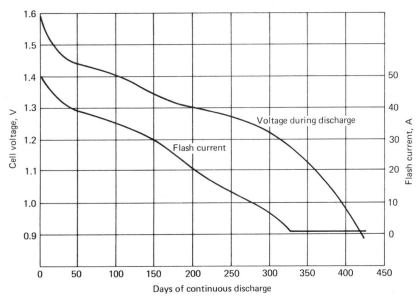

FIG. 5.25 Comparison of voltage under load during discharge and flash current as a function of time. General purpose no. 6 ignition cell discharged through 300-Ω load.

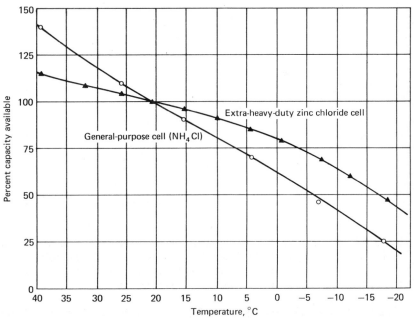

FIG. 5.26 Percentage of capacity available as a function of temperature (moderate-drain discharge).

5-25

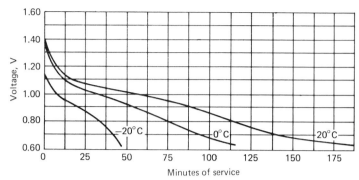

FIG. 5.27 Effect of temperature on voltage characteristics of D size zinc-carbon cell, continuous discharge starting at 667 mA.

TABLE 5.6 Effect of Temperature on a General-Purpose D Size Cell

Temperature, °C	Open-circuit voltage, V	Closed-circuit voltage, V (10-Ω load)	Flash current, A	Internal resistance, Ω
40	1.61	1.57	6.4	0.25
20	1.60	1.56	5.9	0.27
0	1.60	1.54	4.0	0.40
−20	1.59	1.30	0.9	1.80
−30	1.55	0.74	0.3	5.20
−40	0.80	0.0	0.0	

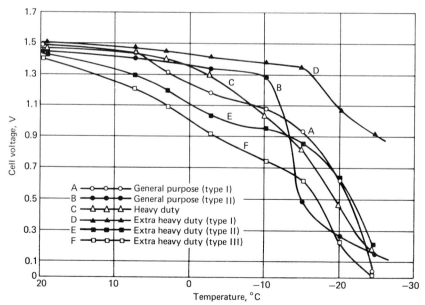

FIG. 5.28 Initial working voltage of various cell formulations as a function of temperature, 2.25-Ω discharge, D size cells.

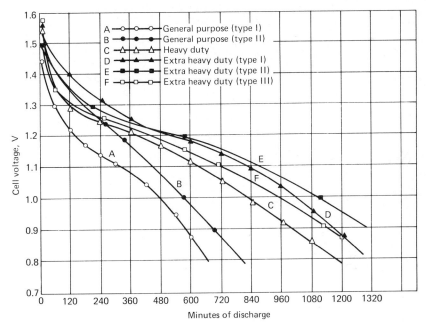

FIG. 5.29   Performance of various cell formulations, D size cells, discharged on the ANSI HIF test at 4 Ω, 20°C.

Several cells show an initial low voltage but recover after a few hours of discharge probably due to internal heat generated during the discharge.

Special low-temperature cells were developed using low-freezing-point electrolytes and a design that minimizes internal cell resistance, but they did not achieve popularity due to the superior overall performance of other types of primary cells. For best operation, at low ambient temperatures, the Leclanche cell should be kept warm by some appropriate means. A vest battery worn under the user's clothing, employing body heat to maintain it at a satisfactory operating temperature, was once used by the military to achieve operation at low temperatures.

5.6.8 Service Life

The service life of the Leclanche cell is summarized in Figs. 5.31 and 5.32, which plot the service life at various loads and temperatures normalized for unit weight (amperes per kilogram) and unit volume (amperes per liter). These curves are based on the performance of a general-purpose cell at the average discharge current under several discharge modes. These data can be used to approximate the service life of a given cell under particular discharge conditions or to estimate the size and weight of a cell required to meet a specific service requirement.

Manufacturers' catalogs should be consulted for specific performance data in view of the many cell formulations and discharge conditions. Table 5.7 presents typical data from a manufacturer for two formulations of the AA size cell.

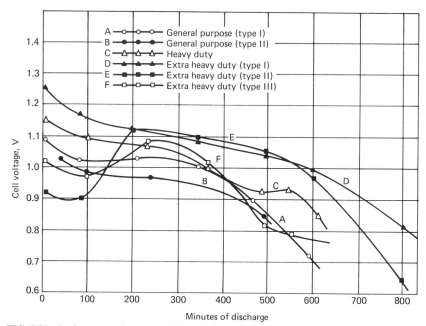

FIG. 5.30 Performance of various cell formulations, D size cells, discharged on the ANSI HIF test at 4 Ω, −20°C.

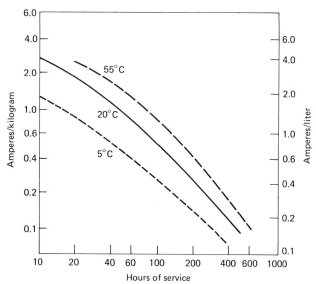

FIG. 5.31 Service hours for a general-purpose zinc-carbon cell discharged 2 h/day to 0.9 V.

5-28

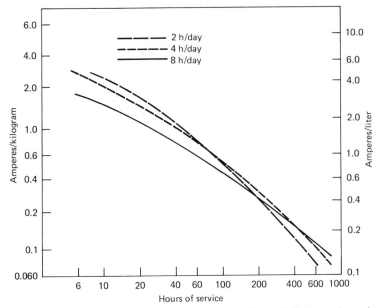

FIG. 5.32   Service hours for a general-purpose zinc-carbon cell discharged intermittently to 0.9 V.

5.6.9 Shelf Life

The Leclanche cell gradually loses capacity while idle. The deterioration, which is greater for partially discharged cells than for unused cells, results from zinc corrosion, chemical side reactions, and moisture loss. The rate of deterioration depends on storage temperature; high temperatures accelerate the loss; at low temperatures the cell activity is reduced, which increases the shelf life. Figure 5.33 shows the capacity retention after storage at 45, 20, and −20°C. The shaded areas give the range of capacity retention as a function of the storage time. A range is necessary because of the many manufacturing variations already mentioned. Cells stored at 21°C for 12 months have 85 to 90% of the initial capacity remaining. After 24 months, anywhere from 58 to 80% can be expected. Cells stored at 45°C will have 85 to 95% of initial capacity after only 3 months of storage. At 6 months, the cells stored at 45°C would be expected to retain only 40 to 70% of initial capacity. Cells stored at −20°C would be expected to retain approximately 80% of their initial capacity at 10 y.

Since low temperatures retard deterioration, storage at low temperatures is an advantageous method for preserving battery capacity. A storage temperature of 0°C is very effective. Freezing is not harmful as long as there is no repeated cycling from high to low temperatures. Use of case materials with widely different coefficients of expansion may lead to cracking. When batteries are removed from cold storage, they should be allowed to reach room temperature in order to provide satisfactory performance. Moisture condensation during warm-up should be prevented as this may cause electrical leakage.

TABLE 5.7 Manufacturer's Data—AA Size Zinc-Carbon Cells*

Eveready No. 915 (General Purpose), Estimated Average Hours Service at 21.1°C

Schedule	Starting drains, mA	Load, Ω	Cutoff voltage					
			0.75 V	0.8 V	0.9 V	1.0 V	1.1 V	1.2 V
2 h/day	5	300		265	255	240	210	145
	10	150		125	120	110	92	60
	20	75		54	50	46	36	22
	30	50		32	28	25	20	11
	50	30		14	12	10	8	3.8
4 h/day	5	300		260	250	230	200	130
	10	150		120	110	98	84	54
	20	75		48	43	37	30	18
	30	50		26	22	18	14	7.6
8 h/day	1	1500		1275	1190	1080	950	680
	2	750		620	560	500	420	300
	5	300		240	225	200	160	105
	10	150		110	94	78	64	42
	20	75		42	35	28	21	14
	30	50		22	18	15	10.5	6.7
12 h/day	1	1500		1420	1280	1210	1120	820
	2	750		700	640	600	510	360
	5	300		230	215	180	140	100
	10	150		100	84	74	60	37
	20	75		40	33	26	19	11
24 h/day	1	1500		1440	1250	1180	1040	700
	2	750		670	580	500	420	300
	5	300		220	200	160	130	90
	10	150		94	80	66	52	35
	20	75		37	30	25	19	11
	30	50		20	16	13	9	6.7
	50	30		10	8.5	6.8	4.8	2.8
4 h/day (radio test)	37.5	40	23		19			
½ h/day (calculator test)	125	12	6		4.5			
5 min/day (general-purpose intermittent test—flashlight)	375	4	130 min					
4 min/h 8 h/day 16 h rest (light-industrial flashlight test)	375	4	90 min		68 min			

*Based on Ref. 11.

Eveready No. 1215 (Zinc Chloride), Estimated Average Hours Service at 21.1°C

Schedule	Starting drains, mA	Load Ω	Cutoff voltage			
			0.75 V	0.8 V	0.9 V	1.0 V
4 h/day (radio test)	37.5	40	33		31	
	60	25	20		18	
½ h/day (calculator test)	125	12	8.6		7.7	
2 h/day (cassette test)	187.5	8	5.8		4.9	
4 min on, 11 min off, 8 h/day 16 h rest (camera cranking test)	250 mA constant current					118 min
5 min/day (general-purpose intermittent test—flashlight)	375	4	190 min			
4 min/h 8 h/day 16 h rest (light-industrial flashlight test	375	4	175 min		139 min	
Continuous (toy test)	375	4		110 min		

Temperature vs. service capacity	Temperature, °C	Percent of (21.1°C) service
Schedule:		
4 min/h	37.8	115
8 h/day	32.2	110
16 h rest	26.7	105
(light-industrial	21.1	100
flashlight test)	15.6	90
Starting drain:	10.0	80
375 mA	4.4	70
Load:	−1.1	60
4 Ω	−6.7	45
Cutoff:	−12.2	30
0.9 V	−17.8	20

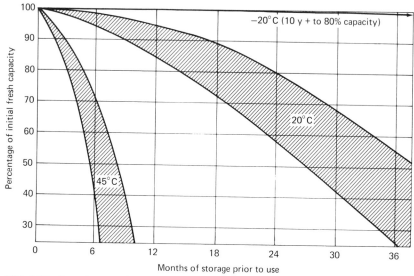

FIG. 5.33 Capacity retention after storage for zinc-carbon cells (initial capacity = 100%).

5.7 SPECIAL DESIGNS

The zinc-carbon system currently is being used in special designs to enhance particular performance characteristics or adapt to new applications.

5.7.1 Flat-Pack or "Wafer Battery" Design

One variation, developed for camera applications requiring a high power capability, is the "flat-pack" design.[12] The active components are similar to other zinc-carbon cells, but the very thin, flat structure provides the necessary active surface area for the high power output. The battery, as illustrated in Fig. 5.34, consists of four cells in a laminated stack. The basic cell is constructed with a duplex electrode which is a conductive sheet, top-coated with manganese dioxide (cathode of one cell) and bottom-coated with zinc (anode of adjacent cell). Three complete duplexes and two half-duplexes (one on each end of the assembly), with the separator wetted with electrolyte solution in gel form, complete the assembly. The adhesive perimeter on each separator seals the entire battery; the vent permits only gases to escape. The entire four-cell, 6-V battery is encased with a paper and polyester overwrap. A summary of the characteristics of this battery is given in Table 5.8. Typical performance characteristics are shown in Figs. 5.35 to 5.37. Particularly significant are the high drain performance and power-to-weight and power-to-volume characteristics. Storage life is similar to other zinc-carbon batteries; storage below 25 to 30°C is recommended.

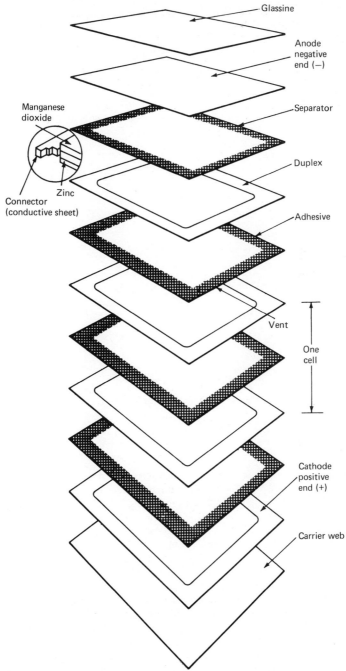

Glassine

Anode
negative
end (−)

Separator

Manganese
dioxide

Duplex

Zinc

Connector
(conductive sheet)

Adhesive

Vent

One
cell

Cathode
positive
end (+)

Carrier web

FIG. 5.34 Exploded view of 6-V Polaroid Polapulse battery. (*Courtesy of Polaroid Corp.*)

TABLE 5.8 Polaroid Polapulse Battery Typical Technical Characteristics*

Voltage	6 V, nominal
	6.4 V, typical open-circuit voltage
Dimensions	9.474 $\times$ 7.722 $\times$ 0.457 cm
Volume	30.2 cm³
Weight	27 g
Use temperature range	-7 to 54°C
High discharge current	26 A instantaneous, 5 A after 30 s,
	2.5 A after 60 s (through 0.05 Ω)
Internal resistance	0.25 Ω at 1-A drain

*Based on Ref. 12.

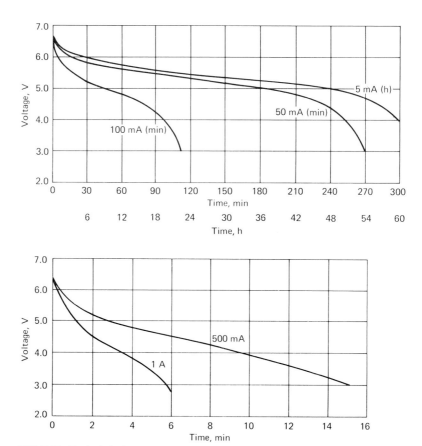

FIG. 5.35 Typical discharge curves for Polapulse cells at 20°C. (5-mA curve uses the hour scale; 50- and 100-mA curves use the minute scale.) (*Courtesy of Polaroid Corp.*)

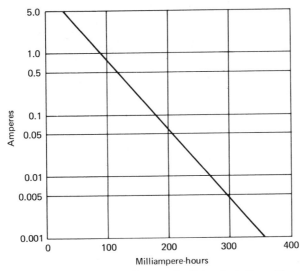

FIG. 5.36 Capacity vs. load current for Polapulse battery. (*Courtesy of Polaroid Corp.*)

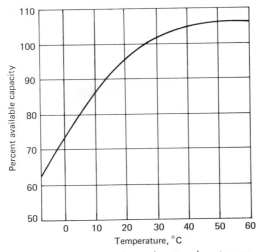

FIG. 5.37 Percentage of capacity at various temperatures, Polapulse cell, discharged at 100 mA. (*Courtesy of Polaroid Corp.*)

5.7.2 Paper Cell

Another new structure is the "paper" battery, so-named because it is made in extremely flat sheets. One such type, illustrated in Fig. 5.38, measures 20 by 70 by 0.8 mm.[13] It is similar to other flat cells but is made as a complete unit cell. It also uses zinc perchlorate $[Zn(ClO_4)_2]$ as the electrolyte. The cell illustrated has a capacity of 60 mAh

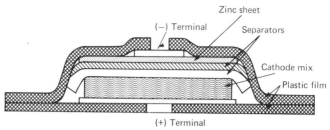

FIG. 5.38 Paper cell. (*Courtesy of Matsushita Battery Industrial Co., Ltd.*)

when discharged at 0.1 mA (38 Wh/kg). The paper cell is constructed of a thin sheet of zinc, separator, and cathode mix in a sandwich fashion. The entire battery is enclosed by a plastic film except where the metallic terminals will be connected. A summary of the characteristics of this cell is given in Table 5.9; typical discharge curves at 20°C are shown in Fig. 5.39 for two levels of discharge. Figure 5.40 summarizes the performance at 20°C, giving the service hours at various resistive loads to two end voltages.

TABLE 5.9 Paper Battery Specifications*

Size	20 × 70 × 0.8 mm
Weight	1.9 g (approx.)
Nominal voltage	1.5 V
Discharge capacity	60 mAh when discharged continuously through 15 kΩ to 1.2 V at 20°C
Suggested current	Less than 2 mA
Impedance (at 1 kHz)	Less than 10 Ω
Short-circuit current	1 A
Terminals	Positive side: stainless steel Negative side: brass or nickel-plated steel
Shelf life	Projected the same as that of ordinary carbon zinc

*Based on Ref. 13.

5.8 CELL AND BATTERY TYPES AND SIZES

Zinc-carbon cells are made in a number of sizes with different formulations to meet a variety of applications. The unit cells and multicell batteries are classified as "industrial," "general purpose," "heavy duty," "photoflash," etc., according to their output capability under specific discharge conditions.

Table 5.10 lists the more popular cell sizes with estimated service hour performance at various loads under a typical 2 h per day intermittent discharge, except for the

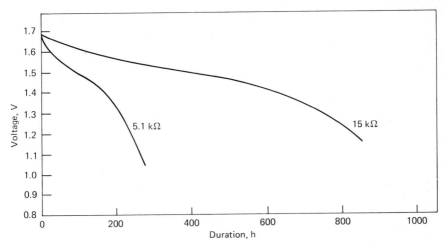

FIG. 5.39 Paper cell, typical discharge curves. (*Courtesy of Matsushita Battery Industrial Co., Ltd.*)

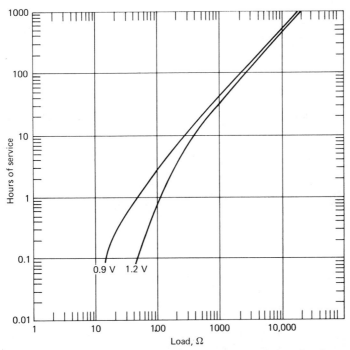

FIG. 5.40 Paper cell, service hours vs. load. (*Courtesy of Matsushita Battery Industrial Co., Ltd.*)

TABLE 5.10 Characteristics of 1.5-V Zinc-Carbon Unit Cells

Cell designation			Weight, g	Maximum dimensions, mm		Capacity: 2 h/day			
ANSI	IEC	NEDA		Diameter	Height	Standard cell		Zinc-chloride cell	
						Starting drain, mA	Service hours	Starting drain, mA	Service hours
N	R-1	910	6.2	12	30.2	1.5 7.5 15	320 60 20		
AAA	R-03	24	8.5	10.5	44.5	2 10 20	350 54 21		
AA	R-6	15	17	14.5	50.5	3 15 30	450 80 32	187.5 375	5.8 1.9 (continuous)
C	R-14	14	40	26.2	50.0	5 25 50	520 115 53	187 375	17 4.6 (continuous)
D	R-20	13	94	34.1	61.5	10 50 100	525 125 57	9 187 667	1100 42 5.3 (continuous)
F	R-25	60	160	34.1	91.4	15 75 150	630 135 60	187 375 667	10 36 18
G	R-26	...	180	32.0*	105*				
J	R-27	...	270	32.0*	150*				
#6	R-40	905 906 911	910	67.0	172	50 50 500	750 210 95		

*Nominal dimensions.

TABLE 5.11 Performance Requirements for Zinc-Carbon Unit Cells*

(a) Requirements for General-Purpose Unit Cell Batteries

NEDA	ANSI	IEC	GP 2¼-Ω Initial, min	GP 2¼-Ω 6-month delay, min	GP 4-Ω Initial, min	GP 4-Ω 6-month delay, min	GP 5-Ω Initial, min	GP 5-Ω 6-month delay, min	LIF 2¼-Ω Initial, min	LIF 2¼-Ω 6-month delay, min	LIF 4-Ω Initial, min	LIF 4-Ω 6-month delay, min
13F	D	R-20	425	380	…	…	…	…	350	315	600	540
14F	C	R-14	…	…	350	300	…	…	…	…	200	…
15F	AA	R-6	…	…	100	85	…	…	…	…	70	…
24F	AAA	R-03	…	…	…	…	60	50	…	…	…	…
910F	N	R-1	…	…	…	…	35	25	…	…	…	…

(GP = General-purpose intermittent tests; LIF = Light industrial flashlight tests)

(b) Performance Requirements for Flashlight Unit Cell Batteries

Light industrial flashlight (LIF) tests

NEDA	ANSI	IEC	2¼-Ω Initial 0.65 V	2¼-Ω 6-month delay 0.65 V	4-Ω Initial 0.9 V	4-Ω Initial 0.75 V	4-Ω 6-month delay 1.1 V	4-Ω 6-month delay 0.9 V	4-Ω 6-month delay 0.75 V
14D	C	R-14	…	…	300	400	…	270	360
13C,13D	D	R-20	600	540	850	500	450	750	…

Heavy industrial flashlight (HIF) tests

NEDA	ANSI	IEC	2¼-Ω Initial 0.65 V	2¼-Ω 6-month delay 0.65 V	4-Ω Initial 1.1 V	4-Ω Initial 0.9 V	4-Ω Initial 0.75 V	4-Ω 6-month delay 1.1 V	4-Ω 6-month delay 0.9 V	4-Ω 6-month delay 0.75 V
13C,13D	D	R-20	400	450	…	450	800	…	400	720

(c) Performance Requirements for No. 6 Cells

NEDA	Types	Light-intermittent test 20 Ω, days	Heavy-intermittent test 2⅓ Ω, Initial, h	Heavy-intermittent test 6-month delay, h	Telephone-battery test 50 Ω, days	Alarm-battery test 300 Ω, days	Alarm-battery test 500 Ω, days
905	No. 6 general purpose	200	60	55	400		
911	No. 6 industrial	310	90	80	500		
906	No. 6 alarm		…	…	…	300	500
906	No. 6 "regular" telephone	300	…	…	470		
906	No. 6 "special" telephone	370	…	…	680		

*Based on Ref. 10.

TABLE 5.12 Multicell Zinc-Carbon Batteries*

Battery designation		Voltage, V	Diameter, mm		Overall height, mm		Length, mm		Width, mm	
NEDA	IEC		Max.	Min.	Max.	Min.	Max.	Min.	Max.	Min.
900	...	1.5	...	...	111.1	107.1	68.2	65.1	68.2	65.1
6	4R25	6.0	...	...	103.2	96.0	68.2	65.1	68.2	65.1
920	4R25	6.0	...	...	103.2	96.0	68.2	65.1	68.2	65.1
908, 924, 915D	4R25	6.0	...	...	101.6	96.0	68.2	65.1	68.2	65.1
915, 925, 908D	4R25	6.0	...	...	111.1	107.1	68.2	65.1	68.2	65.1
918	4R25-2	6.0	...	...	125.4	122.2	136.5	132.5	73.0	69.0
907	4R40Y	6.0	...	...	190.5	181.8	269.9	263.5	71.4	67.5
903	...	7.5	...	...	163.5	160.3	185.7	181.0	103.2	100.0
1611	...	9.0	19.1	...	50.8	...	...	...	...	...
1604	6F22	9.0	...	...	49.2	...	26.5	24.5	17.5	15.5
1600	6F24	9.0	25.4	...	50.0	...	...	...	...	...
1602	...	9.0	...	...	69.9	67.5	35.7	34.1	34.9	32.5
1605	...	9.0	...	...	61.9	60.3	46.0	44.5	46.0	44.5
1603	6F100	9.0	...	...	80.2	78.6	65.1	63.5	51.6	50.0
904	...	9.0	...	...	163.5	...	217.9	...	103.2	...
922	...	12.0	...	...	183.4	...	265.1	...	69.1	...
926	...	12.0	...	...	125.4	122.2	136.5	132.5	73.0	69.0
923	...	12.0	...	...	106.4	102.4	136.5	132.5	73.0	69.0
220	10F15	15.0	...	...	34.9	...	15.9	...	15.1	...
221	15F15	22.5	...	...	50.8	...	15.9	...	15.1	...
215	15F20	22.5	...	...	50.8	...	27.0	...	15.9	...
225	...	22.5	...	...	56.4	...	34.9	...	27.0	...
710	...	22.5	...	...	77.8	...	89.7	...	55.5	...
210	20F20	30.0	...	...	65.1	...	27.0	...	15.9	...
201	30F40	45.0	...	...	95.3	...	63.9	...	24.5	...
205	...	45.0	...	...	106.4	...	79.4	...	60.3	...
200	45F40	67.5	...	...	96.8	...	71.4	...	34.9	...
204	60F40	90.0	...	...	95.3	...	94.6	...	34.9	...

*Based on Ref. 10.

TABLE 5.13 Cross-Reference of Zinc-Carbon Cells and Batteries*

NEDA no.	Berec	Bright Star	Burgess	Duracell	Eveready	Panasonic	Ray-O-Vac	U.S. Army
				Cells				
910F	D23	20	310	M910	904	...	716	
24F	HP16	58	7	M24F	912	...	400	
15F	R6	59	910	M15F	915	UM3	7AA	BA-58
14F	R14	11-M	105	M14F	935	UM2	1C	BA-42
13F	R20	10M	205	M13F	950	UM1	2D	BA-30
905	6	61gn	61gn	M905	1S6	...	61gn	
				Multicell batteries				
900	993PP	464	410	M900	735	...	900	
6	PJ944	646	F4Pl	...	744	...	A6	BA-210/U
920	...	646	F4BW	...	2744N	...	920	
908	996	460	F4M	M908	509	...	941	BA-200/U
915	PJ992	460S	F4BP	M915	510S	...	942	BA-803/U
918	991	158	TW1	M918	731	...	918	
907	...	146	S461	...	1461	...	641	BA-249/U
903	...	155	4F5H	...	715	...	903	BA-804/U
1611	...	...	L6	M1611	206	...	T1611	
1604	BF22(PP3)	0920	2U6	M1604	216	006P	1604	BA-90/U
1600	PP4	...	P6	M1600	226	...	1600	
1602	PP6	...	2N6	M1602	246	...	1602	
1605	PP7	...	M6	M1605	266	...	1605	
1603	PP9	...	D6	M1603	276	...	1603	
904	...	164	4F6H	...	716	...	904	BA-207/U
922	...	187	2G8H	...	1463	...	922	
923	...	...	...	...	2780N			
220	10F15	15P	Y10	M504	504	W10E	220	BA-332/U
221	15F15	22P	Y15	M505	505	MV15E	221	
215	15F20	12P	U15	M215	412	015	215	BA-261/U
225	...	...	K15	...	420			
710	...	...	4156	...	763	...	A710	BA-2
210	...	...	U20	...	413	...	A210	BA-305
201	...	...	XX30	...	455	...	201	BA-56
205	...	30-59	...	...	738	...	205	BA-63
200	...	45N	XX45	...	467	...	200	BA-51
204	...	60N	V60	...	490	...	204	BA-805/U

*Based on Ref. 14.

continuous "toy" battery test. The performance of these unit cells, under several intermittent discharge conditions, is given in Table 5.11. (Table 5.4 gives the discharge schedule for these tests.)

Table 5.12 lists some of the major multicell zinc-carbon batteries that are available commercially. The performances of these batteries can be estimated by using the IEC designation to determine the cell complement (e.g., NEDA 6, IEC 4R25 battery consists of four F size cells connected in series). Table 5.13 gives cross-references to the zinc-carbon cells and batteries of the various manufacturers.

In view of the different cell formulations used for zinc-carbon cells, manufacturers' catalogs should be consulted for specific performance data.

REFERENCES

1. Samuel Ruben, *The Evolution of Electric Batteries in Response to Industrial Needs*, Dorrance, Philadelphia, 1978, chap. 5.

2. George Vinal, *Primary Batteries*, Wiley, New York, 1950.

3. A. Kozawa and R.A. Powers, "Electrochemical Reactions in Batteries," *J. Chem. Ed.* **49**:587 (1972).

4. Richard Huber, in K.V. Kordesch (ed.), *Batteries*, Dekker, New York, 1974, vol. 1, chap. 1.

5. R.J. Brodd, A. Kozawa, and K.V. Kordesch, "Primary Batteries 1951–1976," *J. Electrochem. Soc.* **125**(7)(1978).

6. N. C. Cahoon, in N. C. Cahoon and G. W. Heise (eds.), *The Primary Battery*, Wiley, New York, 1976, vol. 2, chap. 1.

7. M. Bregazzi, *Electrochem. Technol.* **5**:507 (1967).

8. C. L. Mantell, *Industrial Carbon*, 2d ed., Van Nostrand, New York, 1946.

9. C. L. Mantell, *Batteries and Energy Systems*, 2d ed., McGraw-Hill, New York, 1983.

10. "American National Standard Specification for Dry Cells and Batteries," ANSI C18.1–1979, American National Standards Institute, Inc., May 1979.

11. *Eveready Battery Engineering Data*, Union Carbide Corp., New York, 1976.

12. Polaroid Corporation, Cambridge, Mass.

13. Matsushita Battery Industrial Co., Ltd., Osaka, Japan.

14. *NEDA Battery Index*, National Electronic Distributors Association, Chicago, 1980.

Magnesium and Aluminum Cells

by
David Linden

6.1 GENERAL CHARACTERISTICS

Magnesium and aluminum are attractive candidates for use as anode materials in primary cells. As shown in Part 1, Chap. 1, Table 1.1, they have a high standard potential. Their low atomic weight and multivalence change result in a high electrochemical equivalence on both a gravimetric and volumetric basis. Further, they are both abundant and relatively inexpensive.

As an active primary cell, magnesium has been used successfully in a magnesium/manganese dioxide (Mg/MnO_2) cell. This system has two main advantages over the zinc-carbon cell, namely, twice the service life or capacity of a zinc cell of equivalent size and the ability to retain this capacity, during storage, even at elevated temperatures (Table 6.1). This excellent storability is due to a protective film that forms on the surface of the magnesium anode.

Several disadvantages of the magnesium cell are its "voltage delay" and the parasitic corrosion of magnesium that occurs during the discharge once the protective film is removed, generating hydrogen and heat. The magnesium battery also loses its excellent storability after being partially discharged and, hence, is unsatisfactory for long-term intermittent use. For these reasons, the active (nonreserve) magnesium battery, while

TABLE 6.1 Major Advantages and Disadvantages of Magnesium Cells

Advantages	Disadvantages
Good capacity retention, even under high-temperature storage	Delayed action (voltage delay)
	Evolution of hydrogen during discharge
Twice the capacity of corresponding Leclanche cell	Heat generated during use
Higher cell voltage than zinc-carbon cells	Poor storage after partial discharge
Competitive cost	

being used successfully in military applications such as radio transceivers and emergency or standby equipment, has not found wide commercial acceptance. Recent developments, however, have demonstrated the possibility of reducing the voltage delay and of reacting the hydrogen generated inside the cell to regenerate water that is required for the cell reaction.[1] These improvements could significantly increase the acceptability and use of this battery system.

Aluminum, however, has not been used successfully in an active primary cell despite its potential advantages. Like magnesium, a protective film forms on the aluminum, which is detrimental to battery performance, resulting in a cell voltage that is considerably below theoretical and causing a voltage delay that can be significant for partially discharged cells or those that have been stored. While the protective oxide film can be removed by using suitable electrolytes or by amalgamation, gains by such means are accompanied by accelerated corrosion and poor shelf life.

6.2 CHEMISTRY

The magnesium primary cell uses a magnesium alloy for the anode, manganese dioxide as the active cathode material but mixed with acetylene black to provide conductivity, and an aqueous electrolyte consisting of magnesium perchlorate, with barium and lithium chromate as corrosion inhibitors and magnesium hydroxide as a buffering agent to improve storability (pH of about 8.5). The amount of water is critical as water participates in the anode reaction and is consumed during the discharge.

The overall discharge reaction of the magnesium/manganese dioxide cell is

$$Mg + 2MnO_2 + H_2O \rightarrow Mn_2O_3 + Mg(OH)_2$$

The theoretical potential of the cell is over 2.8 V, but this voltage is not realized in practice. The observed values are decreased by about 1.1 V, giving an open-circuit voltage of 1.9 to 2.0 V, still higher than the zinc-carbon cell.

The corrosion of magnesium under storage conditions is slight. A film of $Mg(OH)_2$ that forms on the magnesium provides good protection, and treatment with chromate inhibitors increases this protection. However, when the protective film is broken or removed during discharge, corrosion occurs with the generation of hydrogen

$$Mg + 2H_2O \rightarrow Mg(OH)_2 + H_2$$

This wasteful reaction is a problem, not only because of the need to vent the hydrogen from the cell and to prevent it from accumulating, but also because it uses water which is critical to the cell operation, produces heat, and reduces the efficiency of the anode.

The efficiency of the magnesium anode is about 60 to 70% during a typical continuous discharge and is influenced by such factors as the composition of the magnesium alloy, the cell components, discharge rate, and temperature. On low drains and intermittent service, the anode efficiency can drop to 40 to 50% or less. The anode efficiency also is reduced with decreasing temperature.

Considerable heat is generated during the discharge of a magnesium battery, particularly at high discharge rates, due to the exothermic corrosion reaction (about 82 kcal per gram-mole of magnesium) and the losses resulting from the difference between the theoretical and operating voltage. Proper battery design must allow for the dissipation of this heat to prevent overheating and shortened life. On the other hand, this heat can be used to advantage at low ambient temperatures to maintain the battery at higher, and more efficient, operating temperatures.

The good shelf life of the magnesium battery results from the protective film which forms on the magnesium anode. This film, however, is responsible for a voltage delay[1,2]—a delay in the cell's ability to deliver full output voltage after it has been placed under load (Fig. 6.1). This delay, as shown in Fig. 6.2, is usually less than 1 s but can be longer (up to a minute or more) for discharges at low temperatures and after prolonged storage at high temperatures.

The standard potential for aluminum in the anode reaction

$$Al \rightarrow Al^{3+} + 3e$$

is reported as -1.7 V. A cell with an aluminum anode should have a potential about 0.9 V higher than the corresponding zinc cell. However, this potential is not attained, and the potential of an Al/MnO_2 cell is only about 0.1 to 0.2 V higher than that of a zinc cell. The Al/MnO_2 cell never progressed beyond the experimental stage because of the problems with the oxide film, excessive corrosion when the film was broken, voltage delay, and the tendency for aluminum to corrode unevenly. The experimental cells that were fabricated used a two-layer aluminum anode (to minimize premature failure due to can perforation), an electrolyte of aluminum or chromium chloride, and a manganese dioxide-acetylene black cathode similar to the conventional zinc/manganese dioxide cell.[3] The reaction mechanism is

$$Al + 3MnO_2 + 3H_2O \rightarrow 3MnO \cdot OH + Al(OH)_3$$

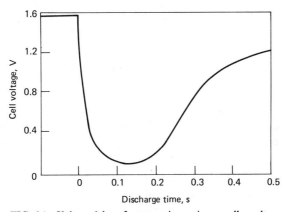

FIG. 6.1   Voltage delay of a magnesium primary cell—voltage profile at 20°C.

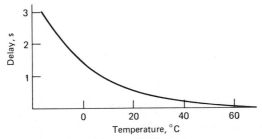

FIG. 6.2   Voltage delay vs. temperature.

6.3 CONSTRUCTION—Mg/MnO₂ CELLS

Magnesium/manganese dioxide (nonreserve) primary cells are generally constructed in a cylindrical configuration.

6.3.1 Standard Construction

The construction of the magnesium cell is similar to the cylindrical zinc-carbon cell. A cross section of a typical cell is shown in Fig. 6.3. A magnesium alloy can, containing small amounts of aluminum and zinc, is used in place of the zinc can. The cathode consists of an extruded mix of manganese dioxide, acetylene black for conductivity and moisture retention, barium chromate, and magnesium hydroxide. The electrolyte is an aqueous solution of magnesium perchlorate or magnesium bromide with lithium chromate. A carbon rod serves as the cathode current collector. The separator is an absorbent kraft paper as in the paper-lined zinc cell structure. Sealing of the magnesium cell is critical, as it must be tight to retain cell moisture during storage but provide a means for the escape of hydrogen gas which forms as the result of the corrosion reaction during the discharge. This is accomplished by a mechanical vent—a small hole in the plastic top seal washer under the retainer ring which is deformed under pressure, releasing the excess gas.

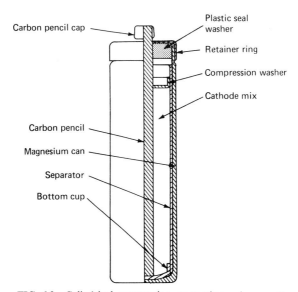

FIG. 6.3 Cylindrical construction, magnesium primary cell.

6.3.2 Inside-Out Construction

The basis of the "inside-out" design (Fig. 6.4) is a highly conductive carbon structure which can be molded readily into complex shapes. The carbon structure is formed in the shape of a cup which serves as the cell container; an integral center rod is incorpo-

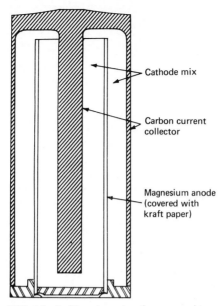

FIG. 6.4   Inside-out construction, magnesium primary cell. (*Courtesy of ACR Electronics, Inc.*)

rated to reduce current paths. The cups are structurally strong, homogeneous, and impervious to the passage of liquids and gases and the corrosive effects of the electrolyte. The cell consists of the carbon cup, a cylindrical magnesium anode, paper separator, and a cathode mix consisting of manganese dioxide, carbon black, and inhibitors with aqueous magnesium bromide or perchlorate as the electrolyte. The cathode mix is packed into the spaces on both sides of the anode and is in intimate contact with the inside and outside surfaces of the anode, the center rod, and the inside surfaces of the cup. This configuration provides larger electrode surface areas. External contacts are made by two metallic end pieces. The positive terminal is bonded during the forming process to the closed end of the carbon cup. The negative terminal, to which the anode is attached, together with a plastic ring, forms the insulated closure and seal for the open end of the cup. The entire cell assembly is enclosed in a crimped tin-plated steel jacket.

6.4 PERFORMANCE CHARACTERISTICS— Mg/MnO₂ CELLS

6.4.1 Discharge Performance

Typical discharge curves for the cylindrical magnesium/manganese dioxide primary cell are shown in Fig. 6.5. The discharge profile is generally flatter than for the zinc-carbon cells; the magnesium battery also is less sensitive to changes in the dis-

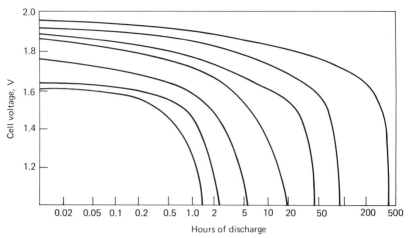

FIG. 6.5   Typical discharge curves, magnesium/manganese dioxide cylindrical cell, 20°C discharge temperature.

charge rate. The average discharge voltage is in the order of 1.6 to 1.8 V, about 0.4 to 0.5 V above that of the zinc-carbon cell; the typical end voltage is 1.2 V.

The performance of the magnesium primary cell at low temperatures is also superior to the zinc-carbon cell, operating to temperatures of −20°C and below. Figure 6.6 shows the performance of the magnesium cell at different temperatures based on the 20-h discharge rate. The low-temperature performance is influenced by the heat generated during discharge and is dependent on the discharge rate, size of battery, battery configuration, and other such factors. Actual discharge tests should be performed if precise performance data are needed.

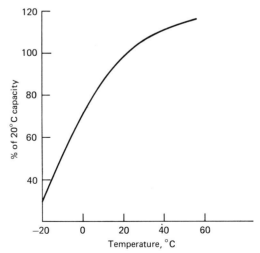

FIG. 6.6   Performance vs. temperature, magnesium/manganese dioxide cylindrical cell.

On extended low-rate discharges, the magnesium cell may split open. This rupture is due to the formation of magnesium hydroxide which occupies about one and one-half the volume of the magnesium. It expands and presses against the cathode mix which has hardened appreciably from the loss of water during the discharge. This opening of the cell can cause the cell voltage to rise about 0.1 V, also increasing cell capacity due to the air that can enter into the cell reaction.

The service life of the magnesium/manganese dioxide primary cell, normalized to unit weight (kilogram) and volume (liter), at various discharge rates and temperatures, is summarized in Fig. 6.7. The data are based on a rated performance of 60 Ah/kg and 120 Ah/L.

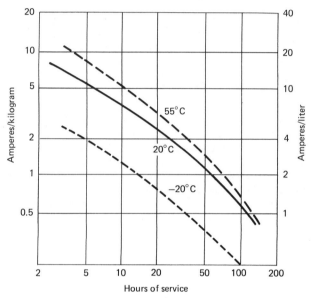

FIG. 6.7   Service life of the magnesium/manganese dioxide primary cell at various discharge rates and temperatures (to 1.2 V per cell end voltage).

6.4.2 Shelf Life

The shelf life of the magnesium/manganese dioxide primary cell at various storage temperatures is shown in Fig. 6.8, with a comparison with the shelf life of the zinc-carbon cell. The magnesium cell is noted for its excellent shelf life. The cell can be stored for periods of 5 y or longer at 20°C with a total capacity loss of 10 to 20% and as high as 55°C with losses of about 20%/y.

6.4.3 Inside-Out Cells

The discharge characteristics of the cylindrical inside-out magnesium primary cells are shown in Fig. 6.9. This structure has better high-rate and low-temperature performance than the conventional structure. These cells can be discharged at temperatures as low

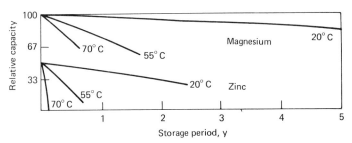

FIG. 6.8 Comparison of service vs. storage of the magnesium/manganese dioxide and zinc-carbon cells.

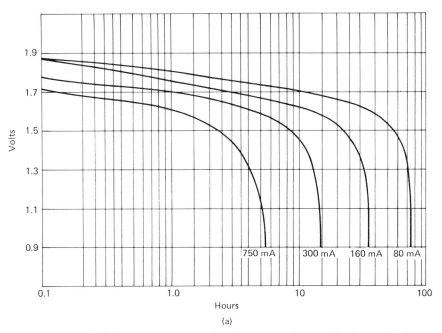

(a)

FIG. 6.9 Typical discharge curves, magnesium inside-out primary cell, D size. (a) 20°C. (*Courtesy of ACR Electronics, Inc.*)

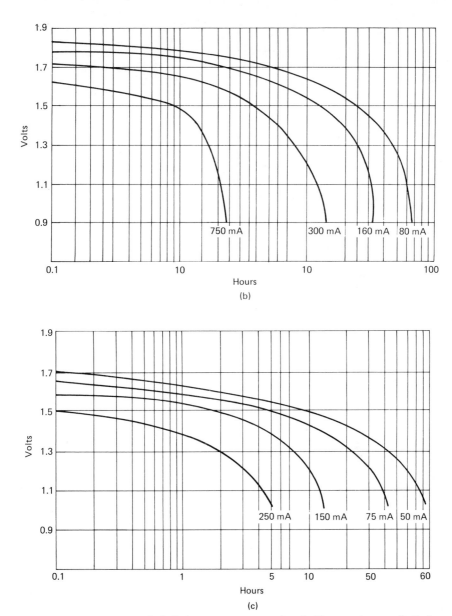

FIG. 6.9 (*Continued*) Typical discharge curves, magnesium inside-out primary cell, D size. (*b*) 0°C; (*c*) −20°C. (*Courtesy of ACR Electronics, Inc.*)

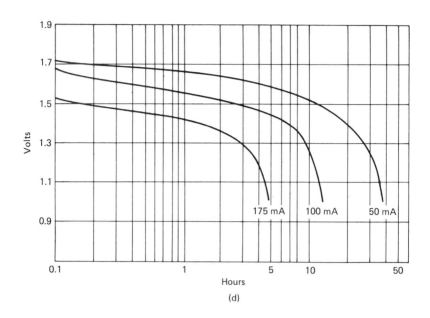

(d)

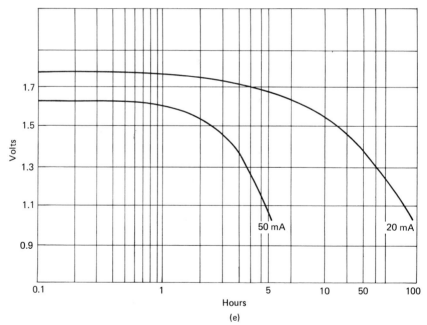

(e)

FIG. 6.9 (*Continued*) Typical discharge curves, magnesium inside-out primary cell, D size. (*d*) 30°C; (*e*) −40°C. (*Courtesy of ACR Electronics, Inc.*)

6-10

as $-40°C$, although at lighter discharge loads at the lower temperatures. Discharge curves are characteristically flat. They also have good and reproducible low-drain, long-term discharge characteristics as they do not split under these discharge conditions. Discharges for a 2½-y duration are realized with a D size cell at a 270-μA drain at 20°C.

6.4.4 Battery Design

Battery configuration has an important influence on the performance of the magnesium/manganese dioxide cell because of the heat generated during the discharge. As discussed in Sec. 6.2, proper battery design must allow for dissipation of this heat to prevent overheating, premature dry-out, and shortened performance—or for using this heat to improve performance at low ambient temperatures. In some low-temperature applications, it is advantageous to insulate the battery against heat loss. Actual discharge tests will be required to obtain precise performance data under the variety of possible conditions and battery design.

The battery and equipment design must also consider the hydrogen that is generated during discharge. The hydrogen must be vented and kept from accumulating because hydrogen-air mixtures are flammable above 4.1% and explosive above 18%.

6.5 CELL SIZES AND TYPES—Mg/MnO₂ CELLS

The cylindrical magnesium/manganese dioxide cells are manufactured in several of the popular standard ANSI sizes, as summarized in Table 6.2. Most of the production of the conventional cell is used for military radio transceiver applications, and cells are not readily available through commercial sources. Inside-out cells are available in the cell sizes listed.

6.6 OTHER TYPES OF MAGNESIUM PRIMARY BATTERIES

Magnesium primary cells have been developed in other structures and with other cathode materials, but these designs have not achieved commercial success. Flat cells,

TABLE 6.2 Cylindrical Magnesium Primary Cells

| | | | | Capacity, Ah* | |
| | | | | Conventional | Inside-out |
Cell type	Diameter, mm	Height, mm	Weight, g	structure	cell†
N	11.0	31.0	5	0.5	
B	19.2	53.1	26.5	2.0	
C	25.4	49.7	45	. . .	3.0
C (long)	23.7	82.0	60	4.0	
D	33.6	60.5	105	. . .	7.0
FD	41.7	49.1	125	. . .	8.0
No. 6	63.5	159.0	1000	. . .	65

*50-h discharge rate.
†Available from ACR Electronics, Inc., Hollywood, Fla.

using a plastic-film envelope, were designed but were never produced commercially. The use of organic depolarizers, such as *meta*-dinitrobenzene (*m*-DNB), in place of manganese dioxide was of interest because of the high capacity that could be realized with the complete reduction of *m*-DNB to *m*-phenylenediamine (2 Ah/g). The discharge of actual cells, while having a flat voltage profile and a higher ampere-hour capacity than the MnO_2 cell, had a low operating voltage of 1.1 to 1.2 V per cell. Watthour capacities were not significantly higher than the magnesium/manganese dioxide cells, and the *m*-DNB cell was inferior at low temperatures and high current drains. Commercial development of these cells never materialized.

Magnesium/air cells were studied, again because of the higher operating voltage than with zinc. These cells, too, were never commercialized. Magnesium, however, is a very useful anode in reserve batteries. Its application in these types of batteries is covered in Part 5.

6.7 ALUMINUM PRIMARY CELLS

Experimental work on Al/MnO_2 primary or "dry" cells was concentrated on the D size cylindrical cell using a construction similar to the one used for the Mg/MnO_2 cell (Fig. 6.3). The most successful anodes were made of a duplex metal sheet consisting of two different aluminum alloys. The inner, thicker layer was more electrochemically active, leaving the outer layer intact in the event of pitting of the inner layer. The cathode bobbin consisted of manganese dioxide and acetylene black, wetted with the electrolyte. Aqueous solutions of aluminum or chromium chloride, containing a chromate inhibitor, were the most satisfactory electrolytes.

The performance of the experimental Al/MnO_2 D size cells, compared with the Zn/MnO_2 cell, is shown in Fig. 6.10.[3] The aluminum cell shows a higher operating voltage and higher energy density on a continuous discharge and, except on long-term or intermittent discharges, will generally outperform the zinc cell. The performance of these aluminum cells has not been characterized at either high or low temperatures,

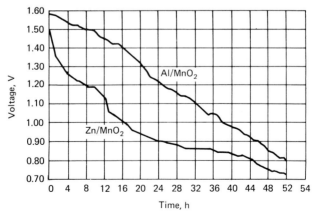

FIG. 6.10 Discharge curves for D size cells; 15-Ω continuous discharge at 20°C. (*From John J. Stokes, "The Aluminum Dry Cell," Electrochem. Tech., vol. 6, no. 1–2, 1968.*)

but indications are that these cells will not perform satisfactorily at low temperatures. Shelf life on room-temperature storage is good; 95% retention was reported after 3 y of storage. Delayed action after storage remains a problem, however, with stored or partially discharged cells.

Aluminum active primary cells were never produced commercially. While the experimental aluminum cells delivered a higher energy output than conventional zinc cells, anode corrosion, causing problems on intermittent and long-term discharges and irregularities in shelf life, and the voltage-delay problem restrained commercial acceptance.

Aluminum has been more successfully used as an anode in aluminum/air systems (Part 4, Chap. 30) and in some reserve batteries (Part 5).

REFERENCES

1. Dhanji, S., "Magnesium—The Safe Power Solution," *SAFE Symposium,* Las Vegas, Nev., Dec. 1982.
2. Kumar, B.V.R. and Sathyanarayana, S., "Delayed Action of Magnesium Anode in Primary Batteries," *J. Power Sources,* vol. 9, Elsevier Sequoia S.A. Lausanne, Switzerland, 1983.
3. Stokes, John J., "The Aluminum Dry Cell," *Electrochem. Technol.* **6** (1-2) (1968).

BIBLIOGRAPHY

Belitskus, D.: "Performance of Aluminum—Manganese Dioxide Dry Cells," *J. Electrochem. Soc.,* **119**:295, 1972.

Robinson, J.L.: "Magnesium Cells," in G. E. Heise and N. C. Cahoon (eds.), *The Primary Battery,* Wiley, New York, 1976, vol. 2, chap. 2.

Stokes, John J. and David Belitskus: "Aluminum Cells," in G. E. Heise and N.C. Cahoon (eds.), in ibid., chap. 3.

Wood, D.B.: "Magnesium Batteries," in K.V. Kordesch (ed.), *Batteries,* vol. 1, Dekker, New York, 1974, chap. 4.

7

Alkaline Manganese Dioxide Cells

by
Karl V. Kordesch

7.1 GENERAL CHARACTERISTICS

The alkaline manganese dioxide (zinc/potassium hydroxide/manganese dioxide) cell has had a remarkable growth as a consumer battery in the past decade, capturing a large share of the primary cell market. The advantages and disadvantages of this battery system are summarized in Table 7.1. The alkaline cell has a higher capacity and an advantage over the zinc-carbon Leclanche cell on low discharge rates and intermittent service. Its performance excels, however, on high discharge rates and continuous

TABLE 7.1   Major Advantages and Disadvantages of the Alkaline Manganese Dioxide Cell

Advantages	Disadvantages
Good high-rate discharge capability	Higher initial unit cost than Leclanche cells
Higher energy output than Leclanche cell (depending on load)	If shorted or abused, cell temperatures could rise to high levels (e.g., 100°C)
Good shelf life	Sloping discharge curve, but less pronounced than Leclanche cell.
Good leakage resistance	
"Rest periods" not necessary—efficient when used continuously	
Good low-temperature performance	
Good shock resistance	
Low gassing rate	

service; under these conditions it can outperform the Leclanche cell by 5 times or more. The alkaline cell is also superior to the other aqueous electrolyte primary cells at low temperatures. Further, the design of the cell, using a steel container which can be sealed effectively, provides the cell with good leakage resistance and long shelf life. The alkaline manganese dioxide cell is more costly than the zinc-carbon cell but for many applications, particularly at high discharge loads and continuous drains, the use of the alkaline cell is cost-effective. A detailed comparison is given in Part 2, Chap. 4.

The history of the alkaline manganese dioxide cell dates back to 1900. The first commercial cell, as produced in 1950,[1] used a manganese dioxide-graphite cathode in the form of a pellet, KOH or NaOH soaked into an absorbent separator, and an increased surface area zinc anode. It was called the "crown cell" because of its bottle-cap-shaped closures, and it was used mainly as a low-current radio battery. The modern high-drain alkaline manganese dioxide cell was developed in the 1960s. With the phenomenal increase in the use of portable electronic equipment (radios, cassette-recorders, etc.), large-volume, worldwide production was achieved in the decade from 1970 to 1980.

7.2 CHEMISTRY

The zinc/alkaline/manganese dioxide cell uses the same electrochemically active materials, zinc and manganese dioxide, as the Leclanche cell but differs in the cell construction and in the use of highly conductive potassium hydroxide electrolyte which results in a lower cell internal resistance. The zinc anode, which does not have to double as the cell container, is formed of zinc powder, which gives it a large surface area, and is amalgamated to suppress hydrogen gassing. Synthetically produced, electrolytic manganese dioxide with increased reactivity and higher capacity than natural ores is used for the cathode.

The simplified reaction mechanisms, based on a one-valence-step reduction for the manganese dioxide, are

Anode: $$Zn + 2OH^- \rightarrow Zn(OH)_2 + 2e$$

$$Zn(OH)_2 + 2OH^- \rightarrow [Zn(OH)_4]^{2-}$$

Cathode: $$2MnO_2 + H_2O + 2e \rightarrow Mn_2O_3 + 2OH^-$$

Overall reaction: $$Zn + 2MnO_2 \rightarrow ZnO + Mn_2O_3$$

The Gibbs free energy is calculated to be between -276 and -284 kJ/mol, depending on the type of MnO_2 chosen, which results in a cell potential of 1.4 to 1.5 V. The electrode potential is determined by the homogeneous distribution of Mn^{3+} and Mn^{4+}, whereby the ratio of the two species determines the state of reduction and also the shape of the discharge curve. It has been found[2] that the expansion of the MnO_2 lattice is small between MnO_2 and $MnO_{1.6}$. Figure 7.1 shows that a large increase in the electrode resistance (swelling) takes place after the one-electron discharge.[3]

Zinc is thermodynamically unstable in contact with alkaline solutions generating hydrogen. The behavior of zinc changes completely if it is amalgamated; in particular, the harmful effects of heavy metal impurities are strongly reduced. In commercial batteries the mercury content is kept between 4 and 8%, but the trend is to lower the concentration.

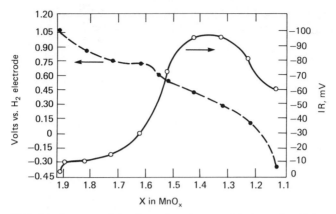

FIG. 7.1 Increase of the electrode resistance with decreasing X in MnO_x. (*From Ref. 3.*)

The zinc passivation phenomenon is important for practical applications. Low current densities, increased temperatures, stirring, etc. prevent passivation. The formed zinc-covering layers may be dense or porous. They may become more or less soluble depending on accompanying ZnO precipitations from oversaturated zincate solutions. In any case, large-surface powdered zinc does not easily passivate, while smooth surfaces do. As a result, powdered-zinc anodes operate satisfactorily with small amounts of electrolyte, while zinc can anodes do not.

For reasons of conductivity and solubility of $Zn(OH)_2$, potassium hydroxide solutions have been chosen as the electrolyte for alkaline MnO_2 cells. KOH is available in a sufficiently pure form as a 45 to 50% solution with a specific gravity of 1.46 to 1.52. Its maximum conductivity (0.5 S/cm) occurs with a 30 wt% KOH solution which is 7 molar in $(OH)^-$ ion concentration. A 30% KOH solution has a freezing point of $-66°C$ (eutectic). The range of electrolytes used in alkaline MnO_2 cells is from 7 to 14 molar (50% KOH).

The electrolyte is either immobilized by the addition of a gelling agent (e.g., sodium carboxymethyl cellulose, usually 4 to 7%), or it is soaked up by the porous separator system or the capillary action of the powdered-zinc anode (compressed type). Zinc oxide is added as corrosion-inhibitor, as paste-former, and also for overcharge protection.

The conductivity of the electrolyte is changed by the dissolution of $Zn(OH)_2$ during operation of the cell, but, on the other hand, the bulk conductivity in a powdered-zinc gel anode is determined by the metallic structure until failure of particle contacts. Separator conductivity is determined by its porosity.

7.3 CELL COMPONENTS AND MATERIALS

7.3.1 Manganese Dioxide Cathode Formulation

Different types of manganese dioxide show very different electrochemical behavior in batteries; seemingly minor structural variations are technologically important. With alkaline batteries the situation is simplified because only electrolytically produced

MnO_2 is used. The industrial process uses $MnCO_3$ ore (rhodochrosite) dissolved in sulfuric acid as the starting material. The electrolysis conditions are carefully controlled with respect to temperature, concentration, and current density. Titanium, lead alloys, or carbon electrodes are used for the deposition of MnO_2. The level of impurities is very critical in view of the expected battery shelf life. A typical chemical analysis is shown in Table 7.2. Extensive literature exists on the production of electrolytic MnO_2 and its properties,[2] testing, and evaluation.[4,5]

TABLE 7.2   Typical Range of Chemical Analysis of Battery Grade EMD Samples

MnO_2	92.0 $\pm$ 0.3%	Sb	0.0001%
Total Mn	60.0 $\pm$ 0.12%	As	0.0003%
H_2O (−)*	1.52 $\pm$ 0.20%	Mo	0.0002%
Fe	0.008%	Na	0.24%
Pb	0.0007%	NH_3	0.001%
Cu	0.0003%	SiO_2	0.02%
Ni	0.0005%	Insoluble to HC1	0.62%
Co	0.0008%		
Cr	0.0008%	SO_4^{2-}	0.79%

*H_2O (−): Adsorbed moisture which can be removed by heating at 110 to 120° C. Balance is structural water which is released by heating above 120° C.

The composition of a typical cathode in the alkaline MnO_2 cell is given in Table 7.3. A D size cell contains 50 g of cathode mix. The theoretical capacity (based on a one-electron change) is 9.76 Ah. On a relatively high-rate discharge (16-h rate) the actual ampere-hour capacity is about 7 Ah, for a MnO_2 utilization of 73%. At a lower rate, e.g., the 100-h rate, the cell will deliver about 9.6 Ah to 0.8 V for a utilization of 98%. There is still service left below 0.8 V, and for some uses 0.65 V may be a suitable cutoff voltage. In this case, for a lower end voltage corresponding to further reduction of the MnO_2, a figure of 0.41 Ah per gram of MnO_2 is used for the determination of MnO_2 capacity. On this basis, the theoretical capacity of the D size cell is nearly 13 Ah.

TABLE 7.3   Composition of a Typical Cathode in Alkaline MnO_2 Cell

70% MnO_2 (with 90% available MnO_2, balance is water, etc.)
10% graphite (higher percentage in high-power cells)
1–2% acetylene black
Balance: binders and electrolyte for molding

Density of mix: 4.0–4.2 g/cm³
Theoretical capacity of cathode mix (based on a one-electron change for MnO_2 or 0.31 Ah/g):
  0.9 × 0.7 × 0.31 Ah/g = 0.195 Ah/g
Volumetric capacity: 0.195 Ah/g × 4 g/cm³ = 0.8 Ah/cm³

7.3.2 Zinc Powder Anode Formulation

For alkaline cells, zinc must be of high purity (99.85 to 99.90% Zn). It is produced commercially by electroplating or by distilling. Sometimes 0.04 to 0.06% Pb is added for corrosion resistance. Powdered zinc is obtained by discharging a thin stream of molten zinc into an air jet where it is "atomized." The specifications for a typical battery grade zinc powder require a certain particle size distribution (0.0075 to 0.8 mm) determined by sieve fractions. The surface area of a typical powdered zinc is 0.2 to 0.4 m^2/g. Two types of powdered-zinc anodes are used in commercial cells: powdered zinc mixed with gelled electrolyte and pressed powdered zinc which soaks up the KOH solution.

Gelled Anodes   A typical formulation contains 76 wt % zinc, 7% mercury (metal), 6% sodium carboxymethyl cellulose, and the rest KOH solution.[2] The mass is very viscous and can be extruded into the cell. In miniature alkaline cells NaOH is used to reduce the caustic creepage around the seal area. It is important to realize that the gel anode is not electronically conductive enough until 30% by volume is metallic zinc.[6] Such a homogeneous gel lacks good zinc utilization at high current densities; therefore, two-phase anodes consisting of more compact zinc-powder gel and a clear gel have been devised. That way 90% zinc usage is achieved. The volume capacity of the gelled anodes is 2 Ah/cm³. The anode capacity of a primary alkaline MnO_2 cell must be balanced against the (1e + 33%) cathode capacity (0.41 Ah/g) in order to avoid gassing of a fully discharged cell, which could occur if the cathode is completely consumed but zinc is left. The potential of the zinc against the exhausted cathode is sufficient to produce hydrogen at the remaining carbon structure if the cell is left on load. The reaction is

$$Zn + 2H_2O = H_2 + Zn(OH)_2$$

In reference to the previously discussed D size cell, a cathodic capacity of 13 Ah must be balanced. A total of 20 g of the gelled anode with 76% zinc results in an anode capacity of 20 × 76% × 0.82 Ah/g = 12.5 Ah or a factor of 0.96. In practical cells this factor can be 1.05, considering that 10% zinc is lost.

Porous Zinc Anodes   These anodes are produced by cold-pressing of zinc powder wetted with mercury; this method welds the particles together.[7] In another process the porosity is controlled by filler materials (for example, NH_4Cl) which can easily be removed later. Plastic binders are also used to fabricate porous zinc electrodes. Bonding zinc oxide and electrolytically reducing the electrode is done only in rechargeable cells. Such anodes can carry very high currents due to the large-surface zinc which is deposited. The capacity figures for porous zinc bodies dependent on current density and temperature are given in Table 7.4.

TABLE 7.4   Capacity of Porous Zinc Bodies*

Temperature, °C	$i^a = 5$ mA/cm²		$i^a = 10$ mA/cm²		$i^a = 20$ mA/cm²	
	Ah/g	Ah/cm³	Ah/g	Ah/cm³	Ah/g	Ah/cm³
20	0.612	2.35	0.538	2.35	0.460	2.20
0	0.406	1.96	0.348	1.93	0.230	1.85
−20	. . .	. . .	0.198	1.19	0.112	0.85
−30	0.140	0.83	0.097	0.69	0.048	0.49
−40	0.098	0.41·	0.042	0.35	0.004	0.25

*Based on Ref 8.

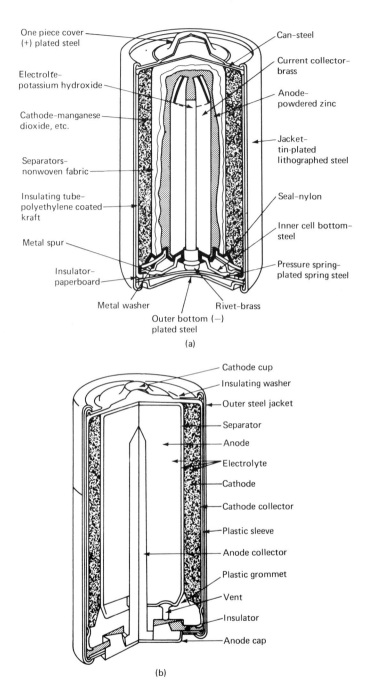

One piece cover — (+) plated steel

Can-steel

Current collector-brass

Electrolte-potassium hydroxide

Anode-powdered zinc

Cathode-manganese dioxide, etc.

Jacket-tin-plated lithographed steel

Separators-nonwoven fabric

Insulating tube-polyethylene coated kraft

Seal-nylon

Inner cell bottom-steel

Metal spur

Insulator-paperboard

Pressure spring-plated spring steel

Metal washer

Rivet-brass

Outer bottom (−) plated steel

(a)

Cathode cup

Insulating washer

Outer steel jacket

Separator

Anode

Electrolyte

Cathode

Cathode collector

Plastic sleeve

Anode collector

Plastic grommet

Vent

Insulator

Anode cap

(b)

FIG. 7.2 Cross section of cylindrical alkaline manganese dioxide cells. (a) Union Carbide Corp. (*reprinted from Ref. 2, by courtesy of Marcel Dekker, Inc.*); (b) Duracell International, Inc.

7-7

7.3.3 Separators[9]

In general, alkaline MnO_2 cells use "macroporous" separators made from woven, bonded, or felted materials. Rechargeable cells must use "microporous" membrane-type materials, like regenerated cellulose (cellophane, sausage casing, grafted membranes, or polymeric films). Inorganic microporous layers with binders are expensive, but synthetic sheets from which a filler has been leached out are not. Pore diameters range from 2.5 to 10 nm (membranes) to several hundred nanometers for porous-sheet separators.

7.4 CONSTRUCTION

7.4.1 Cylindrical Cells

The principal features of the zinc/alkaline/manganese dioxide cell are the manganese dioxide of high density, a zinc anode of high surface area, and the highly conductive potassium hydroxide electrolyte. An inside-out cell design, with the manganese dioxide cathode external to the centrally located zinc anode, was adopted to obtain better mass transport (diffusion) properties and low internal resistance. The zinc can and bobbin cathode construction, which is the usual design of the zinc-carbon cell, was generally not followed for the alkaline cell.

Figure 7.2 shows the construction of the alkaline cylindrical cell used by two manufacturers. The MnO_2 mixed with graphite or carbon black is pressed against the inner surface of the steel can which also serves as the cathode current collector. The anode is centrally located and consists of a mixture of granular or powdered zinc, which is amalgamated to reduce hydrogen evolution, and the electrolyte. In some designs, a gelling agent is used to immobilize the electrolyte and minimize leakage. A highly absorbent chemically inert separator divides the electrodes. A central current collector makes intimate contact with the zinc, leads to the bottom negative terminal, and provides the "right" polarity (positive terminal on top of cell). An external insulating sleeve or tube is used between the steel case and outer steel jacket.

Special attention is given to the seals and safety vents of the cells to provide protection from cell rupturing, caustic burns, etc. in case of mistreatment of the cell by short-circuiting, fast-charging (charging of the primary alkaline cell is not recommended), fire disposal, or other abusive use.

Figure 7.3 shows the manufacturing steps for the alkaline MnO_2 cell.

7.4.2 Miniature Cells (Button Cells)

The cell construction shown in Fig. 7.4 is typical for a small hearing-aid or watch battery. Usually the cathode is formed in the bottom can (steel, nickel-plated), and the zinc-powder gel or pressed porous zinc pellet is molded into the copper-clad top-steel can. The separator is wetted with the electrolyte. The sealing gasket is very important in view of the leakproofness required of such cells.

7.4.3 Special Constructions

Alkaline MnO_2 cells are produced to meet a wide range of performance goals. To achieve those, special batteries have been designed: reserve-type cells which have an

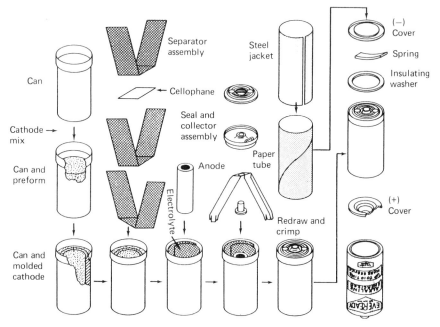

FIG. 7.3 Assembly of cylindrical alkaline manganese dioxide cell (Union Carbide Corp. E-95 cell).

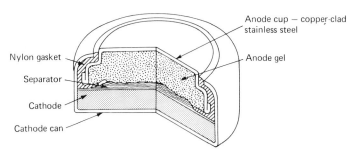

FIG. 7.4 Construction of miniature alkaline manganese dioxide cell. (Reprinted from Ref. 2, by courtesy of Marcel Dekker, Inc.)

unlimited shelf life and are activated (usually by mechanical means) before use; cells with double cathodes; thin-plate cells (e.g., for high pulse loads in photographic equipment); and special (military) batteries. A zinc-alkaline battery, using a flat-cell design, is available in a 6- and 12-V configuration. This battery will deliver most of its capacity to -10°C.[10] Low-temperature applications require different electrolyte compositions from those in cells formulated for use in the tropics. Low-temperature performance benefits very much from large interface designs and thin-plate or rolled-up (spiral wound) cell constructions.[11] Batteries of these designs are experimental.

7.5 PERFORMANCE CHARACTERISTICS

7.5.1 Voltage

The open-circuit voltage of the alkaline MnO_2 cell is about 1.52 V. The nominal operating voltage is 1.25 V, but the actual voltage is dependent on the discharge load and state of charge. The usual end voltage is 0.9 V per cell, but longer service can be obtained to lower end voltages, particularly at high current drains.

7.5.2 Discharge Performance

Figure 7.5 shows the discharge curves of the AA size alkaline MnO_2 cell, rated at 1700 mAh, on continuous discharge at various loads at 20°C. These discharge curves are typical for the alkaline cell performance and will generally be the same for cells of other sizes at similar hourly discharge rates.

The sloping shape of the discharge curve is typical for the alkaline MnO_2 cell, although the slope is not as pronounced as with the zinc-carbon cell. A general sum-

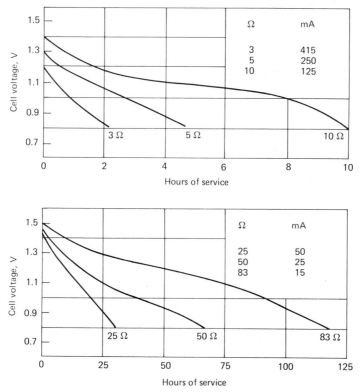

FIG. 7.5 Typical discharge curves, alkaline manganese dioxide cell (AA size) at 20°C. (*Reprinted from Ref. 2, by courtesy of Marcel Dekker, Inc.*)

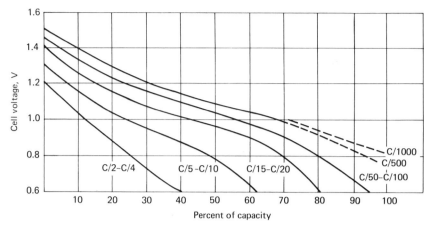

FIG. 7.6   Voltage and capacity of alkaline manganese dioxide cells as a function of discharge rate at 20°C.

mary of the relationship of the capacity with discharge rate can be obtained from Fig. 7.6.

Continuous discharge performance data do not give the complete picture of the capability of the alkaline cell because of the effect of discharge load, intermittency of operation, and duty cycle. For this reason, a tabulation of service data under various discharge conditions is preferred for some applications. Table 7.5 is an ex-

TABLE 7.5   Average Service of AA Size Alkaline Cell at 20°C*

Schedule	Typical drains at 1.2 V, mA	Load, Ω	Cutoff voltage, V						
			0.75	0.8	0.9	1.0	1.1	1.2	1.3
24 h/day	0.8	1500	2800 h	2700 h	2500 h	2000 h	1700 h	1400 h	800 h
24 h/day	8	150	260 h	255 h	235 h	190 h	160 h	127 h	70 h
8 h/day	8	150	260 h	255 h	235 h	190 h	160 h	130 h	74 h
2 h/day	8	150	265 h	255 h	235 h	190 h	160 h	133 h	78 h
24 h/day	80	15	20 h	19 h	18 h	15 h	12 h	7 h	2 h
8 h/day	80	15	20 h	19 h	18 h	16 h	13 h	7 h	2 h
2 h/day	80	15	22 h	21 h	19 h	17 h	14 h	8 h	3 h
24 h/day	800	1.5	65 min	60 min	46 min	24 min	10 min	4 min	2 min

Eveready No. E91, ANSI L-40, IEC LR-6.
SOURCE: Data taken from Eveready Battery Engineering Data, Union Carbide Corp. 1982.

ample, listing the average service of the AA size alkaline cell under different use conditions.

7.5.3 Effect of Temperature

The alkaline MnO_2 cell performs well at low temperatures, outperforming even the best Leclanche cells. The cell can operate at temperatures as low as $-40°C$. Figure 7.7 shows the capacity or service life of typical alkaline MnO_2 cells (C and D size cells) at various temperatures under a high and light discharge load. The advantage of the light discharge load is evident. These data, again, are typical for the alkaline cell and apply to cells of other sizes discharged under similar conditions.

7.5.4 Internal Resistance

The internal resistance of fresh, undischarged alkaline MnO_2 cells (AA and D size cells) at various temperatures is shown in Fig. 7.8.

TABLE 7.5 (*Continued*)
Simulated application tests, estimated average service at 21°C

Schedule	Typical drains at 1.2 V, mA	Load, Ω	Cutoff voltage, V			
			0.75	0.8	0.9	1.0
4 h/day (radio test)	16	75	121 h	117 h	105 h	90 h
4 h/day (radio test)	30	40	63 h	59 h	52 h	44 h
½ h/day (calculator test)	48	25	34 h		28 h	
½ h/day (calculator test)	100	12	17 h	16 h	14 h	11 h
1 h/day (cassette test)	120	10	13 h		11 h	
4 min every 15 min, 8 h/day, 16 h rest (camera cranking test)	250 mA constant current					211 min
4 min/h 8 h/day, 16 h rest (light-industrial flashlight test)	300	4	300 min	275 min	245 min	200 min
Continuous (toy test)	300	4	245 min	240 min	225 min	175 min
1 h/day (toy test)	308	3.9	290 min	270 min	220 min	160 min
5 min/day (general-purpose intermittent flashlight test)	308	3.9	260 min	250 min	220 min	150 min

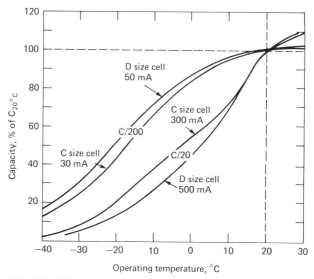

FIG. 7.7 Effect of temperature on the performance of alkaline manganese dioxide cells.

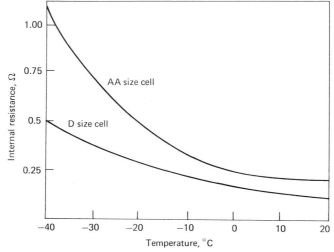

FIG. 7.8 Internal resistance of fresh alkaline manganese dioxide cells.

7.5.5 Service Life

The service life of the alkaline MnO_2 cell at various discharge loads and temperatures, normalized for unit weight (amperes per kilogram) and volume (amperes per liter), and to an end voltage of 0.9 V per cell, is summarized in Fig. 7.9. These curves, based on

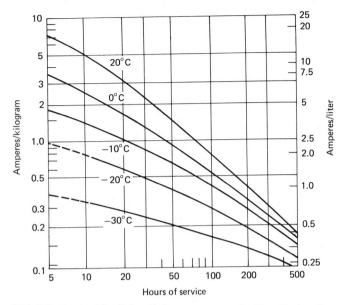

FIG. 7.9   Service life of the alkaline manganese dioxide cell at various discharge rates and temperatures (to 0.9 V per cell end voltage).

an energy density of 75 Ah/kg and 190 Ah/L, can be used to approximate the service life of a given cell under particular discharge conditions or to estimate the size or weight of a cell to meet a particular service requirement.

7.5.6 Shelf Life

The shelf life of the alkaline MnO_2 cell is good; the capacity retention is about 90 to 95% after 1 y of storage at 20°C. Figure 7.10 shows the capacity loss on stand as a function of storage temperature for the alkaline MnO_2 cell compared with the alkaline HgO cell. Points A and A' represent 20 and 10% loss after 3 y at 20°C; points B and B' represent 20 and 10% loss after 3 months at 45°C; point C represents a 20% capacity loss in 20 days. As a rule of thumb, 1 week of storage at 70°C is equivalent to 3 months storage at 45°C.

7.6 CELL AND BATTERY TYPES AND SIZES

The physical and electrical characteristics of the major alkaline MnO_2 cells are listed in Table 7.6. A plot of the capacity vs. the weight and volume of the cells (Fig. 7.11) shows the linearity of the sizes, although the specific energy values, particularly on a volume basis, are higher for the smaller cylindrical cells. This is due to the better utilization of the thin layers of MnO_2 and powdered zinc, especially at the high rates. Cells of other configurations are less efficient.

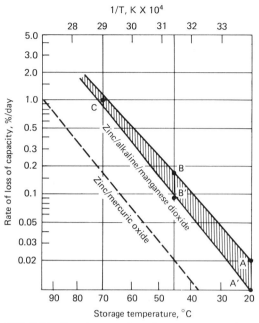

FIG. 7.10 Capacity retention after storage at various temperatures.

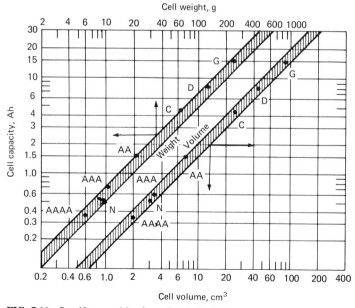

FIG. 7.11 Specific capacities for cylindrical alkaline manganese dioxide cells.

TABLE 7.6 Characteristics of 1.5-V Alkaline MnO$_2$ Cells*

ANSI†	IEC‡	NEDA	Cell size	Capacity, ~Ah	Diameter, mm	Height, mm	Weight, g	Volume, cm³
				Cylindrical cells				
	LR-1		AAAA	0.35	8	40	6	2.0
L-25	LR-7		N	0.5	12	28	10	3.0
L-30	LR-03		⅓A	0.6	16	16	11	3.6
L-40	LR-6	24A	AAA	0.75–0.8	10	44	11.5	3.6
L-70	LR-14	15A	AA	1.5–1.7	14	50	22	7.5
L-90	LR-20	14A	C	4.5–5.0	26	50	65	25
L-100	LR-25	13A	D	8–10	33	60	130	50
	LR-26		F	15–20	33	91	200	80
			G	15–20	33	105	235	95
				Button Cells				
	LR-41			0.030	7.9	3.6	0.6	0.2
	LR-43			0.090	11.6	4.2	1.4	0.3
	LR-44			0.120	11.6	5.4	2.3	0.5
	LR-48			0.060	7.9	5.4	0.9	0.3
L-15	LR-53			0.130	23.2	6.1	6.8	2.5
	LR-54			0.065	11.6	3.1	1.1	0.3
	LR-55			0.030	11.6	2.1	0.9	0.2
		825		0.200	23.0	6.0	7.0	2.0

*The values presented are average figures from different manufacturers. (Conformity with codes is approximative; some measures include special terminals.)

†ANSI C-18.1 (1979): Specifications for dry cells and batteries, American National Standards Institute, Inc., New York.

‡IEC: International Electrochemical Commission. The L denotes alkaline MnO$_2$-Zn. In the Japanese code the same sizes are: AM4, AM3, AM2, AM1. The IEC code is also the German DIN code.

Table 7.7 lists the major multicell alkaline MnO_2 batteries that are available commercially. Table 7.8 gives cross-references to the alkaline cells and batteries of the various manufacturers.

TABLE 7.7 Characteristics of Alkaline MnO_2 Batteries

Voltage, V	Current range, mA	Code	Description	Dimensions, mm	Weight, g	Volume, cm	Capacity, Ah
3	0–10	PX-30	2 · 825	25 D, 12 H	14.3	5.0	0.2
4.5	0–10	PX-31	3 · 825	41 · 17 · 11.4	12.8	6.6	0.2
3	0–70	NEDA 1308 A	2 · L-25	17 D, 33 H	22.7	7.3	0.6
4.5	0–70	NEDA 1306 A	3 · L-25	17 D, 50 H	33.2	11.0	0.6
6	0–10	Varta 7250	4 · button	13 D, 25 H	14.2	3.3	0.1
6	0–100	Eveready 539	4 · AAAA	48 · 35.6 · 9	31.0	16.4	0.35
9	0–100	NEDA 1604 A	6 · AAAA†	50 · 26 · 17.5	45.0	21	0.35
6	0–1300	NEDA 930 A	4 · L-100	141 · 118 · 53	1130	883	15
6	Special	IP-5*	4 cells	97 · 67 · 67	600	435	8.5
6	Special	IP-26*	4 cells	230 · 140 · 75	3500	2400	80

*Imatra Alkaline Batteries, Finland; batteries up to 320 Ah for navigation.
†Some 9-V batteries consist of six flat cells LF-22.

TABLE 7.8 Cross-References of Codes for Alkaline MnO_2 Cells and Batteries

IEC NEDA	LR-07 watch	LR-9 photo	Button	LR-1 910 A	LR-03 24 A	LR-6 15 A	LR-14 14 A	LR-20 13 A
Eveready	675	625	EPX-825	E-90	E-92	E-91	E-93	E-95
Duracell	675	625	PX-825	MN 9100	MN 2400	MN 1500	MN 1400	MN 1300
Ray-O-Vac	675	625	RPX-825	810	824	815	814	813
Burgess	. . .	. . .	APX-825	AL-N	AL-7	AL-9	AL-1	AL-2
Bright Star	. . .	. . .	BPX-825	7528	7526	7524	7522	7520
Varta	. . .	4626	7201	4001	4003	4006	4014	4020
Japan	LR-44	. . .	. . .	AM-5	AM-4	AM-3	AM-2	AM-1
Military	. . .	. . .	. . .	. . .	. . .	BA-3058	BA-3042	BA-3030
NEDA	. . .	. . .	1308-A	1306-A	. . .	. . .	1604-A	930-A
Eveready	EPX-30	538	532	523	537	539	522	520
Duracell	PX-30	7R-31	PX-24	PX-21	. . .	7 K 67	MN 1604	
Ray-O-Vac	RPX-30	RPX-31	RPX-24	RPX-21				
Burgess	. . .	HPX-31	APX-24	AL-523				
Bright Star	. . .	. . .	7531	7530				
Varta	. . .	. . .	. . .	7251	7250	. . .	4022	

REFERENCES

1. W. S. Herbert, *J. Electrochem. Soc.* **99:**190 C (1952).

2. K. V. Kordesch, "Manganese Dioxide," in K.V. Kordesch (ed.), *Batteries,* vol. 1, Dekker, New York, 1974, chap. 2.

2a. A. Kozawa, "Electrolytic Manganese Dioxide," in ibid., chap. 3.

3. D. Boden, C. J. Venuto, D. Wisler, and R. B. Wylie, *J. Electrochem. Soc.,* **114:**415 (1967).

4. G. W. Heise and N. C. Cahoon, *The Primary Battery,* vol. 1, Wiley, New York, 1971.

5. A. Kozawa and R. J. Brodd (eds.), *Proceedings, Manganese Dioxide Symposium,* vol. 1, Electrochemical Society, Cleveland Section, I. C. Sample Office, Cleveland, OH, 44145, 1975.

5a. B. Schumm, H. M. Joseph, and A. Kozawa (eds.): *Proceedings, Manganese Dioxide Symposium,* vol. 2, Electrochemical Society, Cleveland Section, I. C. Sample Office, Cleveland, OH, 44145.

6. J. L. S. Daley, *Annu. Power Sources Conf.,* **15:**96 (1961).

7. J. F. Jammet, U.S. Patent 3,556,861 (1971).

8. F. Przybyla and F. J. Kelly, *Power Sources 2, Proceedings of the 6th International Symposium,* Brighton, Pergamon, New York, 1968.

9. S. U. Falk and A. J. Salkind, *Alkaline Storage Batteries,* Wiley-Interscience, New York, 1969.

10. Imatra Battery, Dry Battery B. V., The Netherlands, and Fremont Battery Co., Fremont, Ohio.

11. K. V. Kordesch and A. Kozawa, U.S. Patent 3,945,847 (1976).

Mercuric Oxide Cells

by
Denis Naylor

8.1 GENERAL CHARACTERISTICS

The alkaline zinc/mercuric oxide battery is noted for its high capacity per unit volume, constant voltage output, and good storage characteristics. The system has been known for over a century, but it was not until World War II that a practical cell was developed by Samuel Ruben in response to a requirement for a cell with a high capacity-to-volume ratio which would withstand storage under tropical conditions. The early cells performed extremely well under these conditions using designs similar to present-day button cells.[1-4]

An early zinc/mercuric oxide cell is shown in Fig. 8.1. The cell anode was made from zinc powder or foil amalgamated with mercury; the electrolyte was a solution of sodium or potassium hydroxide, to which zinc oxide was added; mercuric oxide was used for the cathode with graphite added to provide a conductive matrix to the electrode. These basic ingredients are still used in present-day cells.

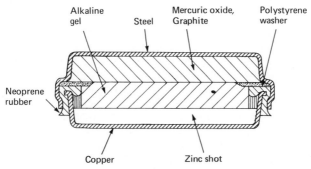

FIG. 8.1   Zinc/mercuric oxide cell—early design.

The use of cadmium in place of zinc results in a very stable battery with excellent storage and performance at extreme temperatures due to the low solubility of cadmium in caustic alkali over a wide range of temperature. Although it was realized some time ago that an anode of cadmium could be used with a mercuric oxide cathode, the cost of the material was high and the cell voltage was low, less than 1.0 V. Hence, the battery is a fairly recent innovation and is being used in special applications where its unique characteristics are advantageous. Cell structures are similar to those used for the zinc/mercuric oxide cells, with only slight modifications at the top/anode interface being necessary.[5-7]

The major characteristics of these two battery systems are summarized in Table 8.1.

8.2 CHEMISTRY

It is generally accepted that the basic cell reaction for the zinc/mercuric oxide cell is

$$Zn + HgO \rightarrow ZnO + Hg$$

For the overall reaction $\triangle G^0 = 259.7$ kJ. This gives a thermodynamic value for E^0 at 25°C of 1.35 V which is in good agreement with the observed values of 1.34 to 1.36 V for the open-circuit voltage of commercial cells.[8,9] From the basic reaction equation it can be calculated that 1 g of zinc provides 819 mAh and 1 g of mercuric oxide provides 247 mAh.

Some types of zinc/mercuric oxide cells exhibit open-circuit voltages between 1.40 and 1.55 V. These cells contain a small percentage of manganese dioxide in the cathode and are used where voltage stability is not of major importance for the application.

The basic cell reaction for the cadmium/mercuric oxide cell is

$$Cd + HgO + H_2O \rightarrow Cd(OH)_2 + Hg$$

For the overall reaction $\triangle G^0 = -174.8$ kJ. This gives a thermodynamic value for E^0 at 25°C of 0.91 V, which is in good agreement with the observed values of 0.89 to 0.93 V. From the basic reaction it can be calculated that 1 g of cadmium should provide 477 mAh.

8.3 CELL COMPONENTS

8.3.1 Electrolyte

Two types of alkaline electrolyte are used in the zinc/mercuric oxide cell, one based on potassium hydroxide and one on sodium hydroxide. Both these bases are very soluble in water, and highly concentrated solutions are used; zinc oxide is also dissolved in varying amounts in the solution to suppress hydrogen generation.

Potassium hydroxide electrolytes generally contain between 30 and 45% w/w KOH and up to 7% w/w zinc oxide. They are more widely used than the sodium hydroxide electrolytes because of their greater operating temperature range and ability to support heavier current drains. For low-temperature operation both the potassium hydroxide and zinc oxide contents are reduced, and this introduces some instability at higher temperatures with respect to hydrogen generation in the cell.

TABLE 8.1 Characteristics of the Zinc/Mercuric Oxide and Cadmium/Mercuric Oxide Cell

Zinc/mercuric oxide cell		Cadmium/mercuric oxide cell	
Advantages	Disadvantages	Advantages	Disadvantages
High energy-to-volume ratio, 400 Wh/L	The cells are expensive; although widely used in miniature sizes, they tend to be used only for special applications in the larger sizes	Long shelf life under adverse storage conditions	The cells are more expensive than zinc/mercuric oxide cells due to the high cost of cadmium
Long shelf life under adverse storage conditions	Disposal of quantities of spent batteries creates ecological problems	Flat discharge curve over a wide range of current drains	Disposal of spent batteries creates an ecological problem, with both cadmium and mercury being toxic
Over a wide range of current drains recuperative periods are not necessary to obtain a high capacity from the cell	After long periods of storage, the cell electrolyte tends to seep out of the seal and is evidenced by a white carbonate deposit at the seal insulation	Ability to operate efficiently over a wide temperature range, even at extreme high and low temperatures	The system has a low output voltage (open-circuit voltage = 0.90 V)
High electrochemical efficiency	Moderate energy-to-weight ratio	Can be hermetically sealed because of inherently low gas evolution level	Moderate energy-to-volume ratio
High resistance to impact, acceleration, and vibration			Low energy-to-weight ratio
Very stable open-circuit voltage (1.35 V)			
Flat discharge curve over a wide range of current drains			

Sodium hydroxide electrolytes are prepared in similar concentration ranges and are used in cells where low-temperature operations and/or high current drains are not required. These electrolytes are suitable for long-term-discharge cells because of the reduced tendency of the electrolyte to seep out of the cell seal after long periods of storage.

Generally only potassium-based alkaline electrolytes are used in the cadmium/mercuric oxide cell. As cadmium is practically insoluble in all concentrations of aqueous potassium hydroxide solutions, the electrolyte can be optimized for low-temperature operation.

The freezing-point curve for caustic potash solutions is shown in Fig. 8.2; it will be seen that the eutectic with a freezing point below $-60°C$ is 31% w/w KOH, which is the electrolyte most frequently used. This allows the system to operate down to $-55°C$. Improvements in low-temperature cell efficiency have been made in some cases by the addition of a small percentage of cesium hydroxide to the electrolyte.

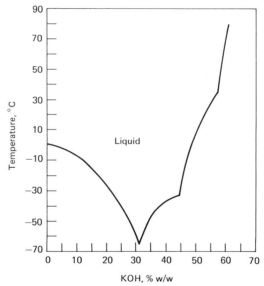

FIG. 8.2 Freezing-point curve for aqueous caustic potash solutions.

8.3.2 Zinc Anode

Alkaline electrolytes act as ion carriers in the cell reactions. The reaction at the zinc negative electrode may be written

$$Zn + 4 OH^- \rightarrow Zn(OH_4)^{2-} + 2e$$

$$Zn(OH_4)^{2-} \rightarrow ZnO + 2OH^- + H_2O$$

These reactions imply the dissolution of the zinc electrode, with the crystallization of zinc oxide from the electrolyte. Once this mechanism is understood, the reaction at the anode can be simplified to

$$Zn + 2OH^- \rightarrow ZnO + H_2O + 2e$$

Direct solution of the zinc electrode in the alkaline solution on open circuit is minimized by dissolving zinc oxide in the electrolyte and by amalgamating the zinc in the electrode. Mercury levels used in zinc electrodes are usually in the range of 5 to 15% w/w. Great attention is also paid to the impurity levels in the zinc since minor cathodic inclusions in the electrode can drive the hydrogen generation reaction despite the precautions indicated above.[10-14]

8.3.3 Cadmium Anode

The reaction at the anode is

$$Cd + 2OH^- \rightarrow Cd(OH)_2 + 2e$$

This implies the removal of water from the electrolyte during discharge, necessitating an adequate quantity of electrolyte in the cell and the desirability of a high percentage of water in the electrolyte. Cadmium has a high hydrogen overvoltage in the electrolyte, and so amalgamation is neither necessary nor desirable, since the electrode potential is some 400 mV less electropositive than zinc.

Cadmium metal powders as produced conventionally are unsuitable for use as electrode materials. Activated cadmium anodes are produced by (1) electroforming the anode, (2) electroforming powder by special process followed by pelleting, or (3) precipitating by special process as a low-nickel alloy and pelleting. All of these processes are used by different manufacturers to give cells with various performance parameters.[15-18]

8.3.4 Mercuric Oxide Cathode

At the cathode the overall reaction may be written

$$HgO + H_2O + 2e \rightarrow Hg + 2(OH)^-$$

Mercuric oxide is stable in alkaline electrolytes and has a very low solubility. It is also a nonconductor, and adding graphite is necessary to provide a conductive matrix. As the discharge proceeds, the ohmic resistance of the cathode falls and the graphite assists in the prevention of mass agglomeration of mercury droplets. Other additives which have been used to prevent agglomeration of the mercury are manganese dioxide, which increases the cell voltage to 1.4 to 1.55 V, lower manganese oxides, and silver powder, which forms a solid-phase amalgam with the cathode product.

Graphite levels usually range from 3 to 10% and manganese dioxide from 2 to 30%. Silver powder is used only in special-purpose cells because of cost considerations but may be up to 20% of the cathode weight. Again, great care is taken to obtain high-purity materials for use in the cathode; trace impurities soluble in the electrolyte are liable to migrate to the anode and initiate hydrogen evolution. An excess of mercuric oxide capacity of 5 to 10% is usually maintained in the cathode to "balance" the cell and prevent hydrogen generation in the cathode at the end of discharge.[19, 20]

8.3.5 Materials of Construction

Materials of construction for the zinc/mercuric oxide cells are limited not only by their ability to survive continuous contact with strong caustic alkali, but also by their electrochemical compatibility with the electrode materials. As far as external contacts for cells are concerned, these are decided by corrosion resistance, compatibility with the equipment interface with respect to galvanic corrosion, and, to some degree, cosmetic appearance. To maintain simplicity of design, many cells have the same material at both external positive terminal and internal cathode contact. Metal parts may be homogeneous, plated metal, or clad metal. Insulating parts may be injection-, compression-, or transfer-molded polymers or rubbers. Table 8.2 lists the materials commonly used in zinc/mercuric oxide cells.

TABLE 8.2   Materials Commonly Used in Zinc/Mercuric Oxide Cells

External cell contacts	Stainless steels, nickel and some nickel alloys, gold
Anode-compatible materials	Copper and some copper alloys, tin, silver, gold
Cathode-compatible materials	Stainless steels, nickel and some nickel alloys, gold
Sealing grommet materials	Polyolefins, some polyamides, neoprene
Internal insulating materials	Polyolefins, some polyamides, polystyrene, polyvinyl chloride, some acrylics
Electrolyte absorbents, barriers, and separators	α-Cellulose and cellulose derivatives, polyvinyl alcohol, polyolefins, polyvinyl chloride, butadiene-styrene rubber

With the exception of the anode contact, materials for the cadmium/mercuric oxide cell are generally the same as for the zinc/mercuric oxide cell. However, because of the wide range of storage and operating conditions of most applications, cellulose and its derivatives should not be used, and low-melting-point polymers are also avoided. Nickel is usually used on the anode side of the cell and also, conveniently, at the cathode.

8.4 CONSTRUCTION

The zinc/mercuric oxide cell is manufactured in three basic structures: button, flat, and cylindrical configurations. There are several design variations within each configuration.

8.4.1 Button Cell

The button configuration of the zinc/mercuric oxide cell is shown in Fig. 8.3.[21] The cell top is copper or copper alloy on the inner face and nickel or stainless steel on the outer face. This part may also be gold plated, depending on the application. Within the top is a dispersed mass of amalgamated zinc powder ("gelled anode"), and the top is insulated from the can by a nylon grommet. The whole top/grommet/anode assembly

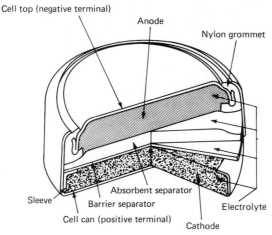

Cell top (negative terminal)

Anode

Nylon grommet

Sleeve

Absorbent separator

Barrier separator

Cell can (positive terminal)

Electrolyte

Cathode

FIG. 8.3   Zinc/mercuric oxide cell—button cell. (*Courtesy of Duracell, Inc.*)

presses down onto an absorbent which contains most of the electrolyte, the remainder being dispersed in the anode and cathode. Below the absorbent is a permeable barrier which prevents any cathode material from migrating to the anode. The cathode of mercuric oxide and graphite is consolidated into the cell can, and a sleeve support of nickel-plated steel prevents collapse of the cathode mass as the cell discharges. The cell can is of nickel-plated steel, and the whole cell is tightly held together by crimping the top edge of the cell can as shown.

Variants of this single top structure are widely used in the construction of button cells; these cells are used in hearing aids, watches, calculators, and photographic devices.

8.4.2 Flat-Pellet Cell

Another form of a larger-sized zinc/mercuric oxide cell is shown in Fig. 8.4.[22-24] In these cells, the zinc powder is amalgamated and pressed into a pellet with sufficient porosity to allow electrolyte impregnation. A double top is used, with an integrally

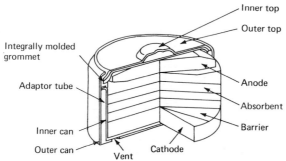

Inner top

Outer top

Integrally molded grommet

Adaptor tube

Inner can

Outer can

Vent

Cathode

Anode

Absorbent

Barrier

FIG. 8.4   Zinc/mercuric oxide cell—flat-pellet cell. (*Courtesy of Duracell, Inc.*)

molded polymer grommet, as a safeguard to relieve excessive gas pressures and maintain a leak-resistant structure. The outer top is of nickel-plated steel, and the inner top is nickel-plated steel but tin plated on its inner face. This cell also uses two nickel-plated steel cans with an adaptor tube between the two, the seal being effected by pressing the top/grommet assembly against the inner can and crimping over the outer can. A vent hole is pierced into the outer can so that if gas is generated within the cell, it can escape between the inner and outer cans, any entrained electrolyte being absorbed by the paper adaptor tube.

8.4.3 Cylindrical Cell

The larger cylindrical zinc/mercuric oxide cell is constructed from annular pressings as shown in Fig. 8.5. The anode pellets are rigid and pressed against the cell top by the neoprene insulator slug. A number of variations of the cylindrical cell are in use with dispersed anodes, where contact with the anode is made either by a nail welded to the inner top or a spring extending from the base insulator to the top.

8.4.4 Wound-Anode Cell

Another design of the zinc/mercuric oxide cell which can operate well at low temperatures is the wound-anode or jelly-roll structure shown in Fig. 8.6.[25-28] Structurally, the cell is similar to the flat cell shown in Fig. 8.4, but the anode and absorbent have been

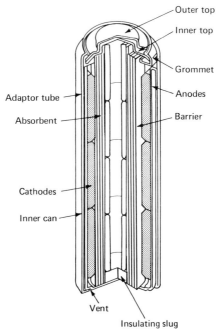

FIG. 8.5   Zinc/mercuric oxide cell—cylindrical cell. (*Courtesy of Duracell, Inc.*)

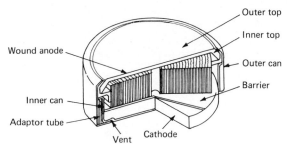

FIG. 8.6   Zinc/mercuric oxide cell—wound-anode cell. (*Courtesy of Duracell, Inc.*)

replaced by a wound anode which consists of a long strip of corrugated zinc interleaved with a strip of absorbent paper. The paper edge protrudes at one side and the zinc strip at the other. This provides a large surface area anode. The roll is held in a plastic sleeve and the zinc is amalgamated in situ. The paper swells in the cell electrolyte and forms a tight structure which is compressed in the cell at the assembly stage with the zinc edge in contact with the cell top.

Electrolyte formulations can be adjusted for low-temperature operation, long storage life at elevated temperature, or a compromise between the two. The performance is optimized by careful adjustment of the anode geometry.

8.4.5 Zinc Can Cell

Good performance down to −20°C can be obtained from a zinc/mercuric oxide cell by suitable designs which provide for a large anode surface area. The Kalium cell is an example and uses zinc-carbon style technology as shown in Fig. 8.7. The cathode mix in the bobbin is mercuric oxide and carbon with a multilayer lap-wound paper separator sealed at the bottom with a plastic disk and at the top with a plastic washer. A zinc can is the anode collector, and the anode itself is a mass of dispersed amalgamated zinc powder providing a high electrode surface area. The zinc powder is also distributed in the outer layers of the bobbin wrap to maximize the surface area and allow for electrolyte penetration of the anode. A conventional aqueous potassium hydroxide electrolyte is used. Kalium cells are available as multicell encased or encapsulated units for a variety of high-rate and/or low-temperature applications.[29]

8.4.6 Low Current Drain Structures

Cells designed for operation at low current drain require modification of the structure to prevent internal electrical discharge paths forming from the conductive materials in both anode and cathode. After partial discharge metallic mercury globules are particularly troublesome in this respect. The problem can be minimized by the use of silver powder in the cathode.

All available passages through which material could form an electrical track need to be blocked if long-term discharge is to be realized. Figure 8.8 shows an RM-1 (R50) size cell structure originally designed for use as a power source for a heart pacemaker. A lap-wound multilayer absorbent/barrier system is used; it is pressed against the welded double caps and at the top is in close contact with the gasket seal

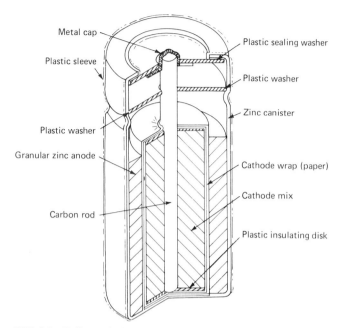

Metal cap

Plastic sleeve

Plastic washer

Granular zinc anode

Carbon rod

Plastic sealing washer

Plastic washer

Zinc canister

Cathode wrap (paper)

Cathode mix

Plastic insulating disk

FIG. 8.7 Kalium cell. (*Courtesy of Crompton-Parkinson, Ltd.*)

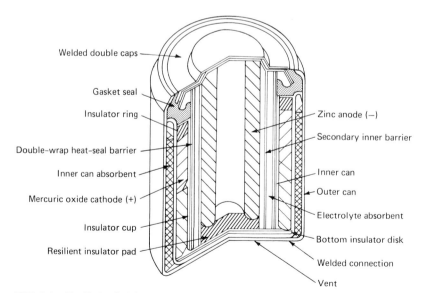

Welded double caps

Gasket seal

Insulator ring

Double-wrap heat-seal barrier

Inner can absorbent

Mercuric oxide cathode (+)

Insulator cup

Resilient insulator pad

Zinc anode (−)

Secondary inner barrier

Inner can

Outer can

Electrolyte absorbent

Bottom insulator disk

Welded connection

Vent

FIG. 8.8 Certified cell, 1.35 V, 1.0 Ah. (*Courtesy of Duracell, Inc.*)

8-10

and at the base is sealed off with a polyethylene insulator cup. A polymer insulator disk is spread across the base of the inner can, and at the top of the cathode pellets an annular insulator ring is pressed in. This structure discharged effectively at the 3- to 5-y rate.

A similar problem has to be faced with cells designed for use in watches. Figure 8.9 shows a typical button watch cell utilizing multiple barrier layers and a polymer insulator washer effectively sealing off the anode from the cathode by compressing these layers against the support ring. The other components shown in Fig. 8.9 are similar to those shown in Fig. 8.3. These structures are expected to discharge at the 1- to 2-y rate.[30,31]

8.4.7 Cadmium/Mercuric Oxide Cells

Cadmium/mercuric oxide cells are not internationally standardized and are produced in a wide range of sizes in button, prismatic, and cylindrical forms, many of which are for special purposes. A button-type cell having the cathode conventionally compacted into the can with a support ring is shown in Fig. 8.10. The barrier is an ion-permeable polymer and the absorbent a polyolefin fiber mat. The grommet is nylon, and all the metal components are nickel-plated steel. In this case the anode is pressed from an electroformed powder.

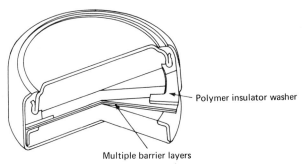

Polymer insulator washer

Multiple barrier layers

FIG. 8.9   Button cell for low-rate discharge.

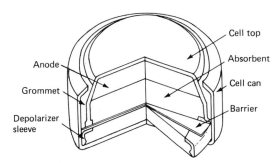

Cell top

Absorbent

Anode

Cell can

Grommet

Barrier

Depolarizer
sleeve

FIG. 8.10   Cadmium/mercuric oxide cell.

8.5 PERFORMANCE CHARACTERISTICS: ZINC/MERCURIC OXIDE CELL

8.5.1 Voltage

The open-circuit voltage of the zinc/mercuric oxide cell is 1.35 V. Its voltage stability under open-circuit or no-load conditions is excellent, and these cells have been widely used for voltage reference purposes. The no-load voltage is nonlinear with respect to both time and temperature. A voltage-time curve is shown in Fig. 8.11; the no-load voltage will remain within 1% of its initial value for several years. A voltage-temperature curve is shown in Fig. 8.12; temperature stability is even better than age stability. From −20 to +50°C the total no-load voltage range is in the region of 2.5 mV.

8.5.2 Discharge Performance

A flat discharge curve is characteristic of the system and is shown in Fig. 8.13 for a pressed-powder anode cell of the RM-1 (R50) size over a current range of 10 to 80 mA at 20°C. The endpoint voltage is generally considered to be 0.9 V, although at higher current drains the cells may discharge usefully below this voltage. At low current drains the discharge profile is very flat and the curve is almost "squared off."

The capacity or service of the zinc/mercuric oxide cell is about the same on either continuous or intermittent discharge regimes. Discharge curves similar to those shown in Fig. 8.14 will normally be obtained over the current drain range recommended by the manufacturer, irrespective of the duty cycle.

Under overload conditions, however, a considerable shift in available capacity can be realized by the use of "rest" periods which may increase service life considerably. Figure 8.15 shows the effect of a high current drain for a particular cell size; the higher on-load voltage on continuous load at the beginning of the discharge curve is due to the heating effect produced by drawing high current from the cell.

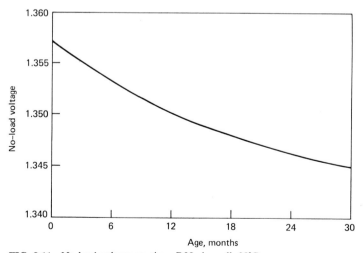

FIG. 8.11   No-load voltage vs. time, R20 size cell, 20°C.

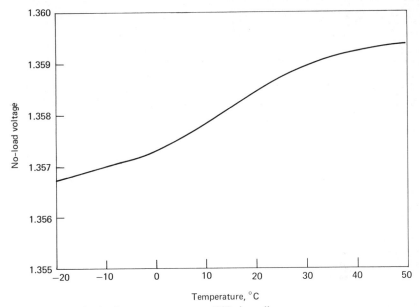

FIG. 8.12   No-load voltage vs. temperature, R20 size cell.

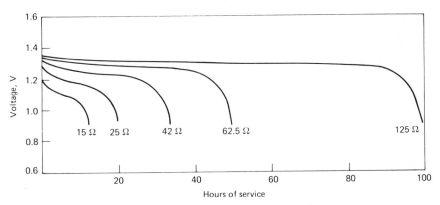

FIG. 8.13   Discharge curves, R50 size zinc/mercuric oxide cell, 20°C.

Problems are not encountered at low rates of discharge with cells designed for the purpose unless a high current drain pulse is superimposed on a continuous low-drain base current; special designs are necessary to cope with this problem.

8.5.3 Effect of Temperature

The zinc/mercuric oxide cell is best suited for use at normal and elevated temperatures from 15 to 45°C. Discharging cells at temperatures up to 70°C is also possible if the

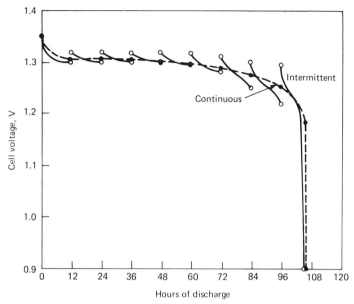

FIG. 8.14 Voltage/time curve for R44 size zinc/mercuric oxide cell, 20°C, 625-Ω load, continuous vs. intermittent discharge (12 h/day).

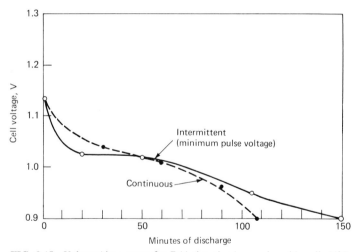

FIG. 8.15 Voltage/time curve for R44 size zinc/mercuric oxide cell, 20°C, 25-Ω load, continuous vs. intermittent (1 s/min).

discharge period is relatively short. The zinc/mercuric oxide cell generally does not perform well at low temperatures. Below 0°C, discharge efficiency is poor unless the current drain is low. Figure 8.16 shows the effect of temperature on the performance of the zinc/mercuric oxide cell.

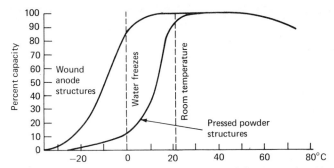

FIG. 8.16   Effect of temperature on performance of zinc/mercuric oxide cell.

The wound-anode or "dispersed"-powder anode structures are better suited to high rates and low temperatures than the pressed-powder anode.[32,33] The performance of a wound-anode cell of 4.5-Ah capacity is shown on a 25-Ω load at temperatures from -20 to $+20°C$ in Fig. 8.17. Below $-20°C$, the system becomes too inefficient to operate unless the current drains are very low.

8.5.4 Impedance

Impedance is usually measured at a frequency of 1 kHz, with or without an external load on the cell. In some hearing-aid circuits it is important for the impedance to have a reasonably low value. The impedance curve is almost a mirror image of the voltage discharge curve, rising very steeply at the end of the useful discharge life, as shown in

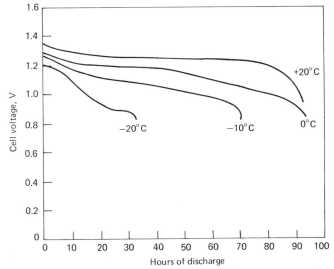

FIG. 8.17   Discharge curve for wound-anode zinc/mercuric oxide cell, 25-Ω load at temperatures between $+20$ and $-20°C$.

Fig. 8.18. The value obtained is frequency-dependent to some degree, particularly above 1 MHz, and a fixed frequency has to be specified. A frequency vs. impedance curve under no-load conditions is shown in Fig. 8.19.

8.5.5 Storage

The zinc/mercuric oxide cell has good storage characteristics. In general, the cells will store for over 2 y at 20°C with a capacity loss of 10 to 20% and 1 y at 45°C with about

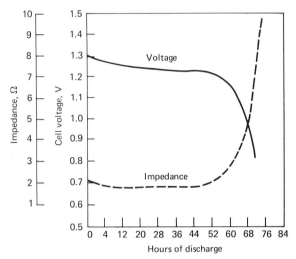

FIG. 8.18 Internal impedance, R9 size zinc/mercuric oxide cell, 350 mAh, 20°C, 1 kHz, 250-Ω load.

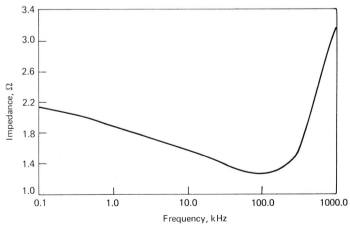

FIG. 8.19 Variation of impedance with frequency for R44 size zinc/mercuric oxide cell at 20°C.

a 20% loss. Storage at lower temperatures, e.g., down to −20°C, will, as with other battery systems, increase storage life.

The storability will depend on the discharge load placed on the cell and also on the cell structure. Failure in storage is usually due to the breakdown of cellulosic compounds within the cell which, at first, results in a reduction of the limiting-current density at the anode. Further breakdown produces low-drain internal electrical paths and a real loss of capacity due to self-discharge. Eventually, complete self-discharge can occur, but these processes at 20°C and below take many years.

Typical storage characteristics for a pressed-powder anode are shown in Fig. 8.20. Long storage lives are within the capabilities of the system as shown in Figs. 8.21 and 7.10. The cell in this case is a wound-anode cell with a noncellulosic barrier which shows a capacity loss in the region of only 13% over 6 y. With cells designed for long-term storage, dissolution of mercuric oxide from the cathode and its transfer to the anode become a significant factor in cell capacity loss. At elevated temperatures storage life is reduced, unless the cell is designed for this treatment; after 9 to 12 months at +45°C the performance will not be maintained, and total cell failure may occur. However, with a suitable cell structure, survival to 2 y and beyond is possible, as shown in Fig. 8.22.

8.5.6 Service Life

The performance of the zinc/mercuric oxide cell at various temperatures and loads is summarized in Figs. 8.23 and 8.24 on a weight and volume basis. These data, based on the performance of an RM-1 (R50) size cell with a dispersed anode, can be used to approximate the performance of a zinc/mercuric oxide cell. However, due to the variety of cell structures, manufacturers' specification sheets should be consulted for specific cell performance.

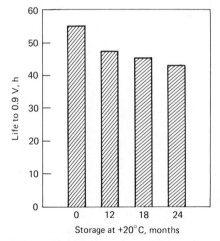

FIG. 8.20 Storage characteristics—zinc/ mercuric oxide cell, R6 pressed-powder anode, 20°C storage (discharge at 20°C, 30-Ω load, discharge endpoint of 0.9 V).

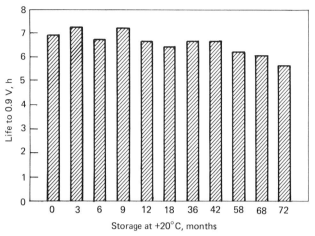

FIG. 8.21 Storage characteristics—zinc/mercuric oxide cell, 3.5-Ah wound-anode cell, 20°C storage (discharge at 20°C, 25-Ω load, discharge endpoint of 0.9 V).

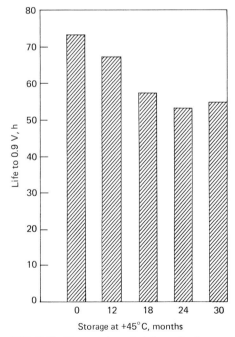

FIG. 8.22 Storage characteristics—zinc/mercuric oxide cell, 3.5-Ah wound-anode cell, 45°C storage (discharge at 20°C, 25-Ω load, discharge endpoint of 9 V).

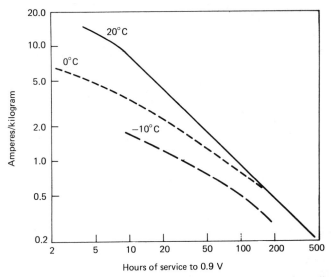

FIG. 8.23 Projected service life of a typical zinc/mercuric oxide cell (weight basis).

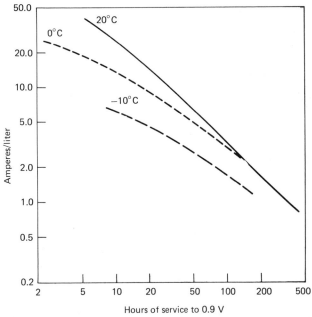

FIG. 8.24 Projected service life of a typical zinc/mercuric oxide cell (volume basis).

8.5.7 Stepped-Voltage Cells

A cell which produces a clearly defined "stepped" voltage during discharge is useful in an alarm device. This can be achieved by using materials in either cathode or anode which discharge at a different potential from the base electrode. In Fig. 8.25, a nine-cell battery discharge is shown with a stepped-curve V, produced by seven zinc/mercuric oxide cells (curve V_2) and two hybrid cells (curve V_1) in series. The hybrid cells have cathodes in which part of the mercuric oxide has been replaced by cadmium oxide in sufficient quantity to leave the cell with the same "balanced" capacity. When all the mercuric oxide has been reduced in the hybrid cells, their voltage falls by 750 mV per cell onto the lower "warning" plateau which can be used to trigger an alarm indicating the need for battery replacement—in this particular case the voltage step was arranged to occur at two-thirds the discharge life of the whole battery of cells. This type of battery has been used in smoke alarm systems.

8.6 PERFORMANCE CHARACTERISTICS: CADMIUM/MERCURIC OXIDE CELL

8.6.1 Discharge

An outstanding feature of the cadmium/mercuric oxide cell is its ability to operate over a wide temperature range. The usual operating range is from -55 to $+80°C$; but with the low gassing rate and thermal stability of the cell, operating temperatures to 180°C have been achieved with special designs.

Figure 8.26 shows the discharge curves for a typical button cell (500-mAh size) at 20°C. Excellent voltage stability and flat discharge curves are characteristic of these cells but at a low operating voltage (open-circuit voltage is only 0.9 V). Figure 8.27 shows the discharge of the same cell at various temperatures. Figure 8.28 shows the

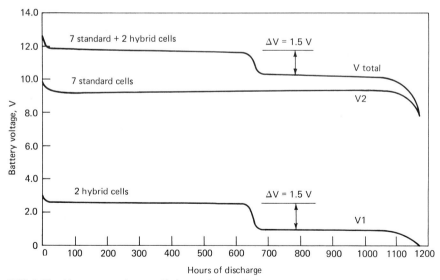

FIG. 8.25 Alarm system battery discharge characteristics.

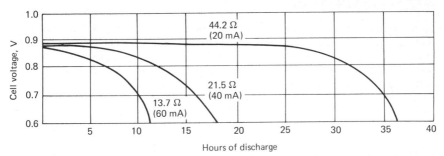

FIG. 8.26   Discharge curves—cadmium/mercuric oxide button cell (500 mAh, 20°C). (*Courtesy of Elca Battery Co.*)

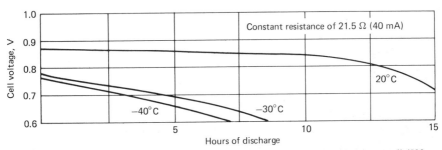

FIG. 8.27   Effect of temperature on cell discharge—cadmium/mercuric oxide button cell (500 mAh), discharge at 40 mA. (*Courtesy of Elca Battery Co.*)

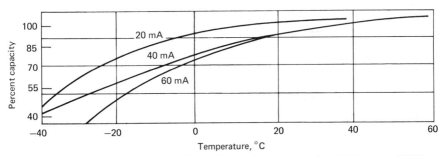

FIG. 8.28   Effect of temperature on cell capacity—cadmium/mercuric oxide button cell (500 mAh). (*Courtesy of Elca Battery Co.*)

effect of temperature on the capacity at various discharge loads. A high percentage of the 20°C capacity is available at the lower temperatures. The endpoint voltage is usually taken as 0.6 V, although at higher current densities and lower temperatures more useful life can be obtained to lower end voltages.

The performance of the cadmium/mercuric oxide cell is summarized in Figs. 8.29 and 8.30 on a weight and volume basis, respectively. The data were derived from the performance of typical button cells. Again, manufacturers' data should be used to determine the performance of specific cells in view of the special designs that are available.[34]

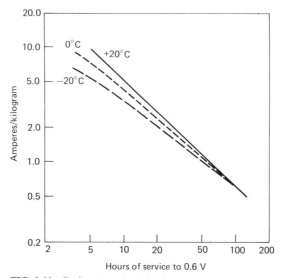

FIG. 8.29 Projected service life of a typical cadmium/mercuric oxide cell (weight basis).

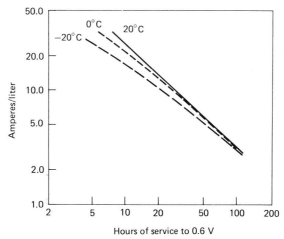

FIG. 8.30 Projected service life of a typical cadmium/mercuric oxide cell (volume basis).

8.6.2 Storage

Storage life over the temperature range -55 to $+80°C$ is remarkably good, and if the barrier-absorbent system is designed to withstand elevated-temperature storage, the major self-discharge mechanism is by dissolution of the mercuric oxide and its transfer to the anode. A shelf life of 10 y at ambient temperatures with less than 20% capacity loss is within the capabilities of the system. Elevated-temperature storage is exceptionally good, and since neither electrode should generate hydrogen, the cells can be hermetically sealed with minimal risk of electrolyte leakage or cell distortion.[31]

8.7 CELL TYPES AND SIZES

The zinc/mercuric oxide cells are manufactured in a wide selection of sizes to fit particular types of applications. A listing of typical cells, with their physical and electrical characteristics, is given in Table 8.3.

In many instances, cells with different internal designs and formulations may be dimensionally interchangeable but are functionally different (e.g., different rate capability, shelf life, capacity, voltage characteristics). Hence, special care must be taken in selecting cells for specific purposes. The zinc/mercuric oxide cell is manufactured often in the same dimensions, but with either a 1.35- or 1.4-V output, for use in different applications. In addition, attention should be given to the electrochemical system because, for example, silver oxide cells are available in identical configurations but with a 1.5-V nominal output voltage.

The industry, in recent years, has begun to standardize on some of the cell sizes and characteristics. Standard designations have been established by such organizations as the International Electrotechnical Commission (IEC), American National Standards Institute (ANSI), and National Electronic Distributors Association (NEDA) (see Part 2, Chap. 4). A list of the sizes used for some of the zinc/mercuric oxide cells is shown in Table 8.3. Numbers with a prefix R indicate the cell size, and a prefix M is used for the 1.35-V cells and N for the 1.4-V cells. Several alternative designations are used by manufacturers; the major ones are also shown in Table 8.3. As listed, these cells are dimensionally interchangeable but may be electrically and functionally different.

Table 8.4 lists typical cadmium/mercuric oxide cells and some of their characteristics. The standard cells cover low to medium discharge rate applications (25 to 1000 h); the high-rate cells are used for the 5- to 100-h rate applications. The high-rate performance is achieved at the expense of a slight reduction of shelf life. In addition to those listed, prismatic cells, ranging in capacity from 1 to 200 Ah, are also manufactured.

8.8 HANDLING, USE, AND DISPOSAL OF MERCURY CELLS

8.8.1 Handling

The cells should be adequately packed for transport and not stacked so high on top of each other as to cause physical damage to the cells. Mercury cells are heavy. Batteries should be stored in a ventilated area, preferably between 10 and 25°C with a maximum relative humidity of 65%. Severe shock, vibration, and physical damage should be avoided. Momentary short-circuiting of a cell will cause no permanent damage, but prolonged short-circuiting must not be allowed to occur. Do not solder leads directly onto cell contacts unless advised to do so by the manufacturer.

TABLE 8.3 Common Zinc/Mercuric Oxide Round-Cell Sizes

IEC design-ation	Approx. diam., mm	Approx. ht., mm	Approx. wt., g	Approx. capacity, mAh	Alternative designations*	
					With variable letter prefix and/or suffix	Other
R1	11	30	12	800	401, 35	H-R, HM-R
R6	13.5	50	30	2400	502, 9	H-T, HM-T
R9	16	6	4	350	625, 53	H-D, HM-D, 7102, 7002
R41	8	3.5	0.5	45	312, 57, 325, 5	325, 507, 7107, H-A
R42	11.6	3.5	1.5	100	575, 56, 343, 10, 31	343, 509, 7109
R43	11.6	4	2	150	41, 54, 354, 11	354, 508, 7108
R44	11.6	5.2	2.5	210	675, 52, 313, 15	313, 501, H-C, HM-C, 7101, 7001, 7103
R48	8	5.2	1	95	13, 58, 323, 6	323, 506, 7106
R50	16	16.5	12	1000	1, 40, 57	H-P, HM-P
R51	16	50	40	3600	12, 77, 59	H-U
R52	16	11	8	500	640, 30	H-N, HM-N

*Alternative designations do not necessarily comply in all respects with the requirements of IEC Publication 86.

TABLE 8.4 Typical Cadmium/Mercuric Oxide Cells

Cell type*	Dimensions, mm		Weight, g	Capacity (25-h rate), mAh
	Diam.	Height		
M500	25.1	6.4	12.5	500
M500H	25.1	6.4	12	500
M1000	34	7.1	26.4	1000
MC3	34	13.7	47.7	3000

*Elca Battery Co., Mountain View, California.

8.8.2 Use

Only the correct size, voltage, and grade of cell for the equipment should be used. The cell contacts should be clean before fitting. Where more than one cell is used in equipment, do not mix cells of different capacities, cells in different stages of discharge, or cells from different manufacturers. A complete set of cells must be replaced at one time. Primary cells should not be recharged.

8.8.3 Disposal

Mercury is poisonous and the cell electrolyte is corrosive; do not dismantle mercury cells. In the event of accidental rupture the entire contents should be placed under water in a suitable container (a polythene or polypropylene bucket).

If electrolyte comes into contact with the skin, neutralize at once with a saturated solution of boric acid or, if that is not available, swill with copious quantities of water. Do not dispose of mercury cells in fire. Spent zinc/mercuric oxide cells should be returned to your dealer or manufacturer for ecological reasons; facilities for the recovery of mercury from discharged cells are available in many countries.

8.9 APPLICATIONS

The zinc/mercuric oxide cells are used where stable voltage, long storage life, or high energy-volume ratios are required. The characteristics of the zinc/mercuric oxide system are particularly advantageous for miniature cells used in hearing aids, photographic applications, and small electronic equipment such as calculators and wrist watches as well as in the original smoke detectors. The mercuric oxide cell was also used in the early cardiac pacemakers until being replaced by the longer-lived solid electrolyte cells which were acceptable for this application requiring very low power. The cell has been used as a voltage reference source and also in electrical instruments and portable electronic equipment, such as sonobuoys, emergency beacons, rescue transceivers, and radio and surveillance sets, but these applications have not become widespread except for military and special uses because of the relatively higher cost of the mercuric oxide systems.

The cadmium/mercuric oxide cells are used mainly in special applications requiring the particular performance capabilities of this system. These include gas and oil well logging, telemetry from engines and other heat sources, and alarm systems and for operation of remote equipment, such as data-monitoring devices, buoys, weather stations, and emergency equipment.[35,36]

REFERENCES

1. C. L. Clarke, U.S. Patent 298,175 (1884).

2. S. Ruben, "Balanced Alkaline Dry Cells," *Proc. Electrochem. Soc. Gen. Meeting,* Boston, October 1947.

3. M. Friedman and C. E. McCauley, "The Ruben Cell—A New Alkaline Primary Dry Cell Battery," *Proc. Electrochem. Soc. Gen. Meeting,* Boston, October 1947.

4. S. Ruben, U.S. Patent 2,422,045 (1947).

5. M. G. Klein and M. Eisenberg, "A New Long Wet-Shelf Life Primary Battery System," *Electrochem. Technol.* **3**:58 (1965).

6. B. Berguss, "Cadmium-Mercuric Oxide Alkaline Cell," *Proc. Electrochem. Soc. Meeting,* Chicago Sect., October 1965.

7. M. Eisenberg, "Shelf Life and Low Temperature Performance of Mercuric Oxide-Cadmium Button Type Cells," *Proc. Electrochem. Soc. Meeting,* New York, October 1974.

8. P. Ruetschi, "The Electrochemical Reactions in Mercuric Oxide-Zinc Cell," in D. H. Collins (ed.), *Power Sources,* vol. 4, 1972, Oriel Press, Newcastle-upon-Tyne, England, 1973, p. 381.

9. W. M. Latimer, *Oxidation Potentials,* 2d ed., Prentice-Hall, New York, 1953.

10. T. P. Dirkse and R. Timmer, "The Corrosion of Zinc in KOH Solutions," *J. Electrochem. Soc.* **116**:162 (1969).

11. F. Mansfeld and S. Gilman, "The Effect of Several Electrode and Electrolyte Additives on the Corrosion and Polarisation Behaviour of the Alkaline Zinc Electrode," *J. Electrochem. Soc.* **117**:1328 (1970).

12. D. P. Boden, R. B. Wylie, and V. J. Spera, "The Electrode Potential of Zinc Amalgam in Alkaline Zincate Solutions," *J. Electrochem. Soc.* **118**:1298 (1971).

13. D. P. Gregory, P. C. Jones, and D. P. Redfearn, "The Corrosion of Zinc Anodes in Aqueous Alkaline Electrolytes," *J. Electrochem. Soc.* **119**:1288 (1972).

14. T. P. Dirkse, "Passivation Studies on the Zinc Electrode," in D. H. Collins (ed.), *Power Sources*, vol. 3, 1970, Oriel Press, Newcastle-upon-Tyne, England, 1971, p. 485.

15. D. Weiss and G. Pearlman, "Characteristics of Prismatic and Button Mercuric Oxide-Cadmium Cells," *Proc. Electrochem. Soc. Meeting*, New York, October 1974.

16. F. Przybyla, G. R. Ramsey, and T. C. O'Nan, "Passivation of a Porous Cadmium Electrode in Potassium Hydroxide Solution," in D. H. Collins (ed.), *Power Sources*, vol. 4, 1972, Oriel Press, Newcastle-upon-Tyne, England, 1973, p. 401.

17. F. Przybyla, U.S. Patent 3,847,784 (1974).

18. B. K. Jochmann, and T. Nervik, U.S. Patent 3,741,749 (1973).

19. S. Ruben, U.S. Patent 2,542,575 (1951).

20. R. J. Dawson, U.S. Patent 3,600,231 (1971).

21. S. Ruben, U.S. Patent 2,481,539 (1949).

22. R. Colton, U.S. Patent 2,636,062 (1953).

23. F. D. Williams, U.S. Patent 2,712,565 (1955).

24. R. R. Clune, U.S. Patent 3,096,217 (1963).

25. S. Ruben, U.S. Patent 2,422,046 (1947).

26. F. D. Williams, U.S. Patent 2,422,606 (1947).

27. D. Naylor, "Wound Anode Mercury Cells," *Proc. Second Intern. Symp. on Batteries*, Bournemouth, Hants., U.K., October 1960.

28. R. R. Clune and R. M. Goodman, U.S. Patent 3,205,097 (1965).

29. G. Matthews, "Mercury Cell—Kalium Version," *Proc. First Intern. Symp. on Batteries*, Christchurch, Hants., U.K., October 1958.

30. P. Ruetschi, U.S. Patent 4,136,236 (1979).

31. P. Ruetschi, "Longest Life Alkaline Primary Cells," in J. Thompson (ed.), *Power Sources*, vol. 7, 1978, Academic, London, 1979, p. 533.

32. R. R. Clune, "Recent Developments in the Mercury Cell," *Proc. 14th Annu. Power Sources Symp.*, United States, May 1960.

33. S. J. Angelovich, R. Di Palma, and C. W. Fleischmann, "Low Temperature Mercury-Zinc Batteries," *Proc. 24th Annu. Power Sources Symp.*, United States, May 1970.

34. Commercial literature, Elca Battery Co., Mountain View, Calif.

35. M. Eisenberg, "The New Mercuric-Oxide-Cadmium (Mercad) Battery System for Medical and Implantation Applications," *Proc. Intersoc. Energy Convers. Eng. Conf.*, United States, 1969.

36. R. G. Barnhart and D. P. Bowden, "A Review of Cadmium-Mercuric Oxide Batteries for Underwater Applications," *Proc. Intersoc. Energy Convers. Eng. Conf.*, United States, 1971.

Silver Oxide Cells

by
Akio Shimizu and Yoshio Uetani

9.1 GENERAL CHARACTERISTICS

The silver-zinc (zinc/alkaline electrolyte/silver oxide) primary battery is an important miniature power source because its energy density is among the highest of all battery systems and it can deliver this energy at a high current rate and at a constant voltage level. The silver oxide cell has good low-temperature performance, delivering about 70% of its 20°C capacity at 0°C and 35% at −20°C. It also has good storability, retaining more than 90% of its capacity after 1 y of storage at 20°C (see Table 9.1).

As a result of these performance advantages, the silver-zinc battery has found wide acceptance in electronic equipment, such as hearing aids, photographic applications, watches, and calculators, which require small and thin batteries with high capacity and

TABLE 9.1 Major Advantages and Disadvantages of Zinc/Silver Oxide Primary Cells

Advantages	Disadvantages
High energy density	Use limited to button and miniature cells because of high cost
Good voltage regulation—high rate capability	
Flat discharge curve—can be used as a reference voltage.	
Comparatively good low-temperature performance	
Leakage and salting negligible	
Good shock and vibration resistance	
Good shelf life	

long service life. In these applications, the higher cost of the silver cell has been tolerated.

The primary zinc/silver oxide cell is manufactured in the miniature flat or button configuration in capacities ranging from 5 to 250 mAh; the more popular sizes are 35 mAh and above.

9.2 CHEMISTRY AND COMPONENTS

The zinc/silver oxide primary cells are composed of compressed silver oxide powder for the cathode, an aqueous solution of potassium hydroxide or sodium hydroxide saturated with zincates for the electrolyte, and powdered zinc for the anode.

9.2.1 Silver Oxide

Although there are three oxidation states of silver oxide,[1] monovalent, divalent, and trivalent, the monovalent form has been the most popular for commercial use because it is the most stable under a variety of conditions.

The overall electrochemical reaction of monovalent silver oxide on discharge is

$$Ag_2O + H_2O + 2e \rightarrow 2\ Ag + 2OH^- \qquad E^0 = +0.345\ V$$

The formation of metallic silver contributes to maintaining a low internal impedance during the discharge.

The theoretical capacity by weight of monovalent silver oxide is 231 mAh/g, and the theoretical capacity per volume is 1650 Ah/L. The practical capacity in a battery is reduced as the silver oxide is blended with graphite to increase the conductivity of the electrode. This results in a lower packing density and a lower silver oxide content.

Monovalent silver oxide is stable as compared with higher-valent silver oxide in alkaline solutions, but a slight decomposition of silver oxide may occur in contact with graphite during storage at elevated temperatures. The decomposition rates depend on the type of graphite and the proportion used in the positive electrode; decomposition is attributed to the reaction of silver oxide with a volatile component of the graphite.[2]

Some positive electrodes in commercial cells contain a small amount of manganese dioxide. These mixtures may be tailored to give a flat discharge curve, a gradual reduction of voltage (to give an indication of end of life), or increased service.

9.2.2 Zinc

Amalgamated zinc powders of high surface area are generally used as anode materials with gelling agents such as sodium carboxymethyl cellulose or sodium polyacrylate. This anode design helps achieve the good low-temperature performance of the zinc/silver oxide battery. The amalgamation concentration of the zinc powder is usually from 2 to 15% by weight.

The overall electrochemical reaction of zinc on discharge is

$$Zn + 2OH^- \rightarrow Zn(OH)_2 + 2e \qquad E^0 = -1.245\ V$$

The theoretical capacity of zinc is 820 mAh/g. The practical capacity by volume is 2870 Ah/L as the apparent density of amalgamated zinc powder is about 3.5 g/cm^3.

9.2.3 Electrolyte

An aqueous solution of potassium hydroxide (KOH) or sodium hydroxide (NaOH) saturated with zincates is used for the electrolyte. Potassium hydroxide electrolyte is generally preferred in batteries for high-drain use and the sodium hydroxide electrolyte for low-drain use. The sodium hydroxide electrolyte is preferable for batteries for long-term use because of the lower incidence of salting or creeping. However, the anti-salting or creeping characteristics of the silver oxide button cells with KOH electrolyte have been progressively improved by better sealing techniques, gasket material, and treatment of the anode cap material.[3]

The conductivity of KOH and NaOH solutions at various concentrations is given in Fig. 9.1.[4,5] The electrolyte concentration used in commercial cells ranges from 20 to 45%.

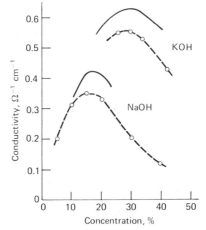

FIG. 9.1 Specific conductivity of alkaline hydroxide solutions. (Solid line, 25°C, Ref. 4; dotted line, 15°C, Ref. 5.

9.2.4 Separators

The properties required for the separator of silver oxide batteries are:

1. Stability in alkaline solution

2. Durability in contact with silver oxide

3. Permeability of hydroxide ions

4. Retardation of soluble silver oxide diffusion

A radiation-grafted polyethylene membrane, a microporous polypropylene film, or cellophane is usually used alone or in a lamination of these films with nonwoven materials made of nylon fiber or vinylon (polyvinyl alcohol fiber) as electrolyte absorbents.

9.3 CELL CONSTRUCTION

A cross-sectional view of a typical zinc/silver oxide button cell is illustrated in Fig. 9.2. The cathode is usually composed of monovalent silver oxide (Ag_2O) as the active material and graphite for conductivity. These are compacted into pellets and inserted into the bottom of the can. In some cases, manganese dioxide is added to the cathode material as an auxiliary active material; in other cases, the cathode is composed only of Ag_2O.

The anode consists of amalgamated zinc powder and a gelling agent which is dissolved in the alkaline electrolyte solution.

The cathode and anode materials are separated by one or more barrier layers of a suitable microporous film, a grafted plastic membrane, specially treated cellophane, and nonwoven absorbent fibers.

The top cup usually consists of two or more layers of laminated copper, tin, stainless steel, and nickel, so that the anode makes contact with the copper or tin.

The bottom cup usually is nickel-plated steel which serves as the positive terminal.

The top cup is insulated from the bottom cup by an insulating and sealing gasket which may be made of any suitable resilient electrolyte-resistant material such as nylon.

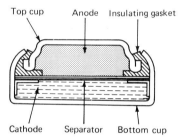

FIG. 9.2 Cutaway view of typical zinc/silver oxide button cell.

9.4 PERFORMANCE CHARACTERISTICS

9.4.1 Voltage

The open-circuit voltage of the zinc/monovalent silver oxide cell is about 1.6 V and depends on the electrolyte concentration, the content of zincate in the electrolyte, and the temperature. The dependence of the open-circuit voltage on the temperature is shown in Fig. 9.3.[6]

9.4.2 Discharge Characteristics

Typical discharge curves for the zinc/silver oxide cell for discharges at various loads at 20°C are presented in Fig. 9.4. The flat discharge curve is characteristic of this system. Although shown for a particular cell size, the curves are applicable to cells of other sizes on discharges giving the same hours of service.

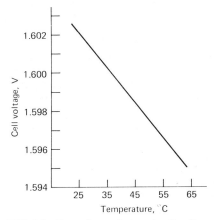

FIG. 9.3   Dependence of open-circuit voltage of zinc/silver oxide cell on temperature.

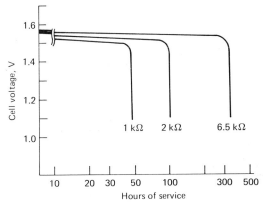

FIG. 9.4   Typical discharge curves of zinc/silver oxide cell at 20°C (size: 11.6 × 3.0 mm).

The relationship of capacity to the discharge rate on discharges at 20°C is illustrated in Fig. 9.5. The data are based on rating the capacity of a KOH-electrolyte cell as 100% at the 150-h rate and the NaOH-electrolyte cell as 100% at the 400-h rate.

9.4.3 Effect of Temperature

The zinc/silver oxide button cell is capable of operating over a wide temperature range. The battery can deliver more than 70% of its 20°C capacity at 0°C and 35% at −20°C, at moderate loads. At heavier loads, the loss is greater. Higher temperatures tend to accelerate capacity deterioration, but temperatures as high as 60°C can be tolerated for several days with no serious effect.

Figures 9.6 to 9.8 show the initial closed-circuit voltage of representative sizes of zinc/silver oxide cells at various loads and temperatures. Again, these data can be used

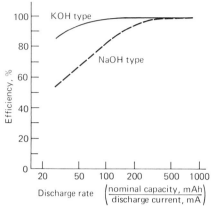

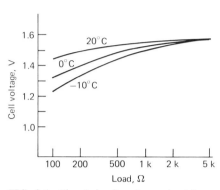

FIG. 9.5 Dependence of discharge efficiency on discharge rate, zinc/silver oxide button cells at 20°C.

FIG. 9.6 Closed-circuit voltage, zinc/silver oxide cell (size: 11.6 × 3.6 mm).

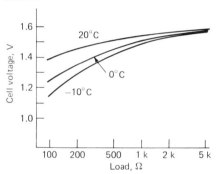

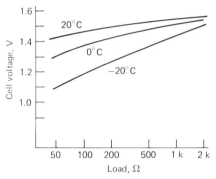

FIG. 9.7 Closed-circuit voltage, zinc/silver oxide cell (size: 7.9 × 3.0 mm).

FIG. 9.8 Closed-circuit voltage, zinc/silver oxide cell (size: 11.6 × 5.4 mm).

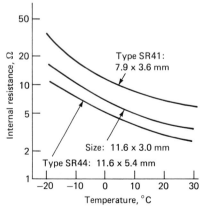

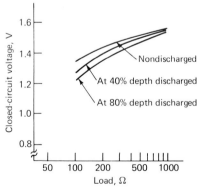

FIG. 9.9 Internal resistance of typical zinc/silver oxide button cells.

FIG. 9.10 Closed-circuit voltages on pulse load discharge for 5 s for Type SR1130: 11.6 × 3.0 mm.

to approximate the closed-circuit voltages of cells of other sizes by using a discharge load that gives the same hours of service.

9.4.4 Internal Resistance

The internal resistance of representative types of zinc/silver oxide button cells at various temperatures is shown in Fig. 9.9. The impedance of these cells is generally constant during the discharge and does not rise appreciably until the cell voltage falls below a useful operating level.

9.4.5 Pulse Loads

In some applications, a heavy-current pulse of short duration is required in addition to the low background current. Figure 9.10 shows the closed-circuit voltages that are obtained at different pulse loads on a zinc/silver oxide cell that had previously discharged to different depths of discharge.

The performance obtained on a discharge simulating the electronic shutter mechanism of a still camera is illustrated in Fig. 9.11. A high voltage level is maintained even after the battery had been previously discharged to a 75% depth.

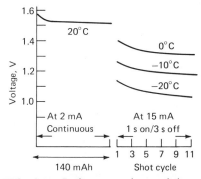

FIG. 9.11 Performance characteristics of zinc/silver oxide S244 cell (size: 11.6 × 5.4 mm) simulating electronic shutter mechanism pulse discharging a partially discharged cell at 15 mA.

9.4.6 Service Life

The service life of the zinc/silver oxide button cell at various discharge loads and temperatures, normalized for unit weight (amperes per kilogram) and volume (amperes per liter) to an end voltage of 1.0 V per cell, is summarized in Fig. 9.12. These curves can be used to approximate the service life of a given cell under a particular discharge condition or to estimate the size or weight of a cell required to meet a

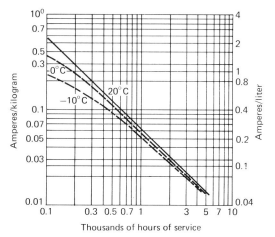

FIG. 9.12   Service life of the zinc/silver oxide button cell at various discharge rates and temperatures (end voltage: 1.0 V).

particular service requirement. The curves are based on an energy density of 60 Ah/kg and 240 Ah/L at the 200-h rate at 20°C. The actual performance will vary with cell design and size.

9.4.7 Shelf Life

The capacity retention of the zinc/silver oxide button cell at various temperatures and storage periods is summarized in Fig. 9.13. The retention of capacity is good compared with other aqueous-electrolyte cells. The capacity retention is also dependent on discharge load as shown in Fig. 9.14.

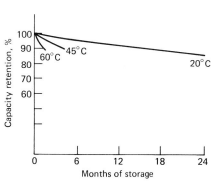

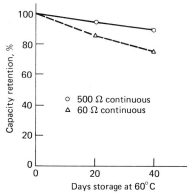

FIG. 9.13   Capacity retention of the zinc/silver oxide cell at various storage temperatures (size: 11.6 × 3.0 mm; discharge conditions: 6.5 kΩ continuous at 20°C).

FIG. 9.14   Capacity retention of zinc/silver oxide cell (size: 11.6 × 5.4 mm; SR44 storage temperature: 60°C; discharge temperature: 20°C).

TABLE 9.2 Zinc/Silver Oxide Primary Cells

Applications	Diam × height, mm	Nominal voltage, V	Average service capacity, mAh	Maxell	Eveready (UCC)	Duracell*	Ray-O-Vac	Seiko	Timex	Varta	Renata	Bulova	Standard IEC	ANSI	JIS
High drain	11.6 × 5.4	1.55	190	SR44W	357	10L14	RW42	SB-B9	J	541	7	228	SR44	S15	SR44(G13)
	11.6 × 4.2	1.55	120	SR43W	386	10L124	RW44	SB-B8	H	548	6	266	SR43	S10	SR43(G12)
	11.6 × 3.05	1.55	80	SR1130W	389	10L122	RW49	SB-BU	...	544	17	317			
	11.6 × 2.05	1.55	45	SR1120W	391	10L130	RW40	SB-BS	...	...	23				
	9.5 × 2.73	1.55	45	SR926W	399			SB-BP							
	7.9 × 3.6	1.55	38	SR41W	392	10L125	RW47	SB-B1	K	547	2	274B	SR41	S4	SR41(G3)
	7.9 × 2.6	1.55	27	SR726W	396										
	7.9 × 2.15	1.55	20	SR721W											
	9.5 × 2.73	1.55	44	XR9527W**											
Watches — Low drain	11.6 × 5.4	1.55	165	SR44SW	301	WS11	RW44	SB-A9	D	528	1	226	SR44	WS15	SR44S(GS13)
	11.6 × 4.2	1.55	100	SR43SW	390	...	RW39	SB-A8	...	534	11	603	SR43	WS10	SR43S(GS12)
	11.6 × 3.05	1.55	82	SR1130SW	381	...	RW30	SB-AU	...	533	23LD				
	11.6 × 2.05	1.55	45	SR1120SW	366			SB-AS							
	11.6 × 1.65	1.55	29	SR1116SW											
	9.5 × 3.6	1.55	70	SR936SW	394		RW33	SB-A4		524	27				
	9.5 × 2.73	1.55	45	SR926SW	395			SB-AP		523	25				
	9.5 × 2.05	1.55	29	SR920SW	371			SB-AN							
	9.5 × 1.65	1.55	18	SR916SW	373			SB-AJ							
	7.9 × 3.6	1.55	38	SR41SW	384	10L15	RW37	SB-A1		527	10	247	SR41	WS4	SR41S(GS3)
	7.9 × 2.6	1.55	28	SR726SW	397			SB-AL		536	26				
	6.8 × 2.6	1.55	23	SR626SW											
	6.8 × 2.15	1.55	15	SR621SW	364			SB-AG		531	...	602			
	6.8 × 1.65	1.55	10	SR616SW											
	9.5 × 2.73	1.55	55	TR927SW†											
Cameras and other applications	11.6 × 5.4	1.55	190	SR44F(G13F)	EP × 76	...	...	...		...	...	...	SR44	S15	SR44(G13)
	11.6 × 5.4	1.55	165	SR44(G13)	...	...	...	...		...	...	...	SR44	S15	SR44(G13)
	11.6 × 4.2	1.55	120	SR43(G12)	S41	...	...	...		...	...	...	SR43	S10	SR43(G12)
	11.6 × 3.05	1.55	80	SR1130	289										
	11.6 × 2.05	1.55	45	SR1120											
	13.0 × 25.2	6.2	165	4SR44(4G13)	544	PX28	RPX28	...		4028	...	...	4SR44	...	4SR44(4G13)

*Duracell may also follow Eveready type numbers.
**Rechargeable type.
†Silver dioxide battery.

9.5 CELL SIZES AND TYPES

The characteristics of commercially available button zinc/silver oxide cells are summarized in Table 9.2.

9.6 DIVALENT SILVER OXIDE CELL

Although the theoretical capacity for the divalent silver oxide (AgO) is about 3200 Ah/L, corresponding to twice the value for monovalent silver oxide, the divalent form has not been used in button cells because AgO is not stable in alkaline solutions and discharges with two voltage plateaus, at 1.85 and 1.58 V.

The stability of AgO has been improved by the addition of PbO, Cd, and Au. The problem of the two plateaus on discharge has been solved by a redesign of the cathode structure. A layer of Ag_2O provides the contact to the bottom can which is the positive terminal, isolating the divalent silver oxide from the terminal.[7] The discharge voltage then remains constant according to the following reactions in the cathode during the discharge

$$Ag_2O + H_2O + 2e^- \rightarrow 2Ag + 2OH^-$$

$$Ag + AgO \rightarrow Ag_2O$$

The capacities of the divalent silver oxide cells are from 20 to 40% higher than the zinc/monovalent silver oxide cells.

REFERENCES

1. A. Fleischer and J. J. Lander (eds.), *Zinc Silver Oxide Batteries,* Wiley, New York, 1971.
2. A. Shimizu and Y. Uetani, The Institute of Electronics and Communication Engineers of Japan, Technical paper CPM79-55, 1979.
3. Y. Uetani, A Shimizu, and K. Kajita, "Electric Power Sources in Horological and Microtechnical Products," *Intern. Symp.,* Mulhouse, France, April 1981.
4. E. J. Rubin and R. Baboian, "A Correlation of the Solution Properties and the Electrochemical Behavior of the Nickel Hydroxide Electrode in Binary Aqueous Alkali Hydroxides," *J. Electrochem. Soc.* **118**:428 (1971).
5. "Kagaku Benran," Maruzen, Tokyo, 1966.
6. S. Hills, "Thermal Coefficients of EMF of the Silver (I) and the Silver (II) Oxide-Zinc-45% Potassium Hydroxide Systems," *J. Electrochem. Soc.* **108**:810 (1961).
7. R. J. Dawson, U.S. Patent 3,484,295 (1969).

Zinc/Air Cells

by
Steven F. Bender and John W. Cretzmeyer

10.1 GENERAL CHARACTERISTICS

Zinc/air batteries* use oxygen directly from the atmosphere to produce electrochemical energy. Oxygen diffuses into the cell and is used as the cathode reactant. The air cathode catalytically promotes the reaction of oxygen with an aqueous alkaline electrolyte and is not consumed or changed during the discharge. As the air cathode is extremely compact yet has an essentially unlimited capacity, very high energy densities are achieved, resulting from the increased volume available for the zinc anode.

For many applications, zinc/air technology offers the highest available energy density of any primary system. Other advantages include a flat discharge voltage, long shelf life, safety and ecological benefits, and low energy cost. Since the cells are open to the ambient, a factor limiting universal application of zinc/air technology is the tradeoff between long service life (high environmental tolerance) and maximum power capability (lower environmental tolerance). The major advantages and disadvantages of this battery type are summarized in Table 10.1.

TABLE 10.1 Major Advantages and Disadvantages of Zinc/Air (Button) Cells

Advantages	Disadvantages
High energy density	Not independent of environmental conditions:
Flat discharge voltage	"Drying out" limits shelf life once opened to air
Long shelf life	
No ecological problems	"Flooding" limits power output
Low cost (on service energy basis)	Limited power output
Capacity independent of load and temperature when within operating range	

*Rechargeable metal/air batteries are covered in Part 4, Chap. 30.

The effect of atmospheric oxygen as a depolarizing agent in electrochemical systems was first noted early in the nineteenth century. However, it was not until 1878 that a cell was designed in which the manganese dioxide of the famous Leclanche cell was replaced by a porous platinized carbon/air electrode. Limitations in technology prevented the commercialization of zinc/air batteries until the 1930s. In 1932 Heise and Schumacher constructed alkaline electrolyte zinc/air cells which had porous carbon air cathodes impregnated with wax to prevent flooding. This design is still used almost unchanged for the manufacture of large industrial zinc/air batteries. These batteries are noted for their very high energy densities but low power output capability. They are used as power sources for remote railway signaling and navigation aid systems. Broader application is precluded by low current capability and bulk.

The thin, efficient air cathode used in today's zinc/air button cells is the result of fuel-cell research for space programs and was made possible through the emergence of fluorocarbons as a polymer class. The unique surface properties of these materials, hydrophobicity combined with gas porosity, made possible the development of thin, high-performance gas electrodes using a Teflon bonded catalyst structure[1] and a hydrophobic cathode composite.[2] A method for continuous fabrication of this composite was developed in the early 1970s.[3]

Early efforts to apply this technology were directed toward power packs for man-pack radios and radar systems. After further development, miniature high-performance button cells were successfully commercialized in 1977. They have been most successful as hearing-aid batteries and are used also in electronic watches, for memory maintenance, and in other low-power applications.

A closed-battery system, using a zinc anode, pure oxygen, and a KOH electrolyte (known as the ZOX system) was designed for spacecraft applications as a means of increasing energy output; this was never fully developed.[4]

10.2 CHEMISTRY

The more familiar types of primary alkaline systems are the zinc/manganese dioxide, the zinc/mercuric oxide, and the zinc/silver oxide cells. These, typically, use potassium or sodium hydroxides, in concentrations from 25 to 40% by weight, as the electrolyte, which functions primarily as an ionic conductor and is not consumed in the discharge process. In simple form, the overall discharge reaction for these metal oxide cells can be stated as follows:

$$MO + Zn \rightarrow M + ZnO$$

During the discharge, the metal oxide (MO) is reduced, either to the metal as shown or to a lower form of an oxide. Zinc is oxidized and, in the alkaline electrolyte, usually reacts to form ZnO. Thus it can be seen that, at 100% efficiency, electrochemically equivalent amounts of metal oxide and zinc must be present. Therefore, an increase in capacity of any cell must be accompanied by an equivalent increase in both cathodic and anodic materials.

In the zinc/oxygen couple, which also uses an alkaline electrolyte, it is necessary to increase only the amount of zinc present to increase cell capacity. The oxygen is supplied from the outside air which diffuses into the cell as it is needed. The air cathode acts only as a reaction site and is not consumed. Theoretically, the air cathode has infinite use life and its physical size and electrochemical properties remain unchanged during cell discharge. The reactions of the air cathode are complex but can be simplified to show the cell reactions as follows:

		E_0
Cathode	$\frac{1}{2}\ O_2 + H_2O + 2e \rightarrow 2OH^-$	0.40 V
Anode	$Zn \rightarrow Zn^{2+} + 2e$	
	$Zn^{2+} + 2OH^- \rightarrow Zn\ (OH)_2$	1.25 V
	$Zn\ (OH)_2 \rightarrow ZnO + H_2O$	
Overall reaction	$Zn + \frac{1}{2}\ O_2 \rightarrow ZnO$	1.65 V

10.3 CONSTRUCTION

10.3.1 Industrial Batteries

A typical industrial-type zinc/air cell, the Edison Carbonaire ST brand, is manufactured in a 1100-Ah size and is available in two- and three-cell configurations as illustrated in Fig. 10.1. The cell case and cover are molded from a tinted transparent acrylic plastic. The construction features are shown in Fig. 10.2. The batteries are composed of large cells with wax-impregnated carbon cathodes and rigid zinc anodes. Connections are made with leads through the plastic or rubber case, similar to the leads in an automotive battery. These systems also normally have a bed of lime flake to absorb carbon dioxide during the long open discharge life and regenerate the electrolyte.

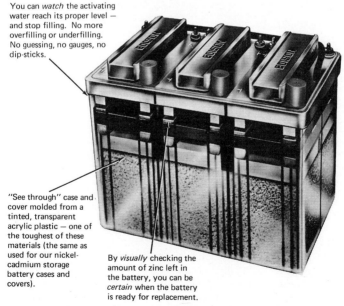

You can *watch* the activating water reach its proper level — and stop filling. No more overfilling or underfilling. No guessing, no gauges, no dip-sticks.

"See through" case and cover molded from a tinted, transparent acrylic plastic — one of the toughest of these materials (the same as used for our nickel-cadmium storage battery cases and covers).

By *visually* checking the amount of zinc left in the battery, you can be *certain* when the battery is ready for replacement.

FIG. 10.1   Edison Carbonaire zinc/air battery. (*Courtesy of McGraw-Edison Co.*)

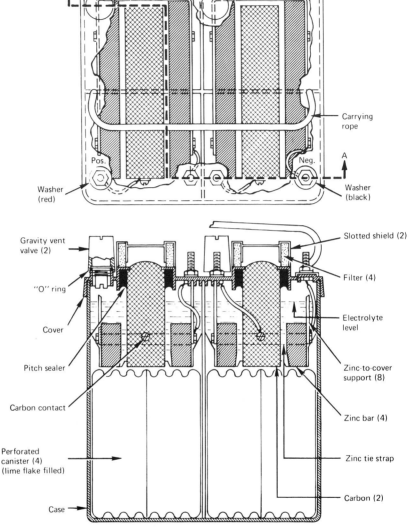

FIG. 10.2 Cross section of Edison Carbonaire zinc/air battery, type ST-22. (*Courtesy of McGraw-Edison Co.*)

The cells and batteries are supplied sealed. The caustic potash (potassium hydroxide), which will act as electrolyte, and the lime flake are present in the dry form. The unit is activated by removing the seals and adding the appropriate amount of water to dissolve the potassium hydroxide, which forms the electrolyte. Periodic inspection and addition of water to achieve proper electrolyte level are the only required maintenance. The use of transparent cases allows visual inspection and simplifies maintenance since the electrolyte level and state of charge can be determined without the need for dipsticks or floats.

Primary zinc/air batteries, ranging in capacity from 320 to 3000 Ah, also have been developed for maritime applications, such as for powering light buoys. Typical batteries are illustrated in Fig. 10.3.

10.3.2 Button Cells

The construction features of zinc/air button cells are quite different from those of the large industrial units. The zinc anode material is generally a loose powder mixed with electrolyte and, in some cases, a gelling agent to immobilize the composite. The containers housing the cathode and anode active materials also act as the cell terminals, insulation between the two containers being provided by a plastic insulator (gasket).

A schematic representation of a typical zinc/air button cell is given in Fig. 10.4. A zinc/metal oxide cell is shown for comparison. The reason for increased energy density in the zinc/air cell is graphically illustrated by comparing the anode compartment volumes. The very thin cathode of the zinc/air cell (about 0.5 mm) permits the use of twice as much zinc in the anode compartment as can be used in the mercuric oxide equivalent. Since the air cathode has theoretically infinite life, the electrical capacity

FIG. 10.3 "Air-zinc" cells. (*Courtesy of Central Laboratory of Electrochemical Power Sources, Bulgarian Academy of Sciences, Sofia, Bulgaria and Mussalla, Works, Samokov, Bulgaria.*)

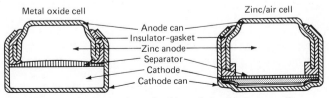

FIG. 10.4 Cross section of metal oxide and zinc/air button cells.

of the cell is determined only by the anode capacity, resulting in at least a doubling of energy density.

A portion of the total volume internally available for the anode must be reserved to accommodate the expansion that occurs when zinc is converted to zinc oxide during cell discharge. This space also provides additional cell tolerance to sustained water gain during operating conditions. Referred to as the anode free volume, it is typically 15 to 25% of the total anode compartment volume.

Figure 10.5 shows a magnified cross-sectional view of the cathode region of Gould's Activair zinc/air cell. The cathode structure includes the separators, catalyst layer, metallic mesh, hydrophobic membrane, diffusion membrane, and air-distribution layer. The catalyst layer contains oxides of manganese in a carbon conducting medium. It is made hydrophobic by the addition of finely dispersed Teflon particles. The metallic mesh provides structural support and acts as the current collector; the hydrophobic membrane maintains the gas-permeable waterproof boundary between the air and the cell's electrolyte; the diffusion membrane regulates gas diffusion rates (not used when air hole controls gas diffusion); and the air distribution layer distributes oxygen evenly over the cathode surface.

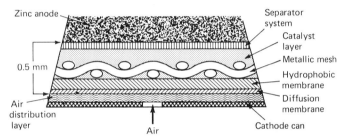

FIG. 10.5 Zinc/air button cell—cathode area cross section. (*Courtesy of Gould, Inc.*)

An air-access hole(s) on the positive terminal of a zinc/air cell provides a path for oxygen to enter the cell and diffuse to the cathode catalyst sites. The rate at which oxygen and other gases transfer into or out of the cell is regulated either by the hole area or by the porosity of the diffusion membrane at the surface of the cathode layer. Regulating oxygen diffusion sets a limit to a zinc/air cell's maximum continuous current capability, because operating current is directly proportional to oxygen consumption (5.81×10^{-5} cm^3 of oxygen per milliampere-second). Accordingly, current capability increases with increasing hole area or membrane porosity until the current density (reaction rate) at the air electrode becomes limiting.

If only oxygen transfer rates mattered, gas diffusion in zinc/air cells would not be regulated, resulting in higher operating current capability. Regulation is necessary because other gases, most importantly water vapor, can enter or leave the cell. If not properly controlled, undesirable gas transfer can cause a degradation in cell power capability and service life.

Water vapor transfer is generally the dominant form of gas transfer performance degradation. This transfer occurs between the cell's electrolyte and the ambient (Fig. 10.6). A zinc/air cell's aqueous electrolyte has a characteristic water vapor pressure. A typical electrolyte consisting of 30% potassium hydroxide by weight is in equilibrium with the ambient at room temperature when the relative humidity is approximately 60%. A cell will lose water from its electrolyte on drier days and gain water on more

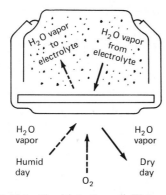

FIG. 10.6   Zinc/air button cell. Water vapor transfer is the dominant form of gas transfer degradation.

humid days. In the extreme, either water gain or water loss can cause a zinc/air cell to fail before delivering full capacity. A smaller hole or lower diffusion membrane porosity yields greater environmental tolerance because water transfer rates are reduced, resulting in a longer practical service life.

The maximum continuous current capability of a zinc/air cell, as determined by gas diffusion regulation, is typically specified as the limiting current, denoted I_L. The relationship between gas transfer regulation, limiting current, and service life is illustrated in Fig. 10.7.

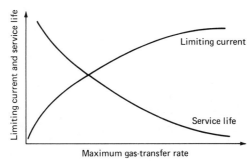

FIG. 10.7   Zinc/air button cell. Gas transfer regulation determines the limiting current and useful service life of a zinc/air cell.

10.4 PERFORMANCE CHARACTERISTICS—BUTTON CELLS

10.4.1 Cell Sizes

Zinc/air button cells are available in several sizes and in high- and low-rate configurations. Capacities range from 80 to 400 mAh. Table 10.2 lists the physical and electrical characteristics of the available cells. In general, the high-rate cells are best suited for applications requiring higher current levels and relatively short service life as in hear-

TABLE 10.2 Activair Zinc/Air Cells

Cell Type	Cell classification	Diameter, mm	Height, mm	Rated capacity, mh	Limiting current I_L	Maximum power output $=I_L \times 1.1$ V	Typical useful service life
A675HP	High rate	11.6	5.4	400	12.0 mA	13.2 mW	≤ 3 months
A675H	High rate	11.6	5.4	400	7.0 mA	7.7 mW	≤ 3 months
A41H	High rate	11.6	4.2	300	5.2 mA	5.7 mW	≤ 3 months
A13H	High rate	7.9	5.4	170	3.8 mA	4.2 mW	≤ 2 months
A312H	High rate	7.9	3.5	80	3.5 mA	3.9 mW	≤ 2 months
A675L	Low rate	11.6	5.4	400	20.0 μA	22.0 μW	≤ 5 y
A41L	Low rate	11.6	4.2	300	20.0 μA	22.0 μW	≤ 5 y
A13L	Low rate	7.9	5.4	170	7.0 μA	7.7 μW	≤ 3.5 y
A312L	Low rate	7.9	3.5	80	7.0 μA	7.7 μW	≤ 3.5 y

SOURCE: Gould, Inc, Portable Battery Division, Eagen, Minn.

ing-aid applications. Low-rate cells are designed for low-current, long-life applications as in electronic watches.

10.4.2 Voltage

The nominal open-circuit voltage of the zinc/air cell is 1.45 V. The initial closed-circuit voltage at 20°C ranges from 1.1 to 1.4 V, depending on the discharge load. The discharge is relatively flat, with 0.9 V the typical end voltage.

10.4.3 Energy Density

The zinc/air button cell has a high volumetric energy density compared with other primary battery systems and is about 920 Wh/L at the 5-μA discharge rate. Table 4.2, Part 2 (Chap. 4) compares the performances of the different primary battery systems.

10.4.4 Discharge Characteristics

Discharge profiles for the A13H and A675H high-rate zinc/air button cells are presented in Fig. 10.8. As the air cathode in the cell is not chemically altered during

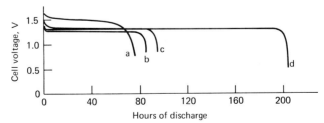

FIG. 10.8 Discharge profiles of primary button cells, 20°C, 620-Ω discharge load. Cell types: (a) zinc/silver oxide, 11.6 × 5.4 mm; (b) zinc/air, 7.9 × 5.4 mm; (c) zinc/mercuric oxide, 11.6 × 5.4 mm; (d) zinc/air, 11.6 × 5.4 mm.

discharge, the voltage remains quite stable. Curves for the zinc/mercuric oxide and zinc/silver oxide cells are provided for comparison. On continuous discharge at the loads shown, the zinc/air system will deliver twice the service of the other batteries.

A set of discharge curves, typical of the performance of the zinc/air button cell at 20°C, is presented in Fig. 10.9. After an almost instantaneous initial voltage drop, the discharge curves are relatively flat. The cell delivers about the same capacity over the current range shown; the operating voltage is lower at the higher discharge currents, as is typical with all battery systems.

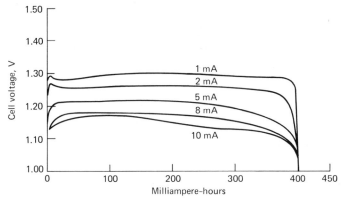

FIG. 10.9   Typical discharge curves of zinc/air button cells at 20°C, type A675.

10.4.5 Voltage-Current Performance

The power performance of a zinc/air cell is typically limited below the system's maximum capability to optimize the tradeoff between power output and shelf life–service life requirements for the targeted set of applications. Many cell design variables impact power vs. life performance, including the choice of separator system, electrolyte, cathode catalyst, and gas diffusion regulation.

Voltage vs. current profiles for Gould's Activair high-rate and low-rate cells are presented in Fig. 10.10 (refer to Table 10.2 for cell dimensions). High-rate cells are designed for applications requiring a high power output and relatively short service life. Low-rate cells are designed for applications requiring low power delivered over a long service life. The level of gas diffusion regulation is designed to optimize the tradeoff between the cell's environmental tolerance (setting service life capability) and power output.

The maximum continuous operating current of a zinc/air cell, as determined by gas diffusion regulation, is referred to as the "limiting current." For a given level of diffusion control, the limiting current will vary directly with the surface area of the cathode, which is proportional to cell diameter. The relationship between limiting current, maximum power capability, and service life for high- and low-rate cells is illustrated in Table 10.2.

As shown in Fig. 10.10, the cell voltage rapidly falls when continuous currents above the limiting current are applied. This occurs because the cell has become oxygen-

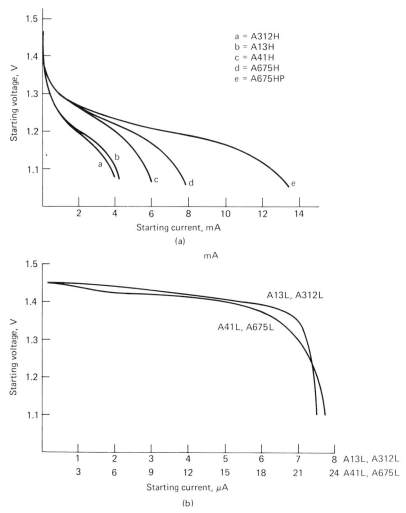

FIG. 10.10   Voltage vs. current profiles for zinc/air button cells at 20°C. (*a*) High-rate cells; (*b*) low-rate cells. (*Courtesy of Gould, Inc.*)

starved. It is consuming oxygen at a faster rate than the rate at which oxygen is entering the cell. Voltage will fall until the load current is reduced to an equilibrium condition.

10.4.6 Cell Internal Impedance

The effects of discharge level and signal frequency on the internal impedance characteristics of the A13H and A675 button cells are presented in Fig. 10.11.

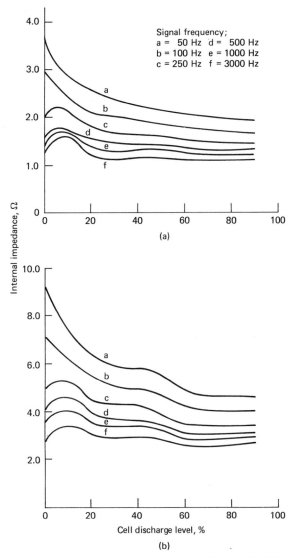

FIG. 10.11   Internal impedance profiles vs. discharge level. (*a*) 675H cell; (*b*) A13H cell. (*Courtesy of Gould, Inc.*)

10.4.7 Pulse Load Performance

Zinc/air cells can handle pulse currents much higher than the limiting current, the current level depending on the nature of the pulse. This capability results from a reservoir of oxygen that builds up within the cell when the load is below the limiting current.

Figure 10.12 illustrates general voltage profiles for various pulse waveforms. In Fig.

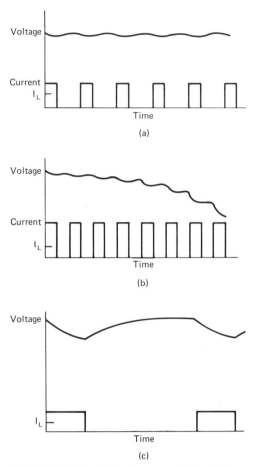

FIG. 10.12 Pulse load performance of zinc/air button cells. (a) $I_{avg} < I_L$, short pulse duration; (b) $I_{avg} > I_L$, short pulse duration; (c) $I_{avg} < I_L$, long pulse duration.

10.12a, a pulse current double the magnitude of I_L is applied. However, the pulses are spaced so that the average current is less than I_L. Therefore, the average rate of oxygen ingress is sufficient to sustain the average load current. Furthermore, the pulses are of short duration, so that the cell does not become oxygen-starved before the end of a single pulse. The cell's voltage profile shows the ripple effect of the pulse current, but the cell maintains a continuous useful average operating voltage. If the peak pulse current is increased so that the average load current is greater than I_L, the cell will eventually become oxygen-starved and voltage will decline, as shown in Fig. 10.12b.

In choosing a cell for a particular application, the designer must ensure that the average current for the duration of the pulse load is less than I_L. The designer must also check for the possibility of the cell becoming oxygen-starved during the application of a single pulse. This condition is illustrated in Fig. 10.12c. In this figure, the average current is less than I_L, but the cell becomes oxygen-starved and the voltage declines sharply during the time that a pulse is applied because of the long pulse duration. The voltage recovers between pulses, maintaining an average close to the average of Fig. 10.12a.

The capability of a zinc/air cell to handle the pulse load requirements of a typical quartz analog watch application is illustrated in Table 10.3. The pulse load is two orders of magnitude higher than the 7-μA limiting current, yet the average load current is only 5.7 μA. The ripple voltage is less than 10 mV, about an average operating voltage of 1.4 V.

Figure 10.13 illustrates actual performance of an A675H cell under sustained current loads greater than the limiting current.

TABLE 10.3 Capability of a Low-Rate Zinc/Air Cell to Meet the Pulse Requirements of a Quartz Analog Watch

Item	Specification/performance
Cell type	A312L
Cell-limiting current	7.0 μA
Watch quiescent current	0.8 μA
Stepping motor pulse	700.0 μA
Pulse duration	0.007 s
Pulse frequency	1.0 pulse/s (1 Hz)
Average load current*	5.7 μA
Cell ripple voltage	10.0 mV
Average cell voltage	1.4 V

$*I_{avg} = (0.007) \times (700 \ \mu A) + (1\text{-}0.007) \times (0.8 \ \mu A) = 5.7 \ \mu A.$

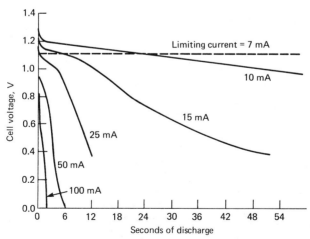

FIG. 10.13 Voltage-time response of zinc/air button cell to continuous loads greater than the limiting current, A675H size cell.

10.4.8 Effect of Temperature

The effect of temperature on discharge performance is illustrated in Fig. 10.14. Alkaline systems in general are not good performers at low temperatures. This degradation is caused primarily by a reduction in ionic diffusion capability through the electrolyte. Rather high concentrations of potassium or sodium hydroxide (25 to 40% by weight) are used to obtain good electrical conductivity. This high concentration is subject to changes in viscosity and a general lowering of ion mobility as the temperature drops.

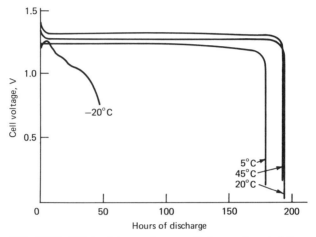

FIG. 10.14   Effect of temperature on discharge characteristics, zinc/air button cell (A675 cell) discharge load of 620 Ω.

Operating voltage also varies with temperature. At lower temperatures and a fixed current, the voltage is lower than when the cell is discharged at room temperature or warmer. Although the optimum discharge temperature is between 10 and 40°C, the cell can still function at temperatures below 10°C, but at lower voltages and energy densities. The operating voltage of the zinc/air cell at various discharge loads and temperatures is shown in Fig. 10.15.

It should be noted that at low discharge rates and moderate temperatures, the effect of temperature is minimal. In selecting a zinc/air system that must perform at low temperatures, it is important to consider the required current density in order to preclude failure due to diffusion limitations.

10.4.9 Storage Life

Four principal mechanisms affect the capacity of zinc/air cells during storage and operating service. One mechanism, self-discharge of the zinc (corrosion), is an internal reaction, and the other three are caused by gas transfer. The gas transfer mechanisms are direct oxidation of the zinc anode, carbonation of the electrolyte, and electrolyte water gain or loss.

During storage the air access hole(s) of a zinc/air cell can be sealed to prevent gas transfer decay. A typical material for sealing a cell is a polyester tape. Note that, unlike conventional cells, one of the zinc/air cell's reactants, oxygen, is sealed outside the cell

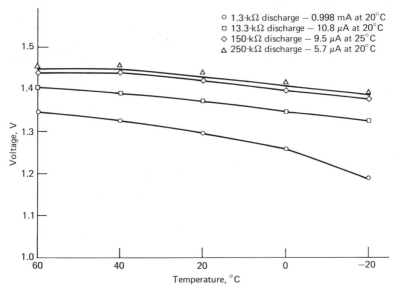

○ 1.3-kΩ discharge − 0.998 mA at 20°C
□ 13.3-kΩ discharge − 10.8 μA at 20°C
◇ 150-kΩ discharge − 9.5 μA at 25°C
△ 250-kΩ discharge − 5.7 μA at 20°C

FIG. 10.15   Effect of temperature on cell voltage, A675 zinc/air button cell.

during storage. This characteristic gives zinc/air cells excellent shelf life performance.

The primary mechanism affecting the shelf life of a zinc/air cell is the self-discharge reaction. Zinc is thermodynamically unstable in an alkaline solution (electrolyte) and reacts to form zinc oxide (discharged zinc) and hydrogen gas (an additional advantage of zinc/air cells is that hydrogen evolved during the reaction is vented through the sealing tape to prevent the pressure buildup that can cause cell deformation in conventional cells). This reaction is controlled by additives, such as mercury, to the zinc. Results of shelf life evaluation of A675H cells at room ambient conditions over a 4½-y storage period are presented in Table 10.4. Capacity retention over this period is 91% of initial capacity, yielding an average capacity loss per year of less than 2%. Similar performance is achieved only by lithium cells.

Elevated temperatures will dramatically increase the rate of the self-discharge reaction. Capacity retention performance of A675H cells after 20 days storage at 60°C presented in Table 10.5 indicates the detrimental effect of high-temperature storage. Storage life at high temperatures can be optimized through trade-offs between other performance parameters and choice of cell design components.

TABLE 10.4   A675H Cell Capacity Retention vs. Storage Time at 20°C

Storage, y	Average capacity,* mAH	% Change from initial capacity	Average % change per annum
0	422	0.00	0.00
2	405	4.03	2.00
4.5	385	8.77	1.89

*Discharge load = 2 mA, 12 hr/day.

TABLE 10.5 A675H Cell Capacity Retention at 60°C

Storage time, days	Average capacity,* mAh	Change from initial capacity, %
0	401	0
20	374	6.73

*Discharge load = 2.5 mA.

10.4.10 Factors Affecting Service Life

The combined effects of self-discharge and gas transfer degradation determine the service life performance of a zinc/air cell. For most applications water transfer is the dominant factor. However, under some conditions electrolyte carbonation and direct oxidation mechanisms can adversely affect performance.

Carbonation of the Electrolyte Carbon dioxide, which is present in the atmosphere at a concentration of approximately 0.04%, reacts with an alkaline solution (electrolyte) to form an alkali metal carbonate and bicarbonate. Zinc/air cells can be satisfactorily discharged using a carbonated electrolyte, but there are two disadvantages of extreme carbonation: (1) vapor pressure of the electrolyte is increased, aggravating water vapor loss in low humidity conditions and (2) crystals of carbonate formed in the cathode structure may impede air access, eventually causing cathode damage with subsequent deterioration of cathode performance. As indicated in Fig. 10.16, carbonation must be extreme to be detrimental to cell performance in most applications.[5]

Direct Oxidation The zinc anode of a zinc/air cell can be oxidized directly by oxygen which enters the cell and dissolves in and diffuses through the electrolyte. Measurements of the effect of direct oxidation on cell capacity have been predicted experimentally[6] using advanced microcalorimetry techniques to predict direct oxidation effects on A675H cells (cells open, not sealed). This research indicates a capacity loss of less than 1.5% of rated capacity per year at 25°C and less than 5%/y at 37.5°C. The loss for low-rate cells will be greater than an order of magnitude less because of stricter gas diffusion regulation.

The combined effect of oxidation and self-discharge on fresh high-rate cells, open to the ambient, yields a predicted capacity loss of less than 4%/y (2%/y for low-rate cells). However, the typical useful service life for a high-rate cell, as presented in Table 10.2, is about 3 months.

Effect of Water Vapor Transfer on Service Life The decay mechanism determining the useful service life is water vapor transfer. It occurs when a partial pressure difference exists between the vapor pressure of the electrolyte and the surrounding environment. As previously indicated, a cell with a typical electrolyte, consisting of 30% concentration of potassium hydroxide, will lose water when the humidity at room temperature is below 60% and will gain water at humidities above 60%. Excessive water loss increases the concentration of the electrolyte and can eventually cause the cell to fail because of inadequate electrolyte to maintain the discharge reaction. Excessive water gain dilutes the electrolyte, which also reduces conductivity. Furthermore, the catalyst layer of the air cathode will flood under sustained water gain conditions, reducing electrochemical activity and eventually causing cell failure.

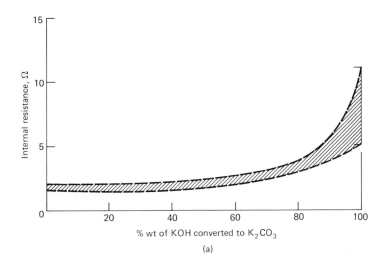

(a)

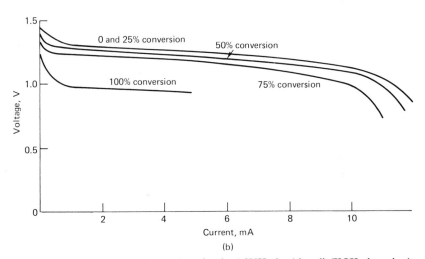

(b)

FIG. 10.16   Effect of electrolyte carbonation in A675H zinc/air cell (KOH electrolyte). (a) On internal impedance at 20°C (4000 Hz); (b) on voltage/current profile at 20°C. (*From Ref. 4.*)

   The design of a zinc/air cell can be optimized to compensate for water transfer for a specific set of operating conditions. Trade-offs can be made between the volume and composition of the electrolyte, the amount of zinc, and the degree of gas diffusion regulation to maximize service life. For most applications and cell types, diffusion control is the most important variable.

   The effect of diffusion control on water vapor transfer rates for two zinc/air cell sizes is presented in Table 10.6. Actual instantaneous water transfer rates at a given environment will vary throughout the life of the cell, dependent on the length of exposure. As

TABLE 10.6   Effect of Gas Diffusion Regulation on Zinc/Air Button Cell Water Vapor Transfer Rates*

Relative humidity, %	Cell type			
	A13H	A13L	A675H	A675L
25	−0.520	−0.0067 †	−0.990	−0.020
60	0	0	0	0
75	+0.270	+0.010	+0.530	+0.030

*Room temperature; data in milligrams per day of weight change; high-rate cell data averaged over 30 days; low-rate cell data averaged over 4 months.
†Projected data.
SOURCE: Gould Inc.

a cell gains or loses water, transfer rates will decrease because the electrolyte equilibrium relative humidity moves toward the ambient relative humidity. The figures presented in Table 10.6 represent average rates computed over the indicated exposure period. Actual average rates over the life of the cell would be lower.

The degree of increased water vapor (gas) transfer tolerance of a low-rate cell vs. its high-rate equivalent is approximated in Table 10.7 by taking the ratios of the water transfer rates given in Table 10.6. Under continuous exposure at the specified conditions, the service life of a low-rate cell can be estimated to be longer than the high-rate cell of the same size by the multiple in the table.

TABLE 10.7   Relative Degree of Environmental Tolerance of Low-Rate vs. High-Rate Cells

Relative humidity, %	A13L*	A675L*
25	78 times A13H	50 times A675H
75	27 times A13H	18 times A675H

*Ratio of high-rate to low-rate water vapor transfer rates from Table 10.6.
SOURCE: Gould, Inc.

Discharge profiles of high-rate cells discharged under sustained high- and low-humidity conditions are presented in Fig. 10.17. Additional data on capacity performance are provided in Table 10.8. Note that the larger cell, A675H, has greater tolerance to the detrimental effects of water gain or loss. This is typical in a comparison of cells of two different sizes. Larger cells have more electrolyte and also a greater anode-free volume, making them more tolerant to water loss and gain conditions, respectively.

Projected performance of low-rate cells at high and low humidities is presented in Table 10.9. The estimates of service life are for the period over which at least 80% of cell initial capacity will be delivered before cell failure. In addition to capacity degradation resulting from gas transfer, capacity loss of 2%/y has been assumed due to self-discharge. It should be noted that low-rate cells can transfer more water (either gain or loss) than high-rate cells before they fail because they are operating at much lower current densities. Therefore, in particular, the 2.2-y projection for an A13L cell operating in a continuous 75% relative humidity environment is a conservative estimate.

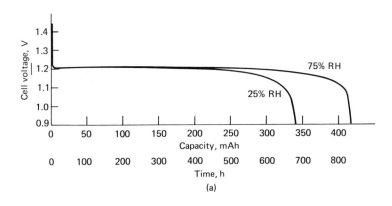

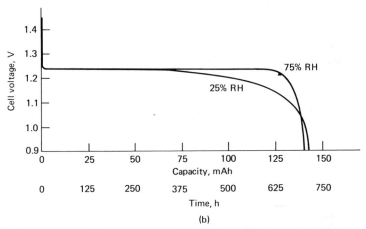

FIG. 10.17 Discharge profiles of zinc/air cells at high and low humidities at 20°C. (*a*) A675H cell, 5 mA, 2.4 h/day discharge; (*b*) A13H cell, 2 mA, 2.4 h/day discharge.

TABLE 10.8 Capacity Performance of High-Rate Zinc/Air Cells after Open Storage at High and Low Humidities*

Storage time	A13H capacity, mAh after storage at:		A675H capacity, mAh after storage at:	
	75% RH	25% RH	75% RH	25% RH
Initial	172	172	429	429
100 h	166	164	NA	NA
500 h	168	158	NA	NA
700 h†	139	NA	NA	NA
1000 h	65	134	423	410
3000 h	NA	NA	402	392

*Conditions: 20°C; load after storage = 1 mA; load during storage = open circuit.
† 2 mA, 10% duty cycle.
SOURCE: Gould, Inc.

TABLE 10.9 Projected Service Life Capability
of Low-Rate Cells at High and Low Humidities*

| Relative | Service life, y† | |
humidity, %	A13L	A675L
25	> 4	> 4
75	> 2.2	> 4

*Results projected from data in Tables 10.8 and 10.9.
†Service life calculated for period over which 80% of rated
capacity will be delivered before cell failure.

10.5 PERFORMANCE CHARACTERISTICS— INDUSTRIAL BATTERIES

10.5.1 Edison Carbonaire Batteries

The Edison Carbonaire ST type zinc/air cell is manufactured in an 1100-Ah size and is available in two- and three-cell batteries, with the cells connected in series or parallel. The physical and electrical characteristics of these batteries are listed in Table 10.10.

The three voltage vs. time curves shown in Fig. 10.18 depict the performance obtained at various discharge rates. Capacities obtained are quite consistent over the range of 0.15 to 1.25 A continuous discharge, although, at the higher rates, voltage variation with temperature must be considered. For example, if tight voltage control is required, a series-parallel assembly of batteries may be needed to reduce the current drain per cell and thus minimize voltage fluctuations with temperature. Maximum discharge rates for the 1100-Ah cell are shown in Table 10.11.

TABLE 10.10 Edison Carbonaire Zinc/Air Batteries

| Type | Dimensions, cm | | | Weight (filled), kg | Connection | Nominal voltage, V | Nominal capacity, Ah |
	Length	Width	Height				
					Two cells		
ST-22-1100	21.9	20.0	28.9	14	Series	2.5	1100
ST-22-2200					Parallel	1.25	2200
					Three cells		
ST-33-1100	32.4	20.0	28.9	21	Series	3.75	1100
ST-33-3300					Parallel	1.25	3300

SOURCE: McGraw-Edison Co., Power Systems Division, Bloomfield, N.J.

TABLE 10.11 Edison Carbonaire ST Type Zinc/Air Batteries*

Duty cycle, °C	10% on (up to 0.5 s on)	20% on (up to 2.0 s on)	50% on (up to 1 s on)	100% on (continuous)
20	3.5	2.8	2.3	1.25
−5	2.4	1.9	1.6	0.75

*Maximum discharge rates (amperes) to 1.0 V per cell.
SOURCE: McGraw-Edison Co., Power Systems Division, Bloomfield, N.J.

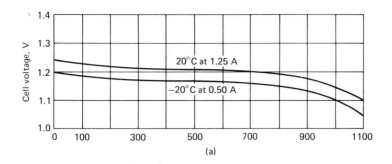

(a)

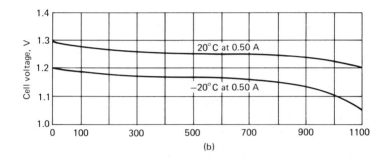

(b)

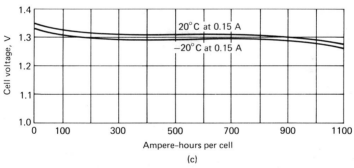

Ampere–hours per cell

(c)

FIG. 10.18 Discharge characteristics of Edison Carbonaire zinc/air cells, 1100-Ah size. (a) Typical discharge voltage at maximum continuous rates; (b) typical discharge voltage at moderate continuous rates; (c) typical discharge voltage at low continuous rates. (*Courtesy of McGraw-Edison Co.*)

10.5.2 "Air-Zinc" Batteries

The "Air-Zinc" cells, illustrated in Fig. 10.3, use a gas-diffusion air cathode electrode based on fuel cell technology which provides a low polarization and stable long life operation and a large surface area zinc sponge anode. Table 10.12 lists the physical and electrical characteristics of three size cells. Typical discharge curves are shown in Fig. 10.19 and the polarization curves at ambient and low temperatures are shown in Fig. 10.20.

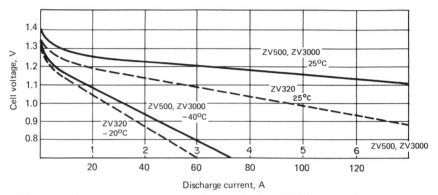

FIG. 10.19   Discharge characteristics of the "air/zinc" cells at 25°C. (*Courtesy of Central Laboratory of Electrochemical Power Sources, Sofia, Bulgaria.*)

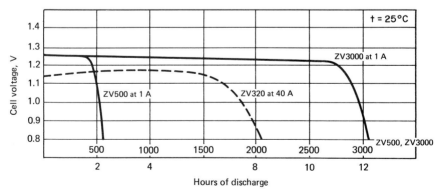

FIG. 10.20   Polarization curves, "air/zinc" cells. (*Courtesy of Central Laboratory of Electrochemical Power Sources, Sofia, Bulgaria.*)

TABLE 10.12 Characteristics of the "Air-Zinc" Cell

| Type | Dimension, mm | | | Weight, kg | | Open circuit voltage, V | Continuous Discharge | | | Energy density | |
	Length	Width	Height	Without electrolyte	With electrolyte		Nominal capacity, Ah	Discharge current, A	Cut-off voltage, V	Wh/kg	Wh/L
ZV 320	135	48	313	1.55	2.15	1.41	320	40	0.8	170	175
ZV 500	135	48	313	1.5	2.45	1.41	500	2	0.8	250	290
ZV 3000	260	170	225	7.0	14.0	1.41	3200	2	0.8	270	395

REFERENCES

1. G. W. Elmore and H. A. Tanner, U.S. Patent 3,419,900.
2. A. M. Moos, U.S. Patent 3,276,909.
3. R. E. Biddick, U.S. Patent 4,129,633.
4. Eagle-Picher Co, Joplin, Missouri.
5. J. W. Cretzmeyer, H. R. Espig, and R. S. Melrose, "Commercial Zinc-Air Batteries," *Power Sources,* vol. 6, Oriel Press, Newcastle-upon-Tyne, England, 1977.
6. S. F. Bender, J. W. Cretzmeyer, and J. C. Hall, "Long Life Zinc-Air Cells as a Power Source for Consumer Electronics," *Progress in Batteries and Solar Cells,* vol. 2, JEC Press, Ohio, 1979.

BIBLIOGRAPHY

Bender, S. F., and D. W. Biegger: "High Energy Density Zinc-Air Cells for Portable Electronics," Wescon 1980 Professional Program Paper, "Power Cells, Energy Crisis in Miniature," Electronic Conventions, Inc., Calif., 1980.

————, J. W. Cretzmeyer, and J. C. Hall: "Long Life Zinc-Air Cells as a Power Source for Consumer Electronics," *Progress in Batteries and Solar Cells,* vol. 2, JEC Press, Ohio, 1979.

Espig, H. R., and D. F. Porter: "Primary Zinc-Air Batteries," *Power Sources,* vol. 4, Oriel Press, Newcastle-upon-Tyne, England, 1973.

Gould Product Literature: *ACTIVAIR Primary Zinc-Air Batteries, Guide For Designers,* 1978.

Heise, G. W., and N. C. Cahoon: *The Primary Battery,* Wiley, New York, 1971, vol. 1.

Meuller, G. A.: *The Gould Battery Handbook,* Gould, Inc., Mendota Heights, Minn. 1973.

Product literature, McGraw-Edison Co., Power Systems Division, Bloomfield, N.J.

11

Lithium Cells

by
David Linden

11.1 GENERAL CHARACTERISTICS

Lithium metal is attractive as a battery anode material because of its light weight, high-voltage, high electrochemical equivalence, and good conductivity. Because of these outstanding features, the use of lithium has predominated in the development of high-performance primary and secondary batteries during the last decade. (Also see Part 2 Chap. 12, Part 3 Chap. 25, Part 4, and Part 5.)

Serious development of high energy density battery systems was started in the 1960s and was concentrated on nonaqueous batteries using lithium as the anode. The lithium cells were first used in the early 1970s in selected military applications, but their use was limited as suitable cell structures, formulations, and safety considerations had to be resolved. More recently, lithium primary cells and batteries have been designed, using a number of different chemistries, in a variety of sizes and configurations.[1] Sizes range from less than 5 mAh to 20,000 Ah; configurations range from small button and cylindrical cells for memory backup and portable applications to large prismatic cells for standby and emergency electrical power.

Lithium cells, with their outstanding performance and characteristics, will see wide application in the next decade with concurrent reduction in cost and increase in production.

11.1.1 Advantages of Lithium Cells

Primary cells using lithium anodes have many advantages over conventional batteries. The advantageous features include:

1. *High voltage:* Lithium cells have voltages as high as 3.9 V, depending on the cathode material, compared with 1.5 V of most primary battery systems. The higher cell voltage reduces the number of cells in a battery pack by a factor of about 2.

2. *High energy density:* The energy output of a lithium cell (over 200 Wh/kg and 400 Wh/L) is 2 to 4 or more times better than conventional zinc anode batteries.

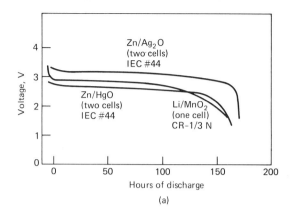

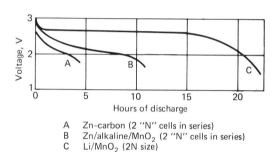

A   Zn-carbon (2 "N" cells in series)
B   Zn/alkaline/MnO$_2$ (2 "N" cells in series)
C   Li/MnO$_2$ (2N size)

(b)

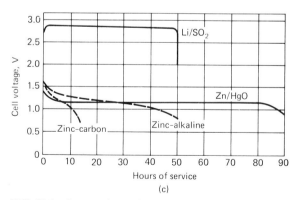

FIG. 11.1 Comparative performance of lithium and conventional primary cells at 20°C. (a) Button cells, discharge load 2700 Ω; (b) N size cylindrical cells, discharge load 50 Ω; (c) D size cells, 200-mA average discharge current.

3. *Operation over a wide temperature range:* Many of the lithium cells will perform over a temperature range from above 70 to −40°C or below.

4. *Good power density:* Some of the lithium cells are designed with the capability to deliver their energy at high current and power levels.

5. *Flat discharge characteristics:* A flat discharge curve (constant voltage and impedance through most of the discharge) is typical for most lithium cells.

6. *Superior shelf life:* Lithium cells can be stored for long periods even at elevated temperatures. Storage of up to 10 y at room temperature is projected; storage of more than 5 y at 20°C and 1 y at 70°C has been demonstrated.

The performance advantages of several types of lithium cells compared with conventional primary and secondary batteries, are shown in Part 1, Sec. 3.5 and Part 2, Sec. 4.2. The advantage of the lithium cell on a volumetric basis is shown graphically in Figs. 11.1 *a, b,* and *c,* which compare the service life of various button and N and D size cylindrical cells. Only the zinc/mercuric oxide and zinc/silver oxide cells, which are noted for their high volumetric energy density, approach the capability of the lithium systems at 20°C. These cells however do not compare as favorably on a gravimetric basis (Figs. 4.7 and 4.8) nor as well at lower temperatures.

11.1.2 Classification of Lithium Primary Cells

Lithium cells use nonaqueous solvents for the electrolyte because of the reactivity of lithium in aqueous solutions. Organic solvents, such as acetonitrile, propylene carbonate, and inorganic solvents, such as thionyl chloride, are typical. A compatible solute is added to provide the necessary electrolyte conductivity. (Solid-state and molten-salt electrolytes are also used in some other primary and reserve lithium cells, Part 2, Chap. 12 and Part 5.) Many different materials are used for the active cathode material; sulfur dioxide, manganese dioxide, and carbon monofluoride are now in popular use. The term "lithium cell," therefore, applies to many different types of cells or chemistries, each using lithium as the anode but differing in cathode material, electrolyte, and chemistry as well as in design and other physical and mechanical features.

Lithium primary cells can be classified into several categories, based on the type of electrolyte (or solvent) and cathode material that is used. These classifications, typical examples, and the major characteristics of each are listed in Table 11.1.

Soluble Cathode Cells These use liquid or gaseous cathode materials, such as sulfur dioxide (SO_2) or thionyl chloride ($SOCl_2$), that dissolve in the electrolyte or are the electrolyte solvent. Their operation depends on the formation of a protective coating on the lithium anode resulting from a reaction between the lithium and the cathode material. This prevents further chemical reaction (self-discharge) between the anode and cathode. These cells are manufactured in many different configurations and designs (e.g., high and low rate) and with a very wide range of capacity. They are generally fabricated in cylindrical configuration in the smaller sizes, up to about 25 Ah, using a bobbin construction for the low-rate cells and a jelly-roll structure for the high-rate designs. Prismatic containers, having flat parallel plates, are generally used for the larger cells up to 20,000 Ah in size. Flat or "pancake-shaped" configurations have also been designed. These soluble cathode lithium cells are used for low to high discharge rates. The high-rate designs, using large electrode surface areas, are noted for their high power density and are capable of delivering about the highest current densities of any primary cell.

TABLE 11.1 Classification of Lithium Primary Cells

Cell classification	Typical electrolyte	Power capability	Size, Ah	Operating range, °C	Shelf life, y	Typical cathodes	Nominal cell voltage, V	Key characteristics
Soluble cathode (liquid or gas)	Organic or inorganic (w/solute)	Moderate to high power, W	0.5 to 20,000	−55 to 70	5–10	SO_2 $SOCl_2$ SO_2Cl_2	3.0 3.6 3.9	High energy output, high power output, low temperature operation, long shelf life
Solid cathode	Organic (w/solute)	Low to moderate power, mW	0.03 to 5	−40 to 50	5–8	V_2O_5 CrO_x Ag_2CrO_4 MnO_2 $(CF)_n$ S CuS FeS_2 FeS CuO $Bi_2Pb_2O_5$ Bi_2O_3	3.3 3.3 3.1 3.0 2.6 2.2 1.7 1.6 1.5 1.5 1.5 1.5	High energy output for moderate power requirements, non-pressurized cells
Solid electrolyte (see Part 2, Chap. 12)	Solid state	Very low power, μW	0.003 to 0.5	0 to 100	10–25	$PbI_2/PbS/Pb$ I_2 (P2VP)	1.9 2.8	Excellent shelf life, solid state—no leakage, long-term microampere discharge

NOTE: For reserve cells, see Part 5, Chaps. 15, 19, and 20.

Solid Cathode Materials   The second type of lithium anode primary cells uses solid rather than soluble gaseous or liquid materials for the cathode. With these solid cathode materials, the cells have the advantage of not being pressurized or requiring a hermetic-type seal, but they do not have the high rate capability of the soluble cathode systems. They are designed, generally, for low- to medium-rate applications (milliamperes) such as memory backup, portable electronic equipment, photographic equipment, watches and calculators, and small lights. Button, flat, and cylindrical-shaped cells are available in low-rate and the higher-rate jelly-roll configurations. A number of different solid cathodes are being used in lithium primary cells as listed in Table 11.1. The discharge of the solid cathode cells is not as flat as that of the soluble cathode cells, but, at the lower discharge rates, their capacity (energy density) may be close to that of the lithium/sulfur dioxide cell.

Solid Electrolyte Cells   These cells are noted for their extremely long storage life, in excess of 20 y, but are capable of only low-rate discharge in the microampere range. They are used in applications such as memory backup, cardiac pacemaker, and similar equipment where current requirements are low but long life is most important (see Part 2, Chap. 12).

In Fig. 11.2, the size or capacity of these three types of lithium cells (up to the 25-Ah size) is plotted against the current levels at which they are typically discharged. The approximate weight of lithium in each of these cells is also shown.

11.2 CHEMISTRY

11.2.1 Lithium

The main requirements for electrode materials for high-performance (high energy density) batteries are a high electrochemical equivalence (high coulombic output for a given weight of material) and a high electrode potential. It is apparent from Table 11.2, which lists the characteristics of metals that are used as battery anodes, that

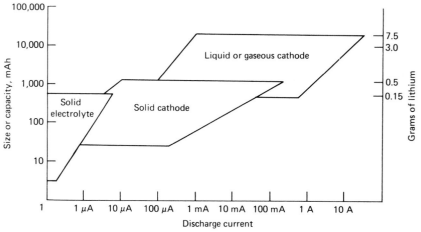

FIG. 11.2   Classification of lithium primary cell types.

TABLE 11.2 Characteristics of Anode Materials

Material	Atomic weight, g	Standard potential at 25°C, V	Density, g/cm³	Melting point, °C	Valence change	Electrochemical equivalence Ah/g	g/Ah	Ah/cm³
Li	6.94	−3.05	0.54	180	1	3.86	0.259	2.08
Na	23.0	−2.7	0.97	97.8	1	1.16	0.858	1.12
Mg	24.3	−2.4	1.74	650	2	2.20	0.454	3.8
Al	26.9	−1.7	2.7	659	3	2.98	0.335	8.1
Ca	40.1	−2.87	1.54	851	2	1.34	0.748	2.06
Fe	55.8	−0.44	7.85	1528	2	0.96	1.04	7.5
Zn	65.4	−0.76	7.1	419	2	0.82	1.22	5.8
Cd	112	−0.40	8.65	321	2	0.48	2.10	4.1
Pb	207	−0.13	11.3	327	2	0.26	3.87	2.9

lithium is an outstanding candidate. Its standard potential and electrochemical equivalence are the highest of the metals; it excels in theoretical gravimetric energy density; and, with its high potential, it is inferior only to aluminum and magnesium on a volumetric energy (watthour per liter) basis. Aluminum, however, has not been used successfully as an anode material because of its poor electrochemical behavior, and magnesium has a low practical operating voltage. Further, lithium is preferred to the other alkali metals because of its better mechanical stability and lower reactivity. Calcium has been investigated as an anode, in place of lithium, because its higher melting point (838°C compared with 181°C for lithium) may result in safer operation, reducing the possibility of thermal runaway if high internal cell temperatures should occur.

Lithium is one of the alkali metals and is the lightest of all the metal elements with a density about half that of water. When first made or freshly cut, lithium has the luster and color of bright silver but tarnishes rapidly in moist air. It is soft and malleable, can be readily extruded into a thin ribbon, and is a good conductor of electricity. Table 11.3 lists some of the physical properties of lithium.[2]

TABLE 11.3 Physical Properties of Lithium

Melting point	180.5°C
Boiling point	1347°C
Density	0.534 g/cm³ (25°C)
Specific heat	0.852 cal/g (25°C)
Specific resistance	9.35 × 10⁶ Ω·cm (20°C)
Hardness	0.6 (Mohs scale)

Lithium reacts vigorously with water, releasing hydrogen and forming lithium hydroxide

$$2 \text{ Li} + 2 \text{ H}_2\text{O} \rightarrow 2 \text{ LiOH} + \text{H}_2$$

This reaction is not as vigorous as that of sodium and water, probably due to the fairly low solubility of and the adherence of LiOH to the metal surface. Because of this reactivity, however, lithium must be handled in a dry atmosphere and, in a battery, be used with nonaqueous electrolytes. (The lithium-water battery, described in Part 5, Chap. 35, is an exception to this condition.)

11.2.2 Cathode Materials

A number of inorganic and organic materials have been examined for use as the cathode in primary lithium batteries.[1,3] The critical requirements for this material to achieve high performance are a high battery voltage, a high energy density, and compatibility with the electrolyte (i.e., being essentially nonreactive or insoluble in the electrolyte). Preferably, the cathode material should be conductive, although there are few such materials available, and solid cathode materials are usually mixed with a conducting material, such as graphite, and applied to a conductive grid to provide the needed conductivity. If the cathode reaction products are a metal and a soluble salt (of the anode metal), this feature can improve cathode conductivity as the discharge proceeds. Other desirable properties for the cathode material are low cost, availability (noncritical material), and favorable physical properties, such as nontoxicity and nonflammability. Table 11.4 lists some of the cathode materials that have been studied for primary lithium batteries and gives their cell reaction mechanisms and the theoretical cell voltages and capacities.

11.2.3 Electrolytes

The reactivity of lithium in aqueous solutions requires the use of nonaqueous electrolytes for lithium anode batteries.[3] Polar organic liquids are the most common electrolyte solvents for the active primary cells, except for the thionyl chloride ($SOCl_2$) and sulfuryl chloride (SO_2Cl_2) cells where these inorganic compounds serve as both the solvent and active cathode material. The important properties of the electrolyte are:

1. It must be aprotic, i.e., have no reactive protons or hydrogen atoms, although hydrogen atoms may be in the molecule.

2. It must be nonreactive with lithium (or form a protective coating on the lithium surface to prevent further reaction) and the cathode.

3. It must be capable of forming an electrolyte of high ionic conductivity.

4. It should be liquid over a broad temperature range.

5. It should have favorable physical characteristics, such as low vapor pressure, stability, nontoxicity, and nonflammability.

A listing of the organic solvents commonly used in lithium batteries is given in Table 11.5. These organic electrolytes, as well as thionyl chloride (mp $-105°C$, bp $78.8°C$) and sulfuryl chloride (mp $-54°C$, bp $69.1°C$), are liquid over a wide temperature range with low freezing points. This characteristic provides the potential for operation over a wide temperature range, particularly at low temperatures.

Lithium salts, such as LiCl, $LiClO_4$, LiBr, and $LiAlCl_4$, are the electrolyte solutes most commonly used to provide ionic conductivity. The solute must be able to form a stable electrolyte which does not react with the active electrode materials. It must be soluble in the organic solvent and dissociate to form a conductive electrolyte solution. Maximum conductivity is normally obtained with a 1-molar solute concentration, but generally the conductivity of these electrolytes is about one-tenth that of the aqueous systems. To accommodate this lower conductivity, close electrode spacing and cell designs to minimize impedance and provide good power density are used.[4,5]

TABLE 11.4 Cathode Materials Used in Lithium Primary Batteries

Cathode material	Mol. wt.	Valence change	Density, g/cm³	Theoretical faradic capacity (cathode only)			Cell reaction mechanism (with lithium anode)	Theoretical cell	
				Ah/g	Ah/cm³	g/Ah		Voltage, V	Capacity, Wh/kg
SO_2	64	1	1.37	0.419	...	2.39	$2Li + 2SO_2 \rightarrow 2Li_2S_2O_4$	3.1	1170
$SOCl_2$	119	2	1.63	0.450	...	2.22	$4Li + 2SOCl_2 \rightarrow 4LiCl+S+SO_2$	3.65	1470
SO_2Cl_2	135	2	1.66	0.397	...	2.52	$2Li + SO_2Cl_2 \rightarrow 2LiCl+SO_2$	3.91	1405
Bi_2O_3	466	6	8.5	0.35	2.97	2.86	$6Li + Bi_2O_3 \rightarrow 3Li_2O+2Bi$	2.0	640
$Bi_2Pb_2O_5$	912	10	9.0	0.29	2.64	3.41	$10Li+Bi_2Pb_2O_5 \rightarrow 5Li_2O+2Bi+2Pb$	2.0	544
$(CF)_n$	$(31)_n$	1	2.7	0.86	2.32	1.16	$nLi + (CF)_n \rightarrow nLiF+nC$	3.1	2180
$CuCl_2$	134.5	2	3.1	0.40	1.22	2.50	$2Li + CuCl_2 \rightarrow 2LiCl+Cu$	3.1	1125
CuF_2	101.6	2	2.9	0.53	1.52	1.87	$2Li + CuF_2 \rightarrow 2LiF+Cu$	3.54	1650
CuO	79.6	2	6.4	0.67	4.26	1.49	$2Li + CuO \rightarrow Li_2O+Cu$	2.24	1280
CuS	95.6	2	4.6	0.56	2.57	1.79	$2Li + CuS \rightarrow Li_2S+Cu$	2.15	1050
FeS	87.9	2	4.8	0.61	2.95	1.64	$2Li + FeS \rightarrow Li_2S+Fe$	1.75	920
FeS_2	119.9	4	4.9	0.89	4.35	1.12	$4Li + FeS_2 \rightarrow 2Li_2S+Fe$	1.8	1304
MnO_2	86.9	1	5.0	0.31	1.54	3.22	$Li + Mn^{IV}O_2 \rightarrow Mn^{III}O_2(Li^+)$	3.5	1005
MoO_3	143	1	4.5	0.19	0.84	5.26	$2Li + MoO_3 \rightarrow Li_2O+Mo_2O_5$	2.9	525
Ni_3S_2	240	4		0.47		2.12	$4Li + Ni_3S_2 \rightarrow 2Li_2S+3Ni$	1.8	755
$AgCl$	143.3	1	5.6	0.19	1.04	5.26	$Li + AgCl \rightarrow LiCl+Ag$	2.85	515
Ag_2CrO_4	331.8	2	5.6	0.16	0.90	6.25	$2Li + Ag_2CrO_4 \rightarrow Li_2CrO_4+2Ag$	3.35	515
V_2O_5	181.9	1	3.6	0.15	0.53	6.66	$Li + V_2O_5 \rightarrow LiV_2O_5$	3.4	490

TABLE 11.5 Properties of Organic Electrolyte Solvents for Lithium Primary Batteries

Solvent	Structure	Boiling point (at 10^5 Pa), °C	Melting point, °C	Flash point, °C	Density (at 25°C), g/cm³	Specific conductivity, with 1 M LiClO₄, Ω^{-1} cm^{-1}
Acetonitrile (AN)	$H_3C-C\equiv N$	81	-45	5	0.78	3.6×10^{-2}
γ-Butyrolactone (BL)		204	-44	99	1.1	1.1×10^{-2}
Dimethylsulfoxide (DMSO)	$H_3C-S-CH_3$	189	18.5	95	1.1	1.4×10^{-2}
Dimethylsulfite (DMSI)	$O=S\langle^{OCH_3}_{OCH_3}$	126	-141		1.2	
1,2-Dimethoxyethane (DME)		83	-60	1	0.87	
Dioxolane (1,3-D)		75	-26	2	1.07	
Methyl formate (MF)	$H-C-O-CH_3$	32	-100	-19	0.98	3.2×10^{-2}
Nitromethane (NM)	H_3C-NO_2	101	-29	35	1.13	1×10^{-2}
Propylene carbonate (PC)		242	-49	135	1.2	7.3×10^{-3}
Tetrahydrofuran (THF)		65	-109	-15	0.89	

11.2.4 Cell Couples and Reaction Mechanisms

The overall discharge reaction mechanism for the various lithium primary batteries is shown in Table 11.4, which also lists the theoretical cell voltage of each cell. The mechanism for the discharge of the lithium anode is the oxidation of lithium to form lithium ions (Li^+) with the release of an electron

$$Li \rightarrow Li^+ + e$$

The electron moves through the external circuit to the cathode where it reacts with the cathode material, which is reduced. At the same time, the Li^+ ion, which is small (0.06 nm in radius) and mobile in both liquid and solid-state electrolytes, moves through the electrolyte to the cathode where it reacts to form a lithium compound.

A more detailed description of the cell reaction mechanism for the different lithium primary batteries is given in the sections on those cell systems[1,6] (Secs. 11.4 to 11.15).

11.3 CHARACTERISTICS OF LITHIUM PRIMARY BATTERIES

11.3.1 Summary of Design and Performance Characteristics

A listing of the major lithium primary cells now in production or advanced development and summarizing their constructional features, key electrical characteristics, and manufacturers is presented in Table 11.6. Because lithium cells are in their early production stage, the types of cells and sizes offered by each manufacturer are subject to change until the technology matures and standardization of chemistry, cell design, size, and performance materializes. The performance characteristics of these systems, under theoretical conditions, is given in Table 11.4. Comparisons of the performance of the lithium cells with comparably sized conventional primary batteries are covered in Part 1, Chap. 3 and Part 2, Chap. 4.

TABLE 11.6 Characteristics of Lithium Primary Cells

| System | Cathode | Electrolyte | | Separator | Container | Construction | Voltage, V | |
		Solvent	Solute				Nominal	Working* (20°C)
				Soluble Cathode Cells				
Lithium/ sulfur dioxide (Li/SO₂)	SO₂ with carbon and binder on Al screen	AN	LiBr	Polypro- pylene	Nickel- plated steel	Spiral "jelly-roll" cylindrical construction; glass-to-metal seal	3.0	2.9–2.7
Lithium/ thionyl chloride Li/SOCl₂)	SOCl₂ with carbon and binder on Ni or SS	SOCl₂ (some with additives)	LiAlCl₄	Glass mat or filter paper	Nickel- plated steel or SS	Glass-to- metal seal	3.6	3.5–3.3
Low rate						"Bobbin" in cylindrical construction		
High capacity						Prismatic with flat plates		
High rate						Spiral "jelly-roll" cylindrical construction or flat disk		

11.3.2 Soluble Cathode Lithium Primary Cells

Two types of soluble cathode lithium primary cells are available (Table 11.1). One uses SO_2 as the active cathode dissolved in an organic electrolyte solvent. The second type uses an inorganic solvent, such as the oxychlorides $SOCl_2$ and SO_2Cl_2, which serves as both the active cathode and the electrolyte solvent. These materials fortuitously form a passivating layer or protective film of reaction products on the lithium surface which protects it from further reaction. Even though the active cathode material is in contact with the lithium anode, self-discharge is prevented by the protective film, and the shelf life of these batteries is excellent. This film, however, causes a voltage delay to occur: a time delay to penetrate or break the film and for the cell voltage to reach the operating level when the discharge load is applied. These lithium cells have a high energy density and, with proper design, such as the use of high surface area electrodes, are capable of delivering this high energy density at high power densities.

These cells generally require a hermetic-type seal. Sulfur dioxide is a gas at 20°C (boiling point is $-10°C$), and the undischarged cell has an internal pressure of 3 to 4×10^5 Pa at 20°C. The oxychlorides are liquids at 20°C, but with boiling points of 78.8°C for $SOCl_2$ and 69.1°C for SO_2Cl_2, a moderate pressure can develop at high operating temperatures. In addition, as SO_2 is a discharge product in the oxychloride cells, the internal cell pressure increases as the cell is discharged.

The lithium/sulfur dioxide (Li/SO_2) cell is the most advanced of these lithium primary cells. These cells are typically manufactured in cylindrical configurations in capacities up to about 30 Ah. They are noted for their high power density (about the highest of the lithium primary cells), high energy density, and good low-temperature performance. They are used in important military and specialized industrial and commercial applications where these performance characteristics are mandatory.

The lithium/thionyl chloride $(Li/SOCl_2)$ cell has one of the highest energy densities of all the practical battery systems. Figures 4.7 and 4.8 illustrate the advantages of the $Li/SOCl_2$ cell over a wide temperature range at moderate discharge rates. Figure 11.3 compares the discharge profile of the $Li/SOCl_2$ cell with the Li/SO_2 cell. At 20°C, at moderate discharge rates, the $Li/SOCl_2$ cell has a higher working voltage and about

Energy density† (approx.)		Power density	Operating temperature, °C	Discharge profile	Storage	Available sizes	Representative manufacturers
Wh/kg	Wh/L						
			Soluble Cathode Cells				
280	440	High	−55 to 70	Very flat	to 75°C; 10 y at 20°C (est.)	Cylindrical cells	Duracell; Power Conversion; Honeywell; Silberkraft; SAFT
			−40 to 70		to 75°C		
300	650	Low to moderate		Flat		Cylindrical cells: 1.0–10 Ah	Union Carbide; Tadiran; GTE Sylvania; Honeywell; SAFT
480	950	Moderate		Flat		2000–15,000 Ah	GTE Sylvania; Honeywell
300	900	Moderate to high		Flat		>1.0– 8000 Ah	W. Greatbach; Altus; GTE Sylvania

TABLE 11.6   Characteristics of Lithium Primary Cells *(Continued)*

System	Cathode	Electrolyte Solvent	Electrolyte Solute	Separator	Container	Construction	Voltage, V Nominal	Working* (20°C)
				Soluble Cathode Cells				
Lithium/ sulfuryl chloride (Li/SO$_2$Cl$_2$)	SO$_2$Cl$_2$ with carbon and binder on SS screen	SO$_2$Cl$_2$ (some with additives)	LiAlCl$_4$	Glass	Nickel-plated steel or SS	Spiral "jelly-roll" cylindrical construction; glass-to-metal seal	3.9	3.5
				Solid Cathode Cells				
Lithium/ lead bismuthate (Li/Bi$_2$Pb$_2$O$_5$)	Bi$_2$Pb$_2$O$_5$ pellet with lead powder and binder	1,3D	LiClO$_4$			"Button" cell construction	1.5	1.6–1.5 then 1.4–1.3
Lithium/ bismuth trioxide (Li/Bi$_2$O$_3$)	Bi$_2$O$_3$	Organic electrolyte				"Button" cell construction	1.5	1.7 to 1.5
Lithium/ carbon monofluoride (Li/(CF)$_n$)	CF with carbon and binder on nickel collector	DMSI BL + THF or PC + D	LiAsF$_6$ LiBF$_4$	Polypropylene	Nickel-plated steel or SS	"Coin" cell; crimped seal Spiral "jelly-roll" cylindrical construction crimped or glass-to-metal seal Rectangular with flat plates	2.6	2.7–2.6
Lithium/ copper oxide (Li/CuO)	CuO pressed in cell can	1,3D	LiClO$_4$	Polypropylene	Nickel-plated steel	"Bobbin" inside-out cylindrical construction; crimped seal	1.5	1.5–1.4
Lithium/ copper sulfide (Li/CuS)	CuS pellet with C, S, Cu	DME + 1,3D or THF + DME	LiClO$_4$	Glass mat	SS	Flat pellets in flat rectangular cell; crimped seal	1.7	2.1 and 1.6 (low loads) 1.6–1.5 (Moderate loads)
Lithium/ iron sulfide (Li/FeS)	FeS pellet with carbon and binder	PC + DME		Polypropylene	Nickel-plated steel	"Button" cell construction; crimped seal	1.5	1.4–1.2
Lithium/ iron sulfide (LiFeS$_2$)	FeS$_2$	PC + DME				"Button" cell construction; crimped seal	1.6	1.6 and 1.4
Lithium/ manganese dioxide (Li/MnO$_2$)	MnO$_2$ with carbon and binder on supporting grid	PC + DME	LiClO$_4$	Polypropylene	Nickel-plated steel or SS	"Coin" cells with flat electrodes; "jelly-roll" cylindrical construction; all crimped seal	3.0	2.8–2.7
Lithium/ silver chromate (Li/Ag$_2$CrO$_4$)	Ag$_2$CrO$_4$ pellet with carbon	PC	LiClO$_4$	Polypropylene	Nickel-plated steel or SS	Button or flat cells; crimped seal; small rectangular or "D" shape	3.0	3.2 and 2.5
Lithium/ chromium oxide (Li/CrO$_x$)	CrO$_x$	Organic electrolyte			SS	Small cylindrical cells	3.3	3.7 to 3.2
Lithium/ vanadium pentoxide (Li/V$_2$O$_5$)	V$_2$O$_5$ pressed with graphite	MF	LiAsF$_6$ and LiBF$_4$	Polypropylene	SS	Button cells; glass-to-metal seal Bobbin-type cylindrical cells; glass-to-metal seal	3.3	3.3 and 2.8–2.5

*Working voltages are typical for discharges at favorable loads.
†Energy densities are for 20°C, under favorable discharge conditions. See details in appropriate sections.

Energy density† (approx.)		Power density	Operating temperature, °C	Discharge profile	Storage	Available sizes	Representative manufacturers
Wh/kg	Wh/L						
			Soluble Cathode Cells				
500	1000	Moderate to high	−40 to 70	Flat	to 75°C	In early production stage	W. Greatbach; GTE Sylvania
			Solid Cathode Cells				
150	400	Low but with pulse capability	−10 to 45	Double-plateau	to 60°C	Button cells: 18–185 mAh	SAFT
90	350	Low to moderate	−20 to 50	Moderately flat		Button cells to 50 mAh	VARTA
200	400	Low to moderate	−20 to 60	Moderately flat	to 45°C; 5 y at 20°C (est.)	Coin cells to 500 mAh	Panasonic; Eagle-Picher
						Cylindrical cells to 5 Ah	
						Rectangular cells	Eagle-Picher
275	650	Low	−10 to 70	High initial voltage drop, then moderately flat	Long shelf life to 70°C	Cylindrical cells: 500–3400 mAh	SAFT
135	335	Low to moderate		Double-plateau at low loads; moderate slope at higher loads		9 V, 1604 size; pacemaker	Ray-O-Vac; Cordis
125	370	Low		Flat, after initial voltage drop		Not in production	Hitachi
130	385	Low		Double-plateau or high initial drop		Button cells: 35–295 mAh	Union Carbide
200	400	Low to moderate	−20 to 55	Moderately flat	to 70°C; 5 y at 20°C (est.)	Coin and cylindrical cells: 65–1100 mAh	Duracell; Sanyo; GE; Varta; SAFT; Ray-O-Vac; Polaroid
275	700	Low	−10 to 55	Double-plateau	Excellent shelf life; 4 y at 45°C; 1 month at 100°C	Button cells: 130–2800 mAh; prismatic and "D" shape: 1.2–2.4 Ah	SAFT
270	675	Low	−30 to 60	Moderately flat at low drains		½AA	VARTA
200	600	Low to moderate	−40 to 60	Double-plateau		Button cylindrical prismatic cells: to 30 Ah	Honeywell

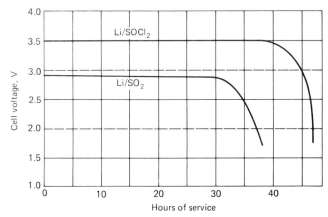

FIG. 11.3 Comparison of performance of Li/SO_2 and $Li/SOCl_2$ C size cells (100-mA discharge load) at 20°C.

a 25% advantage in service life. The Li/SO_2 cell, however, does have better performance at low temperatures and high discharge rates and less of a voltage delay after storage. $Li/SOCl_2$ cells have been fabricated in many sizes and designs ranging from small button and cylindrical cells, with capacities below 1 Ah, to large prismatic cells with capacities as high as 20,000 Ah. Low-rate cells have been used successfully for several years; high-rate cells are under development.

The lithium/sulfuryl chloride (Li/SO_2Cl_2) cell has potential advantages because of its higher voltage and resultant higher energy density. The relative voltage advantage of the Li/SO_2Cl_2 cell is illustrated in Fig. 11.4. Suitable cathode electrode formulations and cell designs are being developed to achieve the full capability of this electrochemical system, and recent work has resulted in significant performance improvements.[7]

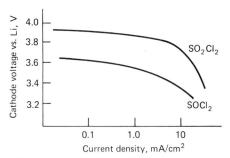

FIG. 11.4 Comparison of cathode polarization curves—Li/SO_2Cl_2 vs. $Li/SOCl_2$. (*From Ref. 7.*)

Halogen additives are being used in both the $SOCl_2$ and SO_2Cl_2 electrolytes as a means to improve performance and provide safer operation under abusive conditions.

Calcium is also being investigated as an anode material in place of lithium in thionyl chloride cells. Safer operation is anticipated with calcium since its melting temperature of 838°C is not likely to be reached by any internally driven cell condition. While the discharge voltage is about 0.4 V lower than the $Li/SOCl_2$ cell (open-circuit voltage is

3.25), the $Ca/SOCl_2$ cell has a flat discharge profile and about the same volumetric ampere-hour capacity. Shelf life characteristics are also similar to those of the lithium anode cell.[8,8a,8b]

11.3.3 Solid Cathode Lithium Primary Cells

The lithium/solid cathode cells are generally used in low- to moderate-drain applications and are manufactured mainly in small flat and/or cylindrical sizes ranging in capacity from 30 mAh to about 5 Ah, depending on the particular electrochemical system. Larger cells have been examined in cylindrical and prismatic configurations, but they have not attained large-production status.

A comparison of the performance of the solid cathode lithium cells and conventional batteries is covered in Part 2, Chap. 4.

The solid cathode cells have the advantage, compared with the soluble cathode lithium primary cells, of being nonpressurized and thus not requiring a hermetic-type seal. A polymer mechanically crimped seal, which can be used effectively for the small, flat cells and is easier to manufacture, is satisfactory for most applications. At light discharge loads, the energy density of some of the solid cathode systems is comparable to the soluble cathode systems. Their disadvantages, again compared with the soluble cathode cells, are a lower rate capability, poorer low-temperature performance, and a sloping discharge profile.

To maximize their high-rate performance and compensate for the lower conductivity of the organic electrolytes, designs are used for these lithium cells to increase electrode area, such as a large-diameter "coin" cell instead of button cells, multiple parallel electrodes, or the jelly-roll construction for the cylindrical cells.

A number of different cathode materials are used in the lithium/solid cathode cells. These are listed in Tables 11.4 and 11.6 which present some of the theoretical and practical performance data of these cells. The major features of the lithium/solid cathode cells are compared in Table 11.7. Many of the characteristics are similar, although an important property is the 3-V cell voltage obtained with several of the cathodes. These cells have a high energy density, and the higher voltage reduces the number of cells required in a battery pack by 50%. The lithium/manganese dioxide (Li/MnO_2) cell was one of the first solid cathode cells to be used commercially. It is relatively inexpensive, has excellent shelf life, good high-rate and low-temperature performance, and is available in flat and cylindrical-shaped cells. The lithium/carbon monofluoride $[Li/(CF)_n]$ cell was another of the first solid cathode cells and is attractive because its theoretical capacity is among the highest of the solid cathode systems. It is manufactured in flat and cylindrical-shaped cells. The higher cost of polycarbon monofluoride may affect the commercial potential for this system. The lithium/silver chromate (Li/Ag_2CrO_4) cell is noted for its high volumetric energy density for low-rate, long-term applications. Its high cost, due to the silver chromate, will limit its use to special applications. The lithium/vanadium pentoxide (Li/V_2O_5) cell has a high volumetric energy density but with a two-step discharge profile. Its main application has been in reserve cells (Part 5, Chap. 39). A low-rate lithium button cell, using sulfur as the active cathode material, has been proposed. This LiIs system (2.2 V) has a high specific energy, in the order of 2600 Wh/kg.[8c]

The lithium/copper sulfide (CuS) cell has a working voltage of about 1.7 V and has been developed as a replacement for conventional cylindrical zinc cells, as it has a 2:1 performance advantage. This system has not been commercialized extensively. The

Table 11.7 Characteristics of Typical Lithium/Solid Cathode Cells

Type of cell	Operating voltage, V	Characteristics
Li/MnO$_2$	3.0	High gravimetric and volumetric energy density. Wide operating temperature range (-20 to 55°C). Performance at relatively high discharge rates. Minimal voltage delay. Relatively low cost. Available in flat (coin) and cylindrical cells.
Li/(CF)$_n$	2.6	Highest theoretical gravimetric energy density. Low to moderate rate capability. Wide operating temperature range (-20 to 60°C). Voltage delay observed. Available in flat (coin) and cylindrical cells.
Li/CuS	1.7	Good volumetric energy density. Limited availability—designed only in 9-V NEDA size 1604. Potential for low cost.
Li/CuO	1.5	Highest theoretical volumetric coulombic capacity (Ah/L). Voltage compatible for conventional 1.5-V cells. Low to moderate rate capability. Operating temperature range up to 125–150°C. No apparent voltage delay. Available in cylindrical cells.
Li/FeS$_x$	1.5	Direct replacement for conventional Zn/HgO and Zn/Ag$_2$O button cells. Lower power capability than conventional cells but better low-temperature performance and storability. Relatively low cost. Available in button cell sizes.
Li/Bi$_2$Pb$_2$O$_5$	1.5	Replacement for conventional button cells. Operating range from -20 to 50°C. Low drain but with pulse load capability. Available in button cell sizes.
Li/Ag$_2$CrO$_4$	3.1	High voltage, high energy density. Low rate capability. High reliability. Used in low-rate, long-term applications. High cost. Available in button, small rectangular, and D-shaped cells.
Li/CrO$_x$	3.3	High voltage. High energy density. Low rate capability. Available in cylindrical cells.
Li/V$_2$O$_5$	3.3	High volumetric energy density. 2-step discharge. Used mainly in reserve cells (Part 5, Chap. 7).

other lithium/solid cathode cells operate in the range of 1.5 V and were developed to replace conventional 1.5-V button or cylindrical cells. The lithium/copper oxide (Li/CuO) cell is noted for its high coulombic energy density and has the advantage of higher capacity or lighter weight and better shelf life when compared with conventional cylindrical cells. The iron sulfide cells, in both the Li/FeS and Li/FeS$_2$ forms, have similar advantages over the conventional cells, and their development also has been directed to the replacement market. Similarly, the bismuth oxide systems have been designed as a replacement for "watch" cells, a low drain cell with a heavier pulse capability.

Typical discharge curves for the major solid cathode cells are shown in Fig. 11.5. The discharge curve of the Li/SO$_2$ cell, showing its flatter discharge profile, also is plotted for comparison purposes.

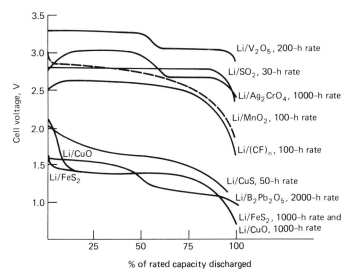

FIG. 11.5   Typical discharge curves of lithium/solid-cathode cells.

A comparison of the performance of several of the solid cathode cells in a low-rate flat-cell configuration and the higher-rate cylindrical configuration is presented in Part 1, Sec. 3.5, and Part 2, Chap. 4. In the flat-cell configuration (Table 4.4), the lithium cells have an advantage in gravimetric energy density (Wh/kg) over many of the conventional cells. This advantage may not be too important in these small cell sizes, but the lithium cells have an advantage of lower cost, particularly when compared with the silver cells, and longer shelf life. In the larger cylindrical sizes (Table 4.5), the lithium cells have an advantage in both volumetric and gravimetric energy density. In some designs this advantage is even more significant at higher discharge loads. Figure 4.3. shows another comparison of the performance of solid cathode and soluble cathode lithium cells. Although the energy densities are similar at low discharge rates, the advantage of the soluble cathode cells at high discharge rates is evident.

This comparison, as well as the need to identify the specific discharge conditions when comparisons are made, is illustrated in Fig. 11.6.[9] In Fig. 11.6a, the different systems are plotted in order of increasing energy density at the medium rate of dis-

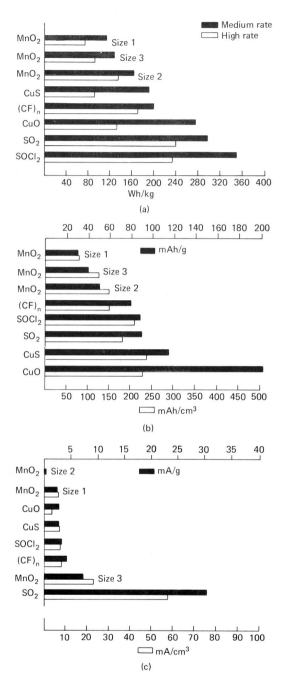

FIG. 11.6 Comparison of lithium primary cells. (*a*) Energy density at high and medium rates. (*b*) Rated capacity on weight and volume basis. (*c*) High rate current on weight and volume basis. (*From Ref. 9.*)

charge; the values at a higher discharge rate are also shown. (The three sizes for the Li/MnO_2 couple are: Size 1, smallest; Size 2 and Size 3 are the same size, but with Size 3 being the high-rate version; Size 1 is also a high-rate version in the smaller cell size.) In this comparison, the soluble cathode systems give the highest values, particularly under the high-rate discharge conditions.

Figure 11.6b is a comparison of rated capacity, in ampere-hours, of the cell divided by its weight and volume, without considering the cell voltage. As may be expected, changes in the rank order take place. The CuO and CuS show a high gravimetric capacity; this advantage is not retained when considered on a volumetric basis. Another change in rank order occurs when the systems are considered on the basis of high rate capability (based on the highest rate quoted by the manufacturer) as plotted in Fig. 11.6c. In this ranking, the high-rate versions of the Li/MnO_2 system show a higher ranking; the relatively low ranking of the $Li/SOCl_2$ couple is due to the fact that the design studied was not of a high-rate "jelly-roll" construction and low in high rate capability. It is also important to note that, although this figure illustrates the general performance characteristics of each system and the effect of several factors on the performance, the values cited are specific for the cells tested. They may not necessarily be representative for all batteries, because of the further effect on performance of design, size, etc.

The selection of a lithium vs. conventional cell thus becomes a trade-off between the lower initial cost of most of the conventional cells, the performance advantages of the lithium cells, and the key requirements of the specific application,

11.4 SAFETY AND HANDLING OF LITHIUM BATTERIES

11.4.1 Factors Affecting Safety and Handling

Attention must be given to the design and use of lithium cells and batteries to ensure safe and reliable operation. As with most battery systems, precautions must be taken to avoid physical and electrical abuse because some batteries can be hazardous if not used properly. This is important in the case of lithium cells since some of the components are toxic and/or inflammable[10] and the relatively low melting point of lithium (180.5°C) suggests that cells be prevented from reaching high internal temperatures.

Because of the variety of lithium cell chemistries, designs, sizes, etc., the procedures for their use and handling are not the same for all cells and batteries and depend on a number of factors such as:

1. *Electrochemical system:* The characteristics of the specific chemicals and cell components influence operational safety.

2. *Size and capacity of cell and battery:* Safety is directly related to the size of the cell and the number of cells in a battery. Small cells and batteries, containing less material and, therefore, less total energy, are "safer" than larger cells of the same design and chemistry.

3. *Amount of lithium used:* The less lithium is used, implying less energetic cells, the safer they should be.

4. *Cell design:* High-rate designs, capable of high discharge rates, vs. low-power designs where discharge rate is limited, use of "balanced" cell chemistry, and other features affect cell performance and operating characteristics.

5. *Safety features:* The safety features incorporated in the cell and battery will obviously influence handling procedures. These features include cell-venting mechanisms to prevent excessive internal cell pressure, thermal cutoff devices to prevent excessive temperatures, electrical fuses, and diode protection. Cells are hermetically or mechanically crimped-sealed, depending on electrochemical system, to effectively contain cell contents if cell integrity is to be maintained.

11.4.2 Safety Considerations

The electrical and physical abuses that may arise during the use of lithium cells are listed in Table 11.8 with some generalized comments on corrective action. The behavior of specific cells is covered in the other sections of this chapter; the manufacturer's data should be consulted for more details on the performance of individual cells.

High-Rate Discharges or Shorting Low-capacity cells, or those designed as low-rate cells, may be self-limiting and not be capable of high-rate discharge. The temperature rise will thus be minimal with no safety problems. Larger and/or high-rate cells can develop high internal temperatures if shorted or operated at excessively high rates. These cells are generally equipped with safety venting mechanisms to avoid a more serious hazard. Such cells or batteries should be fuse-protected (to limit discharge current); thermal fuses or thermal switches should also be used to limit the maximum temperature rise.

Forced Discharge or Voltage Reversal Voltage reversal can occur in a multicell series-connected battery, when the better-performing cells can drive the poorer cell below

TABLE 11.8 Considerations for Use and Handling of Lithium Primary Batteries

Abusive condition	Corrective procedure
High-rate discharging or shorting	Low-capacity or low-rate cells may be self-limiting. Electrical fusing, thermal protection. Limit current drain; apply battery properly.
Forced discharge (cell reversal)	Voltage cutoff. Use low-voltage batteries. Limit current drain. Special designs ("balanced" cell).
Charging	Prohibit charging. Diode protection.
Overheating	Limit current drain. Fusing, thermal cutoff. Design battery properly. Do not incinerate.
Physical abuse	Avoid opening, puncturing, or mutilating cells. Maintain cell integrity.

0 V, into reversal, as the battery is discharged toward 0 V. In some types of lithium cells, this forced discharge can result in cell venting or, in more extreme cases, cell rupture. Precautionary measures include the use of voltage cutoff circuits to prevent a battery from reaching a low voltage, the use of low-voltage batteries (since this phenomenon is unlikely to occur with a battery containing only a few cells in series), and limiting the current drain since the effect of forced discharge is more pronounced on high-rate discharges. Special designs, such as the "balanced" Li/SO_2 cell (Sec. 11.5), also have been developed that are capable of withstanding this discharge condition.

Charging   Lithium cells, as well as the other primary cells, are not designed to be recharged. If they are, they may vent or explode. Cells which are connected in parallel or which may be exposed to a charging source (as in battery-backup CMOS memory retention circuits) should be diode-protected to prevent charging.

Overheating   As discussed above, overheating should be avoided. This can be accomplished by limiting the current drain, using safety devices such as fusing and thermal cutoffs, and designing the battery to provide necessary heat dissipation.

Incineration   Lithium cells are either hermetically or mechanically crimped-sealed. They should not be incinerated without proper protection because they may rupture or explode at high temperatures.

Currently, special procedures govern the transportation and shipment of lithium batteries containing more than 0.5 g of lithium (equivalent to about 1 Ah in capacity).[11] Procedures for the use, storage, and handling of lithium batteries also have been recommended.[12] Disposal of some types of lithium cells also is regulated. The latest issue of these regulations should be consulted for the most recent procedures. Several types and sizes of lithium cells have received Underwriters Laboratories, Inc. (Northbrook, Ill.) "component recognition."

11.5 LITHIUM/SULFUR DIOXIDE (Li/SO₂) CELLS

One of the more advanced of the lithium primary cells is the lithium/sulfur dioxide (Li/SO_2) system. The cell has an energy density of up to 280 Wh/kg and 440 Wh/L compared with 100 Wh/kg and 220 Wh/L for the premium conventional primary cells. The Li/SO_2 cell is particularly noted for its capability to handle high current or high power requirements and for its excellent low-temperature performance.

11.5.1 Chemistry

The Li/SO_2 cell uses lithium as the anode and a porous carbon cathode electrode with sulfur dioxide as the active cathode material. The cell reaction mechanism is

$$2\ Li + 2\ SO_2 \rightarrow Li_2S_2O_4 \quad \text{(lithium dithionite)}$$

As lithium reacts readily with water, a nonaqueous electrolyte, consisting of sulfur dioxide and an organic solvent, typically acetonitrile, with dissolved lithium bromide, is used. The specific conductivity of this electrolyte is relatively high and decreases only moderately with decreasing temperature (Fig. 11.7), thus providing a basis for good

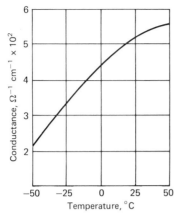

FIG. 11.7 Conductance of acetonitrile–lithium bromide–sulfur dioxide electrolyte (70% SO₂). (*Courtesy of Duracell, Inc.*)

high-rate and low-temperature performance. About 70% of the weight of the electrolyte/depolarizer is SO_2. The internal cell pressure, in an undischarged cell, due to the vapor pressure of the liquid SO_2, is 3 to 4 × 10⁵ Pa at 20°C. The pressure at various temperatures is shown in Fig. 11.8. The mechanical features of the cell are designed to safely contain this pressure without leaking and vent the electrolyte if excessively high temperatures and resulting high internal pressures are encountered.

During discharge, the SO_2 is used and the cell pressure is reduced. The discharge is generally terminated by the full use of available lithium, in designs where the lithium is the stoichiometrically limiting

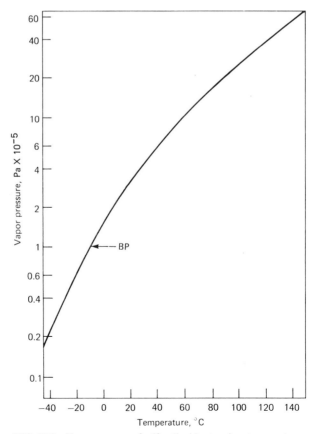

FIG. 11.8 Vapor pressure of sulfur dioxide at various temperatures.

electrode ("balanced cell"), or by blocking of the cathode by precipitation of the discharge product. The good shelf life of the Li/SO$_2$ cell results from the protective lithium dithionite film on the anode formed by the initial reaction of lithium and SO$_2$; it prevents further reaction and loss of capacity during storage.

Most Li/SO$_2$ cells are now fabricated in a "balanced" construction where the lithium-sulfur dioxide stoichiometric ratio is in the range of Li:SO$_2$ = 1:0.9 to 1.1. With the earlier designs, where the ratio was in the order of Li:SO$_2$ = 1.5:1, high temperatures, cell venting, or rupture and fires due to an exothermic reaction between lithium (which deposits on the cathode during reversal) and acetonitrile, in the absence of SO$_2$ could occur on deep or forced discharge. Cyanides and methane can also be generated through this reaction. In the balanced cell, the anode becomes polarized almost at the same time as the cathode with little or no opportunity for lithium deposition. The conditions for the hazardous reaction are eliminated since there is no reservoir of reactable lithium on the cathode and some protective SO$_2$ remains in the electrolyte.[13] A higher negative cell voltage, in reversal, of the balanced cell is also beneficial for diode protection which is used in some designs to bypass the current through the cell and minimize the adverse effects of reversal.

11.5.2 Construction

The Li/SO$_2$ cell is typically fabricated in a cylindrical structure as shown in Fig. 11.9. A jelly-roll construction is used, made by spirally winding rectangular strips of lithium foil, a microporous polypropylene separator, the cathode electrode (a Teflon–carbon black mix pressed on a supporting expanded aluminum screen), and a second separator layer. This design provides the high surface area and low cell resistance to obtain the

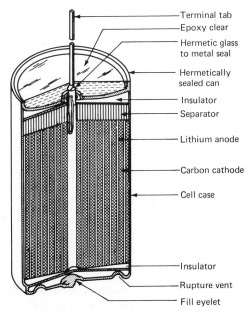

FIG. 11.9   Lithium/sulfur dioxide cell. (*Courtesy of Duracell, Inc.*)

high current and low-temperature performance. The roll is inserted in a nickel-plated steel can, electrical connections are made, the top is sealed in place, and the electrolyte/depolarizer is added. The safety rupture vent releases when the internal cell pressure reaches excessive levels, caused by inadvertent abusive use such as overheating or short circuit, and prevents cell rupture or explosion. The vent activates at approximately 90°C, well above the upper temperature limit for operation and storage, safely relieving the excess pressure and preventing possible cell rupture.

11.5.3 Performance

Voltage The nominal voltage of the Li/SO_2 cell is 3 V. The specific voltage on discharge is dependent on the discharge rate, discharge temperature, and state of charge; typical working voltages range between 2.9 and 2.7 V (see Figs. 11.10, 11.12, and 11.14). The end or cutoff voltage, the voltage by which most of the cell capacity has been exhausted, is 2 V.

Discharge Typical discharge curves for the Li/SO_2 cell at 20°C are given in Fig. 11.10. The high cell voltages and the flat discharge profile are characteristic of the Li/SO_2 cell. Another unique feature is the ability of the Li/SO_2 cell to be efficiently discharged over a wide range of current or power levels, from high-rate short-term or pulse loads to low-drain continuous discharges for periods of 5 y or longer. At least 90% of the cell's rated capacity may be expected on the long-term discharges.

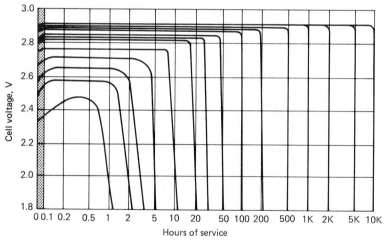

FIG. 11.10 Typical discharge characteristics of Li/SO_2 cell at various discharge loads at 20°C. (*Courtesy of Duracell, Inc.*)

The high rate capability of the Li/SO_2 cell is illustrated in Fig. 11.11, which compares the performance of various D size primary cells (see Fig. 3.23 for comparison with sealed secondary batteries). The Li/SO_2 cell maintains a high capacity almost to the 1-h rate, whereas the zinc and magnesium primary cell performance begins to drop off significantly at the 20- to 50-h rate. The Li/SO_2 cells are capable of higher-rate discharges on pulse loads. For example, a D cell designed in a high-rate construction

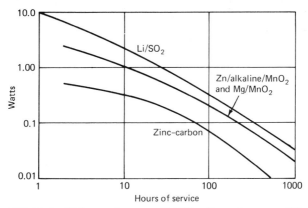

FIG. 11.11   High-rate advantage of Li/SO$_2$ cell—performance of D size cells at 20°C. (*Courtesy of Duracell, Inc.*)

can deliver pulse loads as high as 30 A at a voltage above 2 V. Extended discharges, however, at rates above the 5-h rate may cause overheating. The actual heat rise depends on the battery design, type of discharge, temperature, and voltage, etc. As discussed in Sec. 11.4, the design and use of the battery should be controlled to avoid overheating.

Effect of Temperature   The Li/SO$_2$ cell is noted for its ability to perform over a wide temperature range, from −55 to 70°C. Discharge curves for a Li/SO$_2$ cell at various temperatures are shown in Fig. 11.12. Significant, again, are the flat discharge curves over the wide temperature range, the good voltage regulation, and the high percentage of the 20°C performance available at the temperature extremes. As with all battery systems, the relative performance of the Li/SO$_2$ battery is dependent on the rate of

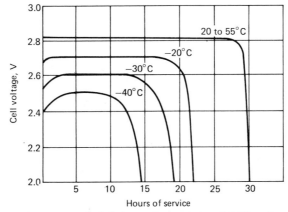

FIG. 11.12   Typical discharge characteristics of Li/SO$_2$ cell at various temperatures, C/30 discharge rate. (*Courtesy of Duracell, Inc.*)

discharge. In Fig. 11.13, the discharge performance is plotted as a function of load and cell temperature (see also Figs. 4.7 and 4.8).

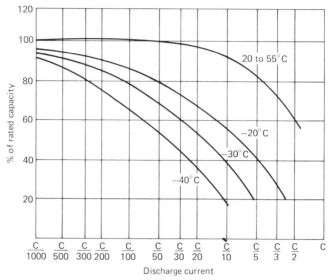

FIG. 11.13 Li/SO$_2$ cells—performance as a function of discharge temperature and load. (*Courtesy of Duracell, Inc.*)

Internal Resistance and Discharge Voltage The Li/SO$_2$ cell has a relatively low internal resistance (about one-tenth that of conventional primary batteries) and good voltage regulation over a wide range of discharge load and temperature. The average voltage of the discharge of the Li/SO$_2$ cell (to an end voltage of 2 V) at various discharge rates and temperatures is plotted in Fig. 11.14.

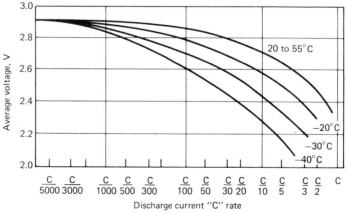

FIG. 11.14 Li/SO$_2$ cells—average discharge voltage. (*Courtesy of Duracell, Inc.*)

Service Life   The capacity or service life of the Li/SO_2 cell at various discharge rates and temperatures is given in Fig. 11.15. The data are normalized for a 1-kg or 1-L size cell and presented in terms of hours of service at various discharge rates. The linear shape of these curves, except for the falloff at high current levels and low temperatures, is again indicative of the capability of the Li/SO_2 battery to be efficiently discharged at these extreme conditions. These data can be used in several ways to calculate the approximate performance of a given cell or to select a Li/SO_2 cell of a suitable size for a particular application, recognizing that the energy density of the larger-size cells is higher than that of the smaller ones.

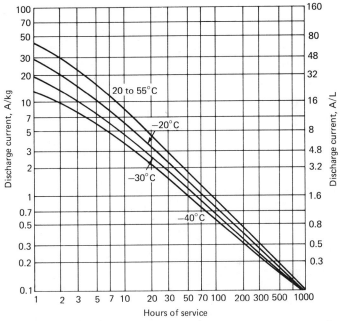

FIG. 11.15   Service life of Li/SO_2 cells (2.0-V end voltage). (*Courtesy of Duracell, Inc.*)

The service life of a cell at a given current load can be estimated by dividing the current (in amperes) by the weight or volume of the cell. This value is located on the ordinate, and the service life, at a specific current and temperature, is read off the abscissa.

The weight or volume of a cell needed to deliver a required number of hours of service at a specified current load can be estimated by locating a point on the curve corresponding to the required service hours and discharge temperature. The cell weight or volume is calculated by dividing the value of the specified current (in amperes) by the value of amperes per kilogram or amperes per liter obtained from the ordinate.

Shelf Life   The Li/SO_2 battery is noted for its excellent storage characteristics, even at temperatures as high as 70°C. Shelf life data for the Li/SO_2 cell at various temperatures are presented at Fig. 11.16. Most primary batteries lose capacity while idle or on

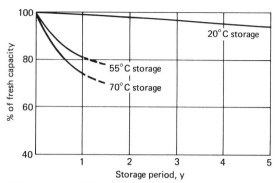

FIG. 11.16 Shelf life characteristics of Li/SO₂ cells. (*Courtesy of Duracell, Inc.*)

stand due to anode corrosion, side chemical reactions, or moisture loss. With the exception of the magnesium cell, most of the conventional primary cells cannot withstand temperatures in excess of 50°C and should be refrigerated if stored for long periods. The Li/SO_2 cell, however, is hermetically sealed and protected during storage by the formation of a film on the anode surface. Capacity losses during stand are minimal. Five-year storage data at 20°C show a capacity loss of less than 10%; a storage capability of 10 y is projected. At elevated temperatures, after 1-y storage, the capacity loss is less than 35% at 70°C and 20% at 55°C, although the specific performance also is dependent on the conditions of discharge. The rate of capacity loss, as shown in Fig. 11.16, is higher initially but decreases as the storage period is extended. A generalized relationship of capacity loss with storage temperature is plotted in Figs. 3.22 and 4.9.

Voltage Delay After extended long-term storage at elevated temperatures, the Li/SO_2 cell may exhibit a delay in reaching its operating voltage (above 2.0. V) when placed on discharge, especially at high current loads and low temperatures. This start-up or voltage delay is caused by the protective film formed on the lithium anode, the characteristic responsible for the excellent shelf life of the cell. The specific delay time for a cell cannot be predicted with accuracy for it depends on the history of the cell, the specific cell design and components, the storage time and temperature, discharge load, and temperature, etc. Typically, the voltage delay is minimal or nonexistent for discharges at moderate to low rates at temperatures above −20° C. No delay is evident on discharge at 20°C even after storage at 70°C for 1 y. On discharge at −30°C, the delay time is less than 200 ms after 8 weeks of storage at 70°C on discharges lower than the 40-h rate. At higher rates, the voltage delay increases with increasing storage temperature and time. At the 2-h discharge rate, for example, the maximum start-up time is about 80 s after 8 weeks of storage at 70°C; it is 7 s after 2 weeks of storage.[14] The start-up voltage delay can be eliminated by preconditioning with a short discharge at a higher rate until the operating voltage is reached, because the delay will return only after another extended storage period.

11.5.4 Cell Types and Sizes

The Li/SO_2 cells are manufactured in a number of cylindrical cell sizes, ranging in capacity from 0.450 to 30 Ah. Some of the cells are manufactured in standard ANSI cell sizes in dimensions of popular conventional zinc primary cells. While these single cells may be physically interchangeable, they are not electrically interchangeable because of the higher cell voltage of the lithium cell (3.0 V for lithium, 1.5 V for the conventional zinc cells). Table 11.9 lists some of the sizes and rated capacities of Li/SO_2 cells currently available.

11.5.5 Use and Handling of Li/SO_2 Cells and Batteries—Safety Considerations

The Li/SO_2 cell is designed as a high-performance system and is capable of delivering a high capacity at high discharge rates. The cell should not be physically or electrically abused, safety features should not be bypassed, and manufacturers' instructions should be followed.

The abusive conditions that could adversely affect the performance of the Li/SO_2 cell and result in cell venting, rupture, explosion, or fire—and corrective measures— are discussed in Sec. 11.4.

The Li/SO_2 cell is pressurized and contains materials that are toxic or flammable. Properly designed cells are hermetically sealed so that there will be no leakage or

TABLE 11.9 Lithium/Sulfur Dioxide Cylindrical Cells

Cell size	Rated capacity, Ah*	Dimensions			Weight, g	Energy density†	
		Diam., mm	Ht., mm	Vol., cm³		Wh/kg	Wh/L
AA Diam.	0.45	14	24	3.7	6.5	193	341
	1.05 (AA size)	14	50	7.6	14.0	210	386
A Diam.	0.55	16.3	24	5.0	9.0	170	308
	0.90	16.3	34.5	7.2	12.0	210	350
	1.20 (A size)	16.3	44.5	9.3	15.0	224	361
	1.75	16.3	57.0	11.8	19.0	258	408
	0.65	24.0	18.5	8.3	13.0	140	218
	3.0	24.0	53.0	24.0	37.0	227	350
C Diam.	1.0	25.7	17.2	8.9	17.0	162	314
	3.5 (C size)	25.7	50	26	41	238	378
	4.4	25.7	60	31	50	245	392
	9.5	25.7	130	68	115	230	395
	5.4	29.1	60	40	62	244	378
D Diam.	8.0 (D size)	33.8	60	54	85	263	417
	16.5	33.8	120	108	165	280	428
	8.5	38.7	50	59	95	249	406
	21.0	38.7	114	134	207	283	440
	9.5	41.6	51	69	105	252	384
	25.0	41.6	118	160	230	302	437
	30.0	41.6	141	191	280	300	440

*Rated capacity at the C/30 discharge rate. Some of the cells listed are available in high rate designs giving better service at the higher discharge rates, but with lower capacity at the C/30 rate.
†Based on 2.8-V average voltage.

outgassing, and they are equipped with safety vents which release if the cells reach excessively high temperatures and pressures, thus preventing explosive damage.

The Li/SO_2 cell can deliver very high currents. Because high internal temperatures can develop from continuous high current drain and short circuit, cells should be protected by electrical fusing and thermal cutoffs. Charging of Li/SO_2 cells may result in venting, rupture, or even explosion and should not be attempted. Cells or groups of cells connected in parallel should be diode-protected to prevent one group from charging another. The balanced Li/SO_2 cell is designed to handle forced discharges or cell reversal and will perform safely within the specified bounds, but design limits should not be exceeded in any application.[13]

Proper battery design, using the Li/SO_2 cell, should follow these guidelines:

Use electrical fusing and/or current limiting devices to prevent high currents or short-circuit

Protect with diodes if cells are paralleled or connected to a possible charging source

Minimize heat buildup by adequate heat dissipation and protect with thermal cutoff devices

Do not restrict cell vents

Do not use flammable materials in construction of batteries

Allow for release of vented gases (or use a battery case that can safely contain them)

Currently, special procedures govern the transportation, shipment, and disposal of Li/SO_2 batteries containing more than 0.5 g of lithium, equivalent to 1.0 Ah in capacity.[10–12] Procedures for the use, storage, and handling of these batteries also have been recommended. The latest issue of these regulations should be consulted for the most recent procedures.

11.5.6 Applications

The desirable characteristics of the Li/SO_2 battery and its ability to deliver a high energy output and operate over a wide range of temperature, discharge load, and storage conditions have resulted in its use in an increasing number of applications. Its light weight and small size, high drain capability, and low-temperature operation have opened up applications for primary batteries that, heretofore, were beyond the capability of primary battery systems (see Part 1, Sec. 3.5).

Major applications for the Li/SO_2 battery are in military equipment, such as radio transceivers and portable surveillance devices, taking advantage of its light weight and wide temperature operation (Fig. 3.25). Other military applications, such as sonobuoys and munitions, have long shelf life requirements, and the active Li/SO_2 primary battery can replace the reserve batteries used earlier. Similar industrial applications are developing, particularly to replace secondary batteries and eliminate the need for recharging. Consumer applications have been limited, to date, because of restrictions in shipment and transportation. The Li/SO_2 battery does have a significant advantage in applications where the battery is required for heavy current loads, low-temperature operation, and long shelf life. With increasing production, user experience, and cost reduction, the Li/SO_2 battery should become cost-competitive with conventional battery systems and receive wider acceptability in military, industrial, and consumer applications.[15]

11.6 LITHIUM/THIONYL CHLORIDE (Li/SOCl₂) CELLS

The lithium/thionyl chloride (Li/SOCl$_2$) cell has one of the highest energy densities of the practical battery systems; energy densities range up to 500 Wh/kg and 900 Wh/L, the highest values being achieved with the large, high-capacity low-rate cells. Figures 4.7, 4.8, and 11.3 illustrate some of the advantageous characteristics of the Li/SOCl$_2$ cell.

Li/SOCl$_2$ cells have been fabricated in a variety of sizes and designs, ranging from small cells, with capacities as low as 500 mAh, to large 20,000-Ah prismatic cells. The thionyl chloride system originally suffered from a chemical instability that led to an explosion hazard, especially on high-rate discharges and overdischarge, and a voltage delay that was most evident on low-temperature discharges after high-temperature storage. Low-rate cells have been used successfully for several years, and recently significant progress is being made to attain safe and reliable performance with high-rate and high-capacity cells.[16]

11.6.1 Chemistry

The Li/SOCl$_2$ cell consists of a lithium anode, a carbon cathode, and a nonaqueous SOCl$_2$:LiAlCl$_4$ electrolyte. Thionyl chloride is both the electrolyte and the active cathode material. The generally accepted overall reaction mechanism is[1]

$$4 \text{ Li} + 2 \text{ SOCl}_2 \rightarrow 4 \text{ LiCl} + \text{S} + \text{SO}_2$$

The sulfur and sulfur dioxide are soluble in the excess thionyl chloride electrolyte, and there is a moderate buildup of pressure due to the generation of sulfur dioxide during the discharge. The lithium chloride, however, is not soluble and precipitates within the porous carbon black cathode as it is formed. In some cell designs and discharge conditions, this blocking of the cathode is the factor that limits the cell's service or capacity. Formation of sulfur as a discharge product can also present a problem because of a possible reaction with lithium which can result in thermal runaway.

The lithium anode is protected from reacting with the thionyl chloride electrolyte during stand by the formation of a protective LiCl film on the anode as soon as it contacts the electrolyte. This passivating film, while contributing to the excellent shelf life of the cell, can cause a voltage delay at the start of a discharge, particularly on low-temperature discharges after long stands at elevated temperatures.

The low freezing point of the electrolyte (below $-110°C$) and its relatively high boiling point (78.8°C) enable the cell to operate over a wide range of temperature. The electrical conductivity of the electrolyte decreases only slightly with decreasing temperature. Some of the components of the Li/SOCl$_2$ system are toxic and inflammable; thus exposure to open or vented cells or cell components should be avoided.

11.6.2 Bobbin Type Cylindrical Cells

An early application of the Li/SOCl$_2$ cell was one of the battery power sources for the cardiac pacemaker.[17] In this application, high capacity and long life were important in order to meet a service life of at least 4 y. The AA size (50 mm high $\times$ 14.3 mm in diameter) bobbin cell used in this application had an energy density in excess of 330 Wh/kg and 800 Wh/L (Fig. 11.17).

Li/SOCl$_2$ bobbin cells are manufactured in a cylindrical configuration in sizes

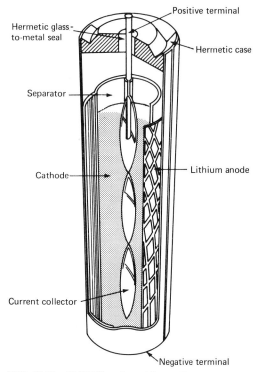

FIG. 11.17 Li/SOCl₂ cell used in cardiac pacemaker. (*From Ref. 17.*)

conforming to ANSI standards. These cells are designed for low- to moderate-rate discharge and are not discharged at rates higher than the C/50 rate. They have a high energy density. For example, the C size cell delivers 5 Ah at 3.4 V compared with 2 Ah and 4.5 to 5 Ah at 1.5 V for the conventional zinc-carbon and zinc-alkaline cells, respectively (Tables 4.5 and 11.10).

Construction Figure 11.18 shows the constructional features of the cylindrical Li/SOCl₂ cell, which is built as a bobbin-type cell. The anode is made of lithium foil which is swagged against the inner wall of a stainless or nickel-plated steel can; the separator is made of nonwoven glass; the cylindrical, highly porous cathode, which takes up most of the cell volume, is made of Teflon-bonded carbon black. The cell is hermetically welded and a glass-to-metal seal is used for the positive terminal lead. These low-rate cells may or may not be provided with a vent, depending on the manufacturer and cell design.

Performance The open-circuit *voltage* of the Li/SOCl₂ cell is 3.66 V; typical operating voltages range between 3.3 and 3.5 V with an end voltage of 3.0 V. Typical *discharge curves* for the Li/SOCl₂ cell at various temperatures are shown in Fig. 11.19 for the D size cell. The Li/SOCl₂ cell discharges are characterized by a flat profile with good performance over a wide range of temperature and low- to moderate-rate discharges. The relationship of capacity with *temperature* is given in Fig. 11.20 showing the

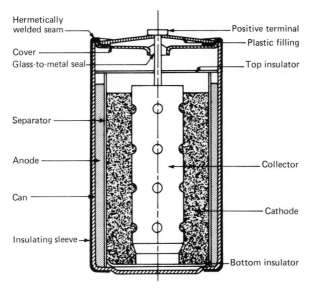

FIG. 11.18   Bobbin-type Li/SOCl₂ cell—cross section. (*From Ref. 18a.*)

performance from −40 to 80°C. The Li/SOCl₂ cell is capable of performance at unusually high temperatures. For example, Fig. 11.21 shows the capacity of an AA cell (rated at 1.7 Ah) at various discharge loads at 120°C. At 145°C (Fig. 11.22), the cells deliver most of their capacity at high rates and up to 70% at low discharge rates (20 days of discharge),[18] but a temperature of 150°C should not be exceeded.

Figure 11.23 shows the behavior of AA cells on continuous low-rate discharge at 45°C. Cells discharged at 0.1 mA show no capacity loss over the 23-month discharge period, delivering close to the rated 1.7-Ah capacity. The discharge curve is very flat at these low-current drains.

The capacity or *service life* of the bobbin-type Li/SOCl₂ cell, normalized for a 1-kg and 1-L size cell, at various discharge temperatures and loads, is summarized in Fig. 11.24.

The long *shelf life* of the Li/SOCl₂ cell is due to the stability of the lithium anode in contact with the electrolyte, as a result of a protective LiCl film that forms on the lithium surface. The long shelf life can also be attributed to the stability of other cell components; e.g., the can and cover are cathodically protected by the lithium, and the carbon, stainless-steel collector, and glass separator are all inert in the electrolyte. Figure 11.25 shows the loss of capacity after 3 y at 20°C, a loss of about 1 to 2%/y. Storage at 70°C results in a capacity loss of about 30%/y, but the rate of loss decreases with increasing storage time. Cells should preferably be stored in an upright position; storage on the side or upside down will result in a capacity loss.

After storage, the Li/SOCl₂ cell may exhibit a *delay* in reaching its operating voltage caused by the formation of the LiCl film on the lithium surface. The voltage delay becomes more pronounced with a heavier discharge load and lower discharge temperature. Figure 11.26 shows the dependence of the closed-circuit voltage on the discharge current of AA cells after 2 y storage at 25°C. Once the cell discharge is started, the

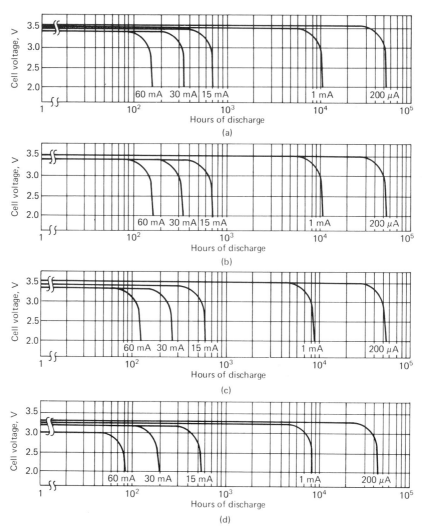

FIG. 11.19   Discharge characteristics of the Li/SOCl₂ cylindrical bobbin cell. (*a*) Discharge curves at +45°C; (*b*) discharge curves at +25°C; (*c*) discharge curves at −10°C; (*d*) discharge curves at −40°C. (*From Ref. 18a.*)

passivation film is dissipated gradually, the internal resistance returns to its normal value, and the plateau voltage is reached. The passivation film may be removed more rapidly by the application of high-current pulses for a short period or, alternatively, by short-circuiting the cells, momentarily, several times.

Cell Sizes   The bobbin-type Li/SOCl₂ cells are manufactured in the standard ANSI cell sizes, as well as in special cell and battery configurations [e.g., for printed circuit board mounting]. Table 11.10 summarizes the properties of some of the cells. Although

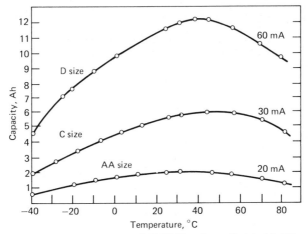

FIG. 11.20 Performance of Li/SOCl₂ cells—cylindrical bobbin cells at various temperatures. (*From Ref. 18a.*)

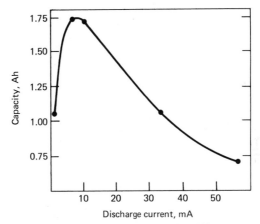

FIG. 11.21 Performance of Li/SOCl₂ bobbin-type cylindrical AA size cell at 120°C. (*From Ref. 18a.*)

these cells may be physically interchangeable, they are not electrically interchangeable with the conventional zinc cells because of their higher voltage.

The characteristics of an alternate design of a low-rate cylindrical Li/SOCl₂ cell, slightly smaller than an AA cell, are also listed in Table 11.10. This cell uses a design with a central anode and a peripheral carbon cathode in contact with the cell case. The cell is closed by compression-crimping a fluorinated hydrocarbon polymer seal between the can wall and the cover. A resealable vent is installed under the top cap. Cells with this seal exhibited no leakage after temperature cycling.[19] The capacity of this cell at various discharge rates and temperatures is summarized in Fig. 11.27. Two of these cells can be used in a battery equivalent in size to the 9-V transistor battery (NEDA size 1604) and give significantly longer service (see Fig. 4.5).

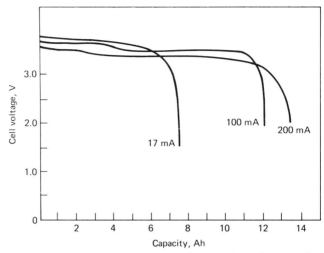

FIG. 11.22 Discharge characteristics of Li/SOCl₂ bobbin-type cylindrical D size cell at 145°C. (*From Ref. 18b.*)

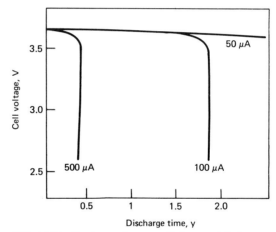

FIG. 11.23 Continuous low-rate discharges, AA size Li/SOCl₂ cylindrical bobbin cell, at 45°C. (*From Ref. 18a.*)

Special Characteristics The bobbin cells are designed to limit the possibility of hazardous operation and to eliminate the need (in some designs) for a safety vent. This is achieved by minimizing the reactive surface area and by increasing the heat dissipation, thus limiting the short-circuit current and a hazardous temperature rise, respectively. These cells also are cathode-limited, a design that was found safer than anode-limited cells for the Li/SOCl₂ system.[20] The cells, reportedly, have withstood short-circuit, forced discharge, and charging under certain conditions with no hazard.[18,19] Cells should not be disposed of in fire or subjected to long-term exposure at temperatures above 180°C because they may explode.

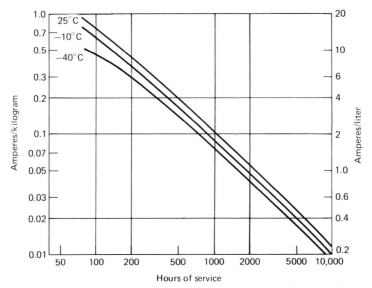

FIG. 11.24 Service life of Li/SOCl₂ bobbin-type cylindrical cells.

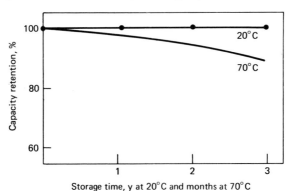

FIG. 11.25 Capacity retention, Li/SOCl₂ bobbin-type cylindrical cells. (*From Ref. 18a.*)

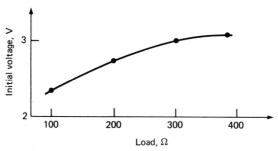

FIG. 11.26 Li/SOCl₂ bobbin-type cylindrical AA size cells, minimum voltage vs. load after 2-y storage at 25°C. (*From Ref. 18a.*)

TABLE 11.10 Characteristics of Cylindrical Bobbin-type Li/SOCl$_2$

Cell designation	Sub* AA	½AA	AA	1/3C	C	1/6D†	D
Rated capacity, Ah, at C/1000 rate	1.25	0.75	1.6	1.8	5.1	1.0*	10.2
Dimensions: Diameter, mm Height, mm Volume, cm^3	12.1 41.6 4.8	14.7 25.5 4.3	14.7 51.0 8.0	26.0 19.0 10	26.0 49.8 26	32.9 10.0 8.2	32.9 61.3 51
Weight, g	10	10	19	21	52	26	100
Maximum current, mA for continuous use:		15	42	30	90	0.7	125
Energy density Wh/kg Wh/cm^3	425 885	250 600	280 680	290 620	330 665	130 415	340 675

*Data from Ref. 19.
†For memory backup, C/10,000 rate.
SOURCE: Tadiran Israel Electronics Industries, Ltd., Tel Aviv, Israel (except for sub AA cell).

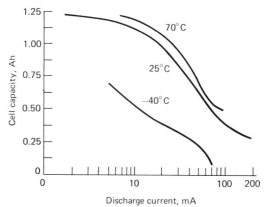

FIG. 11.27 Energy density of the Li/SOCl$_2$, 1.2-cm size, cell at various discharge loads and temperatures. (*From Ref. 19.*)

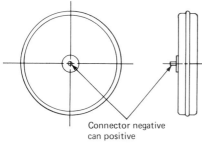

FIG. 11.28 Disk-type Li/SOCl$_2$ cell. (*Courtesy of Altus Corp.*)

11.6.3 Flat or Disk-Type Li/SOCl$_2$ Cells

The Li/SOCl$_2$ system is also manufactured in a flat or disk-shaped cell configuration with a moderate to high discharge rate capability. These batteries are hermetically sealed and incorporate a number of features to safely handle abusive conditions, such as short circuit, reversal, and overheating, within design limits.[21]

The cell consists, as shown in Fig. 11.28, of a single or multiple assembly of disk-shaped lithium anodes, separators, and carbon cathodes sealed in a stainless-steel case containing a ceramic feed-through for the anode and insulation between the positive and negative terminals of the cell.[22]

The cells are manufactured in small and large (43.2-cm diameter) sizes. The characteristics of these cells are summarized in Table 11.11. Discharge curves are shown in Figs. 11.29 and 11.30 for the small and large cells, respectively. Typically, the cells have a high energy density, flat discharge profiles, and the capability of performance over the temperature range of −40 to 70°C. On storage, they can retain 90% of their capacity after storage of 5 y at 20°C, 6 months at 45°C, and 1 month at 70°C.

The cell design features:

Short-circuit protection: Structure of interconnects fuse at high currents providing an open circuit.

Reverse-voltage chemical switch: Upon cell reversal, it allows cell to endure 100% capacity reversal, up to 10-h rate, without venting or pressure increase.

TABLE 11.11   Characteristics of Disk-Type Li/SOCl$_2$ Cells

Capacity, Ah	Dimensions		Weight, g	Maximum continuous current, A	Energy density	
	Diameter, cm	Height, cm			Wh/kg	Wh/L
			Small cells			
0.22	2.24	0.33	5	0.035	150	580
1.40	3.18	0.64	20	0.170	240	935
5.50	6.35	0.90	75	0.7	250	660
18.0	10.16	0.90	225	2.0	275	850
			Large cells			
500	43.2	1.27	7,270	7	240	915
1400	43.2	3.5	1,600	16	350	930
2000	43.2	5.1	17,700	25	385	910
8000	43.2	18.7	56,800	40	475	990

SOURCE: Courtesy Altus Corp., San Jose, Calif.

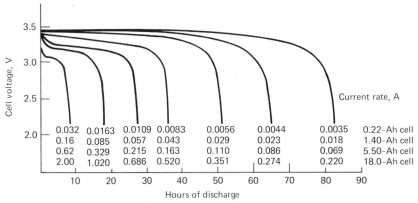

FIG. 11.29   Performance characteristics of disk-type, low-capacity Li/SOCl$_2$ cells. Initial performance capacity to 2.5 V. (*Courtesy of Altus Corp.*)

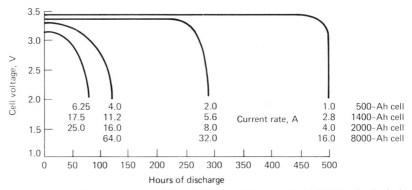

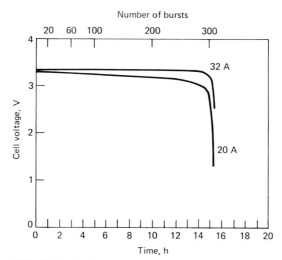

FIG. 11.30 Performance characteristics of disk-type, high-capacity Li/SOCl₂ cells. Typical performance at 0 to 25°C temperature range. (*Courtesy of Altus Corp.*)

Antipassivation (precoat lithium anode): Reduces voltage delay by retarding growth of LiCl film; large cells stored for 2 y reach operating voltage within 20 s.

Self-venting: Ceramic seal is designed to vent cell at predetermined pressures.[21]

Another flat cell design, consisting of a multiassembly of disk-shaped electrodes with glass separators stacked in parallel, has been developed to handle high-rate military applications. The cell, 7.6 cm in diameter × 1.3 cm in height, has a capacity of about 20 Ah. Its performance at 25°C is illustrated in Fig. 11.31; the 300 intermittent pulses at 3.2 and 20 A correspond to a capacity of 21 Ah. The cell impedance is 1.7 mΩ, and short-circuit current is about 1500 A.[23]

FIG. 11.31 Performance characteristics of the flat Li/SOCl₂ cell at 25°C. Cell capacity = 211 Ah. (*From Ref. 23.*)

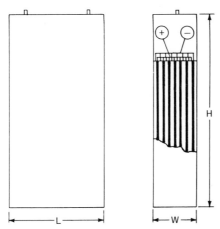

FIG. 11.32   Schematic presentation of the rectangular Li/SOCl$_2$ cell construction. (*From Ref. 24.*)

These cells are used mainly in military applications as multicell batteries. The manufacturer's data should be consulted to obtain the most recent and specific design and performance data and conditions to ensure safe use.

11.6.4 Large Prismatic Li/SOCl$_2$ Cells

The large high-capacity Li/SOCl$_2$ cells are being developed mainly as a standby power source for those military applications requiring a power source that is independent of commercial power and the need for recharging.[24-26] They generally are built in a prismatic configuration as shown schematically in Fig. 11.32. The lithium anodes and Teflon-bonded carbon electrodes are made as rectangular plates with a supporting grid structure, separated by nonwoven glass separators and housed in a hermetically sealed stainless-steel container. The terminals are brought to the outside by glass-to-metal feed-throughs or by a single feed-through isolated from the positive steel case. The cells are filled through an electrolyte filling tube.

The characteristics of several prismatic cells are summarized in Table 11.12. These cells have a very high energy density (950 Wh/L and 480 Wh/kg). They are generally discharged continuously at relatively low rates (200- to 300-h rate), but are capable of heavier discharge loads. A typical discharge curve is shown in Fig. 11.33. The voltage profile is flat, and the cell operates just slightly above the ambient at this discharge load. During the course of the discharge there is a slight buildup of pressure, reaching a pressure of about 2×10^5 Pa at the end of the discharge. A higher-rate pulse discharge is shown in Fig. 11.34. The 2000-Ah cell was discharged continuously at a 5-A load, with 40-A pulses, 16 s in duration, superimposed once every day. A steady discharge voltage was obtained throughout most of the discharge with only a slight reduction in voltage during the pulse. The cells are capable of performance from -40 to 50°C; shelf life losses are estimated at 1%/y. Extensive design evaluation and testing are now being conducted to determine the performance characteristics, reliability, and safety of the cells under use conditions.[25,26]

TABLE 11.12   Characteristics of Large Prismatic Li/SOCl$_2$ Cells

| Capacity, Ah | Dimensions | | | Weight, kg | Energy denisty | |
	Height, cm	Length, cm	Width, cm		Wh/kg	Wh/L
2,000*	44.8	31.6	5.3	15	460	910
10,000*	44.8	31.6	25.5	71	480	950
16,500†	38.7	38.7	38.7	113	495	970

*GTE Sylvania, Needham, Mass.
†Honeywell Power Sources Center, Horsham, Pa.

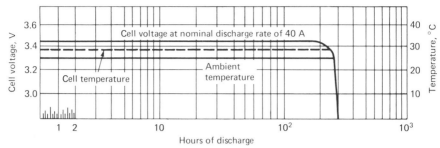

FIG. 11.33 Discharge curves, 10,000-Ah Li/SOCl$_2$ cell. (*Courtesy of GTE Sylvania, Inc.*)

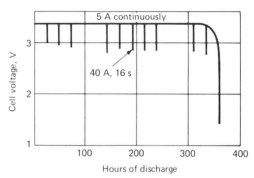

FIG. 11.34 Discharge of the high-capacity 2000-Ah Li/SOCl$_2$ cell. (*From Ref. 24.*)

11.6.5 Applications

The applications of the Li/SOCl$_2$ system take advantage of the high energy density and long shelf life of this battery system. The low-drain cylindrical cells are used as a power source for CMOS memories, instrumentation, and small electronics. Wide application in consumer-oriented applications is still limited because of the relatively high cost and concern with the safety and handling of these types of lithium batteries.

The larger Li/SOCl$_2$ cells are used mainly in military applications where high energy density is needed to fulfill important mission requirements. A significant application for the large 10,000-Ah cells is as the standby power source in missile silos in the event of loss of commercial or other power. This lithium battery is always in a state of readiness and does not require any recharging.[22,26]

11.7 LITHIUM/OXYCHLORIDE CELLS

The lithium/sulfuryl chloride (Li/SO$_2$Cl$_2$) cell is, in addition to the lithium/thionyl chloride cell, the other oxychloride that has been considered for primary lithium batteries. The Li/SO$_2$Cl$_2$ cell has two potential advantages over the Li/SOCl$_2$ cell:

1. A higher energy density as a result of a higher operating voltage (3.9-V open-circuit voltage) as shown in Fig. 11.4 and less solid product formation (which may block the cathode) during the discharge.

2. Inherently greater safety because sulfur, which is a possible cause of thermal runaway in the $Li/SOCl_2$ cell, is not formed during the discharge of the Li/SO_2Cl_2 cell.

Another type of lithium/oxychloride cell involves the use of halogen additives in both the $SOCl_2$ and SO_2Cl_2 electrolytes. These additives give an increase in the cell potential (3.9 V for the $Li/BrCl$ in the $SOCl_2$ system, 3.95 V for the Li/Cl_2 in the SO_2Cl_2 system), energy densities up to 950 Wh/L and 500 Wh/kg, and safer operation under abusive conditions.

11.7.1 Lithium/Sulfuryl Chloride (Li/SO_2Cl_2) Cells

The Li/SO_2Cl_2 cell is similar to the thionyl chloride cell using a lithium anode, a carbon cathode (which is prepared differently than the thionyl chloride cell cathode), and the electrolyte/depolarizer of $LiAlCl_4$ in SO_2Cl_2. The discharge mechanism is

Anode: $Li \rightarrow Li^+ + e$

Cathode: $SO_2Cl_2 + 2e \rightarrow 2Cl^- + SO_2$

Overall: $2Li + SO_2Cl_2 \rightarrow 2LiCl + SO_2$

The open-circuit potential is 3.909 V.

The Li/SO_2Cl_2 system is still in the developmental stage and serious investigation of this system has been started only recently.[1] Optimized cathode electrode formulations have to be developed since the Teflon-bonded Shawinigan carbon black electrodes used in the thionyl chloride cells do not perform satisfactorily. Initial work on optimized electrodes is summarized in Fig. 11.35. It compares the performance of Li/SO_2Cl_2 cells made with improved cathodes with $Li/SOCl_2$ cell performance and shows the potential advantage of the sulfuryl chloride system;[7] but optimization for other discharge conditions, storability, etc. remain to be accomplished. Initial indications are that the protective film formed on the surface of lithium stored in the SO_2Cl_2 electrolyte is less compact than the film formed in the $SOCl_2$ electrolyte and, hence, self-discharge in the Li/SO_2Cl_2 cells may proceed at a higher rate than with the $Li/SOCl_2$ cells.[27] Future work in these areas, as well as cell and battery design for specific applications, will determine the acceptability and deployment of this system.

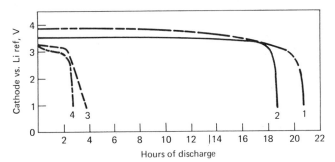

Legend: 1 Cathodic reactant $-$ SO_2Cl_2, I = 5 mA/cm²
         2 Cathodic reactant $-$ $SOCl_2$, I = 5 mA/cm²
         3 Cathodic reactant $-$ SO_2Cl_2, I = 20 mA/cm²
         4 Cathodic reactant $-$ $SOCl_2$, I = 20 mA/cm²

FIG. 11.35 Comparison of cathodic discharge curves for SO_2Cl_2 and $SOCl_2$. *(From Ref. 7.)*

TABLE 11.13 Typical Halogen Additive

	BrCl in $SOCl_2$				Li/Cl in SO_2Cl_2		
Type designation	½AA	AA	C	D	⅔A	D	DD
Rated capacity, Ah (100-h rate)	0.8	2	7	14	1.8	12	30
Dimensions Diameter, mm Height, mm Volume, cm³	14.0 24.6 3.8	14.3 50.0 7.8	26.0 49.8 26	33.5 60 51	16 34.6 7.0	33.5 60 51	33.5 112 100
Weight, g	10	17	55	110	16	110	220
Current capability, maximum, mA	20	100	500	1000	—	—	—
Energy density Wh/kg Wh/L	200 775	430 900	460 950	470 960	415 950	405 870	475 1050

SOURCE: Courtesy Electrochem Industries, Clarence, NY 14031.

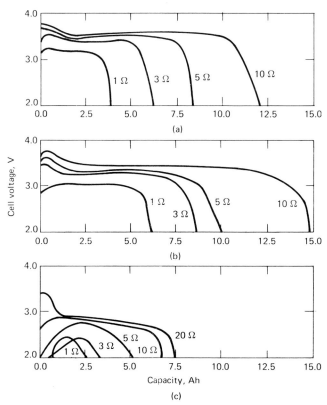

FIG. 11.36 Discharge characteristics of D size Li/BrCl, $SOCl_2$ cells:
(a) 72° ± 2°C; (b) 24° ± 3°C; (c) −40° ± 2°C. (*From Ref. 28.*)

11.7.2 Halogen-Additive Lithium/Oxychloride Cells

Two types of halogen-additive lithium/oxychloride cells have been developed.

Li/SOCl₂ System with BrCl Additive   This cell has an open-circuit voltage of 3.9 V and an energy density of up to 1000 Wh/L on low-rate discharges at 20°C. Table 11.13 summarizes the characteristics of cylindrical cells that are available with this chemistry. The cells are fabricated by winding the lithium anode, the carbon cathode, and two layers of a separator of nonwoven glass into a cylindrical roll and packaging them in a hermetically sealed can with a glass-to-metal feed-through. The discharge curves for the D size cell at various temperatures and discharge rates are shown in Fig. 11.36. The discharge curves are relatively flat with a working voltage of about 3.5 V. The cells are capable of performance at relatively high rates over the temperature range of −40 to 70°C. The effect of storage at 70 and 20°C is illustrated in Fig. 11.37. There is no significant loss in capacity as a result of this storage, although the current-delivering capability was reduced after storage, as shown by the lower discharge voltage. The addition of BrCl to the depolarizer may also prevent the formation of sulfur as a discharge product, at least in the early stage of the discharge, and minimize the hazards of the Li/SOCl₂ attributable to sulfur or discharge intermediates. The cells show abuse resistance when subjected to the typical abuse tests, such as short circuit, forced discharge, and exposure to high temperatures.[28]

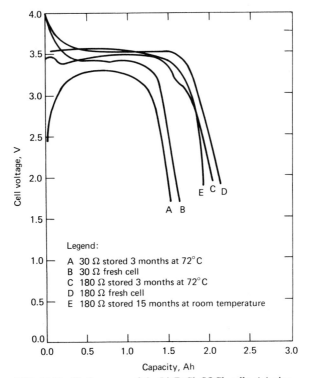

FIG. 11.37   Performance of the Li/BrCl, SOCl₂ cells, AA size, after storage. (*From Ref. 28.*)

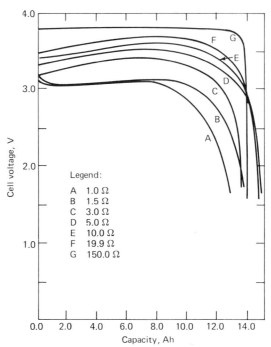

FIG. 11.38 Performance characteristics of the Li/Cl$_2$, SO$_2$Cl$_2$ cell, D size, at 20°C. (*From Ref. 29.*)

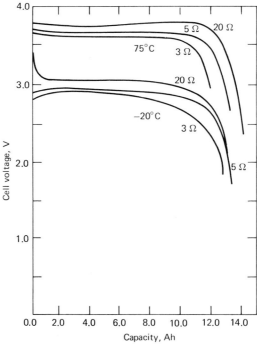

FIG. 11.39 Performance characteristics of the Li/Cl$_2$, SO$_2$Cl$_2$ cell, D size, 75°C and −20°C. (*From Ref. 29.*)

Li/SO₂Cl₂ with Cl₂ Additive This cell has an open-circuit voltage of 3.95 V at 20°C and an energy density of up to 1000 Wh/L even when discharged at rates as high as the C/4 rate. The operating temperature range is -30 to 150°C, although specially designed cells must be employed for temperatures above 90°C. Cells are fabricated in cylindrical sizes, similar to those listed in Table 11.13, using the spirally wound internal construction and hermetically sealed stainless steel cans. The performance of D size cells at 20, 75, and -20°C is given in Fig. 11.38 and 11.39. No significant loss in capacity is reported after storage at temperatures up to 70°C for periods of 1 month. The cells show abuse resistance similar to the Li/BrCl in SOCl₂ cells when subjected to abuse tests.[29]

11.8 LITHIUM/MANGANESE DIOXIDE (Li/MnO₂) CELLS

The Li/MnO₂ cell was one of the first lithium/solid cathode systems to be used commercially because of its attractive performance at moderate and low discharge rates (energy densities above 200 Wh/kg and 400 Ah/L), high cell voltage (nominal voltage 3 V), and cost-competitiveness with some of the conventional batteries. In flat and cylindrical cell constructions, it is being used in watches, cameras, calculators, and similar applications. Higher-capacity cylindrical and rectangular cells are under development. The performance of a flat-type Li/MnO₂ cell is compared with similar-sized mercury and silver oxide cells, and a cylindrical cell is compared with comparable zinc cells, in Fig. 11.1. The higher energy output of the Li/MnO₂ cell is evident. (Other comparative data are presented in Sec. 11.1 and Part 2, Chap. 4.) The Li/MnO₂ cell is also noted for its good performance over a wide temperature range and storability even at elevated temperatures.

11.8.1 Chemistry

The Li/MnO₂ cell uses lithium for the anode, an electrolyte containing lithium perchlorate in a mixed organic solvent of propylene carbonate and 1,2-dimethoxyethane, and a specially prepared heat-treated form of MnO₂ for the active cathode material.

The cell reactions for this system are

Anode reaction: $\quad Li \rightarrow Li^+ + e$

Cathode reaction: $\quad Mn^{IV}O_2 + Li^+ + e \rightarrow Mn^{III}O_2(Li^+)$

Total cell reaction: $\quad Li + Mn^{IV}O_2 \rightarrow Mn^{III}O_2(Li^+)$

Manganese dioxide is reduced from the tetravalent to the trivalent state by lithium; $Mn^{III}O_2(Li^+)$ signifies that the Li^+ ion enters into the MnO₂ crystal lattice.[1,30]

11.8.2 Construction

Two cathode structures have been developed for the Li/MnO₂ cell—a pressed powder cathode for the low-rate cells and a thin, pasted electrode on a supporting grid structure for the higher-rate flat and cylindrical cells. These cell configurations, including the cylindrical cell with its spirally wound electrodes, are shown in Fig. 11.40. A safety vent is incorporated in the larger, high-rate cells and activates under abusive use, such as short circuit, forced discharge, or overheating.

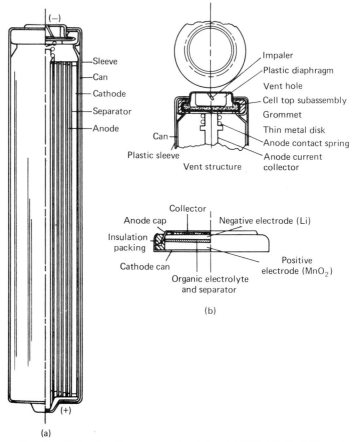

FIG. 11.40 Li/MnO$_2$ cells—cross-sectional view. (*a*) Cylindrical cell (size 2N) and (*b*) flat or coin cell. (*Courtesy of Duracell, Inc.*)

Welded sealed cells are also available, in addition to the crimped-seal design. These cells for memory backup and other low-drain, long-term applications are reported to have a 10-y life.[1]

11.8.3 Performance

Voltage The open-circuit voltage of the Li/MnO$_2$ cell is 3.5 V. The nominal voltage of the cell is 3.0 V, although the initial closed-circuit voltage may be somewhat higher. The end or cutoff voltage, the voltage by which most of the cell capacity has been expended, is 2.0 V.

Discharge Typical discharge curves for the Li/MnO$_2$ cell at 20°C are given in Fig. 11.41 for the different cell constructions. The discharge profile is fairly flat except for

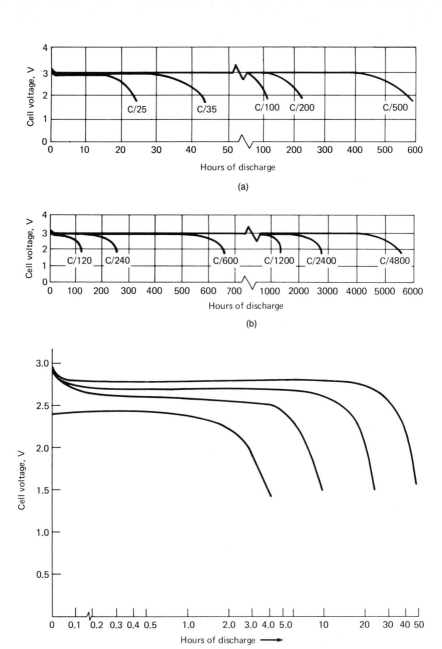

FIG. 11.41 Li/MnO₂ cells, typical discharge curves. (*a*) High-rate flat cell (*b*) low-rate flat cell. (*c*) cylindrical cell. (*Courtesy of Duracell, Inc.*)

the higher-rate discharges. Some applications (e.g., LED watch with backlight) require a high pulse load superimposed on a low background current. The performance of a flat cell under these conditions is shown in Fig. 11.42, illustrating the good voltage regulation even at the high discharge rates. The welded cells can be discharged at very light loads for periods up to 10 y or more, at 20°C, the discharge profile is very flat, at a voltage of about 3 V per cell.[31]

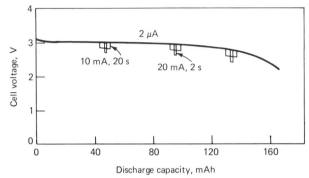

FIG. 11.42   Performance of Li/MnO₂ flat cell under pulse discharge at 20°C, cell type 423OH. (*Courtesy of Duracell, Inc.*)

Effect of Temperature   The Li/MnO₂ cell is capable of performing over a wide temperature range, from about -20 to 55°C. Discharge curves, showing the performance of various cell designs, are presented in Fig. 11.43.

In Fig. 11.44, the performance of the Li/MnO₂ cells is plotted as a function of load and temperature. The advantage of the lower-rate discharges is evident, but the Li/MnO₂ cell will deliver a high percentage of its capacity even at relatively high discharge rates. The average voltage of these discharges is given in Fig. 11.45.

Impedance   The impedance of the Li/MnO₂ cell, as within most battery systems, is dependent on the cell size, design, electrode, separator, etc., as well as the chemistry. Inherently, the conductivity of the organic solvent–based electrolytes is lower than the aqueous electrolytes, and the Li/MnO₂ cell, therefore, has a higher impedance than conventional cells of the same size and construction; designs which increase electrode area and decrease electrode spacing, such as coin-shaped flat cells and jelly-roll configurations, are used. Further, the lithium cells will perform relatively more efficiently at the lower temperatures because the conductivity of the organic solvents is less sensitive to temperature change than the aqueous solvents. Table 11.14 lists typical impedance values for a low-rate button cell, rated at 65 mAh. Impedance values will be lower for the larger and/or higher-rate cells.

Service Life   The capacity or service life of the different types of Li/MnO₂ cells, normalized for a 1-g and 1-cm³ cell, at various discharge loads and temperatures, is summarized in Fig. 11.46. These data can be used to approximate the performance of a given cell or to determine the size and weight of a cell for a particular application.

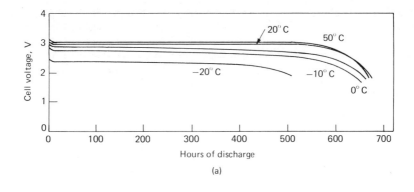

(a)

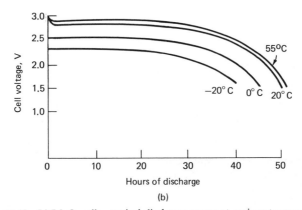

(b)

FIG. 11.43   Li/MnO$_2$ cells—typical discharge curves at various temperatures. (*a*) Low-rate flat cells, C/600 rate; (*b*) cylindrical cells, C/50 rate. (*Courtesy of Duracell, Inc.*)

Shelf Life   The storage characteristics of the Li/MnO$_2$ cell are shown in Fig. 11.47. The system is very stable, and its useful shelf life is limited primarily by the amount of solvent diffusing through the seal. Projections of these data predict that 85% of the fresh capacity will be available even after 6 y of storage at 20°C.

11.8.4   Cell and Battery Sizes

The Li/MnO$_2$ cells are manufactured in a number of flat and cylindrical cells ranging in capacity from about 30 to 1200 mAh; larger-size cells are under development in cylindrical and rectangular configurations. The physical and electrical characteristics of some of these are summarized in Table 11.15. In some instances, interchangeability with other battery systems is provided by doubling the size of the cell to accommodate the 3-V output of the Li/MnO$_2$ cell compared with 1.5 V of the conventional primary cells.[31]

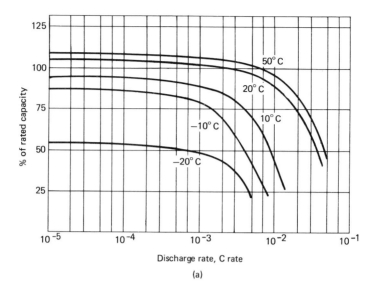

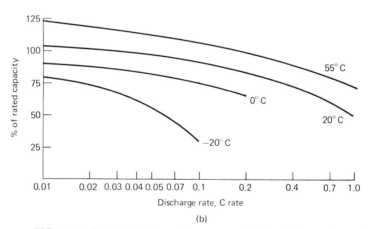

FIG. 11.44   Li/MnO$_2$ cells—performance as a function of temperature and discharge load, 2-V end voltage. (a) Low-rate flat cells; (b) cylindrical cells. (*Courtesy of Duracell, Inc.*)

11.8.5 Special Designs

A flat-pack battery (see Part 2, Chap. 5, Sec. 5.7.1) has also been designed using the Li/MnO$_2$ system. The design is similar to the one illustrated in Fig. 5.34, with identical dimensions, but only two lithium cells are required for the 6-V pack. At a 20-mA drain, the battery will operate for more than 70 h with the advantageous features of the Li/MnO$_2$ system. Figure 11.48 shows the performance of the battery under different load conditions at 20°C.

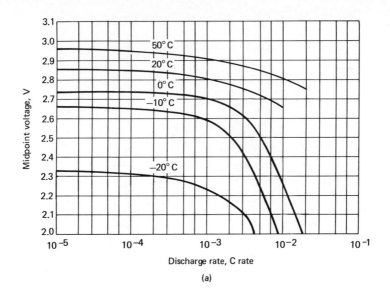

(a)

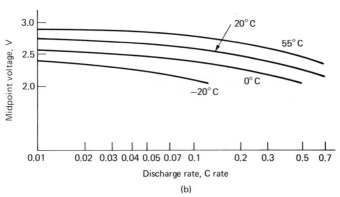

(b)

FIG. 11.45 Li/MnO₂ cells—average voltage during discharge, 2-V end voltage. (*a*) Low-rate flat cells; (*b*) cylindrical cells. (*Courtesy of Duracell, Inc.*)

TABLE 11.14 Impedance of Li/MnO₂ Cells, 65-mAh Capacity*

	Impedance, Ω			
	1000 Hz		40 Hz	
% depth of discharge†	Room temp. (20 to 35°C)	Low temp. (0 to −10°C)	Room temp. (20 to 35°C)	Low temp. (0 to −10°C)
Initial	175–400	300–500	190–410	380–620
10	180–335	380–540	185–330	440–530
30	165–210	330–550	160–210	380–550
50	132–170	400–1400	140–170	420–1460
60	130–170	460–1930	140–170	480–2070
80	220–290	760–2800	220–290	780–3030

*UCC Type Y 2022.
†Discharged at 15,000 Ω, 35°C.

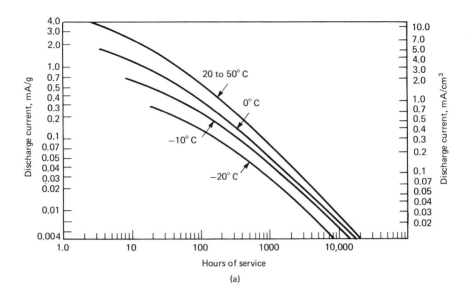

(a)

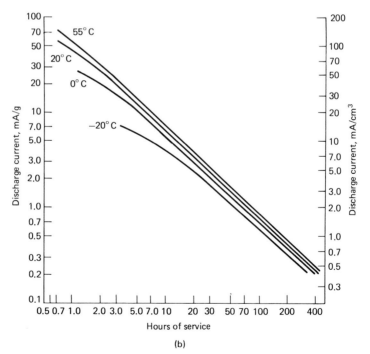

(b)

FIG. 11.46 Li/MnO₂ cells—service life to 2-V end voltage. (*a*) Low-rate flat cells; (*b*) cylindrical cells. (*Courtesy of Duracell, Inc.*)

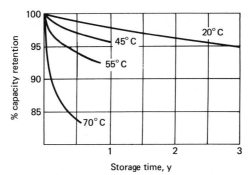

FIG. 11.47   Shelf life of Li/MnO$_2$ cells. (*Courtesy of Duracell, Inc.*)

11.8.6 Applications and Handling

The main applications of the Li/MnO$_2$ system currently are in the small flat and cylindrical cells and batteries ranging up to several ampere-hours in capacity, taking advantage of its higher volumetric energy density, better high-rate capability, and longer shelf life compared with the conventional zinc batteries—and lower cost compared with the zinc/silver oxide system. The Li/MnO$_2$ cells are used in memory applications, watches, calculators, cameras, etc.; at the higher drain rates, motor drives, photoflash, toys, and small electronics are excellent applications. As the Li/MnO$_2$ technology matures, larger cells are produced, and costs are lowered, the Li/MnO$_2$ system will be able to compete more successfully with the conventional batteries in general-purpose consumer and industrial applications.

The low-capacity Li/MnO$_2$ cells can generally be handled without hazard, but, as with the conventional primary battery systems, charging and incineration should be avoided as these conditions could cause a cell to explode.

The higher-capacity cylindrical cells are generally equipped with a venting mechanism to prevent explosion, but the batteries, nevertheless, should be protected to avoid short circuit and cell reversal, as well as charging and incineration. The specific conditions for the use and handling of Li/MnO$_2$ cells are dependent on the size as well as the specific design features, and manufacturers' recommendations should be consulted.

11.9 LITHIUM/CARBON MONOFLUORIDE [Li/(CF)$_n$] CELLS

The lithium/carbon monofluoride [Li/(CF)$_n$] cell is also one of the first lithium/solid-cathode systems to be used commercially. It is attractive as its theoretical gravimetric energy density (over 2000 Wh/kg) is among the highest of the solid-cathode systems. Its open-circuit voltage is 3.1 V, with an operating voltage of about 2.5 V. Its practical energy density ranges up to 250 Wh/kg and 600 Wh/L. The system is used primarily at low to medium discharge rates.

TABLE 11.15 Li/MnO$_2$ Cells and Batteries

International (IEC) type number	Rated capacity, mAh*	Weight, g	Dimensions Diameter, mm	Height, mm	Volume, cm^3	Energy density† Wh/kg	Wh/L
			Low-rate flat or coin cells				
	65	1.7	11.5	4.2	0.44	105	410
CR-1220	30	0.8	12.5	2.0	0.25	105	335
CR-1620	50	1.2	16.0	2.0	0.40	116	350
CR-2016	50	2.0	20.0	1.6	0.50	70	280
CR-2020	90	2.3	20.0	2.0	0.63	110	400
CR-2025	120	2.5	20.0	2.5	0.79	135	425
CR-2420	120	3.0	24.5	2.0	0.94	112	360
CR-2032	170	3.0	20	3.2	1.00	160	475
CR-2325	160	3.8	23	2.5	1.04	120	430
CR-2430	200	4.0	24.5	3.0	1.41	140	400
CR-WM	3500	42.0	45	12.0	19.08	230	515
			High-rate flat or coin cells				
CR-2016H	50	2.0	20.0	1.6	0.50	70	280
CR-2025H	100	2.4	20.0	2.5	0.79	115	355
CR-2420H	100	3.0	24.5	2.0	0.94	94	298
CR-2032H	130	2.8	20.0	3.2	1.00	130	365
CR-2430H	160	4.0	24.5	3.0	1.41	112	320
			Cylindrical cells				
CR-772	30	1.0	7.9	7.2	0.35	84	240
CR-1/3N	160	3.0	11.6	10.8	1.14	150	395
CR-1/2AA	500	8.5	14.5	25.0	4.13	165	340
CR-2N	1000	13.0	12.0	60.0	6.78	215	415
CR-2/3A	1100	15.0	16.4	32.8	7.25	205	425
			Batteries				
2-CR-1/3N (two cells in series)	160	8.8	13.0	25.2	3.3	100	275
Flat-pack (two cells in series)	1200	34	94 × 77 × 4.5		30.2	210	238
PC mounting (3 V)	200	12	27.2 × 28 × 5.08		3.8	50	160

*Low-rate cells: C/200 rate. High-rate and cylindrical cells: C/30 rate.
†Based on average voltage of 2.8 V per cell.

11.9.1 Chemistry

The active components of the cell are lithium for the anode and solid carbon polymonofluoride (CF)$_n$ for the cathode. Carbon monofluoride is an intercalation compound, formed by the reaction between carbon powder and fluorine gas. While electrochemically active, the material is chemically stable in the organic electrolyte and does not thermally decompose up to 400°C, resulting in a long storage life. Different electrolytes have been used; typical electrolytes are lithium hexafluoroarsenate (LiAsF$_4$) in dimethyl sulfite (DMSI), lithium tetrafluoroborate (LiBF$_4$) in butyrolac-

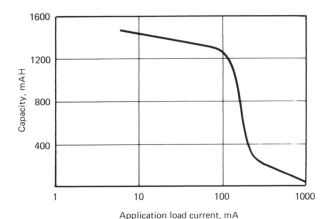

Application load current, mA

FIG. 11.48 Performance of Li/MnO$_2$ 6-V flat-pack battery, capacity to 3-V cutoff. (*Courtesy of Polaroid Corp.*)

tone (BL), and tetrahydrofuran (THF) or propylene carbonate (PC) and dimethoxyethane (DME).

The simplified discharge reactions of the cell are

Anode reaction: $n\mathrm{Li} \rightarrow n\mathrm{Li}^+ + n\mathrm{e}$

Cathode reaction: $(\mathrm{CF})_n + n\mathrm{e} \rightarrow n\mathrm{C} + n\mathrm{F}^-$

Overall cell reaction: $n\mathrm{Li} + (\mathrm{CF})_n \rightarrow n\mathrm{LiF} + n\mathrm{C}$

The polycarbon monofluoride changes into conductive carbon as the discharge progresses, thereby increasing the cell's conductivity, improving the regulation of the discharge voltage, and increasing the discharge efficiency. The LiF precipitates in the cathode structure.[1,32]

11.9.2 Construction

The Li/(CF)$_n$ system is adaptable to a variety of sizes and configurations: cells are available in flat coin or button, cylindrical, and rectangular shapes ranging in capacity from 0.020 to 25 Ah; larger-sized cells are under development.

Figure 11.49 shows the construction of a button-type cell. The Li/(CF)$_n$ cells are

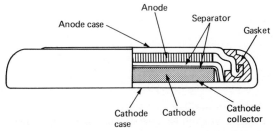

FIG. 11.49 Cross-sectional view of the Li/(CF)$_n$ flat button-type cell. (*From Ref. 32; courtesy of Panasonic Co., Secaucus, N.J.*)

typically constructed with an anode of lithium foil rolled onto a collector and a cathode of Teflon-bonded polycarbon monofluoride and carbon black on a nickel collector. Nickel-plated steel or stainless steel is used for the case material. The button cells are crimped-sealed using a polypropylene gasket. The cylindrical cells use spirally wound (jelly-roll construction) flexible electrodes, and the cells are either crimped or hermetically sealed. The pin-type cells use an inside-out design with a cylindrical cathode and a central anode in an aluminum case. The larger cells are provided with low-pressure safety vents.[32,33]

The $Li/(CF)_n$ system has also been used in a thin "paper" battery structure described in "Paper cell" in Part 2, Sec. 5.7, and illustrated in Fig. 5.38.

11.9.3 Performance

Flat Cells   Figure 11.50 presents the discharge curves at 20°C for a typical $Li/(CF)_n$ button cell rated at 110 mAh. The voltage is constant throughout most of the discharge, and the coulombic utilization is close to 100% under low-rate discharge. Figure 11.51*a* presents the discharge curves for the same cell at various discharge temperatures. The average load voltage (plateau voltage during discharge) at various temperatures and loads is given in Fig. 8.51*b*. Performance data are summarized in Fig. 11.52 for a $Li/(CF)_n$ button cell, normalized for a 1-g and 1-cm³ (mL) cell. These curves can be used to approximate the size or performance of a cell for a particular application.

Cylindrical Cells   The cylindrical cells are generally designed to operate at proportionally heavier discharge loads than the button cells. Figure 11.53 shows the discharge characteristics at 20°C of a cylindrical C size cell (26 mm in diameter by 50 mm high) rated at 5Ah. Figure 11.54 presents the discharge curves at other discharge temperatures. The average load voltages for cylindrical cells at various temperatures and loads are given in Fig. 11.55. The performance data are then summarized in Fig. 11.56 which shows the effect of temperature and load on the service life of a cell normalized to unit weight (grams) and volume (cubic centimeters). The internal impedance of these cells is relatively low. For example, the internal impedance of the BR-⅔A cell ranges between 0.4 Ω at 25°C to 0.6 Ω at −20°C.[32b]

Micropower Cell   The long shelf life and flat discharge characteristics of the $Li/(CF)_n$ system are utilized in prismatic cells designed for mounting on PC boards and used for long-term low-drain or standby applications. The dimensions and performance characteristics of this cell are given in Table 11.16 and Fig. 11.57, respectively. The other $Li/(CF)_n$ cells are also capable of long-term discharge.

Paper Battery   The dimensions and performance of the thin $Li/(CF)_n$ paper battery are given in Table 11.16. The cell is rated at 1500 mAh to a 2.0-V end voltage when discharged at a 500-Ω load. Short-circuit current is 3 A and the internal resistance of the cell is 2 Ω. Shelf life at 20°C is estimated at 3 y.

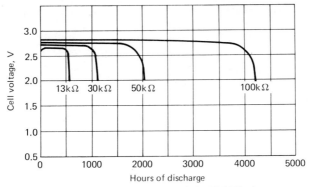

FIG. 11.50   Typical discharge curves of the Li/(CF)$_n$ button-type cell at 20°C (cell rated capacity: 110 mAh). (*Courtesy of Panasonic Co, Secaucus, N.J.*)

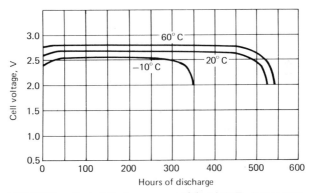

FIG. 11.51*a*   Typical discharge curves of the Li/(CF)$_n$ button-type cell at various temperatures, 13,000-Ω discharge load. (*Courtesy of Panasonic Co., Secaucus, N.J.*)

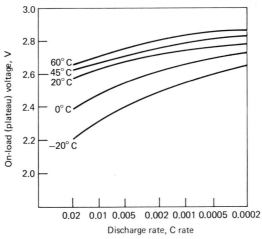

FIG. 11.51*b*   Li/(CF)$_n$ button-type cells—on-load plateau voltage vs. discharge rate.

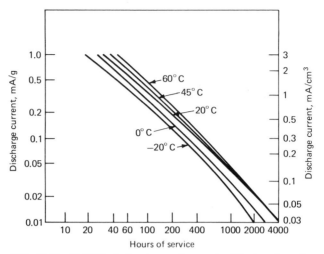

FIG. 11.52 Li/(CF)$_n$ button-type cells—service life at various discharge rates and temperatures, 2.0-V end voltage.

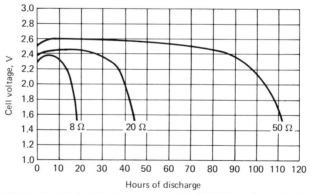

FIG. 11.53 Typical discharge curves of the Li/(CF)$_n$ cylindrical cell at 20°C (cell rated capacity = 5 Ah). (*Courtesy of Panasonic Co., Secaucus, N.J.*)

Shelf Life and Voltage Delay The shelf life of the Li/(CF)$_n$ cell is comparable to the other solid-cathode cells (see Fig. 11.54). Capacity retention of up to 95% after 5 y at 20°C and 6 months at 45°C is projected. An initial low voltage is observed with the Li/(CF)$_n$ cell; i.e., the cell voltage drops initially below the operating level on load and recovers gradually as the discharge progresses. This is attributed to the fact that (CF)$_n$ is an insulator but gradually changes to conductive carbon as the reaction proceeds.

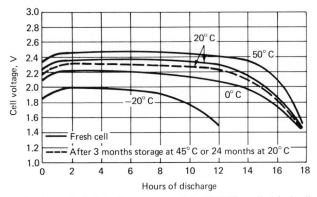

FIG. 11.54 Typical discharge curves of the Li/(CF)$_n$ cylindrical cell at various temperatures; cell rated capacity = 5Ah, 8-Ω discharge load. (*Courtesy of Panasonic Co., Secaucus, N.J.*)

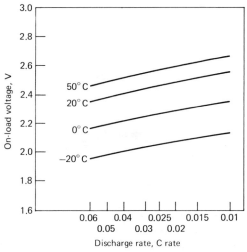

FIG. 11.55 Li/(CF)$_n$ cylindrical cells—on-load plateau voltage vs. discharge rate.

11.9.4 Cell and Battery Types

The Li/(CF)$_n$ cells are manufactured in a number of flat or button, cylindrical, and rectangular sizes. The major electrical and physical characteristics of some of these cells are listed in Table 11.16. Manufacturers' specifications should be consulted for the most recent listing of commercially available cells.[34]

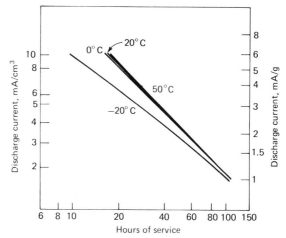

FIG. 11.56 Li/(CF)$_n$ cylindrical cells—service life at various discharge rates and temperatures, 1.8-V end voltage.

11.9.5 Applications and Handling

The applications of the Li/(CF)$_n$ cell are similar to those of the other lithium/solid-cathode cells, again taking advantage of the high energy density and long shelf life of these cells. The flat or button Li/(CF)$_n$ cells are used as a power source for watches, portable calculators, memory applications, etc. The low-capacity miniature pin-type cells have been used as an energy source for LEDs and for microelectronic equipment. The larger cylindrical and rectangular cells can also be used in memory applications, but their higher drain capability also covers applications in radio sets, for telemetry, photographic, and similar general-purpose applications. The BR-⅔ A cell is used in the Kodak disk camera.[32b] New applications will also develop for this battery system as the technology matures.

Handling considerations for the Li/(CF)$_n$ system, too, are similar to those for the other lithium/solid-cathode systems. The limited current capability of the button and low-capacity cells restricts temperature rise during short circuit and reversal. These cells can generally withstand this abusive use even though they are not provided with a safety venting mechanism. The larger cells are provided with a venting device, but short circuit, high discharge rates, and reversal should be avoided as these conditions could cause the cell to vent. Charging and incineration likewise should be avoided for all cells. The manufacturer's recommendations should be obtained for handling of specific cell types.

11.10 LITHIUM/COPPER SULFIDE Li/CuS CELLS

The lithium/copper sulfide (Li/CuS) cell[1,3,35,36] has, at low discharge rates, about twice the capacity of an equivalently sized alkaline Zn/MnO$_2$ cell, is very stable on storage,

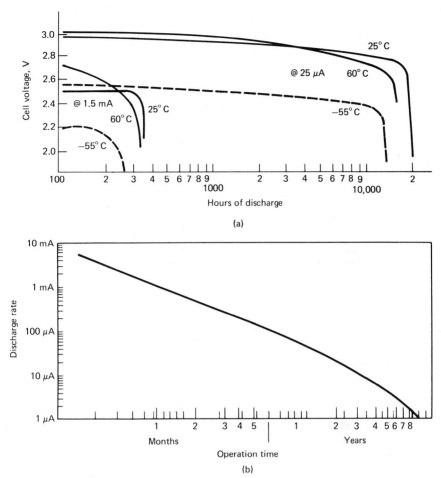

FIG. 11.57 Li/(CF)$_n$ micropower cell. (*a*) Typical discharge characteristics; (*b*) operational life vs. continuous discharge load at 25°C. (*Courtesy of Eagle-Picher Industries, Joplin, Mo.*)

and has good low-temperature performance. Its operating voltage is about 1.7 V, but its relatively high ampere-hour capacity results in an energy density that compares favorably with the higher-voltage systems. The energy density of the Li/CuS cell, in a NEDA size 1604 9-V battery, is about 135 Wh/kg and 335 Wh/L. A comparison of the performance of the NEDA size 1604 Li/CuS battery with an alkaline Zn/MnO$_2$ battery is given in Fig. 11.58.

11.10.1 Chemistry

The active materials of the Li/CuS cell are lithium for the anode and copper sulfide for the cathode. The electrolyte is a mixture of 1,2-dimethoxyethane, 1,3-dioxolane, and 2,5-dimethylisoxazole as a stabilizer with LiClO$_4$ as the solute. This electrolyte was

TABLE 11.16 Lithium/Carbon Monofluoride Li/$(CF)_n$ Cells

International (IEC) type number	Type	Dimensions, mm				Volume, cm^3	Weight, g	Capacity, mAh	Energy/density*	
		Diameter	Height	Length	Width				Wh/kg	Wh/L
BR-425	Pin	4.19	25.9	...	...	0.36	0.6	20	83	140
BR-435	Pin	4.19	35.8	...	...	0.49	0.9	40	110	205
BR-2016	Coin	20	1.6	...	...	0.50	1.5	60	100	300
BR-2025	Coin	20	2.0	...	...	0.63	2.3	90	98	355
BR-2320	Coin	23	2.0	...	...	0.83	2.5	110	110	330
BR-2325	Coin	23	2.5	...	...	1.04	3.1	150	120	360
†	Coin	33	2.6	...	...	2.22	8.1	350	108	395
†	Coin	29	5.1	...	...	3.37	8.3	500	150	370
BR-½A	Cylinder	16.7	22.5	...	...	4.93	10.1	750	185	380
BR-2/3A	Cylinder	16.7	33.3	...	...	7.29	13.5	1200	220	410
BR-C	Cylinder	26	50	...	...	26.5	47	5000	265	470
	Paper battery	...	94	75	1.4	9.8	12	1500	310	385
†	Rectangular	...	35.6	37.6	8.6	11.5	35	3000	215	600
†	Rectangular (high rate)	...	35.3	66.8	21.8	51.4	144	5000	87	243
†	Rectangular (low rate)	...	17.8	30.5	8.1	4.4	9.4	500	135	284

*Based on 2.5 V per cell.
SOURCE: Panasonic Co., Secaucus, N.J. and †Eagle-Picher Industries, Joplin, Mo.

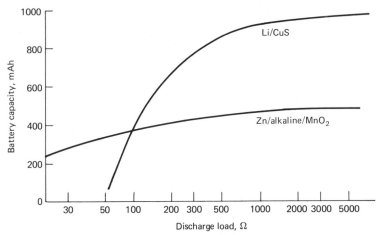

FIG. 11.58   Comparison of Li/CuS and Zn/alkaline/MnO$_2$, 9.0-V 1604 batteries at 20°C (5.4-V end voltage).

formulated to optimize the discharge rate capability. Another electrolyte used for this system is tetrahydrofuran (THF) and dimethoxyethane (DME) binary solvent with LiClO$_4$ for the solute.

The discharge reactions of the Li/CuS cell are not fully clarified under all conditions. The cell reaction occurs in two steps. The first step:

Anode reaction: $Li \rightarrow Li^+ + e$

Cathode reaction: $CuS + xLi^+ + xe \rightarrow Li_xCuS$

On continued discharge, the second step of the cathode discharge occurs:

$$2Li_xCuS + 4e \rightarrow 2Li_xCu + 2S^{2-}$$

11.10.2 Construction

The 9-V, NEDA size 1604 battery is constructed from four flat rectangular cells using the Li/CuS system. The cell design is shown in Fig. 11.59. The anode is cut from lithium foil which is pressure-welded to the stainless steel lid. The continuous weld between the lid and lithium ensures highly efficient discharge of the anode. The cathode is formed from a mixture of copper, sulfur, cupric sulfide, and graphite. After pelleting, the mixture is heated to complete the reaction, thus forming a durable pellet of about 40% porosity. An electrolyte reservoir is placed between the cathode and can to obtain the most efficient use of the cathode. This reservoir retains electrolyte at the cathode surface next to the can and improves higher discharge rate performance. Both the electrolyte reservoir and the separator are made of a short fiber glass mat of about 90% porosity. The can and lid are made of 304 stainless steel and the grommet of polypropylene.

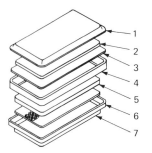

1 Stainless-steel lid
2 Unalloyed lithium foil anode welded to lid
3 Glass fiber mat separator
4 Grommet, polypropylene
5 Copper sulfide cathode
6 Electrolyte reservoir
7 Stainless-steel can

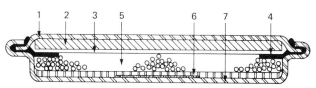

FIG. 11.59 Li/CuS cell for size 1604, 9.0-V battery. (*Courtesy of Ray-O-Vac Corp., Madison, Wis.*)

TABLE 11.17 Characteristics of Li/CuS, 9-V, 1604 Battery

Dimensions	Electrical
Length: 26.2 mm maximum Width: 16.9 mm maximum Height, shoulder: 45.4 mm maximum Height, overall: 48.8 mm maximum Volume: 20.1 cm³ Weight: 50 g	Open-circuit voltage: 8.6 V Initial closed-circuit voltage: 7.9 V Capacity (250 Ω, 4 h per day to 5.4 V: 750 mAh Energy density: 135 Wh/kg 335 Wh/L

SOURCE: Ray-O-Vac Corp., Madison, Wis.

11.10.3 Performance

The characteristics of the 9-V, four-cell battery are summarized in Table 11.17. Figure 11.60 shows the typical discharge curve for the NEDA size 1604 Li/CuS battery at low and high discharge rates. A two-plateau discharge is typical at the low rates. The high-rate discharge causes a mixed potential discharge to occur, although at this rate a slight plateau is seen at the cuprous level.

Typical capacity data at various loads are detailed in Table 11.18. The battery operates more efficiently when discharged intermittently and, at heavier drains, will operate more efficiently when discharged to a lower end voltage.

Impedance typically decreases as frequency increases. Typical data are given in Table 11.19.

Table 11.20 details the effect of temperature on discharge. Discharge at elevated temperatures predictably increases the efficiency of discharge. However, at tempera-

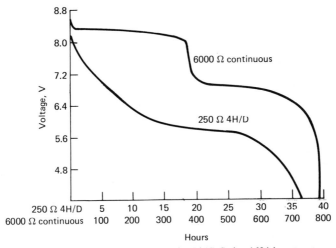

FIG. 11.60 Typical discharge curves for Li/CuS size 1604 battery at 20°C. (*Courtesy of Ray-O-Vac Corp., Madison, Wis.*)

TABLE 11.18 Capacity of Li/CuS, 9-V, 1604 Battery at 20°C*

| | End voltage: 5.4 V | | | |
Load	Average drain, mA	Average voltage, V	mAh	mWh
64 Ω cont.	95.6	6.12	73	447
64 Ω 30 min/h	98.8	6.32	99	626
250 Ω 4 h per day	25.0	6.24	750	4680
500 Ω 4 h per day	13.4	6.72	870	5840
1000 Ω cont.	7.04	7.04	890	6260
6000 Ω cont.	1.26	7.56	970	7330

| | End voltage: 4.2 V | | | |
Load	Average drain, mA	Average voltage, V	mAh	mWh
64 Ω cont.	79.4	5.08	314	1600
64 Ω 30 min/h	78.8	5.04	440	2240
250 Ω 4 h per day	24.3	6.08	860	5200
500 Ω 4 h per day	13.3	6.64	920	6120
1000 Ω cont.	7.04	6.96	930	6440
6000 Ω cont.	1.25	7.52	980	7400

*Theoretical capacity 1150 mAh.
SOURCE: Ray-O-Vac Corp., Madison, Wis.

tures above 70°C, a loss of reliability has been observed. Capacity available at low temperatures is related to the electrolyte used. With the standard electrolyte, significant capacities are obtained at temperatures as low as −30°C at low rates.

Table 11.21 shows the good capacity retention demonstrated by this system. Capacity retention is believed to be directly related to seal quality and the resultant electrolyte loss and water gain. Storage capability of 7 y at 20°C is forecast.

TABLE 11.19 Typical Impedance Values of 1604 Battery at 20°C

Frequency, Hz	Impedance, Ω		
	Li/CuS	Alkaline-MnO$_2$	Zinc-carbon
100	16	6	79
1000	12	2	36
5000	7	2	32

SOURCE: Ray-O-Vac Corp., Madison, Wis.

TABLE 11.20 Effect of Temperature on Discharge, Li/CuS, 9-V, 1604 Battery

Temperature	250 Ω 4 h per day		6000 Ω	
	5.4 V	4.2 V	5.4 V	4.2 V
71°C	110%	103%		
54°C	103%	100%	100%	100%
21°C	100%	100%	100%	100%
0°C	53%	88%	97%	98%
−18°C*	38%	3%		
	(48%)	(73%)		
−29°C			4%	46%

*Special electrolytes, not optimum at 21°C, will improve this performance to the values in parentheses.
SOURCE: Ray-O-Vac Corp., Madison, Wis.

TABLE 11.21 Capacity Retention of Li/CuS, 9-V, 1604 Battery

Storage	250 Ω 4 h per day		6000 Ω	
	To 5.4 V	To 4.2 V	To 5.4 V	To 4.2 V
6 months, 21°C	98%	95%	98%	98%
6 months, 45°C	97%	98%	97%	97%
3 months, 54°C	98%	99%	100%	100%

SOURCE: Ray-O-Vac Corp., Madison, Wis.

11.10.4 Battery Types and Applications

The Li/CuS system has been investigated in flat as well as cylindrical configurations.[36] The NEDA size 1604 9-V battery has been investigated, but it has not been commercialized. Cordis Corp. has manufactured a Li/Cus cell in a D-shaped flat configuration for use in cardiac pacemakers. The cell (5.1 cm × 3.2 cm × 0.51 cm thick) is rated at 1.8 Ah and is designed for the light pacemaker loads.

The handling considerations of the Li/CuS system also are similar to the other lithium/solid cathode cells of the same size and capacity. Cells can withstand short-circuit but should not be forced-discharged or exposed to temperatures as high as 180°C, the melting point of lithium. New designs, replacing the LiClO$_4$ which is very reactive, appear effective in improving the safety of the cell under abusive conditions.[35]

11.11 LITHIUM/COPPER OXIDE (Li/CuO) CELLS

The lithium/copper oxide system is characterized by a high energy density (about 300 Wh/kg and 600 Wh/L) as CuO has one of the highest volumetric faradic capacities of the practical cathode materials (4.26 Ah/cm^3). The cell has an open-circuit voltage of 2.25 V and an operating voltage of 1.2 to 1.5 V, which makes it interchangeable with conventional batteries. Figure 11.61 compares the performance of a Li/CuO AA size cylindrical cell with the alkaline Zn/MnO$_2$. The Li/CuO cell has a 2:1 capacity advantage at low discharge rates but loses this advantage at higher discharge rates.

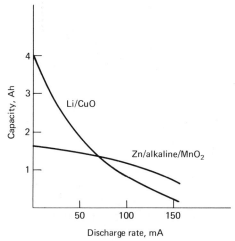

FIG. 11.61 Comparison of Li/CuO and Zn/alkaline/MnO$_2$ AA size cell at 20°C.

Li/CuO cells have been designed in button and cylindrical configurations for low- to medium-rate applications in sizes up to 3.5 Ah. Larger sizes and higher-rate designs are under development.

11.11.1 Chemistry

The discharge reaction of the Li/CuO cell is

$$2Li + CuO \rightarrow Li_2O + Cu$$

The discharge proceeds stepwise, CuO $\rightarrow$ Cu$_2$O $\rightarrow$ Cu, but the detailed mechanism has not been clarified.[1,37] A double-plateau discharge has been observed at high temperature (70°C) discharges at low rates which blends into a single plateau under more normal discharge conditions.[40]

11.11.2 Construction and Performance

Button Cells   The construction of the button-type cell is similar to other conventional and solid cathode cells (Fig. 11.62). The anode is made from lithium sheet; the nickel sponge serves as an electrolyte absorber as well as a conductive lead. The cathode is a pressed pellet consisting of CuO (lithium-doped to optimize conductivity), acetylene black, or graphite as the conductive component, with a binder such as Teflon [polytetrafluorethylene (PTFE)]. The electrolyte is a mixed organic solvent, such as propylene carbonate (PC) and tetrahydrofuran (THF) with lithium perchlorate added as a solute. The cell cases are made from nickel-plated steel; the sealing grommet is polypropylene.

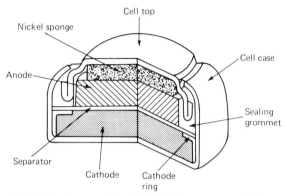

FIG. 11.62   Cross section of Li/CuO button-type cell. (*From Ref. 37.*)

Figure 11.63 shows the typical discharge curves at 20°C of a Li/CuO button cell (11.6 mm in diameter × 5.4 mm high); the performance at different temperatures is shown in Fig. 11.64. These cells also demonstrated good shelf life. No appreciable

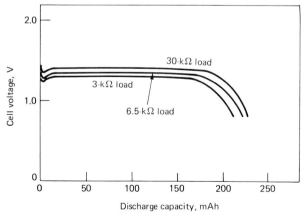

FIG. 11.63   Typical discharge curves under various fixed loads, at 20°C, Li/CuO button cell (11.6 mm diam. × 5.4 mm high). (*From Ref. 37.*)

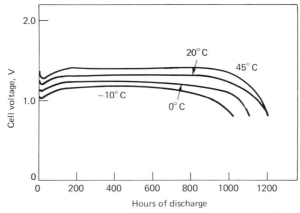

FIG. 11.64   Influence of temperature on Li/CuO button cell (11.6 mm diam. × 5.4 mm high), 6.5-kΩ load. (*From Ref. 37.*)

decrease in cell performance was observed after storage at 20°C for 6 months, 45°C for 3 months, or 60°C for 1 month.[37]

Cylindrical Cells   The cylindrical Li/CuO cell uses the inside-out bobbin construction, similar to the one used for the zinc/alkaline/manganese dioxide cell, as illustrated in Fig. 11.65.[1, 38-41] A nickel-plated steel can is used, and the top is insulated from the

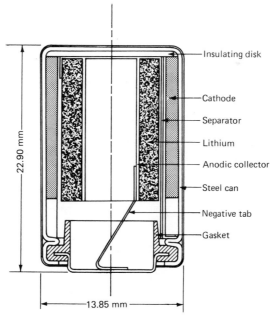

FIG. 11.65   Schematic diagram of Li/CuO 1/2 AA size cell. (*From Ref. 41.*)

can with a polypropylene gasket. The can is connected to the cylindrical copper oxide cathode and the top to the lithium anode. The electrolyte is dioxolane with $LiClO_4$ as the solute.

Table 11.22 lists the physical and electrical characteristics of some of the cylindrical Li/CuO cells. The discharge curves for the AA size cell at 20°C are shown in Fig. 11.66. After a high initial load voltage, the discharge profile is flat at the relatively light loads; the bobbin construction does not lend itself to high-rate discharges, and the cell capacity is significantly lowered with increasing discharge rates. The Li/CuO cylindrical cell operates over a wide temperature range, typically from 70 to −20°C, although the cell can operate outside these limits but with changes in the discharge profile and/or load capability. Discharge curves at several different temperatures are shown in Fig. 11.67. The performance of the cell at temperatures from 70 to −40°C and at various loads is summarized in Fig. 11.68.[41] The high capacity of the cell at the lighter loads falls off sharply with increasing load and decreasing temperature.

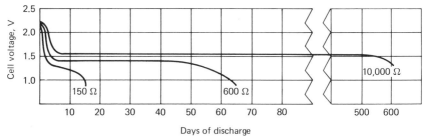

FIG. 11.66 Typical discharge curves for Li/CuO AA size cell at 20°C. (*Courtesy of SAFT Corp. of America.*)

TABLE 11.22 Cylindrical Lithium/Copper Oxide Cells

IEC designation	¼R6	R3	R6
Size (ANSI)	¼AA	½AA	AA
Nominal voltage, V	1.5	1.5	1.5
Dimensions:			
Diameter, mm	14.1	14.1	14.1
Height, mm	12	24.5	49.5
Volume, cm³	1.9	3.9	7.8
Weight, g	4.5	7.3	17.4
Rated capacity, Ah	0.445*	1.8†	3.4‡
Energy density:			
Wh/kg	140	345	275
Wh/L	330	645	6.10
Impedance (from cell), Ω	100	37	15

*4.8-kΩ load.
†1.6-kΩ load.
‡120-Ω load.
SOURCE: SAFT Corp. of America.

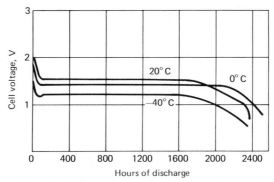

FIG. 11.67 Effect of temperature on Li/CuO AA size cell, 1.5-mA load. (*Courtesy of SAFT Corp. of America.*)

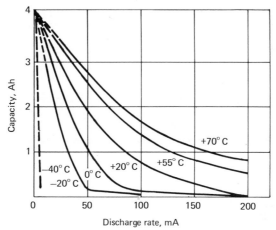

FIG. 11.68 Capacity as a function of discharge load and temperature, Li/CuO AA size cell. (*Courtesy of SAFT Corp. of America.*)

The Li/CuO cell also has good storage characteristics; a shelf life of 10 y at 20°C is predicted. Cells stored for 2 y at 20°C show no loss in capacity; performance after storage at high temperatures is plotted in Fig. 11.69. Retention of residual capacity in partially discharged cells was shown to be equivalent to that of fully charged cells. The Li/CuO cell also is free from the voltage delay phenomenon generally experienced with lithium anode cells, showing a rapid response after application of the load even after storage for 2 y at 20°C.[40]

11.11.3 Cell Types and Applications

Li/CuO cells are available in the small cylindrical sizes listed in Table 11.22. These cells are manufactured in the bobbin construction for use mainly in low-rate applications.

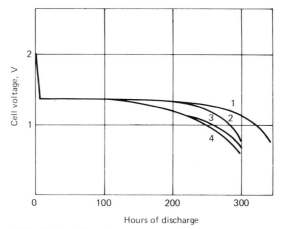

FIG. 11.69   Effect of storage on performance of Li/CuO cylindrical cells, discharges at 20°C after 70°C storage. Curve 1, no storage; curve 2, 6 months of storage; curve 3, 12 months of storage; curve 4, 18 months of storage. (*Courtesy of SAFT Corp. of America.*)

Under these low-drain conditions, the cell has a 2:1 capacity advantage over the conventional premium batteries. Combined with its excellent storability and operation over a wide temperature range, the Li/CuO system is a useful power source for applications such as memory preservation, clocks, and telemetry and in high-temperature environments.

11.12   LITHIUM/IRON SULFIDE Li/FeS$_2$ AND Li/FeS CELLS

Iron sulfide, in both the FeS$_2$ and FeS forms, is being used in solid cathode lithium cells. These systems have a nominal voltage of about 1.5 V and can, therefore, be used as a direct replacement for aqueous cells having a similar voltage. The lithium/iron sulfide cells are being manufactured primarily in flat or button-type configurations. Their performance is similar to that of similar-sized Zn/Ag$_2$O cells with higher impedance and slightly lower power capability, but with better low-temperature performance and storability and lower cost. The performances of these two systems are compared in Fig. 11.70 in a button cell application (11.5 mm in diameter $\times$ 5.6 mm high).

11.12.1   Chemistry

The discharge reactions are

$$FeS_2 + 4Li \rightarrow Fe + 2Li_2S$$

$$FeS + 2Li \rightarrow Fe + Li_2S$$

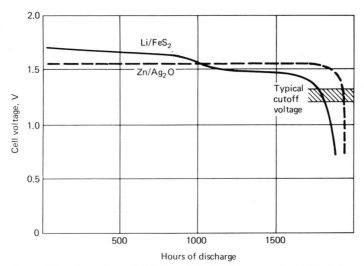

FIG. 11.70   Comparison of Li/FeS$_2$ and Zn/Ag$_2$O button cells, 15,000-Ω discharge load at 20°C (cell size: 11.5 mm diam. $\times$ 5.6 mm high). (*Courtesy of Union Carbide Corp.*)

The FeS$_2$ electrode has a performance advantage over the FeS electrode because of a higher sulfur content and higher cell voltage. The FeS electrode has the advantages of reduced corrosion, longer life, and a single voltage plateau compared with the FeS$_2$ electrode, which discharges in two steps.

11.12.2 Construction

The construction of the lithium/iron sulfide cells is very similar to that of the zinc/silver oxide button cell, as shown in Fig. 11.71. An organic solvent, such as propylene carbonate (PC) and dimethoxyethane (DME), is used for the electrolyte.[1,42]

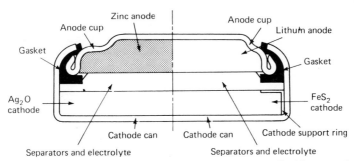

FIG. 11.71   Cross section: Zn/Ag$_2$O and Li/FeS$_2$ button cells. (*Courtesy of Union Carbide Corp.*)

11.12.3 Physical and Electrical Characteristics— Performance

Li/FeS₂ Cell[42]   Table 11.23 lists the physical and electrical characteristics of several of the button-type cells manufactured with the Li/FeS_2 system. The discharge at various temperatures of a typical cell (11.5 mm in diameter $\times$ 4.2 mm high) rated at 100 mAh is shown in Fig. 11.72. The impedance of the cell during these discharges is listed in Table 11.24. Discharges conducted after various periods of high-temperature storage (Fig. 11.73) illustrate the storage characteristics of this cell. Only a small loss of capacity is observed even after storage at elevated temperatures.

Li/FeS Cell[43,44]   The construction of the Li/PC + DME (LiClO₄)/FeS button cell is similar to the one illustrated in Fig. 11.71. Its dimensions are 11.6 mm in diameter $\times$ 3.6 mm high, 1.1 g weight. Lithium metal, 1.2 mm thick, is used as the anode; the

TABLE 11.23 Lithium/Iron Sulfide (Li/FeS₂) Cells

Cell no.	Eveready no.	Nominal voltage, V	Dimensions			Rated capacity, mAh	Energy density, Wh/L
			Diam., mm	Height, mm	Vol., cm³		
Y 1868	895	1.5	8.4	2.7	0.15	35*	325
Y 1841	801	1.5	11.5	4.2	0.44	100†	325
Y 2020	803	1.5	11.5	5.6	0.58	160†	385
. . .	. . .	1.5	23.0	2.3	0.95	295	435

*Rated at 30,000 Ω at 35°C.
†Rated at 15,000 Ω at 21°C.
SOURCE: Union Carbide Corp., Danbury, Conn.

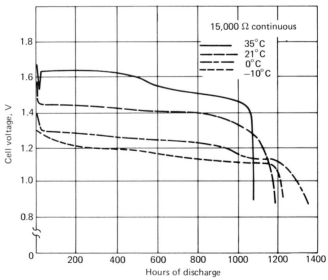

FIG. 11.72 Typical discharge curves, Li/FeS₂ button cells. (*Courtesy of Union Carbide Corp.*)

TABLE 11.24   Typical Impedance Range of Li/FeS₂ Cell, 100 mAh

| | Impedance, Ω | | | |
| | 1000 Hz | | 40 Hz | |
% depth of discharge*	Room temp. (20 to 35°C)	Low temp. (0 to −10°C)	Room temp. (20 to 35°C)	Low temp. (0 to −10°C)
Initial	140–160	350–375	125–150	375–425
10	85–115	275–300	85–115	300–350
30	75–100	240–260	75–100	275–325
50	75–100	215–245	65–90	235–285
60	85–115	215–245	75–100	250–300
80	125–150	250–275	135–160	275–325

*Discharged at 15,000 Ω, 35°C.
SOURCE: Union Carbide Corp., Danbury, Conn.

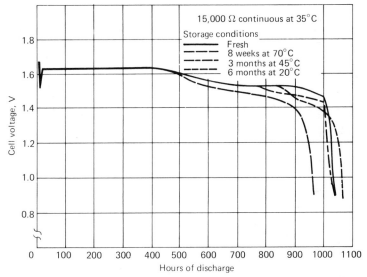

FIG. 11.73   Effect of storage on performance of Li/FeS₂ button cell. (*Courtesy of Union Carbide Corp.*)

cathode consists of FeS and graphite with a PTFE binder pressed to 0.78 mm. A nonwoven polypropylene separator is used. The container is made from nickel-plated steel with a polypropylene gasket.

Figure 11.74 shows the discharge characteristics of this cell at various loads at 30°C, corresponding to energy densities of about 120 Wh/kg and 250 Wh/L at the light loads. The working voltage was 1.4 V for the 10-kΩ discharge and 1.25 V for the 2.5 kΩ discharge. Figure 11.75 illustrates the behavior of the cell at various temperatures at a 2.5-kΩ load. Performance to 0°C was obtained at this moderate discharge rate.

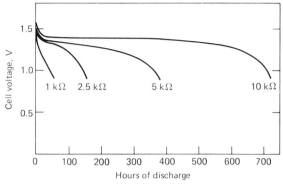

FIG. 11.74 Typical discharge curve of a Li/FeS button cell at 30°C. (*From Ref. 43.*)

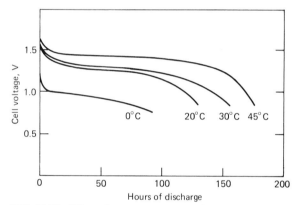

FIG. 11.75 Effect of temperature on performance of Li/FeS button cell. (*From Ref. 43.*)

11.12.4 Cell Types and Applications

The major applications of the 1.5-V lithium/iron sulfide cells are as a direct replacement for the zinc/silver oxide and zinc/mercuric oxide miniature button cells used in long-life low-power applications such as watches, calculators, and photographic devices. The advantages of the lithium cells are their better shelf life, low-temperature performance, leakage resistance, and lower cost. The aqueous cells have the advantage of lower internal resistance and better performance at higher-rate discharges. There is no significant difference in capacity or service life.

Table 11.23 lists the Li/FeS$_2$ cells that are available. Cells using the Li/FeS system have not been manufactured for commercial use.

These low-power low-capacity lithium cells can be handled in the same manner as the aqueous button cells; charging and incineration should be avoided.

11.13 LITHIUM/BISMUTH OXIDE (Li/Bi$_2$Pb$_2$O$_5$ and Li/Bi$_2$O$_3$) CELLS

The lithium/lead bismuthate and the lead/bismuth trioxide cells are other lithium/ solid cathode cells with a working voltage of about 1.5 V, similar to conventional systems. The cells are manufactured in a button-type configuration and are interchangeable with the zinc/silver oxide and zinc/mercuric oxide cells, providing advantages in longer shelf life, lighter weight, and lower cost. The Li/Bi$_2$ Pb$_2$O$_5$ and Li/Bi$_2$O$_3$ button cells are designed for low-power applications, but their internal resistance is low enough to support pulsing loads required in analog watches and other electronic devices.[1,45]

11.13.1 Chemistry

The Li/Bi$_2$Pb$_2$O$_5$ cell uses a lithium anode, a cathode of the active material, Bi$_2$Pb$_2$O$_5$, mixed with lead powder to provide electronic conductivity, and a Teflon binder. The electrolyte is a solution of LiClO$_4$ in 1,3-dioxolane. The reaction mechanism is given as

$$10 \text{ Li} + \text{Bi}_2\text{Pb}_2\text{O}_5 \rightarrow 5 \text{ Li}_2\text{O} + 2 \text{ Bi} + 2 \text{ Pb}$$

and involves a 10-electron transfer per mole of Bi$_2$Pb$_2$O$_5$. The theoretical cell voltage, as computed by assuming the mixture Bi$_2$O$_3$ + 2 PbO, is approximately 2 V. In an actual cell, the observed open-circuit voltage ranges from 2.2 to 2.5 V; the higher voltage is attributed to the presence of oxygen or higher oxide states of lead or bismuth. On discharge, the high voltage rapidly falls to approximately 1.8 V at typical discharge rates.

The reaction mechanism of the Li/B$_2$O$_3$ cell involves a 6-electron transfer:

$$6\text{Li} + \text{Bi}_2\text{O}_3 \rightarrow 3\text{Li}_2\text{O} + 2\text{Bi}$$

The theoretical voltage is about 2.04 V, the observed open-circuit voltage is approximately 2.2 V. Even under very light loads, the operating voltage falls to 1.7 V.

11.13.2 Construction

The cell is constructed in a typical button cell configuration, similar to conventional cells. The cathode is shaped into a pellet by cold-pressing a mixture of the active material, lead powder, and a Teflon binder.

11.13.3 Performance

Typical discharge curves at 20°C for the 185-mAh Li/Bi$_2$Pb$_2$O$_5$ cell are shown in Fig. 11.76. The effect of discharge under varying current load conditions on the delivered capacity and average discharge voltage is shown in Fig. 11.77. The operation of the cell under pulse load conditions is shown in Fig. 11.78. The operating temperature range is 45 to -10°C. On storage, the self-discharge rate is about 3%; a 6% loss of capacity occurs at 20°C, 6% after 3 months storage at 60°C.

The performance of the Li/Bi$_2$O$_3$ cells on discharge is shown in Fig. 11.79. The resistance of the 45-mAh cell is about 150 Ω; the 35-mAh cell about 200 Ω.

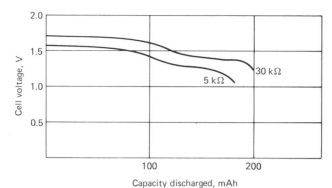

FIG. 11.76 Lithium/lead bismuthate cells—typical discharge curves, 20°C, 185-mAh cell. (*Courtesy of SAFT Corp. of America.*)

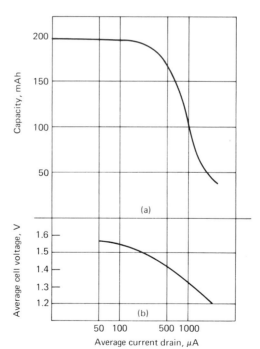

FIG. 11.77 Lithium/lead bismuthate cells—characteristics as a function of current drain, 20°C, 185-mAh cell (*a*) Capacity to 1.2-V end voltage; (*b*) average discharge voltage to 1.2-V end voltage. (*Courtesy of SAFT Corp. of America.*)

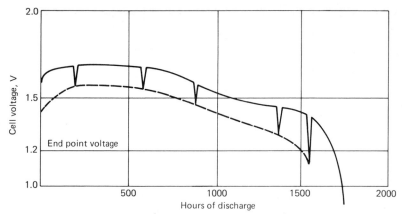

FIG. 11.78 Lithium/lead bismuthate cells—operation under pulse load at 20°C, 185-mAh cell. Background load, 15 kΩ; pulse load, 1 mA of 1-ms duration. (*Courtesy of SAFT Corp. of America.*)

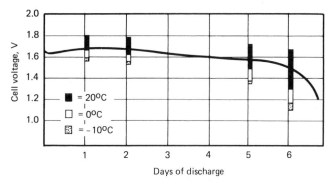

FIG. 11.79 Discharge characteristics of the lithium/bismuth trioxide cell. (*Courtesy of VARTA.*)

11.13.4 Cell Types and Sizes

A listing of the the $Li/Bi_2Pb_2O_5$ and Li/Bi_2O_3 cells is given in Table 11.25.[46]

11.14 LITHIUM/SILVER CHROMATE (Li/Ag_2CrO_4) CELL

The lithium/silver chromate cell is noted for its high volumetric energy density and long storability. It is intended to be used in low-rate applications requiring high reliability, long shelf life, and operating periods which may exceed 10 y, such as cardiac pacemakers, human implant devices, and memory preservation, where its high cost is acceptable.[1] The energy density of the cell at the low discharge rate is about 275 Wh/kg and 700 to 750 Wh/L. Figure 11.80 compares the energy output of the Li/Ag_2CrO_4

TABLE 11.25   Lithium/Bismuth Oxide and Lithium/Lead Bismuthate Cells

| Type | IEC size | Dimensions | | | Weight, g | Rated capacity, mAh | Energy density | |
		Diam., mm	Ht., mm	Vol., cm³			Wh/g	Wh/L
1154	44	11.6	5.4	0.57	1.85	185	150	485
1136	42	11.6	3.6	0.38	1.25	100	120	395
1121	55	11.6	2.1	0.22	0.85	50	88	340
926*		9.5	2.6	0.18	0.7	45	95	375
921*	60	9.5	2.1	0.15	0.55	35	90	340
721	58	7.9	2.1	0.10	0.40	18	70	270

SOURCE: SAFT Corp. of America, Valdosta, Ga. ($Li/Bi_2Pb_2O_5$) and *VARTA, Hannover, West Germany (Li/Bi_2O_3).

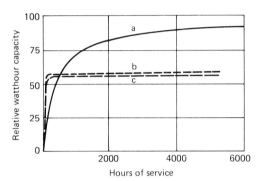

FIG. 11.80   Comparison of performance of Li/Ag_2CrO_4 with conventional alkaline button cells. (*a*) Li/Ag_2CrO_4; (*b*) Zn/HgO; (*c*) Zn/Ag_2O.

cell with that of conventional cells of similar size; the advantage of the lithium cell at low power levels is evident.

11.14.1 Chemistry

The lithium/silver chromate cell uses a lithium anode on an expanded nickel collector grid, a cathode of silver chromate and graphite, and an electrolyte of lithium perchlorate in propylene carbonate. The discharge reaction is given as

$$2Li + Ag_2CrO_4 \rightarrow Li_2CrO_4 + 2Ag$$

The open-circuit voltage is about 3.3 V. The discharge data show two voltage plateaus; the first occurs at about 3.2 V due to the reduction of the silver ion and which covers about 70% of the total discharge duration; the second occurs at about 2.5 V and is due to the reduction of the chromate ion. The second plateau can serve as an end-of-discharge indicator. The stability of the pure silver chromate and the lithium in the propylene carbonate electrolyte is responsible for the cell's good storage and longevity.

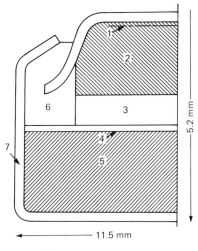

Legend:

1 Expanded nickel grid
2 Lithium anode
3 Nonwoven polypropylene separator
4 Barrier layer
5 Silver chromate cathode
6 Polypropylene grommet
7 Nickel plated steel

FIG. 11.81 Cross section of Li/Ag$_2$CrO$_4$ button cell. (*Courtesy of SAFT Corp. of America.*)

11.14.2 Construction

The lithium/silver chromate cell is manufactured in several configurations: flat or button, rectangular, and D shapes. A typical button cell, sealed with a polypropylene gasket, is shown in Fig. 11.81 and a hermetically sealed rectangular cell in Fig. 11.82.

11.14.3 Performance

A typical discharge curve of the Li/Ag$_2$-CrO$_4$ cell at 37°C, illustrating the two-step discharge, is shown in Fig. 11.83. The change of the cell's impedance during the discharge is also shown. The operating range of the cell is from -10 to 55°C.

The lithium/silver chromate cell shows negligible self-discharge. Performance of the cell after storage is shown in Fig. 11.84; performance is unaffected after 4 y of storage at 45°C. Other tests showed that the cell can be stored at much higher temperatures, e.g., 1 month at 100°C, without affecting the capacity. Under extremely light pacemaker loads, the service life of the cell should reach 6 to 10 y.[47,48]

11.14.4 Cell Types and Sizes

A listing of typical lithium/silver chromate cells is given in Table 11.26.[46]

11.15 LITHIUM/CHROMIUM OXIDE (Li/CrO$_x$) CELLS

11.15.1 General Characteristics

The lithium/chromium oxide system has also been manufactured in small cylindrical cell configurations for low drain applications. The cell has an open-circuit voltage of 3.8 V; the nominal voltage is 3 V.

11.15.2 Construction

The cylindrical Li/CrO$_x$ cell uses an inside-out bobbin construction, similar to the one used for the Li/CuO cell, as illustrated in Fig. 11.85. A stainless-steel nail is used for the anode current collector.

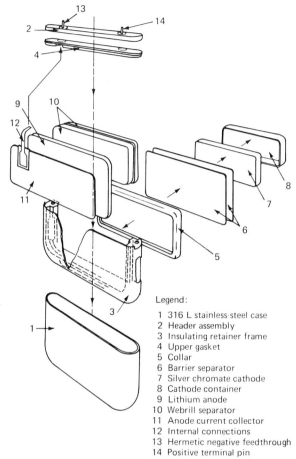

Legend:
1 316 L stainless-steel case
2 Header assembly
3 Insulating retainer frame
4 Upper gasket
5 Collar
6 Barrier separator
7 Silver chromate cathode
8 Cathode container
9 Lithium anode
10 Webrill separator
11 Anode current collector
12 Internal connections
13 Hermetic negative feedthrough
14 Positive terminal pin

FIG. 11.82   Cross section of Li/Ag$_2$CrO$_4$ rectangular cell. (*Courtesy of SAFT Corp. of America.*)

11.15.3 Cell Types and Performance

The Li/CrO$_x$ cell has been manufactured in a ½ AA size cell with the characteristics listed in Table 11.27. Typical discharge curves for the ½ AA size cell, rated at 900 mAh when discharged at a load of 5.6 kΩ, are shown in Fig. 11.86. The cell is designed mainly for low drain applications and has an internal resistance of about 50 Ω.

REFERENCES

1. J.P. Gabano, *Lithium Batteries,* Academic Press, London, 1983.

2a. Technical data, Foote Mineral Co., Exton, Pa.; Lithium Corp. of America, Gastonia, N. C.

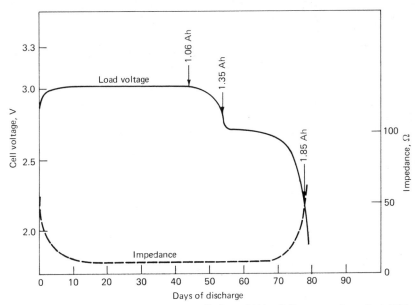

FIG. 11.83 Discharge and impedance profile for a Li/Ag₂CrO₄ rectangular cell at 37°C. (*Courtesy of SAFT Corp. of America.*)

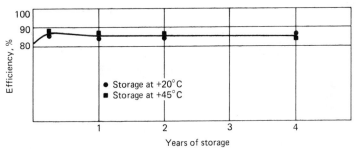

FIG. 11.84 Effect of storage on discharge efficiency of Li/AgCrO₄ cell. (*Courtesy of SAFT Corp. of America.*)

2b. H. R. Grady, "Lithium Metal for the Battery Industry," *J. Power Sources* **5**:127 (1980), Elsevier Sequoia S.A., Lausanne, Switzerland.

3. G. Eichinger and J. O. Besenhard, "High Energy Density Lithium Cells," pt. II, "Cathodes and Complete Cells," *J. Electroanal. Chem.* **72**:1 (1980), Elsevier Sequoia S.A., Lausanne, Switzerland.

4. J. T. Nelson and C. F. Green, "Organic Electrolyte Battery Systems," U.S. Army Material Command report no. HDL-TR-1588, Washington, March 1972.

5. J. O. Besenhard and G. Eichinger, "High Energy Density Lithium Cells," pt. I, "Electrolytes and Anodes," *J. Electroanal. Chem.* **68**:1 (1976), Elsevier Sequoia S.A., Lausanne, Switzerland.

TABLE 11.26 Typical Lithium/Silver Chromate Cells

Cell type number		Rated capacity, mAh	Dimensions, mm				Weight, g	Internal impedance, Ω		Energy density	
SAFT	IEC		Height	Diameter	Length	Width		BOL*	EOL**	Wh/kg	Wh/L
					Button cell						
114	R44	130	5.4	11.4	…	…	1.7	800	<2000	230	395
210		810	9.2	21.0	…	…	8.9	400	<2000	270	760
273		1220	7.9	27.3	…	…	13	300	<1500	275	790
355/6		1750	6.0	35.5	…	…	19.5	100	<1000	270	880
355/10		2870	10.0	35.5	…	…	29.4	150	<1000	290	870
					Rectangular cells						
LiR123		2090	23	…	45	8.9	26	50	<1000	240	677
LiR132		2450	32	…	45	8.9	35			210	575
					D-shaped cells						
LiD123		1180	22	radius 22.5 mm	45	7.0	20	500	<1000	180	525
LiD128		1800	28	radius 22.5 mm	45	8.0	25	50	<1000	215	530

*Beginning of life.
†End of life.
SOURCE: SAFT, Poitiers, France, and SAFT Corp. of America, Valdosta, Ga., 31601

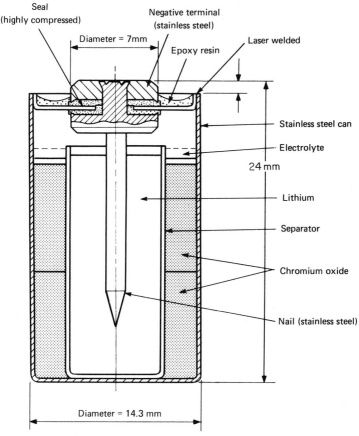

FIG. 11.85 Schematic diagram of the lithium/chromium oxide cell, 1/2 AA size. (*Courtesy of VARTA.*)

TABLE 11.27 Characteristics of Lithium/Chromium Oxide Cell

Cell type	Dimensions Diam., mm	Ht., mm	Vol., cm³	Weight, g	Rated capacity, mAh	Energy density Wh/kg	Wh/L
½AA	14.5	24	4	10	900 at 5.6 kΩ	270	675

6. A. N. Dey, "Lithium Anode Film and Organic and Inorganic Electrolyte Batteries," *Thin Solid Films*, Elsevier Sequoia S.A., Lausanne, Switzerland, 1977, vol. 43, p. 131.

7. S. Gilman and W. Wade, "The Reduction of Sulfuryl Chloride at Teflon-Bonded Carbon Cathodes," *J. Electrochem. Soc.* **127**:1427 (1980).

8. A. Meitav and E. Peled, "Calcium-Ca(AlCl₄)₂-Thionyl Chloride Cell: Performance and Safety," *J. Electrochem. Soc.* **129**:3 (1982).

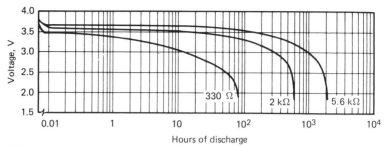

FIG. 11.86   Discharge characteristics of the lithium/chromium oxide cell, 20°C, 1/2 AA size [load 330 Ω (~82 mA), 2 kΩ (~155 mA), 5.6 kΩ (~0.58 mA)]. (*Courtesy of VARTA.*)

8a. R. L. Higgins and J. S. Cloyd, "Development of the Calcium-Thionyl Chloride System," *Proc. 29th Power Sources Conf.,* Electrochemical Society, Pennington, N.J., June 1980.

8b. M. Binder, S. Gilman, and W. Wade, "Calcium-Sulfuryl Chloride Primary Cell," *J. Electrochem. Soc.* **129**:4 (1982).

8c. H. Yamin and E. Peled, "Electrochemistry of a Non-Aqueous Lithium/Sulfur Cell," *J. Power Sources* 9:281(1983), Elsevier Sequoia, S.A. Lausanne, Switzerland.

 9. L. W. Langrish, "A Comparison of the Primary Lithium Systems," *Proc. 29th Power Sources Conf.,* Electrochemical Society, Pennington, N.J., June 1980.

10. N. I. Sax, *Dangerous Properties of Industrial Materials,* Van Nostrand Reinhold Co., New York, 1979.

11a. *Transportation,* Code of Federal Regulations CFR 49, U.S. Government Printing Office, Washington, D.C.

11b. Exemption DOT-E-7052, Department of Transportation, Washington, D.C.

11c. "Technical Instructions for the Safe Transport of Dangerous Goods by Air," International Civil Aviation Organization, DOC 9284-AN/905, Montreal, Quebec, Canada.

12. E. H. Reiss, "Considerations in the Use and Handling of Lithium-Sulfur Dioxide Batteries," *Proc. 29th Power Sources Conf.,* Electrochemical Society, Pennington, N.J., June 1980.

13. R. E. Ralston, "Recent Advances in LiSO₂ Battery Technology," *Proc. 29th Power Sources Conf.,* Electrochemical Society, Pennington, N.J., June 1980.

14. H. Taylor, "The Storability of Li/SO₂ Cells," *Proc. 12th Intersociety Energy Conversion Engineering Conf.,* American Nuclear Society, LaGrange Park, Ill., 1977.

15. D. Linden and B. McDonald, "The Lithium-Sulfur Dioxide Primary Battery—Its Characteristics, Performance and Applications," *J. Power Sources* 5:35 (1980), Elsevier Sequoia S.A., Lausanne, Switzerland.

16. R.C. McDonald et al., "Investigation of Lithium Thionyl Chloride Battery Safety Hazard," Technical Report N60921-81-C-0229, Naval Surface Weapons Center, Silver Spring, Md., January 1983.

17. D. Morley and R. J. Solar, "The Li/SOCl₂ Cell as an Effective Power Source," *Proc. 28th Power Sources Conf.,* Electrochemical Society, Pennington, N.J., June 1980.

18a. "Tadiran Lithium Inorganic Cells," Tadiran Israel Electronics Industries, Ltd., publication LBR-1507, Tel Aviv, Israel, August 1981.

18b. M. Babai and U. Zak, "Safety Aspects of Low-Rate Li/SOCl₂ Batteries," *Proc. 29th Power Sources Conf.,* Electrochemical Society, Pennington, N.J., June 1980.

19. R. L. Zupancic, "Performance and Safety Characteristics of Small Cylindrical Li/SOCl₂ Cells," *Proc. 29th Power Sources Conf.,* Electrochemical Society, Pennington, N.J., June 1980.

20. K. M. Abraham and R. M. Mank, "Some Safety Related Chemistry of Li/SOCl₂ Cells," *Proc. 29th Power Sources Conf.,* Electrochemical Society, Pennington, N.J., June 1980.

21. Manufacturer's data, Altus Corp., San Jose, CA 95112.

22. J. F. McCartney, A. H. Willis, and W. J. Sturgeon, "Development of a 200 kWh Li/SOCl₂ Battery for Undersea Applications," *Proc. 29th Power Sources Conf.,* Electrochemical Society, Pennington, N.J., June 1980.

23. A. N. Dey, N. Hamilton, and W. Bowden, "Primary Li/SOCl₂ Cells—Development of Batteries for Laser Designator Applications," *Proc. 29th Power Sources Conf.,* Electrochemical Society, Pennington, N.J., June 1980.

24. N. Marincic and F. Goebel, "Performance Characteristics of Large Primary Li/SOCl₂ Cells," *Proc. 28th Power Sources Symp.,* Electrochemical Society, Pennington, N.J., June 1978.

25. K. F. Garoutte and D. L. Chua, "Safety Performance of Large Li/SOCl₂ Cells," *Proc. 29th Power Sources Conf.,* Electrochemical Society, Pennington, N.J., June 1980.

26. F. Goebel, R. C. McDonald, and N. Marincic, "Performance Characteristics of the Minuteman Lithium Power Source," *Proc. 29th Power Sources Conf.,* Electrochemical Society, Pennington, N.J., June 1980.

27. K. A. Klinedinst, "Lithium/Sulfuryl Chloride Electrochemical Cells," Electrochemical Society, fall meeting, Pennington, N.J., October 1981.

28. C. C. Liang, P. W. Krehl, and D. A. Danner, "Bromine Chloride as a Cathode Component in Lithium Inorganic Cells," *J. Appl. Electrochem.* 1981.

29. C. C. Liang, M. E. Bolster, and R. M. Murphy, "The Li/Cl₂ in SO₂Cl₂ Inorganic Battery System," *J. Electrochem. Soc.,* vols. 128, 130, Pennington, N.J., August 1981, June 1983.

30. H. Ikeda, S. Narukawa, and S. Nakaido, "Characteristics of Cylindrical and Rectangular Li/MnO₂ Batteries," *Proc. 29th Power Sources Conf.,* Electrochemical Society, Pennington, N.J., 1980.

31. Manufacturers' data sheets, Duracell, Inc., Tarrytown, N.Y.; Sanyo Electric Co., Osaka, Japan; Union Carbide Corp., Danbury, Conn.; Ray-O-Vac Corp., Madison, Wis.

32a. A. Morita, T. Iijima, T. Fujii, and H. Ogawa, "Evaluation of Cathode Materials for the Lithium/Carbon Monofluoride Battery," *J. Power Sources* 5:111 (1980), Elsevier Sequoia, Lausanne, Switzerland, 1980.

32b. P.S. Clark, *Design News,* May 23, 1983.

33. R. L. Higgins and L. R. Erisman, "Applications of the Lithium/Carbon Monofluoride Battery," *Proc. 28th Power Sources Symp.,* Electrochemical Society, Pennington, N.J., June 1978.

34. Manufacturer's data sheets, Eagle-Picher Industries, Joplin, Mo.; Panasonic Co., Secaucus, N. J.

35. A. Bredland, T. Messing, and J. Paulson, "Performance and Safety Characteristics of a Li/CuS NEDA 1604 Battery," *Proc. 29th Power Sources Conf.,* Electrochemical Society, Pennington, N.J., June 1980.

36. D. Linden, N. Wilburn, and E. Brooks, "Organic Electrolyte Batteries," *Power Sources,* vol. 4, Oriel Press, Newcastle-upon-Tyne, England, 1972.

37. T. Iijima, Y. Toyoguchi, J. Nishimura, and H. Ogawa, "Button-Type Lithium Battery Using Copper Oxide as a Cathode," *J. Power Sources* 5:1 (1980), Elsevier Sequoia, Lausanne, Switzerland.

38. G. Lehmann, G. Gerbier, A. Brych, and J. P. Gabano, "The Copper Oxide-Lithium Cell," *Power Sources,* vol. 5, Academic, London 1974.

39. Y. Jumel, M. Broussely, A. Thunder, and G. W. Allvey, "Properties of the Li-CuO Couple," *Proc. 28th Power Sources Symp.,* Electrochemical Society, Pennington, N.J., June 1978.

40. J. Turner, et al., "Further Studies on the High Energy Density Li/CuO Organic Electrolyte System," *Proc. 29th Power Sources Conf.,* Electrochemical Society, Pennington, N.J., June 1980.

41. Manufacturers' data sheets, SAFT, Poitiers, France; SAFT Corp. of America, Valdosta, Ga.; SAFT-Sogea Batteries, Ltd., Hampton, England.

42. Manufacturer's technical data, Union Carbide Corp., Danbury, Conn.

43. Y. Uetani, K. Yokoyama, and T. Kawai, "FeS/Li Organic Electrolyte Cell," *Proc. 28th Power Sources Symp.,* Electrochemical Society, Pennington, N.J., June 1978.

44. Y. Uetani, K. Yokoyama, and O. Okamoto, "Preparation of Iron Sulfides and the Study of Their Electrochemical Characteristics for Use in a Non-Aqueous Lithium Battery," *J. Power Sources* 5:1 (1980), Elsevier Sequoia, Lausanne, Switzerland.

45. M. Broussely, Y. Jumel, and J. P. Gabano, "Lithium Batteries with Voltage Compatibility with Conventional Systems," *J. Power Sources,* 5:1 (1980), Elsevier Sequoia, Lausanne, Switzerland.

46. Manufacturers' technical data, SAFT, Poitiers, France; SAFT Corp. of America, Valdosta, Ga.

47. G. Lehmann, T. Rassinoux, and J. P. Gabano, "The Silver Chromate-Lithium Cell," *Power Sources,* vol. 4, *Proc. 8th Intern. Symp.,* Oriel Press, Newcastle-upon-Tyne, England, 1972.

48. G. Gerbier, G. Lehmann, P. Lenfant, and J. P. Rivault, "Reliability of Lithium-Silver Chromate Cells," *Power Sources,* vol. 6, *Proc. 10th Intern. Symp.,* Academic Press, London, 1976.

Solid-Electrolyte Cells

by
Boone B. Owens, Paul M. Skarstad,
and Darrel F. Untereker

12.1 GENERAL CHARACTERISTICS

Any battery consists of an electrochemically reactive couple separated by an ion-transport medium or electrolyte. In most familiar batteries, the electrolyte is a liquid. However, the availability of solids capable of being fabricated into electronically insulating elements with fairly low overall ionic resistance has stimulated the development of solid-electrolyte batteries. Several of these types of batteries have become commercially available and are important power sources at normal ambient temperatures ($\approx$ 25°C) for heart pacemakers, for preserving volatile computer memory, and for other low-power applications requiring long shelf and service life. The characteristics of the major types of ambient-temperature solid-electrolyte batteries are summarized in Table 12.1 and covered in detail in this chapter. Review articles, which cover all solid-electrolyte battery systems, including those not commercially available, are found in Refs. 1 to 5.

TABLE 12.1 Characteristics of Solid-Electrolyte Cells

System	Cell voltage, V	Energy density at 1-y rate	
		Wh/L	Wh/kg
$Ag/RbAg_4I_5/Me_4NI_5$, C	0.66	40–80	15–25
$Li/LiI(Al_2O_3)/PbI_2,PbS,Pb$	1.9	300–600	75–150
$Li/LiI/I_2(P2VP)$	2.8	350–700	120–200
$Li/LiI(SiO_2)/Me_4NI_5$	2.75	400	125

Commercially available solid-electrolyte batteries use lithium anodes. Lithium is attractive as an anode material for several reasons; first, it has a high specific capacity on both a weight and a volume basis. Second, it is strongly electropositive; this leads to high voltages when coupled with typical cathode materials. Finally, suitable lithium ion conductors are available for use as solid electrolytes. Table 12.2 compares theoretical values of the equivalent weight, equivalent volume, voltage, specific capacity, and specific energy for battery systems forming the metal iodide as the discharge product. Among the alkali metals, the small equivalent volume of lithium more than compensates for the slightly higher voltages obtained with the heavier members of the group in determining the theoretical energy density. On the other hand, the high voltage with lithium compensates for the capacity advantage of the polyvalent metals. Only metal/I_2 cells of Ca, Sr, and Ba have theoretical specific energies approaching that of Li. Moreover, no suitable conductive solid electrolytes are known for the polyvalent metals in Table 12.2. Therefore, lithium is the anode material of choice for commercial cells at the present time.

TABLE 12.2 Theoretical Values for the Volumetric Specific Capacities and Energies of Balanced Metal/I_2 Cells*

Anode Metal	Anode equivalent weight, g/eq	Anode equivalent volume, cm³/eq	E^0, V	Cell specific capacity, Ah/cm³	Cell specific energy, Wh/cm³
Monovalent metals					
Li	6.9	13.0	2.8	0.69	1.9
Na	23.0	23.7	3.0	0.54	1.6
K	39.1	44.9	3.4	0.38	1.3
Rb	85.5	55.9	3.4	0.33	1.1
Cs	132.9	71.1	3.5	0.28	1.0
Cu	63.5	7.1	0.7	0.82	0.59
Ag	107.9	10.3	0.7	0.75	0.51
Tl	204.4	17.2	1.3	0.63	0.81
Polyvalent metals					
Be	4.5	2.4	1.1	0.96	1.1
Mg	12.2	7.0	1.9	0.82	1.5
Ca	20.0	13.0	2.8	0.69	1.9
Sr	43.8	16.9	2.9	0.63	1.9
Ba	68.7	19.6	3.1	0.59	1.9
Zn	32.7	4.6	1.1	0.93	1.0
Cd	56.2	6.5	1.1	0.89	1.0
Al	9.0	3.3	1.0	0.83	0.9

*Equivalent weights, densities, and cell voltages were obtained from data in Ref. 6.

Solid-state cells offer several advantages over solid-electrolyte cells with fluid electrode components. They generally exhibit high thermal stability, low rates of self-discharge (shelf life of 5 to 10 y or better), the ability to operate over a wide range of environmental conditions (temperature, pressure, and acceleration), and high energy densities (300 to 700 Wh/L). On the other hand, limitations associated with a complete solid-state battery include relatively low power capability (microwatt range) due to the high impedance of most solid electrolytes at normal ambient temperature, possible mechanical stresses due to volume changes associated with electrode discharge reactions, and reduced electrode efficiencies at high discharge rates (see Table 12.3).

TABLE 12.3   Major Advantages and Disadvantages of Lithium Solid-Electrolyte Cells

Advantages	Disadvantages
Excellent storage stability—shelf life of 10 y or better	Low current drains (microamperes)
	Power output reduced at low temperatures
High energy densities	Care must be exercised to prevent shorting
Hermetically sealed—no gassing or leakage	or shunting of cell (which could be a
Wide operating temperature range, up to 200°C	relatively high drain on cell)
Shock- and vibration-resistant	

The two commercial solid-electrolyte battery systems are based on the solid electrolyte lithium iodide, either formed in situ during cell manufacture or dispersed with alumina and formed as a discrete $LiI(Al_2O_3)$ separator.

In the battery system

$$Li/LiI/I_2(P2VP)$$

the solid electrolyte LiI is formed in situ as the discharge product of the cell reaction. The cathode is a mixture of solid iodine and a saturated viscous liquid solution containing poly-2-vinylpyridine (P2VP) and iodine. The $Li/LiI/I_2(P2VP)$ battery can be regarded as a quasi-solid-state system because of the high viscosity of the polymer-containing liquid phase and the preponderance of solid iodine in the material. However, the viscous liquid phase does impart a plasticity to the $I_2(P2VP)$ cathode which makes these solid-state cells better able to adapt to volumetric changes during cell discharge.

The other commercial battery

$$Li/LiI(Al_2O_3)/metal salts$$

is a true solid-state battery; all components are solids with no significant amount of liquid or vapor component coexisting.

12.2 Li/LiI(Al₂O₃)/METAL SALT CELLS

12.2.1 General Characteristics

Solid-state batteries based on this system have the following properties: (1) long shelf life; (2) low power capability; (3) hermetically sealed, no gassing; (4) broad operating temperature range: −40 to 170°C for pure lithium anodes; up to and beyond 300°C with compound anodes; (5) high volumetric energy, density up to 0.6 Wh/cm³ in the system presently in production[7,8] and up to 1.0 Wh/cm³ in systems under development.[9,10]

These batteries are recommended for low-rate applications. They are particularly suited for applications requiring long life at low-drain or open-circuit conditions.

Possible ambient-temperature applications include watches, pacemakers, and monitoring devices. The power characteristics and long shelf life are well-suited to providing backup power for preserving volatile computer memory.

12.2.2 Cell Chemistry

Two different cathodes have been used in these commercial solid-state cells. The original battery system used a mixture of PbI_2 and Pb for the cathode.[7] This has been replaced by a mixture of PbI_2, PbS, and Pb.[8] A new system under development, with increased energy density, uses a mixture of TiS_2 and S as the cathode.[9] Properties of these battery systems are summarized in Table 12.4. Other cathode materials have also been investigated in cells of this type. They include As_2S_3 and various other metal sulfides.[10]

TABLE 12.4 Practical Solid-State Lithium Battery Systems*

Battery system	Cell	Practical energy density, Wh/cm³	Cell reactions	$E,^0$ V
1X	Li/LiI(Al₂O₃)/PbI₂, Pb	0.1–0.2	$2 Li + PbI_2 \rightarrow 2 LiI + Pb$	1.9
3X	Li/LiI(Al₂O₃)/PbI₂, PbS, Pb	0.3–0.6	$2 Li + PbI_2 \rightarrow 2 LiI + Pb$	1.9
			$2 Li + PbS \rightarrow Li_2S + Pb$	1.8
5X	Li/LiI(Al₂O₃)/TiS₂, S	0.9–1.0	$Li + TiS_2 \rightarrow Li TiS_2$	2.5–1.9
			$2 Li + S \rightarrow Li_2S$	2.3

*Ref. 17.
MANUFACTURER: Duracell, Inc. Bethel, Conn.

The solid electrolyte used in these solid-state cells is a dispersion of lithium iodide and lithium hydroxide with alumina. Pure lithium iodide has a conductivity of about 10^{-7} $\Omega^{-1} \cdot cm^{-1}$ at room temperature.[11–13] The conductivity of lithium iodide can be enhanced by several orders of magnitude by dispersal of substantial amounts (≈ 50 m/o) of high surface-area alumina in the lithium iodide. Lithium ion conductivities as high as $10^{-4} \Omega^{-1} \cdot cm^{-1}$ have been reported in such dispersions at 25°C.[10,14]

The dispersed-phase lithium iodide with enhanced lithium ion conductivity remains a good electronic insulator with the partial electronic (hole) conductivity approximately 10^{-40} $\Omega^{-1} \cdot cm^{-1}$.[10] Thus, this material is well-suited for application as a solid electrolyte. The mechanism by which the high surface-area alumina, itself a poor conductor, enhances the conductivity of lithium iodide is not well understood. Enhanced interfacial conduction has been suggested.[15] The Arrhenius plot of conductivity for LiI (Al_2O_3) (Fig. 12.1) in the range -50 to 300°C is characterized by a single activation energy (0.4 eV).[10] At 300°C the conductivity reaches 0.1 $\Omega^{-1} \cdot cm^{-1}$, allowing the construction of batteries with substantial rate capability. High-temperature solid-state secondary batteries using this electrolyte have been investigated for load-leveling and vehicle-propulsion applications.[10,16]

12.2.3 Cell Construction

The dispersed-phase ionic conductor $LiI(Al_2O_3)$ can be fabricated as a dense, conductive solid electrolyte element by uniaxial compression at room temperature.[8] No high-

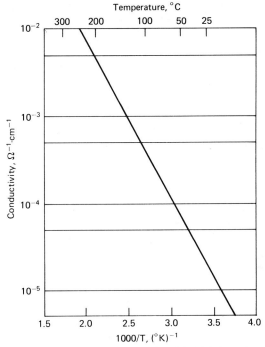

FIG. 12.1   Ionic conductivity of the $Li^{\cdot}(Al_2O_3)$ electrolyte. (*Courtesy of Duracell, Inc.*)

temperature sintering is required. The minimum thickness of an electrolyte element in a practical cell construction is about 0.2 mm. The cell is designed in a button or circular disk configuration as shown in Fig. 12.2. The cell is fabricated by sequentially pressing the cell components (cathode, electrolyte, and anode) at appropriate pressures into the cell in the form of a thin circular disk. This cell can be incorporated in series-parallel combinations to give batteries with desired voltage and current characteristics. The battery is fabricated by placing the cells in a stainless-steel container, holding the cells under spring pressure. There is no need to encapsulate or isolate each cell as required in conventional batteries. The energy density achieved in the cell and battery described in Table 12.5 is 0.4 Wh/cm³. Energy densities up to 0.6 Wh/cm³ have been achieved in practical cells with this chemical system.

TABLE 12.5   Properties of Commercial $Li/LiI/PbI_2,PbS$ Batteries

Type number	E^0, V	Capacity, mAh	Diameter, cm	Height, cm	Volume, cm³	Weight, g	Energy density Wh/cm³	Wh/g
305127	1.9	350	2.90	0.25	1.5	7.25	0.445	0.92
305159	3.8	350	2.97	0.58	4.0	15.86	0.332	0.84

MANUFACTURER: Duracell, Inc., Bethel, Conn.

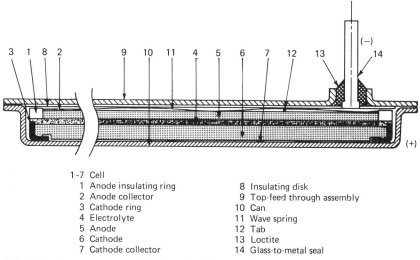

1-7 Cell
1 Anode insulating ring
2 Anode collector
3 Cathode ring
4 Electrolyte
5 Anode
6 Cathode
7 Cathode collector

8 Insulating disk
9 Top-feed through assembly
10 Can
11 Wave spring
12 Tab
13 Loctite
14 Glass-to-metal seal

FIG. 12.2 Cross section of solid-state cell. (*Courtesy of Duracell, Inc.*)

12.2.4 Performance Characteristics

The discharge properties of these solid-state batteries are characterized by high practical energy densities, low rate capability, and low rates of self-discharge. From the linear polarization curve (Fig. 12.3)[17] it is apparent that the resistance is essentially ohmic. Bridge measurements of cell resistance at 1 kHz are in agreement with values determined from polarization curves.

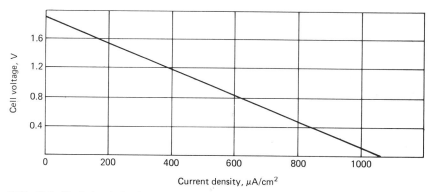

FIG. 12.3 Typical polarization curve (IX cell, 25°C, cell surface area: 1 cm²). (*Courtesy of Duracell, Inc.*)

Figure 12.4 shows the discharge curves for the 350-mAh cell at various discharge rates at 20°C. At discharge rates below 6 μA, the discharge efficiency is close to 100% to a 1-V cutoff. The discharge curves for the same cell at various temperatures are shown in Fig. 12.5 for discharges at 3 μA (0.55 μA/cm² of electrolyte area) and 12 μA. The improved performance attained at the higher discharge temperatures is evi-

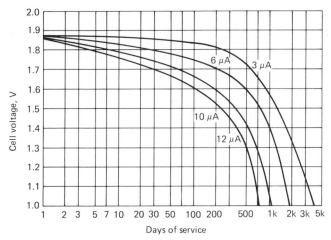

FIG. 12.4 Discharge curves, solid-state cell, 350 mAh, 20°C. (*Courtesy of Duracell, Inc.*)

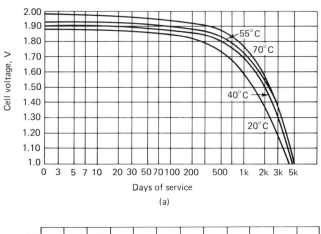

(a)

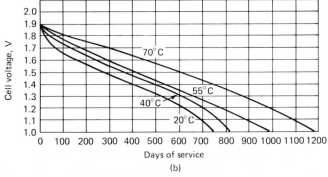

(b)

FIG. 12.5 Discharge curves, solid-state cell, 350 mAh. (*a*) 3-μA discharge; (*b*) 12-μA discharge. (*Courtesy of Duracell, Inc.*)

dent. These cells can be discharged at higher temperatures than shown; however, for operation above 125°C, a modified electrode structure is required. At lower temperatures, the current capability of the solid-state cell is reduced. In typical CMOS memory applications, however, the load current requirement usually is similarly reduced and the cell output tracks the CMOS power requirements. The service life delivered by the 350-mAh cell at various discharge temperatures and loads is given in Fig. 12.6. These data can be used to calculate performance of the cell under various conditions of use.[17, 17a]

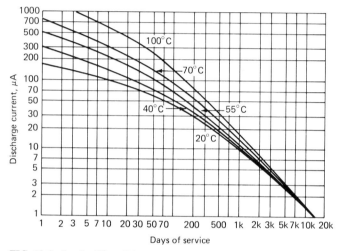

FIG. 12.6 Service life, solid-state cell, 350 mAh (to 1-V cutoff). (*Courtesy of Duracell, Inc.*)

12.2.5 Storage

Long-term storage tests show that there is no loss of capacity after storage periods up to 1 y at 20, 45, and 60°C (Fig. 12.7). The cell can withstand storage to 200°C, but for best performance such high temperatures should be avoided. Temperatures above 200°C may cause bulging and failure of the seal. The long shelf life results from the chemical compatibility of the cell components, the absence of chemical reaction between the electrodes and the electrolyte, and the low electronic conductivity of the solid electrolyte which minimizes self-discharge. Longer-term tests show that there is no measurable loss of capacity after storage periods of 4 y at 20°C and up to 1 y at temperatures as high as 100°C. On the basis of these tests, it is projected that the shelf life of these cells is at least 15 to 20 y under normal storage conditions.[7]

12.2.6 Handling

The solid-state batteries are designed primarily for low power and long service life. These cells can withstand short-circuit and voltage reversal, although the conditions should be avoided. No explosion due to pressure buildup or chemical reaction is known

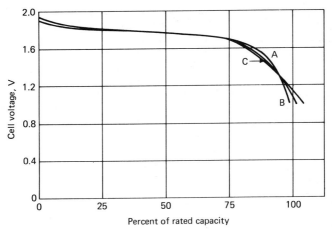

FIG. 12.7 Storage tests, solid-state cell, discharge after (*a*) no storage; (*b*) 1 y at 45°C; (*c*) 1 y at 60°C. (*Courtesy of Duracell, Inc.*)

to occur under recommended operating temperatures. Prolonged short-circuiting will result in a separation between the electrode and the electrolyte, making the cell inoperative.

12.3 THE LITHIUM/IODINE CELL

12.3.1 General Characteristics

Lithium/iodine batteries are based upon the reaction

$$Li + 1/2I_2 \rightarrow LiI$$

The specific reactions for the cell using poly-2-vinylpyridine (P2VP) in the cathode are

Anode $2\,Li \rightarrow 2\,Li^+ + 2e$

Cathode $2\,Li^+ + 2e + P2VP \cdot n\,I_2 \rightarrow P2VP \cdot (n-1)I_2 + 2LiI$

Overall $2\,Li + P2VP \cdot n\,I_2 \rightarrow P2VP \cdot (n-1)I_2 + 2\,LiI$

Lithium and iodine are consumed, and their reaction product, LiI, precipitates in the region between the two reactants. The LiI not only is the cell discharge product, but also serves as the cell separator. The theoretical specific energy for this cell is 1.9 Wh/cm^3 (see Table 12.2). Practical values approaching 1 Wh/cm^3 can be obtained at discharge rates of 1 to 2 μA/cm^2. Commercially available lithium/iodine batteries have a solid anode of lithium and a polyphase cathode which is largely iodine. The iodine is made conductive by the addition of an organic material. Pyridine-containing polymers are most often used for this purpose, the additive in all present commercial batteries being P2VP. At ambient temperatures the iodine/P2VP mixtures are two-phase in undischarged batteries, liquid plus excess solid iodine.[18] The iodine content of the cathode decreases during discharge of the battery, and the remaining cathode material becomes hard as the battery nears depletion. As discharge proceeds, the layer of lithium iodide becomes thicker. The resistance of the battery also increases because of the growing amount of discharge product.

The volume change accompanying the cell discharge is negative. The theoretical value for this volume change is -15% for complete discharge of a balanced mixture of pure iodine and lithium.[18] It is somewhat less when the chemical cathode is not pure iodine. For example, a volume change of -12% is expected if the cathode is 91% iodine by weight.[19] The volume change may be accommodated by the formation of a porous discharge product or by the formation of macroscopic voids in the cell.

Because the discharge product also functions as the electrolyte, this battery requires no discrete separator between the anode and the cathode material. Cells are formed by contacting the iodine-containing cathode directly with the lithium anode. The chemical reaction between these two materials immediately forms a thin layer of lithium iodide between the anode and cathode. This layer serves to electronically separate the two electroactive materials and prevents failure due to internal shorting of the anode and the cathode. This makes them especially suitable for applications requiring very high reliability.

Features of this cell system include low self-discharge, high reliability, and no gassing during discharge. Shelf life is 10 y or longer, and the cells can take a considerable amount of abuse without any catastrophic effects. Cells of this type have found commercial applications powering various low-power devices such as cardiac pacemakers, solid-state memories, and digital watches. Power sources for portable monitoring and recording instruments and the like are also possible applications.

All the currently available Li/I_2 batteries have a nominal capacity of 15 Ah or less, and most have deliverable capacities under 5 Ah. All the Li/I_2 batteries intended for medical applications are designed to be cathode-limited.

12.3.2 Battery Construction

Five types of Li/I_2 batteries are presently produced, three of which are used for medical applications such as cardiac pacemakers. Figure 12.8 shows the first type. This battery has a case-neutral design. It consists of a stainless-steel housing with a plastic insulator that lines the inside of the case. A lithium envelope (the anode) fits inside the plastic and contains the $I_2(P2VP)$ depolarizer. The cathode current collector is located in the center of the cell. Current collector leads from both the anode and the cathode go through hermetic feedthroughs in the case. The cell is formed by pouring molten iodine depolarizer into the lithium envelope. After the cathode material solidifies, the top of the lithium envelope is closed, the plastic cup is added, and the final assembly is completed. The construction used in this battery eliminates any contact between the case and the iodine depolarizer.

A second construction uses a case-positive design for batteries of similar size. A cutaway view of such a battery is shown in Fig. 12.9. This cell is manufactured in a slightly different manner from that in Fig. 12.8. It contains a central lithium anode and uses the stainless-steel case of the cell as the positive current collector. Most models are completely assembled with their header welded to the can before the cathode is added to the cell. Hot depolarizer is poured into the battery can through a small fill port which is later welded shut. The anode current collector is brought out via a glass and metal feedthrough.

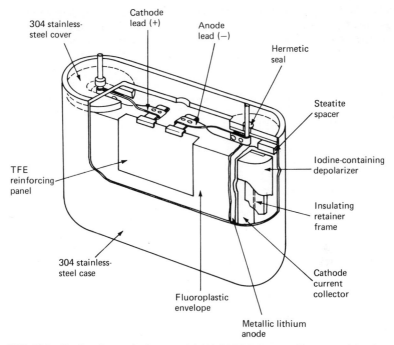

FIG. 12.8   Catalyst Research Corp. model 802/35 Li/I₂ battery. (*Courtesy of Catalyst Research Corp.*)

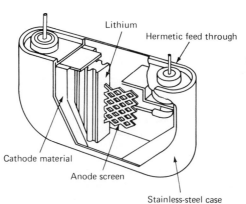

FIG. 12.9   Cutaway view of a typical can-positive Li/I₂ battery. (*Courtesy of Wilson Greatbatch, Ltd.*)

Manufacturers of these case-positive designs also precoat their anode assembly with a layer of pure P2VP prior to assembly. This coating is designed to protect the anode from the environment before assembly, but it also alters the electrical discharge behavior (see Sec. 12.3.4 below).

Another case-positive medical application battery has recently been introduced.

This battery is very similar to the other case-positive designs, but the cathode is not poured into the battery can. The iodine and P2VP are pelletized and then pressed onto the central anode assembly. After the pressing operation, the entire unit is slipped into a nickel battery can. An exploded view of this cell is shown in Fig. 12.10.

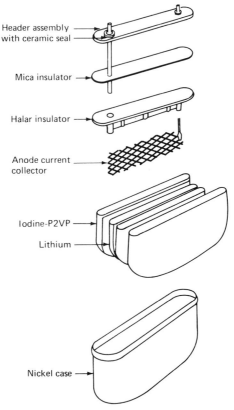

Header assembly with ceramic seal

Mica insulator

Halar insulator

Anode current collector

Iodine-P2VP

Lithium

Nickel case

FIG. 12.10 Case-positive Li/I₂ battery—exploded view. (*Courtesy of Catalyst Research Corp.*)

Case-neutral batteries were developed to prevent corrosion of the exterior case and to minimize leakage to the feedthrough by the iodine depolarizer. However, 5 y of real-time data have shown that no significant corrosion of stainless steel in contact with the cathode depolarizer or its vapor takes place in sealed can-positive cells. Tests show that corrosion occurs during the first few months after assembly and is limited to a 50-μ layer. Even at 60°C, corrosion of stainless steel by the iodine depolarizer has not proved to be a problem in the dry environment of the battery.

Li/I₂ medical batteries are produced in a variety of sizes and shapes to meet specific applications. Their profiles range from rectangular to semicircular or a combination. All of them are made quite thin and have flat sides because their primary application is in cardiac pacemakers. Cell thickness is typically 10 mm or less. The area of the lithium anodes ranges between 10 and 20 cm^2 in current batteries. Some cells use a ribbed anode (see Fig. 12.9) to increase the amount of active anode surface area in the battery.

The nonmedical batteries are made in more conventional button and cylindrical configurations. The hermetically sealed button cells are intended for powering digital watches and serving as backup power for computer memories. These batteries are made by pressing iodine cathode and lithium anode layers into a stainless-steel cup. The cup is the positive-current collector. A glass and metal feedthrough brings the negative terminal to the exterior. Figure 12.11 shows a view of this cell type. The cylindrical (D size) Li/I$_2$ cell is hermetically welded like the other batteries described above. It is case-positive; the negative connection is a button on the end of the cell. It is designed to withstand substantial shock, vibration, and abuse without venting, swelling, leaking, or exploding.

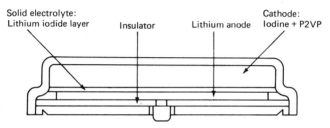

Solid electrolyte: Lithium iodide layer   Insulator   Lithium anode   Cathode: Iodine + P2VP

FIG. 12.11 Button-type Li/I$_2$ battery. (*Courtesy of Catalyst Research Corp.*)

Manufacturing of all Li/I$_2$ batteries is done in a dry environment (typically less than 1% RH); in addition, all these batteries are hermetically sealed to prevent exchange of any material with the environment. Good sealing is required in order to maintain the desired electrical characteristics.

Connection to the medical-grade batteries is made by soldering or spot welding. The case-positive varieties usually have a pin or wire welded to the case to facilitate making the positive connection (see Figs. 12.9 and 12.10).

Manufacturers keep detailed records of the construction and manufacture of each cell intended for medical applications. Each of the batteries is individually serialized. This procedure allows the systematic tracing of the history and behavior of every battery, should the need arise.

12.3.3 Commercially Available Cells

Table 12.6 summarizes the manufacturers' specifications for some typical commercially available cells. Batteries for medical use are relatively expensive because of the low manufacturing volume and demands of high reliability. Unit prices are between $75 and $125. However, the inherent simplicity of the Li/I$_2$ system is amenable to high-volume production and lower cost. The small button cells are in the $2 to $3 range; the cylindrical D size cell sells for $50 to $75.

TABLE 12.6 Manufacturers' Specifications of Typical Commercial Li/I$_2$ Cells

Manufacturer*	Type	Manufacturer's rated capacity, Ah	Wt. g	Vol., cm³	Size, mm (length × width × diam.)	Energy density Wh/cm³	Wh/g
CRC†	802/35	3.8	54	18.7	45 × 13.5 × 35	0.53	0.186
CRC†	901/23	2.5	26	7.4	45 × 9.3 × 23	0.92	0.254
WGL†	761/15	1.3	17	4.6	45 × 8.6 × 15	0.71	0.202
WGL†	761/23	2.5	27	7.6	45 × 8.6 × 23	0.82	0.245
WGL†	762M	2.5	29	8.2	45 × 8.0 × 28	0.82	0.228
CRC‡	S23P-15	0.12	3.8	0.83	23 diam. × 1.8	0.39	0.084
CRC‡	S27P-15	0.17	5.3	1.0	27 diam. × 1.8	0.46	0.085
CRC‡	S19P-20	0.12	2.8	0.57	19 diam. × 2.0	0.57	0.114
CRC	LID	14	140	38.6	33 diam. × 57	0.94	0.265
CRC	2736	0.46	7.1	2.2	27 diam. × 3.8	0.59	0.168
CRC	3740	0.870	17.8	4.2	37 diam. × 4.0	0.56	0.130

*Code: CRC, Catalyst Research Corp., Baltimore, Md.; WGL, Wilson Greatbatch, Ltd., Clarence, N.Y.
†Medical application battery.
‡Button design.

12.3.4 Discharge Performance Characteristics

The open-circuit voltage of a Li/I$_2$ battery is very near 2.80 V. This value is maintained through the useful life of the battery. The resistance of the lithium iodide discharge product controls the load voltage for most of the discharge. The morphology of the lithium iodide discharge product, in turn, depends upon the way the battery is constructed. Only at the end of the discharge cycle is the resistance of the iodine depolarizer significant in the total internal resistance of the battery. Because the battery resistance is quite high and increases throughout life, the discharge curves are not flat, even at moderate current drains. Two characteristic types of discharge curves are observed.

Batteries made by putting the depolarizer against bare lithium anode metal discharge by growing a nearly planar layer of lithium iodide. This layer becomes thicker in proportion to the amount of discharge; hence, the load voltage during this first phase of the discharge decreases linearly with the amount of charge removed from the battery, until all the available iodine is consumed. At that time the voltage drops at a steeper rate, indicative of the increasingly significant resistance of the iodine-deficient depolarizer. In the final, third phase, the open-circuit as well as the load voltage drops as iodine is no longer available at unit thermodynamic activity. The cells, for medical use, are made cathode-limited so their end-of-life voltage change is much less abrupt than if they were lithium-limited. The normalized slope (volts per ampere-hour) for the straight-line portion of the discharge curves is approximated by

$$\frac{\Delta E}{\Delta Q} = \frac{3600\,M}{F\sigma d}\cdot\frac{i}{A^2} = K\frac{i}{A^2}$$

where $\Delta E/\Delta Q$ = slope of discharge curve, V/Ah
M = molecular weight of LiI
d = density of LiI
σ = lithium ion conductivity of LiI

F = Faraday's constant
i = discharge current
A = area of anode, cm^2

Literature values of the Li^+ conductivity in LiI range from 0.2 to 0.8 $\times$ 10^{-6} $\Omega^{-1} \cdot cm^{-1}$ at 37°C.[11-13] K then lies between 1.5 and 6.0 $\times$ 10^6. Experimental values of K measured from discharge data have been found to be between 1.0 and 1.5 $\times$ 10^6. At high discharge rates fracturing of the LiI layer apparently occurs, and this equation holds only over small portions of the discharge curve.

The resistance of these batteries behaves consistently with the load voltage curves. It builds up linearly with discharge until the cells are depleted of iodine. At discharge rates much above 10 $\mu A/cm^2$ a nonohmic component of the polarization begins to become increasingly important.[20] The nonohmic component appears to originate near the discharge product–cathode interface. The effects of this polarization may last a long time after high-rate discharge. Thus, extrapolation of performance between rates is not simple.

A different discharge curve is obtained from lithium/iodine batteries which are constructed with a coating of pure P2VP applied to the anode before the iodine depolarizer is added to the battery. This coating alters the discharge behavior of these cells. The discharge product no longer grows in a planar fashion but rather grows in columnar-type groupings. This results in many channels between the crystallites of the lithium iodide. Hence, the lithium iodide is effectively thinner for any given state of discharge than for an equivalent battery which has no coating. Since the effective thickness of the lithium iodide is much less, coated-anode batteries may be used at a higher drain rate than the typical maximum for the uncoated batteries.

Plots of discharge voltage vs. state of discharge for these batteries are not linear. The battery resistance increases exponentially with the state of discharge. This dependence extends from the beginning of life throughout the useful voltage range of the batteries. The resulting discharge curves do not have a sharp knee, which indicates that there is a gentle transition between regions where the battery resistance is dependent upon the resistance of the lithium iodide and where the resistance of the depleted cathode depolarizer dominates.[20]

Figure 12.12 shows a typical plot of load voltage vs. state of discharge for both coated- and uncoated-anode groups of otherwise identical batteries at 37°C. The voltage data are shown between the initial voltage of 2.8 V and the cutoff voltage of 2.0 V. All data were obtained at a current density of 6.7 $\mu A/cm^2$. The voltage of the uncoated batteries decreases linearly until iodine depletion starts to occur. At this time the slope of the curve changes. The same batteries with P2VP-coated anodes exhibit a higher load voltage at this current density. However, as the current density decreases, the differences between the discharge curves for the coated- and uncoated-anode batteries become smaller. Figure 12.13 shows the discharge voltage data for the coated-anode battery in Fig. 12.12 to 0 volts with the corresponding resistance data. Log R vs. Q is linear, and R varies between 100 and 8000 Ω during discharge. Typical discharge curves for a Catalyst Research Corp. (CRC) 800 series cell, as well as changes in cell resistance, are shown in Fig. 12.14.[21] Catalyst Research Corp. projects their 900 series batteries (manufactured by pressing the cathode depolarizer onto a central lithium anode assembly) to have discharge behavior very similar to that of the coated-anode cells.[21]

The button and D size batteries made by Catalyst Research Corp. are expected to

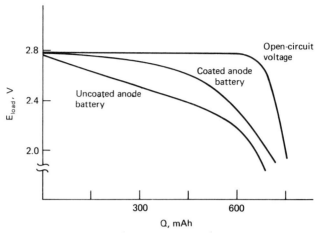

FIG. 12.12 Load voltage vs. discharge state for uncoated and P2VP-coated anode Li/I_2 batteries discharged at 6.7 $\mu A/cm^2$, 37°C. (*Courtesy of Energy Technology, Medtronic, Inc.*)

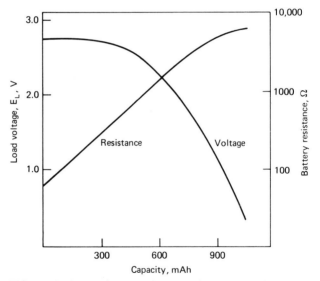

FIG. 12.13 Load voltage and battery resistance vs. Q for coated-anode, Li/I_2 battery discharged at 100 μA, 37°C. (*Courtesy of Energy Technology, Medtronic, Inc.*)

have discharge curves like the 900 series batteries. Figures 12.15 and 12.16 show projected discharge curves for two types of button cells.[21] Figure 12.17 is a similar discharge plot for the D battery at 25°C.[21] This cell delivered 7 Ah (0.45 Wh/cm³) at the 2-month (5-mA) rate; capacities and energy densities at lower rates are expected to be larger. The projected discharge curves for the 1- and 2-mA rates are also shown.

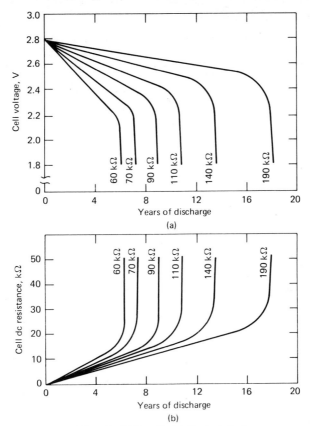

FIG. 12.14 800 series Li/I$_2$ cell. (a) Typical discharge curves; (b) changes in cell resistance during discharge. (*Courtesy of Catalyst Research Corp.*)[20]

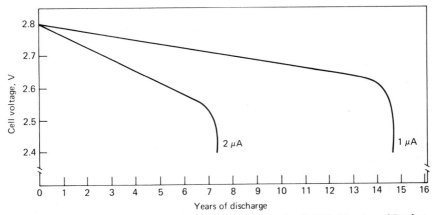

FIG. 12.15 Projected discharge curves for S19P-20 button watch cell, 25°C. (*Courtesy of Catalyst Research Corp.*)

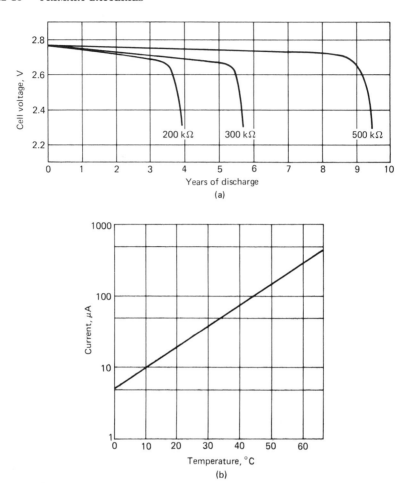

FIG. 12.16 Discharge characteristics for 2736 button cell. (*a*) Projected discharge at 25°C; (*b*) maximum continuous discharge current vs. temperature.[20] (*Courtesy of Catalyst Research Corp.*)

12.3.5 Self-Discharge

In addition to the ohmic and nonohmic polarization losses, there are two other major losses of capacity in the Li/I_2 battery system. The first is self-discharge. It occurs by the direct combination of lithium and iodine which has diffused through the lithium iodide layer to reach the lithium anode material. The amount of self-discharge observed in these batteries is dependent upon the effective thickness of the lithium iodide discharge layer. For this reason the largest percentage of self-discharge occurs early in the battery's life when the lithium iodide layer is very thin. In fact, at a discharge rate of 1 to 2 $\mu A/cm^2$ virtually all self-discharge loss will occur by the time the battery is 25% depleted. Figure 12.18 shows a plot of power lost to self-discharge vs. state of discharge for a 2000-mAh stoichiometric capacity Li/I_2 medical battery at

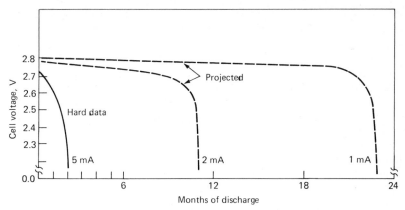

FIG. 12.17 Discharge data for D size Li/I$_2$ cell, 25°C. (*Courtesy of Catalyst Research Corp.*)

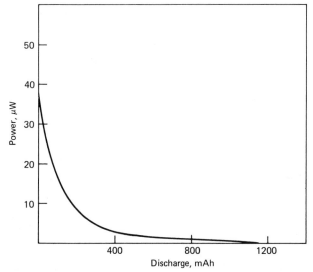

FIG. 12.18 Power (heat) loss due to self-discharge vs. discharge state for typical Li/I$_2$ battery (calorimetric measurements made at open-circuit voltage). (*Courtesy of Energy Technology, Medtronic, Inc.*)

37°C. These measurements were made using a microcalorimeter. The amount of self-discharge observed in the batteries precoated with a layer of P2VP is a little greater than in the batteries which do not receive this treatment. This is because the diffusion path between the cathode depolarizer and the lithium anode is thinner than in the uncoated batteries. However, the amount of self-discharge in any of these batteries is 10% or less (dependent upon discharge rate) of the rated capacity of the cell over a 10- to 15-y lifetime and is therefore small compared with many other battery systems. Calorimetric measurements have shown that self-discharge losses are lessened while current is being drawn from the battery. Most estimates (including that above) are made at open circuit; therefore, they overestimate self-discharge losses. The open-circuit estimates of loss may be 100% high for some application rates.

12.3.6 Other Performance Losses

The total iodine utilization in the Li/I_2 battery depends upon the initial ratio of iodine to donor material in the cathode depolarizer as well as loss to self-discharge. The iodine and amine nitrogen form a very strong 1:1 complex which is not dischargeable at a useful voltage. For this reason the amount of iodine that can be discharged is limited to that between the initial starting ratio of the battery material depolarizer and that of the 1:1 complex. In addition, the P2VP used as a donor material in these batteries is chemically attacked by the iodine in the battery. The result is that approximately another one-half mole of iodine per mole of nitrogen atoms in the donor materials becomes unavailable for discharge.[20] Thus, the maximum amount of discharge capacity is equal to the iodine that lies between the initial cathode ratio and about a 1.5:1 final ratio. This value does not include the self-discharge and polarization losses discussed above.

The effect of all losses upon the available capacity as a function of rate is summarized in Fig. 12.19 for a battery at 37°C. Here the deliverable capacity is estimated by starting with the stoichiometric capacity and subtracting the losses and limitations which arise from the various processes described above. This battery had an initial cathode-to-donor ratio of 6.2 moles iodine per pyridine ring (15:1 weight ratio) in the cathode complex. The maximum utilization is predicted to be 60% of stoichiometric capacity for this ratio of iodine to P2VP. The mole ratio of iodine to donor material ranges from 6 to 8 in currently available batteries. The achievable utilization will depend upon this ratio. (Most literature for commercial batteries gives the initial ratio in terms of weight ratio of iodine to donor material.)

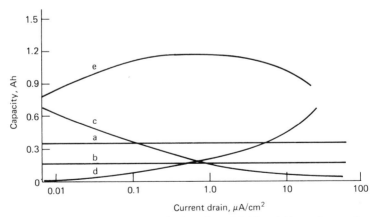

FIG. 12.19 Capacity and loss projections for a 2.0-Ah stoichiometric capacity Li/I_2 battery. (*a*) Loss to 1:1 adduct formation; (*b*) loss to iodination reaction; (*c*) projected loss to self-discharge; (*d*) projected loss to polarization; (*e*) net utilizable battery capacity. (*Courtesy of Energy Technology, Medtronic, Inc.*)

12.3.7 Effect of Temperature

Since the majority of these batteries have been used as power sources for heart pacemakers, most experience with them is at body temperature (37°C). However, testing has been done at temperatures between -10 and 60°C. These results are more important for the nonmedical batteries. Figure 12.20 shows the variation of maximum current and expected capacity for the Catalyst Research Corp. D cell as a function of temperature. Hard data exist between -10 and 50°C. Performance at the lower temperature is reduced due to the low conductivity of the depolarizer and LiI. At 60°C self-discharge is increased along with the rate of other parasitic reactions, although the battery is usable at this temperature. Manufacturers give operating temperature ranges for nonmedical batteries between -20 or 0 and 50°C. The temperature at which best performance can be expected is between room temperature and 40°C.

12.4 OTHER SOLID-ELECTROLYTE CELLS

Three other solid-electrolyte systems have reached an advanced development stage.

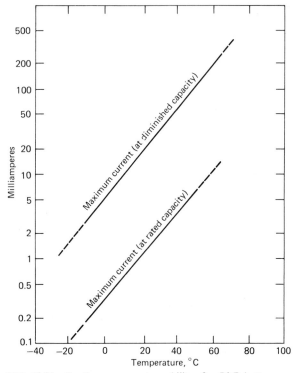

FIG. 12.20 Continuous current capability of an Li/I$_2$ battery as a function of temperature. (*Courtesy of Catalyst Research Corp.*)

12.4.1 Li/LiI(SiO₂,H₂O)/Me₄NI₅·C

This cell was evaluated for use in cardiac pulse generators. The cells exhibited voltages of 2.75 V and were projected to have energy densities of 0.4 Wh/cm³.[22] The cathode is a mixture of carbon and tetramethylammonium pentaiodide (Me_4NI_5).

12.4.2 Ag/RbAg₄I₅/Me₄NI$_n$·C

These solid electrolyte cells are based on rubidium silver iodide ($RbAg_4I_5$) as the electrolyte. This material exhibits an unusually high ionic conductivity of $0.26 \, (\Omega \cdot cm)^{-1}$ at 25°C.[1] This permits cell discharge at much higher current drains than those available with LiI-based cells. The initial characteristics of small 40-mAh, five-cell batteries are shown in Table 12.7; both tetramethylammonium pentaiodide and enneaiodide were evaluated as active cathode agents.[23] These batteries demonstrated efficient discharge capability at temperatures ranging from −40 to 71°C. Constant-load discharge curves are shown in Figs. 12.21 and 12.22.[2] The two cathodes are similar. The Me_4NI_9 cathode has the higher iodine content and activity, which results in higher energy density and rate capability. The lower iodine activity of the Me_4NI_5 cathode gives rise to a lower rate of self-discharge and, therefore, better long-term storage characteristics.

TABLE 12.7 Solid-State Ag/I₂ Batteries*

	Ag/Me₄NI₉	Ag/Me₄NI₅
Single-cell voltage, V	0.662	0.650
Battery (OCV) voltage, V	3.31	3.25
Internal resistance, Ω	30	35

*1.27 cm diam. × 1.9 cm ht, capacity = 40 mAh.

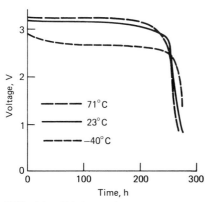 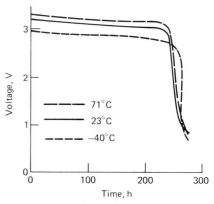

FIG. 12.21 Discharge of Ag/Me₄NI₅ battery, 21.5-kΩ constant load. (*From Topics in Applied Physics, vol. 21, reprinted by permission of the publisher, Springer-Verlag.*)[2]

FIG. 12.22 Discharge of Ag/Me₄NI₉ battery. (*From Topics in Applied Physics, vol. 21, reprinted by permission of the publisher, Springer-Verlag.*)[2]

Table 12.8 summarizes the findings for the batteries made with the Me_4NI_9 cathode. The batteries were discharged after 5-y storage. The data are reported in terms of percent of original capacity.[23] Disproportion of the $RbAg_4I_5$ occurred after prolonged storage at $-15°C$, causing the battery resistance to increase. A temperature soak at 40 to 60°C solved this problem.

TABLE 12.8   Percent Deliverable Capacity after 5-Y Storage

Storage temperature, °C	Discharge temperature, °C		
	−40	23	71
−15	96	69	88
23	91	81	74
71	70	55	53

The Ag/I_2 batteries were inherently deficient in two significant areas, voltage and size. The low cell voltage and high equivalent weights and volumes restricted these batteries to very low energy densities (0.08 Wh/cm³ or 13 Wh/kg). Consequently no commercial application resulted.

12.4.3 Li/LiBr/Br₂(P2VP)

The Li/Br_2 solid-electrolyte cell was designed in a manner analogous to the Li/I_2 battery, but due to the higher reactivity of the bromine and the high impedance of LiBr, batteries of this type do not appear promising.[24]

REFERENCES

1. B. B. Owens, "Solid Electrolyte Batteries," in C. W. Tobias (ed.), *Advances in Electrochemistry and Electrochemical Engineering,* vol. 8, Wiley, New York, 1971, chap. 1, pp. 1–62.

2. B. B. Owens, J. E. Oxley, and A. F. Sammells, "Applications of Halogenide Solid Electrolytes," in S. Geller (ed.), *Solid Electrolytes, Topics in Applied Physics,* vol. 21, Springer-Verlag, New York, 1977, chap. 4, pp. 67–104.

3. C. C. Liang, "Solid-State Batteries," *Appl. Solid State Sci.* (4): 95–135 (1974).

4. T. Takahaski, *Denki Kagaku* **36**:402–412, 481–490 (1968).

5. B. B. Owens and P. M. Skarstad, "Ambient Temperature Solid-State Batteries," in P. Vashishta, J. N. Mundy, and G. K. Shenoy (eds.), *Fast Ion Transport in Solids,* Elsevier-North Holland, New York, 1979, pp. 61–68.

6. R. C. Weast (ed.), *Handbook of Chemistry and Physics,* 59th ed., Chemical Rubber Publishing Company, West Palm Beach, Fla., 1978–1979.

7. C. C. Liang, *J. Electrochem. Soc.* **120**:1289 (1973).

8. C. C. Liang and L. H. Barnette, *J. Electrochem. Soc.* **123**:453–458 (1976).

9. J. R. Rea, L. H. Barnette, C. C. Liang, and A. V. Joshi, "A New High Energy Density Battery System," in B. B. Owens and N. Margalit (eds.), *Proceedings of the Symposium on Biomedical Implantable Applications and Ambient Temperature Lithium Batteries,* Proceedings vol. 80–4, Electrochemical Society, Princeton, N.J., 1980, pp. 245–253.

10. C. C. Liang, A. V. Joshi, and N. E. Hamilton, *J. Appl. Electrochem.* **8**:445–454 (1978).

11. D. E. Ginnings and T. E. Phipps, *J. Am. Chem. Soc.* **52**:1340 (1930).

12. B. J. H. Jackson and D. A. Young, *J. Phys. Chem. Solids* **30**:1973–1976 (1969).

13. C. R. Schlaijker and C. C. Liang, "Solid-State Batteries and Devices," in W. Van Gool (ed.), *Fast Ion Transport in Solids,* North Holland, Amsterdam, 1973, pp. 689–694.

14. C. C. Liang, U.S. Patent 3,713,897 (1973).

15. A. M. Stoneham, E. Wade, and J. A. Kilner, *Mater. Res. Bull.* **14**:661–666 (1979).

16. J. R. Rea, *DOE Battery and Electrochemical Contractors' Conf.,* Arlington, Va., vol. II, session VII, December 1979.

17. Commercial Literature, Duracell, Inc., Bethel, Conn.

17a. S. J. Garlock, "Characteristics of Lithium Solid State Batteries in Memory Circuits," Wescon 81, San Francisco, Calif., Professional Program Session Recod No. 29, Electronic Conventions, Inc., El Segundo, Calif., September 1981.

18. G. M. Phillips and D. F. Untereker, "Phase Diagram for the Poly-2-vinylpyridine and Iodine System," in B. B. Owens and N. Margalit (eds.), *Proceedings of the Symposia on Biomedical Implantable Applications and Ambient Temperature Lithium Batteries,* Proceedings vol. 80–4, Electrochemical Society, Princeton, N.J., 1980.

19. P. M. Skarstad, B. B. Owens, and D. F. Untereker, "Volume Changes on Discharge in Lithium-Iodine (Poly-2-vinylpyridine) Cells," *152d Nat. Electrochem. Soc. Meeting,* Atlanta, October 1977, abstract no. 23.

20. K. R. Brennen and D. F. Untereker, "Iodine Utilization in Li/I$_2$ (Poly-2-vinylpyridine) Batteries," in B. B. Owens and N. Margalit (eds.), *Proceedings of Symposia on Biomedical Implantable Applications and Ambient Temperature Lithium Batteries,* Proceedings vol. 80–4, Electrochemical Society, Princeton, N.J., 1980.

21. Commercial literature, Catalyst Research Corp., Baltimore, Md.

22. B. B. Owens, K. Fester, and T. Kuder, extended abstracts, vol. 77–2, Electrochemical Society, October 9–14, 1977, Atlanta, Ga., abstract no. 25.

23. D. L. Warburton, R. F. Bis, and B. B. Owens, "Five-Year Storage Tests of Solid-State Ag/I$_2$ Batteries," *28th Power Sources Symp.,* Atlantic City, N.J., June 12–15, 1978.

24. W. Greatbatch, R. T. Mead, R. L. McLean, F. Rudolph, and N. W. Frenz, U.S. Patent 3,994,747 (1976).

PART 3
SECONDARY BATTERIES

13

Introduction

by
David Linden

13.1 GENERAL CHARACTERISTICS AND APPLICATIONS OF SECONDARY BATTERIES

Secondary or rechargeable batteries are widely used in many applications. The most familiar are the ones used in starting, lighting, and ignition (SLI) automotive applications, in industrial truck materials-handling equipment, and the stationary batteries used for emergency and standby power. Small secondary batteries are also used to power portable equipment, such as tools, toys, lighting, and photographic, radio and similar devices. More recently, secondary batteries have received renewed interest as a power source for electric vehicles and electric utility load leveling. Major development programs have been initiated toward improving the performance of existing battery systems and developing new systems to meet the stringent specifications of these new applications.

The applications of secondary batteries fall into two major categories:

1. Those applications in which the secondary battery is discharged (similar in use to a primary battery) and recharged after use, either in the equipment in which it was discharged or separately. Secondary batteries are used in this manner for convenience (in hand-held calculators, electronic flash units, etc.), for cost savings (as they can be recharged rather than replaced), or for power drains beyond the capability of primary batteries. Electric vehicle, traction, industrial truck, and some stationary battery applications fall in this category.

2. Those applications in which the secondary battery is used as an energy storage device, being charged by a prime energy source and delivering its energy to the load on demand, when the prime energy source is not available or is inadequate to handle the load requirement. Examples are automotive and aircraft systems, emergency no-fail and standby power sources, and hybrid applications.

Secondary batteries are characterized, in addition to their ability to be recharged, by high power density, flat discharge profiles, and good low-temperature performance. Their energy densities, however, are usually lower and their charge retention is poorer than those of the primary battery systems (see Sec. 3.5 in Part 1, Chap. 3).

Secondary batteries have been in existence for over 100 y. The lead-acid battery was developed in 1859 by Planté. It is still the most widely used battery, albeit with many design changes and improvements, with the automotive SLI battery by far the dominant one. The nickel-iron alkaline battery was introduced by Edison in 1908 as a power source for the early, but short-lived, electric automobile; it eventually saw service in industrial trucks, underground work vehicles, railway cars, and stationary applications. Its advantages were durability and long life, but it gradually lost its market share because of its high cost and lower specific energy.

The pocket-plate nickel-cadmium battery has been manufactured since 1909 and was used primarily for heavy-duty industrial applications. The sintered-plate designs, which led to increases in power capability and energy density, opened the market for aircraft engine starting and communications applications during the 1950s. Later the sealed nickel-cadmium cells were developed and led to new consumer applications, which they share with the sealed lead-acid cells. As with the primary battery systems, significant performance improvements have been made with the older secondary battery systems, and a number of new types, such as the silver-zinc, the nickel-zinc, the hydrogen, lithium and halogen batteries, and the high-temperature systems, have been introduced into commercial use or serious development. Much of the new development work has been supported by the need for high-performance batteries for electric vehicles and utility power load leveling.

The worldwide secondary battery market is now in excess of $6 billion annually. A world perspective of the use of secondary batteries by application is presented in Table 13.1. The lead-acid battery is by far the most popular, with the SLI battery accounting for a major share of the market. This share is declining gradually, due to increasing applications for other types of batteries. The market share of the alkaline battery systems is about 10 to 15%. A major growth area has been the nonautomotive consumer applications for small secondary batteries.[1]

TABLE 13.1   Worldwide Secondary Battery Market—1979 (in millions of dollars)

	Battery system	
Market segment	Lead-acid	Alkaline and others
Vehicle SLI	3900	
Industrial: Standby and UPS	400	300
Traction: Including fork-lift trucks	400	100
Consumer and instruments		
Small sealed cells (power size)	60	200
Electronic size (button and specialty)	. . .	10
Energy storage		
Solar	15	< 1
Load leveling	< 1	< 1
Military and space	30 (incl. submarines)	70
Vehicular propulsion, electric, including:		
Golf	46	< 1
Hybrid	1	1
Total	≈ 5000	≈ 700*

*Approximately $500 million worth of nickel-iron and nickel-cadmium cells are fabricated in the U.S.S.R. (not included in these values).

The typical characteristics and applications of secondary batteries are summarized in Table 13.2.

13.2 TYPES AND CHARACTERISTICS OF SECONDARY BATTERIES

The important characteristics of the secondary or rechargeable cell are that the charge and discharge—the transformation of electrical energy to chemical energy and back again to electrical energy—should proceed nearly reversibly, should be energy efficient, and should have minimal physical changes that can limit the cell's cycle life. Chemical action, which may cause deterioration of the cell's components, loss of life, or loss of energy, should be absent, and the cell should possess the usual characteristics (such as high specific energy, low resistance, and good performance over a wide temperature range) desired of a battery. These requirements limit the number of materials that can be successfully employed in a rechargeable battery system.

The lead-acid battery system has many of these characteristics. The charge/discharge process is essentially reversible, the system does not suffer from deleterious chemical action, and, while its energy density may not be as high as desired, the lead-acid battery performs reliably over a wide temperature range. A key factor in its popularity and dominant position is its combination of relatively low cost with good performance and cycle life characteristics.

The lead-acid battery is designed in many configurations, as listed in Table 13.2 (also see Chap. 14, Table 14.2), from small, sealed cells with a capacity of 2.5 Ah to large cells, up to 12,000 Ah. The automotive SLI battery is, by far, the most popular and the one in widest use. Table 13.3 summarizes some of the advances in SLI battery design and performance over the past 25 y. Most significant are the use of lighter-weight plastic containers, the improvement in shelf life, the "dry-charge" process, and the "maintenance-free" design. The latter, using calcium-lead or low-antimony grids, has practically eliminated water loss during charging (minimizing the need to add water) and has reduced the self-discharge rate so that batteries can be shipped or stored in a wet, charged state for relatively long periods.

The lead-acid industrial storage batteries are generally larger than the SLI batteries, with stronger, higher-quality construction. Applications of the industrial batteries fall in several categories. The motive power traction types are used in materials-handling trucks, tractors, mining vehicles, automobiles (mostly experimental) and, to a limited extent, golf carts and personnel carriers (although the majority in use are automotive-type batteries). A second category is diesel locomotive engine starting and the rapid-transit batteries, replacing the nickel-iron battery in the latter application. Significant advances are the use of lighter-weight plastic containers in place of the hard-rubber containers, better seals, and changes in the tubular positive-plate designs. Another category is stationary service: telecommunications systems, electric utilities for operating power distribution controls, emergency power and standby power systems, uninterruptible power systems (UPS), and in railroads for signaling and car power systems.

The industrial batteries use three different types of positive plates: tubular and pasted flat plates for motive power, diesel engine cranking, and stationary applications, and Planté designs, forming the active materials from pure lead, mainly in the stationary batteries. The flat-plate batteries use either lead-antimony or lead-calcium grid alloys. A relatively recent development for the telephone industry has been the "round

TABLE 13.2 Major Characteristics and Applications of Secondary Batteries

System	Characteristic	Application
Lead-acid		
Automotive	Popular, low-cost secondary battery; moderate specific energy, high-rate and low-temperature performance	Automotive SLI, lawn mowers, tractors, aircraft, marine
Traction (motive power)	Designed for deep 6- to 9-h discharge, cycling service	Industrial trucks, materials handling, electric vehicles; special types for submarine power
Stationary	Designed for standby float service, long life	Emergency power, utilities, telephone, UPS, load leveling, energy storage
Portable	Sealed, maintenance-free, low cost, good float capability	Portable tools, small appliances and devices, TV and portable electronic equipment
Nickel-cadmium		
Vented	Good high rate, low-temperature capability; flat voltage, excellent cycle life	Aircraft batteries, industrial and emergency power applications, communication equipment
Portable	Sealed, maintenance-free; good high-rate low-temperature performance, excellent cycle life	Photographic, portable tools, appliances and electronic equipment; standby power, memory backup
Nickel-iron	Durable, rugged construction, long life, low specific energy	Materials handling, stationary applications, railroad cars (now being considered for electric vehicles)
Nickel-zinc	High specific energy, moderate life	In development stage for electric vehicles, energy storage, emergency power, portable-type applications
Silver-zinc	Highest specific energy, good high rate capability; low cycle life, high cost	Lightweight portable electronic and other equipment; torpedo propulsion, drones, submarines, and other military equipment
Silver-cadmium	High specific energy, good charge retention, moderate cycle life; high cost	Portable equipment requiring a lightweight, high-capacity battery
Nickel-hydrogen Silver-hydrogen	High energy density, long cycle life under deep discharge	In development stage, primarily for aerospace applications
Ambient temperature rechargeable "primary" types (Zn/HgO, Zn/AgO, lithium systems)	Low cost, good capacity retention, sealed and maintenance-free; limited cycle life	Small button and cylindrical cell applications, watches and calculators, small solar cell power sources
Advanced systems	High specific energy and power, competitive cost	In development for electric vehicles and energy storage (see Part 4)

TABLE 13.3   Comparison of SLI Batteries, 1952—1977

Component or characteristic	1952	1977
Container and cover	Rubber	Polypropylene
Seal	Asphalt	Heat sealed
Intercell connector	Over cover	Through partition
Separators	Wood, treated paper, microporous rubber	Microporous plastic, treated paper, microporous rubber
Grid alloy	> 6% Sb	< 4% Sb or Ca
Specific gravity of electrolyte	1.285	1.260
Watering interval	7–30 days	6 months to several years
Weight, kg	18	15
Density, kg/L	2.2	1.9
Voltage, V	6	12
Capacity, Ah	100	48*
Charger	Generator	Alternator

*The 20-h discharge rating which was in use in 1952 is no longer used for SLI batteries. The rating shown for 1977 is for comparative purposes.
SOURCE: Based on Ref. 1.

cell," designed for trouble-free, long-life service. This battery uses plates, circular in shape with pure lead grids, which are stacked one above the other in a cylindrical cell container, rather than the normal prismatic structure with flat, parallel plates.

The small "sealed" lead-acid batteries are constructed in two configurations, prismatic cells with parallel plates, ranging in capacity from 1 to 30 Ah, and cylindrical cells similar in appearance to the popular primary dry cells and ranging in capacity up to 25 Ah. The electrolyte in these cells is either a gel or a solution absorbed in the plates and in highly porous separators so they can be operated in any position without the danger of leakage. The grids generally are of lead-calcium alloy; some use grids of pure lead. Another feature is an internal design that provides for the recombination of gases that might form on charge. Sealed lead-acid batteries are used in emergency lighting equipment, portable instruments and tools, cordless TV, and other such consumer-type applications requiring small, portable batteries.

Lead-acid batteries also are used in other types of applications, such as in submarine service, for reserve power in marine applications, and in areas where engine-generators cannot be used, e.g., indoors and in mining equipment. New applications, to take advantage of the cost effectiveness of this battery, in addition to the electric vehicle, include load leveling for utilities and solar photovoltaic systems. These applications will require improvements in the energy and power density of the lead-acid battery.

All the other conventional types of secondary batteries use an aqueous alkaline solution (KOH or NaOH) as the electrolyte. Electrode materials are less reactive with alkaline electrolytes than with acid electrolytes. Furthermore, the charge/discharge mechanism in the alkaline electrolyte involves only the transport of oxygen or hydroxyl ions from one electrode to the other; hence, the composition or concentration of the electrolyte does not change.

The nickel-cadmium secondary battery is the most popular alkaline secondary battery and is available in several cell designs and in a wide range of sizes. The original cell design used the pocket-plate construction. The vented pocket-type cells are very rugged and can withstand both electrical and mechanical abuse. They have very long life and require little maintenance beyond occasional topping with water. This type of battery is used in heavy-duty industrial applications, such as materials-handling trucks, mining vehicles, railway signaling, emergency or standby power, and diesel engine starting. The sintered-plate construction is a later development, having higher energy density. It gives better performance than the pocket-plate type at high discharge rates and low temperatures but is more expensive. It is used in applications, such as aircraft engine starting and communications and electronics equipment, where the lighter weight and superior performance are required. The smaller "sealed" cell is a third design. These cells incorporate features to prevent the buildup of pressure caused by gassing during charge. They are available in prismatic, button, and cylindrical configurations and are used in consumer and small industrial applications.

The nickel-iron battery was important from its introduction in 1908 until the 1970s, when it lost its market share to the industrial lead-acid battery. It was used in materials-handling trucks, mining and underground vehicles, railroad and rapid-transit cars, and in stationary applications. The main advantages of the nickel-iron battery, with major cell components of nickel-plated steel, are extremely rugged construction, long life, and durability. Its limitations (low specific energy, poor charge retention, and poor low-temperature performance) and its high cost of manufacture, compared with the lead-acid battery, led to the decline in usage. Recently, however, interest in electric vehicles has led to new approaches to the manufacture of nickel-iron batteries which may be more competitive in cost and performance.

The silver-zinc (zinc/silver oxide) battery is noted for its high energy density, the highest of any battery presently available commercially, low internal resistance desirable for high-rate discharges, and a flat discharge profile. This battery system is useful in applications where high energy density is a prime requisite, such as lightweight portable electronic equipment, submarine and torpedo propulsion, and other military and space uses. It is not employed for general storage battery applications because its cost is high, its cycle life and activated life are limited, and its performance at low temperatures falls off more markedly than do those of other secondary battery systems. The silver-cadmium (cadmium/silver oxide) battery has better cycle life and low-temperature performance than the silver-zinc battery but is inferior in these characteristics compared with the nickel-cadmium battery. Its energy density, too, is between that of the nickel-cadmium and silver-zinc batteries. The battery is also very expensive, using two of the more costly electrode structures. As a result, the silver-cadmium battery was never developed commercially and was used only in special applications. A third silver system, the silver-iron (iron/silver oxide), has been considered for special applications. It has a high energy and power capability; however, due to its high cost and limited life, wide use of this battery system is not anticipated.

The nickel-zinc (zinc/nickel oxide) battery also has characteristics midway between the nickel-cadmium and the silver-zinc battery systems. Its energy density is about twice that of the nickel-cadmium battery, but the cycle life is still limited due to the tendency of the zinc electrode to form dendrites, which cause internal shorting. The nickel-zinc battery has not yet achieved commercial success but is a promising candidate because of its high energy density and potentially competitive cost, if its cycle life can be increased. The interest in electric vehicles had provided increased support for nickel-zinc battery development, but this effort has been deemphasized.

Another secondary battery system uses hydrogen for the active anode material (with a fuel cell–type electrode) and conventional cathodes, such as nickel oxide or silver oxide. These batteries are being developed mainly for the aerospace programs which require high energy densities, deep discharge, and long cycle life. The high cost of these batteries is a disadvantage which could limit their application. A further extension is the hydrogen/oxygen secondary battery (or electrically regenerative H_2/O_2 fuel cell). This system is basically a combination of a hydrogen/oxygen primary fuel cell, with hydrogen as the anode and oxygen as the active cathode material, and a water electrolysis cell which charges or regenerates the discharge product water to the original reactants. This secondary system has not been developed to the point of practical application.[2]

The energy crisis, and the need for secondary batteries for electric vehicles, electric utility load leveling, and similar applications, have created an unparalleled interest in the development of new, more advanced battery systems. These fall into two general groups: one consists of ambient-temperature systems that make use of active materials which are stable in aqueous electrolytes, such as iron, zinc, aluminum, and nickel; the second includes systems which use alkali metals as active materials and require nonaqueous electrolytes for operation. These use organic solvents or molten salt or ceramic electrolytes which operate at high temperatures ranging from 200 to 600°C. These systems are covered in Part 4.

Finally, several of the conventional primary battery systems, for example, the zinc/alkaline/manganese dioxide and zinc/silver oxide batteries, are also manufactured as rechargeable batteries. Their main advantages are their low initial cost and good energy density, but their cycle life is limited. The major application of these cells is in "rechargeable" button cells for small electronic devices; they are also manufactured in several of the common small cylindrical sizes. Ambient-temperature rechargeable cells using lithium as the anode are also attractive because of the high energy densities that are possible. The cycle life of these lithium batteries is still low, and commercial applications have not yet been developed.

13.3 COMPARISON OF THE PERFORMANCE CHARACTERISTICS OF THE SECONDARY BATTERY SYSTEMS

13.3.1 General

The characteristics of the major secondary systems are summarized in Table 13.4. This table is supplemented by Tables 1.2 and 3.3 (in Part 1), which list several theoretical and practical electrical characteristics of these secondary battery systems. A graphical comparison of the theoretical and practical performances of various battery systems is given in Fig. 3.1 (Part 1). This shows that only about 25% of the theoretical capacity of a battery system is attained under practical conditions as a result of design and the discharge requirements.

It should be noted, as discussed in detail in Part 1, Chap. 3, that these types of data and comparisons (as well as the performance characteristics shown in this section) are necessarily approximations, with each system presented under favorable discharge conditions. The specific performance of a battery system is very dependent on the cell design and all the detailed and specific conditions of the use and discharge/charge of the battery.

A qualitative comparison of the various secondary battery systems is given in Table 13.5. The different ratings given to the various designs of the same electrochemical

TABLE 13.4 Characteristics of the Major Secondary Battery Systems

	Lead-acid				Nickel-cadmium		
Common name	SLI	Traction	Stationary	Portable	Vented pocket plate	Vented sintered plate	Sealed
Chemistry:							
Anode	Pb	→	→	→	Cd	→	→
Cathode	PbO_2	→	→	→	NiOOH	→	→
Electrolyte	H_2SO_4 (aqueous solution)	→	→	→	KOH (aqueous solution)	→	→
Cell voltage (typical), V:							
Nominal	2.0	2.0	2.0	2.0	1.2	→	→
Open circuit	2.1	2.1	2.1	2.1	1.29	→	→
Operating	2.0–1.8	2.0–1.8	2.0–1.8	2.0–1.8	1.25–1.10	→	→
End	1.75 (lower operating and end voltage during cranking operation)	1.75	1.75 (except when on float service)	1.75 (when cycled)	1.0	→	→
Operating temperature, °C	−40 to 55	−20 to 40	−10 to 40^c	−40 to 60	−20 to 45	−40 to 50	−40 to 45
Energy density (at 20°C):							
Wh/kg	35	25	10–20	30	20	37	30
Wh/L	70	80	50–70	80	40	90	80
Discharge profile (relative)	Flat	Flat	Flat	Flat	Flat	Very flat	Very flat
Power density	High	Moderately high	Moderately high	High	High	High	Moderate to high
Self-discharge rate (at 20° C), % loss per month^d	20–30 (Sb-Pb) 2–3 (maintenance-free)	4–6	——	4–8	5	10	15–20
Calendar life, y	3–6	6	18–25	2–8	8–25	3–10	2–5
Cycle life, y^f	200–700	1500	——	250–500	500–2000	500–2000	300–700
Advantages	Low cost, ready availability; good high-rate, high- and low-temperature operation (good cranking service); good float service; new maintenance-free designs	Lowest cost of competitive systems (also see SLI)	Designed for "float" service; lowest cost of competitive systems (also see SLI)	Maintenance-free; long life on float service; low- and high-temperature performance; no "memory" effect; operates in any position	Very rugged, can withstand physical and electrical abuse; good charge retention, storage, and cycle life; lowest cost of alkaline batteries	Rugged; excellent storage; good specific energy and high-rate and low-temperature performance	Sealed, no maintenance; good low-temperature and high-rate performance; long life cycle; operates in any position
Limitations	Relatively low cycle life; limited energy density; poor charge retention and storability; hydrogen evolution	Low energy density; less rugged than competitive systems; hydrogen evolution	Hydrogen evolution	Cannot be stored in discharged condition; lower cycle life than sealed nickel-cadmium; difficult to manufacture in very small sizes	Low energy density	High cost; "memory" effect; thermal runaway	Sealed lead-acid battery better at high temperature and float service; "memory" effect
Status	Most widely produced secondary battery	In production	In production	In production	In production	In production	In production
Major cell types available	Prismatic cells: 30–200 Ah at 20-h rate	Based on positive plate design: 45–200 Ah per positive plate	Based on positive plate design: 5–400 Ah per positive plate	Sealed cylindrical cells: 2.5–25 Ah; prismatic cells: 0.9–35 Ah	Prismatic cells: 5–1300 Ah	Prismatic cells: 10–100 Ah	Button cells to 0.5 Ah; cylindrical cells to 10 Ah
Approximate cost, $/kWh	50	70	100	200–500	400–1000	600–1000	1000 (cylindrical cells) 3000 (button cells)
Representative manufacturers	Gould, Globe, Exide, Delco, General, VARTA, Prestolite, Chloride, Lucas, Japan Storage Battery, Matsushita, Yuasa, and others	Globe, Exide, C&D, VARTA, Chloride, Gould Oldham, Eagle-Picher Industries	Globe, Exide, C&D, Gould, Chloride, VARTA	Gates, Globe, Elpower, Eagle-Picher Industries, Yuasa, SAFT, Chloride, Matsushita, Exide	SAB NIFE, NIFE, Inc., McGraw-Edison, SAFT, VARTA, Yuasa, Chloride	GE, Marathon, SAFT, Gould	GE, UCC, SAFT, VARTA, Sanyo, Panasonic

^{a}In small cylindrical or button cell sizes.
^{b}Ambient temperature, organic electrolyte type (small sizes); also see Part 4.
^{c}Normally used indoors at room temperature.
^{d}Self-discharge rate usually decreases with increasing storage time.
^{e}Low-rate Zn/AgO cell.
^{f}Dependent on depth of discharge (DOD).
^{g}Experimental.

13-10

Nickel-iron (conventional)	Nickel-zinc	Zinc/silver oxide (silver-zinc)	Cadmium/silver oxide (silver-cadmium)	Nickel-hydrogen	Silver-hydrogen	Rechargeable "primary" types MnO_2	AgO^a	Lithium systems (ambient temperature)[b]
Fe NiOOH KOH (aqueous solution)	Zn NiOOH KOH (aqueous solution)	Zn AgO KOH (aqueous solution)	Cd AgO KOH (aqueous solution)	H_2 NiOOH KOH (aqueous solution)	H_2 AgO KOH (aqueous solution)	MnO_2 Zn AgO KOH (aqueous solution)		Li (Al) TiS_2 Organic solvent
1.2 1.37 1.25–1.05 1.0	1.6 1.73 1.6–1.4 1.2	1.5 1.85 1.7–1.3 1.0	1.2 1.4 1.4–1.0 0.7	1.4 1.32 1.3–1.15 1.0	1.4 1.4 1.15–1.05 0.9	1.5 1.5 1.3–1.0 1.0	1.55 1.6 1.55 1.2	2.0 2.4 2.0–1.4 1.4
−10 to 45	−20 to 60	−20 to 60	−25 to 70	0 to 50	0 to 50	−20 to 40	0 to 40	−20 to 55
27 55	60 120	90 180	55 110	55 60	80 90	30 60	70 250	85 145
Moderately flat Moderate to low	Flat High	Double plateau High (for high rate designs)	Double plateau Moderate to high	Moderately flat Moderate	Moderately flat Moderate	Sloping Low	Sloping Low	
20–40	10	3	3	60	——	~1	<1	<1
8–25 2000–4000	—— 50–200	1–3[c] 100–150[c]	2–3 150–600	—— 1500–6000[g]	—— 500–3000[g]	20–50 (limited DOD)		15–25
Very rugged, can withstand physical and electrical abuse; long life (cycling or stand)	High energy density; relatively low cost; good low-temperature performance	Highest energy density; high discharge rate; low self-discharge	High energy density; low self-discharge, good cycle life	High energy density; long cycle life at high DOD; can tolerate overcharge	High energy density; long cycle life at high DOD; high ampere-hour efficiency	Good shelf life; low cost		Potential for high specific energy; good shelf life
Low power and energy density; high self-discharge; hydrogen evolution; high cost	Poor cycle life	High cost; low cycle life; decreased performances at low temperatures	High cost; decreased performance at low temperatures	High initial cost; self-discharge proportional to H_2 pressure	High cost (may be limited to special military and aerospace applications)	Limited cycle life; used only to moderate DOD; low drain applications; small sizes only		Limit cycle life; present designs are low rate
Older designs becoming obsolete	In development	In production	In production	In development mainly for aerospace applications (up to 100 Ah)	In development (25- to 100-Ah sizes)	In production		In development
New designs in development for EV and mobile traction	Not commercially available	Prismatic cells: <1 to 800 Ah; special types to 6000 Ah	Prismatic cells: <1 to 500 Ah	Not commercially available	Not commercially available	Cylindrical cell to 5 Ah	Button cells: 50 mAh	Not commercially available
——	200 (projected)	800– >1500	1000–>2000	2000	>2000	1000^d	4000^a	——
NIFE, VARTA	Energy Research Corp., Exide	Yardney Electric Corp., Eagle-Picher Industries	Yardney Electric Corp., Eagle-Picher Industries	Energy Research Corp., Eagle-Picher Industries	Energy Research Corp., Eagle-Picher Industries	UCC	Duracell, Hitachi-Maxell	

TABLE 13.5 Comparison of Secondary Batteries

System	Energy density	Power density	Flat discharge profile	Low temperature operation	Charge retention	Charge acceptance	Efficiency	Life	Mechanical properties	Cost
Lead-acid										
Pasted	4	4	3	3	4	3	2	3	5	1
Tubular	4	5	4	3	3	3	2	2	3	2
Planté	5	5	4	3	3	3	2	2	4	2
Sealed	4	3	3	2	3	3	2	3	5	2
Nickel-cadmium										
Pocket	5	3	2	1	2	1	4	2	1	3
Sintered	4	1	1	1	4	1	3	2	1	3
Sealed	4	1	2	1	4	2	3	3	4	2
Nickel-iron	5	5	4	5	5	2	5	1	1	3
Nickel-zinc	2	3	2	3	4	3	3	4	3	3
Silver-zinc	1	2	5	3	1	5	1	5	3	4
Silver-cadmium	2	3	5	4	1	5	1	4	3	4
Nickel-hydrogen	2	3	4	4	5	3	5	2	3	5
Silver-hydrogen	2	3	4	4	5	3	5	2	3	5
Zinc/manganese dioxide, silver oxide (rechargeable "primary" type)	1	5	5	3	1	4	4	5	4	2

NOTE: Rating: 1 to 5, best to poorest.

system are an indication of the effect of design on the performance characteristics of a battery.

13.3.2 Voltage and Discharge Profile

The discharge curves of the conventional secondary battery systems, at the C/5 rate, are compared in Fig. 13.1. The lead-acid battery has the highest cell voltage of the conventional systems. The average voltage of the alkaline systems ranges from about 1.65 V for the nickel-zinc system to about 1.1 V. At the C/5 discharge rate at 20°C, there is relatively little difference in the shape of discharge curve for the various designs of a given system. However, the difference under other discharge rates and temperatures could be significant.

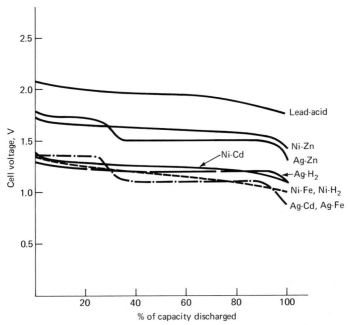

FIG. 13.1   Discharge profiles of the conventional secondary battery systems, at approximately C/5 discharge rate.

Most of the secondary battery systems have a flat discharge profile, except for the silver oxide systems which show the double plateau due to the discharge of the silver oxide electrode.

13.3.3 Energy Density

The delivered energy densities, both gravimetric and volumetric, of the various secondary battery systems are shown graphically in Fig. 13.2. The values can also be compared with the theoretical values for these systems given in Table 1.2 (Part 1); the energy densities of practical cells are only 10 to 30% of the theoretical values.

The silver oxide systems deliver the highest capacities, followed by the nickel-zinc and nickel-hydrogen cells and finally by the nickel-cadmium and lead-acid systems. These data show the significance of the cell and battery design on the energy output as well as the conditions under which the battery is discharged. Accordingly, there is a wide range and even an overlap in performance, indicating the need to consider all of the conditions of operation before making a choice of battery system and design.

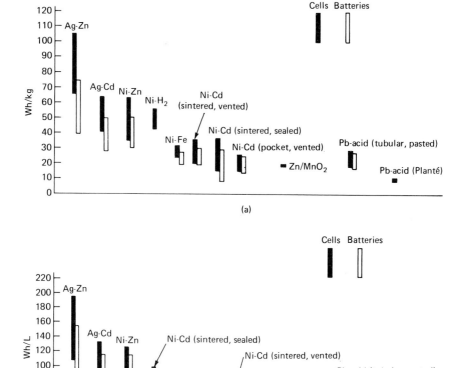

FIG. 13.2 Energy density of the secondary battery systems at 20°C: (a) gravimetric energy density; (b) volumetric energy density. (*Adapted from Falk and Salkind.*)

13.3.4 Effect of Discharge Rate on Energy Density

In Fig. 13.3, the effect of the discharge rate on the performance of each battery system is compared, based on the 10-h-rate watthour capacity for each system. With the exception of the conventional nickel-iron battery, which was not designed for high rate capability, the capacity of the alkaline batteries decreases relatively little with increasing discharge rate compared with the lead-acid system. This is due to the limited time available for the diffusion of the sulfuric acid electrolyte during the discharge of the lead-acid battery. The sintered-plate nickel-cadmium system shows the best performance.

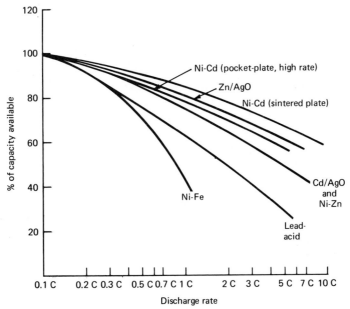

FIG. 13.3   Effect of discharge rate on capacity of secondary battery systems at 20°C.

These characteristics are compared again in Fig. 13.4. This figure shows the hours of service each battery type (unitized to 1 kg battery weight) will deliver at various power (discharge current × midpoint voltage) levels. The sharper slope is indicative of superior retention of capacity with increasing discharge load. The energy density can be calculated by

$$\text{Energy density} = \text{power density} \times \text{hours of service}$$

or
$$\text{Wh/kg} = \text{W/kg} \times \text{h} = \frac{A \times V \times h}{kg}$$

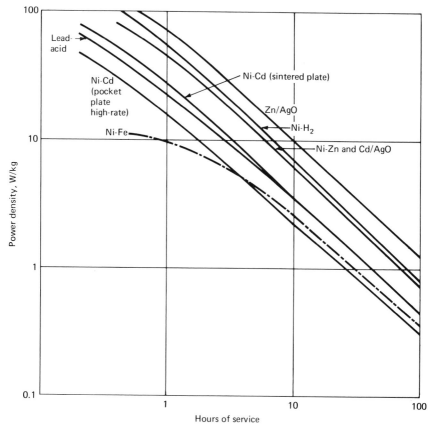

FIG. 13.4   Comparison of performance of secondary battery systems—power density vs. hours of service at 20°C.

13.3.5 Effect of Temperature

The performance of the various secondary batteries over a wide temperature range is shown in Fig. 13.5 on a gravimetric basis. In this figure, the gravimetric energy density for each battery system is plotted from -40 to $60°C$ at about the $C/5$ discharge rate. The zinc/silver oxide system has the highest energy density throughout most of the range, but the sintered-plate nickel-cadmium cell shows the highest percentage retention, similar to its superior performance at high discharge rates (Fig. 13.3). In general, the low-temperature performance of the alkaline batteries is better than the performance of the lead-acid batteries, again with the exception of the nickel-iron system. The lead-acid system shows better characteristics at the higher temperatures. These data are necessarily generalized for the purposes of comparison and present each system under favorable discharge conditions; performance is strongly influenced by the specific discharge conditions.

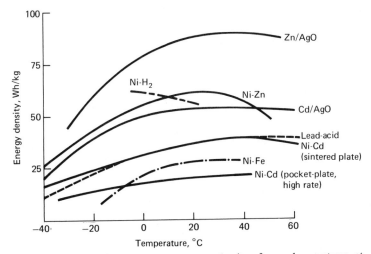

FIG. 13.5 Effect of temperature on energy density of secondary systems, at approximately C/5 discharge rate.

13.3.6 Charge Retention

The charge retention of most secondary batteries is poor compared with that of primary battery systems (see Fig. 3.22 in Part 1). Normally, secondary batteries are recharged on a periodic basis or maintained on "float" charge if they are to be in a state of readiness. Most alkaline batteries, especially the nickel-cadmium and nickel-iron batteries, can be stored for long periods of time even in a discharged state without permanent damage and can be recharged when required for use. The lead-acid batteries, however, can not be stored in a discharged state because sulfation of the plates, which is detrimental to battery performance, will occur.

Figure 13.6 shows the charge retention properties of the different secondary battery systems. These data are also generalized for the purpose of comparison. There are wide variations of performance depending on design and many other factors, with the variability increasing with increasing storage temperature. Typically, too, the rate of loss of capacity decreases with increasing storage time; thus storage time is another influencing factor (illustrated, for example, in Figs. 17.15 and 18.13 in Part 3).

The silver secondary batteries have the best charge retention characteristics of the secondary battery systems (not including the rechargeable "primary" battery types). Low-rate silver cells may lose as little as 10 to 20% per year, but the loss with high-rate cells with large surface areas could be 5 to 10 times higher. The vented pocket- and sintered-plate nickel-cadmium batteries and the nickel-zinc systems are next; the sealed cells and the nickel-iron batteries have the poorest charge retention properties of the alkaline systems.

The charge retention of the lead-acid batteries is dependent on the design, electrolyte concentration, and formulation of the grid alloy as well as other factors. The charge retention of the standard automotive SLI batteries, using the standard antimonial-lead grid, is poor, and these batteries have little capacity remaining after 6 months' storage at room temperature. The low antimonial-lead designs and the maintenance-free batteries have much better charge retention with losses in the order of 20 to 40%/y.

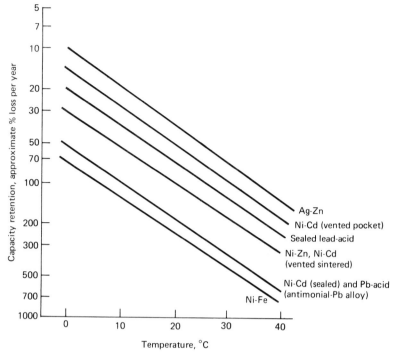

FIG. 13.6   Capacity retention of secondary battery systems.

13.3.7 Life

The lives (cycle life and calendar life) of the different secondary battery systems are listed in Table 13.4. Again, these data are approximate because specific performance is dependent on the particular design and the conditions under which the battery is used. The depth of discharge (DOD), for example as illustrated in Fig. 13.7, and the charging regime strongly influence the battery's life.[4]

Of the conventional secondary systems, the nickel-iron and the vented pocket-type nickel-cadmium batteries are best with regard to cycle life and total lifetime. The nickel-hydrogen battery, being developed for aerospace applications, has already demonstrated good cycle life under deep depth of discharge. The silver-zinc batteries and the "primary" rechargeable batteries have the poorest life characteristics. The lead-acid batteries do not match the performance of the best alkaline batteries. The pasted cells have the shortest life of the lead-acid cells; the best cycle life is obtained with the tubular design; and the Planté design has the best lifetime.

13.3.8 Charge Characteristics

The typical charge curves of the various secondary systems at normal constant-current charge rates are shown in Fig. 13.8. All the batteries can be charged under constant-current conditions, which is usually the preferred method of charging, although, in practice, constant-potential or modified constant-potential methods are used. Some of the sealed cells, however, may not be charged by constant-potential methods because

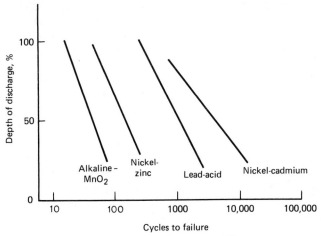

FIG. 13.7 Effect of depth-of-discharge on cycle life.

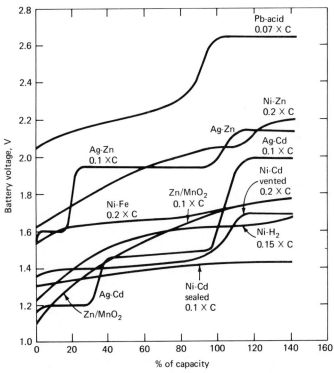

FIG. 13.8 Typical charge characteristics of secondary battery systems, constant-current charge at 20°C. (*Adapted from Falk and Salkind.*)

of the possibility of thermal runaway. Table 13.6 summarizes the typical conditions for charging the different systems. Generally, the vented nickel-cadmium cell has the most favorable charge properties and can be charged by a number of methods and in a short time. These cells can be charged over a wide temperature range or can be overcharged without damage. Nickel-iron and the sealed nickel-cadmium cells have good charge properties, but the temperature range is narrower for these systems. The lead-acid cells also have good charge characteristics, but care must be taken to prevent excessive overcharging. The silver-zinc system is the most sensitive one; the permissible charge current is low, and overcharging is very detrimental to cell life. Data on the efficiency of charging are also listed in Table 13.6.

13.3.9 Cost

The cost of a secondary battery may be evaluated on several bases, depending on the mode of operation. The initial cost is one of the bases for consideration. Other factors are the number of charge/discharge cycles that are available, or the number delivered in an application, during a battery's lifetime and the cost determined on a dollar-per-cycle or dollar-per-total-kilowatthour basis. The cost of charging and maintenance may also have to be considered in this evaluation (see Sec. 3.5.2 in Part 1, Chap. 3). In an emergency standby service or SLI-type application, the important factors may be the calendar life of the battery (rather than cycle life) and the cost as evaluated on a dollar-per-operating-year basis.

Table 13.4 lists the initial cost of each secondary battery in terms of dollars per kilowatthour based on the rated capacity. The lead-acid battery system is, by far, the least costly of the secondary batteries, particularly the SLI types. The lead-acid traction and stationary batteries, having more expensive constructional features and not as broad a production base, are several times more costly but are less expensive than the other secondary batteries. The nickel-cadmium batteries are the next lowest in cost. As shown in the table, the cost is very dependent on the cell size or capacity, the smaller button cells being considerably more expensive than the larger cylindrical and prismatic cells. The nickel-iron battery is more expensive and, for this reason among others, lost out to the less expensive battery systems.

The most expensive of the conventional-type secondary batteries are the silver batteries. Their higher cost and low cycle life have limited their use to special applications, mostly in the military and in portable equipment, which require their high energy density. The hydrogen systems are more expensive due to the use of a fuel cell–type negative electrode and to relatively limited production. However, their excellent cycle life under conditions of deep discharge make them most attractive for aerospace applications.

An important objective of the program for the development of secondary batteries for electric vehicles and energy storage is to reduce the cost of these battery systems. The near-term objective is a production cost of $70 per kilowatthour by 1990.

13.4 STANDARDIZATION OF SECONDARY BATTERIES

Standardization of secondary batteries has focused mainly on the lead-acid SLI batteries. No international standards exist, and those in widest use have been prepared by

TABLE 13.6 Charging Characteristics of Secondary Batteries

System	Charging methods*		Recommended constant-current charge rate C, A	Over-chargeability	Temperature range for charging, °C	Ah and Wh efficiencies†	
	Preferred	Not recommended				Ah efficiency, %	Wh efficiency, %
Lead-acid	cc, cp		0.07	Fair	−40 to 50		
Pasted, Planté						90	75
Tubular						80	70
Nickel-cadmium							
Pocket vented	cc, cp		0.2	Very good	−50 to 40	70	60
Sintered vented	cc, cp		0.2	Very good	−55 to 75	70–80	60–75
Sealed	cc	cp	0.1–0.3	Very good	0 to 40	65–70	60–70
Nickel-iron	cc	cp	0.2	Very good	0 to 45	80	60
Nickel-zinc	cc, cp		0.1–0.4	Fair	−20 to 40	85	70
Silver-zinc	cc,		0.05–0.1	Poor	0 to 50	90	75
Silver-cadmium	cc		0.01–0.2	Fair	−40 to 50	90	70

*Constant current (cc) includes two-rate charging, and constant potential (cp) includes modified constant-potential charging.
†All data are related to normal rates of charge and discharge and to room temperature operation.
SOURCE: Based on Ref. 3.

the automotive industry through the Society of Automotive Engineers (SAE), Warrendale, Pa., and by the battery industry through the Battery Council International (BCI), Chicago, Ill. The BCI nomenclature follows the standards adopted by its predecessor, the American Association of Battery Manufacturers (AABM). These specifications cover standard sizes and test procedures. In Europe and Japan, standards have been developed by several national standards groups. The standards and test procedures for SLI batteries are given in Part 3, Chap. 14.

International standards for various types of secondary batteries are being prepared by the International Electrotechnical Commission (IEC), Geneva; the American National Standards Institute, Inc., New York, also is preparing standards for certain types of secondary batteries which will be published as ANSI C18.2.

The battery manufacturers and important users, such as the Bell System and the military services, have adopted specifications for lead-acid and other secondary battery systems, but these have not been adopted for general use.

REFERENCES

1. A. J. Salkind, D. T. Ferrell, and A. J. Hedges, "Secondary Batteries 1952–1977," *J. Electrochem. Soc.* **125**(8), August 1978.

2. E. Findl and M. Klein, "Electrolytic Regenerative Hydrogen-Oxygen Fuel Cell Battery," *Proc. 20th Annu. Power Sources Conf.,* Electrochemical Society, Pennington, N.J., May 1966.

3. S. U. Falk and A. J. Salkind, *Alkaline Storage Batteries,* Wiley, New York, 1969.

4. L. H. Thaller, "Expected Cycle Life vs. Depth of Discharge Relationships of Well-Behaved Single Cells and Cell Strings," *J. Electrochem. Soc.* **130**(5), May 1983.

Lead-Acid Batteries

by
Alvin J. Salkind, George E. Mayer, and David Linden

14.1 GENERAL CHARACTERISTICS

The lead-acid battery has been a successful article of commerce for over a century. Its production and use continue to grow because of new applications for battery power in energy storage, emergency power, and electric vehicles (including material-handling equipment) and because of the continued growth of automobiles, boats, and planes for which it provides the energy for engine starting, vehicle lighting, and engine ignition (SLI). Its sales represent approximately 60% of the sales of all batteries in the world, or a market value, in 1980, of about $5 billion at manufacturers' levels, and 2 to 3 times this value at retail levels (see Table 13.1). (These values do not include those countries, e.g., U.S.S.R. and China, for which market data are not available.) This battery is also used extensively in telephone systems, power tools, communication devices, emergency lighting systems, and as the power source for mining and material-handling equipment. The wide use of the lead-acid battery in many designs, sizes, and system voltages is accounted for by the low price and ease of manufacture on a local basis of this battery system. The lead-acid battery is almost always the least expensive storage battery for any application, while still providing good performance and life characteristics.

The overall advantages and disadvantages of the lead-acid battery, compared with other systems, are listed in Table 14.1.

The lead-acid battery is manufactured in a variety of sizes and designs, ranging in size from less than 1 to over 10,000 Ah. Table 14.2 lists the various types of lead-acid batteries that are available.

14.1.1 History

Practical lead-acid batteries began with the research and inventions of Raymond Gaston Planté in 1860, although batteries containing sulfuric acid or lead compo-

14-1

TABLE 14.1   Major Advantages and Disadvantages of Lead-Acid Batteries

Advantages	Disadvantages
Popular low-cost secondary battery—capable of manufacture on a local basis, worldwide, from low to high rates of production.	Relatively low cycle life.
	Limited energy density—typically 30 to 40 Wh/kg; other systems are capable of 60 Wh/kg or higher.
Available in large quantities and in a variety of sizes and designs—manufactured in sizes from smaller than 1 Ah to several thousand ampere-hours.	Poor charge retention—sulfation.
Good high-rate performance—suitable for engine starting (but outperformed by some nickel-cadmium batteries).	Long-term storage in a discharged condition can lead to irreversible polarization of electrodes (sulfation).
Good low- and high-temperature performance.	Difficult to manufacture in small sizes (it is easier to make nickel-cadmium button cells in the smaller than 500-mAh size).
Electrically efficient—turnaround efficiency of over 60%, comparing discharge energy out with charge energy in.	Hydrogen evolution in some designs can be an explosion hazard.
High cell voltage—the open-circuit voltage of 2.2 V is the highest of all aqueous electrolyte battery systems.	Stibene and arsine evolution in some designs and applications can be a health hazard.
Good float service.	
Easy state-of-charge indication.	
Good charge retention for intermittent charge applications (if grids are made with high overvoltage alloys).	
Capable of being fabricated in maintenance-free designs.	
Low cost compared with other secondary batteries.	

nents were discussed earlier.[1] Table 14.3 lists the events in the technical development of the lead-acid battery. In Planté's fabrication method, two long strips of lead foil and intermediate layers of coarse cloth were wound spirally and immersed in a solution of about 10% sulfuric acid. The early Planté cells had little capacity, since the amount of stored energy depended on the corrosion of one lead foil to lead dioxide to form the positive active material, and similarly the negative electrode was formed by roughening of the other foil (on cycling) to form an extended surface. Primary cells were used as the power sources for this formation. The capacity of Planté cells was increased on repeated cycling as corrosion of the substrate foil created more active material and increased the surface area. In the 1870s, magnetoelectric generators became available to Planté, and about this time the Siemens dynamo began to be installed in central electric plants. Lead-acid batteries found an early market to provide load leveling and to average out the demand peaks; they were charged at night, similar to the procedure now planned for modern load-leveling, energy-storage systems.

TABLE 14.2   Types and Characteristics of Lead-Acid Batteries

Type	Construction	Typical applications
SLI (starting, lighting, ignition)	Flat-pasted plates (option: maintenance-free construction)	Automotive, marine, aircraft, diesel engines in vehicles and for stationary power
Traction	Flat-pasted plates Tubular and gauntlet plates	Industrial trucks (material handling)
Vehicular propulsion	Flat-pasted plates Tubular and gauntlet plates; also, composite construction	Electric vehicles, golf carts, hybrid vehicles, mine cars
Submarine	Tubular plates Flat-pasted plates	Submarines
Stationary (including energy storage types such as charge retention, solar photovoltaic, load leveling)	Planté Manchester Tubular and gauntlet plates Flat-pasted plate Circular flat plate	Standby emergency power Telephone exchange Uninterruptible power systems (UPS) Load leveling Signaling
Portable (sealed, nonspill)	Flat-pasted plates: gelled electrolyte, electrolyte absorbed in separator Spirally wound electrodes: sealed lead-acid (SLA) Tubular plates	Consumer and instrument applications: toys, portable tools, appliances, lighting, emergency lighting, radio, TV, alarm systems

Subsequent to Planté's first developments, numerous experiments were done on accelerating the formation process and coating lead foil with lead oxides on a lead plate pretreated by the Planté method. Attention then turned to other methods for retaining active material, and two main technology paths evolved:

1. Coating a lead oxide paste on cast or expanded grids, rather than foil, in which the active material developed structural strength and retention properties by a "cementation" process (interlocked crystalline lattice) through the grid. This is generally referred to as flat-plate design.

2. The tubular electrode design, in which a central conducting wire or rod is surrounded by active material and the assembly encased in an electrolyte porous insulated tube, which can be either square, round, or oval.

Simultaneous with the advances in developing and retaining active material was work in strengthening the grid, by casting it from lead alloys such as lead-antimony (e.g., Sellon, 1881) or lead-calcium (e.g., Haring and Thomas, 1935).[2] The technical knowledge for economical manufacture of reliable lead-acid batteries was in place by the end of the nineteenth century, and subsequent growth of the industry was rapid. Improvements in design, manufacturing equipment and methods, recovery methods, active material utilization and production, supporting structures and components, and

TABLE 14.3 Events in the Technical Development of the Lead-Acid Battery

		Precursor systems
1836	Daniell	Two-fluid cell Copper/copper sulfate/sulfuric acid/zinc
1840	Grove	Two-fluid cell Carbon/fuming nitric acid/sulfuric acid/zinc
1854	Sindesten	Polarized lead electrodes with an external source

		Lead-acid battery developments
1860	Planté	First practical lead-acid battery, corroded lead foils to form active material
1881	Faure	Coated lead foils with lead oxide-sulfuric acid paste for positive electrode, to increase capacity
1881	Sellon	Lead-antimony alloy grid
1881	Volckmar	Perforated lead plates to provide pockets for support of oxide
1882	Brush	Mechanically bonded lead oxide to lead plates
1882	Gladstone and Tribe	Double-sulfate theory of reaction in lead-acid battery: $PbO_2 + Pb + 2H_2SO_4 \rightleftharpoons 2PbSO_4 + 2H_2O$
1883	Tudor	Pasted a mixture of lead oxides on a grid pretreated by Planté method
1886	Lucas	Formed lead plates in solutions of chlorates and perchlorates
1890	Phillipart	Early tubular construction—individual rings
1890	Woodward	Early tubular construction
1910	Smith	Slotted rubber tube, Exide tubular construction
1920 to present		Materials and construction research: especially expanders, oxides, and fabrication techniques
1935	Haring and Thomas	Lead-calcium alloy grid
1935	Hamer and Harned	Experimental proof of double-sulfate theory of reaction
1956–1960	Bode and Voss Ruetschi and Cahan J. Burbank W. Feitknecht	Clarification of properties of the two crystalline forms of PbO_2 (alpha and beta)
1970 to present		Expanded metal grid technology Composite plastic/metal grids Sealed and maintenance-free lead-acid batteries Glass fiber and improved separators Through-the-partition intercell connectors Heat-sealed plastic case-to-cover assemblies High energy density batteries (above 40 Wh/kg)

nonactive components such as separators, cases, and seals continue to better the economic and performance characteristics of lead-acid batteries. Recently, more intensive development work has started for new applications, especially for energy storage and electric vehicles (see Sec. 14.10).

14.1.2 Manufacturing Statistics and Lead Use

The major present-day use of lead-acid storage batteries is in vehicle starting (SLI) applications. Most vehicles use a 12-V battery with a capacity in the range of 40 to 60 Ah. This battery, weighing about 14.5 kg, has sufficient high-rate capacity to deliver the 300 to 400 A necessary to start an automobile engine.

Approximately 65% of the battery's weight is lead or lead components. Batteries represent the most important use of lead in the world. The yearly consumption of lead for the manufacture of batteries was about 2 million metric tonnes in 1980, approximately half of the total world output of primary and secondary lead. There is a direct correlation between the number of motor vehicles registered and the number of SLI battery units sold annually, as shown in Table 14.4.[3] Although SLI units represent over 70% of the monetary value of the total lead battery market at present, this percentage is declining as the other applications for storage batteries continue to grow. A world perspective on the manufacture and use of lead-acid batteries is presented in Table 13.1 in Part 3.[3]

An important aspect of lead is that it is a recoverable resource. It has been estimated that much more than 90% of the batteries sold in the United States are ultimately recycled, and it takes considerably less energy to recycle lead, a low-melting metal (327.4°C), than to produce the metals used in other storage battery systems (nickel, iron, zinc, silver, and cadmium) in battery grade quality.

TABLE 14.4 Market Growth of the Lead-Acid Battery—United States*

	1940	1950	1960	1969	1976	1978	1979	1980
MV registration	. . .	55	74	105	134	149	151	158
SLI units (original equipment and replacement)	20	32	34	47	60	72	69	62
Car-battery ratio‡	. . .	1.7	2.2	2.2	2.2	2.1	2.1	2.5
SLI sales	$. . .	. . .	330	510	1040	1600†	1700†	1675†
Industrial	$. . .	. . .	70	105	200	260	360	380
Consumer	$. . .	. . .	1	3	15	35	50	55
Total	$. . .	. . .	400	620	1250	1900	2100	2110

*All units in millions, except car-battery ratio.
†Battery prices increase 50 cents per kilowatthour for each 1-cent increase in the price of lead. Lead prices varied from $0.40 to $1.40/kg between 1978 and 1982.
‡A considerable number of SLI-type batteries are used in boats and other vehicles; therefore, this ratio is not an exact representation of the useful life of an automotive battery.
NOTE: The Canadian market is approximately 10% of the U.S. market.

14.2 CHEMISTRY

14.2.1 General Characteristics

The lead-acid battery uses lead dioxide as the active material of the positive electrode and metallic lead, in a highly reactive porous structure, as the negative active material. The physical and chemical properties of these materials are listed in Table 14.5. Typically, a charged positive electrode contains both variations, α-PbO$_2$ (orthorhombic) and β-PbO$_2$ (tetragonal). The equilibrium potential of the α-PbO$_2$ is more positive than that of β-PbO$_2$ by 0.01 V. The α form also is a larger, more compact crystal which is less active electrochemically and slightly lower in capacity per unit weight; it does, however, promote longer cycle life. Neither of the two forms is fully stoichiometric. Their composition can be represented by PbO$_x$, with x varying between 1.85 and 2.05. The introduction of antimony, even at low concentrations, in the preparation or cycling of these species leads to a considerable increase in their performance. The preparation of the active material precursor consists of a series of mixing and curing operations using leady lead oxide (PbO + Pb), sulfuric acid, and water. The ratios of the reactants and curing conditions (temperature, humidity, and time) affect the development of crystallinity and pore structure. The positive active material, which is formed electrochemically from the cured plate, is a major factor influencing the performance and life of the lead-acid battery.

The electrolyte is a sulfuric acid solution, about 1.28 specific gravity or 37% acid by weight in a fully charged condition.

TABLE 14.5 Physical and Chemical Properties of Lead and Lead Oxides (PbO$_2$)*

Property	Lead	α-PbO$_2$	β-PbO$_2$
Molecular weight	207.2	239.19	239.19
Composition	. . .	PbO$_{:1.94-2.03}$	PbO$_{:1.87-2.03}$
Crystalline form	Face centered cubic	Rhombic (columbite)	Tetragonal (rutile)
Lattice parameters, nm	$a = 0.4949$	$a = 0.4977$ $b = 0.5948$ $c = 0.5444$	$a = 0.491-0.497$ $c = 0.3367-0.340$
X-ray density, g/cm^3	11.34	9.80	~9.80
Practical density at 20°C (depends on purity), g/cm^3	11.34	9.1-9.4	9.1-9.4
Heat capacity, cal/deg/mol	6.80	14.87	14.87
Specific heat, cal/g	0.0306	0.062	0.062
Electrical resistivity, (20°C,$\mu\Omega$/cm)	20	~100 $\times 10^3$	
Electrochemical potential, V, in 4.4 M H$_2$SO$_4$ at 31.8°C	0.356	~1.709	~1.692
Melting point, °C	327.4	–	–

*Based on Ref. 4 and others.

As the cell discharges, both electrodes are converted to lead sulfate; the process reverses on charge.

Negative electrode: $Pb \underset{charge}{\overset{discharge}{\rightleftarrows}} Pb^{2+} + 2e$

$Pb^{2+} + SO_4^{2-} \underset{charge}{\overset{discharge}{\rightleftarrows}} PbSO_4$

Positive electrode: $PbO_2 + 4H^+ + 2e \underset{charge}{\overset{discharge}{\rightleftarrows}} Pb^{2+} + H_2O$

$Pb^{2+} + SO_4^{2-} \underset{charge}{\overset{discharge}{\rightleftarrows}} PbSO_4$

Overall reaction: $Pb + PbO_2 + 2H_2SO_4 \underset{charge}{\overset{discharge}{\rightleftarrows}} 2PbSO_4 + 2 H_2O$

As shown, the basic electrode processes in the positive and negative electrodes involve a dissolution-precipitation mechanism and not some sort of solid-state ion transport or film formation mechanism.[4] The discharge/charge mechanism, known as the "double-sulfate" reaction, is also shown graphically in Fig. 14.1. As the sulfuric acid in the electrolyte is consumed during discharge, producing water, the electrolyte can be considered an "active material" and in certain battery designs can be the limiting material.

As the cell approaches full charge and the majority of the $PbSO_4$ has been converted to Pb or PbO_2, the cell voltage on charge becomes greater than the gassing voltage (about 2.39 V per cell) and the overcharge reactions begin, resulting in the production of hydrogen and oxygen (gassing) and the resultant loss of water.

Negative electrode: $2H^+ + 2e \rightarrow H_2$

Positive electrode: $H_2O - 2e \rightarrow \frac{1}{2} O_2 + 2H^+$

Overall reaction: $H_2O \rightarrow H_2 + \frac{1}{2} O_2$

In the sealed lead-acid cells, this reaction is controlled to prevent hydrogen evolution and the loss of water by recombination of the evolved oxygen with the negative plate (see Sec. 15.2).

The general performance characteristics of the lead-acid cell, during charge and discharge, are shown in Fig. 14.2. As the cell is discharged, the voltage decreases due to depletion of material, internal resistance losses, and polarization. If the discharge current is constant, the voltage under load decreases smoothly to the cutoff voltage and the specific gravity decreases in proportion to the ampere-hours discharged.

An analysis of the behavior of the positive and negative plates can be done by measuring the voltage between each electrode and a reference electrode, the "half-cell" voltage. Figure 14.3 illustrates this analysis, using a cadmium reference electrode.

Voltage   The nominal voltage of the lead-acid cell is 2 V; the voltage on open circuit is a direct function of the electrolyte concentration ranging from 2.125 V for a cell with 1.28 specific gravity electrolyte to 2.05 V with 1.21 specific gravity (see Sec. 14.2.2). The end or cutoff voltage on moderate-rate discharges is 1.75 V per cell but may range to as low as 1.0 V per cell at extremely high discharge rates at low temperatures.

Specific Gravity   The selection of specific gravity used for the electrolyte depends on the application and service requirements (see Table 14.11). The electrolyte concentration must be high enough for good ionic conductivity and to fulfill electrochemical requirements, but not so high as to cause separator deterioration or corrosion of other

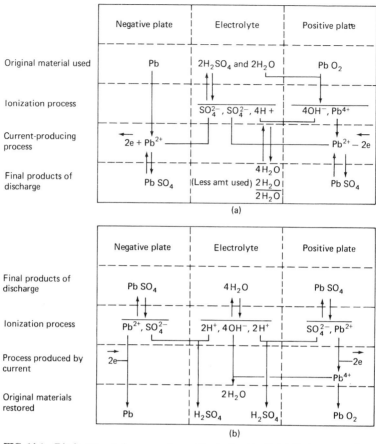

FIG. 14.1   Discharge and charge reactions of the lead-acid cell: (*a*) discharge reactions and (*b*) charge reactions. *(From C. Mantell, Batteries and Energy Systems, 2d ed., McGraw-Hill, New York, 1983.)*

parts of the cell, which would shorten life and increase self-discharge. The electrolyte concentration is deliberately reduced in high-temperature climates. During discharge, the specific gravity decreases from a fully charged to a fully discharged condition (see Table 14.6) in proportion to the ampere-hours discharged. The specific gravity is, thus, an excellent means for checking the state of charge of the battery. On charge, the change in specific gravity should similarly be proportional to the ampere-hour charge accepted by the cell, but there is a lag, because complete mixing of the electrolyte does not occur until gassing commences near the end of the charge.

14.2.2 Open-Circuit Voltage Characteristics

The open-circuit voltage for a battery system is a function of temperature and electrolyte concentration as expressed in the Nernst equation for the lead-acid cell (also see Part 1, Chap. 2, Sec. 2.3).

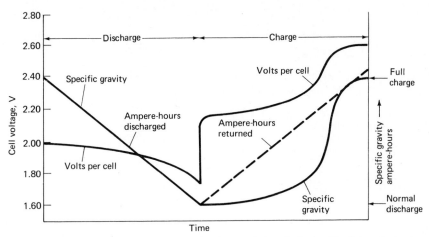

FIG. 14.2 Typical voltage and specific gravity characteristics of a lead-acid cell (constant-rate discharge and charge).

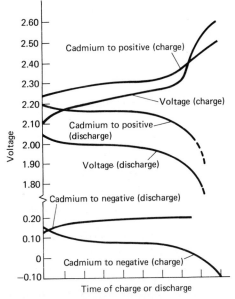

FIG. 14.3 Typical charge-discharge curves of a lead-acid cell. *(From C. Mantell, Batteries and Energy Systems, 2d ed., McGraw-Hill, New York, 1983.)*

TABLE 14.6   Specific Gravity of Lead-Acid Battery
Electrolyte at Different States of Charge*

State of charge	Specific gravity			
	A	B	C	D
100% (full charge)	1.330	1.280	1.265	1.225
75%	1.300	1.250	1.225	1.185
50%	1.270	1.220	1.190	1.150
25%	1.240	1.190	1.155	1.115
Discharged	1.210	1.160	1.120	1.080

*Assumes flooded cell design.
NOTE: Specific gravity may range from 100 to 150 points between full
charge and discharge, depending on cell design:
A is an electric vehicle (EV) battery
B is a traction battery
C is an SLI battery
D is a stationary battery

$$E = 2.047 + \frac{RT}{F} \ln \left(\frac{\alpha \, H_2SO_4}{\alpha \, H_2O} \right)$$

Since the concentration of the electrolyte is varied, the relative activities of H_2SO_4
and H_2O in the Nernst equation change. A graph of the open-circuit voltage vs.
electrolyte concentration at 25°C is given in Fig. 14.4. The plot is fairly linear above
1.20 specific gravity but shows strong deviations at lower concentrations. Open-circuit
voltage is also affected by temperature; the temperature coefficient of the open-circuit
voltage of the lead-acid battery is shown in Fig. 14.5. Where dE/dT is positive, e.g.,
above 0.5 m, the reversible potential of the system increases with increasing temperature. Below 0.5 m, the temperature coefficient is negative. Most lead-acid batteries
operate above 2 m (1.120 sp gr) and have a thermal coefficient of about $+0.2$ mV/°C.

14.2.3 Polarization and Resistive Losses

When a battery is being discharged, the voltage under load is lower than the open-
circuit voltage at the same concentrations of H_2SO_4 and H_2O in the electrolyte and Pb
or PbO_2 and $PbSO_4$ in the plates. The thermodynamically stable state for batteries is
the discharged state. Work (charging) must be done to cause the equilibria of the
electrochemical reactions to go toward PbO_2 in the positive and Pb in the negative.
Thus, the voltage of the power source for recharging the lead-acid battery must be
higher than the Nernst voltage of the battery on open circuit.

These deviations from the open-circuit voltage during charge or discharge are due,
in part, to resistive losses in the battery and, in part, to polarization. These losses can
be measured by use of an interrupted discharge, where the IR losses can be estimated
by Ohm's law ($\Delta E/\Delta I = R$) within a few seconds to a few minutes after the discharge
is stopped; the effect of polarization can take several hours to measure in order for
diffusion to allow the plate interiors to reequilibrate. Polarization is more easily measured by use of a reference electrode. The standard reference of hydrogen on platinum
is not practical for most measurements on lead-acid batteries, and several other sulfate-
based reference electrode systems are used. A review of reference electrodes[6a] neglects
several very practical sulfate electrodes. The most commonly used electrode for battery

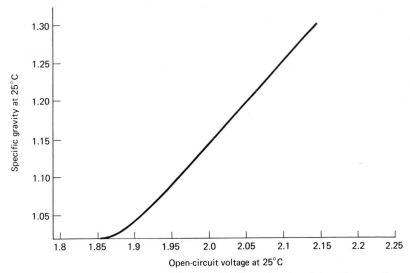

FIG. 14.4   Open-circuit voltage of the lead-acid cell as a function of electrolyte specific gravity.

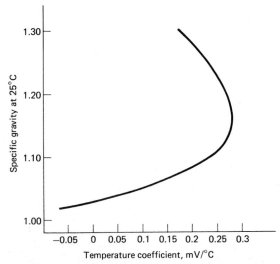

FIG. 14.5   Temperature coefficient of open-circuit voltage of the lead-acid cell as a function of specific gravity of electrolyte.

maintenance is the cadmium "stick," but this is not especially stable ($\pm$20 mV/day). Mercury-mercurous sulfate reference electrodes are stable and are available from several vendors. A novel $Pb/H_2SO_4/PbO_2$ reference electrode has been patented.[6b] This electrode measures the polarization on charge or discharge directly, without need for correction of different thermal coefficients of emf. The change in polarization between the start and the end of discharge is typically 50 mV to several hundred, and the cell capacity is limited by the group (positive or negative) that has the largest change in polarization during discharge. When both groups in a cell change about equally, the capacity limitation is more likely the depletion of H_2SO_4 in the electrolyte than depletion of Pb or PbO_2 in the plates. On charge, the polarization is a good measure that both positives and negatives have been recharged: the plate polarizations change by more than 60 mV between the start and end of recharge. Polarization voltages stabilize at some value when plates are recharged and are gassing freely.

14.2.4 Self-Discharge

The equilibria of the electrode reactions are normally in the discharge direction; e.g., thermodynamically, the discharged state is most stable. The rate of self-discharge (loss of capacity or charge when no external load is applied) of the lead-acid cell is fairly rapid, but it can be reduced significantly by incorporating certain design features.

The rate of self-discharge depends on several factors. Lead and lead dioxide are thermodynamically unstable in sulfuric acid solutions, and, on open circuit, they react with the electrolyte. Oxygen is evolved at the positive electrode and hydrogen at the negative, at a rate dependent on the overvoltage and acid concentration (the gassing rate increases with increasing acid concentration) as follows:

$$PbO_2 + H_2SO_4 \rightarrow PbSO_4 + H_2O + \tfrac{1}{2}\,O_2$$

$$Pb + H_2SO_4 \rightarrow PbSO_4 + H_2$$

For most positives, the formation of $PbSO_4$ by self-discharge is slow, typically much less than 0.5%/day at 25°C. (Some positives which have been made with nonantimonial grids can fail by a different mechanism on open circuit, namely, the development of a grid-active material barrier layer.) The self-discharge of the negative is generally more rapid, especially when the cell is contaminated with various metallic ions; for example, antimony lost from the positive grids by corrosion can diffuse to the negative, where it is deposited, resulting in a "local action" discharge cell which converts some lead active material to $PbSO_4$. New batteries with lead-antimony grids lose about 1% of charge per day at 25°C, but the charge loss increases by a factor of 2 to 5 as the battery ages. Batteries with nonantimonial lead grids lose less than 0.5% of charge per day regardless of age. This is illustrated in Fig. 14.6.[7] Maintenance-free and charge retention–type batteries, where the self-discharge rate must be minimized, use low-antimony or antimony-free alloy (such as calcium-lead) grids. However, because of other beneficial effects of antimony, its complete elimination may not be desirable, and low antimony–lead alloys are a useful compromise.

Self-discharge is temperature-dependent, as shown in Fig. 14.7. The graph shows the fall in specific gravity per day of a new, fully charged battery with 6% antimonial-lead grids. Self-discharge can, thus, be minimized by storing batteries in cool areas between 5 and 15°C.

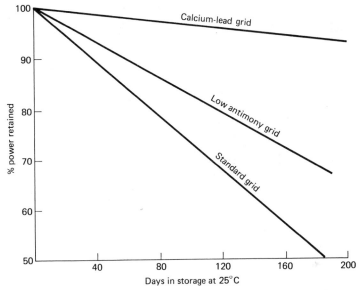

FIG. 14.6   Capacity or power loss during stand or storage at 25°C. *(From Ref. 7.)*

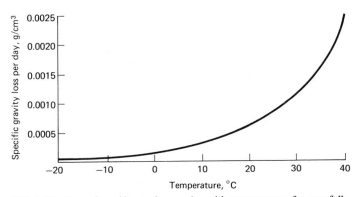

FIG. 14.7   Loss of specific gravity per day with temperature of a new fully charged lead-acid battery with 6% antimonial-lead grids. *(From Ref. 8.)*

14.2.5 Characteristics and Properties of Sulfuric Acid

The major characteristics and properties of the sulfuric acid electrolyte, as they apply to the operation of the lead-acid battery, are listed in Appendix C, Table C.2. The freezing points of sulfuric acid solutions at various concentrations also are plotted in Appendix C, Fig. C.1. The freezing point of aqueous sulfuric acid solutions varies significantly with concentration. Batteries must be designed, therefore, so that the electrolyte concentration is above the value at which the electrolyte would freeze when exposed to the anticipated cold. Alternately, the battery can be insulated and/or heated so that it remains above the electrolyte freezing temperature.

The specific gravities for several types of lead-acid batteries are given in Table 14.11; the change in concentration at different states of charge is shown in Table 14.6. A comparison with freezing point data will show that battery type A will freeze at −30°C when fully discharged while battery type D will freeze at about −5°C, a factor which must be considered in the design of the battery and/or the design of the battery housing. The acid concentration for most lead-acid batteries for use in temperate climates is usually between 1.26 and 1.28 specific gravity. Higher-concentration electrolytes tend to attack the separators and other components; lower concentrations tend to be insufficiently conductive in a partially charged cell and freeze at low temperatures. In high-temperature climates, a lower concentration is used, and in stationary cells with larger proportional electrolyte volumes and no high rate demands, electrolytes with as low as 1.21 sp gr are used (see Table 14.11).

Figure C.3, Appendix C, shows the proportions of concentrated sulfuric acid needed to prepare electrolyte of a required concentration.[5] Additional data on the properties of sulfuric acid and the other component materials of the lead-acid battery are available in Refs. 1, 5, and 8.

14.3 CONSTRUCTIONAL FEATURES, MATERIALS, AND MANUFACTURING METHODS

Lead-acid batteries consist of several major components as shown in a cutaway view in Fig. 14.8. This figure shows the construction of an automotive SLI battery. Batteries for other applications have analogous components as illustrated and described in Secs. 14.4 to 14.7. The applications of the various cells and batteries dictate the design, size, quantities, and types of materials that are used.

The active components of a typical lead-acid battery constitute less than one-half of its total weight. A breakout of the weights of the components of several types of lead-acid batteries is shown in Fig. 14.9.

The battery components are fabricated and processed as shown in the flowsheet (Fig. 14.10). The major starting material is highly purified lead.[9] The lead is used for production of alloys (for subsequent conversion to grids) and for production of lead oxides for subsequent conversion first to paste and ultimately to the lead dioxide positive active material and sponge lead negative active material.

14.3.1 Alloy Production

Pure lead is generally too soft to use as a grid material. Three exceptions, which use pure lead plates, are some special, very thick plate Planté or pasted-plate batteries, some small "jelly roll" wound batteries, and a cylindrical-cell telephone standby battery designed and commercialized by the Bell Telephone System.

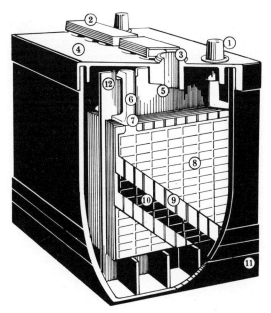

FIG. 14.8   Lead-acid battery—cutaway view of an SLI battery.

1. Terminal post
2. Gang-vent plug
3. Vent
4. One-piece cover
5. Electrolyte mark level
6. Intercell connector welded to the plate strap

7. Plate strap
8. Negative plate
9. Separator
10. Positive plate
11. Container
12. Cell partition

The pure lead has been hardened, traditionally, by the addition of antimony metal. The amount of antimony has varied between 12 and 5% by weight, generally dependent on the availability and cost of antimony. Typical modern alloys, especially for deep-cycling applications, contain 4 to 6% antimony. The trend in grid alloys is to go to even lower antimony contents, in the range of 1.5 to 3% antimony, in order to ultimately reduce the maintenance (water addition) that the battery will require. As the antimony content is decreased below 4%, the addition of small amounts of other elements is necessary to prevent grid fabrication defects and grid brittleness. These elements, such as sulfur, copper, arsenic, selenium, and tellurium, and various combinations of these elements act as grain refiners to decrease the lead grain size.[10-12]

Some of the alloying elements, not previously ascribed as grain refiners, fall into two broad classes of elements that are beneficial or detrimental to grid production or battery performance. Beneficial elements include tin, which operates synergistically with antimony and arsenic to improve metal fluidity and castability; silver is reported to improve corrosion resistance; cobalt is also thought to improve corrosion resistance. Detrimental elements include iron, which increases drossing[1]; nickel, which affects battery operation; and manganese, which attacks the separator. Cadmium has been

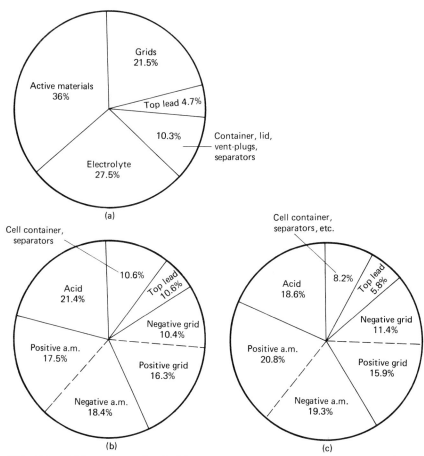

FIG. 14.9 Weight analyses of typical lead-acid batteries. (*a*) SLI battery. (*b*) Tubular cell. (*c*) Flat-plate traction cell. *(From Ref. 8.)*

used in grid alloys to enhance processability but is not popular because of its toxicity and difficulty in removal during lead recovery (recycling) operations. Bismuth exists in many lead ore feedstocks and has been reported to both increase and decrease grid corrosion rates.

A second class of lead alloys has been developed which uses calcium or other alkaline earth elements for stiffening. These alloys were developed originally for telephone service applications.[2,13] Antimony from the grids is dissolved during battery operation and migrates to the negative plates where it redeposits and upsets the cell operating voltage. One of the manifestations of this antimony transfer is greatly increased water consumption as an antimonial-lead battery ages. For telephone applications, more stable battery operations and less frequent watering were desired. The composition of the alloy depends somewhat on the grid processing that will be used. Calcium is used in the range of 0.03 to 0.10%; a variation has been to substitute

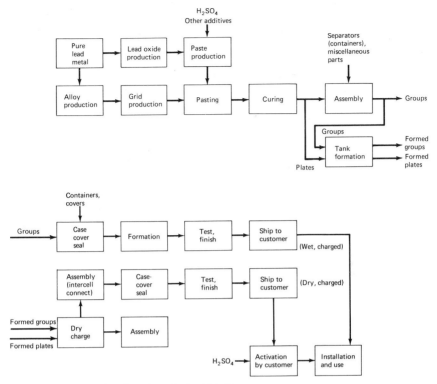

FIG. 14.10   Production flow sheet, lead-acid battery.

strontium for calcium. Barium has been investigated but is generally felt to be detrimental to performance. Tin has been used to enhance the mechanical and corrosion-resistant properties of the Pb-Ca alloys and is usually used in the range of 0.5 to 1.0% by weight. The trend in nonantimonial alloy development is toward ternary alloys (Pb-Ca-Sn) containing a minimal amount of tin because of the expense and relative scarcity of this element. Some batteries have been produced with a quaternary alloy—the fourth element being aluminum—to stabilize the drossing loss of the alkaline earth element (calcium or strontium) from the molten alloy. Grain refining is done by the alkaline earth metal, and no other elements (impurities) are desired. The properties of the alloys are summarized in Table 14.7.[10]

14.3.2 Grid Production

Two general classes of grid production methods virtually describe all modern production, but two other classes of production techniques might become widespread in the future. These are listed in Table 14.8.

The purposes of the grid are to mechanically hold the active material and to conduct

TABLE 14.7 Properties of Lead Alloys

			Alloys of the 1970s				
Property	Conventional antimony	Low antimony	Cast lead-calcium-tin 0.1 Ca 0.3 Sn	0.1 Ca 0.7 Sn	Lead-strontium tin-aluminum	Lead-cadmium antimony	Wrought lead-calcium-tin 0.065 Ca 0.7 Sn
Ultimate tensile strength, Pa $\times 10^{-6}$	38–46	33–40	40–43	47–50	53	33–40	60
Percent elongation	20–25	10–15	25–35	20–30	15	25	10–15

Property	Cast conventional Sb	Cast low Sb	Cast lead calcium	Cast lead strontium	Cast Pb-Cd-Sb	Wrought Pb-Ca-Sn (1st generation)
Ease of grid manufacture	Good	Fair	Fair	Fair—good	Fair	Good
Mechanical	Good	Fair	Fair—good	Fair—good	Fair	Good
Corrosion	Fair	Fair	Good	Good	Fair	Good
Battery performance	Poor	Fair	Good	Good	Good	Fair—good
Economics	Good	Good	Fair	Poor	Fair	Fair—good

Alloys of the 1980s

Property	Cast alloys				Wrought alloys			
	Lead-calcium-tin 0.1 Ca 0.3 Sn	Lead-calcium 0.1 Ca	Lead-calcium-tin with aluminum	Lead-calcium with aluminum	Lead-calcium-tin 0.065 Ca 0.3 Sn	Lead-calcium-tin 0.065 Ca 0.5 Sn	Lead-calcium 0.075 Ca	Low antimony 2.5–3.0% Sb
Ultimate tensile strength, Pa $\times 10^{-6}$	40–43	37–39	40-43	37–39	43–47	47	43	37–40
Percent elongation	25–35	30–45	25–35	30–45	15	15	25	25–40

Property	Cast alloys		Wrought alloys	
	Low antimony	Lead-calcium	Wrought lead-calcium-tin (2d generation)	Wrought low antimony
Ease of grid manufacture	Fair—good	Good (aluminum)	Good	Good
Mechanical	Fair	Fair—good	Good	Good
Corrosion	Fair	Good	Good	Fair—good
Battery performance	Fair—good	Good	Fair—good	Fair
Economics	Good	Good (lower tin)	Good	Good

NOTE: Alloy constituents given in weight percent.

TABLE 14.8 Grid Production Methods

Book mold cast
  Gravity cast
  Injection molded (die cast)

Mechanically worked (Planté, Manchester)

Continuous cast

Continuous cast, wrought-expanded
  Casting
  Working
  Expansion
    Progressive die expansion
    Precision expanded
    Rotary expansion
    Diagonal/slit expansion
  Punching

Composite

electricity between the active material and the cell terminals. The mechanical support can be provided by nonmetallic materials (polymer, ceramic, rubber, etc.) inside the plate, but these are not electrically conductive. Additional mechanical support is sometimes gained by the construction method or by various wrappings on the outside of the plate. Metals other than lead alloys have been investigated to provide electrical conductivity, and some (copper, aluminum, silver) are more conductive than lead. These alternate conductors are not corrosion-resistant in the sulfuric acid electrolyte and are often more expensive than lead alloys. Titanium has been evaluated as a grid material; it is not corroded after special surface treatments but is very expensive.

The grid design is generally a rectangular framework with a tab or lug for connection to the post strap. For cast grids, the framework features a heavy external frame and a lighter internal structure of horizontal and vertical bars. In some grid designs the frame tapers with the greater width near the lug; the internal bars may also be tapered. A recent advance in grid design is the "radial" grid, with the vertical wires displaced along the frame, pointing directly toward the tab area in order to increase grid conductivity (Fig. 14.11). The radial design has been further refined to a composite of a lead alloy radial conductor arrangement cast into a rectangular plastic frame. An example of this composite grid is shown in Fig. 14.12.

"Book mold" casting accounts for most grid production. Permanent molds are made from steel (Meehanite) blocks by machining grooves to form the grid frames and internal lattice structure. The molds are filled when closed with an amount of lead sufficient to fill out the grid and leave an excess gate or sprue which is subsequently trimmed off by a cutting or stamping operation. The grid alloy is entered into the mold by dump from a ladle in a recirculation lead alloy stream, dump from a metering valve in a nonrecirculation lead stream, or dump from a hand-filled ladle. A variation on book mold casting is injection molding or die casting of battery grids. Here, the lead alloy is forced into a clamped mold by a high injection pressure. Depending on the alloy characteristics, injection molding can be capable of very high production rates.

The second major method of grid manufacture is a mechanical working operation of a strip or slab of lead alloy. The traditional operations (Planté-type plate) have been to cut grooves into a thick lead plate, thereby increasing its surface area, or to crimp and roll up lead strips into rosettes which are inserted into round holes in a cast plate.

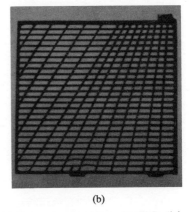

(a)   (b)

FIG. 14.11   Examples of lead-acid battery cast grids. (*a*) Conventional cast flat grid. (*b*) Radial-design grid.

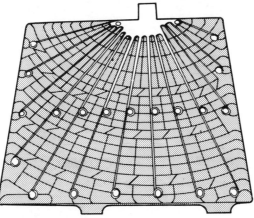

FIG. 14.12   Composite grid, radial conductor. Grid combines diagonal conducting members with light robust plastic frame.

The resultant plates are formed electrolytically into positives in the classic Planté and "Manchester" designs, respectively (Fig. 14.13).

The third major grid production method is circumferential continuous casting on a mold cut into the surface of a drum. Successful high-speed production of up to 150 grids per minute has been reported. Continuous-cast grids are not symmetrical about a central planar axis and need to be overpasted to hold the active material in place.

The fourth major grid production method, expansion from wrought strip, is rapidly supplanting book mold casting as the preferred method for the manufacture of SLI battery grids. The advantages of this method are lower grid weight per unit of battery electrical performance, the capability to manufacture a wide variety of sizes with a minimum investment in tooling, a very high rate production capability (up to 600 plates per minute), and very uniform grid and plate sizes. Most development and commercialization has been done on nonantimonial lead-calcium tin alloys (Fig. 14.14). Strip is produced from cast slabs by a variety of proprietary metal-working processes, and the

FIG. 14.13　(*a*) Planté plate.

FIG. 14.13　(*b*) Manchester plate.

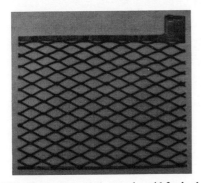

FIG. 14.14 Expanded wrought grid for lead-acid batteries.

thin strip is slit to the width specified by the battery manufacturer. The worked metal increases in strength as it decreases in thickness during this processing.

Machinery to produce grids from wrought strip has been developed and put into production by several manufacturers. Four types of machinery are involved: progressive die expansion, precision expansion, rotary expansion, and diagonal slit expansion. Progressive die expansion (PDE) has been utilized most extensively of the four methods, and PDE machinery has been commercially available for several years.

Whatever grid production method is used, there is often the need for small cast parts for plate and cell interconnections and connection to external equipment. These parts have traditionally been cast in fixed molds, sometimes with mold inserts to allow a variety of similar parts to be made in each mold. Newer battery production methods often produce these various interconnections automatically in the course of battery assembly.

14.3.3 Lead Oxide Production

Lead is used to make the active materials as well as the grids. The lead must be highly refined (usually virgin or primary lead) to preclude contamination of the battery and is described as "corroding-grade" lead in ASTM specification B29.[9] Lead is oxidized by either of two processes: the Barton pot and the ball mill.[14] In the Barton pot process, a fine stream of molten lead is swept around inside a heated, pot-shaped vessel, and oxygen from the air reacts with fine droplets and/or particles to produce an oxide coating around each droplet. Typical Barton pot oxides contain 15 to 30% free lead which usually exists as the core of each fine leady oxide spherical-shaped particle. Barton pots are available in a variety of sizes up to 1000 kg/h output.

Ball milling describes a larger variety of processes. Lead pieces are put into a rotary mechanical mill, and the attrition of the pieces causes fine metallic flakes to form. These are oxidized by an airflow, and the airflow also serves to remove the leady oxide particles to collection in a baghouse. The feedstock for ball mills can range from small cast slugs weighing less than 30 g to full pigs of lead weighing approximately 30 kg. Typical ball mill oxides also contain 15 to 30% free lead in the shape of a flattened platelet core surrounded by an oxide coating.

Some battery positive plates use an additive of red lead (Pb_3O_4) which is more conductive than PbO to facilitate the electrochemical formation of PbO_2. Red lead is produced from leady oxide by roasting this material in an airflow until the desired conversion is complete; such processing reduces the free lead content and generally increases the oxide particle size. A variety of other oxides and lead-containing materials have been used to produce battery plates but are of only historical interest.[14] Some positive plates for the Bell telephone system batteries are made with tetrabasic lead sulfate ($4PbO \cdot PbSO_4$) which is a precursor for α-PbO_2.

14.3.4 Paste Production

Lead oxide is converted to a plastic, doughlike material so that it can be affixed to the grids. Leady oxide is mixed with water and sulfuric acid in a mechanical mixer. Three types of mixers are commonly used: the "change can" or "pony" mixer, the muller, and a vertical muller.

The pony mixer is the traditional unit. A preweighed amount of leady oxide is placed into the mixing tub, and this is wetted first with water and then with sulfuric acid solution. Dry paste additives, if any, are premixed into the leady oxide before water addition. These additives can be plastic fibers to enhance mechanical strength of the dried paste, "expanders" to maintain negative-plate porosity in operation, and various other proprietary additives which ease processing or are believed to improve battery performance. Muller mixers are usually filled first with the water component, then the oxide, then the acid.

As mixing proceeds, the paste viscosity increases, then decreases, as measured by the amount of power consumed by the mixer motor. The paste becomes hot from the mechanical mixing and from the reaction of H_2SO_4 with the leady oxide. Paste temperature is controlled by cooling jackets on the mixer or by evaporation of water from the paste. The amounts of water and acid for a given amount of oxide will be different for the two mixer types and will also depend on the intended use of the plates: SLI plates are generally made at a low $PbO:H_2SO_4$ ratio and deep-cycling plates at a high $PbO:H_2SO_4$ ratio. Sulfuric acid acts as a bulking agent—the more acid used, the lower the plate density will be. The total amount of liquids and the type of mixer used will affect final paste consistency (viscosity). Paste mixing is controlled by measurement of paste density using a cup with a hemispherical cavity and by measurement of paste consistency with a penetrometer.

14.3.5 Pasting

Pasting is the process by which the paste is actually integrated with the grid to produce a battery plate. This process is a form of extrusion, and the paste is pressed by hand trowel or by machine into the grid interstices. Two types of pasting machines are used: a fixed-orifice paster that pushes paste into both sides of the plate simultaneously and a belt paster in which paste is pressed into the open side of a grid that is being conveyed past a paste hopper on a porous belt. The amount of paste applied to a plate by a belt paster is regulated by the spacing of the hopper above the grid on the belt and the type of troweling (roller or rubber "squeegee") used at the hopper exit. Using identical paste and grids, a trowel roller machine packs the paste both thicker and more densely than a rubber squeegee machine. As plates are pasted on either belt pasting machine, water is forced out of the paste, into the belt, and ultimately to a sump on or near the machine. The sump material can be used in place of some of the liquids for subsequent batches of negative paste.

Grids are automatically or manually placed onto the belt before being moved under the paste hopper. Most plates are made as "doubles" joined at the feet (cast) or joined at the tab edge (wrought expanded). Typically, the belt is 35 to 50 cm wide and can handle such doubles. Industrial stationary or traction plates (being larger) are pasted by lengthwise feed into the machine or are hand-pasted.

After pasting, plates are racked or stacked for curing. Stacked plates contain enough moisture to stick together, and so before stacking the plate surfaces are dried somewhat by a rapid passage through a high-temperature drier or over heated platens. Some carbon dioxide from the combustion process might be absorbed on the surface, such that the surface is made "harder." The "flash" drying process may also help start the curing reactions. Thicker plates are usually placed long-edge upward in racks rather than being stacked horizontally on pallets, after flash drying.

Wrought expanded plates and some cast plates are cut into discrete plate portions by a slitter machine in the pasting line. Some manufacturers also have the plate lugs brushed clean of paste and surface oxidation on the same machinery.

In Europe, and less commonly in the United States, a great portion of the heavy-duty battery positive plates are made in porous tubular sheaths. The grid is cast or injection-molded with long-finned spines attached to a header bar and a connection lug. Individual woven fiberglass-plastic sheaths, or a multitube gauntlet, are placed on the spines. These plates are filled with powder or with a slurried paste until the tubes are full. A plastic cap plugs the open sheath ends and becomes the bottom of the plate (Fig. 14.15).

14.3.6 Curing

The curing process is used to make the paste into a cohesive, porous mass and to help to produce a mechanical bond between the paste and the grid. Several different curing processes are used for lead oxides, dependent on paste formulation and the intended use of the battery.[12]

Typical cure for SLI plates is "hydroset," low temperature and low humidity for 24 to 72 h. The temperature is preferably between 25 and 40°C; the humidity is that contained in the flash-dried plates, typically 8 to 20% H_2O by weight. The plates are usually covered by canvas, plastic, or other materials to help retain both temperature and moisture. Some manufacturers use enclosed rooms for the hydroset, and these rooms may be heated where required by climatic conditions. As the plates cure, they reach a peak temperature and the temperature and humidity decrease. Hydroset typically produces tribasic lead sulfate, which gives high energy density and short cycle life.

Some manufacturers use high-temperature, high-humidity ovens for curing plates. This controls the peak temperature and makes sure that sufficient moisture is available to oxidize the remaining free lead in the paste. Peak temperatures in the range of 65 to 90°C are used; at higher temperatures and/or longer times, significant amounts of tetrabasic lead sulfate are produced in the plate, and the plates generally have low energy density and long cycle life. At the end of curing, the free lead content of the paste should be below 4%, preferably as low as possible. (Insufficient paste cure is observed as "soft," easily broken pasted plates, usually pale in color.) If the plates have not cured, they can be rewetted and reheated to force the cure. Another process to force completion of curing is to dip the partially cured plates into dilute sulfuric acid. This latter process ("pickling") is also used for cure of powder-filled tubular positive plates.

Cured plates are stored until use. Shelf life is not critical, but the high cost of inventory usually makes storage time minimal.

14.3.7 Assembly

The simplest cell consists of one negative, one positive, and one separator between them. Most practical cells contain about 3 to 30 plates with separators in between. Individual or "leaf" separators are generally used; some maintenance-free SLI and

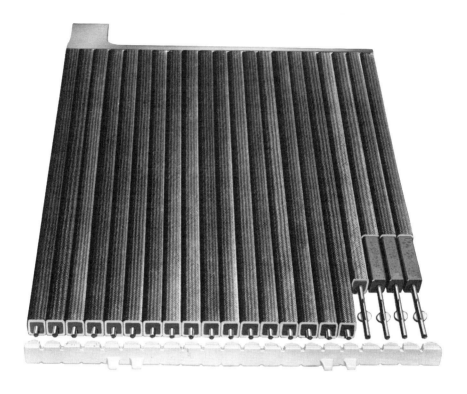

FIG. 14.15 Tubular and gauntlet plates: (*a*) Tubular and (*b*) gauntlet.

small sealed cell batteries use "envelope" separators that surround either the positive or the negative plates or both.

Separators are used to electrically insulate each plate from its nearest counterelectrode neighbors but must be porous enough to allow acid transport into or out of the plates. The properties of typical separators are given in Table 14.9.

SLI batteries typically have phenolic/cellulosic separators or sintered PVC separators, with average pore diameters in the 5- to 30-μm range. Heavy-duty batteries usually have microporous rubber or polyolefin separators which have smaller pore size and give longer life at a higher cost. Glass fiber mats are used in some sealed lead-acid designs to absorb all the electrolyte.

Industrial traction pasted positive plates are usually wrapped with several layers of fibrous glass matting to help retain the active material during use. The inner layer is usually very fine, parallel strands of "slyver" glass, and subsequent layers are randomly oriented glass fibers held together by a plastic (styrene or acrylic) binder. The matting is held in place by a heat-sealed, perforated rubber retainer. In lieu of plate wrapping, many heavy-duty batteries use a layer of glass matting on the ribs of the separator to help retain the positive active material.

Plates and separators are stacked manually or by a stacking machine. The latter is

TABLE 14.9 Properties of Separators

	Cellulose	PVC	Rubber	Microporous polyethylene Standard	Thin	Nonwoven polypropylene
Pore size, μm	25	20	1	0.03		12
Purity	Fair	Good	Good	Good		Excellent
Corrosion resistance	Fair	Excellent	Excellent	Excellent		Good
Mechanical strength	Fair	Good	Fair	Excellent		Good
Backweb thickness, mm	0.5	0.35–0.5	0.5	0.45–0.5	0.25–0.3	0.5
Porosity, %	60	45	60	60	60	60
Electrical resistance, $m\Omega\ cm^{-2}$	3.72	3.25	3.72	3.72	1.86	2.79
Cold start voltage, 5 s, 280 A, −18 °C	7.7	7.8	7.7	7.7	8.3	8.0
Flexibility	Brittle	Slightly brittle	Brittle	Excellent		Excellent
Sealing	No	Excellent	No	Good		Excellent

SOURCE: N. I. Palmer, BCI Convention, April 1975, p. 108.

typically used for SLI plates and rigid separators. Stacked elements are staged on roller conveyors or carts as input to the interplate welding operation. Welding is done by two general methods: melting of the lugs in a mold, with the lugs facing upward, or immersion of lugs facing downward into pools of molten lead alloy contained in a preheated mold. The first method is the traditional assembly method for lead-acid batteries. In this method, the plate lugs fit up through slots in a mold "comb"; the shape and size of the group strap is delineated by the "dam" and "back iron" portions of the tooling. Some battery manufacturers use slotted "crowfoot" posts to fit over the plate lugs to speed the welding process. The second welding method is called the "cast-on strap" process and is typically used for SLI cells. Stacked elements are loaded into slots of the cast-on machine. A mold which has cutouts corresponding to the desired straps and posts is preheated and filled with molten lead (usually approximately Pb-3%Sb) alloy. The mold and/or the stacked elements are moved until the plate lugs are immersed in the strap cutouts. External cooling solidifies the strap onto and around each lug, and the elements are moved to a point where they can be dropped into a battery case. Visual examination can differentiate between the two welding methods: fixture-welded plate straps are usually thicker and smoother than cast-on straps; cast-on also will usually show a convex meniscus of metal between adjacent plate lugs on the underside of the strap. A good weld is required between each plate lug and the strap so that high-rate discharge performance is maximized. The resultant assemblage of plates and separators is known as an "element," and the welded subelements are known as "groups." Electrical testing for short circuits is usually done on elements before further assembly.

Cast-on battery elements are either continuously connected or are made in discrete one-cell modules. The first method requires that long intercell connections be used, which travel over the intercell partition and are seated in a slot in this partition; this is known as the loop-over-partition design. In the second cast-on method, tabs on the ends of the plate straps are positioned over holes that have been prepunched into the intercell partitions of a battery case. These tabs are welded together manually with a very small torch or automatically by a resistance welding machine. The latter also squeezes the tabs and the intercell partition to provide a leak-proof seal.

Industrial traction cells, and old style SLI cells, have been connected into batteries after the cell cases and covers are sealed together. Traction batteries are needed in thousands of different sizes for various applications, and the standard unit of construction is a cell, not a quantity of plates and separators. A heavy steel tray is fabricated and coated with an acid-resistant coating (urethane, epoxy, etc.). Traction cells are placed into the tray and shimmed as necessary, and intercell connections are welded on. Heavy flexible wires (made from welding cable) are welded to the endmost cells for connection to the external circuit.

14.3.8 Case-to-Cover Seal

Four different processes have been used to seal battery cases and covers together. Enclosed cells are necessary to minimize safety hazards related to the acidic electrolyte, to the potentially explosive gases produced on overcharge, and to electrical shock. Most SLI batteries and many modern traction cells are sealed with fusion of the case and cover. The fusion comes from preheating each on a platen, then mechanically forcing the two together, or from ultrasonic vibration of the case and cover against each other. Fusion-sealed batteries are virtually impossible to repair; at best, the elements can be salvaged, but the cover and usually the case are discarded and replaced. Other SLI batteries are sealed by an epoxy cement which fills a groove in the cover; the battery is inserted and positioned so the case and intercell partition lips fit into the epoxy-filled groove. Heat is used to catalyze the setting of the epoxy.

Some small deep-cycling batteries feature tar-sealed cases and individual cell covers. Here, the tar seal allows easy repair to the battery. Traditionally, all batteries were made this way before about 1960, but epoxy and heat seals are typical for SLI batteries today. Molten tar is dispensed from a heated kettle to fill a groove between the cover and the case. The tar must be hot enough to flow easily but cool and viscous enough to solidify before running down into the cell.

Stationary batteries in plastic cases are sealed with epoxy glues, with solvent glues, or (for PVC copolymer cases and covers) with a thermal seal. Terminals are cast or welded on. Some very large stationary and traction cells are made so that coolant can be circulated through the terminals, and others are made with terminals with copper inserts for increased conductivity and mechanical strength.

14.3.9 Tank Formation

Plates or assembled groups can be electrically "formed" or charged before assembly into the case. When SLI plates are formed, these are usually formed as "doubles," with two to five panels stacked together in a slotted plastic formation tank, spaced an inch or less from stacks of counterelectrode pasted panels in adjacent slots. The stacks are arranged so that all positive lugs protrude out of one side of the tank top and all negative lugs protrude out of the other side of the tank top. All lugs with the same polarity are connected by welding to a heavy lead bar, and the two bars are connected to a low-voltage, constant-current power supply. The tank is filled with electrolyte, and current is passed until the plates have been formed: the positives are converted to a deep brownish-black and the negatives to a soft gray which show a bright metallic streak when scratched. Industrial plates are usually formed singly; sometimes these are also formed against "dummy" plates or grids. A variety of tank materials have been used, but the most common are polyvinyl chloride (PVC), polyethylene, or lead. The tanks are arranged so that the acid can be drained and refilled because formation increases the electrolyte concentration.

A variety of formation conditions are used, with variations in electrolyte density, charging rate (current), and temperature. Electrolyte is typically dilute, in the range of 1.050 to 1.150 sp gr. Charging rate is usually fixed, but some manufacturers use a sequence of two or three different charging rates for different periods of time.

Tank-formed groups or plates are somewhat unstable (negatives will spontaneously oxidize in air) and therefore are "dry charged" before use.

Modern electronic chargers for formation are made to operate in a constant-current mode, either by a saturable reactor control or by use of silicon-controlled rectifiers (SCRs). The most recent development in formation chargers is to control the current and time by use of a microcomputer; some formation schedules include charge at three or more different currents which start at low current(s), go to higher current(s), and then revert to lower current(s). Current adjustment during formation minimizes damage to the cells by high temperatures and minimizes the need for cell cooling by water spray or forced air.

14.3.10 Case Formation

The more usual method of formation is to assemble the battery completely, fill with electrolyte, and then apply the formation charge. This method is used for SLI and most stationary and traction batteries. A variety of formation conditions are used, similar to those for tank formation. The two major formation processes are the "two-shot" formation process (used for stationary and traction batteries) and the "one-shot" formation process (used for most SLI batteries). In the two-shot formation, the electrolyte is dumped to remove the low-density initial electrolyte and refilled with more concentrated electrolyte, chosen so that when this is mixed with the dilute initial acid residue which is absorbed in the elements or trapped in the case, the cell electrolyte will equilibrate at the desired density (Table 14.10). Typical values of the electrolyte specific gravity at full charge after formation are given in Table 14.11.

14.3.11 Dry Charge

The performance of wet batteries degrades with long periods of inactivity, especially when stored at warm temperatures. A loss of 1 to 3% of capacity each day is possible with SLI batteries that contain antimonial lead grids. The loss on stand is much less for maintenance-free batteries (0.1 to 0.3%/day). When lead-acid batteries must be stored for a long time, especially in high ambient temperatures, or when batteries are shipped for export, their performance can be stabilized by removal of the electrolyte by one of several methods.

TABLE 14.10   Formation Processes

	"One-shot"	"Two-shot"
Typical application	SLI	All others
Electrolyte concentration, sp gr		
Initial	1.200	1.005 to 1.150
Final	1.280	1.150 to 1.230
Subsequent processing	None	Dump and refill with 1.280 to 1.330 sp gr electrolyte; continue charge for several hours

TABLE 14.11   Specific Gravity of Electrolytes at Full Charge at 25°C

	Specific gravity	
Type of battery	Temperate climates	Tropical climates
SLI	1.260–1.280	1.210–1.230
Heavy duty	1.260–1.280	1.210–1.240
Golf cart	1.260–1.290	1.240–1.260
Golf cart (electric vehicle)	1.275–1.325	1.240–1.275
Traction	1.275–1.325	1.240–1.275
Stationary	1.210–1.225	1.200–1.220
Diesel starting (railroad)	1.250	
Aircraft	1.260–1.285	1.260–1.285

When the electrolyte is removed, the battery is called "dry-charged" (i.e., "charged and dry") or "charged and moist." The first process is done before the battery elements are assembled inside the case and cover. The plates can be tank-formed, water-washed, then dried in an inert gas before the element-welding portion of assembly. Alternately the welded element can be tank-formed, washed, and then dried in an inert gas. The latter process is simpler to carry out, but it is necessary that the separators can be rewetted easily after being washed and dried. The assembly (case, elements, cover) is completed and the battery is sealed. The battery can be stored in this dry-charged state for up to several years before reactivation and use.

Several processing innovations have been commercialized in the past 10 y to convert wet-charged batteries into moist or semidry batteries. In one process, most of the electrolyte is removed by centrifugation. Another process uses an inorganic salt (sodium sulfate) in the electrolyte which minimizes degradation during storage and assists in an eventual reactivation. A battery is formed, dumped, refilled with electrolyte which contains the additive, high-rate-discharge tested, and then finally dumped. The high-rate electrical discharge (to simulate engine cranking) allows test of an assembled "damp-dry" battery but this also probably blocks the plate surfaces with a thin layer of small lead sulfate crystals; these crystals then minimize plate or separator degradation during storage. During storage the batteries are sealed.

14.3.12 Testing and Finishing

Electrical tests are used to check the performance of batteries before they are sold and often before they are put into use. The type of test employed depends on the intended use for the battery: SLI batteries are tested by brief discharges at very high currents (200 to 1500 A) to simulate engine-cranking performance. Stationary and traction cells are discharged at a rate specified by the user usually in the range of 1 to 10 h for stationary and 4 to 8 h for traction. The discharge for SLI batteries is usually done by dissipation through a fixed low-ohmage resistor or by a brief, high-rate electrical discharge driven by a power supply. Heavy-duty batteries are discharged through a resistor, a transistorized load, or an inverter.

The final manufacturing steps consist of improving the battery appearance by washing, drying, painting, installing vent plugs, and labeling as desired. Rubber-cased batteries are usually painted; plastic-cased batteries are not. A large variety of plaques and labels are available which can describe the battery, its performance, and use. Product liability requirements in many countries mandate that the user be warned of the hazardous nature of the battery, especially that the electrolyte is corrosive and that gases are formed which can be explosive.

Traction batteries are physically sized to fit a myriad of different forklift trucks, and so the final assembly for a traction battery consists of inserting preformed and pretested cells into a sturdy metal box ("tray"), making intercell connections, making cable connections, and sometimes adding a tar or plastic (urethane) material in the spaces between cell covers.

14.3.13 Shipping

Small batteries (SLI and golf-cart types) are palletized several layers high for long-distance shipment. The batteries are cushioned by five-sided (slipover) or six-sided

TABLE 14.12   Standards on Lead Batteries*

Description	Indian standards	British standards	Australian standards	American standards	Japanese standards	German standards	International Electro-technical Commission standards
Lead-acid batteries for motor vehicles	7372-1974	3911-1965	D2-1968	SAE J 537h SAE J 240a SAE J 930a	JIS-D-5301-1973 and C 8701-1969		95-1 (1972) 95-2 (1965) 95-3 (1963)
Lead-acid traction batteries	5154-1969	2550-1971	—	WB-133 NEMA IB3 NEMA IB2	JIS-D5303-1969		254-1967
Lead-acid batteries for motorcycles, auto-rickshaws, and similar vehicles	1145-1962	—	—	—	—	—	—
Stationary lead-acid cells and batteries with tubular positive plates	1651-1970	—	—	—	C 8704-1973	DIN 40736 DIN 40737	—
Stationary lead-acid cells and batteries with Planté positive plates	1652-1972	440-1964	—	—	—	DIN 40731 DIN 40732 DIN 40733 DIN 40738	—
Stationary lead-acid cells and batteries with pasted plates	6304-1971	—	—	IEEE STD 484-1975 NEMA-IB4	—	—	—

Lead-acid batteries for aircraft	1846-1961	2G 205-pt.I (1975)	U2-1952	—	—	—	—
Miner's cap lamp lead-acid batteries	2512-1963	4945-1973	—	—	—	—	—
Lead-acid batteries for train lighting and air conditioning	6848-1972	—	—	—	—	DIN 43579 DIN 43567	—
Cycling batteries (EV)	—	—	—	BCI Standards (in development)	SBA-3013-1974	—	IEC Standards (in development)
Rubber and plastic containers for lead-acid batteries	1146-1972	—	D23-1973	ASTM D639-60T	—	DIN 43577	—
Separators for lead-acid batteries	6071-1970 (synthetic)	—	AS 1320-1975	F.S.O.-S801b F 4-14-65	—	—	—
Sulfuric acid for battery use	266-1961	3031-1972	C 60-1961	—	—	—	—
Water for battery use	1069-1964	4974-1975	C 59-1961	—	—	—	—
Battery terminals and terminal lugs	—	—	C 300-1947	—	—	DIN 40746 DIN 40748 DIN 43569	—

*Based on Ref. 8.

cardboard boxes with cardboard or wood sheet between layers. The batteries are held laterally by banding or by plastic sheet which is shrink- or stretch-wrapped around a full pallet. Pallets need to be very sturdy to withstand the battery weight and handling abuse. Larger batteries are palletized, banded, and cushioned as appropriate.

Batteries have traditionally been shipped only minimal distances because of their fragile nature, their weight, and their corrosive contents. The latter cause common carriers to charge a significant premium for battery shipment and usually preclude shipment by air. As the number of small, localized-sales battery manufacturers continues to decrease in the United States, the remaining large manufacturers now usually ship to their distribution chain on their own trucks.

14.3.14 Activation of Dry-Charged Batteries

When batteries have been dry-charged, they have to be reactivated before use. Activation consists of unpacking the battery, filling the cells with electrolyte (which sometimes is shipped with the battery in a separate package), charging the battery (if time is available), and testing the battery performance. When dry-charged batteries are activated, the materials which had been used to seal the vent holes must be removed and discarded.

14.3.15 Standards

Table 14.12 is a listing of some of the standards for lead-acid batteries and components.

14.4 SLI (AUTOMOTIVE) BATTERIES: CONSTRUCTION AND PERFORMANCE

14.4.1 General Characteristics

The design of lead-acid batteries is varied in order to maximize the desired type of performance. Tradeoffs exist for optimization among such parameters as power density, energy density, cycle life, "float-service" life, and cost.

High power density requires that the internal resistance of the battery be minimal; this affects grid design, the porosity, thickness and type of separator, and the method of intercell connection. High power and energy density also requires that plates and separators be thin and very porous and, usually, that paste density be very low. High cycle life requires premium separators, high paste density, the presence of α-PbO_2, modest depth of discharge, good maintenance and, usually, the use of high-antimony (5 to 7% Sb) grid alloy. Low cost requires both minimum fixed and variable costs, high-speed automated processing, and no premium materials for the grid, paste, separator, and other cell and battery components.[1,5,11,15,16]

14.4.2 Construction

An SLI battery, whose main function is to start an internal combustion engine, discharges briefly but at a high current. Once the engine is running, a generator or alternator system recharges the battery and then maintains it on "float" at full charge or slight overcharge. In recent automobile designs, the parasitic electrical load of lights, motors, and electronics causes a gradual discharge of the battery when the engine is not in operation. This factor, coupled with normal self-discharge, introduces a significant cycling component into the normal cranking/floating duty cycle. Studies of SLI battery life and failure modes reported by the Battery Council International in 1970 and 1982[17] show that average SLI battery life is now shorter and failure modes are different as a result of these changes in service conditions (see Secs. 14.4.3 and 14.9.4).

The cranking ability of an SLI battery is directly proportional to the geometric area of plate surface, with the proportionality factor typically 0.155 to 0.186 cold-crank amperes (CCA) at $-17.8°C$ (0°F) per square centimeter of positive-plate surface. Cranking performance is generally limited by the positive plate at higher temperatures ($> 18°C$) and by the negative plate at lower temperatures ($< 5°C$); the ratio of positive surface to negative-plate surface is fixed by design. To maximize cranking capacity, SLI battery designs emphasize grids with minimum electrical resistance (using a variety of radial and expanded grid designs), thin plates, and a higher concentration of electrolyte than motive-power or stationary batteries.

The details of the construction of a typical 12-V SLI battery are shown in Fig. 14.8. Usually, an "outside-negative" (n +1 negative plates interspersed with $2n$ separators and n positive plates) design is used. However, in order to balance the cranking rating with the requirement for electrical load, as well as to facilitate automatic assembly, SLI batteries with an even number of plates or "outside-positive" designs are widely produced in the United States.

A major advance is the "maintenance-free" SLI battery which has several characteristics that distinguish it from the conventional battery. It requires no addition of water during its life; it has significantly improved capacity retention during storage; and it has minimal terminal corrosion. The construction of a typical maintenance-free SLI battery is shown in Fig. 14.16.

The maintenance-free battery has a large acid reservoir, made possible by use of smaller plates and placement of the element directly on the bottom of the container eliminating the sludge space. The positive plates are usually enveloped in a microporous separator that prevents active material from falling to the bottom of the container and creating a short. An important feature of the maintenance-free battery is the use of nonantimonial (e.g., calcium-lead) or low-antimonial lead grids. The use of these grids significantly reduces the overcharge current, reducing the rate at which water is lost during overcharge, as well as improving the stand characteristics (see Sec. 14.8). The use of the expanded grid, produced from wrought lead-calcium strip, is also shown in the figure.

Another refinement of the SLI battery is illustrated in Fig. 14.17. In this design, the plates are approximately one-fifth the width of conventional SLI battery plates and are inserted parallel, rather than perpendicular, to the length of the battery case. This design reduces the internal impedance of the cell and gives very high CCA ratings.

Heavy-duty SLI batteries for trucks, buses, and construction equipment are designed similar to the passenger vehicle SLI batteries but use heavier and thicker plates with high-density paste, premium separators often with glass mats, anchor-bonding of the element to the bottom of the case, rubber cases, and other such features to enhance longer life. This is necessary to provide maximum mechanical strength for the physically large (up to 530- by 285-mm) case dimensions. Because the thick plates provide

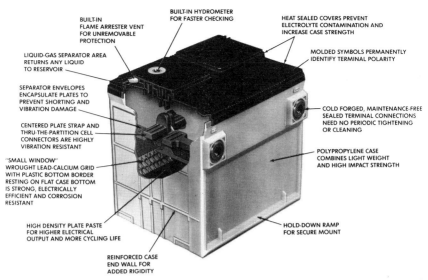

BUILT-IN HYDROMETER
FOR FASTER CHECKING

BUILT-IN
FLAME ARRESTER VENT
FOR UNREMOVABLE
PROTECTION

HEAT SEALED COVERS PREVENT
ELECTROLYTE CONTAMINATION AND
INCREASE CASE STRENGTH

LIQUID-GAS SEPARATOR AREA
RETURNS ANY LIQUID
TO RESERVOIR

MOLDED SYMBOLS PERMANENTLY
IDENTIFY TERMINAL POLARITY

SEPARATOR ENVELOPES
ENCAPSULATE PLATES TO
PREVENT SHORTING AND
VIBRATION DAMAGE

COLD FORGED, MAINTENANCE-FREE
SEALED TERMINAL CONNECTIONS
NEED NO PERIODIC TIGHTENING
OR CLEANING

CENTERED PLATE STRAP AND
THRU-THE-PARTITION CELL
CONNECTORS ARE HIGHLY
VIBRATION RESISTANT

POLYPROPYLENE CASE
COMBINES LIGHT WEIGHT
AND HIGH IMPACT STRENGTH

"SMALL WINDOW"
WROUGHT LEAD-CALCIUM GRID
WITH PLASTIC BOTTOM BORDER
RESTING ON FLAT CASE BOTTOM
IS STRONG, ELECTRICALLY
EFFICIENT AND CORROSION
RESISTANT

HIGH DENSITY PLATE PASTE
FOR HIGHER ELECTRICAL
OUTPUT AND MORE CYCLING LIFE

HOLD-DOWN RAMP
FOR SECURE MOUNT

REINFORCED CASE
END WALL FOR
ADDED RIGIDITY

FIG. 14.16 Maintenance-free lead-acid SLI battery. *(Courtesy of Delco-Remy Div., General Motors Corp., Anderson, Ind.)*

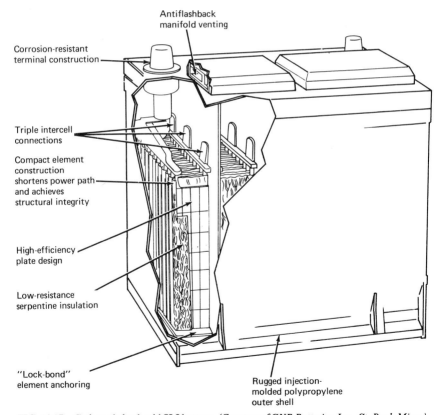

Antiflashback
manifold venting

Corrosion-resistant
terminal construction

Triple intercell
connections

Compact element
construction
shortens power path
and achieves
structural integrity

High-efficiency
plate design

Low-resistance
serpentine insulation

"Lock-bond"
element anchoring

Rugged injection-
molded polypropylene
outer shell

FIG. 14.17 Cathanode lead-acid SLI battery. *(Courtesy of GNB Batteries, Inc., St. Paul, Minn.)*

less cranking current than the thinner plates (since fewer can be included in a cell of given size), series or series-parallel connections of batteries are used. Typically, the 12-V monoblocs are connected in series for cranking at 24 V and in parallel for running and recharging at 12 V. A few sizes of maintenance-free heavy-duty batteries have also been produced.

SLI-type batteries are also used on motorcycles and boats. Batteries for recreational marine use generally have thicker plates (to give more capacity) and higher-density paste. They have the same Battery Council International–type designations as automotive batteries. Marine batteries are also manufactured in four-cell, 8-V monoblocs.

Aircraft use SLI-type batteries with special spill-proof vent caps which preclude loss of electrolyte during flight (Fig. 14.18).

FIG. 14.18   Lead-acid aircraft battery. (*Courtesy of Prestolite Battery, Toledo, Ohio.*)

14.4.3 Performance Characteristics

Discharge Performance   Discharge curves, showing the discharge profile of SLI-type batteries at several constant-current discharge rates, are presented in Fig. 14.19. The typical final or end voltages at these discharge rates are also shown. Higher service capacity is obtained at the lower discharge rates. At the higher discharge rates, the electrolyte in the pore structure of the plates becomes depleted and the electrolyte cannot diffuse rapidly enough to maintain the cell voltage. Intermittent discharge, which allows time for the electrolyte to recirculate, or forced circulation of the electrolyte will improve high-rate performance. In general the lead-acid cell may be discharged without harm at any rate of current it will deliver, but the discharge should not be continued beyond the point where the cell approaches exhaustion or where the voltage falls below a useful value.

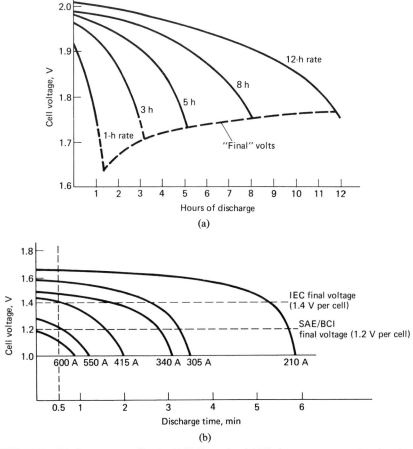

FIG. 14.19   Discharge curves of lead-acid SLI batteries. (*a*) Discharge curves at various hourly rates, 25°C. (*b*) Discharge curves at various high rates at −17.8°C. (Battery rated at 70 Ah at 20-h rate at 25°C.)

Effect of Temperature on Performance   Figure 14.20 shows typical discharge curves
for the lead-acid cell at several discharge temperatures. Higher discharge voltages and
capacities are obtained at the higher temperatures and lower discharge rates.

The effect of discharge rate and temperature on the capacity of the lead-acid cell
is summarized in Fig. 14.21, which shows the percentage of the 20-h rate capacity
delivered under the different discharge conditions. Although the battery will operate
over a wide temperature range, continuous operation at high temperatures may reduce
life as a result of an increase in the rate of corrosion (see Sec. 14.9.1).

The performance of the lead-acid SLI-type cell at different temperatures and loads
is given in another form in Fig. 14.22; the logarithm of the current drain is plotted
against the logarithm of the service hours, in accordance with Peukert's relationship
(see Part 1, Chap. 3, Sec. 3.2.6). The linear relationship is maintained over a wide range
with divergences appearing on the high-rate and low-temperature discharges. In this

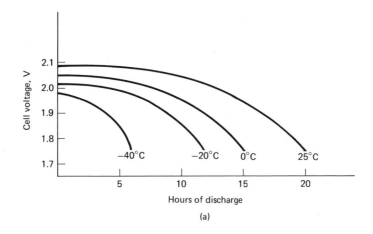

(a)

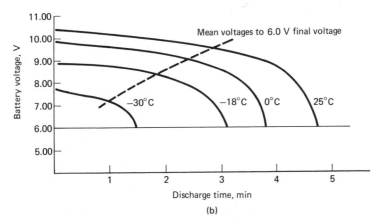

(b)

FIG. 14.20   Discharge curves of lead-acid SLI batteries at various temperatures.
(*a*) Discharge at C/20 rate. (*b*) Discharge at 340 A; 12-V battery, nominal capacity
60 Ah at 20-h rate at 25°C.

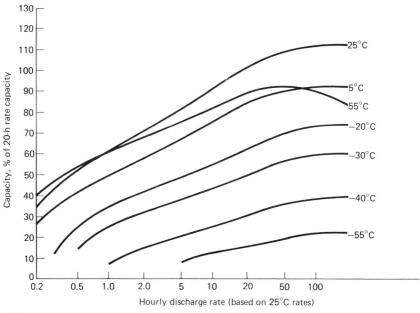

FIG. 14.21 Performance of SLI lead-acid batteries at various temperatures and discharge rates (1.75 V per cell end voltage).

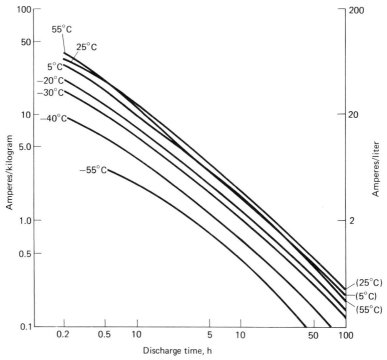

FIG. 14.22 Service life of SLI-type lead-acid cells (to 1.75 V end voltage).

14-40

figure, the data have been normalized to unit cell weight (kilograms) and unit cell volume (liters). This figure can be used to approximate the performance of various size cells over the operating conditions shown or to determine the size and weight of a battery to meet a particular service requirement.

Internal Impedance   The high current requirement for engine cranking demands that the batteries be designed with low resistance, e.g., that conductors have large cross sections and minimal lengths, that separators have maximum porosity and minimum backweb thickness, and that the electrolyte be in the range of low resistance (see Appendix C). The relationship between plate surface area and CCA suggests the involvement of the electrochemical double layer of the porous active materials. Low frequencies are generally necessary to evaluate the capacitive reactance component of battery impedance: strictly, the resistance impedance component can be evaluated by Ohm's law by determining the voltage difference at two levels of discharge current. The resistance of a lead-acid battery increases during a discharge almost linearly with the decrease of the specific gravity of the electrolyte as shown in Fig. 14.23*a*. The difference in resistance between full charge and discharge is in the order of 40%. The effect of temperature on the resistance of the battery is shown in Fig. 14.23*b;* the battery resistance increases by about 50% between 30 and − 18°C. Figure 14.23*c* shows that the battery resistance is much more stable than battery voltage after end of charge. The

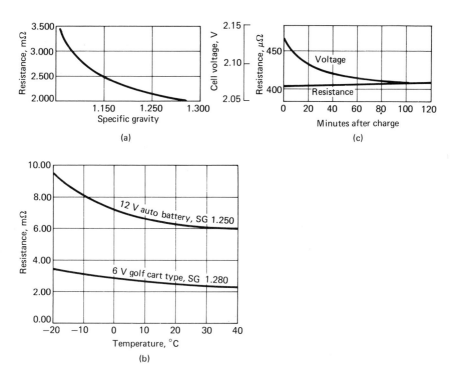

FIG. 14.23   Resistance of SLI-type lead-acid batteries. (*a*) Battery resistance during discharge. (*b*) Effect of temperature on battery resistance. (*c*) Battery voltage and resistance after end of charge. *(Courtesy Palico Instrument Laboratories, Circle Pines, Minn.)*

voltage decays slightly as the surface charge is dissipated, but the resistance rises only slightly, probably because of the consolidation of gas microbubbles in the pores of the active material.

Self-Discharge A lead-acid battery loses capacity during open-circuit stand (self-discharge). This loss is more severe with batteries which use antimonial-lead grid alloys in the positive plates. A comparison of the open-circuit stand loss of conventional antimonial-lead ($> 4\%$ Sb), low antimonial-lead ($< 3\%$ Sb), and nonantimonial lead grids is shown in Fig. 14.6. This loss is most easily detected by a drop in the terminal voltage of the battery and/or the specific gravity. The sulfuric acid reacts, primarily on the surface of the negative plate, in small local self-discharge "cells" where antimony and lead are in contact, becoming a small particle of lead sulfate.

Life and Failure Modes The life of SLI batteries is affected by the design, the processing, and operational environment of the battery. Because of the automated assembly methods used today, SLI batteries are fairly consistent in life under ideal operating conditions, but the wide variety of operating conditions tends to spread the failure distribution. Warranty coverage for a failed battery is often more dependent on marketing strategy than on the statistical expectations of failure rate.

SLI battery design, materials, and operation have changed markedly in the past two decades; life and failure mode distribution have also changed. In Fig. 14.24a, the average age of failed batteries is plotted. Possible explanations for the shorter life in 1982 may be the reduction in the size of the battery and the more demanding performance requirements.[17] Figure 14.24b shows the failure modes for these batteries, which are described in more detail in Table 14.13. A higher incidence of short-circuited batteries used in warmer climates suggests that grid corrosion is still a major failure mode. The "worn-out" category includes low electrolyte level, and it should be noted

TABLE 14.13 Summary of Failure Modes of Lead-Acid SLI Batteries*

1. Open circuits	**4.** Worn out
a. Terminal	*a.* Worn out
b. Cell to terminal	*b.* Undercharge
c. Cell to cell	*c.* Low level (electrolyte)
d. Broken straps	*d.* Terminal corrosion
e. Plates off	*e.* Underformed
2. Short circuits	**5.** Serviceable
a. Plate to strap	*a.* Serviceable
b. Plate to plate (plate fault)	*b.* Discharged only
c. Plate to plate (separator fault)	**6.** Broken
d. Plate to plate (sediment/moss)	*a.* Broken container
e. Vibration short	*b.* Broken cover
3. Poor plates	*c.* Damaged terminal
a. Overcharge/overheat	*d.* Internal damage
b. Grid corrosion	*e.* Other
c. Paste adhesion	
d. Paste sulfation	
e. Paste underformed	

*Based on Ref. 17.

that many maintenance-free SLI batteries are sealed so that water, lost to evaporation and electrolysis, cannot be replaced.

SLI batteries are not designed for deep-cycling service, and very short lives are generally obtained with such operation. The deep-cycling capability of SLI batteries is covered in Sec. 14.5.2.

Standard Tests for Rating SLI Batteries   Several standard tests have been devised to evaluate and rate the performance of SLI batteries under conditions simulating the major requirements of their applications (SAE Standard J 537).

The cold-cranking amperes (CCA) test evaluates the capability of the battery to deliver power to crank an engine at cold temperatures. The cranking-test rating is the

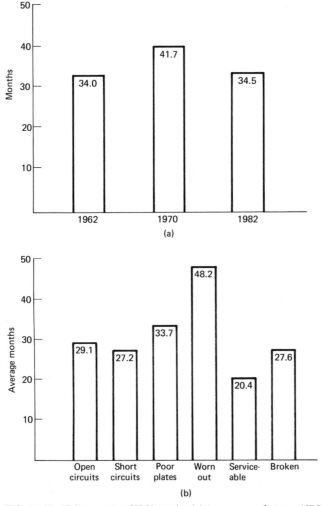

FIG. 14.24   Failure modes of SLI batteries. (*a*) Average age of returned SLI batteries. (*b*) Failure mode of returned batteries, 1982. *(From Ref. 17.)*

current that a fully charged battery can deliver at $-17.8°C$ for 30 s to a voltage of 1.2 V per cell. If the measured voltage is above or below this value at 30 s, the CCA value can be calculated by multiplying the discharge current by the correction factor shown in Fig. 14.25. This test is replacing an earlier test which determined the time a battery could be discharged at 300 A at $-17.8°C$ to 1.0 V per cell. Figure 14.19*b* illustrates the performance of a 70-Ah cell with a CCA rating of 550 A.

Reserve capacity is measured in a test of the battery's ability to provide power for lights, ignition, and the auxiliaries. The reserve capacity is defined as the number of minutes a fully charged battery can maintain a current of 25 A to 1.75 V per cell at 25°C. The 20-h capacity test previously used for SLI batteries is no longer specified.

Other SLI tests are included in the SAE battery test standard J 537 on charge rate acceptance, overcharge life, and vibration resistance. The standard SLI life test is specified in SAE J240A. This test consists of a shallow discharge at 25 A followed by a brief charge at voltage and current limits for 10 min. Although the test procedure indicates a discharge time of 2 min, the test is generally done instead with a discharge time of 4 min in order to give more realistic life and failure modes.

14.4.4 Cell and Battery Types and Sizes

SLI battery sizes have been standardized by both the automotive industry through the Society of Automotive Engineers (SAE), Warrendale, Pa., and the battery industry through the Battery Council International (BCI), Chicago. The BCI nomenclature follows the standards adopted by its predecessor, the American Association of Battery Manufacturers (AABM). The latest standards are published in the annual issue of the SAE Handbook and the Yearbook of BCI.[18] The International Electrotechnical Commission (IEC) has standardized on the BCI sizes.

Tables 14.14 and 14.15 list the standard SLI battery sizes in the United States and Japan, respectively. Japanese and European cars exported to North America use standard BCI/SAE sizes, mostly group 24.

Table 14.16 lists the standard sizes of small SLI batteries used in motorcycles, scooters, and snowmobiles.

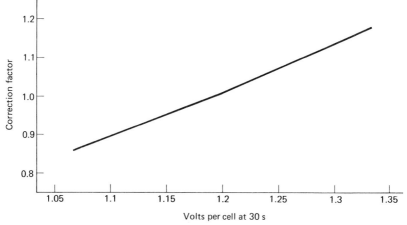

FIG. 14.25 Correction factor for calculating cold-cranking ampere (CCA) rating.

TABLE 14.14 SLI Lead-Acid Batteries

BCI Group Numbers, Dimensional Specifications, and Ratings[a]

BCI Dimensional group size	Maximum overall dimensions, mm			Assembly figure number[b]	Approximate performance ranges	
					Cranking performance,[c] A, at 17.8°C	Reserve capacity, min, at 26.7°C
	L	W	H			
Passenger car and light commercial batteries 12 V (six cells)						
21	208	173	222	10	305–325	68
22F	241	175	211	11F	210–360	45–65
22HF	241	175	229	11F	290–350	69
22NF	240	140	227	11F	180–260	50
22NL	235	133	198	11(Y)	170–200	30
22R	229	175	211	11	290–350	45–65
23	273	175	214	11L	315–410	105
24	260	173	225	10	165–500	69–100
24F	273	173	229	11F	260–475	69–99
24H	260	173	238	10	Listed only as a dimensional reference	
24R	260	173	229	11	435–475	69–99
24T	260	173	248	10	Listed only as a dimensional reference	
25	230	175	225	10	430	100
26	208	173	197	10	310	60
27	306	173	225	10	300–620	102
27F	318	173	227	11F	360–435	105–114
27H	298	173	235	10	Listed only as a dimensional reference	
27HF	318	173	233	11F	Listed only as a dimensional reference	
27SL	298	170	181	11	265	70
29HR	333	173	232	11	435–525	156
29NF	330	140	227	11F	270–430	65–97
41	293	175	175	15	300–395	75–95
42	243	173	173	15	235–330	59
43	334	175	205	15	Listed only as a dimensional reference	
44	418	175	205	15	Listed only as a dimensional reference	
45	240	140	227	10F	250	60
46	273	173	229	10F	315–370	75–90
53	330	119	210	14	210–315	40
54	186	154	212	19	305–310	60
55	218	154	212	19	380	75
56	254	154	212	19	450	90
57	204	183	177	21	310	60
58	239	183	177	21	380	75
60	332	160	225	12	305–385	65–105
61	192	162	225	20	310	60
62	226	162	225	20	380	75
63	258	162	225	20	450	90
64	296	162	225	20	475–535	105–120
71	208	178	216	17	275–350	60–80
72	230	178	210	17	275–430	60–80
73	230	178	216	17	430	80–100
74	260	184	222	17	350–540	80–138
77	306	184	222	17	360–465	107–110

TABLE 14.14 SLI Lead-Acid Batteries (Continued)

Passenger car and light commercial batteries 6 V (six cells)						
1	232	181	238	2	440–485	134–185
2	264	181	238	2	485–650	134–230
2E	492	105	232	5	370–615	140
2L	270	184	206	1	Listed only as a dimensional reference	
2N	254	141	227	1	360–450	138
17HF[d]	187	175	229	2B	270–315	100
19L	216	178	191	2	345–405	110

Heavy-duty commercial batteries 12 V (six cells)[e]						
4D	527	222	276	8	580–850	278
6D	527	254	276	8	750–830	310
8D	527	283	276	8	725–1115	390
30H	343	173	235	10	370–580	147
30HR	343	173	235	11	285–345	90
31	330	173	239	18	475–625	130
32N	362	140	227	11	340–360	102

Heavy-duty commercial batteries 6 V (three cells)[e]						
3	298	181	238	2	525–660	235
4	333	181	238	2	550–975	225–255
5D	349	181	238	2	720–850	310
7D	413	181	238	2	850–965	370

Heavy-duty motor coach and bus batteries 12 V (six cells)[c,e]						
4B	540	283	276	8	430	250

Special tractor batteries 6 V (three cells) and 12 V (six cells)						
3EE (12 V)	491	111	225	9	260–450	75
3EH (6 V)	491	111	249	5	740–830	220
3ET (12 V)	491	111	249	9	385–435	130
4DLT (12 V)	508	208	202	16L	695–750	280
4EH (6 V)	491	127	249	5	880–975	420
12T (12 V)	303	176	278	10	455	160
16TF (12 V)	421	181	283	10F	605–640	240
17TF (12 V)	433	176	202	11L	510	175
20H (12 V)	198	171	237	10	235	50

General utility batteries 12 V (six cells)						
U1	197	132	186	10(X)	155–220	25–30
U1R	197	132	186	11(X)	220	25–30
U2	160	132	181	10(X)	120	17

Electric vehicle batteries 6 V (three cells)						
GC2[f]	264	183	270	2	[g]	[g]
GC2H[f]	264	183	295	2	[g]	[g]

[a] Based on Ref. 18.

[b] See terminals at end of table.

[c] Ratings for batteries recommended for motor coach and bus service are for double insulation. When double insulation is used in other types, deduct 15% from the rating values for cranking performance.

[d] Rod end types—extended top ledge with holes for hold-down bolts.

[e] Length and height dimensions are over handles.

[f] Special-use battery not shown in application section.

[g] Capacity test—75 A for 75 min to 5.25 V at 26.7°C. Ratings may be expressed in the form of a log-log curve (discharge rate vs. time) as supplied by the manufacturers.

TABLE 14.14 SLI Lead-Acid Batteries (*Continued*)

6-V assemblies—terminal position and cell layouts

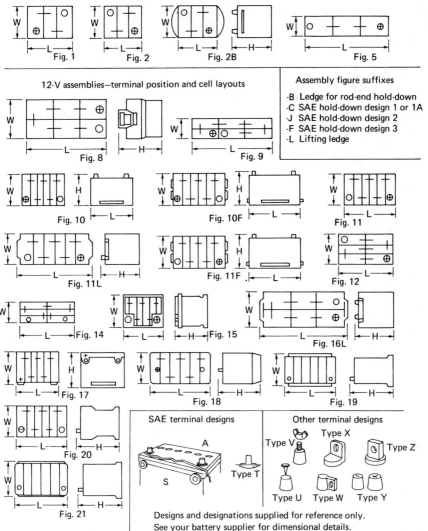

12-V assemblies—terminal position and cell layouts

Assembly figure suffixes

-B Ledge for rod-end hold-down
-C SAE hold-down design 1 or 1A
-J SAE hold-down design 2
-F SAE hold-down design 3
-L Lifting ledge

SAE terminal designs

Other terminal designs

Designs and designations supplied for reference only.
See your battery supplier for dimensional details.

TABLE 14.15 SLI Battery Specifications
Storage Battery Association of Japan, NS type 12 V (Six Cells)

Type	Capacity, Ah 20-h rate	Maximum overall dimensions, mm			Terminal type	Assembly figure no.	Minimum electrical values at −15°C		
		L	W	H			Discharge current, A	Minutes to 6 V	5-s voltage
12N24-3	26	187	127	187	X	2	150	1.6	8.0
12N24-4						1			
NS40	32	197	129	227	Y	1	150	2.5	8.4
NS40L						2			
NS40Z	35	197	129	227	Y	1	150	3.5	8.9
NS40ZL						2			
NS60	45	238	129	227	Y	1	300	1.4	7.3
NS60L						2			
NS70	65	260	173	225	Z	1	300	2.3	7.7
NS70L						2			
N40	40	238	135	232	Z	1	150	2.6	7.8
N40L						2			
N50	50	260	173	225	Z	1	150	3.6	8.2
N50L						2			
N50Z	60	260	173	225	Z	1	300	1.6	7.1
N50ZL						2			
N70	70	306	173	225	Z	1	300	2.3	7.7
N70L						2			
N70Z	70	306	173	225	Z	1	300	2.6	7.9[*]
N70ZL						2			
N100A	80	410	176	233	Z	4	300	3.0	7.9[*]
N100	100	410	176	233	Z	4	300	3.6	8.0[*]
N100Z	100	410	176	233	Z	4	300	4.1	8.2[*]
N120A	110	505	182	257	Z	5	500	1.9	7.3[*]
N120	120	505	182	257	Z	5	500	2.3	7.6[*]
N150	150	508	222	257	Z	5	500	3.1	7.9[*]
N200	200	521	278	270	Z	6	500	4.3	8.2[*]

[*]30-s voltage.
SOURCE: The Storage Battery Association of Japan 6/80, from BCI Yearbook, 1980–1981.

12-V assembly numbers and cell layouts

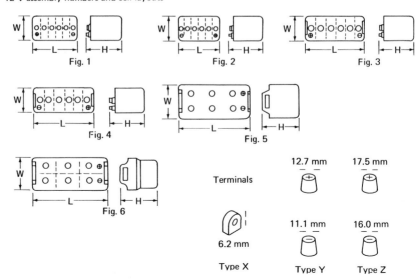

Fig. 1   Fig. 2   Fig. 3

Fig. 4   Fig. 5

Fig. 6

Terminals

12.7 mm   17.5 mm

11.1 mm   16.0 mm

6.2 mm

Type X   Type Y   Type Z

TABLE 14.16 Motorcycle Batteries

Type	10-h rating, Ah	Maximum overall dimensions (except elbow), mm			Terminal type	Assembly figure number	Application *
		L	W	H			
6 V (three cells)							
6N2-2A	2	71	48	99	Leading cable	1	L
6N2A-2C	2	70	47	106	Leading cable	2	L
6N4-2A	4	72	72	97	Leading cable	3	L
6N4A-4D	4	62	58	132	Leading cable	4	L
6N4B-2A	4	102	48	96	Leading cable	6	L
6N6-3B	6	100	58	112	Bolt and nut M5	5	L
6N11-2D	11	151	71	101	Bolt and nut M6	7	SLI
6N12A-2D	12	156	57	116	Bolt and nut M6	8	SLI
12 V (six cells)							
12N5-3B	5	121	62	132	Bolt and nut M5	9	L
12N5.5-3B	5.5	138	62	132	Bolt and nut M5	9	SLI
12N5.5A-3B	5.5	104	91	115	Bolt and nut M5	10	SLI
12N7-3B	7	137	77	135	Bolt and nut M6	9	SLI
12N7C-3D	7	131	91	115	Bolt and nut M6	11	SLI
12N9-3B	9	137	77	142	Bolt and nut M6	9	SLI
12N10-3B	10	138	92	147	Bolt and nut M6	9	SLI
12N12-3B	12	203	77	135	Bolt and nut M6	9	SLI
12N12A-4A	12	135	81	161	Bolt and nut M6	13	SLI
12N14-3A	14	135	90	169	Bolt and nut M6	12	SLI
12N16-3B	16	171	101	156	Bolt and nut M6	9	SLI

*L is lighting application; SLI is starting, lighting, and ignition application.
SOURCE: The Storage Battery Association of Japan 6/80, from BCI Yearbook, 1980–1981.

TABLE 14.16 Motorcycle Batteries (*Continued*)

Assembly numbers

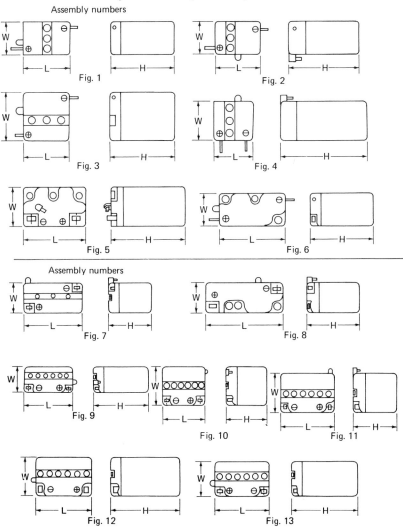

Fig. 1

Fig. 2

Fig. 3

Fig. 4

Fig. 5

Fig. 6

Assembly numbers

Fig. 7

Fig. 8

Fig. 9

Fig. 10

Fig. 11

Fig. 12

Fig. 13

14.5 DEEP-CYCLE AND TRACTION BATTERIES: CONSTRUCTION AND PERFORMANCE

14.5.1 Construction

The prime requirement for deep-cycling batteries for traction applications is maximum cycle life, then, if possible, high energy density and low cost. In an electric fork-lift application, in fact, light weight may not be advantageous because the battery's weight usually is needed to counterbalance the payload. The life of these batteries is improved by use of thick plates with high paste density, usually high-temperature/high-humidity cure, low electrolyte density formation, premium separators, and one or more layers of glass fiber matting (to retain the active material in the positive plates). The major modes of failure are disintegration of the PbO_2 positive active mass and corrosion of the positive grids. The deep-cycling battery is usually designed to be capacity-limited when new by the amount of electrolyte and not by the material in the plates. This serves to protect the plates and maximize their life. Both negatives and positives are degraded during use, but at end of life, battery capacity is generally limited by the positive plate. Battery failure, for cycle life rating purposes, is considered to occur when the battery will no longer produce 60 to 80% of its initial or rated discharge capacity.

A typical traction battery, using flat-pasted plates, is shown in Fig. 14.26. Cells are always made with an "outside-negative" design (e.g., n positive plates, $n + 1$ negative plates). Deep-cycling traction batteries are built as an assemblage of individual cells. If the battery's performance is limited by a catastrophic failure of one (or a few) cell(s), then those cells can be repaired or replaced in a cost-effective manner. Power requirements vary widely with the load, distance traveled, and lifting and/or climbing requirements. Battery sizes are determined by the forklift truck manufacturer and can be "calibrated" in the actual application by the use of an ampere-hour meter. A rough indication of the suitability of a traction battery for an application is the change in the specific gravity of the electrolyte during use. A larger battery size (or battery replacement or repair) is indicated when full operation cannot be achieved.

Although the flat-pasted (Faure) positive plate is typical for deep-cycling batteries in the United States, some cycling batteries in the United States and most cycling batteries in the rest of the world are built with tubular or gauntlet-type positives (Fig. 14.27). The tubular construction minimizes both grid corrosion and shedding, and long life is characteristic of these designs, but at a higher initial cost. Flat-pasted negative plates are used in conjunction with these positives, and the cells are of the outside-negative design.

Small traction batteries (e.g., for golf carts) are designed to be intermediate between full-sized traction batteries and SLI batteries. Traction design concepts sometimes utilized include high paste density, cure and formation to maximize α-PbO_2 content of the positive plates, glass matting against the positive, and tubular positives. SLI concepts sometimes utilized for golf cart and other electric vehicle (EV) batteries include thin cast radial grids, minimum separator resistance, through-the-partition intercell connection, and heat or epoxy-sealed plastic cases and covers. Cost is also an important factor.

For on-the-road EV applications, the major criterion has been high energy density, which results in maximum range, and SLI battery design has prevailed over traction battery design. In traditional cycling batteries with a few widely spaced plates, good electrolyte homogeneity occurs by gentle convectional flow. When plates are made thinner and more closely spaced for high discharge rate applications, such as for electric vehicle propulsion, the electrolyte has been found to become stratified during operation. A variety of electrolyte mixing devices has been designed not only to offset stratification but also to increase the discharge efficiency of the battery.

Military submarines of the diesel-electric type require cycling batteries for propul-

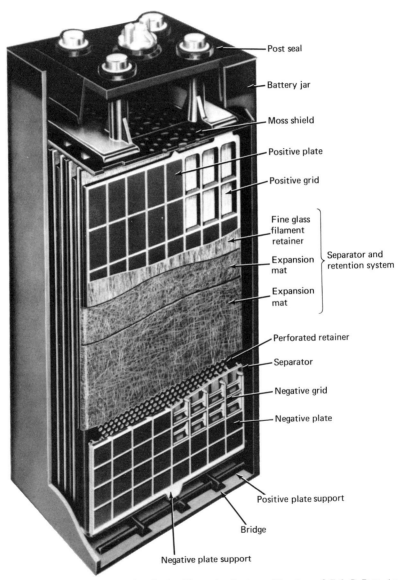

Post seal

Battery jar

Moss shield

Positive plate

Positive grid

Fine glass filament retainer

Expansion mat ⎤ Separator and
 ⎬ retention system
Expansion mat ⎦

Perforated retainer

Separator

Negative grid

Negative plate

Positive plate support

Bridge

Negative plate support

FIG. 14.26   Flat-pasted plate lead-acid traction battery. *(Courtesy of C & D Batteries, Plymouth Meeting, Pa.)*

sion. These batteries are made with nonantimonial lead grids because stibine and arsine produced on charge are unacceptable for personnel health in closed environment. The plates are much larger than most traction cells—up to 600 cm wide and up to 1500 cm tall. Both flat-pasted and tubular positive plates are used.

14.5.2 Performance Characteristics

Batteries for traction and deep-cycle applications use cells with either pasted or tubular positive plates. In general, the performance of the two types of plates is similar, but the tubular or gauntlet plates show lower polarization losses because of the larger active surface area, better retention of the positive active material, and reduced loss on stand.

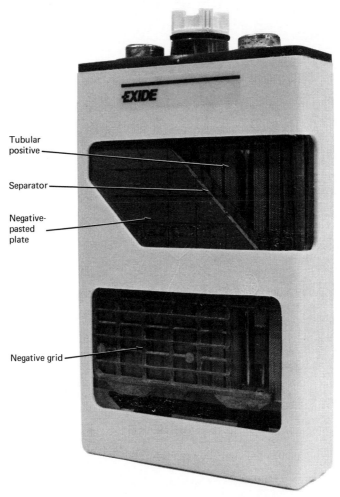

Tubular positive

Separator

Negative-pasted plate

Negative grid

FIG. 14.27   Lead-acid cell with tubular positive plates. *(Courtesy of Exide Corp., Horsham, Pa.)*

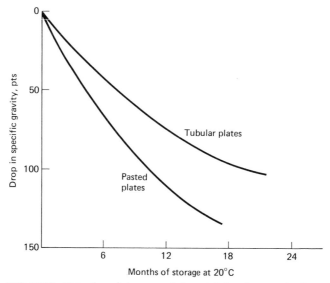

FIG. 14.28 Retention of charge, pasted- vs. tubular-plate traction batteries.

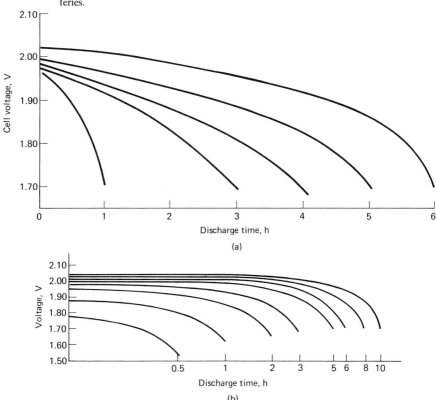

(a)

(b)

FIG. 14.29 Discharge characteristics of traction batteries, 25°C. (*a*) Flat-pasted plate batteries. (*b*) Tubular positive batteries.

The loss of capacity on stand at room temperature for the two plate structures, as measured by the drop in specific gravity, is shown in Fig. 14.28.

Typical discharge curves for the two types of traction cells are shown in Fig. 14.29. The relationship of discharge current to ampere-hour capacity, to various end voltages, is shown in Fig. 14.30. These data are presented on the basis of the positive plate since cell design and performance data of traction batteries are generally based on the number and size of positive plates that are in the cell. As is typical with most batteries, the capacity decreases with increasing discharge load and increasing end voltage.

The same relationship and comparison of the performance of the pasted and tubular plates are plotted in Fig. 14.31. These data show the superiority of the tubular plate as the discharge rate is increased.

Figure 14.32 shows the increase in available service on intermittent discharge, carried out over different periods, as compared with a continuous discharge. The gain is more pronounced at the heavier discharge loads and when the intermittent discharge is spread out over a longer period, thus allowing more time for recovery.

The effect of temperature on the discharge performance of traction-type batteries is given in Fig. 14.33.

The cycle life characteristics of traction batteries is presented in Fig. 14.34. This figure shows the relationship of cycle life to depth of discharge at the 6-h discharge rate, cycle life being defined as the number of cycles to 80% of the rated capacity. It is quite evident that the deeper the cells are discharged, the shorter their useful life and that 80% DOD should not be exceeded if full life expectancy is to be attained. Figure 14.30

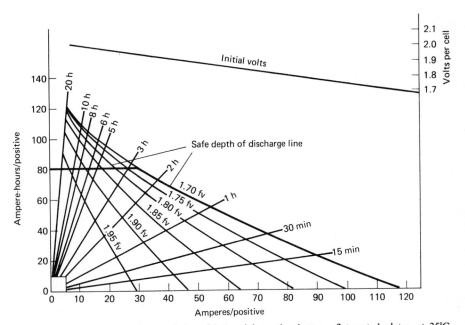

FIG. 14.30 Performance characteristics of industrial traction battery, flat-pasted plates, at 25°C (based on positive plate rated at 100 Ah at 6-h rate) to various final voltages (fv). *(Courtesy of C & D Batteries, Plymouth Meeting, Pa.)*

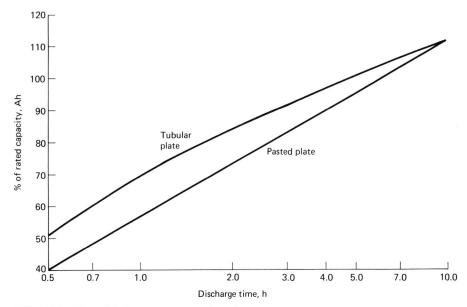

FIG. 14.31 Effect of discharge rate on capacity, 25°C, traction batteries. Comparison of performance of flat-pasted plate vs. tubular-plate batteries.

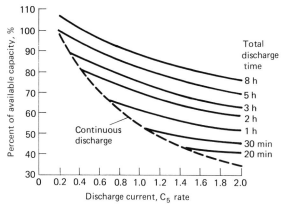

FIG. 14.32 Capacity available on intermittent discharge of traction batteries at 25°C.

shows the safe DOD for other discharge rates. At low rates, the discharge should be terminated at the higher end voltages as shown, until the 1.70-V line is intercepted; then the discharges at the higher rates can be run to 1.7 V per cell, final voltage. Typical cycle life expectancy is 1500 cycles (approximately 6 y).

The relationship of discharge current and service time for several small deep-cycling

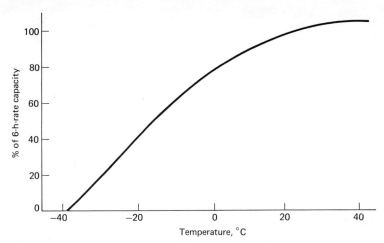

FIG. 14.33 Effect of temperature on the capacity of traction batteries, typical flat-plate design. (*From Ref. 19.*)

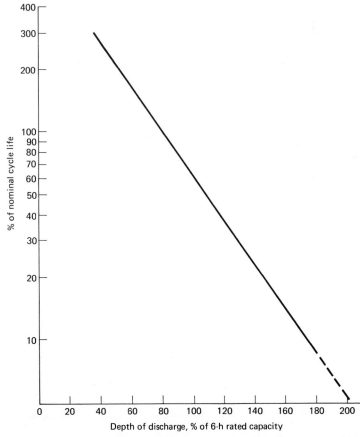

FIG. 14.34 Traction batteries, cycle life vs. depth of discharge. (*Courtesy of C & D Batteries, Plymouth Meeting, Pa.*)

batteries is shown in Fig. 14.35. At very high discharge rates, the Peukert's relationship does not hold as well as for the SLI types, and the performance deviates toward shorter discharge times; nevertheless, such deep-cycling batteries can be used for cranking service and may be preferred if the battery will be deeply or repeatedly discharged in operation. Conversely, an SLI battery generally makes a poor deep-cycling battery; SLI batteries are usually made with nonantimonial lead grids (U.S. practice), and cycling generally causes the development of a grid-active material barrier which shortens cycle life. A comparison of the cycle life at a low discharge rate (25 A) of an SLI battery compared with a deep-cycle design of the same physical size is shown in Fig. 14.36.

14.5.3 Cell and Battery Types and Sizes

Traction or motive-power batteries are made in many different sizes limited only by the battery compartment size and the required electrical service. The basic rating unit is the

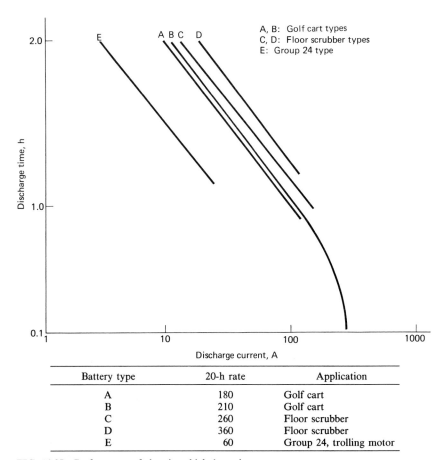

Battery type	20-h rate	Application
A	180	Golf cart
B	210	Golf cart
C	260	Floor scrubber
D	360	Floor scrubber
E	60	Group 24, trolling motor

FIG. 14.35 Performance of electric vehicle batteries.

positive-plate capacity, given in ampere-hours at the 5- or 6-h rate. Table 14.17 lists the typical U.S. traction plate sizes using flat-pasted plates; between 5 and 33 plates are used to assemble traction cells, as also is shown in the table. Ratings of the cell are the product of the capacity of a single positive plate multiplied by the number of positive plates. The cells, in turn, are assembled in a variety of battery layouts, with typical voltages in 6-V increments (e.g., 6, 12,18 to 96 V) resulting in almost 1000 battery sizes. Popular traction battery sizes are the 6-cell, 11 plates-per-cell, 75-Ah positive-plate (375-Ah cell) and the 6-cell, 13 plates-per-cell, 85-Ah positive-plate (510-Ah cell) batteries. Table 14.18 presents similar information on the tubular positive-plate batteries.

Several SLI group sizes have been used for deep-cycling applications, especially taller versions of otherwise SLI lengths and widths. Table 14.19 is a listing of some of these deep-cycling batteries.

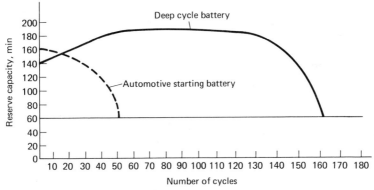

FIG. 14.36 Cycle life characteristics at a low discharge rate (25 A), deep-cycle vs. SLI-type batteries. (*Courtesy of GNB Batteries, St. Paul, Minn.*)

TABLE 14.17 Typical Traction Batteries (United States), Flat-Pasted Plates

Positive-plate capacity, Ah at 6-h rate	Height	Plate dimensions, mm				Cell size,* † (positive plates per cell)
		Width		Thickness		
		Positive	Negative	Positive	Negative	
45	275	143	138	6.5	4.6	5–16
55	311	143	138	6.5	4.6	5–16
60	330	143	138	6.5	4.6	5–16
75	418	143	138	6.5	4.6	2–16
85	438	146	146	7.4	4.6	3–16
90	489	138	143	6.5	4.6	3–16
110	610	143	143	7.4	4.6	4–12
145	599	200	200	6.5	4.7	4–10,12,15
160	610	203	203	7.2	4.7	4–10,12,15

*All cells have n positive plates and $n + 1$ outside negative plates.
†Typical cell characteristics: Six positive, 85-Ah plates (510-Ah cell); weight: 45 kg; size: L 127 mm, W 159 mm, H 616 mm.
SOURCE: C & D Batteries, Plymouth Meeting, Pa.

TABLE 14.18   Typical Traction Batteries, Tubular Plates

Positive-plate capacity, Ah at 6-h rate	Height	Dimensions, mm				Cell size † ‡ (positive plates per cell)
		Width		Thickness		
		Positive	Negative	Positive	Negative	
49	249	147	144	9.1	*	4–10
55	258	147	144	9.1	*	4–10
57	300	147	144	9.1	*	4–10
75	344	147	144	9.1	*	4–10
85	418	147	144	9.1	*	4–10
100	445	147	144	9.1	*	4–10
110	565	147	144	9.1	*	4–10
120	560	147	144	9.1	*	4–10
170	560	204	203	9.1	*	3–8

*Varies from 5 to 8 mm depending on manufacturer.
†All cells have n positive and $n + 1$ outside negative plates. Negatives are flat-pasted plates.
‡Typical cell characteristics: six positive, 85-Ah plates (510-Ah cell); weight: 36 kg; size: L 127 mm, W 157 mm, H 549 mm.
SOURCE: Exide Corp., Horsham, Pa.

TABLE 14.19   Small Deep-Cycling Batteries

BCI type	Volts	Dimensions, mm			Ratings	Typical operational current, A	Applications
		L	W	H			
U1	12	196	130	184	30–45 Ah at 20 h	25	Trolling motors Wheelchairs
24	12	273	171	225	75–90 Ah at 20 h	25	
27	12	306	172	225	90–105 Ah at 20 h	25	
GC2	6	260	181	260	75 min at 75 A	75 (GC)	Golf carts Electric vehicles
(GC2H)	6	260	181	260	85–90 min at 75 A	300 (EV)	
Not assigned	6	260	181	260	100–100 min at 75 A	300 (EV)	
Not assigned	6	260	181	260	120–130 min at 75 A	300 (EV)	
Not assigned	12	261	181	279	105 Ah at 20 h	150 (EV)	
Not assigned	6	295	178	276	200–230 Ah at 20 h	50–75	Floor maintenance machinery
Not assigned	6	295	178	311	230–260 Ah at 20 h	50–75	
Not assigned	6	311	178	362	250–310 Ah at 20 h	50–75	
Not assigned	6	298	178	412	330–360 Ah at 20 h	50–75	
Not assigned	12	241	166	239	50–70 Ah at 20 h	50–75	
Not assigned	12	300	166	251	80–100 Ah at 20 h	50–75	
Not assigned	12	342	175	235	120–130 Ah at 20 h	50–75	
Not assigned	12	394	178	362	170–200 Ah at 20 h	50–75	
Not assigned	12	518	276	279	200–250 Ah at 20 h	50–75	
Not assigned	12	518	276	356	280–300 Ah at 20 h	50–75	
Not assigned	12	457	190	241	100 Ah at 20 h	50–75	
Not assigned	12	518	276	445	350–400 Ah at 20 h	30–50	Mine cars

SOURCE: BCI Technical Committee, Battery Council International, Chicago, Ill.

A variation of the fork-lift battery design is used for some on-the-road electric vehicles. Tables 14.20 and 14.21 list the characteristics of typical EV batteries. Manufacturers' catalogs and data should be consulted for specific information on sizes and performance.

TABLE 14.20   Typical Electric Vehicle (EV) Batteries

Positive-plate capacity, Ah at 3-h rate*	Plate dimensions, mm†		Cell sizes‡ (positive plates per cell)
	Height	Width	
24.5	325	159	⎧ 5,7,8,10,12,
35	432	159	⎨ 13,15,17,19,
42	495	159	⎩ 20,22

*For 6-h rate, multiply by 1.13.
†Positive-plate dimensions.
‡All cells have n positive and $n + 1$ outside negative plates.
Typical cell characteristics: 5 positive, 42-Ah plates (210-Ah cell); weight: 13.6 kg; size: L 67 mm, W 159 mm, H 495 mm.
SOURCE: C & D Batteries, Plymouth Meeting, Pa.

14.6 STATIONARY BATTERIES: CONSTRUCTION AND PERFORMANCE

14.6.1 Construction

Stationary battery designs have changed much more slowly than those for SLI and traction batteries. This is not surprising in light of the much longer service life of the stationary batteries. Heavy, thick plates (including Planté as well as pasted Faure and tubular positives), high paste density, and two-shot formation are generally used to maximize the α-PbO_2 content of the positives.[20] Curing is very important, and pasted plates are usually carefully dried to prevent cracks and degradation of the grid-paste interface.

The stationary battery is designed with excess electrolyte (highly flooded) to minimize maintenance and the watering interval and is generally positive plate–limited in capacity (compared with traction batteries, e.g., which are electrolyte- or acid-limited). The stationary batteries are capable of being floated and modestly overcharged.

The thick-plate design of stationary batteries reflects the fact that high energy and power density are not as necessary as is the case for SLI and traction batteries. The overcharge operation of stationary batteries requires a large electrolyte volume (which can be accommodated as the batteries are mounted in fixed positions) and usually nonantimonial-lead grids (to maximize the intervals between watering). The overcharge causes some positive-grid corrosion, and this is manifested as "growth" or expansion of the grid. The dimensions from the positive plate to the inside container are scaled so that the positives can grow by up to 10% before the plates touch the container walls. If the growth is greater than 10%, the active material is sufficiently loose from the grid so that the capacity becomes severely positive-limited and the battery must be replaced.

TABLE 14.21 Typical Electric Vehicle (EV) Batteries

| Battery model | BCI group | Volts | Plates per battery | Maximum overall dimensions, mm | | | Weight, kg | Ah at 2 h | Ah at 3 h | 75 A (min) | Wh/kg at 3-h rate |
				Length	Width	Height					
EV-250	U1	12	54	197	132	186	95	20	22	15	26
EV-675	24	12	78	260	173	225	22	55	59	39	31
EV-750	GC2	6	57	264	183	270	26	126	135	100	29
EV-800	27	12	90	306	173	225	24	62	68	45	32
EV-950	GC2	6	57	264	183	270	30	150	171	120	33
EV-1000	GC2	6	39	264	183	280	27	158	174	140	37

SOURCE: Globe Battery Div., Johnson Controls, Milwaukee, Wisc.

The positives are usually supported from hanging lugs, or nonconductive rods, that are borne by the tops of the negatives. The containers are usually transparent thermoplastics (acrylonitrile-butadiene-styrene, styrene-acrylonitrile resin, polycarbonate, PVC), but some small stationary batteries are built in translucent polyolefin containers similar to those used for SLI batteries. The stationary batteries were the first application for flame-retardant vent caps which are now also standard on most SLI batteries.

The positive plate has the greater influence on the performance and life of the battery. A variety of positives are used, depending as much on tradition and custom as on the performance characteristics. The flat-pasted stationary batteries are popular in the United States because of their lower cost. For most of the standby, emergency applications the grids are cast in lead-calcium alloy. Planté and tubular designs are popular in Europe because of their longer life. All stationary batteries today use pasted negative plates generally with n positive plates and $n + 1$ negative plates (outside-negative design). This is done because of the need for proper support of the positives which tend to grow or expand during their life. One design used by some manufacturers is to make the two outermost negative plates thinner than the inside negative plates because the outermost surfaces are not easily recharged.

Figure 14.37 is an illustration of a stationary battery system installation.

A significantly different approach to stationary battery design is the cylindrical-shaped battery of the Bell Telephone System and the Western Electric Co.[21,22] Traditionally, prismatic-shaped stationary batteries had failed after 5 to 20 y service in telephone systems. The new battery, illustrated in Fig. 14.38, incorporates a number of innovations in order to achieve a battery life predicted to be 30 y or longer. These include lattice-type circular-shaped pure lead grids (cupped at a 10° angle), plates stacked horizontally one above the other instead of in the conventional vertical construction, chemically produced tetrabasic lead sulfate (TTB) positive paste, positives welded around the external plate circumference, negatives welded to a central conductor core, and heat-sealed copolymer container and cover. The use of pure lead in place of lead-calcium is to reduce grid growth; the circular and slightly concave shape of the plates is to counter the effect of growth and ensure good contact of the active material and grid during the life of the battery.

Some stationary batteries are designed to cycle rather than "float." For these applications, the design criteria of the traction batteries are applicable (see Sec. 14.5.1). Applications of cycling stationary batteries include load leveling or utility peak-power shaving and photovoltaic energy storage systems.

14.6.2 Performance Characteristics

Batteries for stationary applications may use cells with flat-pasted, tubular, Planté, or Manchester positive plates.

Typical discharge curves for the flat-pasted-type stationary cell at various discharge rates at 25°C are shown in Fig. 14.39, and the effect of discharge rate on the capacity of the cell is summarized in Fig. 14.40. Generally, the discharge rate for stationary cells is identified as the "hourly rate" (the current in amperes that the battery will deliver for the rated hours) rather than the "C rate" used for other types of batteries.

Figure 14.41a to d is a series of curves showing the specific performance characteristics of the four types of stationary batteries at 25°C based on positive-plate design. An electrolyte with a specific gravity of 1.215 is used in all these batteries. The format used

FIG. 14.37 Stationary battery system installation *(Courtesy of C&D Batteries Div., Plymouth Meeting, Pa.)*

14-64

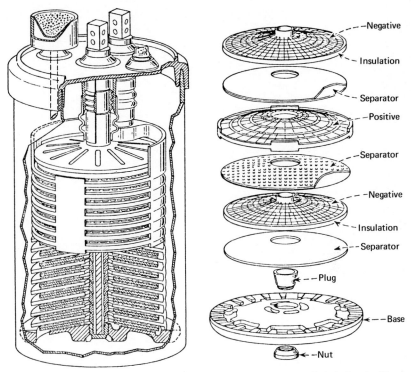

FIG. 14.38 The Bell System lead-acid battery (cutaway and exploded views). *(From Ref. 21.)*

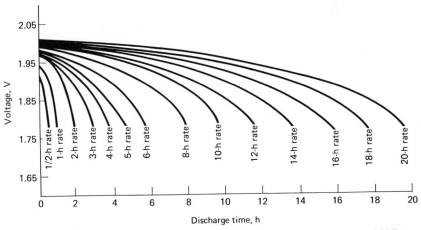

FIG. 14.39 Discharge curves of flat-pasted lead-acid stationary batteries (sp. gr. = 1.215) at various discharge rates at 25°C.

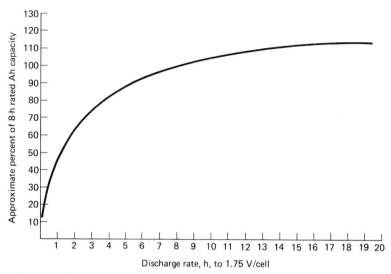

FIG. 14.40   Effect of discharge rate on cell capacity at 25°C, flat-pasted lead-acid stationary batteries (sp. gr. = 1.215) to 1.75-V end voltage.

in these figures consists of two sections. The lower log-log section shows the capacity (expressed in discharge time) the particular positive plate will deliver at the specified current (expressed in amperes per positive plate) to various voltages including a final voltage. The upper semilog section shows the cell voltage at various stages of the discharge at various discharge rates (also expressed in amperes per positive plate). The final voltage is the voltage at which the cell can no longer supply useful energy.

The energy density of the flat-pasted positive-plate and the tubular positive-plate batteries is similar and lower for the Planté positive-plate batteries. The high-rate performance of the flat-pasted positive cells is better because these plates can be made thinner than the tubular or Planté plates.

The optimal temperature range for the use of stationary batteries is from 20 to 30°C, although temperatures from −40 to 50°C can be tolerated. The effect of temperature on the capacity of stationary batteries at different discharge loads is shown in Fig. 14.42. High-temperature operation, however, increases self-discharge, reduces cycle life, and causes other adverse effects, as discussed in Sec. 14.9.1.

The rates of self-discharge of the various types of stationary batteries are compared in Fig. 14.43, which shows the relative float current at a specified float voltage. The float current under these conditions is a measure of self-discharge or local action. The float current is lowest for the calcium-lead grid pasted positives, and the float current remains low throughout the life. The float current is progressively higher for the tubular antimony-lead positives, the pasted antimony-lead positives, and the Manchester-type positives—at the beginning and throughout the battery's life. If the float current is not periodically increased, the antimonial cells will all become progressively self-discharged and sulfated.

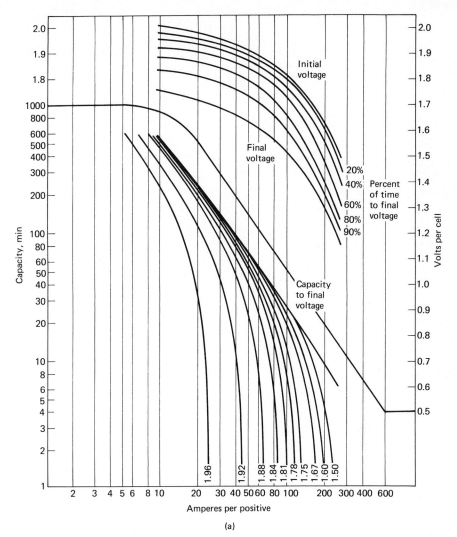

FIG. 14.41 Performance curves of lead-acid stationary batteries, 25°C (S-shaped curves, based on positive-plate performance). (*a*) Antimony flat-pasted plate, 125 Ah at 8-h rate, height, 290 mm, width, 239 mm, thickness, 8.6 mm. (*Courtesy of Exide Corp., Horsham, Pa.*)

For fully charged cells, the self-discharge rate at 25°C for the calcium-lead positive-plate cells is about 1%/month, 3% for the Planté, and about 7 to 15% for the antimonial-lead positive cells. At higher temperatures, the self-discharge rate increases significantly, doubling for each 10°C rise in temperature.

The float current for the calcium-lead and antimonial-lead batteries is shown in Fig.

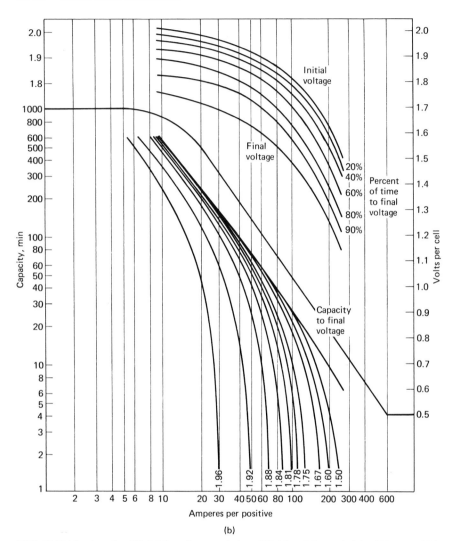

FIG. 14.41 (*Continued*)   (*b*) Calcium flat-pasted plate, 125 Ah at 8-h rate, height, 290 mm, width, 239 mm, thickness, 8.6 mm. (*Courtesy of Exide Corp., Horsham, Pa.*)

14.44 under float charge at voltages between 2.15 and 2.40 V per cell. It has been found that more than 50 mV positive and negative overpotential is necessary to prevent self-discharge so that 0.005-A float current per 100 Ah of battery capacity is required for the lead-calcium batteries. Antimonial-lead batteries initially require at least 0.06 A per 100 Ah, but this increases to 0.6 A per 100 Ah as the battery ages. The higher float current also increases the rate of water consumption and evolution of hydrogen gas.

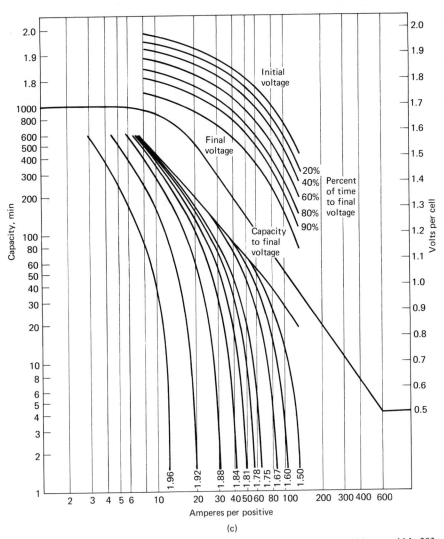

FIG. 14.41 (*Continued*)   (*c*) Ironclad tubular plate, 70 Ah at 8-h rate, height, 274 mm, width, 203 mm, thickness, 8.9 mm. (*Courtesy of Exide Corp., Horsham, Pa.*)

Various, and at times conflicting, claims about the life of stationary battery designs are made by the different manufacturers worldwide. Generally, the flat-pasted antimonial-lead batteries have the shortest life (5 to 18 y) followed by the flat-pasted calcium-lead batteries (15 to 25 y), the tubular batteries (20 to 25 y), and the Planté batteries (25 y).

Life on float service has been found to be related to temperature (Arrhenius-type behavior) as plotted in Fig. 14.45; the growth rate constant k is plotted for several different types of grid alloys used for the Bell System. At 25°C, the time to reach 4%

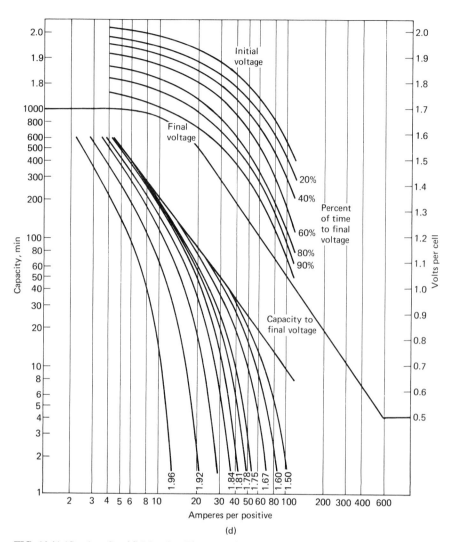

FIG. 14.41 (*Continued*) (*d*) Manchex Plate, 40 Ah at 8-h rate; height, 197 mm; width, 197 mm; thickness, 11.2 mm. (*Courtesy of Exide Corp., Horsham, Pa.*)

growth, an upper limit before the battery's integrity is impaired, is calculated to be 13.8 y for PbSb, 16.8 y for PbCa, and 82 y for pure lead.[21]

14.6.3 Cell and Battery Types and Sizes

Stationary batteries, like traction batteries, are available in a variety of plate and cell sizes. Stationary battery systems are assembled on insulated metal racks with groups

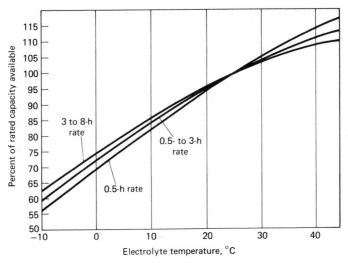

FIG. 14.42 Performance of flat-pasted lead-acid stationary-type batteries at various temperatures and discharge rates. *(Courtesy of Globe Battery Div., Johnson Controls, Milwaukee, Wisc.)*

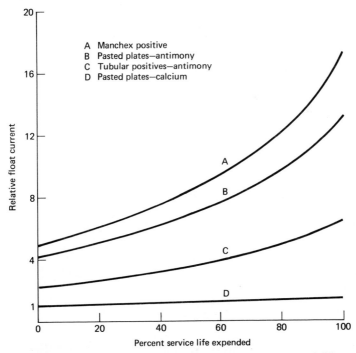

FIG. 14.43 Relative self-discharge of lead-acid stationary batteries of different construction.

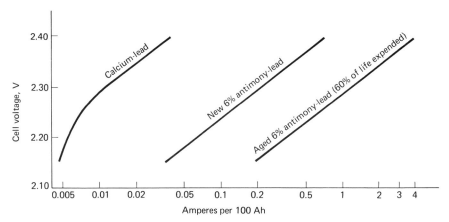

FIG. 14.44   Float current characteristics of stationary batteries at 25°C, 100-Ah cells, fully charged, 1.210 specific gravity.

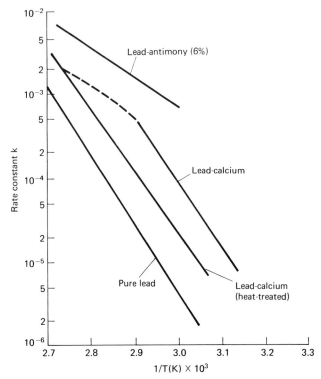

FIG. 14.45   Corrosion rate constant k on log scale vs. i/t (K) for different lead-alloy grids. *(From Ref. 21.)*

TABLE 14.22   Typical Stationary Batteries, Flat-Pasted Plates

Positive-plate capacity, Ah at 8-h rate	Plate dimensions, mm				Cell size* † (positive plates per cell)
	Height	Width	Thickness		
			Positive	Negative	
5	89	63.5	6.6	4.3	2, 4
25	149	143	6.6	4.3	1–8
90–95	290	222	7.9	5.3	2–12
150–155	381	304	6.4	4.6	2–17
168	381	304	7.9	5.3	5–16
195	457	338	6.9	5.3	13–18
412	1816	338	7.6	5.5	17–19

*Typical cell construction: n positive and $n + 1$ outside negative plates per cell. Some smaller cell sizes are assembled in multiples of two, three, or four cells in a monobloc container.
†Typical cell characteristics: 10 positive, 168-Ah plates (1680-Ah cell); weight: 140 kg; size: L 270 mm, W 359 mm, H 575 mm.
SOURCE: C & D Batteries, Plymouth Meeting, Pa.

TABLE 14.23   Typical Stationary Batteries, Tubular Plates

Positive-plate capacity, Ah		Plate dimensions, mm				Cell size* † (positive plates per cell)
At 4-h rate	At 8-h rate	Height	Width	Thickness		
				Positive	Negative	
26	31.25	157	203	8.9	5.6	4
76	96	277P	234P	8.9		
		290N	239N		6.1	3–10
88	105	277P	234P	8.9		
		290N	239N		6.1	3–10
124	152	366	307	8.9	4.8	5–14

*Typical cell construction: n positive and $n + 1$ outside negative plates per cell; used with flat-pasted negative plates.
†Typical cell characteristics: 11 positive, 152-Ah plates (1672-Ah cell); weight: 128 kg; size: L 272 mm W 368 mm, H 577 mm.
SOURCE: Exide Corp., Horsham, Pa. and Tudor AB, Nol, Sweden.

of cells in series, with nominal system voltages between 12 and 160 V. Some battery installations are made with several such series cell strings connected in parallel for greater storage capacity. Most stationary batteries in the United States are made with pasted positive plates. In Europe, Planté and tubular positives are more popular.

Listings of stationary batteries with flat-pasted, tubular, and Planté plates are given in Tables 14.22 to 14.24, respectively. The basic rating unit is the positive plate, given in ampere-hours at the 8-h rate unless specified otherwise; ratings of the cell are the product of the capacity of a single positive plate multiplied by the number of positive plates. A popular stationary battery size is the 1680-Ah cell (168-Ah positive, 10 positive or 21 plates per cell) for use in telephone exchanges. The Bell System has designed a special cylindrical cell in this capacity rating (Western Electric specification KS-20472, list 1).

TABLE 14.24 Typical Stationary Batteries, Planté Plates

| Positive-plate capacity, Ah at 8-h rate | Plate dimensions, mm | | | | Cell sizes* (positive plates per cell) |
| | Height | Width | Thickness | | |
			Positive	Negative	
		Planté type† ‡			
8	140	140	9.5	4.7	3,5,7
20			9.5	4.7	5–17
40			11.1		9–25
80	286	233	9.5	4.7	2–7
83			11.1		13–25
		Manchester type† §			
20	155	149	9.7	4.6	2,3,4
40	197	197	11.2	4.6	2–9
83	282	292	11.2	4.6	5–12

*Typical cell construction: n positive and $n+1$ negative outside plates per cell.
†Used with flat-pasted negative plates.
‡Typical Planté cell characteristics: two positive, 80-Ah plates (160-Ah cell). Two-cell battery: size: L 283 mm, W 159 mm, H 463 mm.
SOURCE: Gould, Inc.
§Typical Manchester cell characteristics: four positive, 40-Ah plates (160-Ah cell). One-cell battery: weight: 40 kg; size: L 131 mm, W 257 mm, H 455 mm.
SOURCE: Exide Corp.

14.7 PORTABLE LEAD-ACID BATTERIES: CONSTRUCTION AND PERFORMANCE

14.7.1 Construction

Small lead-acid batteries for fixed or portable applications are made in two designs: with spirally wound plates (jelly-roll construction) in a cylindrical metal container or with flat plates in a prismatic (usually plastic) container. Both types of batteries are acid-limited, having a minimum amount of electrolyte which is absorbed in the separator material or in a gel. The cells are also positive-limited. The starved-electrolyte and the excess of negative active material facilitate the recombination of oxygen produced in the cell during overcharge or during "float" operation to prevent gassing (also see Part 3, Sec. 15.2). The cells are sealed but have a means for venting in the event of overpressure. These batteries are generally termed sealed lead-acid (SLA) batteries.

An exploded view of a prismatic "gel" cell is shown in Fig. 14.46; an illustration of a cylindrical SLA cell is shown in Fig. 15.1.

The grids for the portable batteries are usually nonantimonial because these grids have a lower self-discharge rate and reduced gassing. However, the nonantimonial lead grids tend to reduce the deep-cyclability of the batteries unless additives are incorporated into the cell. Most of these are proprietary; the most common is phosphoric acid. The balance between cyclability, capacity, and float life is controlled by the ratio of α-PbO_2 to β-PbO_2, the paste density, the amount and concentration of electrolyte, the composition of the grid alloy, and the type and amount of additive.

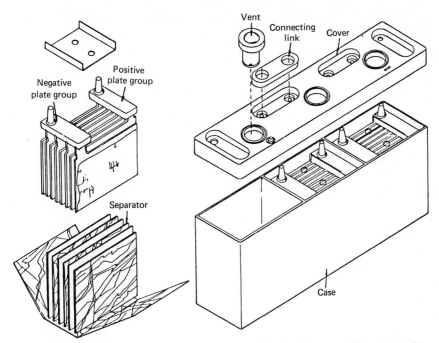

FIG. 14.46 Typical gel cell lead-acid battery (exploded view). *(Courtesy of Globe Battery Div., Johnson Controls, Milwaukee Wisc.)*

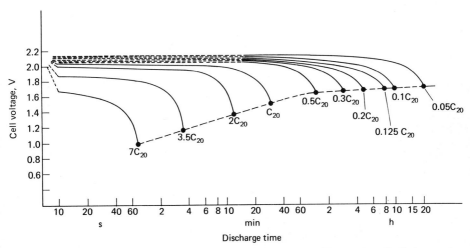

FIG. 14.47 Typical discharge curves of the gel cell lead-acid battery, 20°C, at various C_{20} discharge rates. *(Courtesy of Globe Battery Div., Johnson Controls, Milwaukee, Wisc.)*

In the SLA batteries, the electrolyte is absorbed in blotter-like glass fiber separators or is gelled. In this way, the battery is "nonspill" and can be used in any position without the danger of leakage. Many acid-resistant materials, such as burnt clay, pumice, sand, fuller's earth, plaster of paris, and asbestos, had been used for gelling. Recent practice has been to use fumed silica, sodium silicate, boron phosphate, or hollow polymer microspheres for this purpose. These latter materials are more consistent than the natural products and are less likely to introduce contamination into the battery. When gelling agents are used, the battery is usually formed with dilute acid for the first step of two-step formation, and the gel is introduced only for the second formation step.

The cylindrical SLA cells and batteries are covered in Part 3, Chap. 15.

Mine lamp batteries are the only portable batteries that have flooded (excess) electrolyte and are not sealed. These batteries have antimonial lead grids and tubular or flat-pasted positives.

4.7.2 Performance Characteristics

Typical discharge curves of the gelled lead-acid battery at 20°C are shown in Fig. 14.47, which illustrates the high rate capability and the flat discharge profile of this battery system. The cells are usually rated at the 20-h rate, and the figure shows the performance of the battery at discharge currents, expressed in terms of the C_{20} rate, to various end voltages. For example, for a cell rated at 5 Ah at the 20-h rate, we have

$$\frac{C_{20}}{20} = 0.05\ C_{20} = 0.250\ A$$

The $C_{20}/20$ rate is 0.250 A. The C_{20} rate is 5 A. The capacity at the C_{20} discharge rate is about 50% of the rated 20-h $C_{20}/20$ capacity.

Figure 14.48 shows the effect of discharge rate and temperature on the delivered

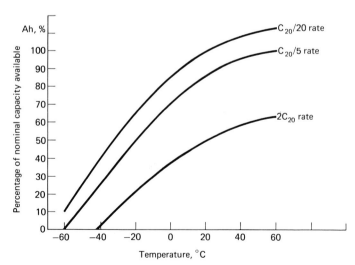

FIG. 14.48   Effect of temperature and discharge rate on capacity of a gel cell lead-acid battery.

capacity of the battery. A fully charged battery can operate over a very wide temperature range. Figure 14.49 summarizes these data based on a cell of unit weight (kilograms) and unit size (liter).

The change in resistance during a discharge of a gelled-electrolyte lead-acid battery is shown in Fig. 14.50.

The self-discharge characteristics of the gelled lead-acid cell are shown in Fig. 14.51. The gelled cell loses relatively little capacity in storage compared with the more conventional antimonial-lead grid battery; the rate of self-discharge is about 4%/month. Once the battery has self-discharged to the level where it has about a 50% state of charge, recharging the battery is advisable.

The service life characteristics of the gelled lead-acid cell on "float" service are shown in Fig. 14.52. The cycle life characteristics, to different depths of discharge (DOD), are shown in Fig. 14.53. A characteristic of the cell is that the capacity increases in the early stages of life and reaches a maximum at about 10 to 30 cycles; it will also gain in capacity while on extended charge as in float service where discharges are infrequent. The cell also does not exhibit a "memory" effect as is the case with the sealed nickel-cadmium cell which may become conditioned, when used for short periods, and may not be able to deliver full capacity when required.

The current-limited, constant-voltage method produces the best results for recharging a gelled lead-acid cell. For cycle service, the battery can be charged at a maximum of 2.4 V per cell, with the current limited to a maximum of C/5. The battery can be assumed to be fully charged and should be removed from the charger when its current

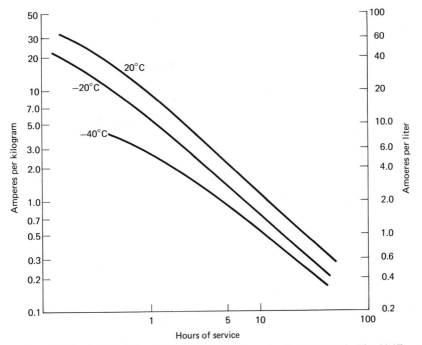

FIG. 14.49 Service life of the gelled lead-acid cell (to end voltages shown in Fig. 14.47).

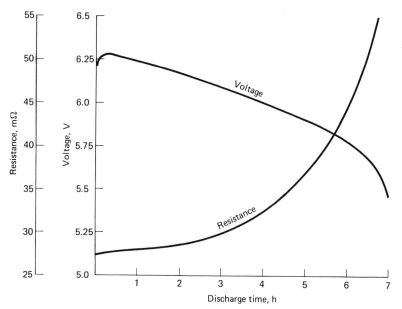

FIG. 14.50   Resistance of a gelled lead-acid 6-V battery during discharge.

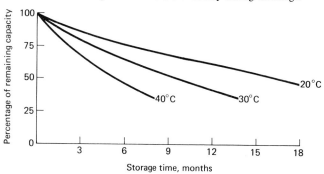

FIG. 14.51   Self-discharge characteristics of a gel cell lead-acid battery at various temperatures.

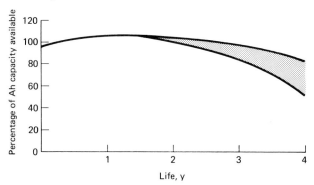

FIG. 14.52   Performance of a gel cell lead-acid battery on float service (20°C). Float voltage = 2.25 to 2.3 V per cell.

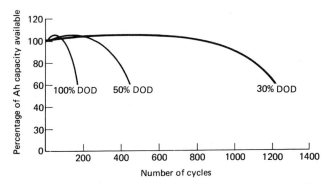

FIG. 14.53   Cycle service life of a gel cell lead-acid battery in relation to depth of discharge (DOD) (20°C).

acceptance (at 2.4 V per cell) drops to about 10 mA per ampere-hour of rated capacity. For float-charging, a voltage of 2.25 to 2.3 V per cell should be used. At this level, the battery can be left on charge continuously (also see Sec. 14.8 and the manufacturer's instructions for more details on charging).

The major advantages of the sealed lead-acid cell, over the competing sealed nickel-cadmium cell, are the absence of the "memory" effect, longer float or standby service, lower rate of self-discharge, and lower cost.

14.7.3 Cell and Battery Types and Sizes

A listing of typical gelled lead-acid cells and batteries is given in Table 14.25, including the battery voltage and manufacturers' part number. Unlike the standardization of other types of secondary batteries prepared by NEDA and the battery industry, no standardized list of sizes has been prepared for gelled lead-acid cells. Hence, sizes, weights, and capacity ratings may vary from manufacturer to manufacturer. A plot of the discharge performance at 20°C for several of these cells is presented in Fig. 14.54.

14.8 CHARGING AND CHARGING EQUIPMENT

14.8.1 General Considerations

In the charging process, direct-current electrical power is used to reform the active chemicals of the battery system to their high-energy charge state. In the case of the lead-acid battery, this involves, as shown in Sec. 14.2, the conversion of lead sulfate in the positive electrodes to lead oxide (PbO_2), the conversion of lead sulfate of the negative electrode to metallic lead (sponge lead), and the restoration of the electrolyte from a low-concentration sulfuric acid solution to the higher concentration of approximately 1.28 sp. gr. Since a change of phase from solid to solution is involved with the sulfate ion, charging lead-acid batteries has special diffusional considerations and is temperature-sensitive. During charge and discharge the solid materials which go into

TABLE 14.25 Typical Sealed Lead-Acid Batteries (Gelled or Absorbed Electrolyte Types)*

Voltage, V	Capacity, Ah at 20-h rate	Dimensions, mm			Weight, g	Part designation by manufacturer					
		Length	Width	Height		Globe	Eagle-Picher	Elpower	Gould	Yuasa	Sonnenschein
2	2.5	59	27	75	227		CF2V2.5				
	2.6										
	5.0						CF2V5				
	5.2	59	27	110	353						
	7.5	53	51	102	545	GC280					1Fx5
	28.0	85	56	166	1,910		CF2V30				
	30.0										
4	0.9	42	35	57.5	168	GC410					
	2.6	90	34	67	409	GC426					2Gx3
6	0.9	51	42	57.5	250	GC610					
	1.0	51	43	51	250	GC620					3Ax2
	1.8	75	56	60	454			EP610-16		NP1-6	3Bx3
	2.6	134	34	60	600	GC626					
	2.6	135	34	67	613			EP626-16	PB626	NP2.6-6	3Gx3
	4.0	152	34	101	1,044	GC640					3Fx3
	4.5	67	67	109	900						
	4.5	152	34	101	1,044	GC645		EP640-21	PB645L		
	6.0	117	51	98	1,180	GC660		EP650-16			
	7.5	152	51	102	1,497	GC680					3Fx5
	8.0	152	51	100	1,417		CF6V8				

Volts	A·h					GC	CF	EP	NP	
	10.0	151	50	96	2,200			EP6120-1	NP10-6	
	12.0	108	70	141	2,500			EP6200-2		3Mx6
	20.0	159	85	165	4,090	GC6200		EP6300-24		
	20.0	178	87	134	3,992					
	30.0	159	85	165	4,860					
	40.0	194	174	174	12,500					
	44.0	165	125	170	7,360		CFCV40			
8	2.6	179	34	67	794	GC826				4Gx3
12	1.2	97	48	51	590				NP1.2-12	
	1.5	179	34	68	681	GC1215				
	4.5	152	65	102	2,042	GC1245				
	6.0	152	80	102	2,587	GC1260				
	8.0	213	70	140	3,720			EP1280-1		
	12.0	213	70	140	5,000			EP12120-1		
	20.0	170	159	165	8,170			EP12200-4		
	20.0	165	124	171	7,360	GC12200	CF12V20			6Mz6
	20.0	177	167	125	7,711					
	30.0	170	159	165	9,750			EP12300-2		
	38.0	197	165	170	14,000				NP38-12	
	40.0	194	174	174	12,500					
24	20.0	249	165	171	15,000		CF24V20			
	28.0	317	170	171	21,135		CF24V30			

*This is a listing of typical batteries; many more sizes than listed are manufactured as size, weight, connector, and capacity ratings may vary from manufacturer to manufacturer. Also see Table 15.2 for characteristics of cylindrical SLA cells.

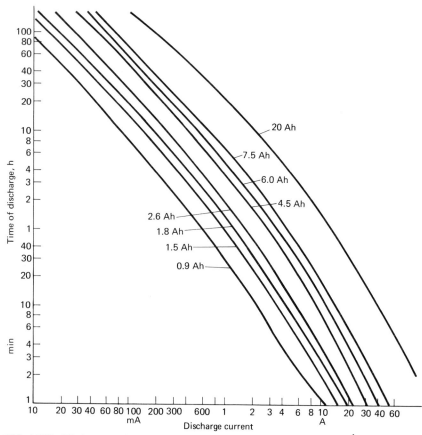

FIG. 14.54   Discharge performance of gel cell lead-acid batteries at 20°C. (*Courtesy of Globe Battery Div., Johnson Controls, Milwaukee, Wisc.*)

the solution as ions are reprecipitated as a different solid compound. This also causes a redistribution of the active material. The rearrangement will tend to make the active material possess a more perfect crystal structure, which results in less chemical and electrochemical activity. Therefore the lead-acid battery is not as reversible physically as it is chemically.[23] This physical degradation can be minimized by proper charging, and often batteries discarded as dead can be restored with a long, slow recharge (3 to 4 days at 2 to 3 A for SLI batteries).

A lead-acid battery can generally be charged at any rate that does not produce excessive gassing, overcharging, or high temperatures. The battery can absorb a very high current during the early part of the charge, but there is a limit to the safe current as the battery becomes charged. This is shown in Fig. 14.55, which is a graphical representation of the ampere-hour rule

$$I = Ae^{-t}$$

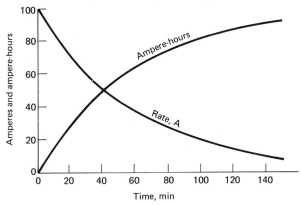

FIG. 14.55 Graphic illustration of the ampere-hour law. *(From Ref. 5.)*

where I is the charging current and A is the number of ampere-hours previously discharged from the battery. Because there is considerable latitude, there are a number of charging regimes and the selection of the appropriate method depends on a number of considerations, such as the type and design of the battery, service conditions, time available for charging, number of cells or batteries to be charged, and charging facilities. Figure 14.56 shows the relation of cell voltage to the state of charge and the charging current. The figure shows that a discharged battery can absorb high currents at relatively low battery voltages. However, as the battery is recharged, the voltage increases to excessively high values at the high charge rates, leading to overcharge and gassing, unless the charge current is controlled to reasonable levels.

In automotive, marine, and other vehicle applications, the dc electric power is usually provided by an on-board generator or alternator driven from the prime engine. These devices have a voltage and current limiter to prevent overcharging. The proper limit is dependent upon the chemistry and physical construction of the cell or battery. For the traditional automotive batteries which used antimonial lead alloy as grid material, voltage limits in the range of 14.1 to 14.6 V for a nominal 12-V battery are usual. With the newer maintenance-free batteries, which use a calcium-lead alloy grid or other grid material with high hydrogen overvoltage, higher charging voltages, in the range of 14.5 to 15.0 V, can be used without danger of overcharge. Batteries in automobile and similar applications today see what is close to cycling rather than float service, but the charging controls are such that very little gas is evolved on charge. This minimizes the requirement for watering but makes accurate control of the charge necessary. Comparisons of the water loss and charging rates for the different types of SLI batteries are shown in Figs. 14.57 and 14.58.[7] The calcium-lead maintenance-free battery is less affected by high settings in the voltage regulator than the other batteries.

In many nonautomotive applications, charging is done separately from the system using the battery. The direct current necessary for charging is usually obtained by rectifying alternating current. These chargers include wall-hung units and mobile units as well as floor-mounted units. Newer charger designs have microprocessor controls, can sense battery condition, temperature, voltage, charge current, etc., and are capable of changing charging rates during the charge. Most rectifiers produce some ac ripple

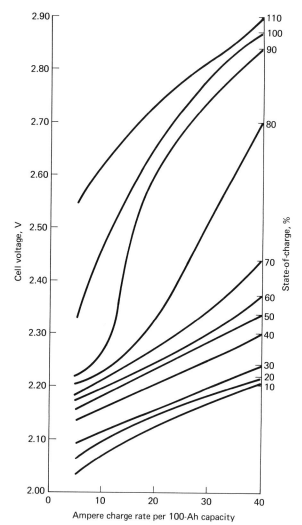

FIG. 14.56   Charging voltage of the lead-acid battery at various states–of–charge. (*From Ref. 19.*)

with the direct current, which causes additional heating of the battery. This should be minimized, especially near the end of the charge when batteries tend to get hot. Pulse charging and the use of asymmetric alternating current have been proposed as a means to overcome this problem, but practical lead-acid batteries have such large capacitances that the pulses are smoothed out and the effects negated.[23]

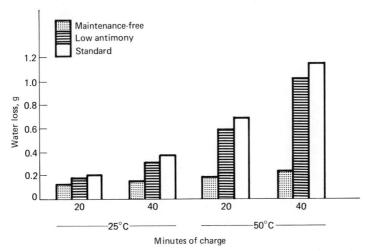

FIG. 14.57 Comparison of water loss on recharge (400-A discharge on new batteries for 9 s). (*From Ref. 7.*)

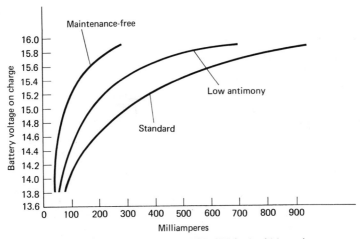

FIG. 14.58 Charging characteristics at 25°C, SLI lead-acid batteries.

14.8.2 Methods of Charging Lead-Acid Batteries

Proper recharging is important to obtain optimum life from any lead-acid battery under any condition of use.

Some of the rules for proper charging are given below and apply to all types of lead-acid batteries.

1. The charge current at the start of recharge can be any value that does not produce an average cell voltage in the battery string greater than the gassing voltage (2.39 V per cell).

2. During the recharge and until 100% of the previous discharge capacity has been returned, the current should be controlled to maintain a voltage lower than the gassing voltage. To minimize charge time, this voltage can be just below the gassing voltage.

3. When 100% of the discharged capacity has been returned under the above voltage control, the charge rate will have normally decayed to the charge "finishing" rate. The charge should be finished at a constant current no higher than this rate, normally 5 A per 100 Ah of rated capacity (at the 5-h rate).

A number of methods for charging lead-acid batteries have evolved to meet these conditions. These charging methods are commonly known as:

1. Constant-current, one-current rate
2. Constant-current, multiple decreasing-current steps
3. Modified constant current
4. Constant potential
5. Modified constant potential with constant initial current
6. Modified constant potential with a constant finish rate
7. Modified constant potential with a constant start and finish rate
8. Taper charge (ferroresonant-type chargers)
9. Pulse charging
10. Trickle charging
11. Float charging

Constant-Current Charging Constant-current recharging, at one or more current rates, is not widely used for lead-acid batteries. This is because of the need for current adjustment unless the charging current is kept at a low level throughout the charge (ampere-hour rule) which will result in long charge times of 12 h or longer. Typical charger and battery characteristics for the constant-current charge, for single and two-step charging, are shown in Fig. 14.59.

Constant-current charging is used for some small lead-acid batteries (see Part 3, Chap. 17, Sec. 17.5). The use of a constant-current charge during the initial battery "formation" charge has been described in Sec. 14.3; constant-current charging is also used at times in the laboratory because of the convenience of calculating ampere-hour input and because constant-current charging can be done with simple, inexpensive equipment. Constant-current charging at half of the 20-h rate can be used in the field to decrease the sulfation in batteries which have been overdischarged and/or undercharged. This treatment, however, may diminish battery life and should be used only with the advice of the battery manufacturer.

Constant-Potential Charging The characteristics of constant-potential and modified constant-potential charging are illustrated in Fig. 14.60. In normal industrial applications, modified constant-potential charging methods are used (methods no. 5, 7, and 8). Modified constant-potential charging (no. 5) is used for on-the-road vehicles and

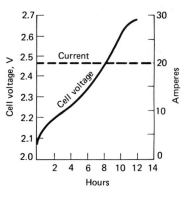

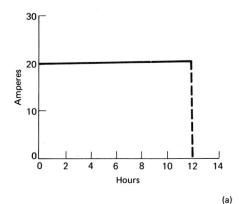

(a)

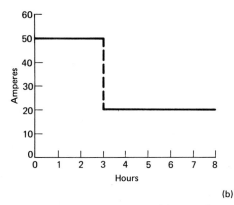

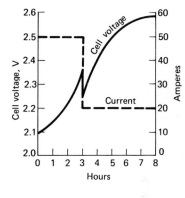

(b)

FIG. 14.59 Typical charger and battery characteristics for constant-current charging of lead-acid batteries. (*a*) Single-step constant-current charging. (*b*) Two-step constant-current charging. (*From Ref. 8.*)

utility, telephone, and uninterruptible power system applications where the charging circuit is tied to the battery. In this case the charging circuit has a current limit, and this value is maintained until a predetermined voltage is reached. Then the voltage is maintained constant until the battery is called on to discharge. Decisions must be made regarding the current limit and the constant-voltage value. This is influenced by the time interval when the battery is at the constant voltage and in a 100% state of charge. For this "float"-type operation with the battery always on charge, a low charge current is desirable to minimize overcharge, grid corrosion associated with overcharge, water loss by electrolysis of the electrolyte, and maintenance to replace this water. To achieve a full recharge with a low constant potential requires the proper selection of the starting current, which is based on the manufacturer's specifications.

The modified constant-potential charge, with constant start and finish rates, is common for deep cycling batteries which are typically discharged at the 6-h rate to a

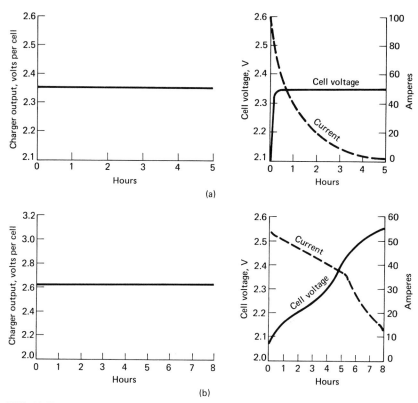

FIG. 14.60 Typical charger and battery characteristics for constant-potential charging of lead-acid batteries. (*a*) Constant-potential charging. (*b*) Modified-constant-potential charging. (*From Ref. 8.*)

depth of 80%; the recharge is normally completed in an 8-h period. The charger is set for the constant potential of 2.39 V per cell (the gassing voltage), and the starting current is limited to 16 to 20 A per 100 Ah of the rated 6-h ampere-hour capacity by means of a series resistor in the charger circuit. This initial current is maintained constant until the average cell voltage in the battery reaches 2.39 V. The current decays at constant voltage to the finishing rate of 4.5 to 5 A per 100 Ah, which is then maintained to the end of the charge. Total charge time is controlled by a timer. The time of charge is selected to ensure a recharge input capacity of a predetermined percent of the ampere-hour output of the previous discharge, normally 110 to 120%, or 10 to 20% overcharge. The 8-h charging time can be reduced by increasing the initial current limit rate.

Taper Charging   Taper charging is a variation of the modified constant-potential method using less sophisticated controls to reduce equipment cost. The characteristics of taper charging are illustrated in Fig. 14.61. The initial rate is limited, but the taper of voltage and current is such that the 2.39 V per cell at 25°C is exceeded prior to the 100% return of the discharge ampere-hours. This method does result in gassing at the critical point of recharge, and cell temperature is increased. The degree of gassing and temperature rise is a variable depending upon the charger design, and battery life can be degraded from excessive battery temperature and overcharge gassing (see Sec. 14.9.3). The gassing voltage decreases with increasing temperature; correction factors given in Table 14.26 provide the voltage correction factors at temperatures other than 25°C.

The end of the charge is often controlled by a fixed voltage rather than a fixed current. Therefore, when a new battery has a high counter-emf, this final charge rate is low and the battery often does not receive sufficient charge within the time period allotted to maintain the optimum charge state. During the latter part of life when the counter-emf is low, the charging rate is higher than the normal finishing rate, and so

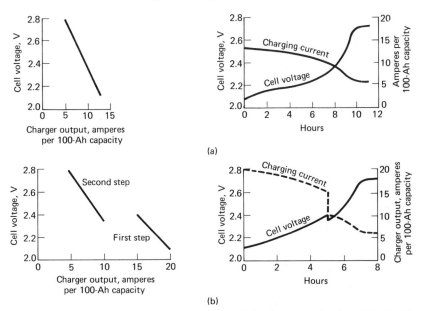

FIG. 14.61 Typical charger and battery characteristics for taper charging of lead-acid batteries. (*a*) Single-step taper charge. (*b*) Two-step taper charge. (*From Ref. 8.*)

TABLE 14.26 Correction Factors for Cell Gassing Voltage

Electrolyte temperature, °C	Cell gassing voltage, V	Correction factor, V
50	2.300	−0.090
40	2.330	−0.060
30	2.365	−0.025
25	2.390	0
20	2.415	+0.025
10	2.470	+0.080
0	2.540	+0.150
−10	2.650	+0.260
−20	2.970	+0.508

the battery receives excessive charge which degrades life. Thus the taper charge does degrade battery life which must be justified by the use of less expensive equipment.

For photovoltaic battery systems and other systems designed for optimum life, charging control and regulation circuits should produce a pattern of voltage and current equivalent to the best industrial circuits. Modified constant-potential charging methods with constant initial currents (methods no. 5 and 7) are preferred. Optimum control to maximize life and energy output from the battery is best achieved when the depth of discharge and the time for recharge are predetermined and repetitive, a condition not always realized in solar photovoltaic applications.

Pulse Charging   Pulse charging is also used for traction applications, particularly in Europe. In this case, the charger is periodically isolated from the battery terminals and the open-circuit voltage of the battery is automatically measured (an impedance-free measurement of the battery voltage). If the open-circuit voltage is above a preset value, depending on a reference temperature, the charger does not deliver energy. When the open-circuit voltage decays below that limit, the charger delivers a direct current pulse for a fixed time period. When the battery state of charge is very low, charging current is connected almost 100% of the time because the open-circuit voltage is below the preset level or rapidly decays to it. The duration of the open-circuit and charge pulses are chosen so that when the battery is fully charged the time for the open-circuit voltage to decay is exactly the same as the pulse duration. When the charger controls sense this condition, the charger is automatically switched over to the finish rate current and short charging pulses are delivered periodically to the battery to maintain it at full charge. In many industrial applications, high-voltage batteries may be used and difficulty can be encountered in keeping the cells in a balanced condition. This is particularly true when the cells have long periods of standby use with different rates of self-decay. In these applications, the batteries are completely discharged and recharged periodically (usually semiannually) in what is called an equalizing charge, which brings the whole string of cells back to the complete charge state. On completion of this process, the liquid levels in the cells must be checked and water added as required to depleted cells. With the newer types of maintenance-free cells, which are semisealed, such equalizing charges and differential watering of the cells may not be possible, and special precautions are taken in the charger design to keep the cells at an even state of charge.

Trickle Charging   A trickle charge is a continuous constant-current charge at a low (about C/100) rate which is used to maintain the battery in a fully charged condition, recharging it for losses due to self-discharge as well as to restore the energy discharged during intermittent use of the battery. This method is typically used for SLI and similar type batteries when the battery is removed from the vehicle or its regular source of charging energy for charging.

Float Charging   Float charging is a low-rate constant-potential charge also used to maintain the battery in a fully charged condition. This method is used mainly for stationary batteries which can be charged from a dc bus. The float voltage for a 1.210 specific gravity nonantimonial grid battery containing electrolyte and having an open-circuit voltage of 2.059 V per cell at 25°C is 2.17 to 2.25 V per cell.

14.8.3 Charging Circuits and Equipment

The battery charger should have the following characteristics:

1. Light weight, small size
2. Low cost
3. Resistance to the environment, particularly temperature extremes, corrosion, fungus growth, vibration, and mishandling
4. Ease of maintenance and repair
5. Safety against electrical shock and other equipment malfunction
6. Quiet as regards audible noise and radio frequency transmission
7. Compatibility with battery configurations and specifications
8. Compatibility with the available power supply (ac)
9. Compatibility with the battery as far as connectors and conductors
10. Ready availability

Not all these specifications necessarily apply to every application, and some are more appropriate than others. For electric vehicle applications and other new uses of lead-acid batteries, the weight and size limitations are particularly important. The present generation of battery chargers for standby power systems features silicon-controlled rectifiers for maximum efficiency and maintenance-free operation. They also have solid-state microprocessing diagnostic circuitry to evaluate battery condition constantly, displaying the state of charge, time remaining to complete charge, charging current, and battery condition. They employ a profile charging technique which results in the least water loss, gassing, and temperature rise. This also minimizes grid corrosion and provides a charging efficiency of up to 90% with a power factor of 0.94. Attention is presently being directed to methods which use changes in electrical impedance of a battery during the charging process to assess its instantaneous current acceptance. The latest development in battery chargers employs high-frequency circuits to improve the power factor and reduce the transformer costs in the charger itself.

14.8.4 Voltage Regulators

The voltage regulator is an essential component in systems using batteries for energy storage. Its function is to regulate the current from the charging source, either rotating equipment, solar panels, or rectified alternating current, to the battery to provide optimum current control during charge. Voltage regulators can vary from a simple manually controlled resistor between the battery and the source of dc energy to a complex temperature-compensating electronic circuit. Cost vs. performance trade-offs must consider such factors as battery life and capacity, efficiency, and power density, reliability, maintainability, size, and weight. The voltage regulator protects against excessive overcharge while at the same time allowing the battery to be charged rapidly.

The basic voltage regulator consists of four elements:

1. Stable reference voltage

2. Voltage-sampling element

3. Voltage comparator

4. Power-dissipating control device

This basic system is shown in the block diagram in Fig. 14.62. The voltage-sampling element shown is a device to translate the output voltage to a level where it can be compared with the reference voltage for that system. The difference between the two signals is amplified and directs the power-dissipating control element to perform the desired regulation function. The control element has normally one of the four types of circuitry also shown in Fig. 14.62.

14.9 MAINTENANCE, SAFETY, AND OPERATIONAL FEATURES

14.9.1 Maintenance

It is common for industrial lead-acid batteries to function for periods of 10 y or longer; proper maintenance can ensure this extended useful life. Five basic rules of proper maintenance are:

1. Match the charger to the battery charging requirements.

2. Avoid overdischarging the battery.

3. Maintain the electrolyte at the proper level (add water as required).

4. Keep the battery clean.

5. Avoid overheating the battery.

In addition to these basic rules, as the battery is made of individual cells connected in series, the cells must be properly balanced periodically.

Charging Practice   Poor charging practice is responsible for short battery life more than any other cause. Fortunately, the inherent physical and chemical characteristics of lead-acid batteries make control of charging quite simple. If the battery is supplied with dc energy at the the proper charging voltage, the battery will draw only the amount of current that it can accept efficiently and this current will reduce as the battery approaches full charge. Several types of devices can be used to ensure that the charge will terminate at the proper time. The specific gravity of the electrolyte should also be checked periodically for those batteries that have a removable vent and adjusted to the specified value (see Tables 14.6 and 14.11).

Overdischarge   Overdischarging the battery should be avoided. The capacity of large batteries, such as those used in industrial trucks, is generally rated in ampere-hours at the 6-h discharge rate to a final voltage of 1.75 V per cell. These batteries can usually deliver more than rated capacity, but this should be done only in an emergency and not on a regular basis. Discharging cells below the specified voltage reduces the electrolyte to a low concentration which has a deleterious effect on the pore structure of the battery. Battery life has been shown to be a direct function of depth of discharge as illustrated in Fig. 14.63.[24]

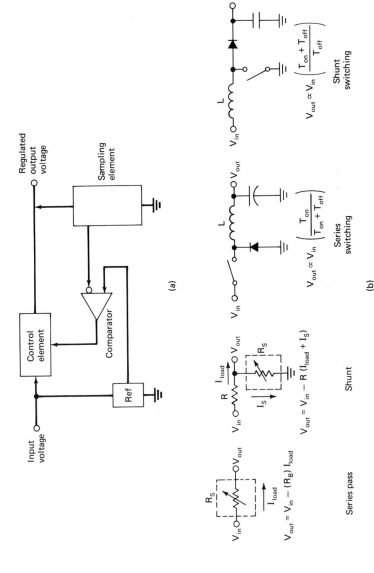

FIG. 14.62 Voltage regulators. (*a*) Block diagram of basic voltage regulator circuit. (*b*) Control element configurations for voltage regulators. (*From Ref. 24.*)

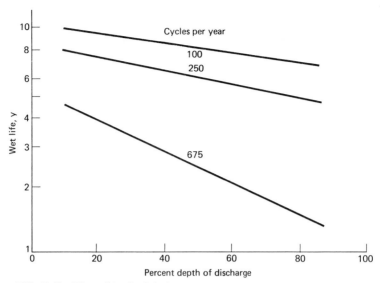

FIG. 14.63   Effect of depth of discharge and number of cycles per year on wet life, 25°C. *(From Ref. 24.)*

Electrolyte Level   During normal operation, water is lost from a battery as the result of evaporation and electrolysis into hydrogen and oxygen, which escape into the atmosphere. Evaporation is a relatively small part of the loss except in very hot, dry climates. With a fully charged battery, electrolysis consumes water at a rate of 0.336 cm³ per ampere-hour overcharge. A 500-Ah cell overcharged 10% can thus lose 16.8 cm³, or about 0.3%, of its water each cycle. It is important that the electrolyte be maintained at the proper level in the battery. The electrolyte not only serves as the conductor of electricity but is a major factor in the transfer of heat from the plates. If the electrolyte is below the plate level, then an area of the plate is not electrochemically efficient; this causes a concentration of heat in other parts of the cell. Periodic checking of water consumption can also serve as a rough check on charging efficiency and may warn when adjustment of the charger is required.

Since replacing water can be a major maintenance cost, water loss can be reduced by controlling the amount of overcharge and by using hydrogen and oxygen recombining devices in each cell where possible. Addition of water is best accomplished after recharge and before an equalization charge. Water is added at the end of the charge to reach the high acid level line. Gassing during charge will stir the water into the acid uniformly. In freezing weather, water should not be added without mixing as it may freeze before gassing occurs. Water added must be either distilled water, demineralized water, or local water which has been approved for use in batteries. Automatic watering devices and reliability testing can reduce maintenance labor costs further. Overfilling must be avoided because the resultant overflow of acid electrolyte will cause tray corrosion, ground paths, and loss of cell capacity. A final check of specific gravity should be made after water is added to ensure correct acid concentration at the end of charge. A helpful approximation is

$$\text{Specific gravity} = \text{cell open-circuit voltage} - 0.845$$

which permits electrical monitoring of specific gravity on an occasional basis (also see Fig. 14.4). Although distilled water is no longer specified by most battery manufacturers, good quality water, low in minerals and heavy metal ions such as iron, will help prolong battery life.

Cleanliness   Keeping the battery clean will minimize corrosion of cell post connectors and steel trays and avoid expensive repairs. Batteries commonly pick up dry dirt which can be readily blown off or brushed away. This dirt should be removed before moisture makes it a conductor of stray currents. One problem is that the top of the battery can become wet with electrolyte any time a cell is overfilled. The acid in this electrolyte does not evaporate and should be neutralized by washing the battery with a solution of baking soda and hot water, approximately 1 kg of baking soda to 4 L of water. After application of such a solution, the area should be rinsed thoroughly with water.

High Temperature—Overheating   One of the most detrimental conditions for a battery is high temperature, particularly above 55°C, because the rates of corrosion, solubility of metal components, and self-discharge increase with increasing temperature. High operating temperature during cycle service requires higher charge input to restore discharge capacity and local action (self-discharge) losses. More of the charge input is consumed by the electrolysis reaction because of the reduction in the gassing voltage at the higher temperature (see Table 14.26). While 10% overcharge per cycle maintains the state of charge at 25 to 35°C, 35 to 40% overcharge may be required to maintain state of charge at the higher (60 to 70°C) operating temperatures. On float service, float currents increase at the higher temperatures, resulting in reduced life. Eleven days float at 75°C is equivalent in life to 365 days at 25°C.

Batteries intended for high-temperature applications should use a lower initial specific gravity electrolyte than those intended for use at normal temperatures (see Table 14.11). Other design features, such as the use of more expander in the negative plate, are also important to improve operation at high temperatures.

Cell Balancing   During cycling, a high-voltage battery having many cells in a series string can become unbalanced, with certain cells limiting charge and discharge. Limiting cells receive more overcharge than other cells in the string, have greater water consumption, and thus require more maintenance. The equalization charge has the function of balancing cells in the string at the top of charge. In an equalization charge, the normal recharge is extended for 3 to 6 h at the finishing rate of 5 A per 100 Ah, 5-h rated capacity, allowing the battery voltage to rise uncontrolled. The equalization charge should be continued until cell voltages and specific gravities rise to a constant, acceptable value. Frequency of equalization charge is normally a function of the accumulative discharge output and will be specified by the manufacturer for each battery design and application.

14.9.2 Safety

Safety problems associated with lead-acid batteries include spills of sulfuric acid, potential explosions from the generation of hydrogen and oxygen, and the generation of toxic gases such as arsine and stibine. All these problems can be satisfactorily handled with proper precautions. Wearing of face shields and plastic or rubber aprons and gloves when handling acid is recommended to avoid chemical burns from sulfuric acid. Flush immediately and thoroughly with clean water if acid gets into the eyes, skin, or clothing and obtain medical attention when eyes are affected. A bicarbonate of soda solution (100 g per liter of water) is commonly used to neutralize any acid accidentally spilled. After neutralization the area should be rinsed with clear water.

Precautions must be routinely practiced to prevent explosions from ignition of the flammable gas mixture of hydrogen and oxygen formed during overcharge of lead-acid cells. The maximum rate of formation is 0.42 L of hydrogen and 0.21 L of oxygen per ampere-hour overcharge at standard temperature and pressure. The gas mixture is explosive when hydrogen in air exceeds 4% by volume. A standard practice is to set warning devices to ring alarms at 20 to 25% of this lower explosive limit (LEL). Low-cost hydrogen detectors are available commercially for this purpose.

With good air circulation around a battery, hydrogen accumulation is normally not a problem. However, if relatively large batteries are confined in small rooms, exhaust fans should be installed to vent the room constantly or to be turned on automatically when hydrogen accumulation exceeds 20% of the lower explosive limit. Battery boxes should also be vented to the atmosphere. Sparks or flame can ignite these hydrogen atmospheres above the LEL. To prevent ignition, electrical sources of arcs, sparks, or flame must be mounted in explosion-proof metal boxes. Battery cells can similarly be equipped with flame arrestors in the vents to prevent outside sparks from igniting explosive gases inside the cell cases. It is good practice to refrain from smoking, using open flames, or creating sparks in the vicinity of the battery. A considerable number of the reported explosions of batteries come from uncontrolled charging in nonautomotive applications. Often, batteries will be charged, off the vehicle, for long periods of time with an unregulated charger. In spite of the fact that the charge currents can be low, fair volumes of gas can accumulate. When the battery is then moved, this gas vents, and if a spark is present, explosions have been known to occur. The introduction of the calcium alloy grids has minimized this problem, but the possibility of explosion is still present.

Some types of batteries can release small quantities of the toxic gases, stibine and arsine. These batteries have positive or negative plates which contain small quantities of the metals antimony and arsenic in the grid alloy to harden the grid and to reduce the rate of corrosion of the grid during cycling. Arsine (AsH_3) and stibine (SbH_3) are generally formed when the arsenic or antimony alloy material comes into contact with nascent hydrogen, usually during overcharge of the battery, which then combines to form these colorless and essentially odorless gases. They are extremely dangerous and can cause serious illness and death. The OSHA 1978 concentration limits for SbH_3 and AsH_3 are 0.1 and 0.05 ppm, respectively, as a maximum allowable weighted average for any 8-h period. Ventilation of the battery area is very important. Indications are that ventilation designed to maintain hydrogen below 20% LEL (approximately 1% hydrogen) will also maintain stibine and arsine below their toxic limits.

The ordinary 12-V SLI automotive battery is a minor shock hazard. The hazard level increases with higher-voltage systems, and systems in the range of 84 to 120 V are being used for electric vehicles. Systems as high as 1000 V are under study for fixed-location energy storage systems for load leveling. Batteries are electrically alive even in the discharged state, and the following precautions should be practiced:

1. Keep the top of the battery clean and dry to prevent ground shorts and corrosion.
2. Do not lay metallic objects on the battery. Insulate all tools used in working on batteries.

3. Remove jewelry and any other electrical conductor before inspecting or servicing batteries.

4. When lifting batteries, use insulated lifting tools to avoid risks of short circuits between cell terminals by lifting chains or hooks.

5. Make sure gases do not accumulate in batteries before they are moved.

14.9.3 Effect of Operating Parameters on Battery Life

Operating parameters which have a strong influence on battery life are depth of discharge, number of cycles used each year, charging control, type of storage, and operating temperature. In some cases, the battery design features which increase life tend to decrease the initial capacity, power, and energy output. It is important, therefore, that the design features of the battery be selected to match the operating and life requirements of the application.

1. Increasing the depth of discharge decreases cycle life, as illustrated in Figs. 14.64 and 14.34.

2. Increasing the number of cycles performed per year decreases the wet life (Sec. 14.9.1 and Fig. 14.63).

3. Excessive overcharging leads to increasing positive grid corrosion, active material shedding, and shorter wet life.

4. Storing wet cells in a discharged condition promotes sulfation and decreases capacity and life.

5. Proper charging operations with good equipment maintain the desired state of charge with a minimum of overcharge and lead to optimum battery life.

6. Stratification of the electrolyte in large cells into levels of varying concentration can limit charge acceptance, discharge output, and life unless controlled during the charge process. During a recharge sulfuric acid of higher concentration than the bulk electrolyte forms in the pores of the plates. This higher-density acid settles to the bottom of the cell, giving higher specific gravity acid near the bottom of the plates and lower specific gravity acid near the top of the plates. This stratification accumulates during the nongassing periods of charge. During the gassing periods of overcharge, partial stirring is accomplished by gas bubbles formed at and rising along the surfaces of the plates and in the separator system. During discharge, acid in the pores of the plates and near their surface is diluted; however, concentration gradients set up by longer charge periods are seldom compensated entirely, particularly if the discharge periods are shorter, as is usually the case. Diffusion processes to eliminate these concentration gradients are very slow, and stratification during repetitive cycling can become progressively greater. Two methods for stratification control are by deliberate gassing of the plates during overcharge at the finishing rate and by stirring of cell electrolyte by pumps (usually air-lift pumps). The degree of success in eliminating stratification is a function of cell design, the design of the pump accessory system, and cell operating procedures.

14.9.4 Failure Modes

The failure modes of lead-acid batteries depend on the type of application and the particular battery design. This is the rationale for the manufacture of different batter-

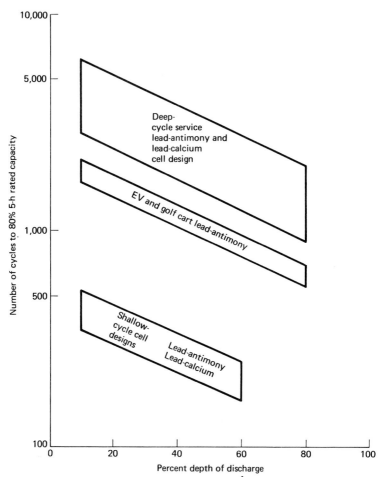

FIG. 14.64   Effect of cell design and depth of discharge on the cycle life of various types of lead-acid batteries (25°C). *(From Ref. 24.)*

ies since each one is designed to give optimum performance in a specific type of use. The more prevalent failure modes for the different types of lead-acid batteries are listed in Table 14.27.[25] Significantly, if a battery is properly maintained, most of the inherent failures are due to the degradation of the positive plate, either through grid corrosion or paste shedding. These failures are irreversible, and when they occur the battery must be replaced. Details of the failure modes of SLI batteries are given in Sec. 14.4.3.

The failure mode and/or the time to failure can be modified by changes in the inner parameters (I), such as battery materials, processing, and design, or by the conditions of use, designated as the outer parameters (O). Some of these are listed in Table 14.28.[25]

TABLE 14.27   Failure Modes of Lead-Acid Batteries

Battery type	Normal life	Normal failure mode
SLI	Several years	Grid corrosion
SLI (maintenance-free)	Several years	Lack of water, damage to positive plates
Golf cart	300–600 cycles	Positive shedding and grid corrosion, sulfation
Stationary (industrial)	6–25 y	Grid corrosion
Traction (industrial)	Minimum 1500 cycles	Shedding, grid corrosion

TABLE 14.28   Modification of Lead-Acid Battery Failure Rate

Failure mechanism	Rate of failure modification*
Shedding positives	I: active mass structure, battery design O: number of cycles, depth of discharge, charge factor
Sulfation/leading of negatives	I: active mass additives O: temperature, charge factor, maintenance
Positive grid corrosion (overall, localized, or positive grid growth)	I: grid alloy, casting conditions, active mass
Separators	I: electrolyte concentration, battery design O: temperature, charge factor, maintenance
Case, cover, vents, external battery connections	I: battery materials and design O: maintenance, abuse

*I: inner parameters; O: outer parameters.

14.10 APPLICATIONS AND NEW MARKETS

The lead-acid battery is used in a wide variety of applications, and in the past few years many new applications have arisen. The various types of lead-acid batteries and their applications are listed in Table 14.2. The new uses of lead-acid batteries are mainly associated with the smaller "sealed" maintenance-free cells used in electronic and portable devices and with the advanced designs for energy storage and electric vehicles.

14.10.1 Automotive Applications

Traditionally, the most common use of the lead-acid battery is for starting, lighting, and ignition in automobiles and other vehicles with internal combustion engines. Almost all of these now have 12-V nominal electrical systems; in the last decade most of the generators have been replaced by alternators and electromechanical regulators by electronic/solid state controls. High cranking ability at low temperatures is still the major design factor, but SLI batteries today see more cycling-type service (compared with float service) because of the electrical load of the auxiliaries. Size and weight reduction have also become important as well as the battery geometry. Batteries are normally located in the cool air stream ahead of the engine to prevent their overheating;

thus their geometry can affect the profile of the vehicle. These factors led to the redesign of the lead-acid battery for SLI application. The most important changes were:

Change from high-antimony (4 to 5%) lead alloy grid to a low-antimony (1 to 2%) or nonantimonial lead alloy grid, thus reducing hydrogen evolution

Use of thinner electrodes

Better separators with lower electrical resistance

Plate tabs located in from corners, and grids redesigned for high conductivity

Semisealed, maintenance-free construction

Automotive-type batteries are also used on trucks, aircraft, industrial equipment, and motorcycles as well as in many other applications. They are used in off-road vehicles, such as snowmobiles, in boats to crank inboard and outboard engines, and in various farm and construction equipment. Military vehicles in the United States and NATO countries have standardized on a 24-V electrical system that is provided by a series connection of two 12-V batteries.

14.10.2 Small Sealed Lead-Acid Cells

In recent years, there has been been a significant increase in the use of battery-operated consumer equipment such as portable tools, lighting devices, instruments, photographic equipment, calculators, radio and TV, toys, and appliances. Batteries for these applications are generally of low capacity, up to 25 Ah. Storage batteries are used frequently because of their high power capability and rechargeability, but they have to be sealed or of the nonspill type in order to function in all positions. Vented lead-acid batteries of the electrolyte-retaining (ER) type and nonvented round (see Chap. 15) or prismatic cells are used in competition with the sealed nickel-cadmium cells. The lead-acid batteries offer lower initial cost, better float service, higher cell voltage, and the absence of memory effect (loss of capacity on shallow cycling). The nickel-cadmium cell has longer life and better cycling service. Both types of batteries have about the same energy density.

The small sealed or semisealed lead-acid cells are available as single 2-V units or as multiple-cell units, usually in 6-V monobloc constructions. They are an outgrowth of the earlier ER-type batteries in which the electrolyte was absorbed in wood pulp separators. The ER-type cells, while spill-proof, contained more electrolyte, did not recombine oxygen on overcharge, and were vented.

A related small deep-discharge lead-acid battery is the one used for miners' lamps and similar equipment. These are 4-V units which are vented and can be watered. They are designed to deliver 1 A for 12 h between charges.

14.10.3 Industrial Applications

Applications for lead-acid batteries, other than the SLI and small sealed power units, fall into two categories, as shown in Table 14.29: those based on automotive-type constructions and those based on industrial-type constructions. Often several designs can be used for a single type of application.

14.10.4 Electric Vehicles

Lead-acid batteries are used as the power source in off-the-road vehicles such as golf carts, forklift trucks, mining vehicles, and special construction and industrial equipment. One million 6-V units, of approximately 750-Wh capacity, are used annually in golf carts in the United States.

An intensive worldwide program was started in the 1970s to improve electric vehicles in response to the oil shortage (see Part 4, Chap. 26). Although many new electrochemical systems are under study for this application, the lead-acid battery is still a leading candidate.[26] The batteries are often used in a tray, which rolls out from below the vehicle to facilitate servicing, as illustrated in Fig. 14.65. The effort has led to the development of batteries with improved energy density, 40 Wh/kg at the 3-h rate, and higher power density. The objective for the advanced lead-acid battery is 50 to 60 Wh/kg, a cycle life of 1000 cycles, and a cost below $50 per kilowatthour (see Part 4, Chap. 26, Sec. 26.3).

TABLE 14.29 Major Applications of Lead-Acid Batteries (Non-SLI Types)

Automotive and small energy storage designs		Industrial designs		
Traction	Special	Stationary	Traction (motive power)	Special
Golf cart	Emergency	Switch gear	Mine	Submarine
Off-road	lighting	Emergency	locomotives	Ocean buoys
vehicles	Alarm signals	lighting	Industrial truck	
On-road	Photovoltaic	Telephone	Large electric	
vehicles	Sealed cells (for	Railway signals	vehicles	
	tools,	Uninterrupted		
	instruments,	power supply		
	electronic	Photovoltaic		
	devices, etc.)	Load leveling		
		and energy		
		storage		

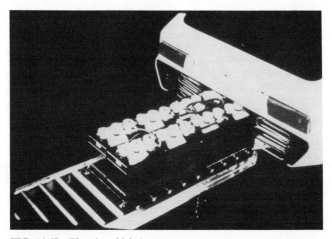

FIG. 14.65 Electric vehicle battery.

In addition to the all-electric vehicles, hybrid vehicles, using an internal combustion engine in combination with a secondary battery, have also been designed. The hybrid designs should ameliorate the problems of the all-electrics with respect to range, speed, acceleration, and recharging. However, initial cost and operation of the hybrid vehicle will be more expensive than the all-electric one, which uses utility power for recharging and avoids the dependence on petroleum fuels.

14.10.5 Energy Storage Systems

Secondary batteries are now being considered for load leveling in electric utility systems as an alternative to meet peak power demands currently provided with energy-expensive oil- or gas-fueled turbines (see Part 4, Chap. 26). Large batteries, in the order of 50 MWh at 1000 V, are required. The lead-acid battery, again, is a major candidate for a near-term solution for this application. A 1.7-MWh standard lead-acid battery and a 1.5-MW power conversion system served as the base line facility for proving in the Battery Energy Storage Test (BEST) Facility located in New Jersey. An optimized lead-acid battery, as well as other advanced secondary battery systems, will be evaluated at BEST during the 1980s. The goal is to obtain in excess of 2000 cycles or 10 y of operation at a cost of about $120 per kilowatthour.

Smaller-sized batteries are used for energy storage in systems employing renewable but interruptible energy sources, such as wind and solar (photovoltaic) energy. These systems are usually located on the customer side of the utility power grid. The system generally handles the following functions:

1. Converts solar, wind, or other such prime energy source to direct electrical power

2. Regulates the electrical power output

3. Feeds the electrical energy into an external load circuit to perform work

4. Stores the electrical energy in a battery subsystem for later use

A block diagram of a typical photovoltaic system is shown in Fig. 14.66.[24]

The selection of the proper battery for these types of applications requires a complete analysis of the battery's charge and discharge requirements, including the load, the output and pattern of the solar or alternative energy source, the operating temperature, the efficiency of the charger and other system components, etc. Golf cart–type lead-acid batteries and modified EV designs are widely used in these small stationary energy storage systems because they are the least expensive design in commercial production. Maintenance-free SLA cells are also being developed for these applications.

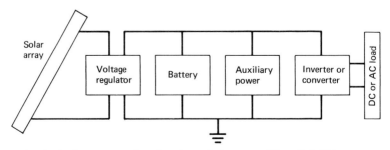

FIG. 14.66 Components of a solar photovoltaic system. (*From Ref. 24.*)

These batteries (100-Ah size) can give maintenance-free unattended operation, a self-discharge rate less than 5%/month, a recharge in less than 8 h, and 1000 to 2000 cycles to an 80% DOD.[27,28]

14.10.6 Power Conditioning and Uninterrupted Power Systems

DC Power Systems   A new concept in standby power is the dc power system with battery backup. These power systems include a battery charger (rectifier/charger) which has a sufficient capacity to recharge the batteries at the proper voltage while simultaneously supplying power to the dc load. Additionally, equipment protection and isolation of the line voltage from the secondary windings of the special power transformer are designed into the system.

Static Uninterruptible AC Power Systems (UPS)   In this power system, a storage battery is linked to the utility power to provide a continuity of service in the event of an interruption of the utility power. The continuous UPS system, Fig. 14.67, is illustrative of this type of device. During normal operation, the ac line supplies power to the static battery charger (rectifier/charger) which, in turn, "float" charges the battery and, at the same time, supplies dc power to the static inverter. The inverter, in turn, supplies power to the ac load. A synchronizing signal (if used) from the ac power line can maintain the phase and frequency of the inverter output the same as in the power line. This maintains the accuracy of timing devices such as clocks and recorder charts.

The voltage regulator within the inverter maintains the ac load voltage constant throughout the load range as well as during periods of "equalizing" charge on the storage battery. Transient and steady-state power line variations are isolated from the load by the regulating action of the battery charger in conjunction with the filtering action of the battery and the inverter.

Upon ac line power failure, the battery charger ceases to operate and the battery instantly supplies power to the inverter and sustains the ac load without interruption. The synchronizing signal is also lost during the power failure; the inverter, therefore, continues to operate on its own internal frequency reference ($\pm$ 1% for standard units).

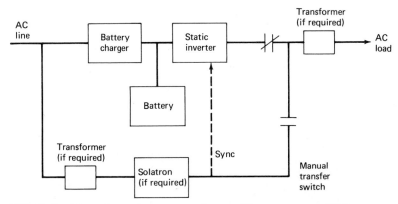

FIG. 14.67   Schematic, continuous-type uninterruptible power system (UPS).

The ampere-hour capacity of the battery determines the protection time of the system. The inverter is designed to maintain a constant voltage output as the battery voltage drops during the discharge. When ac power returns, the charger restores the battery energy and at the same time supplies power to the inverter.

REFERENCES

1. H. Bode, *Lead-Acid Batteries,* Wiley, New York, 1977.

2. H. E. Haring and U. B. Thomas, *Trans. Electrochem. Soc.* **68**:293 (1935).

3. A. J. Salkind, D. T. Ferrell, and A. J. Hedges, "Secondary Batteries: 1952–1977," *J. Electrochem. Soc.* **125**:311C (1978).

4. P. Ruetschi, "Review of the Lead-Acid Battery Science and Technology," *J. Power Sources* **2**:3 (1977/78).

5. G. W. Vinal, *Storage Batteries,* 4th ed., Wiley, New York, 1955.

6a. D. J. G. Ives and G. J. Janz, *Reference Electrodes,* Academic, New York, 1961.

6b. E. A. Willihnganz, U.S. Patent 3,657,639.

7. A. Sabatino, *Maintenance-free Batteries, Heavy Duty Equipment Maintenance,* Irving-Cloud Publishing Co., Lincolnwood, Ill., 1976.

8. Special Issue on Lead-Acid Batteries, *J. Power Sources* **2**(1) (1977/78).

9. *Pig Lead Specifications,* ASTM Specification B29, American Society for Testing and Materials, Philadelphia, 1959.

10. A. T. Balcerzak, *Alloys for the 1980's,* St. Joe Lead Co., Clayton, Mo., 1980.

11. *Grid Metal Manual,* Independent Battery Manufacturers Association (IBMA), Largo, Fla., 1973.

12. N. E. Hehner, *Storage Battery Manufacturing Manual,* Independent Battery Manufacturers Association (IBMA), Largo, Fla., 1976.

13. U. B. Thomas, F. T. Foster, and H. E. Haring, *Trans. Electrochem. Soc.* **92**:313 (1947).

14. N. E. Hehner and E. Ritchie, *Lead Oxides,* Independent Battery Manufacturers Association (IBMA), Largo, Fla.

15. M. Barak, *Electrochemical Power Sources,* Peter Peregrinus, Ltd., Stevenage, U.K., 1980.

16. *Battery Service Manual,* 9th ed., Battery Council International, Chicago, 1982.

17. J. Hoover, *1982 Convention,* Battery Council International, New York, May 1982.

18. *The Storage Battery Manufacturing Industry 1980–1981 Yearbook,* Battery Council International, Chicago, 1980.

18a. *SAE Handbook,* Society of Automotive Engineers, Warrendale, Pa.

19. *Gould Battery Handbook*, Gould Inc., Mendota Heights, Minn., 1973.

20. E. J. Friedman et al., *Electrotechnology,* vol. 3: *Stationary Lead-Acid Batteries,* Ann Arbor Science Publishers, Ann Arbor, Mich., 1980.

21. *Bell Syst. Tech. J.* **49**(7) American Telephone and Telegraph Co., New York, September 1970.

22. R. V. Biagetti and H. J. Luer, "A Cylindrical, Pure Lead, Lead-Acid Cell for Float Service," *J. Power Sources* **4** (1979).

23. E. Ritchie, International Lead-Zinc Research Organization Project LE-82-84, Final Report, New York, December 1971.

24. *Handbook of Secondary Storage Batteries and Charge Regulators in Photovoltaic Systems,* Exide Management and Technology Co., Report 1-7135, Sandia National Laboratories, Albuquerque, N. M., August 1981.

25. G. E. Mayer, "Critical Review of Battery Cycle Life Testing Methods," *5th Int. Electric Vehicle Symp.,* Philadelphia, October 1978.

26. N. P. Yao et al., "Improved Lead-Acid Batteries—the Promising Candidate for Near-Term Electric Vehicles," *Proc. 29th Power Sources Conf.,* Electrochemical Society, Pennington, N.J., June 1980.

27. James R. Willhite, "Development of a Sealed Lead-Acid Battery for Photovoltaic Applications," *Proc. 5th U.S. Dept. Energy Battery and Electrochem. Contractors' Conf.,* Washington, D.C., December 1982.

28. *Maintenance-Free, 100 Ah Lead-Acid Battery for Deep Discharge, Photovoltaic Applications,* Report SAND82-7060, Sandia National Laboratories, Albuquerque, N.M., June 1982.

15

Sealed
Lead-Acid
Batteries

by
Ronald O. Hammel

15.1 GENERAL CHARACTERISTICS

A relatively new design of the lead-acid battery is the cylindrical sealed lead-acid (SLA) rechargeable cell. The cells are sealed to about 4×10^5Pa pressure with a resealable pressure valve and are maintenance-free because there is no need to replace electrolyte or water. They also differ from the conventional lead-acid cell by using a minimum amount of electrolyte ("starved electrolyte") and the gas recombination principle by which, at recommended rates of overcharge, the oxygen and some of the hydrogen generated are recombined within the cell, eliminating the need to add water.

Other advantageous characteristics include the following (see also Table 15.1):

TABLE 15.1 Major Advantages and Disadvantages of Sealed Lead-Acid Batteries

Advantages	Disadvantages
Maintenance-free	Cannot be stored in discharged condition
Long life on float service	Relatively low energy density
High rate capability	Lower cycle life than sealed nickel-cadmium battery
Low- and high-temperature performance	
No "memory" effect or thermal runway (compared to nickel-cadmium battery)	
"State of charge" can be determined by measuring voltage	
Low cost	

Low-Temperature Performance The good low-temperature performance of the lead-acid system has been maintained by the use of a separator system which minimizes diffusion and resistance. This results in good utilization of active material and excellent voltage regulation over a wide temperature range.

High Rate Capability The thin-plate construction of the SLA cell contributes to high utilization of the active plate materials, low internal impedance, and minimal polarization. Thus, the cell can be discharged at high rates with power densities up to 600 W/kg. Voltage regulation during discharge and during changes in discharge current is also excellent. Another advantage of this special construction is the fast recharge capability of the cell. It is possible to recharge the cell to full capacity in less than 1 h.

Long Life in Float Applications The high purity of the lead grids used in the SLA cell results in long life on float charge while the spiral-wound plate design maintains structural resistance to shock and vibration.

Long Cycle Life in Deep-Discharge Applications The spiral-wound plate design also reduces the rate of shedding of the positive plate, thereby increasing the cycle life at deep discharge levels.

Structural Resistance The rugged metal case on the cell further enhances resistance to shock, crushing, or dropping.

Energy Density The SLA cell has the highest energy and power density per unit volume of any lead-acid battery presently available. Most lead-acid cells have about 70 Wh/L, whereas the SLA cell has greater than 85 Wh/L at the C/10 discharge rate.

15.2 CHEMISTRY

Although the design and construction of the SLA cell are different, its chemistry is that of the traditional lead-acid battery. The basic "double sulfate" reaction applies to the overall reaction:

$$PbO_2 + Pb + 2H_2SO_4 \underset{charge}{\overset{discharge}{\rightleftarrows}} 2PbSO_4 + 2H_2O$$

The reaction at the positive electrode

$$PbO_2 + 3H^+ + HSO_4^- + 2e \underset{charge}{\overset{discharge}{\rightleftarrows}} 2H_2O + PbSO_4$$

At the negative electrode

$$Pb + HSO_4^- \underset{charge}{\overset{discharge}{\rightleftarrows}} PbSO_4 + H^+ + 2e$$

When the cell is recharged, finely divided particles of $PbSO_4$ are electrochemically converted to sponge lead at the negative electrode and PbO_2 at the positive by the charging source driving current through the battery. As the cell approaches complete recharge, when the majority of the $PbSO_4$ has been converted to Pb and PbO_2, the overcharge reactions begin. For the typical lead-acid cells, the result of these reactions is the production of hydrogen and oxygen gas and subsequent loss of water.

A unique aspect of the SLA cell is that the majority of the oxygen generated within the cell on overcharge (up to the C/3 rate) is recombined within the cell. The pure lead grids used in the construction minimize the evolution of hydrogen on overcharge. Although most of the hydrogen is recombined within the cell, some is released to the atmosphere.

Oxygen will react with lead at the negative plate in the presence of H_2SO_4 as quickly as it can diffuse to the lead surface. Hydrogen will be oxidized at the PbO_2 surface of the positive plate at a somewhat lower rate, as shown in these overcharge recombination reactions

$$Pb + HSO_4^- + H^+ + 1/2O_2 \underset{\text{charge}}{\overset{\text{discharge}}{\rightleftarrows}} PbSO_4 + H_2O$$

$$PbO_2 + HSO_4^- + H^+ + H_2 \underset{\text{charge}}{\overset{\text{discharge}}{\rightleftarrows}} PbSO_4 + 2H_2O$$

In a flooded lead-acid cell, this diffusion of gases is a slow process, and virtually all the H_2 and O_2 escape from the cell rather than recombine. In the SLA cell, the closely spaced plates are separated by a glass mat which is composed of fine glass strands in a porous structure. The cell is filled with only enough electrolyte to coat the surfaces of the plates and the individual glass strands in the separator, thus creating the starved-electrolyte condition. This condition allows for the homogeneous gas transfer between the plates, necessary to promote the recombination reactions.

The pressure release (Bunsen) valve maintains an internal pressure of about 4×10^5 Pa. This condition aids recombination by keeping the gases within the cell long enough for diffusion to take place. The net result is that water, rather than being released from the cell, is electrochemically cycled to take up the excess overcharge current beyond what is used for conversion of active material. Thus the cell can be overcharged sufficiently to convert virtually all the active material without loss of water, particularly at recommended recharge rates. At continuous high overcharge rates (e.g., C/3 and above), gas buildup becomes so rapid that the recombination process is not as highly efficient, and O_2 as well as H_2 gas is released from the cell.

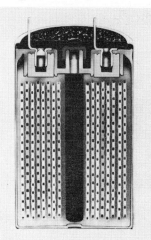

FIG. 15.1 Cross section of the sealed lead-acid cell; components identified in Fig. 15.2. (*Courtesy of Gates Energy Products.*)

15.3 CELL CONSTRUCTION

A cross section of the SLA cell and a breakdown of the basic components contained in the cell are shown in Figs. 15.1 and 15.2. Both the positive and negative plates are made of pure lead and are extremely thin. The plates are pasted with lead oxides, separated by an absorbing glass mat, and spirally wound to form the basic cell element. Lead posts are then

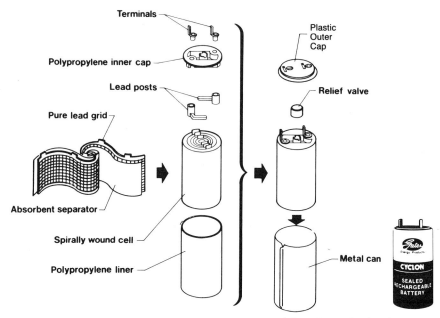

FIG. 15.2   Components of sealed lead-acid cell. (*Courtesy of Gates Energy Products.*)

welded to the exposed positive- and negative-plate tabs. The terminals are inserted through the polypropylene inner top and are effectively sealed by expansion into the lead posts. The element is then stuffed into the liner and the top and liner are bonded together. At this state of construction, the cell is sealed except for the open vent hole. Sulfuric acid is then added and the relief valve is placed over the vent hole. The sealed element is then inserted into the metal can, the outer plastic top added, and crimping completes the assembly. The metal case provides mechanical strength and does not affect the operation of the resealable vent. The cell is then charged for the first time or electrochemically formed.

15.4   PERFORMANCE CHARACTERISTICS

15.4.1   Voltage

The nominal voltage of the SLA cell is 2.0 V, and it is typically discharged to 1.75 V per cell under load. The open-circuit voltage of the cell depends on the state of charge as plotted in Fig. 15.3, based on a C/10 discharge rate. The open-circuit voltage can be used, therefore, to approximate the state of charge. The curve is accurate to within 20% if the cell has not been charged or discharged within 24 h; it is accurate within 5% if the cell has not been used for 5 days. The measurement of the open-circuit voltage to determine the state of charge is based on the relationship between the electromotive force (OCV) and the concentration of the sulfuric acid in the battery.

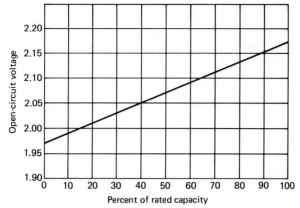

FIG. 15.3 Open-circuit voltage vs. state of charge.

15.4.2 Discharge Characteristics

The discharge voltage profiles of the SLA cell, at various temperatures ranging from −40 to 65°C for various discharge rates, are shown in Fig. 15.4 *a* to *e*. The discharge voltage is very flat at medium to low rates. These curves are based on the smaller 2.5- and 5-Ah cells. Discharge curves for the larger 25-Ah cell are slightly different from those of the smaller cells because of the greater distance from the center of the plate to the external stud. This gives a higher effective internal impedance per unit of capacity and results in a slightly lower performance at higher rates and lower temperatures.

Figure 15.5 shows a set of discharge voltage curves for a 2.5-Ah cell at 25°C, which further illustrates the good voltage regulation of the cell at high rates of discharge.

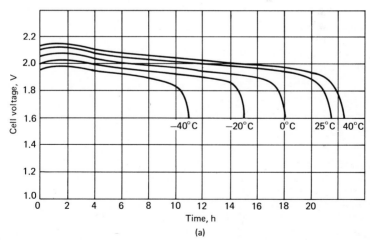

FIG. 15.4 Discharge curves—sealed lead-acid cell for D and X cells: (*a*) discharge rate C/20.

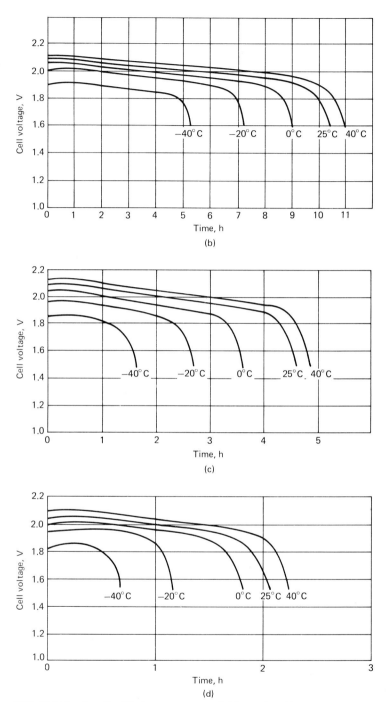

FIG. 15.4 (*Continued*) Discharge curves—sealed lead-acid cells: (*b*) discharge rate C/10; (*c*) discharge rate C/5; (*d*) discharge rate C/2.5.

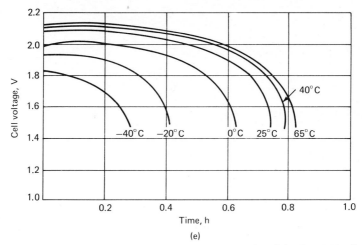

FIG. 15.4 (*Continued*)   Discharge curves—sealed lead-acid cell for D and X cells: (*e*) discharge rate 1C.

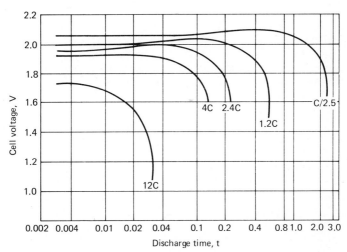

FIG. 15.5   Discharge curves—2.5-Ah sealed lead-acid cell—25°C at high discharge rates.

15.4.3 Effect of Temperature and Discharge Rate

The capacity of the SLA cell, as with most batteries, is dependent on discharge rate and temperature, the capacity decreasing with decreasing temperature and increasing discharge rate. The effect of temperature at the C/10- and 1.2C rate is shown in Fig. 15.6 for the 2.5- and 5-Ah cells. The larger 25-Ah cell gives a lower percentage of the 25°C performance at lower temperatures and higher discharge rates.

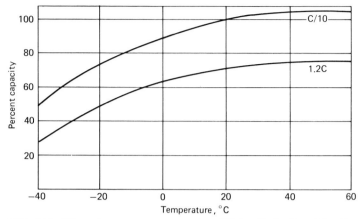

FIG. 15.6 Effect of temperature on capacity of SLA cells for D and X cells.

15.4.4 High-Rate Pulse Discharge

The SLA cell is effective in applications which require a high-rate pulse discharge, such as in engine starting. The voltage-time curves for the cell at room temperature at the 10C discharge rate, both on continuous discharge and for a 16.7% duty cycle (10-s pulse, 50-s rest), are shown in Fig. 15.7 at 25 and $-20°C$.

It is apparent from these data that the capacity of the SLA cell is increased greatly when a pulse discharge is used. This is true because of the phenomenon known as "concentration polarization." As a discharge current is drawn from the cell, the sulfuric acid in the electrolyte reacts with the active materials in the electrodes. This reaction reduces the concentration of the acid at the electrode-electrolyte interfaces. Consequently, the cell voltage drops. During the rest period, the acid in the bulk of the solution diffuses into the electrode pores to replace the acid which has been used up. The cell voltage then increases as acid equilibrium is established. Since during a pulse discharge the acid is allowed to equilibrate between pulses, it is not depleted as quickly and the total cell capacity is increased.

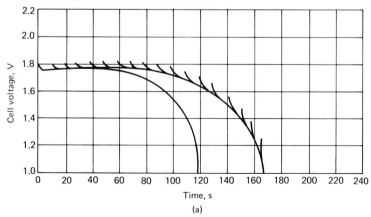

(a)

FIG. 15.7 High-rate pulse performance—10C discharge rate for D and X cells: (*a*) 25°C.

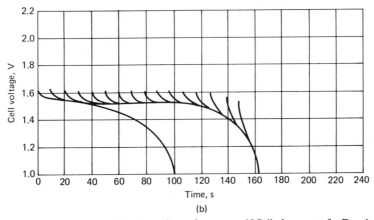

FIG. 15.7 (*Continued*)   High-rate pulse performance—10C discharge rate for D and X cells: (*b*) −20°C.

The ability of the cell to provide high discharge currents and maintain usable voltage is illustrated in Fig. 15.8. The effect of discharge rate on the cell midpoint voltage is shown at both 22 and −20°C. (The voltage indicated was measured midway through the cell discharge.)

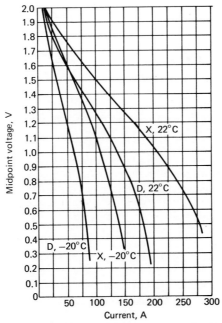

FIG. **15.8**   Effect of discharge rate on cell midpoint voltage.

The curves in Fig. 15.9 illustrate the power of the cell as a function of discharge rate at room temperature and −20°C. Maximum power obtainable from the cell increases as the temperature increases.

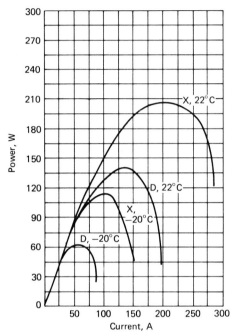

FIG. 15.9 Instantaneous maximum peak power at 22 and −20°C.

15.4.5 Discharge Level

As with all rechargeable batteries, discharging the SLA cell beyond the point at which 100% of the capacity has been removed can shorten the life of the cell or impair its ability to accept a charge.

The voltage point at which 100% of the usable capacity of the cell has been removed is a function of the discharge rate as shown in the upper curve of Fig. 15.10. The lower curve shows the minimum voltage level to which the cell may be discharged with no effect on recharging capability. For optimum life and charge capability the cell should be disconnected from the load at the voltages within the gray area between the two curves.

Under these "overdischarge" conditions, the sulfuric acid electrolyte can be depleted of the sulfate ion and become essentially water, which can create several problems. A lack of sulfate ions as charge conductors will cause the cell impedance to appear high and little charge current to flow. Longer charge time or alteration of charge voltage may be required before normal charging may resume.

Another potential problem is solubility of lead sulfate in water. In a severe deep-discharge condition, the lead sulfate present at the plate surfaces can go into solution

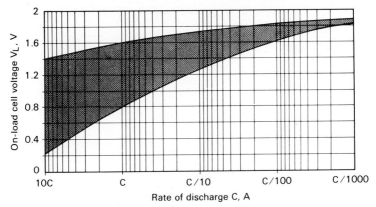

FIG. 15.10   Acceptable voltage discharge levels of SLA cells.

in the water electrolyte. Upon recharge, the water and sulfate ion in the lead sulfate convert to sulfuric acid, leaving a precipitate of lead metal in the separator. This lead metal can result in dendritic shorts between the plates and resultant cell failure.

It is important to note that when the load is removed from the cell, the cell voltage will recover up to approximately 2 V. Figure 15.11 shows an example of a C/16 discharge rate, removal of the load at a cell voltage of 1.6 V, and subsequent recovery of the cell voltage. Because of this phenomenon, some hysteresis must be designed into the battery-disconnect circuitry so that the load is not continuously reapplied to the battery as the battery voltage recovers.

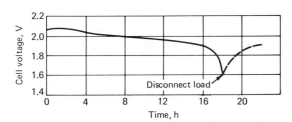

FIG. 15.11   Voltage recovery after removal of discharge load.

15.4.6 Service Life

The performance of the SLA cell is summarized in Fig. 15.12, which shows the service life of the SLA cell at various discharge loads and temperatures on a weight and volume basis. These curves are based on the 2.5- and 5-Ah cells; as discussed in Sec. 15.4.2 above, the larger cells will give poorer performance at higher rates and lower temperatures because of their higher internal impedance.

15.4.7 Storage Characteristics

Most batteries lose their stored energy when allowed to stand on open circuit due to the fact that the active materials are in a thermally unstable state. The rate of self-

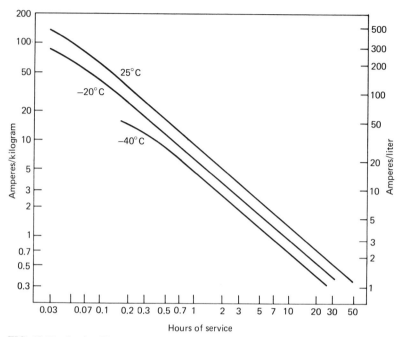

FIG. 15.12   Service life of SLA cell.

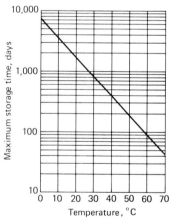

FIG. 15.13   Storage characteristics of SLA cell.

discharge is dependent on the chemistry of the system and the temperature at which it is stored. The SLA cell is capable of long storage without damage, as can be seen in Fig. 15.13, which plots the maximum storage time against storage temperature. This curve shows the maximum number of days at any given temperature from 0 to 70°C for the cell to self-discharge from 2.18 down to 1.81 V open circuit. The cell should not be allowed to self-discharge below 1.81 V, because the recharge characteristics of the cell change appreciably and the cycle life cannot be accurately predicted. The capacity of these cells can be restored by recharging; however, the first charge on a cell that has been allowed to self-discharge down to 1.81 V will take longer than normal, and the first discharge will generally not deliver rated capacity. Subsequent cycles, however, will show an increase in the cell capacity to rated value.

It is important to recognize that the self-discharge rate of the sealed lead-acid cell is nonlinear; thus the rate of self-discharge changes as the state of charge of the cell changes. When the cell is in a high state of charge, i.e., 80% or greater, the self-discharge is very rapid. The cell may discharge from 100 to 90% at room temperature in a matter of a week or two. Conversely, at the same temperature, it may take 10 weeks or longer for the same cell to self-discharge from 20% state of charge down to 10% state of charge. Figure 15.14 is a curve of open-circuit voltage vs. percent remaining storage time that shows the nonlinearity of the self-discharge reaction.

By the use of Figs. 15.13 and 15.14, the number of days of storage which remain before a cell must be recharged can be calculated. As an example, if a cell has an OCV of 2.05 V, the state of charge, as determined from Fig. 15.14, is 35%. From Fig. 15.14 (again at OCV of 2.05 V), the remaining storage time is 82%. Figure 15.13 shows that at 20°C, the cell can be stored for a total of 1200 days before it must be recharged. Therefore, the remaining storage time is 82% of 1200, or 984 days. This is the number of days that a cell at 2.05 V OCV can be stored before it will reach 1.81 V and must be recharged.

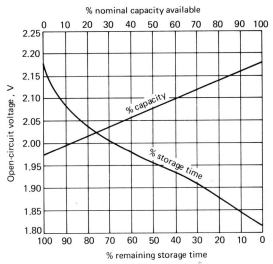

FIG. 15.14 Open-circuit voltage vs. percent remaining storage time.

Figure 15.15 is a curve of the remaining usable capacity in an SLA cell vs. months of storage at various temperatures. This curve is convenient in determining the approximate remaining capacity after a given storage time at a particular temperature.

15.4.8 Life

The life of all rechargeable battery systems is variable, depending upon the type of cycle, environment, and charge to which the cell is subjected during its life. Two basic types of life are considered: cycle life and calendar life.

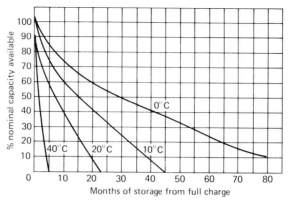

FIG. 15.15   Remaining usable capacity after storage.

Cycle Life   Figures 15.16 and 15.17 show the effect of several of the factors which control cycle life. Figure 15.16 shows the effect of charging voltage and demonstrates the need to select the proper charging voltage for a particular cycle regime. The figure is somewhat misleading, however, in that it would indicate that a low charging voltage,

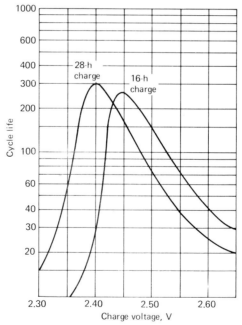

FIG. 15.16   Effect of charge voltage on cycle life at various charge times at 25°C and approx. 100% DOD (end of cycle life: 80% of rated capacity, discharge rate: C/5 to 1.6 V).

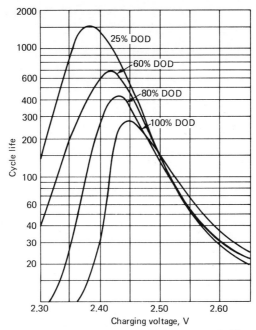

FIG. 15.17   Effect of depth of discharge on cycle life as function of charging voltage, 25°C, 16-h charge.

say 2.35 V, would yield a low cycle life, and this is not true. For example, for an application where the cells would be used about 3 times a week and left on charge the rest of the time, 2.35 V would be quite adequate. Most of the capacity can be returned to the cells within 16 h. The occasional long charge periods would maintain capacity and thus optimize the total cycle life. Generally, 2.45 V per cell is a better charging voltage for regimes of about one cycle per day.

Figure 15.17 shows the general effect of depth of discharge (DOD) on cycle life; typically at 100% DOD, 200 cycles are characteristic. It demonstrates that high cycle life can be achieved by slight oversizing of the battery for the application to reduce the depth of discharge. Figure 15.18 is a curve of capacity vs. cycles for a 2.5-Ah cell (2.35Ah at 5-h rate) cycled at 1 cycle per day at a C/5 discharge rate to 1.6 V per cell and an 18-h charge at 2.5 V constant voltage. The cell takes from 20 to 25 cycles to achieve rated capacity, exceeds rated capacity, and then begins to fall off slowly. The initial increase in capacity is a function of forming the cell.

Float Life   The expected float life of the SLA cell is greater than 8 y at room temperature, arrived at by using accelerated testing methods, specifically, high temperatures.

The primary failure mode of the sealed lead-acid cell can be defined as growth of the positive plate. Because this growth is the result of chemical reactions within the cell, the rate of growth increases with increasing temperature as expressed by the widely accepted Arrhenius equation. In Fig. 15.19, the float life is plotted against temperature. The solid lines represent data from float life tests performed at two float voltages, 2.3

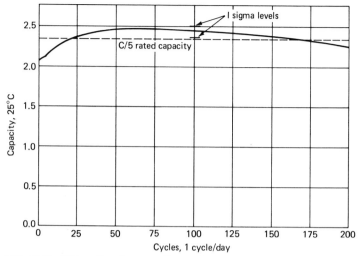

FIG. 15.18 Effect of cycling on cell capacity, 2.5-Ah cell, C/5 discharge rate (rated at 2.35 Ah at C/5 rate).

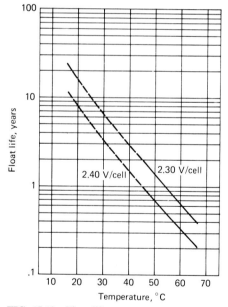

FIG. 15.19 Float life.

and 2.4 V per cell. The lines have been extended through lower temperatures in accordance with the Arrhenius equation. This graph can be used to determine the expected float life at various temperatures. End of life is defined as the failure of the cell to deliver 80% of rated capacity.

15.5 CHARGING CHARACTERISTICS

15.5.1 General Considerations

Charging the sealed lead-acid cell, like charging other secondary cells, is a matter of replacing the energy depleted during discharge. Because this process is somewhat inefficient, it is necessary to return more than 100% of the energy removed during discharge. The amount of energy necessary for recharge depends upon how deeply the cell was discharged, the method of recharge, the recharge time, and the temperature. The overcharge required in the lead-acid cell is associated with generation of gases and corrosion of the positive-grid materials. In conventional lead-acid cells, the gases generated are released from the system, resulting in a loss of water or drying out of the system and subsequent loss of capacity. The SLA cell incorporates the gas recombination principle which allows up to 100% of the oxygen generated at up to the C/3 overcharge rate to be reduced at the negative plate, eliminating oxygen outgassing. Hydrogen gas generation has been substantially reduced by the use of pure lead grid material, which has a high hydrogen overvoltage. The corrosion of the positive grid has been reduced by the use of pure lead. Also, the effects of corrosion of the positive grid have been minimized by the element construction.

Charging can be accomplished by various methods. Constant-voltage charging is the conventional method for lead-acid cells and is also acceptable for the SLA cell. However, constant current, taper current, and variations thereof can also be used.

15.5.2 Constant-Voltage Charging

Constant-voltage charging is the most efficient and the fastest method of charging an SLA cell. Figure 15.20 shows the recharge time at various charge voltages for a cell discharged to 100% of capacity. The charger required to furnish these charge times at given voltages must be capable of at least the 2C rate. If the constant-voltage charger that is used has less than the 2C rate of charge capability, the charge times should be lengthened by the hourly rate at which the charger is limited; i.e., if the charger is limited to the C/10 rate, then 10 h should be added to each of the charge voltage-time relationships; if the charger is limited to the C/5 rate, then 5 h should be added, etc. There are no limitations on maximum current imposed by the charging characteristics of the SLA cell.

Figure 15.21 is a set of curves of charge current vs. time for 2.5-Ah cells charged by constant voltage of 2.45 V with chargers limited to 2-, 1-, and 0.3-A currents. As shown, the only difference in these three chargers is the length of time necessary to recharge the cell.

Figure 15.22 shows the charge rate vs. time and percent of previous discharge capacity returned to the cell vs. time at 2.35-V constant voltage for a cell discharged at the C/2.5 rate to 80% DOD. The time necessary to recharge the cell to 100% of the previous discharge capacity is 1.5 h. The charger used was capable of an output in the 4C range and voltage regulation at the cell terminals of better than 0.1%. The initial high inrush of current caused internal cell heating which enhanced charge acceptance. If the cell had been more deeply discharged, the length of time to reach 100% capacity on charge would have increased. The current decays exponentially with time, and, after 3 h on charge, the current has decayed to a low level. At 2.35 V, the cell will accept whatever current is necessary to maintain capacity.

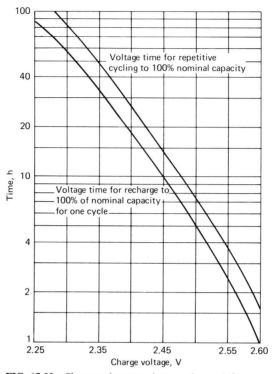

FIG. 15.20 Charge voltage vs. time on charge, 25°C.

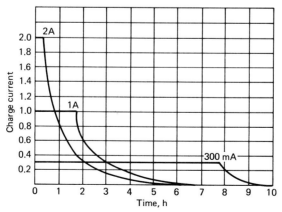

FIG. 15.21 Charge current vs. time at 2.45-V constant voltage with various current limits (2.5-Ah cell, C/10 rate).

15-18

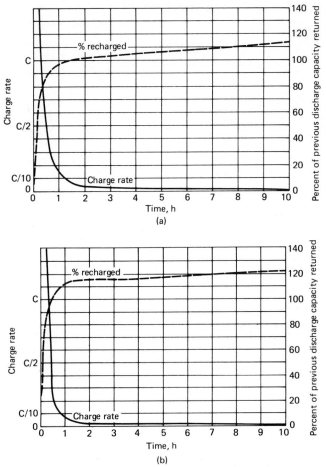

FIG. 15.22   Charge rate and percent recharged vs. time during constant-voltage charge at 25°C. (*a*) 2.35 V; (*b*) 2.50 V.

Figure 15.22*b* is a similar plot but for a cell charged at 2.50-V constant voltage; 100% of the capacity taken out on the previous discharge is returned to the cell in approximately one-half hour.

15.5.3 Fast Charging

A fast charge is defined as a method of charge which will return the full capacity of a cell in less than 4 h. However, many applications require 1 h or less.

Unlike classic parallel-plate lead-acid cells, the SLA cell uses a starved electrolyte system where most of the electrolyte is contained within a highly retentive separator, which then creates the starved plates necessary for homogeneous gas phase transfer. The gassing problem inherent in classic lead-acid cells is not evident with this system, as the oxygen given off on overcharge is able to recombine within the sealed cell. The large surface area of the thin plates used in the SLA system reduces the current density to a level far lower than normally seen in fast charge of classic lead-acid cells, thereby enhancing the fast-charge capabilities.

To study the fast-charge capabilities, a series of tests were run on standard production 6-V, 2.5-Ah and 6-V, 5-Ah Gates batteries. The chargers used were well-regulated, constant-voltage chargers capable of 12 A for charging the 2.5-Ah cells and 25 A for the 5-Ah cells. The cycle regime used was a 3-h cycle: charge for 1 h, rest for 0.5 h, discharge at the 1C rate to 1.6 V, open-circuit the cell, then recharge at the beginning of the third hour. This cycle was carried out through 25 cycles at three different constant-voltage charge voltages, 2.70, 2.55, and 2.35 V per cell.

Figure 15.23 shows the charge rate or the current the SLA cell can accept for the 1-h charge at the three different voltages. The cell has a high charge acceptance at the beginning of the charge time; in fact, in the case of the 2.55-V-per-cell charge, the cell accepted the full current capability of the charger for the first 3 to 4 min. In the case of the 2.7-V-per-cell charge, there was a considerable amount of overcharging starting at 30 min, which caused internal heating and a consequent increase in charge current.

Figure 15.24 is a set of curves of normalized charge efficiency vs. time in minutes for the three different voltages. This efficiency figure was calculated by dividing the total ampere-hour capacity returned by the previous discharge capacity removed. On the 2.55-V charge 100% of the capacity removed on the previous cycle was returned in 15 min. With the 2.7-V charge, a 60% overcharge was returned at the end of the 60-min charge.

Figure 15.25 is a curve of cycles vs. discharge time in minutes for the three charge voltages. Also, a set of reference cells was charged at 2.5-V constant voltage for 16 h and discharged at the 1C rate. This reference curve is displayed to show the expected capacity at the 1C rate. It can be seen from these data that the 2.55-V per cell curve most closely approximates the reference line. The cell charged at 2.7 V per cell received too much overcharge and, therefore, the degradation in capacity after 15 cycles. The cell charged at 2.35 V achieved a value of approximately 75% of the reference and continued to cycle at that level.

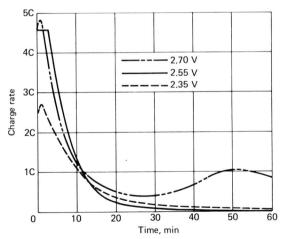

FIG. 15.23 Charge rate vs. time for three charge voltages.

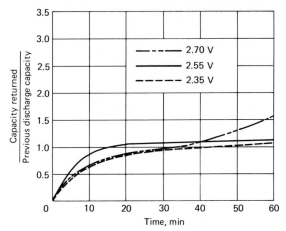

FIG. 15.24   Charge efficiency vs. time for three charge voltages.

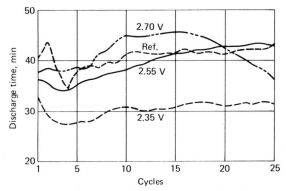

FIG. 15.25   Effect of cycling on discharge time for three charge voltages.

These tests show that the SLA cell can be fast-charged to 100% of rated capacity in less than 1 h. A constant-voltage charger set at 2.5 to 2.55 V per cell and capable of the 3C to 4C rate of charge is preferred.

15.5.4 Float Charging

When the sealed lead-acid cell is to be float-charged in a standby application, the constant-voltage charger should be maintained between 2.3 and 2.4 V for maximum float life. Continuous charging at greater than 2.4 V per cell is not recommended because of accelerated grid corrosion. Figure 15.26 gives the approximate values of voltage a cell will attain when float-charged at 25 and 65°C, or the charge rate a cell will accept, if it has been charged for a sufficient period of time that it is in a state of overcharge equilibrium before these curves can be considered accurate. These curves

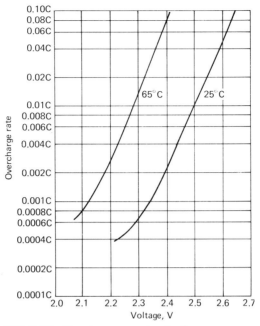

FIG. 15.26 Overcharge current and voltage.

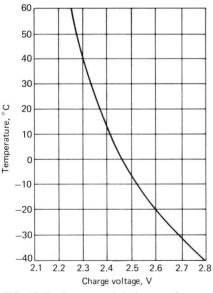

FIG. 15.27 Recommended charge voltage at various temperatures (temperature compensation).

can also be used to determine the approximate value of continuous constant current (trickle charge) which will maintain the proper float voltage. As an example, if a cell were trickle-charged at the 0.001C rate, its average voltage on overcharge would be 2.35 V. Conversely, if a cell were constant-voltage charged at 2.35 V, its overcharge rate would be 0.001C.

High temperatures accelerate the rate of the reactions which reduce the life of a cell. At increased temperatures, the voltage necessary for returning full capacity to a cell in a given time is reduced because of the increased reaction rates within the battery. To maximize life, a negative charging temperature coefficient of approximately -2.5 mV/°C per cell is used at temperatures significantly different from 25°C. Figure 15.27 shows the recommended charging voltage at various temperatures for an SLA cell float-charged at 2.35 V at 25°C. It is obvious from this curve that at extremely low temperatures, a significantly greater temperature coefficient than -2.5 mV/°C is required to achieve full recharge of the cell.

When trickle-charging, it may be necessary to increase the charge rate at higher temperatures to maintain the proper float voltage. From Fig. 15.26, it can be shown that if a cell were trickle-charged at the 0.001C rate at 25°C, then the float voltage would be 2.35 V. However, at the same rate at 65°C, the cell float voltage would be approximately 2.12 V, which is below the open-circuit voltage of the cell. At 65°C, the trickle-charge current would need to be increased to approximately 0.01C to maintain the proper cell float voltage.

15.5.5 Constant-Current Charging

Constant current is another efficient method of charging the SLA cell. Constant-current charging of a cell or battery is accomplished by the application of a non-varying constant-current source. This charge method is especially effective when several cells are charged in series, since it tends to eliminate any charge imbalance in a battery. Constant-current charging charges all cells equally because it is independent of the charging voltage of each cell in the battery. Figure 15.28 is a family of curves of cell voltage vs. percent of capacity of previous discharge returned at different constant-current charging rates. As shown by these curves at different charge rates, the voltage of the cell increases sharply as the full charge state is approached. This increase in voltage is caused by the plates going into overcharge when most of the active material on the plates has been converted from lead sulfate to lead on the negative plate and lead dioxide on the positive plate. The voltage increase will occur at lower states of charge when the cell is being charged at higher rates. This is because at the higher constant-current charge rates the charging efficiency is reduced. The voltage curves in Fig. 15.28 are somewhat different from those for a conventional lead-acid cell due to the effect of the recombination of gases on overcharge within the system. The SLA cell is capable of recombining the oxygen produced on overcharge up to the C/3 rate of constant-current charge. At higher rates, the recombination reaction is unable to continue at the same rate as the gas generation.

While constant-current charging is an efficient method, continued application at rates above C/500, after the cell is fully charged, can be detrimental to the life of the cell. At overnight charge rates (C/10 to C/20), the large increase in voltage at the nearly fully charged state is a useful indicator for terminating or reducing the rates for a constant-current charger. If the rate is reduced to C/500, the cell can be left connected continuously and give 8 to 10 y of life at 25°C. Figure 15.29 is a plot of voltage vs. time for a cell charging at the C/15 rate of constant current at 25°C. This cell had previously been discharged to 100% depth of discharge at the C/5 rate. This curve shows that the cell is not fully charged at the time the voltage increase occurs and must receive additional charging. If a cell is to be charged with constant current at higher than room temperature, then some temperature compensation must be built into the voltage-sensing network. As explained under "Float-charging" above, at higher temperatures and given charging rates the cell voltage on overcharge is reduced. Therefore, the rise in voltage at close to full charge will be somewhat depressed.

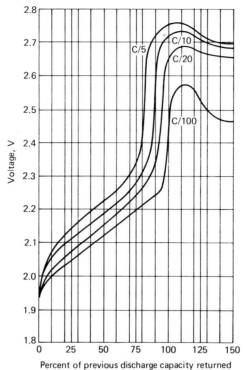

Percent of previous discharge capacity returned

FIG. 15.28 Voltage curves for cells charged at various constant-current rates at 25°C.

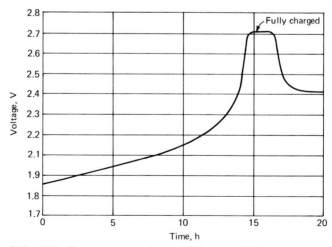

FIG. 15.29 Constant-current charge, C/15 rate at 25°C.

15.5.6 Taper-Current Charging

Although taper-current chargers are among the least expensive types of chargers, their lack of voltage regulation can be detrimental to the cycle life of any type of cell. The SLA cell can withstand charge voltage variations, but some caution in using taper-current chargers is recommended. A taper-current charger contains a transformer for voltage reduction and a half- or full-wave rectifier for converting from alternating to direct current. The output characteristics are such that as the voltage of the battery increases during charge, the charging current decreases. This effect is achieved by use of the proper wire size and the turns ratio. Basically, the turns ratio from primary to secondary determines the output voltage at no load, and the wire size in the secondary determines the current at a given voltage. The transformer is essentially a constant-voltage transformer which depends entirely on the ac line voltage regulation for its output-voltage regulation. Because of this method of voltage regulation, any changes in input line voltage directly affect the output of the charger. Depending on the charger design, the output-to-input voltage change can be more than a direct ratio; e.g., a 10% line voltage change can produce a 13% output-voltage change.

When considering the cost advantage of using a half-wave rectifier vs. a full-wave rectifier in a taper-current charger, it should be noted that the half-wave rectifier supplies a 50% higher peak-to-average-voltage ratio than the full-wave rectifier. Therefore, the total life of the battery for a given average charge voltage can be reduced for the half-wave type of charger because of the higher peak voltages.

There are several charging parameters which must be met. The parameter of main concern is the recharge time to 100% nominal capacity for cycling applications. This parameter can primarily be defined as the charge rate available to the cell when the cell is at 2.2 and 2.5 V. The charge voltage at which approximately 50% of the charge has been returned to the cell at normal charge rates of C/10 to C/20 is 2.2V; the 2.5-volt point represents the voltage at which the cell is in overcharge. Given the charge rate at 2.2 V, the recharge time for a taper-current charger can be defined by the following equation:

$$\text{Recharge time} = \frac{1.2 \times \text{capacity discharged previously}}{\text{charge rate at 2.2 V}}$$

It is recommended that the charge rate at 2.5 V be between C/50 maximum and C/100 minimum to ensure that the battery will be recharged at normal rates and will not be severely overcharged if the charger is left connected for extended time periods.

Figure 15.30 is a set of output voltage vs. current curves for a typical 2.5-Ah cell taper-current charger. The three curves show the change in output with a variation in input voltage from 105 to 130 V ac. This particular charger, at 120-V ac input, will charge a D cell which had been previously discharged to 100% depth of discharge in 30 h by the following equation:

$$\text{Recharge time} = \frac{1.2 \times 2.5}{0.100 \text{ A}} = 30 \text{ h}$$

15.5.7 Parallel/Series Charging

SLA batteries can be charged or discharged in parallel. When more than four strings of cells are paralleled, it is advisable to use steering diodes in both the charge and discharge path. The discharge diodes prevent a battery from discharging into a paral-

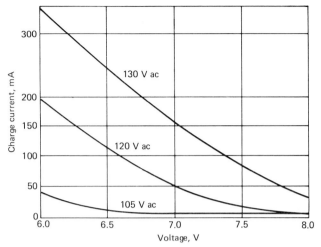

FIG. 15.30   Taper-current charger characteristics.

leled battery should a cell short out in the battery. The charge diodes, in conjunction with the fuse, will prevent a battery with a shorted cell from accepting all the charge current from the charger and subsequently prevent the other paralleled batteries from being fully charged. The fuse should be sized by dividing the maximum charge current by the number of batteries in parallel and multiplying this value times 2. This should result in the fuse opening on charge in a parallel string which has a shorted cell.

When float-charging many cells in series, 12 or more, for example, it is advantageous to use a trickle charge of C/500 maximum in parallel with the float charger. This trickle charge will tend to balance all cells in the battery by driving a continuous trickle charge equally through all cells.

15.5.8 Charge Current Efficiency

Charge current efficiency is the ratio of current which is actually used for electrochemical conversion of the active material from lead sulfate to lead and lead dioxide vs. the total current supplied to the cell on recharge. The current which is not used for charging is consumed in parasitic reactions within the cell such as self-discharge and gas production.

The charging efficiency is excellent for an SLA cell. The distinctly high ratio of plate surface area to ampere-hour capacity allows for higher charging rates and, therefore, efficient charging.

Charge current efficiency is a direct function of state of charge. The charge efficiency of a cell is high until it approaches full charge, at which time the overcharge reactions begin and the charge efficiency decreases. Obviously, past the point of full recharge, the efficiency falls to zero.

Figure 15.31 is a curve of charge current efficiency vs. voltage at various constant voltages. Increasing voltage decreases the efficiency because of increased parasitic currents. The efficiency shows a marked decrease at the open-circuit voltage of 2.18 V, because the charge voltage is not high enough to support the charge reaction and, therefore, the current is only supplying the parasitic currents.

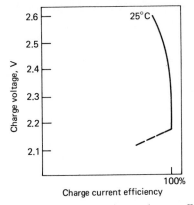

FIG. 15.31 Constant-voltage charge efficiency.

Figure 15.32 is a curve of efficiency vs. log rate at various constant-current charge rates. As can be seen from the curve, at rates up to C/10, the efficiency approaches 100%. At higher rates, there is a decrease in efficiency because, as the cell approaches the fully charged state, the surfaces of the plates become fully charged; this increases the charging reaction rates and results in increased voltages and increased gassing. At low charge rates, the efficiency drops because the charge current is equivalent to the parasitic currents and cell voltage approaches the open-circuit value.

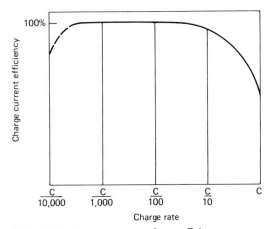

Charge rate

FIG. 15.32 Constant-current charge efficiency.

15.6 SAFETY AND HANDLING

Two primary considerations relative to the application of SLA cells and batteries should be recognized to ensure that their usage is safe and proper: gassing and shorting.

15.6.1 Gassing

Lead-acid batteries produce hydrogen and oxygen gases internally during charging and overcharging. These gases are released in an explosive mixture from conventional lead-acid batteries and therefore must not be allowed to accumulate in a gas-tight container. An explosion could occur if a spark were introduced.

The SLA battery, however, operates on 100% recombination of the oxygen gas produced at recommended rates of charging and overcharging, and so there is no oxygen outgassing. During normal operation, some hydrogen gas and also some carbon dioxide gas are given off. The hydrogen outgassing is essential with each cycle to ensure continued internal chemical balance. The pure-lead grid construction of the SLA battery minimizes the amount of hydrogen gas produced. Carbon dioxide is produced by oxidation of organic compounds in the cell.

The minute quantities of gases which are released from the SLA cell with recommended rates of charge and overcharge will normally dissipate rapidly into the atmosphere. Hydrogen gas is difficult to contain in anything but a metal or glass enclosure; i.e., it can permeate a plastic container at a relatively rapid rate. Because of the characteristics of gases and the relative difficulty in containing them, most applications will allow for their release into the atmosphere. However, if the SLA battery is being designed into a gas-tight container, precautions must be taken so that the gases produced can be released to the atmosphere. If hydrogen is allowed to accumulate and mix with the atmosphere at a concentration of between 4 and 79% (by volume at standard temperature and pressure), an explosive mixture would be present which would be ignited in the presence of a spark or flame.

Another consideration is the potential failure of the charger. If the charger malfunctions, causing higher-than-recommended charge rates, substantial volumes of hydrogen and oxygen will be vented from the cell. This mixture is explosive and should not be allowed to accumulate. Adequate ventilation is required. Therefore, the SLA battery should never be operated in a gas-tight container. The cells should never be totally encased in a potting compound since this prevents the proper operation of the venting mechanism and free release of gas. Furthermore, considerable pressure can build up in a gas-tight container. This can occur during storage because of the continuous generation of carbon dioxide gas. Such pressure is further compounded during charging by the generation of hydrogen.

15.6.7 Shorting

The cells have low internal impedance and thus are capable of delivering high currents if externally shorted. The resultant heat can cause severe burns and is a potential fire hazard. Particular caution should be used when any person working near the open terminals of cells or batteries is wearing metal rings or watchbands. Inadvertently placing these metal articles across the terminals could result in severe skin burns.

15.7 CELL TYPES AND SIZES

A listing of the available SLA cylindrical cells is given in Table 15.2. The performance of these cells, at various current drains at 25°C, is given in Fig. 15.33. A number of multicell batteries are available, using these cells in various series-parallel configurations.

TABLE 15.2   Sealed Lead-Acid (SLA) Cylindrical Cells*

Cell type	Diameter, mm	Length, mm	Weight, g	Capacity, Ah 20-h rate	Capacity, Ah 10-h rate
D	34.2	60.9	180	2.7	2.5
X	44.3	72.4	370	5.2	5.0
J	51.7	123.2	840	13.5	12.5
BC	65.8	163.5	1700	26.0	25.0

*Manufactured by Gates Energy Products, Denver, Colo.

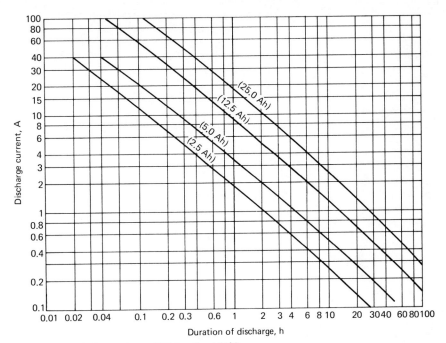

FIG. 15.33 Discharge times of SLA cells at 25°C.

BIBLIOGRAPHY

Battery Applications Manual, Gates Energy Products, Inc., Denver, Colo., 1980.

Bullock, Kathryn R., and Donald H. McClelland: "The Kinetics of the Self-Discharge Reaction in a Sealed Lead-Acid Cell," *J. Electrochem. Soc.* **123**: 327 (March 1976).

Hammel, Ronald O.: "Charging Sealed Lead Acid Batteries," *Proc. 27th Annu. Power Sources Symp.,* Atlantic City, 1976.

———: "Fast Charging Sealed Lead Acid Batteries," Electrochem. Soc. Meeting, Las Vegas, Nev., 1976.

McClelland, Donald H. et al.: U.S. Patent 3,704,173.

———et al.: U.S. Patent 3,862,861, Gates Energy Products, Inc., Denver, Colo.

Mahato, B.K., E.Y Weissman, and E.C. Laird: *J. Electrochem. Soc.* **121**:13 (1974).

Orsino, J.A., H.E. Jensen, John R. Thomas, Donald R. Wolter, and J.P. Mallory: "Maintenance-Free Lead Acid Cells," *Proc. 21st Annu. Power Sources Conf.,* Atlantic City, 1967.

The Sealed Lead Battery Handbook, General Electric Co., Gainesville, Fla., 1979.

Vinal, George Wood: *Storage Batteries,* Wiley, New York, 1955.

Willihnganz, E.: "Accelerated Life Testing of Stationary Batteries," *Electrochem. Tech.,* September/October 1968.

Also see references and bibliography in Part 3, Chap. 14.

16

Vented Pocket Plate Nickel-Cadmium Batteries

by
S. Uno Falk

16.1 GENERAL CHARACTERISTICS

The vented pocket plate battery is the oldest and most mature of the various designs of nickel-cadmium batteries available. It is a very reliable, sturdy, long-life battery which can be operated effectively at high discharge rates and in a wide temperature range. It has very good charge retention properties, and it can be stored for long periods of time in any condition without deterioration. The pocket plate battery can stand both severe mechanical abuse and electrical maltreatment such as overcharging, reversal, and short-circuiting. Little maintenance is needed on this battery. The cost is lower than for any other kind of alkaline storage battery; still, it is higher than that of a lead-acid battery per watthour. The major advantages and disadvantages of this type battery are listed in Table 16.1.

The pocket plate battery is manufactured in a wide capacity range, 5 to more than 1200 Ah, and it is used in a large number of applications. Most of these are of an industrial nature, such as railroad service, switchgear operation, telecommunication, uninterruptible power supply, and emergency lighting. The pocket plate battery is also used in military and space applications. The annual world production of vented pocket plate batteries is of the order of $180 million (1981).

The pocket plate batteries are available in three plate thicknesses to suit the variety of applications. The high-rate designs use thin plates for maximum exposed plate surface per volume of active material. They are used for the highest-rate discharge. The low-rate designs use thick plates to obtain maximum volume of active material per

16-1

TABLE 16.1   Major Advantages and Disadvantages of Vented Pocket Plate Nickel-Cadmium Batteries

Advantages	Disadvantages
Long cycle life	Low energy density
Rugged—can withstand electrical and physical abuse	Higher cost than lead-acid batteries
Reliable—little maintenance required	
Good charge retention	
Excellent long-term storage	
Lowest cost of all alkaline secondary batteries	

exposed plate surface. These types are used for long-term discharge. The medium-rate designs use plates of middle thickness and are suited for applications between, or combinations of, high-rate and long-term discharge.

16.2 CHEMISTRY

The basic idea of the pocket plate design is that of enclosing powdered active material in flat pockets of perforated steel strips and folding these pockets together to form an electrode. The vented pocket plate cell consists of such steel-encased positive nickel hydroxide electrodes, separators or spacers, and similarly steel-encased negative cadmium electrodes, all submerged in a solution of potassium hydroxide in pure water and mounted in a vented cell container of plastic or nickel-plated steel.

The basic electrochemistry is the same as for the vented sintered-plate type as well as other variations of the nickel-cadmium system. The reactions of charge and discharge can be illustrated by the following simplified equation:

$$2NiOOH + 2H_2O + Cd \underset{\text{charge}}{\overset{\text{discharge}}{\rightleftarrows}} 2Ni(OH)_2 + Cd(OH)_2$$

On discharge trivalent nickel hydroxide is reduced to divalent nickel hydroxide under consumption of water. Metallic cadmium is oxidized to form cadmium hydroxide. On charge the opposite reactions take place. The electromotive force (emf) is 1.29 V.

The potassium hydroxide electrolyte is not significantly changed with regard to density or composition during charge and discharge in contrast to the sulfuric acid in lead-acid batteries. The electrolyte density is generally approximately 1.2 g/mL. Lithium hydroxide is often added to the electrolyte for improved cycle life and high-temperature operation.

16.3 CONSTRUCTION

A cutaway view of a modern pocket plate cell is shown in Fig. 16.1. The active material for the positive electrodes consists of nickel hydroxide mixed with graphite for conductivity and additives such as barium or cobalt compounds for improved life and capacity. The active material for the negative electrodes is prepared from cadmium hydroxide or cadmium oxide mixed with iron or iron compounds and sometimes also with nickel.

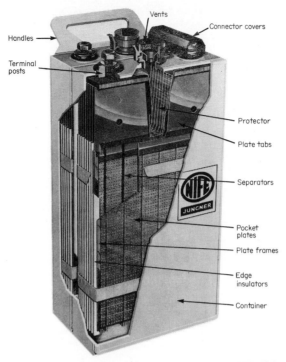

Vents

Connector covers

Handles

Terminal posts

Protector

Plate tabs

Separators

Pocket plates

Plate frames

Edge insulators

Container

FIG. 16.1   Pocket plate cells. (*Courtesy of SAB NIFE AB.*)

The iron and nickel materials are added to stabilize the cadmium and prevent crystal growth and agglomeration. Typical active material compositions are shown in Table 16.2.

Positive and negative electrodes are of the same basic design and are built up of flat pockets of perforated steel strips holding the active materials. The thin steel strips are perforated by hardened steel needles or by a technique using profiled roller dies. The specific hole area is between 15 and 30%. The strips for positive electrodes are nickel-plated to prevent "iron poisoning" of the positive active material.

The active mass is either pressed into briquettes which are fed into the preshaped perforated strip or fed into the preshaped strip as a powder. The upper and lower steel strips are folded together by rollers. A number of these folded strips are arranged to

TABLE 16.2   Typical Composition of Active Materials for Vented Pocket Plate Nickel-Cadmium Batteries in the Discharged State

Positive active material		Negative active material	
Substance	Weight %	Substance	Weight %
Nickel (II) hydroxide	80	Cadmium hydroxide	78
Cobalt (II) hydroxide	2	Iron	18
Graphite	18	Nickel	1
		Graphite	3

interlock with each other to form long electrode sheets, which are then cut to electrode blanks. Electrodes are made from these blanks by providing them with steel frames for mechanical stability and for current takeoff. The electrodes are made with different thicknesses (1.5 to 5 mm) to provide cells for high, medium, and low discharge rates. The negative plate is always thinner (30 to 40%) than the positive.

The electrodes are bolted or welded to electrode groups. Plate groups of opposite polarity are intermeshed and electrically separated from each other by plastic pins and plate edge insulators. Sometimes separators of perforated plastic sheets or plastic ladders are used between the electrodes. The distance between plates of different polarity in an element may vary from less than 1 mm for high-rate cells to 3 mm for low-rate cells.

The elements are inserted into cell containers of plastic or of nickel-plated steel. Plastic containers are made from polystyrene or, more recently, from polypropylene. Important advantages of plastic containers over steel containers are that they allow visual control of the electrolyte level and that they require no protection against corrosion. Also, they have lower weight and they can be more closely packed in the battery. The main drawbacks are that they are more sensitive to high temperatures and they require somewhat more space than steel containers. A pocket plate cell in a steel container is shown in Fig. 16.2.

FIG. 16.2   Pocket plate cell in steel container. (*Courtesy of Chloride-Alcad Ltd.*)

Cells are assembled into batteries in many different ways. Often 2 to 10 cells are mounted to a separate battery unit, several of which may be used to form the complete battery. A typical battery is shown in Fig. 16.3. Cells in plastic containers are also assembled into batteries by putting the single cells close together on a rack or a stand and connecting them with intercell connectors. This is especially the case for stationary applications (Fig. 16.4). Cells in steel containers can be assembled in a similar way; however, here the cells must be spaced from each other and insulated from the rack.

16.4 PERFORMANCE CHARACTERISTICS

16.4.1 Energy Density

Typical energy density values for pocket plate cells are 20 Wh/kg and 40 Wh/L, with the best values for commercially available cells reaching 27 Wh/kg and 55 Wh/L.

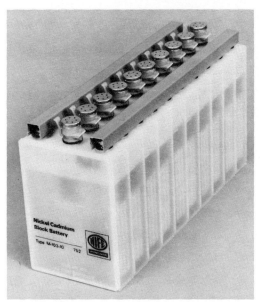

FIG. 16.3 Ten-cell welded polypropylene unit. (*Courtesy of SAB NIFE AB.*)

FIG. 16.4 Typical rack assembly of cells in plastic containers. (*Courtesy of McGraw-Edison Co.*)

Corresponding values for complete pocket plate batteries are 19 Wh/kg and 32 Wh/L and 27 Wh/kg and 44 Wh/L, respectively. These energy density data are based on the nominal capacity of the cells and the average discharge voltage at the 5-h rate.

16.4.2 Discharge Properties

The nominal voltage of a nickel-cadmium cell is 1.2V.

Although discharge rate and temperature are of importance for the discharge characteristics of all electrochemical cells, these parameters have a much smaller effect on the nickel-cadmium cell than on, for instance, the lead-acid battery. Thus, pocket plate nickel-cadmium batteries can be effectively discharged at high discharge rates without losing very much of the rated capacity. They can also be operated over a wide temperature range.

Typical discharge curves at room temperature for high-rate and medium-rate cells at various constant discharge rates are shown in Fig. 16.5. Even at a discharge current as high as $5 \times C$ (where C is the numerical value of the capacity in Ah), a high-rate pocket plate cell can deliver 60% of the rated capacity. Cell capacities as a function of discharge rate and cutoff voltage are given in Fig. 16.6.

Pocket plate nickel-cadmium cells can be used at temperatures down to $-20°C$ with the standard electrolyte. Cells filled with a more concentrated electrolyte can be used down to $-50°C$. Figure 16.7 shows the effect of temperature on relative performance of a high-rate cell. Pocket plate cells can also be used at elevated temperatures. Although occasional operation at very high temperatures is not detrimental, 45 to 50°C is generally considered as the maximum permissible temperature for extended periods of operation.

Figure 16.8 shows typical so-called starter curves for high-rate cells. The cells can deliver as much current as $20 \times C$ A during 1 s to a final voltage of 0.6 V. Figure 16.9 shows the corresponding curves at 0 and $-18°C$.

Occasional overdischarge or reversal of pocket plate cells is not detrimental, nor is complete freezing of the cells. After warming up, the cells will function normally again.

16.4.3 Internal Resistance

Pocket plate cells generally have a low internal resistance. Typical dc resistance values are 0.4, 1, and 2 mΩ, respectively, for a charged 100-Ah high-, medium-, and low-rate cell. The internal resistance is largely inversely proportional to the cell size in a given cell series. Decreasing temperature and decreasing state of charge of a cell will result in an increase of the internal resistance.

16.4.4 Hours of Service

The hours of service on discharge that a vented pocket plate nickel-cadmium cell (high-rate type), normalized to unit weight (kilogram) and size (liter), will deliver at various discharge rates and temperatures, is summarized in Fig. 16.10. The curves are based on a capacity, at the C/5 rate at 20°C, of 16.5 Ah/kg and 33 Ah/L.

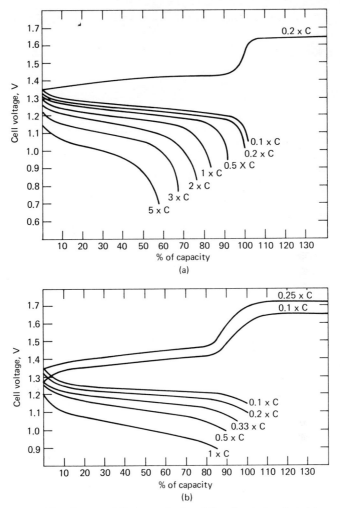

FIG. 16.5   Charge and discharge characteristics of vented pocket plate nickel-cadmium cells at various constant rates at 25°C. (*a*) High-rate cells (SAB NIFE type H); (*Part a courtesy of SAB NIFE AB.*) (*b*) medium-rate cells (Chloride-Alcad types RV and RVP).

16.4.5 Charge Retention

Charge retention characteristics of vented pocket plate cells at 25°C are shown in Fig. 16.11. Charge retention is temperature dependent, the capacity loss at 45°C being about 3 times higher than at 25°C. There is virtually no self-discharge at temperatures lower than −20°C.

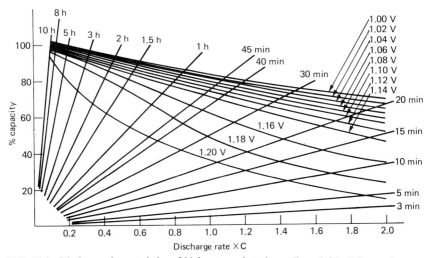

FIG. 16.6   Discharge characteristics of high-rate pocket plate cells at 25°C. Cell capacity as a function of discharge rate and cutoff voltage (SAB NIFE type H). (*Courtesy of SAB NIFE AB.*)

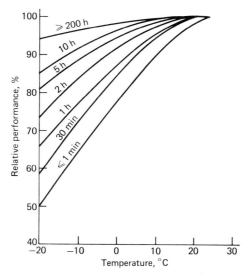

FIG. 16.7   Effect of temperature on relative performance of high-rate pocket plate cells. (*Courtesy of NIFE Inc.*)

16.4.6   Life

The life of a battery can be given either as the number of charge and discharge cycles that can be delivered or as the total lifetime in years. Under normal conditions, a pocket plate battery can reach more than 2000 cycles. The total lifetime may vary between 8 and 25 y or more, depending on the application and the operating conditions. Batteries

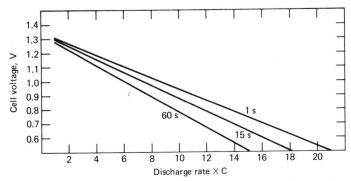

FIG. 16.8   Voltage-current curves for high-rate pocket plate cells at 25°C.

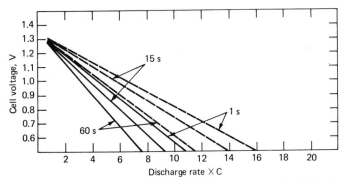

FIG. 16.9   Voltage-current curves for high-rate pocket plate cells, 0°C (- - -) and −18°C (—).

for diesel engine cranking normally last about 15 y. Batteries for train lighting have a normal life of 10 to 15 y, and stationary standby batteries have a life of 15 to 25 y.

Factors which affect the battery life are the operating temperature, the discharge depth, and the charging regime. Low or moderate operating temperatures should always be preferred. Cells operating at elevated temperatures or in cycling applications should be filled with electrolyte to which lithium hydroxide has been added.

The factors behind the excellent reliability and very long life of the pocket plate batteries are the mechanically strong design and the absence of corrosive attack of the electrolyte on the electrodes and other components in the cell and, furthermore, the ability of the battery to stand electrical abuse such as reversal or overcharging and to stand longtime storage in any state of charge.

16.4.7   Mechanical and Thermal Stability

Pocket plate cells and batteries are mechanically very robust and can stand severe mechanical abuse and rough handling in general. The electrode groups are carefully bolted or, in more recent designs, welded together. The cell containers are made of steel

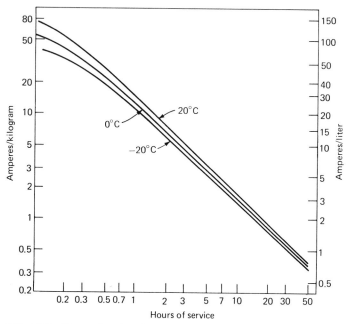

FIG. 16.10   Hours of service of the vented pocket plate nickel-cadmium cell (high-rate type).

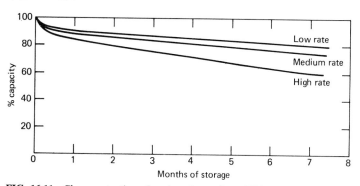

FIG. 16.11   Charge retention of pocket plate cells at 25°C.

or high-impact plastics. The cells in steel containers are mounted in wooden crates or in steel cradles.

The electrolyte does not attack any of the components in the cell, and, accordingly, there is no risk of decreased strength during the lifetime of the battery. Cases of so-called sudden death due to corroded lugs or pole bolts cannot occur.

The thermal resistance of the pocket plate batteries is also very good. These batteries can stand temperatures up to 70°C or more without mechanical damage. Cells in polypropylene or steel containers are the best in this respect. Saline or corrosive environments present no problems for cells in plastic containers.

16.4.8 Memory Effect

The so-called memory effect—the tendency of a battery to adjust its electrical properties to a certain duty cycle to which it has been subjected for an extended period of time —has been a problem with nickel-cadmium batteries in some applications. However, pocket plate batteries do not develop this memory effect under any conditions. Only sintered-type nickel-cadmium cells suffer from this phenomenon.

16.5 CHARGING CHARACTERISTICS

Pocket plate nickel-cadmium cells may be charged at constant current, constant potential, or modified constant potential. Constant-current charge characteristics are shown in Fig. 16.5. Charging is normally carried out at the 5-h rate for 7 h for a fully discharged battery. Overcharging is not detrimental but should be avoided for it leads to increased gassing and decomposition of water. Charging can be carried out in the temperature range -50 to $+45°C$. However, at the extreme temperatures the charging efficiency is lower.

Constant-potential charging characteristics with current limitation are shown in Fig. 16.12. The current is often limited to 0.1 to $0.4 \times C$ amperes, and charging is normally carried out in the voltage range of 1.50 to 1.65 V per cell. The charging time may vary from 5 to more than 25 h, depending on current limitation value and cell type.

In some applications such as emergency and standby, it is necessary to keep the battery in a high state of charge. A convenient way is to connect the battery in parallel with the ordinary current source and the load and to float the battery at 1.40 to 1.45 V per cell. The floating may be combined with a supplementary charge at fixed intervals or after each discharge.

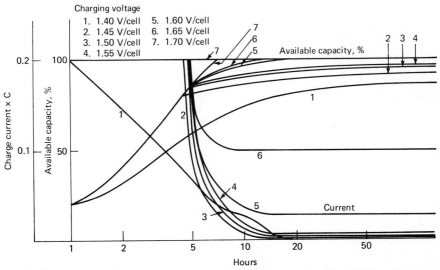

FIG. 16.12   Constant-potential charging with current limitation $0.2 \times C$ of medium-rate pocket plate cells at 25°C.

The ampere-hour efficiency of the pocket plate cell is 72% when going from the discharged to the fully charged state. The watthour efficiency is approximately 60%.

16.6 MANUFACTURERS AND CELL SIZES

Table 16.3 contains data regarding prominent manufacturers of pocket plate nickel-cadmium batteries, trademarks, capacity ranges, and performance types. Table 16.4 lists the characteristics of typical pocket plate cells.

TABLE 16.3 Prominent Manufacturers of Vented Pocket Plate Nickel-Cadmium Batteries and Cell Characteristics

		Cell characteristics		
Manufacturer	Trademark	Performance type	Capacity range, Ah	Cell container material
NIFE Inc. Lincoln, R. I.	NIFE, NIFE JUNGNER	High rate	8.5–570	Plastic, steel
		Medium rate	8.5–1180	Plastic, steel
		Low rate	7.5–1245	Plastic, steel
McGraw-Edison Company, Edison Battery Division, Bloomfield, N.J.	EDISON, Americad	High rate	10–180	Plastic
		Medium rate	10–400	Plastic
		Low rate	16–250	Plastic
SAB NIFE AB, Oskarshamn, Sweden	NIFE, NIFE JUNGNER	High rate	8.5–570	Plastic, steel
		Medium rate	8.5–1180	Plastic, steel
		Low rate	7.5–1245	Plastic, steel
Chloride-Alcad, Ltd. Redditch, England	CHLORIDE ALCAD	High rate	11–900	Plastic, steel
		Medium rate	16–1040	Plastic, steel
		Low rate	5–285	Plastic
VARTA Batterie AG, Hannover, Germany (BDR)	VARTA	High rate	7.5–650	Plastic, steel
		Medium rate	5–1250	Plastic, steel
Société des Accumulateurs Fixes et de Traction (SAFT), Romainville, France	SAFT	High rate	14–310	Plastic, steel
		Medium rate	10–520	Plastic, steel
		Low rate	80–360	Steel
Yuasa Battery Co., Ltd., Tokyo, Japan	YUASA	High rate	20–250	Plastic
		Medium rate	10–900	Plastic
		Low rate	10–800	Plastic

TABLE 16.4 Typical Vented Pocket Plate Nickel-Cadmium Cells, Low Rate

	Dimensions, cm			
Capacity, Ah	Width	Height	Length	Weight, kg
80	16.2	31.1	8.6	5.9
120	16.2	31.1	11.9	8.6
160	16.2	46.4	8.6	9.5
240	16.2	46.4	11.9	13.6
340	16.2	46.4	17.1	20.0
400	16.2	46.4	17.1	20.9

16.7 APPLICATIONS

Because of their favorable electrical properties, long life, excellent reliability, and low maintenance requirements, vented pocket plate nickel-cadmium batteries are used in a large variety of applications. Many of these applications are of an industrial nature, although this type of battery is also used for commercial and military applications.

Stationary use is perhaps the largest application for pocket plate batteries. The battery functions as a reserve or emergency power source and operates when there is a power failure. The battery will take over the load immediately and maintain the service until the normal power supply is restored.

Pocket plate batteries of high- or medium-rate type are often used to take care of short peak loads when closing and tripping breakers in switchgear in electric power and transformer stations. These stations very often have considerable auxiliary electrical equipment for control and protection.

Another important stationary use is emergency lighting. This is needed in many public buildings such as department stores, restaurants, hospitals, and theaters as well as in some outdoor locations. A minimum of maintenance is required for emergency lighting applications. Here pocket plate cells of the low- or medium-rate type are very often used. For these applications special types of batteries have been developed for very long maintenance periods (up to 10 y).

In some cases the energy required is too high for batteries; then an emergency generator, driven by a diesel engine, will be used. High-rate pocket plate batteries are used for the automatic starting of such an engine in the event of power failure.

A novel application is inverter service. Inverters are increasingly required for emergency power supply for electronic equipment such as computers or microwave stations. Here the battery is the power source in case of failure of the primary power supply.

Other stationary applications are telephone exchanges, amplification stations, and solar generator systems.

A very large field for pocket plate batteries is railway service, where these batteries are used for train lighting, air conditioning, diesel engine cranking, railcars, and signaling. The trend here is to use plastic monoblock batteries. These monoblocks have a very strong plastic casing. They are often equipped with flame-arresting vents, they are corrosion-free, and they are provided with a so-called dead top. A related application is in trams, subways, and trolley buses. Pocket plate batteries are also used on board ships for feeding vital equipment in case of primary power failure.

Examples of portable applications are signal lamps, hand lamps, telecommunication sets, searchlights, and portable instruments. Here, pocket plate batteries are used in the larger devices, whereas sealed nickel-cadmium batteries dominate the smaller ones.

The *cost* of the pocket plate nickel-cadmium battery is the lowest of all alkaline storage batteries; but in comparison with a lead-acid battery on a cost per watthour basis, the nickel-cadmium battery is considerably more expensive. However, a pocket plate nickel-cadmium battery is often used in applications where reliability, long life, and low maintenance are of paramount importance. In many uses it is possible to employ a much smaller pocket plate nickel-cadmium battery than a lead-acid battery due to the superior high-rate properties of the pocket plate battery. A typical example is locomotive diesel engine cranking where a pocket plate battery with only half of the capacity of a lead-acid battery can be used. Therefore it is important to calculate the cost not for the watthours of the battery but for the actual service that the battery will give.

Typically the cost of a low-rate pocket plate battery is 30 to 40 cents per watthour. The corresponding figure for a high-rate battery is 70 to 90 cents per watthour (1979).

BIBLIOGRAPHY

Barak, M (ed.): *Electrochemical Power Sources,* Peter Peregrinus, London, 1980.

Falk, S. U., and A. J. Salkind: *Alkaline Storage Batteries,* Wiley, New York, 1969.

Kinzelbach, R.: *Stahlakkumulatoren,* Varta, Hannover, Germany, 1968.

Miyake, Y., and A. Kozawa: *Rechargeable Batteries in Japan,* JEC Press, Cleveland, Ohio, 1977.

17

Vented Sintered-Plate Nickel-Cadmium Batteries

by
John M. Evjen and Arthur J. Catotti

17.1 GENERAL CHARACTERISTICS

The sintered-plate nickel-cadmium battery is a more recent development of the nickel-cadmium system having a higher energy density, up to 50% greater than the pocket-type construction. In addition, as the sintered plate can be constructed in a much thinner form than the pocket plate, the cell has a much lower internal resistance and gives superior high-rate and low-temperature performance. A flat discharge curve is characteristic of the battery, and its performance is less sensitive than other battery systems to changes in discharge load and temperature. The sintered-plate battery has most of the favorable characteristics of the pocket-type battery, although it is generally more expensive. It is electrically and mechanically rugged, is very reliable, requires little maintenance, can be stored for long periods of time in a charged or uncharged condition, and has good charge retention. Cells losing capacity through self-discharge can be restored to full service with normal charge. The major advantages and disadvantages of this battery type are given in Table 17.1.

For these reasons, vented sintered-plate nickel-cadmium batteries are used in applications requiring high-power discharge service such as aircraft turbine engine and diesel engine starting as well as other mobile and military equipment. The battery provides outstanding performance where high peak power and fast recharging are required. In many applications, the vented sintered-plate battery is used because it leads to a reduction in size, weight, and maintenance as compared with other battery systems.

TABLE 17.1   Major Advantages and Disadvantages of Vented Sintered-Plate Nickel-Cadmium Batteries

Advantages	Disadvantages
Flat discharge profile	High cost
Higher energy density (50% greater than pocket-plate)	"Memory" effect
	Controlled charging system required to prevent "thermal runaway"
Superior high-rate and low-temperature performance	
Rugged, reliable, little maintenance required	
Excellent long-term storage	
Good charge retention—capacity can be restored after self-discharge	

This is particularly true in systems subject to low-temperature operation. The rise in battery voltage, at the end of charge, of the vented cell also provides a useful characteristic for controlling the charge.

17.2 CHEMISTRY

Vented sintered-plate nickel-cadmium cells consist of flat positive nickel hydroxide and negative cadmium plates, separated by materials which act as a gas barrier and as an electrical separator. The electrolyte, normally a 31% potassium hydroxide solution, completely covers the plates and separators; for this reason vented cells are sometimes referred to as "flooded cells."

In the sintered-plate design, the active materials are held within the pores of a sintered-nickel structure; nickel hydroxide is the active material of the positive plate while cadmium hydroxide is the active material of the negative plate.

During charge, the nickel hydroxide in the positive electrode is oxidized to the higher-valence NiOOH

$$2Ni(OH)_2 + 2OH^- \rightarrow 2NiOOH + 2H_2O + 2e$$

The cadmium hydroxide in the negative electrode is reduced to metallic cadmium during charge

$$Cd(OH)_2 + 2e \rightarrow Cd + 2OH^-$$

The overall charge-discharge reaction is

$$2Ni(OH)_2 + Cd(OH)_2 \underset{discharge}{\overset{charge}{\rightleftarrows}} 2NiOOH + Cd + 2H_2O$$

The electrolyte concentration does not change during discharge; specific gravity cannot be used as a measure of the state of charge.

The positive electrode accepts charge at close to the theoretical rate up to about 80% of full charge. As full charge is approached, oxygen is also produced at the positive electrode

$$4OH^- \rightarrow 2H_2O + O_2 + 4e$$

As shown in Fig. 17.1, the cell voltage increases about 40 mV from 20 to 80% of full charge at C/10, due to the positive electrode becoming more positive. During overcharge all the charge current is utilized to generate oxygen at the positive electrode.

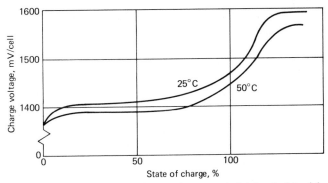

FIG. 17.1   Constant-current charge voltage—vented sintered-plate nickel-cadmium cell, C/10 charge rate at 25°C.

The negative electrode accepts charge at the theoretical rate until it is essentially 100% charged, at which time hydrogen is evolved

$$4H_2O + 4e \rightarrow 2H_2 + 4OH^-$$

The hydrogen overvoltage on the cadmium electrode is quite high, about 110 mV at C/10 rate; consequently there is a sharp rise in voltage as the negative electrode goes into overcharge. This rise in voltage is used in various charging schemes to control or terminate charging.

During overcharge all the current is used to electrolyze water to hydrogen and oxygen as shown in the overall reaction

$$2H_2O \rightarrow 2H_2 + O_2$$

This overcharge reaction consumes water and thereby decreases the level of electrolyte in the cell. The water loss can be limited by controlling the amount of overcharge, so as to maximize the interval between needed water replenishments.

17.3 CONSTRUCTION

Vented cells are designed so that both electrodes reach full charge at about the same time. The positive electrode, as noted above, will begin to evolve oxygen before it is fully charged. If this gas is allowed to reach the negative electrode, it will recombine and generate heat. This will not only prevent the negative from reaching a full state of charge but also result in reduced voltage due to depolarization of the cadmium electrode. To maintain the fullest capability, adequate precautions must be taken to prevent oxygen recombination at the negative plate. This is accomplished by providing a gas barrier between the positive and negative plates and by flooding the plates with excess electrolyte.

Figure 17.2 shows details of a typical vented sintered-plate nickel-cadmium cell.

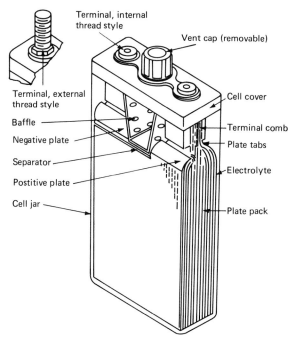

FIG. 17.2   Cross section of a vented sintered-plate nickel-cadmium cell.

17.3.1 Plates and Processes

A variety of plate constructions are in common use in vented sintered nickel-cadmium cells produced by different manufacturers. The plates differ according to the manufacturer in the nature of the substrate, method of sintering, impregnation process, formation, and termination technique.

Substrate   The substrate serves as a mechanical support for the sintered structure and as a current collector for the electrochemical reactions which occur throughout the plate. It also provides mechanical strength and continuity during the manufacturing processes. Two types of substrate are typically used: (1) perforated nickel-plated steel or pure nickel strip in continuous lengths, and (2) woven screens of nickel or nickel-clad steel wire.

A common perforated type may be 0.1 mm thick with 2-mm holes and a void area of about 40%. A typical screen may use 0.18-mm diameter wire with 1.0-mm openings.

Plaque   The sintered structure before impregnation is generally referred to as "plaque." It usually has a porosity of from 80 to 85% and ranges in thickness from 0.40 to 1.0 mm. Two generic sintering processes are used: (1) the slurry coating process and (2) the dry-powder process. Both processes employ special low-density battery grades of carbonyl nickel powder.

In the slurry coating process the nickel powder is suspended in a viscous, aqueous solution containing a low percentage of a thixotropic agent. The nickel-plated strip with the desired perforated pattern is pulled through the suspension. The thickness is controlled by passing it through doctor blades, while wiping the edges free of slurry. The continuous strip is then dried before sintering in a reducing atmosphere at about 1000°C.

The dry-powder processes generally employ wire screen precut to the so-called master plaque dimension. The screens are placed into molds with loose powder on each side. They are then typically sintered in a belt furnace in a reducing atmosphere at 800 to 1000°C.

Impregnation   Various impregnation processes[1-5] are used to load the porous sintered-nickel structure of the positive with nickel hydroxide and of the negative with cadmium hydroxide. Most processes involve a chemical precipitation and with minor variations follow in principle the original process described in 1948.[6] The plaque is impregnated with a concentrated solution of the nitrate, dried to remove most of the water, the hydroxide is precipitated with caustic, and the plaque is washed and dried. This sequence of steps is repeated a number of times so as to fill about 40 to 60% of the pore volume.

Plate Formation   Following impregnation, the plates are mechanically brushed and electrochemically cleaned and formed by charging and discharging the electrode. In the master-plaque process they are formed against inert counterelectrodes or in temporary cells. In the case of the continuous-strip process the formation is done on a machine similar in appearance to a continuous-strip electroplating machine.

Plates blanked from the continuous strip have a clean, wiped area at the top which serves as attachment points for nickel or nickel-plated steel current-collector tabs. In the case of the master-plaque process, a coined or densified area is provided for attachment of these collector tabs.

17.3.2 Separator

The separator system is a thin, multilayered combination of a cloth that electrically separates the positive and negative plates and an ion-permeable plastic membrane that serves as the gas barrier.

Electrical and mechanical separation of the plates is typically provided by either woven or felted nylon material. This material is relatively porous in order to provide sufficient cross section of ionic conduction path through the electrolyte.

The ion-permeable plastic membrane, typically cellophane, is utilized as the gas barrier while at the same time it offers minimum ionic resistance. This thin gas barrier, which becomes relatively soft when wetted, is frequently placed between two layers of the nylon separator and receives significant mechanical support from them.

17.3.3 Plate Pack Cell Assembly

Plate packs are assembled by alternating positive and negative plates with the separator–gas barrier system interleaved between them. The cell terminals are bolted or welded to the current-collector plate tabs.

17.3.4 Electrolyte

Potassium hydroxide electrolyte is used so as to yield a finished concentration of approximately 31% at full charge (specific gravity of 1.30). Performance of the cell, particularly at low temperatures, is significantly dependent on this concentration.

17.3.5 Cell Container

The plate pack is placed into the cell container with the cell terminals extending through the cover. The cell container is usually made of a low-moisture-absorbent nylon and consists of the cell jar and matching cover that are permanently joined together at assembly by solvent sealing, thermal fusion, or ultrasonic bonding. The container is designed to provide a sealed enclosure for the cell, thus preventing electrolyte leakage or contamination, as well as providing physical support for the cell components. The terminal seal is generally provided by means of O rings with Belleville washers and retaining clips.

17.3.6 Vent Cap and Check Valve

The vent cap serves both as a removable plug to provide the access required for replenishment of water to the electrolyte and also as a check valve to release gases generated when water is consumed during overcharge. The check valve prevents atmospheric contamination of the electrolyte. It consists of a nylon body with a hollow center post, around which is placed an elastomeric sleeve which functions as a Bunsen valve to allow gas to escape from the cell but not to enter.

17.4 PERFORMANCE CHARACTERISTICS

17.4.1 Discharge Properties

The discharge curves for a typical vented sintered-plate nickel-cadmium cell at various constant-discharge loads and temperatures are shown in Fig. 17.3. The curves for this cell are characterized by a flat discharge, even at relatively high discharge rates and at low temperatures. Cell voltages at various constant-current discharge loads and state of discharge are given in Fig. 17.4.

The battery, because of its low internal resistance, is capable of delivering pulse currents as high as the 20- to 40-C rate; for this reason, it can be used successfully for very high power applications, such as engine starting (see Sec. 17.4.3).

17.4.2 Factors Affecting Capacity

The total capacity which is deliverable by the fully charged sintered-plate vented cell is dependent on both discharge rate and cell temperature, although the sintered-plate cell is less sensitive to these variables than most other battery systems. The relationships of capacity with discharge load and temperature are shown in Figs. 17.5 and 17.6, respectively. The capacity derating is observed to be a relatively linear relationship with the logarithm of increasing discharge rate, provided that the capacity is measured to an end of discharge voltage (EODV) which is below the "knee" of the curve.

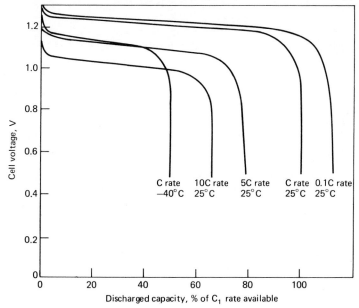

FIG. 17.3 Typical discharge curves.

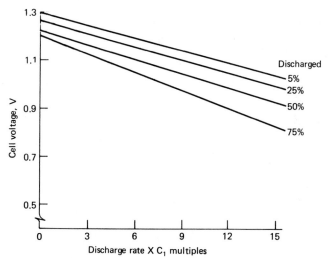

FIG. 17.4 Voltage as a function of discharge load and at various states of charge.

Low-temperature performance is enhanced by the use of eutectic 31% KOH (1.30 specific gravity) electrolyte which freezes at $-66°C$. Lower concentrations will freeze at higher temperatures; e.g., 26% KOH freezes at $-42°C$. As shown in Fig. 17.6, more than 60% of the 25°C capacity is available at $-35°C$, with the temperature having an increasingly significant effect as it is reduced toward $-50°C$. At high discharge rates,

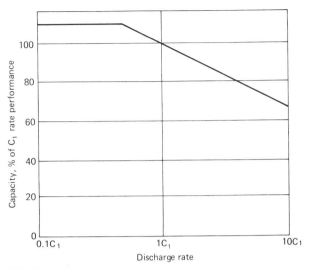

FIG. 17.5 Capacity derating as a function of discharge rate at 25°C.

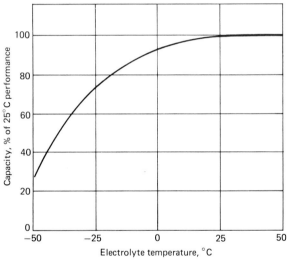

FIG. 17.6 Capacity derating as a function of discharge temperature at C rate of discharge.

heat that is generated may cause the battery to warm up, giving higher performance than would be expected under the ambient conditions.

Vented sintered-plate batteries can be discharged at any elevated temperature. They should not be charged, however, at temperatures above 70°C. As with most chemically based devices, exposure to elevated temperatures for extended periods of time will detract from the life of the battery (see Sec. 17.7.3).

The combined effects of both increased discharge rate and low temperature may be approximated by multiplying the two derating factors.

17.4.3 Variable-Load Engine-Start Power

The most common and demanding use of the vented sintered-plate nickel-cadmium battery is as the power source for starting turbine engines on board aircraft. The discharge in this application occurs at relatively high rates for periods of 15 to 45 s. Typically the load resistance when the start is initiated, particularly in low-temperature marginal-start situations, is of the same order of magnitude as the effective internal resistance of the battery (R_e). The apparent load resistance increases as the engine rotor gradually comes up to speed. This results in a typical discharge current which slowly decreases from some high initial value while the voltage recovers from an initial drop of perhaps 50% or more, back toward 1.2 V per cell effective zero load voltage. The general form of the battery-starter voltage and current, expressed as a function of time for this application, is shown in Fig. 17.7.

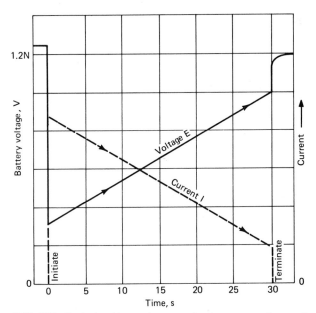

FIG. 17.7   Typical turbine engine start—battery-starter voltage and current as a function of time (N = number of cells).

Expressed as the voltage-current output characteristic of a typical battery, the turbine-starting application is shown as in Fig. 17.8. It will be noted from the two dashed lines that the discharge voltage and current may be considered to always lie on some "instantaneous load regulation line" originating at approximately 1.2N V at zero load current (where N is the number of cells in the battery). Since this instantaneous load regulation line may be considered as linear, only one point, other than the fixed 1.2N V point, is required to describe it. It is convenient to use, for this descrip-

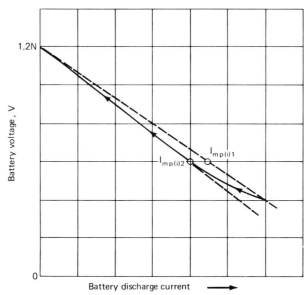

FIG. 17.8 Typical turbine engine start—battery voltage as a function of load current (N = number of cells).

tor point, that value of load current along the straight line at which the battery voltage would be 0.6N V, or one-half of the effective OCV. This value of load current, following application of load to the battery, is designated as instantaneous maximum power current or $I_{mp}(i)$, and its value tends to behave exponentially as shown in Fig. 17.9.

If the effects of the initial transient are disregarded as conservatively negligible, only the value of $I_{mp}(i)$ following 10 s of simulated engine start discharge need be considered. This "steady-state" value is designated I_{mp}. Simulation of it is shown in Fig. 17.10 wherein the load is adjusted continuously to produce the voltage profile shown as a function of time. The voltage in this simulation is controlled so as to represent the most severe load that might be usefully applied to a specific battery at a specific temperature condition. The 0.6N V condition is scheduled to occur at approximately 10 s into the turbine-starting sequence—the simulated light-off point.

The discharge current under this controlled load condition, at that moment when the discharge voltage equals 0.6N V, may thus be conservatively considered as the characteristic I_{mp} available from a specific battery at specific conditions of electrolyte temperature and state of charge. An approximation of I_{mp} may also be measured by performing a "constant potential" discharge at 0.6N V for 15 s.

The maximum power delivery P_{mp} and the effective internal resistance R_e are related to the value of I_{mp} as follows:

$$P_{mp} = 0.6N\, I_{mp}$$

and

$$R_e = \frac{0.6N}{I_{mp}}$$

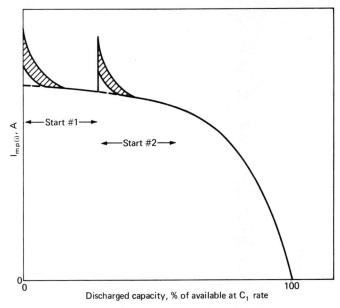

FIG. 17.9 Maximum power current $[I_{mp}(i)]$ as a function of discharged capacity.

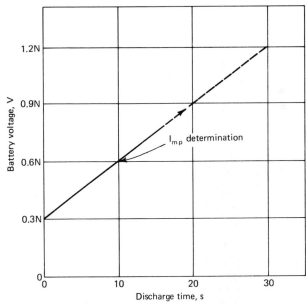

FIG. 17.10 Simulated turbine engine start—battery voltage controlled as a function of discharge time.

17.4.4 Factors Affecting Maximum Power Current

The value of I_{mp}, which a cell (battery) is capable of delivering, is maximum at full charge and at 25°C electrolyte temperature. Derating effects due to state-of-charge and electrolyte temperature factors are shown in Figs. 17.11 and 17.12. It will be noted that both relationships are nonlinear in that the effects on maximum power delivery, per unit of change, increase with decreasing state of charge and with decreasing temperature.

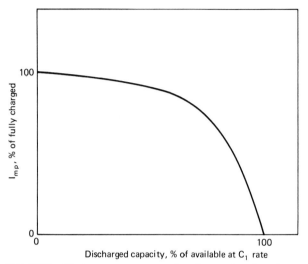

FIG. 17.11   Maximum power current derating as a function of state of charge at 25°C.

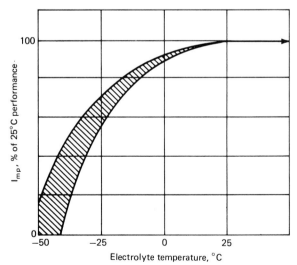

FIG. 17.12   Maximum power current derating as a function of discharge temperature (fully charged cell).

As with capacity, the approximate effect of combined low electrolyte temperature and decreased state of charge may be determined by multiplying the individual factors. It should be noted, however, that high-rate discharge at low temperature may increase the battery temperature to produce one effect, which partially offsets the second effect resulting from the reduction of state of charge, on a subsequent discharge. A negligible effect on I_{mp} occurs with increases in electrolyte temperature above 25°C.

17.4.5 Energy/Power Density

Typical average values for the energy and power densities of the vented sintered-plate nickel-cadmium cell at 23°C are shown in Table 17.2.

TABLE 17.2 Energy and Power Density of Vented Sintered-Plate Nickel-Cadmium Cells

Capacity density (single cell, 1-h rate)	25–31 Ah/kg 48–80 Ah/L
Energy density (1-h rate)	30–37 Wh/kg 58–96 Wh/L
Power density (at maximum power)	330–460 W/kg 730–1250 W/L

17.4.6 Service Life

The service life of the vented sintered-plate nickel-cadmium cell, normalized to unit weight (kilogram) and size (liter) at various discharge rates at 23°C, is approximated in Figs. 17.13 and 17.14. Accurate values of service life may be determined by multiplying the actual battery capacity at C_1 discharge rate and 23°C by the derating factors in Figs. 17.5 and 17.6 and then dividing by the actual discharge rate in amperes.

17.4.7 Charge Retention

"Charge retention" refers to the amount of dischargeable capacity remaining in a battery following prolonged storage under open-circuit conditions. Two mechanisms are responsible for the loss of charge, namely, self-discharge and electrical leakage between cells.

Self-discharge rates are an intrinsic property of cells. Typically, experimental results for the charge retained as a function of open-circuit storage time best fit a semilogarithmic relationship such as that shown in Fig. 17.15. The self-discharge rate of a cell is affected by impurity levels and the electrochemical stability of the electrodes.

The effect of temperature is shown in Fig. 17.16, where the exponential "time constant" (tc), the time to 36.8% capacity retained, is plotted against temperature. Storage temperature is probably the most important factor affecting self-discharge rate.

The second mechanism, the loss of charge due to electrical leakage, is influenced by the history of the battery's use and maintenance. Charge retention usually improves with cycling of the battery, and this will be true unless this cycling history has been abusive. The maintenance factor influencing charge retention is primarily battery clean-

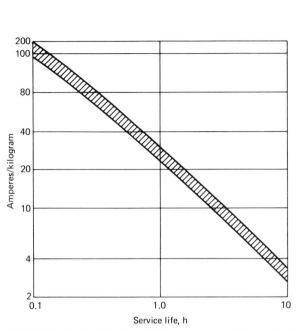

FIG. 17.13 Service life of a typical vented sintered-plate nickel-cadmium cell (gravimetric) at 25°C.

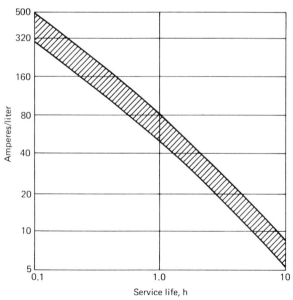

FIG. 17.14 Service life of a typical vented sintered-plate nickel-cadmium cell (volumetric) at 25°C.

17-14

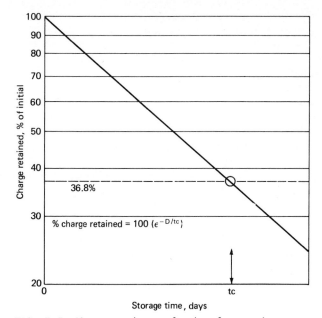

FIG. 17.15   Charge retention as a function of storage time.

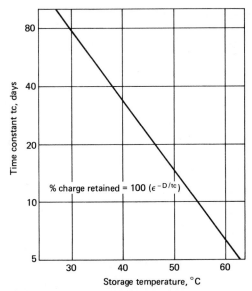

FIG. 17.16   Charge retention time constant as a function of storage temperature.

liness. Battery charge can leak from the terminals of one cell to the terminals of other cells across the cell tops if they are wet with potassium hydroxide. Loss of charge from this cause is relatively unpredictable, but it can usually be prevented by good housekeeping practices. Although surface leakage may affect only a portion of the cells in the battery, it is important since the capacity of the battery is limited to that of the lowest capacity cell.

It should be noted that loss of charge through these mechanisms is not permanent since the battery charge can be completely restored through recharging and comprehensive maintenance practices.

17.4.8 Storage

The sintered-plate cell can be stored in any state of charge and over a very broad temperature range (-60 to $60°C$). Best practice, however, is to store the cells upright with proper electrolyte level at a temperature between 0 and $30°C$. Following storage, the cells are charged to restore them to the operating condition.

17.4.9 Life

The life of the battery is strongly influenced by the care with which it is maintained and reconditioned as well as the way it is used; hence, to generalize on battery life expectancy is difficult. Best life performance is obtained with operation at normal temperatures and minimum reconditioning.

17.5 CHARGING CHARACTERISTICS

The functional design of the vented cell differs from that of the sealed cell primarily by the inclusion of a gas barrier between the positive and negative electrodes. This gas barrier has one principal function, which is to prevent, as discussed in Sec. 17.3, the recombination of generated gases within the cell, thereby allowing both positive and negative plates to return to full charge. This results in an overvoltage during initiation of overcharge, which is used as the feedback signal to control the charging device. As a consequence of this characteristic, however, the vented cell consumes water in overcharge which must be replenished during maintenance.

Charging of the vented sintered nickel-cadmium cell, following its discharge in cyclic use, has four significant objectives. These may be stated as follows:

1. Restore the charge used during discharge as quickly as possible
2. Maintain the "fully charged" capacity as high as possible during the use intervals between maintenance removals
3. Minimize the amount of water usage during overcharge
4. Minimize the damaging effects of overcharge

Fulfillment of the first objective is the principal reason for the design and use of vented cells, since the gas barrier provides the "voltage signal" which may be utilized in several different ways to terminate the fast recharge. The charge may thus be accomplished at the desired high rate without compromising the battery by continuing that rate in overcharge. Objective 2 must inherently be balanced against objectives 3 and 4 in the design and control of charging method. Generally, a continued good capacity between reconditionings is enhanced by providing more overcharge, while more overcharge inherently utilizes more water and, if sufficiently high, may result in damage to the battery. A compromise must therefore be struck; usually about 101 to 105% of the ampere-hours removed on discharge are replaced on the subsequent charge.

All charging techniques which are used in "on-board" systems utilize the "signal" provided by the onset of overvoltage of the vented cell in overcharge. This overvoltage signal is shown in Fig. 17.1. This significant voltage rise is present at all charge rates, and its sharpness actually improves as the cell is cycled in high-rate discharge and recharge. The corollary to this curve, which shows voltage response at constant current, is the current response at constant voltage, which may be expected to look somewhat the reverse of Fig. 17.1 and is illustrated in Fig. 17.17.

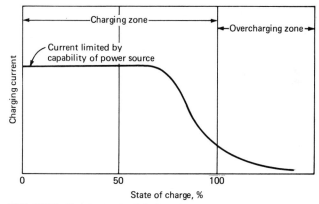

FIG. 17.17   Constant potential charge current.

17.5.1 Constant-Potential Recharging

Constant-potential, or CP, charging is the oldest of the methods still in use and is typically utilized in general aviation aircraft. Similar to an automobile battery charging system, CP charging utilizes a regulated voltage output from the aircraft dc generator, which is mechanically coupled to the engine. The voltage is typically regulated at 1.40 to 1.50 V per cell. Figure 17.17 illustrates the form of charge current as a function of state of charge of the vented cells during CP recharging. Although the initial current could be quite high if limited only by the voltage response of the battery, it most frequently is limited by the capability of the source as shown. As the battery approaches full charge, however, the "back emf" of the battery, illustrated for the constant-current case in Fig. 17.1, reduces the current to that required by the battery to provide an overvoltage equal to the regulated voltage of the charging source. CP charging requires very careful consideration of the selection of charge voltage and its proper maintenance, in order to achieve the balance between objectives 2 and 3. This is particularly difficult to achieve when the battery temperature experiences significant variation, since overvoltage is dependent on battery temperature. This balance may be made essentially independent of battery temperature effects by means of temperature compensation of the CP voltage as discussed in Sec. 17.5.4 below.

17.5.2 Constant-Current, Voltage-Controlled Recharging

A number of commercially available chargers based in general on constant-current charging with voltage cutoff control are utilized in modern aircraft. One of the simplest and most effective of these chargers applies an approximately C-rate constant-current charge to the battery and then terminates it when the battery voltage reaches a predetermined voltage cutoff (VCO) value such as 1.50 V per cell. The control also reinitiates the constant-charge current whenever the open-circuit battery voltage falls to a predetermined lower level, such as 1.36 V per cell. The net result is that the charger will recharge the capacity used during an engine start, typically 10% of the battery's total, in approximately 6 min and then cut the charger "off" due to the sharp rise in voltage as the cells go into overcharge, as illustrated in Fig. 17.1. Shortly thereafter, as the voltage falls below the turn-on voltage, the charge is reinitiated for a short period of time until the battery voltage again reaches the cutoff voltage. This simple on-off action continues at decreasing frequency and decreasing lengths of "on time," thereby maintaining the battery in a "float" condition at a completely full state of charge.

The battery voltage reduction, due to a discharge, inherently initiates the recharging of the battery without additional controls, thus automatically providing the recharge signal function regardless of the discharge rate or the reason. Adjustment of the "cutoff" and "turn-on" voltages as a function of battery temperature, which is described in Sec. 17.5.4, matches this mode of charging to the temperature characteristics of the vented sintered nickel-cadmium battery, thereby maintaining the desired balance of objectives. Both cutoff and turn-on voltages are compensated at the same rate, thereby maintaining a constant differential between turn-on and cutoff. A diagram of the function of this simple basic charge control scheme is shown in Fig. 17.18.

Several other proprietary chargers, based in part on the simple techniques outlined above, may also be found in commercial use. Many of these chargers also provide auxiliary functions such as upper and lower battery temperature charge discontinuation, detection of malfunctioning cells in the battery by detecting half-battery imbalance, and signaling the user in the event of either of these conditions.

17.5.3 Other Charging Methods

The charging methods outlined above are those used in order to achieve fast recharging of a battery which has been discharged in normal use. Periodic maintenance of the vented sintered plate nickel-cadmium battery, however, is generally accomplished by a full and complete discharge of each cell followed by a thorough recharge well into overcharge, in order to place both positive and negative plates in each cell of the battery in full and complete overcharge. The battery may then be returned to service with all plates of all cells in the same full-charged condition, thus enabling the battery to work from the "top down."

The simplest maintenance charge method, requiring the least complex equipment, to ensure this fully balanced, fully overcharged condition, is the low-rate charge. This technique utilizes a constant-current, approximately C/10, charge-overcharge current without voltage feedback control. At this low rate, the charge may be continued safely into overcharge without compromising the physical integrity of the components of the cell. This charge current should be maintained until at least 2 times the rated capacity

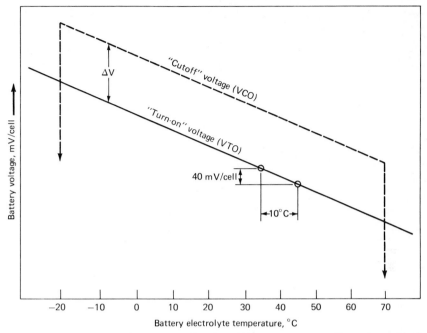

FIG. 17.18   Charger control voltages as a function of battery temperature; C-rate charge (nominal values).

of the battery has been replaced. Since this will inherently result in water usage, water level replenishment is best performed on a fully charged battery just prior to placing the battery back into service at the conclusion of this maintenance charging routine.

Batteries in standby service can be maintained in a fully charged condition by a float or trickle charge similar to pocket plate batteries. The float voltage for vented sintered-plate batteries is 1.36 to 1.38 V per cell.

17.5.4 Temperature Compensation of Charge Voltage

In both the constant-potential and the constant-current VCO charging methods used in repetitive cycling, it has been pointed out that the selection of voltage is always a compromise between the minimization of water usage and the maintenance of a high state of ready charge. The inherent variability of overcharge voltage as a function of battery temperature increases the difficulty of this compromise significantly. This voltage variability is shown on the Tafel curves in Fig. 17.19. The relationship between the Tafel curves at various temperatures indicates a temperature coefficient of $(-)$ 4 mV/°C at constant-current conditions. In other words, overcharge voltage, at constant current, decreases approximately 4 mV per cell for each 1°C rise in cell temperature.

As shown by the slope of these Tafel curves, overcharge voltage is also a linear function of the logarithm of overcharge current. That slope for vented sintered-plate

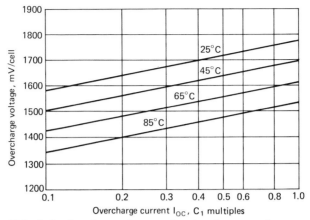

FIG. 17.19   Overcharge voltage as a function of current and temperature (Tafel curves).

nickel-cadmium is approximately 200 mV per cell per decade of change in overcharge current. Thus with constant-potential charging, the overcharge current, and therefore both water usage and overcharge damaging effect, will increase approximately 60% for each 10°C increase in electrolyte temperature.

A convenient technique for avoiding the detrimental effects of an increasing electrolyte temperature is to compensate the "constant" potential voltage, or the constant-current cutoff/turn-on voltages, for this change in battery temperature, at the rate of (−) 4 mV per cell per degree Celsius. This may be accurately accomplished through the use of thermistors or other temperature-sensitive electrical devices installed in the battery case to signal cell temperature. This function for the constant-current charging system is also shown in Fig. 17.18. The selection and use of the proper value of temperature compensation permits the battery charging system to function as though the battery were maintained at a constant temperature. Care must be exercised in the design and manufacture of these devices, since they must operate in an environment of potassium hydroxide which is both electrically conductive and has a propensity to wet and creep on most surfaces. High-grade potting compounds are therefore necessarily used to insulate and protect all auxiliary electronic components and wiring placed inside the battery case.

17.6 MAINTENANCE PROCEDURES

17.6.1 Electrical Reconditioning

The periodic maintenance of the vented sintered plate nickel-cadmium battery has six specific objectives. They are:

1. Assess the timeliness of the preselected maintenance period schedule

2. Recondition the electrical performance, both capacity and power

3. Detect and isolate cell failures and facilitate their replacement

4. Physically clean the battery

5. Replenish the water in the electrolyte

6. Maintain the charging-system voltage calibration

A relatively simple electrical exercise which fulfills the first objective consists of a single discharge, initially at a relatively high rate to simulate engine start and secondly at a relatively low rate approximating that of the 1-h value. This split-rate discharge, when performed on the battery in the condition as removed from the aircraft, is a measure of what its readiness was to perform either function while it was in the aircraft. The 15-s high-rate portion, at approximately the I_{mp} rate of discharge, while measuring the voltage and the capacity removed, determines the relative engine start power capability. The low-rate portion of the discharge at approximately the C rate, while measuring the remainder of the total capacity, determines the status of the available emergency energy capacity. The battery, when performing this measurement, should be in the "fully" charged condition that would typically be encountered when it is in the normal installation. Comparison of this power delivery, and the total available capacity as removed from the aircraft, with the requirements of the application will allow the user to determine whether the maintenance schedule interval may be increased or needs to be decreased.

The restoration of the electrical characteristics of the battery, known as the electrical reconditioning, objective 2, may be accomplished in two additional simple steps. The first is a thorough and complete discharge of each cell in the battery in order to discharge all the active material in each cell. The second step is the thorough and complete recharge of each plate of each cell into full gassing overcharge. The first step is accomplished by the shorting out of each individual cell for 8 to 16 h or more, following the measurement discharge described above. The second step is accomplished by charging the battery with constant current at a value of one-tenth of its ampere-hour rating (C/10) for at least 20 h. Since the capacity for some of the cells in the battery may be as much as 30 to 40% greater than the rated value, the total charge of 2.0C Ah is sufficient to ensure that both plates of all cells reach the full overcharge required.

There are other procedures used, and recommended by the manufacturers of specialized proprietary equipment, to recondition cells in a shorter period of time. The efficacy of each should be verified and the added expense and complexity entailed in the use of these proprietary reconditioning devices in specific applications justified. Care must always be exercised in their use, however, to avoid sustained high-rate overcharge, which may damage the gas barrier material.

Evaluation of cell-to-battery case leakage current, part of objective 3, when the battery is first received for maintenance, will determine the electrical need for cell cleanup as well as the presence of cracked or leaking cell jars. This procedure may be conveniently carried out by the simple expedient of completing a circuit from each cell terminal to battery case with a fused ammeter. A significant amount of leakage current through the ammeter to the case from anywhere in the cell electrical circuit indicates the presence of a conductive path, through potassium hydroxide, on the external surfaces of the cell jars. Such a conductive path may result from spewing of the electrolyte during overcharge, which may indicate either overfilling or the existence of a cracked or leaky cell jar. Isolation of the exact cause can be accomplished by repeating the measurements after physical cleanup of the battery.

The detection of the failure of the gas barrier, a very important part of objective 3, in any of the cells of the battery may be conveniently accomplished by extending the C/10 charge to 24 h. This overcharge will indicate accurately the failure of the gas barrier by either or both of two principal measurements near the end of that overcharge. First, the overcharge gassing rate is extremely sensitive to gas barrier condition and gas recombination. When measured with a simple ball flowmeter, the 24-h gas rate will be less than 80% of normal if the barrier has failed significantly in the cell. The normal value is 11 mL/min for each ampere of C/10 overcharge rate. The second indicator of gas barrier failure will be a 24-h overcharge voltage of less than 1.5 V if the cell is being charged in a 23°C ambient. Some downward adjustment of this voltage criterion may be made at the rate of 4 mV/°C if the battery is being charged in a higher ambient temperature.

17.6.2 Mechanical Maintenance

The replenishment of water in the electrolyte to return the level of the electrolyte to that recommended, objective 5, is the most important routine mechanical procedure employed during battery reconditioning. It is most accurately accomplished near the end of the 24 h of C/10 rate charge, by replenishing with deionized water until the electrolyte reaches the recommended level for a battery in overcharge. A record of the amount of water usage in each cell should be maintained and compared with the battery manufacturer's statement of reserve electrolyte level in each cell. If the total water usage between maintenance fillings, after deducting the amount used during the maintenance procedure, exceeds the reserve available in that cell design, the maintenance interval must be shortened to prevent plate dry-out during use and resultant cell failure. Note that the 24-h C/10 reconditioning procedure will itself use approximately 0.4 mL of water for each ampere hour of rated capacity during the 24-h reconditioning period. For example, 12 mL of water would be used during the reconditioning period for a 30-Ah rated cell, and this must be subtracted from the amount added to determine the amount actually used in service. It should also be noted that a cell with a damaged gas barrier may use less water.

The interval in the maintenance procedure, following the thorough short-down of each cell, may be conveniently utilized to perform battery-cleaning operations and to replace any cell found defective during the electrical measurement. Cleanup generally consists of a thorough rinsing with clear water followed by warm-air drying of the battery. This will dissolve and remove any accumulation of potassium hydroxide and carbonates from the outside of the cell jars. Vent caps should also be washed in warm water, warm water forced through the vent, and then dried with warm air. Power delivery of the battery may be enhanced by removing all intercell link hardware, buffing all contact surfaces, then replacing and retorquing all connectors. This is most safely accomplished with the cells in the completely discharged state. Replacement of any cells found defective at the conclusion of C/10 overcharge requires discharging the cells a second time.

17.6.3 System Inspection Criteria

Reinstallation of the reconditioned and fully charged battery into the aircraft system presents the opportunity for performing the system voltage calibration check in fulfillment of objective 6. The only measurement required for this on a CP charging system is to record the value of battery voltage, after reactivating the system, and following stabilization of that voltage. This stable value is the regulated float voltage that the battery will be subjected to during extended overcharge in use. This voltage measurement should be made with the engine running at a sufficiently high speed to produce a representative and stable value.

The battery voltage measurement, on constant-current VCO systems, should be made just as the system reaches cutoff voltage. The regulated voltage, on either of the two systems, should then be adjusted to the manufacturer's recommended value, if necessary.

17.7 RELIABILITY

17.7.1 Failure Mode

Due to its structural integrity, the sintered plate is practically indestructable in all normal, and most abnormal, operating situations. The cell component which accounts for the overwhelming majority of all vented sintered nickel-cadmium cell failures is the cellophane gas barrier. Failure of the gas barrier results in an effective chemical "short-circuiting" of the overcharge current through the action of oxygen recombination. The symptoms of gas barrier failure are, first, a reduction in the out-gassing rate during low-rate fixed-current overcharge, followed by a reduction in the overcharge voltage. Since gas is not being released by the overcharge current, water is therefore not being used at normal rates, and the energy used in electrical overcharging manifests itself in the form of heat. This results in a potential temperature increase in those cells which have failed in this manner, in addition to nearby cells. Temperature increases due to failed gas barrier, however, are difficult to separate from temperature increases due to many other reasons.

If cells with gas barrier failure are continued in operation for enough cycles, they will eventually lose capacity and the maximum power capability will then be reduced. Extended use of these failed cells may also result in the failure of the nylon electrical separator caused by the local temperature generated at the oxygen recombination sites, resulting in internal shorts.

Several other miscellaneous failure modes, which account for a small portion of cell failures, are as follows: (1) Internally shorted cells resulting from cut-through of the electrical separator by burrs and other plate irregularities, aggravated by cell interplate pressure and vibration. (2) Cracked and leaking cell jars which may result from excessive internal cell pressures caused by insufficient vent valve cleaning. Leaking cell jars may also result from abusive handling of the cells during cell replacement procedures and maintenance, or by defective jar manufacturing or sealing procedure. (3) Burned terminal contacts resulting from faulty cleaning and buffing procedures during maintenance, by insufficient link assembly torques, by terminal screw failure, or by conductive articles being dropped on the intercell connectors of a charged battery.

17.7.2 Memory Effect

In addition to these permanent failure modes, there exists a more common temporary effect which may significantly but gradually reduce both power and energy (capacity) performance. This effect, sometimes referred to as "memory" or "fading," results from continued charging without discharge of some portion of the active materials in the cell, such as in typical engine start use.

This effect, however, is completely reversible by a maintenance cycle consisting of a thorough discharge followed by a full and complete charge-overcharge as described in Sec. 17.6.1.

17.7.3 Factors Influencing Gas Barrier Failure

Gas barrier failure is generally acknowledged to be caused or aggravated by excessive overcharge current, excessive overcharge temperatures, and discharge at high rates with low electrolyte levels. Because gas barrier failure may not be detectable during reconditioning with other than the low-rate constant-current procedures, many of these failures may actually occur during maintenance and then manifest themselves some time after reinstallation in the aircraft. This possibility emphasizes the importance of an accurate assessment of the condition of the gas barrier at the end of the reconditioning period just prior to reinstallation. This assessment is accurately made by the measurement of overcharge gas flow following the extended C/10 overcharge, as described in Sec. 17.6.1.

One indication of the significant importance of the two factors of (1) temperature compensation of charger voltage and (2) effective maintenance practices is the existence of large-scale field data which document real-time failure differences of up to 100 to 1. These life differences exist between well-maintained batteries in temperature-compensated systems on the one hand and the same battery designs in uncompensated CP systems with frequent but abusive maintenance procedures on the other.

17.7.4 Thermal Runaway

Thermal runaway is a system condition peculiar to the use of CP charging on cells with failed gas barrier. The primary symptom of thermal runaway is a continuing increase in the overcharge current on the CP system following recharge. The charge current may exhibit a less than normal decrease at the end of charge and then, after reaching a minimum of some magnitude, will gradually increase again. Due to the heat generated by oxygen recombination, the cell temperature may eventually reach the boiling temperature of the electrolyte, somewhat above 100°C. As previously detailed, this temperature rise is due to the conversion to heat of an ever-increasing portion of the overcharging energy by the mechanism of oxygen recombination. The resulting temperature increase has a significant thermal time lag and may take 2 to 4 h of consecutive charging operation in order to manifest itself. Thus it may actually go undetected in aircraft operation for a significant number of flight hours following initiation of gas barrier failure, if the use of the system is not consecutive and/or sustained.

The thermal runaway condition is first evidenced by the aforementioned failure of charge current to decrease to normal stable levels and/or the subsequent increase of charge current. Next it is evidenced by a gradually increasing temperature, generally after the cells have already reached high overcharge rates, a process which takes several successive hours of operation. Third, if the process is allowed to continue long enough and the water is completely boiled out of one of the cells due to high temperature, inequities in cell voltages may appear. The voltage across the cell which has been boiled dry will increase significantly, thereby decreasing the charge current and the voltage across those cells which are still wetted with electrolyte. The next event will probably be internal shorting of the dried-out cell due to thermal destruction of the electrical separator. The charge current then increases sharply and the process repeats itself.

17.7.5 Potential Hazards

Potential hazards which may be present during the use and maintenance of the vented sintered nickel-cadmium battery fall into five general categories, as described below.

Gas Fire and/or Explosion   Since all functional vented batteries generate a stoichiometric mixture of hydrogen and oxygen gases during overcharge and expel them normally from the cell into the battery container, an explosion of these gases is always a possibility. Two conditions are required for such an explosion, however, and both are recognized and accounted for in the design of a typical system. The first condition is the accumulation of a sufficient quantity of this gas mixture. This condition is prevented in all system designs by the incorporation of adequate battery case ventilation. Some designs rely on either supplying a modest quantity of air to purging tubes on the battery from overboard vents in the aircraft or on natural convection of the gases from a louvered battery case into a ventilated compartment. The large volume of air used in air-cooled designs inherently accomplishes the required ventilation.

Unusual circumstances, however, may defeat any of these gas-purging techniques. It should also be remembered that batteries generate a significant amount of explosive gas during the maintenance procedure, and therefore maintenance should be performed in a well-ventilated shop.

The second condition required for explosion of the gases is the presence of a source of ignition. Although normally there are no ignition sources inside the battery case, several abnormal possibilities do exist. One of these is the internal shorting of a relatively dry cell in overcharge, resulting in an explosion inside the cell with a subsequent ejection of flames into the battery case. A second and more likely source of ignition may exist at an improperly maintained cell terminal due to the high temperatures generated during high-rate discharge. A third source of ignition may occur at the site of stray leakage currents.

Although the coincidence of both a sufficient amount of gas accumulation and a source of ignition is quite rare, it can and has happened. Because of this possibility, batteries for some installations are also designed to be physically capable of absorbing a hydrogen and/or oxygen explosion and containing the effects entirely within the battery case. Generally these batteries will also be electrically functional for at least one C-rate discharge following an explosion.

Arcing and Burning   This potential hazard concerns excessive leakage currents through electrolyte paths outside the cells. Such currents can occur between cells which are physically adjacent but with wide voltage separation in the cell circuit or more probably from cell to grounded metallic case. Arcing is more likely to occur, however, in the circuit of an inappropriately protected auxiliary device located inside the battery case in the environment containing potassium hydroxide. Some examples of these devices are battery heaters, thermal detectors, and voltage sensors. The proper design of these auxiliary appliances must recognize the conductivity of KOH and the ability of that electrolyte to creep along wires and even into mechanically "sealed" insulation. Devices of this type should be tested with high dielectric voltages while submerged in a water-detergent mix before they are installed in the battery case environment.

The result of a sustained leakage current through relatively localized conducting paths may be the ignition of the explosive environment by arcing, as previously discussed. The result might also be the carbonization of insulating materials and the subsequent burning of those materials within the battery case.

Electrical Power   One of the essential functional capabilities required of the vented nickel-cadmium cells is the ability to deliver high-power rates for engine starting. This very capability, however, presents a potential hazard in the form of hot spots on improperly torqued cell terminals during high-rate discharge. It is also a potential hazard to the unwary maintenance technician operating with metallic tools or other objects, such as jewelry, in a careless manner in the vicinity of charged batteries. Since the short-circuit current of these cells (batteries) may range from 1000 to 4000 A, it should be obvious that the exposed conductors of a charged battery should be treated with respectful caution. Very severe burns may occur if, for example, a ring should accidentally make contact between two adjacent cell terminals. Although one of the most obvious, this hazard is one of the most frequently encountered. Insulated cell hardware does provide a partial solution; however, caution and respect for the available power must always be exercised.

Corrosive KOH   Because of the corrosive nature of the KOH used as an electrolyte, all material employed in construction of the battery and its accessory appliances must be KOH immune. Materials such as nylon, polypropylene, nickel-plated steel, steel, and stainless steel are therefore used. The potential hazard of KOH corrosiveness, however, is principally encountered during maintenance. The use of safety glasses and/or safety face shields should be mandatory while performing maintenance on these batteries. A very small amount of KOH in the eye, for example, without prompt, continued, and adequate flushing followed by medical treatment, can result in the loss of eyesight. KOH is also corrosive to the skin, and the affected area should be thoroughly washed and rinsed with water, thereby minimizing any effect.

Electric Shock   Most vented sintered nickel-cadmium batteries are arrayed in groups of 10 to 30 cells, presenting maximum voltages of 15 to 45 V. There are applications, however, in which batteries are connected electrically in series strings of 90 to 200 cells or more. It should be apparent that the voltages presented by this number of cells in series may be lethal to anyone exposed to them. Personnel should also be cautioned, because of the high probability of electrolyte being present between the cell circuit and the conductive external surface of the battery, to exercise care by disconnecting series-connected batteries prior to handling them. Significant shock currents may be carried by relatively small amounts of KOH. The voltage available for the arcing and burning described in "Arcing and Burning" above is, of course, also proportional to the number of cells connected together in the series circuit.

17.8 CELL AND BATTERY DESIGNS

17.8.1 Typical Vented Sintered-Plate Nickel Cadmium Cells

A listing of several typical vented sintered-plate nickel-cadmium cells is given in Table 17.3. The 14-, 22-, and 36-Ah sizes are those typically employed in aircraft batteries. Other cells are available in sizes up to about 350 Ah. The larger cells are generally constructed in steel containers rather than the plastic containers now used for the aircraft size cells.

TABLE 17.3   Typical Vented Sintered-Plate Nickel-Cadmium Cells

Rated capacity (1-h rate), Ah	Maximum power at 25°C, W	Dimensions, cm			Weight, kg
		Length	Width	Height (container)	
14	260–330	6.15	2.73	16.0	0.61
22	360–500	8.06	2.7	19.2	1.0
36	450–840	7.9	3.47	21.9	1.5
50	640–1080	7.9	4.8	22.4	2.0
80	1250	7.56	6.06	21.7	3.1

17.8.2 Typical Battery Designs

A typical arrangement of vented sintered nickel-cadmium cells into a battery configuration is the conventional aircraft battery. An example of this use is shown in Fig. 17.20 and in detail in Fig. 17.21. This arrangement generally consists of a completely enclosing battery case and cover made of either stainless steel or steel with a KOH-resistant finish of epoxy or paint. The cover is typically secured with over-center type latches. The battery case is provided with overcharge gas-purging vents or with freely convective gas-diffusion openings for dilution. The cells are encased in nylon-molded cell jars with terminals extending through a nylon cover sealed to the cell jar. The cells are electrically connected in series with nickel-plated copper links from cell terminal to cell terminal and from the first and last cell to the battery termination and disconnect device. This battery termination extends through the battery case wall and is present

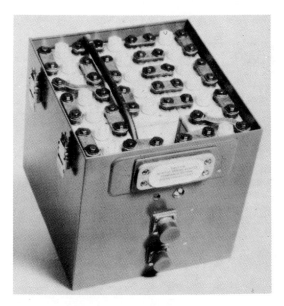

FIG. 17.20   Vented sintered-plate nickel-cadmium battery.

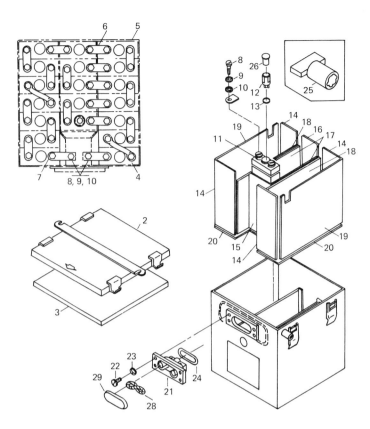

Index no.	Description	Units per assembly	Index no.	Description	Units per assembly
	Battery, nickel-cadmium	1	14	Insulator	4
1	Case assembly	1	15	Insulator	1
2	Cover assembly	1	16	Insulator	1
3	Gasket cover	1	17	Insulator	2
4	Link, cross	2	18	Insulator	2
5	Link	10	19	Insulator	2
6	Link	6	20	Insulator	2
7	Link	2	21	Receptacle and terminal	1
	"Attaching parts"			"Attaching parts"	
8	Screw, slotted hex. hd.	40	22	Screw, oval head	4
9	Washer, flat	40	23	Lockwasher	4
10	Washer, Belleville	40	24	"O" ring	1
11	Cell	19	25	Wrench, vent cap	1
12	Vent cap assembly	19	26	Spray cap	19
13	"O" ring	19	27	Cement (for cover gasket)	
			28	Spring	1
			29	Dust cap	1

FIG. 17.21 Typical assembly—vented sintered-plate nickel-cadmium aircraft battery.

on the outside surface of the battery case as a recessed double male, polarized, high-current receptacle. Functional requirements of aircraft batteries are specified in SAE standard AS 8033.[7]

17.8.3 Air Cooling

The major battery manufacturers produce some battery designs with provision for forced-air cooling. These designs generally take the form of plenum spaces below and above the cells, with cooling passages between the cells connecting the two plenums. The construction provides a means for the external connection of a high-volume, low-pressure, air source. Supplying 23°C air will not only effectively cool an overheating battery, it will also rapidly warm a cold battery. The thermal time constant for a battery with this feature may be as short as 10 to 20% of that of a standard non-air-cooled battery.

17.8.4 Temperature Sensors

Sensors may be provided inside the battery case which sense either typical or average cell temperature. They are equipped with a provision for external electrical connections. These devices may be of the on-off type, such as thermostats, or they may be the continuous type, such as thermistor assemblies. Continuous types have the capability of providing continuous modulation of, for example, the regulated charging voltage of a CP charging source or of the cutoff/turn-on voltages of a constant-current VCO charging system.

17.8.5 Battery Cases

Although the corrosive effect of KOH on bare steel is minimal, and cosmetic in effect, additional KOH resistance in the case material is desirable. In addition to stainless steel and steel with a protective finish, some special applications have used KOH-resistant plastic materials. It must be emphasized, however, that battery cases must withstand significant rough treatment in shock and vibration without losing their KOH-containment capability.

17.8.6 Battery Electrical Termination

Aircraft batteries are normally terminated on the front case surface by a connector of the type shown in Fig. 17.20. Special applications, however, may utilize direct cable connection to the first and last cell terminals, as well as various other special configurations capable of handling the high-power rates available in the event of a short circuit of the external battery circuit.

17.8.7 Battery Heaters

In addition to the airflow heating, heater blankets may also be provided on the inside or outside of the battery case. These blankets may be energized with any available electrical energy source. Primarily this source will be either the dc bus of the same voltage as the battery or an aircraft ac bus of higher potential. Heaters may also be energized from an auxiliary ground supply.

REFERENCES

1. E. J. Casey, "Canadian Developments—Impregnation by Thermal Decomposition," *Proc. Annu. Battery Research Develop. Conf.* **10**:47 (1956).

2. P. L. Bourgault et al., *Can. J. Technol.* **34**:495 (1957).

3. L. Kandler, British Patent 917,291 (1959).

4. E. J. Casey et al., British Patent 882,631 (1960).

5. R. L. Beauchamp, U.S. Patent 3,573,101 (1971), U.S. Patent 3,653,967 (1972).

6. A. Fleischer, *J. Electrochem. Soc.* **94**:289 (1948).

7. "Nickel Cadmium Vented Rechargeable Aircraft Batteries," Aerospace Standard AS 8033, Society of Automotive Engineers, Inc., Feb. 15, 1981.

18

Sealed Nickel-Cadmium Batteries

by
J. A. Wiseman

18.1 GENERAL CHARACTERISTICS

Sealed nickel-cadmium batteries incorporate specific cell design features to prevent buildup of pressure in the cell caused by gassing during charge. As a result, cells can be sealed and require no servicing or maintenance other than recharging. These unique characteristics for a secondary battery have created an acceptance for use in a wide range of applications requiring lightweight portable (e.g., tools, photography, toys, housewares) or standby (e.g., emergency lighting, alarms, computer memory) power.

The major advantages and disadvantages of the sealed nickel-cadmium battery are summarized in Table 18.1. The important characteristics are described below.

Maintenance-Free Operation The cells are sealed, contain no free electrolyte, and require no servicing or maintenance other than recharging.

TABLE 18.1 Major Advantages and Disadvantages of Sealed Nickel-Cadmium Batteries

Advantages	Disadvantages
Cells are sealed; no maintenance required	Sealed lead-acid battery better at high temperature and float service
Long cycle life	
Good low-temperature and high-rate performance capability	"Memory" effect
	Higher cost than sealed lead-acid battery
Long shelf life in any state of charge	

High-Rate Charging   Sealed nickel-cadmium cells are capable of recharge at high rates under controlled conditions. Many cells can be charged in 3 to 5 h without special controls, and all can be recharged within 14 h.

High-Rate Discharge   Low internal resistance and constant-discharge voltage make the nickel-cadmium battery especially suited for high-discharge or pulse current applications.

Wide Temperature Range   Sealed nickel-cadmium cells can operate over the range from about -40 to $50°C$ and are particularly noted for their low-temperature performance. Premium performance cells extend this range to $70°C$.

Long Service Life   Over 500 cycles of discharge, or up to 5 y of standby power, are common to sealed nickel-cadmium cells.

18.2 CHEMISTRY

The active materials of the sealed nickel-cadmium cell are the same as for other types of nickel-cadmium cells, namely, in the charged state, cadmium for the negative, nickel oxide for the positive, and a solution of potassium hydroxide for the electrolyte.

In the discharged state, nickel hydroxide is the active material of the positive electrode, and cadmium hydroxide that of the negative. During charge, nickel hydroxide, $Ni(OH)_2$, is converted to a higher valence oxide

$$Ni(OH)_2 + OH^- \rightarrow NiOOH + e$$

At the negative electrode, cadmium hydroxide, $Cd(OH)_2$, is reduced to cadmium

$$Cd(OH)_2 + 2e \rightarrow Cd + 2OH^-$$

The overall reaction is

$$Cd + 2NiOOH + 2H_2O \underset{charge}{\overset{discharge}{\rightleftharpoons}} Cd(OH)_2 + 2Ni(OH)_2$$

During operation, the active materials undergo changes in their oxidation state but little change in their physical state. Similarly, there is little if any change in the electrolyte concentration. The active materials of both electrodes, in both charged and discharged states, are relatively insoluble in the alkaline electrolyte, remain as solids, and do not dissolve while undergoing changes in their oxidation state. Because of these and other properties, nickel-cadmium cells are characterized by long life in both cyclic and standby operation and by a relatively flat voltage profile over a wide range of discharge current.

The operation of the sealed cell is based on the use of a negative electrode having a higher effective capacity than the positive. During charge, the positive plate reaches full charge before the negative and begins to evolve oxygen. The oxygen migrates to the negative electrode where it reacts with and oxidizes or discharges the cadmium to produce cadmium hydroxide

$$Cd + \tfrac{1}{2} O_2 + H_2O \rightarrow Cd(OH)_2$$

A separator permeable to oxygen is used so oxygen can pass through the separator to the negative; also, a limited amount of electrolyte is used (starved electrolyte system) as this facilitates the transfer of oxygen.

At a steady state, the recombination reaction rate during overcharge must be proportional to the current to prevent buildup of pressure. The internal pressure is sensitive to current, the reactivity of the negative electrode, the electrolyte level, and the temperature. Solid cadmium, gaseous oxygen, and liquid water must coexist in mutual contact for the recombination reaction to occur. If, for example, the electrolyte level is too high (the electrodes are in a flooded state), the oxygen is prevented from contacting the electrode, and the reaction rate at a given temperature and pressure is substantially lowered.

A safety venting mechanism is used in the cell design to prevent rupture in case of excessive pressure buildup due to a malfunction, high charge rate, or abuse.

18.3 CELL CONSTRUCTION

Sealed nickel-cadmium cells are available in several constructions. The most common types are the cylindrical cells, ranging in capacity from about 0.07 to 10 Ah, and the smaller button cells with capacities from 0.02 to 0.50 Ah. Rectangular and oval-shaped cells are also manufactured but in limited sizes.

18.3.1 Cylindrical Cells

The cylindrical cell is the most widely used type because the cylindrical design lends itself readily to mass production and because excellent mechanical and electrical characteristics are achieved with this design. Figure 18.1 shows a cross section of the cylindrical cell.

The positive electrode in cylindrical cells is a highly porous sintered-nickel structure into which the active material is introduced by impregnation with a molten nickel salt, followed by the precipitation of nickel hydroxide by immersion in an alkaline solution. The negative electrodes are made by several methods: using a sintered-nickel substrate similar to the positive, by pasting or pressing the negative active material onto a substrate, or by a continuous electrodeposition process.

After processing, the positive and negative electrodes are cut to size and wound together in a jelly-roll fashion with a separator material between them. The separator material, usually unwoven nylon or polypropylene, is highly absorbent to the potassium hydroxide electrolyte and permeable to oxygen. The roll is inserted into a rugged nickel-plated steel can, and the electrical connections are made; the negative is welded to the can and the positive is welded to the top cover. The very small amount of the electrolyte, enough for efficient operation, is absorbed by the separator. There is no free liquid electrolyte. The cover assembly of cylindrical cells incorporates a fail-safe vent mechanism to prevent cell rupture in case of excessive pressure buildup which could result from extreme overcharge or discharge rates.

18.3.2 Button Cells

Button cells usually have electrodes made from "pressed" plates. The active materials are compressed in molds into disks or plates, and the electrodes are assembled in a "sandwich" configuration as shown in Fig. 18.2.

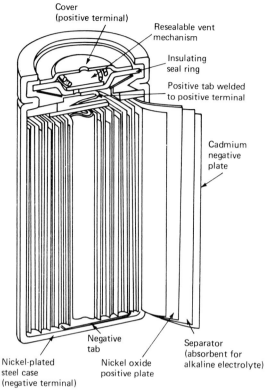

Cover
(positive terminal)

Resealable vent
mechanism

Insulating
seal ring

Positive tab welded
to positive terminal

Cadmium
negative
plate

Negative
tab

Separator
(absorbent for
alkaline electrolyte)

Nickel-plated
steel case
(negative terminal)

Nickel oxide
positive plate

FIG. 18.1 Sealed cylindrical cell construction. (*Courtesy of Gould, Inc.*)

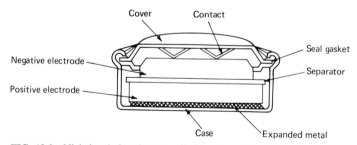

Cover   Contact

Seal gasket

Negative electrode

Separator

Positive electrode

Case   Expanded metal

FIG. 18.2 Nickel-cadmium button cell. (*Courtesy of Gould, Inc.*)

In some cases the electrodes are backed with expanded metal, or screen, to enhance electrical conductivity and mechanical strength. The button cell does not have a failsafe device, but its construction allows the cell to expand, either breaking the electrical continuity or opening the seal to relieve excess pressure caused by an abnormal circumstance.

Button cells are very suitable for low-current, low-overcharge-rate applications.

18.3.3 Rectangular and Oval Cells

The rectangular cell is housed in a nickel-plated steel can using a construction similar to the vented cell but incorporating the features needed for sealed-cell operation. This construction is particularly suited for high discharge rates because of the large electrode area. The oval cell is a special configuration of the cylindrical cell.

18.4 PERFORMANCE CHARACTERISTICS

18.4.1 General Characteristics

A typical charge-discharge duty cycle for the sealed nickel-cadmium cylindrical cell, at 25°C, is shown in Fig. 18.3. The voltage increases slowly, but steadily, at the C/10 charge rate to a steady-state condition, decays slightly during the 1-h rest, is relatively flat during the 1-h discharge to 1.0 V. The cell voltage recovers rapidly over the next hour, while at rest, to near 1.2 V.

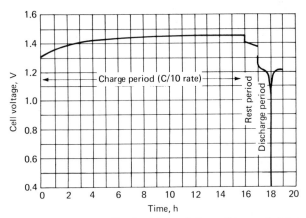

FIG. 18.3 Voltage profile of a nickel-cadmium cell in a typical duty cycle.

18.4.2 Discharge Characteristics

Typical discharge curves for the cylindrical cell at 25°C at various discharge loads are shown in Fig. 18.4. The flat voltage profile, after the initial voltage drop, is characteristic.

The capacity that can be obtained from a cell or battery is dependent on the rate of discharge, the voltage at which discharge is terminated, the discharge temperature, and the previous history of the cell or battery. Figure 18.5 shows the relationship of percent of rated capacity available and the discharge rate of a typical cell, assuming a C/10 charge, 25°C, and a 1.0-V cutoff. The average voltage during discharge decreases as the discharge rate increases, as shown in Fig. 18.6. If the cell or battery were allowed to discharge to a lower cutoff voltage, a greater percentage of the C/5 rate capacity could be obtained.

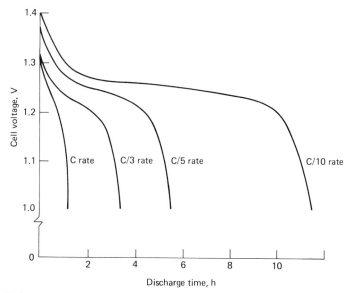

FIG. 18.4   Constant-current discharge curves at 25°C, sealed nickel-cadmium cell.

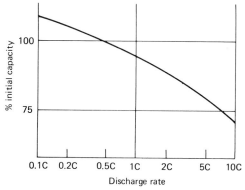

FIG. 18.5   Percent of C/5 rate capacity vs. discharge rate at 25°C, 1.0-V cutoff, typical sealed nickel-cadmium cell.

18.4.3 Effect of Temperature

The sealed nickel-cadmium cell is capable of good performance over a wide temperature range. Best operation is between −20 and +30°C, although usable performance can be obtained beyond this range. The low-temperature performance, particularly at high rates, is generally better than that of the lead-acid battery but usually inferior to that of the vented sintered-plate battery. The reduction in performance at low temperatures is due to an increase in the internal resistance; at high temperatures, the loss can be due to a depressed operating voltage or to self-discharge.

Figure 18.7 shows some typical discharge curves of the sealed cell at −20°C. A flat

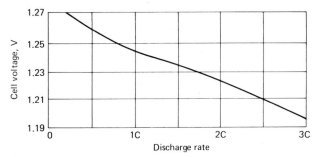

FIG. 18.6 Average voltage vs. discharge rate of a sealed nickel-cadmium cell at 25°C, 1.0-V cutoff.

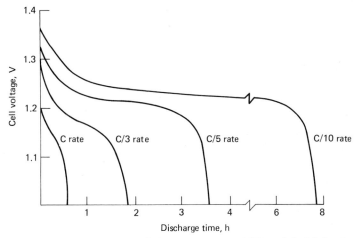

FIG. 18.7 Constant-current discharge curves at −20°C, sealed nickel-cadmium cell.

discharge is still characteristic, but at a lower operating voltage than at room temperature. Figure 18.8 shows the effect of temperature on the average voltage. Ambient temperatures significantly above or below 25°C have a depressing effect on the average discharge voltage.

The effect of temperature and discharge rate on the capacity of a sealed nickel-cadmium cell is shown in Fig. 18.9.

18.4.4 Internal Impedance

The internal resistance is dependent upon the internal impedance, resistances due to activation and concentration polarization, capacitive reactance, and the condition of the cell.

In most applications the effects of capacitive reactance can be ignored. Resistance effects due to activation and concentration polarization decrease with increasing temperature, while they increase at reduced temperatures. The measured impedance follows this same general relationship, as shown in Fig. 18.10.

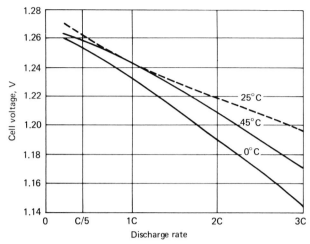

FIG. 18.8 Discharge voltage vs. rate at various temperatures, sealed nickel-cadmium cell.

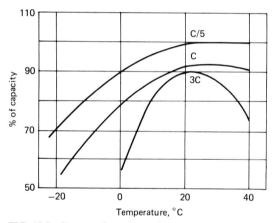

FIG. 18.9 Percent of rated capacity vs. temperature at different discharge rates (1-V cutoff), typical sealed nickel-cadmium cell.

When discharge is at moderate rates, the activation and concentration polarization effects are negligible in a nickel-cadmium cell, and the voltage remains relatively constant during discharge. This is typical of electrochemical systems in which the internal impedance is relatively insensitive to the state of charge. Figure 18.11 shows the typical measured impedance vs. state of charge for a 1.2-Ah cell.

With use over time, a nickel-cadmium cell gradually loses capacity, resulting in a gradual increase in internal resistance. This is caused by gradual deterioration of the separator and by loss of liquid through the seals, which changes the electrolyte concentration and level. The net effect is an increase in internal impedance.

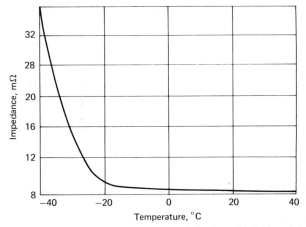

FIG. 18.10   Impedance vs. temperature, fully charged 1.2-Ah sealed nickel-cadmium cell.

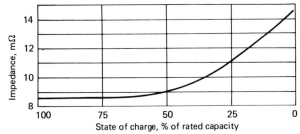

FIG. 18.11   Impedance vs. state of charge, 1.2-Ah sealed nickel-cadmium cell.

18.4.5 Service Life

The service life of a sealed nickel-cadmium cell, normalized to unit weight (kilogram) and size (liter), at various discharge rates and temperatures, is summarized in Fig. 18.12. The curves are based on a capacity, at the C/5 discharge rate at 25°C, of 25 Ah/kg and 70 Ah/L. (Also see Fig. 18.23.)

18.4.6 Cell Reversal

When three or more cells are series-connected, the lowest-capacity cell can be driven into reversal by the others. The larger the number of cells in series, the greater the possibility of this occurring. During reversal, hydrogen may evolve from the positive electrode and oxygen from the negative. Frequent and extended reversal will lead to higher cell pressure and opening of the safety vent and, hence, should be avoided.

Some cell designs provide a limited amount of built-in protection against deep reversal by adding a small amount of cadmium hydroxide to the positive electrode. The term used for the material added to the positive for reversal protection is "antipolar mass"

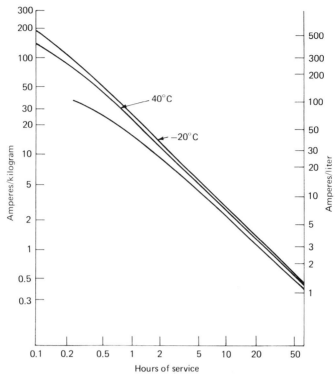

FIG. 18.12 Service life of the sealed nickel-cadmium cell at various discharge rates and temperature.

(APM). When the positive electrode is completely discharged, the cadmium hydroxide in that electrode is converted to cadmium, which, combining with the oxygen generated from the negative electrode, depolarizes the positive, preventing hydrogen generation for a time. This reaction occurs at about -0.2 V. This reaction can sustain for only a limited time after which hydrogen is evolved from the positive. Because hydrogen combines with the cell materials to only a limited extent, repetitive cell reversal will gradually increase a cell's internal pressure, ultimately causing the cell to vent.

In applications of multicell batteries where it is likely that the battery will be frequently discharged below 1.0 V per cell, a voltage-limiting device is recommended to avoid cell reversal.

18.4.7 Shelf and Cycle Life

Shelf Life As with all battery systems, temperature plays an important part in both charge retention and battery life. Table 18.2 lists temperature limits for nickel-cadmium cells.

The rate of self-discharge during storage is a function of the storage temperature and the cell design. Figure 18.13 serves as a guide for the shelf life, or charge retention, of sealed nickel-cadmium cells. In general, the lower-resistant, high discharge rate cylindrical cells do not have as good charge retention as the button cells.

TABLE 18.2 Operating and Storage Limits for
Sealed Nickel-Cadmium Cells

Cell type	Temperature, °C	
	Storage	Operating
Button	−40 to 50	−20 to 50
Standard cylindrical	−40 to 50	−40 to 50
Premium cylindrical	−40 to 70	−40 to 70

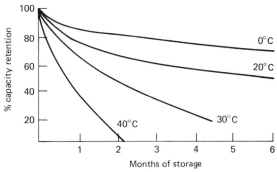

FIG. 18.13 Capacity retention (shelf life) of sealed nickel-cadmium cells.

Storage   The sealed nickel-cadmium cell can be stored in the charged or discharged state without damage. It can be restored for service by recharging (one or two charge/discharge cycles).

Cycle Life   Sealed cells have a long cycle life. Under controlled service and charge conditions, over 500 cycles can be expected on a full discharge. On shallow discharge, considerably higher cycle life is obtained, as shown in Fig. 18.14. Cycle life is also very dependent on the charge and discharge rates, frequency of cycling, and temperature as well as the cell design and components. Figure 18.15 shows the improvement in cycle

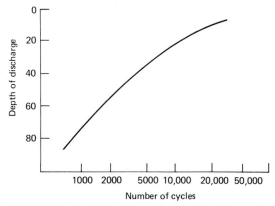

FIG. 18.14 Cycle life of the sealed nickel-cadmium cell.

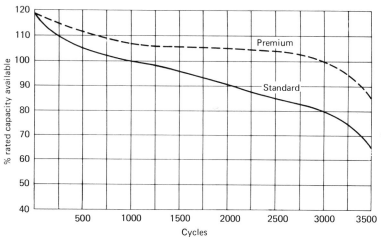

FIG. 18.15   Cycle life at 25°C, 1.2-Ah sealed nickel-cadmium cell (charge 5 h at 400 mA; discharge 3 h at 400 mA; capacity measured every 100 cycles—400 mA to 1.0 V).

life with "premium" cells using a polypropylene separator instead of the standard nylon. The polypropylene separator gives longer life, particularly at elevated temperatures, because it is much more resistant to oxidation, permitting satisfactory cell life at continuous temperatures up to 70°C. Premium designs are available in most cell sizes.

18.4.8 Life Expectancy and Cell Failure

The useful life of a nickel-cadmium battery can be measured in terms of either the number of cycles before failure or in units of time. It is virtually impossible to know all the detailed information necessary to make any kind of accurate prediction of battery life in a given application. The best that can be provided is an estimate based on laboratory test data and field experience or extrapolation of accelerated test data.

Basically, failure of a battery occurs when it fails to operate the device, for whatever reason, at the prescribed performance level, despite the possibility that the battery may still be useful in another application with less demanding requirements.

Failure of a nickel-cadmium cell can be classified in two general categories: reversible and irreversible failure. When a battery fails to meet the specified performance requirements but can, by appropriate reconditioning, be brought back to an acceptable condition, it is considered to have suffered a reversible failure. Permanent or irreversible failure occurs when the battery cannot be returned to an acceptable performance level by reconditioning or any other means.

Reversible Failures   A battery can suffer reversible loss of capacity when it is cycled repetitively at consistent rates of charge, discharge, and time. This effect is frequently referred to as the "memory" or "hysteresis" effect. The effect is the same whether due to high current drains to a low-voltage cutoff or light drains to a higher-voltage cutoff. The basic source of the capacity degradation is the shallowness of the discharge, with the coefficient of overcharge exerting only a minor influence. Operation at high temperatures accelerates this type of capacity loss.

The reduction in capacity that a cell undergoes when repetitively cycled in shallow discharges results from a lowering of the average discharge voltage. For example, if a cell is repeatedly discharged at the C/5 rate for 1 h, the voltage profile will gradually change, as shown in Fig. 18.16. The graph shows that the fifth-cycle discharge has a voltage plateau of about 1.25 V, and after 500 cycles the plateau has shifted to about 1.15 V. Most of this discharge voltage loss can be restored with a few deep reconditioning cycles.

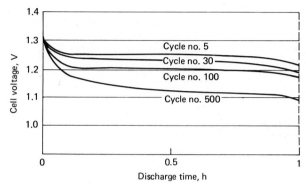

FIG. 18.16 Typical effects on the voltage profile of a sealed nickel-cadmium cell when repeatedly discharged for 1 h at the C/5 rate.

A similar reversible failure can occur with long-term overcharging, particularly at elevated temperatures. Figure 18.17 shows the voltage "step" near the end of discharge that can be induced by long-term overcharging. The capacity is still available but at a lower voltage than when it was freshly cycled. Again, this is a reversible failure; a few charge and discharge cycles will restore normal voltage and expected capacity.

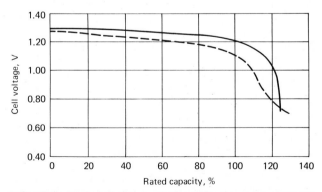

FIG. 18.17 Typical discharge voltage profile of a sealed nickel-cadmium cell after long-term overcharge at C/10 (dotted line) vs. 16-h charge at C/10 rate.

Irreversible Failures   Permanent failure in nickel-cadmium cells results from essentially two causes: short-circuiting and loss of electrolyte. An internal short circuit may be of relatively high resistance and will be evidenced by an abnormally low on-charge voltage and by a drop of voltage as the cell's energy is dissipated through the internal short. A short may also be of such a low resistance that virtually all the charge current travels through it, or it is a "dead" short.

Any loss of electrolyte will cause degradation in capacity. Charging at high rates relative to the battery temperature, repeated cell reversals, and direct shorting are all ways that can cause loss of electrolyte through the pressure relief device. Electrolyte can also be lost over a long period of time through the cell seals, and capacity is lost in proportion to the reduction in electrolyte volume. Capacity degradation caused by electrolyte losses is more pronounced at high discharge rates.

High temperature degrades cell performance and life. A nickel-cadmium cell gives optimum performance and life at temperatures between 18 and 30°C. Higher temperatures reduce life by promoting separator deterioration and increasing the probability of shorting. Higher temperatures also cause more rapid evaporation of moisture through the seals. These effects are all long-term; but the higher the temperature, the more rapid the deterioration.

18.5 CHARGING CHARACTERISTICS

18.5.1 General Considerations

Charging of sealed nickel-cadmium batteries is usually done by the constant-current method. Normally, the 10-h rate is used and the battery is charged for 12 to 16 h (140%). At this rate, the cell can withstand overcharging without harm, although most sealed cells can be safely charged from the C/100 to C/3 rate. At higher charge rates, care must be taken not to overcharge the battery excessively or develop high cell temperatures. The voltage profile of a sealed nickel-cadmium cell during charge is shown in Fig. 18.18; the sharp rise in voltage at the higher charge rate is clearly evident. The voltage profile of the sealed cell is different from the one for the vented cell. The end-of-charge voltage for the sealed cell is lower since, because of the oxygen recombination, the negative plate does not reach as high a state of charge as it does in the vented construction.

Overcharging, at rates beyond the ability of oxygen recombination or heat dissipation, can result in cell failure. "Fast" charging methods, as high as the 5- to 20-C rate, have been used successfully, but a means must be provided for monitoring and terminating the charge before the overcharge state is reached. Temperature rise, voltage, and/or pressure can be monitored and used effectively as a cutoff.

18.5.2 Pressure, Temperature, and Voltage Relationships

Figure 18.19 shows the relationships of pressure, temperature, and voltage of a typical cell during charging at C/10 and C/3 rates. The internal pressure of a nickel-cadmium cell remains low until oxygen is generated at the positive electrode, which occurs when the state of charge reaches about the 75% level. As oxygen accumulates, the pressure

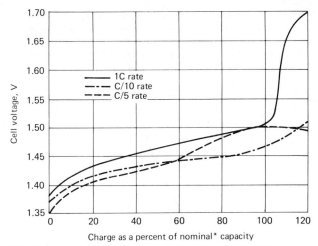

*Nominal capacity determined by charging for 16 h at C/10 rate, then discharging at the C/5 rate to 1.0 V.

FIG. 18.18   Voltage profile of a sealed nickel-cadmium cell during charge (to 120% of capacity of previous discharge).

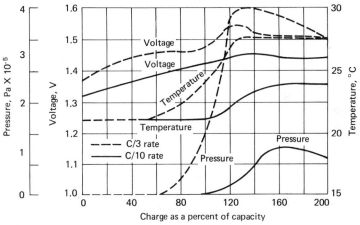

FIG. 18.19   Typical pressure, temperature, and voltage relationships of a sealed nickel-cadmium cell during charge.

gradually rises, and when the cell goes into the overcharge condition and most of the current is used to produce oxygen, the pressure rises more steeply. For continuous overcharge, the current must be lower than the value at which the safety vent operates. When a cell is overcharged at an acceptable level, the pressure stabilizes to a constant whose value is determined not only by the overcharge current and the temperature but by certain of the cell's design parameters, the most important of which are the separator, activity of the negative electrode, and electrolyte level.

When the cell goes into overcharge, most of the current generates oxygen which produces heat when it reacts with the cadmium of the negative plate. The amount of heat produced during overcharge equals the product of the voltage and current. The steady-state temperature is primarily governed by the overcharge rate (and voltage), the heat transfer characteristics of the cell or battery, and the ambient temperature.

Figure 18.20 shows the comparative temperature rise of a typical cell during C/10-, C/3-, and C-rate charging. Temperature rises dramatically in overcharge at rates above C/3. Figure 18.19 shows that the cell voltage is higher for the higher C/3 charge rate at the outset and it increases at about the same rate as the C/10 rate, but the pressure and temperature start rising sooner and at a faster rate than at the C/10 charge rate. The voltage at charge rates in excess of C/3 also rises sharply in overcharge, as was shown in Fig. 18.18.

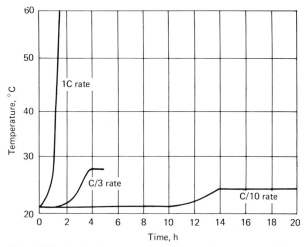

FIG. 18.20   Typical temperature rise during charge of a sealed cylindrical nickel-cadmium cell.

Charging at temperatures between 0 and 30°C is best for sealed cells. At lower temperatures, the cell voltage increases more, and, because recombination of oxygen is slower at these lower temperatures, charging rates must be reduced. Above 40°C the charging efficiency is low and the higher temperatures cause cell deterioration. Figure 18.21 shows the end-of-charge voltage at different temperatures after 16 h charge at the C/5 rate.

Constant-potential charging is not recommended for it can lead to thermal runaway. It can be used if precautions are taken to limit current toward the end of the charge. Similar precautions must be taken with float charging. Trickle charging, at a low constant-current rate, can be used to maintain the cell in a state of full charge. A periodic discharge every 6 months followed by a charge is advisable to ensure optimum performance.

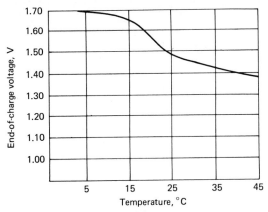

FIG. 18.21   Typical end-of-charge voltage vs. temperature, sealed nickel-cadmium cell (after 16-h charge at C/5 rate).

18.6 CELL SIZES AND MANUFACTURERS

Multicell batteries are available in a variety of voltages and configurations. Some of these are illustrated in Fig. 18.22.

Table 18.3 lists typical sealed nickel-cadmium cells and their major electrical and physical specifications. Figure 18.23 is a guide to determine the approximate cell size required for a given performance requirement or application. These data are based on performance at 20 to 25°C, and necessary allowance must be factored into the estimate to determine the performance under other discharge conditions.

Table 18.4 lists the National Electronic Distributors Association (NEDA) standard cylindrical sealed nickel-cadmium cells and manufacturers' type numbers.

TABLE 18.3 Typical Sizes and Ratings of Available Cells

Cell size	Capacity, Ah at 5-h rate	Diameter, mm	Height, mm	Weight, g	Resistance, mΩ, ± 20%
		Cylindrical cells			
½ AAA	0.070	10.21	22.12	4.0	220
AAA	0.180	10.21	44.45	9.0	120
⅓ AA	0.100	14.27	16.76	6.5	80
½ AA	0.250	14.27	30.28	10.8	60
AA	0.475	14.27	49.96	19.8	28
⅓ A	0.150	15.62	17.58	7.9	100
A	0.600	15.62	48.77	24.0	26
⅓ A_f	0.225	17.30	18.54	10.8	30
⅔ A_f	0.450	17.30	29.59	15.6	48
A_f	0.800	17.30	50.93	30.0	22
C_s	1.2	22.94	43.71	45.4	15
C_s	1.4	22.94	43.71	46.8	15
½ C	0.750	26.26	25.27	31.0	23
⅗ C	1.0	26.26	30.86	39.7	20
C	2.0	26.26	50.01	68.0	17
½ D	2.2	32.89	38.94	80.0	10
D	4.0	32.89	61.11	130.0	9
F	7.0	32.89	91.29	210.0	7
. . . . .	10.0	43.0	91.0	400.0	
		Button cells			
20 B	0.020	11.61	5.41	1.1	5
50 B	0.050	15.60	6.05	2.6	2
100 B	0.100	25.20	6.40	7.9	1
150 B	0.150	25.20	6.91	8.5	0.7
225 B	0.225	25.20	8.92	11.6	0.6
225 BH	0.225	25.20	9.09	12.5	0.4
450 B	0.450	43.10	8.00	31.2	0.3
500 BH	0.500	34.39	9.98	25.5	0.2

FIG. 18.22 Typical sealed nickel-cadmium cells and batteries. (*Courtesy of Gould, Inc.*)

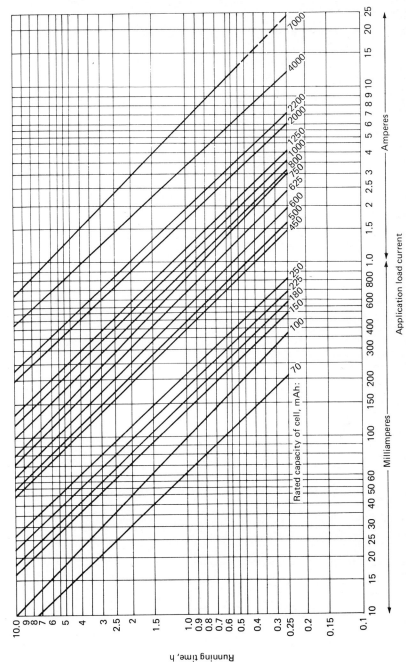

FIG. 18.23 Cylindrical cell selector guide. [Guide can be used to determine the approximate required cell size, given the load and desired run (service) time. Data based on a fully charged battery and 20°C operating temperature.] (*Courtesy of Gould, Inc.*)

TABLE 18.4   Cross-Reference of Sealed Nickel-Cadmium Cell and Batteries

Size	Neda number	Berec	Burgess	Duracell	Eveready	General Electric
D (1.2 V)*	13NC	RX20	CD10	NC13	CH50	GC3
C (1.2 V)*	14NC	RX14	CD14	NC14	CH35	GC2
AA (1.2 V)*	15NC	RX6	CD6	NC15	CH15	GC1
D (1.2 V)*	16NCD	NCC400			CH4	HD1000
7.2 V†	1604NC			NC1604	CH22	GC9
8.4 V†	1605NC	RX22				

Size	Neda number	Gould	Panasonic	Ray-O-Vac	Sanyo
D (1.2 V)*	13NC	CSD	NRD	613	N4000D
C (1.2 V)*	14NC	CSC	NRC	614	N1800C
AA (1.2 V)*	15NC	CSAA	NRAA	615	N475AA
D (1.2 V)*	16NCD				
7.2 V†	1604NC	CS919		NC1604	6N-75P
8.4 V†	1605NC	CS9T			

*For 1.5-V applications (as replacement for zinc primary cells).
†For 9-V applications (as replacement for zinc primary cells).
MANUFACTURERS:

General Electric Co.
Battery Business Dept.
Gainesville, Fla.

Union Carbide Corporation
Battery Products Division
Danbury, Conn.

Sanyo Electric Trading Co., Ltd.
Osaka, Japan

Panasonic
Matsushita Electric Industrial Co., Ltd.
Dry Battery Division
Osaka, Japan

Varta Batterie AG
Hannover, Germany

Société des Accumulateurs Fixes et de Traction (SAFT)
Romainville, France

SAFT America, Inc.
Cockeysville, Md.

SOURCE: National Electric Distributors Association, Park Ridge, Ill. Reprinted with permission.

BIBLIOGRAPHY

Eveready Battery Engineering Data, Union Carbide Corp., Battery Products Division, Danbury, Conn., 1976.

The Gould Battery Handbook, 1st ed., Gould Inc., St. Paul, Minn., 1973.

Nickel-Cadmium Battery Application Engineering Handbook, 2d ed., General Electric Co., Gainesville, Fla., 1975.

Rechargeable CADNICA Battery Engineering Handbook, Sanyo, Osaka, Japan, 1976.

Nickel-Zinc Batteries

by
Martin Klein

19.1 GENERAL CHARACTERISTICS

The nickel-zinc (zinc/nickel oxide) battery system is created by the marriage of the nickel electrode as used in nickel-cadmium batteries and the zinc electrode as used in silver-zinc batteries. The combination sought to achieve the long-life characteristic of the nickel-cadmium battery with the capacity advantage of the zinc anode. Initial investigations of the nickel-zinc battery date back to a series of Drumm patents in the 1930s.[1] Despite this history, the nickel-zinc battery system has not achieved commercial importance to date, due primarily to the still-limited life of the zinc electrode. On the basis of new interest, however, and advances being made in the improvement of its cycle life, it promises to have major commercial significance because of its high energy density (about twice that of the nickel-cadmium battery) and attractive materials cost. Table 19.1 summarizes the major advantages and disadvantages of the nickel-zinc system.

TABLE 19.1 Advantages and Disadvantages of Nickel-Zinc System

Energy density	55–80 Wh/kg	Exceeded only by the Zn/AgO system
Cost, projected	50–200 $/kWh	Relatively low compared with other secondary batteries
Power density	100$^+$ W/kg	High power capability
Operating temperature	−20 to 60°C	Good low-temperature performance
Cycle life	50–200	Poor life, improvements under way

19.2 CHEMISTRY

The zinc/nickel oxide battery system uses zinc as the negative active material and nickel oxide as the positive. The electrolyte is an alkaline potassium hydroxide solution, normally in the 25 to 35% concentration range. The charge and discharge reactions in simplified form are

Charge E^0
Negative electrode: $Zn(OH)_2 + 2e \rightarrow Zn + 2OH^-$ 1.24 V
Positive electrode: $2Ni(OH)_2 + 2OH^- \rightarrow 2NiOOH + 2H_2O$ 0.49 V
Overall reaction: $Zn(OH)_2 + 2Ni(OH)_2 \rightarrow 2NiOOH + Zn + 2H_2O$ 1.73 V

Discharge
Overall reaction: $2NiOOH + Zn + 2H_2O \rightarrow 2Ni(OH)_2 + Zn(OH)_2$

Overcharge
Negative electrode: $2H_2O + 2e \rightarrow H_2 + 2OH^-$
Positive Electrode: $2OH^- \rightarrow \frac{1}{2} O_2 + H_2O + 2e$
Overall reaction: $H_2O \rightarrow H_2 + \frac{1}{2} O_2$

As noted in the equations, the nickel electrode operates between two oxidation states, $+3$ charged to $+2$ discharged. The standard free energy E^0 of the couple is 1.730 V, and, on the basis of 100% utilization of active materials, the system is theoretically capable of 334 Wh/kg:

1. $Ni(OH)_2 \cdot \frac{1}{4} H_2O$ 3.63 g/Ah
    ZnO 1.55 g/Ah
    Total   5.18 g/Ah
2. Theoretical energy density $= 1.73 \times 1000/5.18$ g/Ah $= 334$ Wh/kg

In actual practice, the battery operates at 1.55 to 1.65 V and delivers 55 to 77 Wh/kg. Based on the standard free energy voltage and theoretical reactants, the battery requires 1.3 kg of metallic nickel and 0.7 kg of metallic zinc per kilowatthour.

A critical factor that determines battery usefulness—cycle life—has limited the utility of the nickel-zinc battery system. On repetitive cycling, the zinc electrode is subject to deterioration, which results in capacity decay. The zinc electrode is partially soluble in the electrolyte, and on cycling, dissolution and redeposition of the zinc do not occur at the same location and with the same morphology. This results in a loss of usable active material. Extensive investigation is under way to manage this problem and achieve acceptable performance.

19.3 CELL COMPONENTS

19.3.1 Nickel Electrode

A number of different types of nickel electrodes are used in nickel-zinc cells.

Pocket Plate Electrode   This is the same type of electrode used in pocket plate nickel-cadmium and nickel-iron batteries.[2] Electrodes are prepared by loading nickel hydroxide hydrate active material and a conductive additive (graphite and/or nickel flake) into tubular flat pockets which are then assembled into electrodes. Little interest currently exists in using this type of electrode in nickel-zinc cells since modern cells attempt to utilize lightweight electrode construction.

Sintered-Nickel Electrode   This type of electrode is prepared by sintering high bulk density carbonyl nickel powder into a porous plaque (75 to 80% porous).[3] The plaque usually contains a woven nickel screen or nickel-plated screen. The active nickel hydroxide is loaded into the pores of the plaque. The plaque serves to give the electrode structural integrity and to provide a conductive matrix for contact with the active material but does not participate in the electrochemistry of the cell.

A number of techniques are utilized to prepare the porous plaque and load the pores with active material. The impregnation process can be accomplished by successive dipping or vacuum impregnation of the plaque with concentrated nickel nitrate solution and subsequently converting the nitrate solution in the pores of the plaque to nickel hydrate by precipitation with alkali.[4] An alternate process is the thermal decomposition of the nickel nitrate and subsequent reaction with alkali to form a nickel hydrate.[5]

Both of the above processes require multiple steps to fill the pores with a sufficient amount of active material.

A third impregnation approach consists of the electro-chemical deposition of the nickel hydroxide into the pores of the plaque.[6] This is accomplished by cathodizing the nickel plaque in a molten or high-temperature solution of nitrate against counteractive or inert anodes. Typical sintered-nickel electrodes have an inactive to active nickel ratio of 1:1 to 1.4:1. These electrodes, similar to those used in the nickel-cadmium cell, have demonstrated excellent cycle life and stability.

Nonsintered Nickel Electrodes   To reduce the total nickel content and simplify manufacturing techniques, emphasis is being placed on development of nonsintered-nickel electrodes.[7] A typical electrode is prepared by kneading and calendering an electrode strip consisting of nickel hydroxide hydrate, graphite as a conductive additive, and Teflon as a plastic binder. The electrode strips are laminated on either side of a nickel or nickel-plated steel current collector and pressed into final electrode shape. Figure 19.1 shows the steps in the process.

Controlled Microgeometry (CMG)   The controlled microgeometry foil electrode[8] is manufactured by preparing a thin electroformed nickel foil sheet with a selected hole

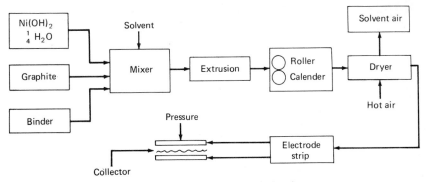

FIG. 19.1   Rolling process. (*Courtesy of Energy Research Corp.*)

pattern. A typical design consists of 4-μm foil with 20-mm holes and 40% open area. A layer of nickel hydroxide hydrate active material, typically 40 μm thick, is coated onto the nickel foil. Multiple layers of the coated foil are then stacked to make the composite electrode. In the stacking process, the holes line up to allow electrolyte access to all areas of the electrode. The nickel foil serves as a current collector throughout the electrode structure and is contacted at one edge for terminal connections.

Table 19.2 summarizes various types of nickel electrodes, listing their nickel content, energy density capabilities, and a relative ranking of life and electrode cost.

TABLE 19.2   Summary of Nickel Electrode Types

Type	Total electrode, g/Ah	Total electrode, cm³/Ah	Nickel content, g/Ah	Life rank	Cost rank
Pocket	12–15	4.2	4	Good	Moderate
Sintered	9–10	3	5	Excellent	High
Pressed	6–6.5	3.6	3	Good	Low
CMG	5–5.5	3.3	3	Fair	High

19.3.2 Zinc Electrode

Zinc electrode fabrication techniques are less varied. Most electrodes are prepared by starting with zinc oxide, possibly zinc metal powder, additives, and a plastic binder. Electrodes can be prepared by rolling electrode strips as shown in Fig. 19.1 or by dry powder pressing.

Starting with zinc oxide as the active material results in an electrode constructed in the fully discharged condition. In the initial formation cycle, the zinc oxide is charged to metallic zinc, which imparts good conductivity and structural integrity to the electrode. Therefore the physical characteristics of zinc electrodes are acceptable from various manufacturing processes.

Zinc electrodes are normally designed to be 60 to 70% porous in their all-metallic state. This level of porosity allows for the active material growth encountered during the discharge, due to the lower density of the resultant zinc hydroxide reaction product. If sufficient porosity is not provided, the discharge product has the effect of coating over the passivating unused active material and rendering it nonfunctional.

Zinc electrode problems manifest themselves during repeated cycling due to the partial solubility of the discharge product in the electrolyte. A portion of the discharge product dissolves during discharge in limited electrolyte cells and redeposits back on the electrode structure on charge. Due to a number of effects (thermal, concentration gradients, current density distribution, and electrolyte inventory variations), the dissolution and redeposition process is not mirrored. As the cycling proceeds in normal prismatic, limited electrolyte cells, this nonuniformity has the net effect of moving active material from the electrode edges toward the center, as shown in Fig. 19.2. For static electrolyte cell designs, a number of approaches are currently under investigation to reduce the extent of this material redistribution. Such approaches include the complexing of the zinc reaction product with an additive such as calcium hydroxide to limit the zinc from going into solution,[9] additives to the electrolyte to reduce the solubility of zinc in the electrolyte,[10] and inert additives to the zinc electrode to create an electrode substructure for more uniform redeposition of zinc in solution.[11]

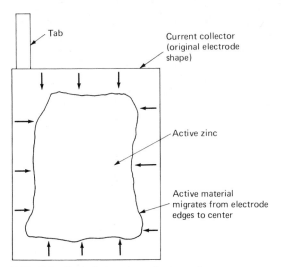

FIG. 19.2   Typical zinc electrode shape change.

Independent of the material redistribution problems, the zinc electrode can also be subject to a dendrite problem. At critical current densities and zincate concentration levels, the zinc redeposited on charge can grow into needlelike dendrites. These dendrites have a tendency to grow toward the counterelectrode and can penetrate the separator, causing cell shorting. The dendrite-forming process can be managed by additives to the electrolyte that have plating leveling effects, the use of membrane-type separator materials to retard zinc dendrite penetration, and controlled-charging techniques that minimize the occurrence of the critical zincate concentration and current density.

The zinc electrode is also prone to corrosion in the alkaline electrolyte, which results in the evolution of hydrogen gas. The thermodynamic zinc potential is approximately 0.4 V above hydrogen evolution. The zinc exists in aqueous electrolytes because of its high hydrogen overvoltage characteristic. Additives or impurities to the zinc electrode which reduce its hydrogen overvoltage contribute to accelerated corrosion and hydrogen gassing. Other additives can reduce its corrosion rate by raising the hydrogen overvoltage. Mercury is typically used for this purpose but is being replaced with other additives.[12]

19.3.3 Separator

The nickel-zinc battery separator system has the multiple function of providing an electrolyte inventory for the respective electrodes and the retarding zinc dendrite shorting. This is usually accomplished with multiple material absorbers on the surface of the positive and negative electrodes and a microporous or membrane-type material to retard zinc dendrite penetration. It is in the area of this microporous or membrane material that most development work is now concentrated. Typical membrane materials are listed in Table 19.3. There are also variations of most of these materials. The

TABLE 19.3   Candidate Separator Materials for Nickel-Zinc cells

Type	Thickness, in Dry	Wet	Resistivity, $m\Omega/in^2$	Chemical stability	Source
Cellophane	0.001	0.003	15–18	Poor	Du Pont
Sausage casing	0.003	0.008	22	Poor	Union Carbide
PVA	0.0015	0.003	35	Poor	Mono Sol
NASA-astropower coated asbestos[13]	0.012	0.012	40	Good	Kimberly Clark
Electroporous[14]	0.002	0.0023	40	Good	Gould, Inc.
Celgard[15]	0.001	0.001	40	Good	Celanese
RAI	0.001	0.001	40–60	Good	Radiation Applications, Inc.
SAC	0.002	0.0022	12	Good	Surface Activation Corp.
Microporous	0.008	0.008	120	Good	W. R. Grace
	0.004	0.004	35	Good	W. R. Grace
Nylon base	0.002	0.002	40	Good	Energy Research Corp.
PVA cross-linked[16]	0.002	0.0025	35	Fair	Energy Research Corp.

commercially available standards are cellophane, fibrous sausage casing, and PVA. These materials are attacked slowly in the cell environment by alkaline solutions and oxygen; therefore current development is directed at more chemically stable materials.

19.3.4 Electrolyte

Most nickel-zinc cells employ potassium hydroxide electrolyte in the 25 to 35% concentration range. Some experimental cells use approximately 1% lithium hydroxide as an additive. This is used in most nickel batteries and possibly serves to improve the semiconductive properties of the nickel active material. Other additives to the electrolyte such as borates have been reported to reduce zinc solubility.[17]

19.4 CONSTRUCTION

19.4.1 Prismatic Cell

Cells of traditional design consist of parallel-plate, prismatic construction as pictured in Fig. 19.3. Cells are assembled with multiple positive and negative electrodes with separator layers between electrodes. Individual positive and negative electrodes are all connected to common positive and negative terminals.

19.4.2 Vibrocel

A vibrating zinc version of the nickel-zinc battery is being developed to overcome the deficiencies of the rechargeable zinc electrode.[18,19] In this concept, the negative electrode is vibrated in a plane parallel to the electrode surface during charge, as shown in Fig. 19.4. The vibrating motion of the negative electrode creates a turbulence in the electrolyte which has a dual effect. First, the microturbulence in the bulk of the electrolyte uniformly disperses any dissolved zinc oxide and ensures a uniform plating of zinc.

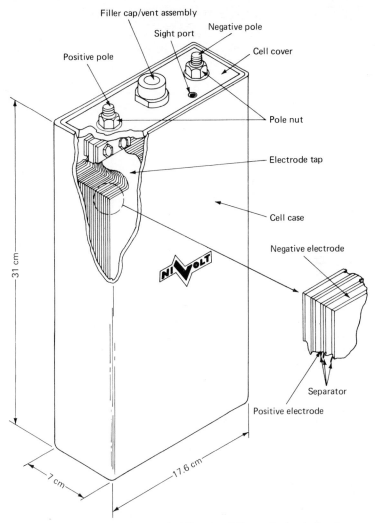

FIG. 19.3   Nickel-zinc cell, 250 Ah. (*Courtesy of Energy Research Corp.*)

Second, the turbulence reduces the zincate concentration gradients and discourages formation of dendrites. If a dendrite should form, it is brushed off by a plastic mesh separator between the positive and negative electrodes.

The vibration is stopped when the charge is completed and is not used during the battery discharge. The cell design is such that a sufficient excess of electrolyte exists within the cell, and under normal conditions all the discharge product goes into solution. Excellent cycle life capability has been reported for this concept—one cycle a day for 80% DOD over a 4-y period. However, the design adds complexity and the need to vibrate every zinc electrode. Debits are potential seal problems, extra inventory of electrolyte resulting in a somewhat bulkier battery, larger interelectrode spacings

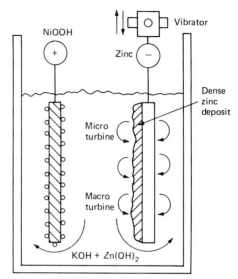

FIG. 19.4 Principle of operation of nickel-zinc Vibrocel. (*Courtesy of ESB Technology Co.*)

required for the zinc electrode vibration, and the need for a structurally independent nickel electrode since it is unsupported in this construction.

Over the years there have been a number of attempts at dynamic solutions to the zinc problem—slurry electrolyte,[19] rotating electrodes,[20] and windshield wipers.[21] None of these dynamic solutions has found acceptance to date, especially because the added complexities have traded away the advantages of the nickel-zinc system.

19.4.3 Sealed Cells

Various designs of sealed nickel-zinc cells have been reported.[19,20] For a sealed cell to function properly, gases generated during charge must be recombined within the cell. A finite quantity of ampere-hour overcharge is required to replace capacity removed during discharge. This is a consequence of the charge acceptance efficiency of the nickel positive electrode. The nickel electrode charging potential exists at a voltage very close to the oxygen evolution potential on the nickel electrode structure. Therefore during battery charging, the oxygen evolution process competes with the nickel electrode charging process. The split in charging current between charging active material and evolving oxygen is a function of state of charge of the nickel electrode, charging current density, temperature, and electrode structure. Figure 19.5 shows a typical nickel electrode charge acceptance characteristic. The zinc electrode, unlike the nickel electrode, accepts charge efficiently with virtually no hydrogen evolution until all the active material is charged. It then exhibits a voltage jump corresponding to the hydrogen overvoltage potential and hydrogen evolution on the zinc electrode. Figure 19.6 shows individual electrode potentials vs. an Hg/HgO reference and the total cell voltage.

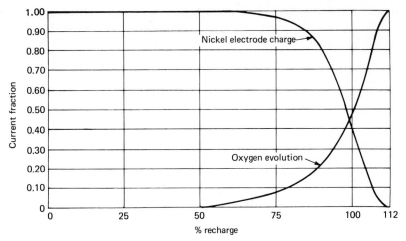

FIG. 19.5 Typical nickel electrode charge acceptance.

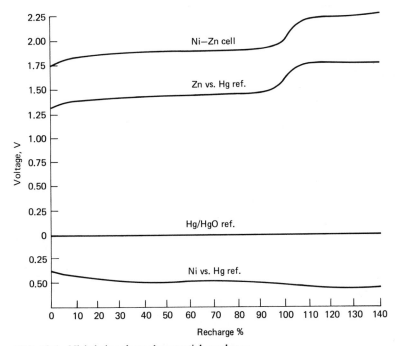

FIG. 19.6 Nickel-zinc electrode potential on charge.

In a classic sealed-cell design, the cell is constructed with an excess of uncharged zinc. During the charging process, oxygen evolution occurs before the zinc becomes fully charged. If the cell is constructed to allow for the oxygen to react with the zinc electrode, the zinc electrode state of charge can be maintained at less than the fully converted form, and hydrogen evolution will not occur. The sealed cell is then capable of a steady-state overcharge without the occurrence of hydrogen evolution or a buildup of oxygen pressure. This occurs if the overcharge (oxygen generation) rate is equal to or less than the rate at which oxygen is recombined with the zinc electrode. This is the same sealed-cell concept utilized in sealed nickel-cadmium secondary batteries. Unlike nickel-cadmium batteries, the use of membrane separators in nickel-zinc cells limits the transport and oxygen access to the zinc electrode for direct recombination. Therefore rates of recombination and rates of steady-state overcharge of nickel-zinc cells are lower than in sealed nickel-cadmium cells.

To enhance the recombination process, sealed cells are generally designed with reduced quantities of electrolyte, which has a deleterious effect on active material utilization and cell heat transfer characteristics. To enhance the recombination, some studies have focused on the use of a third electrode (fuel cell gas diffusion type) constructed of either activated carbon or silver catalyst to serve as the reaction site for oxygen reduction.[25]

In sealed nickel-zinc cells, some provision must also be made to manage the small amount of hydrogen created by zinc electrode corrosion. The direct recombination of hydrogen with the active nickel electrode is quite slow and not sufficient to avoid hydrogen pressure buildup, except in cells which use highly amalgamated zinc electrodes. Some cells have incorporated a third electrode connected to the positive or a catalytic recombination device to recombine the hydrogen with oxygen generated during overcharge.

Advantages of the sealed-cell concept are reduced or no maintenance, retention of the state of charge of the zinc electrode in a region that minimizes zinc dendrite formation, and a reduced rate of shape change. Balancing these factors are the need to construct gas-tight cells, the need for cell containers that can withstand the oxygen pressure levels generated, increased attack on the separator material due to the oxygen environment, and reduced capacity due to starved electrolyte operation. It is, therefore, probable that sealed nickel-zinc cells will be the second-generation version of the nickel-zinc cells discussed.

19.5 PERFORMANCE CHARACTERISTICS

19.5.1 Discharge Performance

The open-circuit voltage of the nickel-zinc cell is 1.73 V. A typical discharge is illustrated in Fig. 19.7, which shows the discharge of a four-cell, nominal 250-Ah battery module, constructed with nonsintered-nickel electrodes. The discharge curve is flat with a plateau voltage of about 1.6 V per cell. The battery actually delivered 310 Ah and, on the basis of a cell assembly weight of 31 kg, 64 Wh/kg. Figure 19.8 gives the characteristics of a typical 90-Ah cell discharged at 25°C at various loads, and Fig. 19.9 shows the performance of this cell at different temperatures. As with other types of batteries, the nickel-zinc cell exhibits a decrease in operating voltage and capacity as the discharge rate is increased and the operating temperature lowered. The decrease in operating voltage is caused by internal resistance losses and the capacity decrease by poorer active material utilization. These effects are less pronounced in alkaline batteries than in lead-acid types. The nickel-zinc battery system has a relatively low internal resistance (0.01 to 0.02 mΩ/Ah) and is capable of high rates of charge and

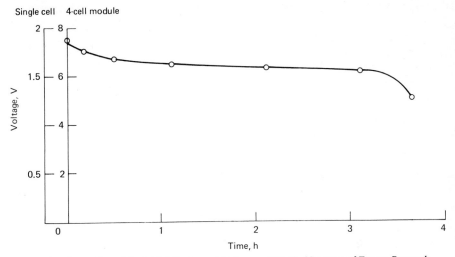

FIG. 19.7   Four-cell module, initial discharge (discharge of 85 A). (*Courtesy of Energy Research Corp.*)

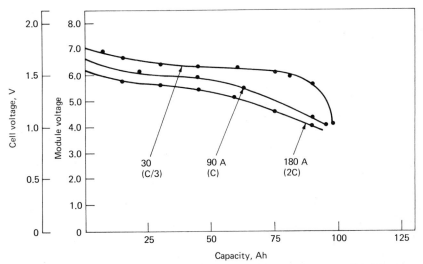

FIG. 19.8   Effect of discharge rate on performance (charge and discharge at 25°C). (*Courtesy of Yardney Electric Corp.*)

discharge. The performance data are summarized in Fig. 19.10, which shows the service life of a nickel-zinc cell at various discharge loads and temperatures on a weight and volume basis.

Figure 19.11 shows the typical performance of a 350-Ah nickel-zinc Vibrocel cell.

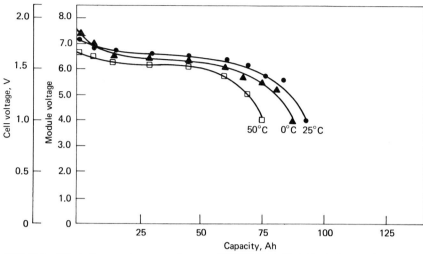

FIG. 19.9 Effect of temperature on performance [4XNZEV6-1; 30-A (C/3) discharge]. (*Courtesy of Yardney Electric Corp.*)

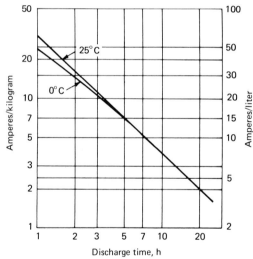

FIG. 19.10 Service life of nickel-zinc cell.

19.5.2 Charge Retention and Cycle Life

The retention of charge of a nickel-zinc battery, as a function of temperature after 3 months of storage, is shown in Fig. 19.12. In most secondary battery applications, the energy removed in each charge-discharge cycle (depth of discharge) varies from cycle to cycle. The battery capacity is selected for the maximum discharge duration, but the battery is normally recharged after each use or mission regardless of the energy

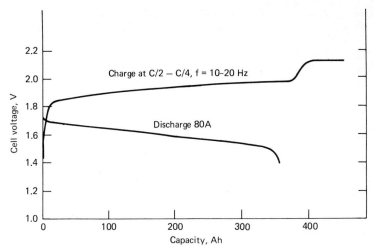

FIG. 19.11   Typical performance of 350-Ah nickel-zinc Vibrocel. (*Courtesy of ESB Technology Co.*)

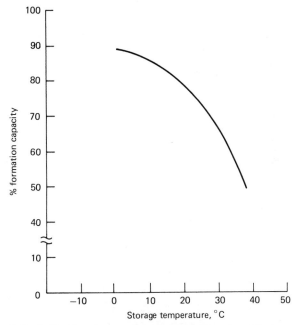

FIG. 19.12   Capacity retention vs. temperature, 3-month storage. (*Courtesy of Yardney Electric Corp.*)

removed. This results in a large number of applications in which the battery is subjected to many shallow depth cycles and a few deep discharge cycles. For characterization purposes, most cycle life testing is conducted at fixed deep depth of discharge. In any application, as cycling proceeds, the usable capacity diminishes, and, hence, it is necessary to define the end of life (e.g., 60% of original capacity). Therefore, cycle life can vary significantly depending on the conditions of use. Figure 19.13 presents data showing the effect of DOD on the cycle life of a nickel-zinc cell. Figure 19.14 shows the capacity retention vs. cycle life for an experimental 400-Ah cell.

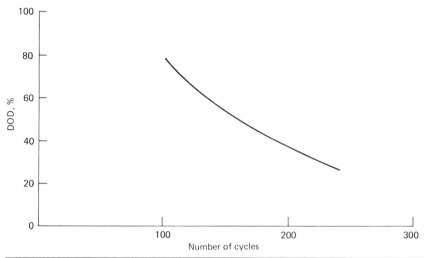

	Charge		Discharge		
Depth of discharge	Time, h	Current, A	Time, h	Current, A	Cycle time, h
75	2.25	83	5.75	36	8
50	1.5	83	4.5	31	6.0
20	0.75	83	2.0	31	3.0

FIG. 19.13   Effect of depth of discharge on cycle life; first-generation electric vehicle nickel-zinc cells. (*Courtesy of Energy Research Corp.*)

19.6 CHARGING CHARACTERISTICS

The nickel-zinc battery requires an overcharge to replace the nickel electrode's capacity removed from the previous discharge. Classic vented cells are normally constructed with an excess of zinc-active material. Charging schemes to be considered are modified constant potential in which an initial current limit is established, constant potential, and constant current or two-step constant current. Figure 19.15 shows the characteristics of a typical 100-Ah cell charged by the three methods. In all cases, the proper amount of overcharge is required to bring the battery back to the fully charged state. This overcharge fraction will vary with the different charge methods due to the changing characteristics of the nickel electrode's charge acceptance efficiency. All methods require some determination of the moment when the fully charged state is achieved.

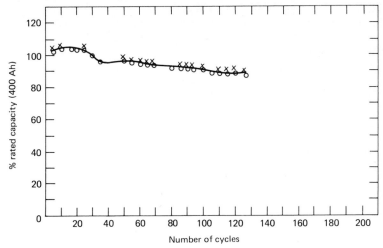

Design capacity (C/3)   400 Ah
Maximum delivered capacity   431 Ah*
Specific energy density   76.7 Wh/kg
Volumetric energy density   138 Wh/L
Peak power 80% DOD (C/3)   118 W/kg
Specific power 58% DOD (C/3)   45 W/kg
*All values average of two cells.

FIG. 19.14   Capacity–cycle life relationship for nickel-zinc cell. (*Courtesy of Gould, Inc.*)

Unfortunately the nickel-zinc battery does not exhibit any characteristic change in voltage to indicate the fully charged state. Consideration has been given to the measurement of quantities of gas evolved as an indication of a state of charge, or in some cases a 10 to 15% overcharge fraction is selected. A coulometric pilot cell, constructed with a matched active nickel electrode and hydrogen electrode in tandem with the main battery, can be used for this purpose. The hydrogen pressure of the pilot cell indicates the proper amount of overcharge and when to terminate the charge. Typical data for this system are shown in Fig. 19.16.

In any of the above schemes, the cumulative ampere-hour overcharge over a successive number of cycles will bring the zinc electrode up to a point where all its active material is fully charged and hydrogen evolution occurs. With constant-current charging, a second plateau charging step will occur when hydrogen evolution starts. It is not practical to use this voltage step as a point for charge termination since it represents virtually a return of 100% ampere-hour removed from previous discharge and does not provide for the necessary overcharge to bring the nickel electrode to its fully charged state. Charging beyond the hydrogen breakpoint is the region in which zinc dendrite formation is favored, and some manufacturers may recommend periodic battery conditioning to avoid reaching this state of charge of the zinc electrode. Conditioning consists of discharging the cell normally and then short-circuiting the positive and negative electrodes to discharge all the reserve zinc in the electrode.

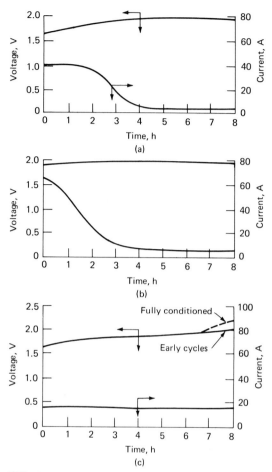

FIG. 19.15   Charge characteristics for 100-Ah nickel-zinc cell: (*a*) constant potential current limited (40 A); (*b*) constant potential; (*c*) constant current.

19.7 APPLICATIONS

At present, significant interest in the nickel-zinc battery is directed at electric vehicle applications (see Part 4). For typical small delivery vans or passenger cars, batteries of approximately 100-V, 250-Ah size are needed. Figure 19.17 shows the layout of a 60-cell electric vehicle battery under development at Energy Research Corporation. The battery weight fraction of a passenger electric vehicle is 30 to 40%. Battery energy consumption is dependent on vehicle design and driving pattern. For a well-designed

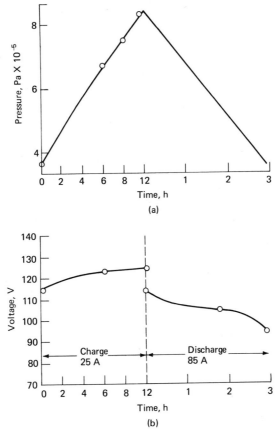

FIG. 19.16 Charge-discharge of 66-cell battery; (*a*) fuel gauge and (*b*) battery. (*Courtesy of Energy Research Corp.*)

vehicle, average battery energy consumption is 150 Wh/km. Therefore, a 25-kWh nickel-zinc battery-powered vehicle is capable of a 160-km range under most driving conditions. Since the load profile that a battery experiences in an electric vehicle varies considerably, a number of standard tests have been established. The most popular test is the J227a, D profile to simulate stop-start city driving as shown in Fig. 19.18. Each 120-s sequence is equivalent to 0.6 km of driving, and the sequence is continually repeated until the battery cannot meet the profile. Figure 19.19 shows the results of a 250-Ah nickel-zinc cell tested in this manner. The commercial success of electric vehicles, yet to be demonstrated, will be dependent on the competitive cost and availability of gasoline and battery life cycle cost.

Nickel-zinc batteries are also being considered for standby load service. Certain

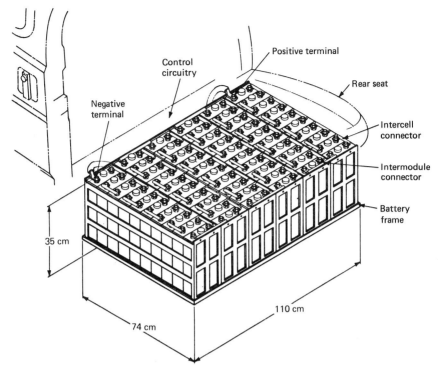

FIG. 19.17   Vehicle battery, 60 cells, 96 V, 250 Ah. (*Courtesy of Energy Research Corp.*)

military and commercial installations require backup power for commercial power failures. Large-capacity nickel cells are under development for this purpose. Sizes as large as 2000 Ah have been built at Energy Research Corporation. Other potential applications are small appliances, mine lamps, military aircraft, military electronics, torpedoes, off-the-road electric vehicles, electric wheelchairs, and other medical power supplies.

19.8 SPECIAL HANDLING

Nickel-zinc batteries contain concentrated alkaline electrolyte that can be hazardous if handled improperly. Particular care should be taken to avoid contact with eyes since severe burns can occur. Eye protection should be worn at all times by anyone working with alkaline batteries, and if electrolyte-eye contact occurs, the eyes should be rinsed with a large quantity of water or diluted boric acid and a physician consulted.

Hydrogen and oxygen gases are generated during open-circuit charge and over-charge of vented batteries, and provision must be made to vent these gases. Accumulation of hydrogen gas in a confined space can result in a potentially dangerous explosive mixture. Furthermore a battery should never be placed in a sealed secondary outer container without pressure relief since the gases generated can develop significant pressure.

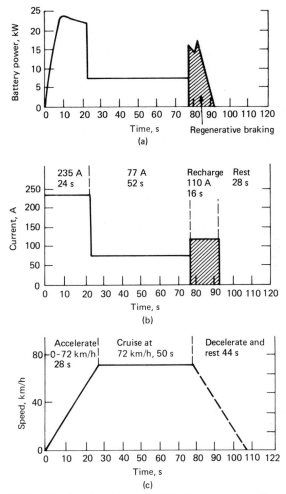

FIG. 19.18 Simulated electric vehicle J227a, D test profile: (a) electric vehicle battery power profile; (b) simplified test currents for nominal 100-V battery; (c) test standard.

REFERENCES

1. J. J. Drumm, to Drumm Battery Co., Ltd., British Patent 365,125, July 17, 1930; J. J. Drumm, to Drumm Battery Co., Ltd., U.S. Patent 1,955,115, April 17, 1934; J. J. Drumm and A. G. Burnell, to Drumm Battery Co., Ltd., U.S. Patent 2,277,636, March 24, 1942.

2. S. U. Falk and A. J. Salkind, *Alkaline Storage Batteries,* Wiley, New York, 1969, chap. 1.

3. P. E. Plehn, *Nickel-Cadmium Storage Batteries,* German FIAT Report no. 800, 1947.

4. A. Fleischer, "Sintered Plates for Nickel Cadmium Batteries," *J. Electrochem. Soc.* **94**:289, 1948.

5. P. L. Baugault, P. E. Lake, and E. J. Casey, U.S. Patent 2,831,044.

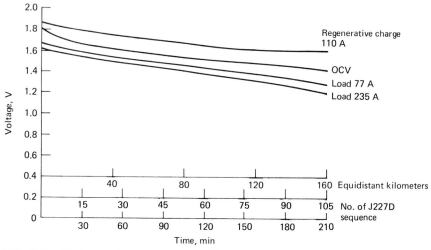

FIG. 19.19   Discharge voltage as a function of J227AD discharge sequence (250-Ah nickel-zinc cell).

6. R. L. Beauchamp, Extended Abstracts of Electrochemical Society 138th National Meeting, Paper no. 65, fall 1970; Bell Telephone Laboratories, Inc., U.S. Patent 3,653,967, April 4, 1972.

7. M. Klein, to Energy Research Corp., U.S. Patent 3,898,099, August 5, 1975.

8. J. Edwards and J. Whittle, to International Nickel Co., Inc., U.S. Patent 3,785,867, Jan. 15, 1974.

9. Hitachi Maxwell, Japanese Patent 51-12642, 1976.

10. R. Thornton and E. Carlson, "Properties of Alternate Electrolytes for Secondary Zinc Batteries," *J. Electrochem Soc.* **127** (7):1448, 1980.

11. A. Charkey, U.S. Patent 4,022,953, May 10, 1977.

12. A. Himy, *Proc. 28th Power Sources Conf.,* Electrochemical Society, Pennington, N.J., 1978.

13. J. M. Parry, C. S. Leung, and R. Wells, "Structure and Operating Mechanisms of Inorganic Separators," NASA Report CR-134692, March 1974.

14. G. D. Bucci et al., *DOE Electrochemical Contractors Conf.* Arlington, Va., December 1979.

15. H. Bierbaum and R. Isaacson, "Micro Porous Polymeric Films," *Prod. Eng.* February 1974.

16. Energy Research Corporation, Annual Report, 1978, R&D and Demo of Ni-Zn Batteries for EV Propulsion ANL/OEPM-78-10, October 1979.

17. F. Schneider et al., "A Bipolar NiO (OH) - $K_3 BO_3$ - 2n Accumulator," *4th Int. Power Sources Symp.,* Brighton, England, September 1972.

18. O. Von Krusenstierna, "High-Energy Long Life Zinc Battery for Electric Vehicles," *6th Int. Power Sources Symp.,* Brighton, England, September 1976.

19. A. J. Salkind and E. Pearlman, "Batteries and Electric Cells, Secondary Cells Alkaline," in *Kirk-Othmer Encyclopedia of Chemical Technology,* Wiley, New York, 1978, pp. 618–619.

20. P. R. Shipps, "Metal-Air Systems," *Proc. 20th Power Sources Conf.,* Electrochemical Society, Pennington, N.J., 1966.

21. Z. Stachurski, U.S. Patent 3,440,098, April 22, 1969.

22. L. R. McCoy, "Zinc-Nickel Oxide Secondary Battery," *10th IECEC Conf.*, Newark, Del., 1975, p. 1131.

23. A. Charkey, "Sealed Nickel-Zinc Cells," *Proc. 25th Annu. Power Sources Conf.*, Electrochemical Society, Pennington, N.J., 1972.

24. T. Shirogani, "Development of Sealed Cylindrical Ni-Zn Secondary Battery," Toshiba Research & Development Center, Kawasaki, Japan.

25. M. Klein, "Nickel-Zinc Development," *DOE Electrochemical Contractors Conf.* Germantown, Md., June 1978.

20

Iron Electrode Batteries

by
Ralph J. Brodd

20.1 GENERAL CHARACTERISTICS

Iron electrodes have been used in rechargeable battery systems since the introduction of the nickel-iron rechargeable battery at the turn of the century by Junger in Europe and Edison in the United States.[1] Even today the batteries are produced in a fashion similar to that of the original construction. New constructions are under development which should give better high-rate performance and have lower manufacturing costs. Today the nickel-iron battery is the most important commercial rechargeable system using iron electrodes. Iron-silver batteries have been used in special electronic applications, and iron-air batteries show promise as motive power systems. The characteristics of the iron battery systems are summarized in Tables 20.1 and 20.2.

As designed by Edison, the nickel-iron battery was and is almost indestructible. It has a very rugged physical structure and can withstand electrical abuse such as overcharge, overdischarge, discharged stand for extended periods, and short-circuiting. The battery is best applied where high cycle life at repeated deep discharges is required (e.g., traction applications) and as a standby power source with 10 to 20 y life. Its limitations are low power density, poor low-temperature performance, poor charge retention, and gas evolution on stand. The cost of the nickel-iron battery lies between the lower-cost lead-acid and the higher-cost nickel-cadmium battery. In recent years the nickel-iron battery has been losing its market to the nickel-cadmium battery in many applications, but the new generation of nickel-iron batteries, now under development and demonstration, is expected to find its major applications in electric automobiles and mobile industrial equipment.

TABLE 20.1 Iron Electrode Battery Systems

System	Uses	Advantages	Disadvantages

Iron/nickel oxide (tubular)	Material handling vehicles, underground mining vehicles, miners' lamps, railway cars and signal systems, emergency lighting	Physically almost indestructible
		Not damaged by discharge stand
		Long life, cycling or stand
		Withstands electrical abuse:
		  Overcharge
		  Overdischarged
		  Short-circuiting
			High self-discharge
			Hydrogen evolution on charge and discharge
			Low power density
			Lower energy density than competitive systems
			Poor low-temperature performance
			Damaged by high temperatures
			Higher cost than lead-acid
			Low cell voltage

Iron/air	Motive power	Good energy density Uses readily available materials Low self-discharge	Low efficiency Hydrogen evolution on charge Poor low-temperature performance Low cell voltage
Iron/silver oxide	Electronic	High energy density High cycle life	High cost

TABLE 20.2 System Characteristics

System	Nominal voltage, V		Energy density			Cycle life, 100% DOD
	Open-circuit	Discharge	Wh/kg	Wh/L	W/kg	
Iron/nickel oxide						
Tubular	1.4	1.2	30	60	25	4000
Developmental	1.4	1.2	65	130	145	> 1200
Iron/air	1.2	0.82	80	. . .	40	1000
Iron/silver oxide	1.48	1.26	105	160	. . .	500

20.2 CHEMISTRY

The active materials of the nickel-iron battery are metallic iron for the negative electrode, nickel oxide for the positive, and a potassium hydroxide solution with lithium hydroxide for the electrolyte. The nickel-iron battery is unique in many respects. The overall electrode reactions result in the transfer of oxygen from one electrode to the other. The intimate details of the reaction can be very complex and include many species of transitory existence.[2-4] The electrolyte apparently plays no part in the overall reaction, as noted in the overall reaction given below:

$$Fe + 2NiOOH + 2H_2O \underset{charge}{\overset{discharge}{\rightleftarrows}} 2Ni(OH)_2 + Fe(OH)_2 \quad \text{(first plateau)}$$

$$3Fe(OH)_2 + 2NiOOH \underset{charge}{\overset{discharge}{\rightleftarrows}} 2Ni(OH)_2 + Fe_3O_4 + 2H_2O \quad \text{(second plateau)}$$

The overall reaction is

$$3Fe + 8NiOOH + 4H_2O \underset{charge}{\overset{discharge}{\rightleftarrows}} 8Ni(OH)_2 + Fe_3O_4$$

The electrolyte remains essentially nonvariant during charge and discharge. It is not possible to use the specific gravity of the electrolyte to determine the state of charge as for the lead-acid battery. However, the individual electrode reactions do involve an intimate reaction with the electrolyte.

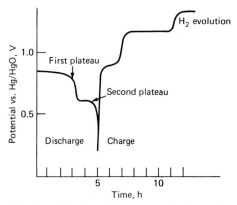

FIG. 20.1   Discharge-charge curve of an iron electrode. (*From Anderson and Ojefors, 1979.*)

A typical charge-discharge curve of an iron electrode is shown in Fig. 20.1. The two plateaus on discharge correspond to the formation of the stable $+2$ and $+3$ valent forms of the iron reaction products. The reaction at the iron electrode can be written as

$$\mathrm{Fe} + n\mathrm{OH}^- \rightarrow \mathrm{Fe(OH)}_n^{2-n} + 2e \qquad \text{(first plateau)}$$

and

$$\mathrm{Fe(OH)}_n^{2-n} \rightarrow \mathrm{Fe(OH)}_2 + n\mathrm{OH}^-$$

$$2\mathrm{Fe(OH)}_2 + 2\mathrm{OH}^- \rightarrow 2\mathrm{Fe(OH)}_3 + 2e \qquad \text{(second plateau)}$$

then

$$2\mathrm{Fe(OH)}_3 + \mathrm{Fe(OH)}_2 \rightarrow \mathrm{Fe}_3\mathrm{O}_4 + 4\mathrm{H}_2\mathrm{O}$$

Iron dissolves initially as the $+2$ species in alkaline media. The divalent iron complexes with the electrolyte to form the $\mathrm{Fe(OH)}_n^{2-n}$ complex, perhaps a ferrate, of low solubility. The tendency to supersaturate plays an important role in the operation of the electrode and accounts for many important aspects of the electrode performance characteristics. Continued discharge forms the $+3$ valent iron which, in turn, interacts with $+2$ valent iron to form $\mathrm{Fe}_3\mathrm{O}_4$.

The superior cycling characteristics of the iron electrode result from the low solubility of the reaction intermediates and oxidized species. The supersaturation on discharge results in the oxidized material forming small crystallites near the reaction site. On charge the low solubility also slows the crystal growth of the iron, thereby helping ensure formation of the original active high-surface-area structure. The low solubility also accounts for poor high-rate and low-temperature performance as the discharged (oxidized) species precipitate at or near the reaction site and block the active surface.

The performance characteristics are substantially improved, however, in the advanced nickel-iron batteries by the use of a superior electrode grid structure, such as fiber-metal, which provides intimate contact with the iron active material throughout the volume of porous structure.

Sulfide addition to the iron electrode radically changes the electrocrystallization kinetics. It increases the supersaturation and makes reaction more reversible. Sulfide also absorbs on the surface to block crystallization sites and raises the hydrogen evolution reaction on charge. Lithium additions seem to make the electrode perform more reversibly, perhaps by enhancing the solubility of the reaction intermediates.

Nickel electrode reactions[6,7] are generally thought to be solid-state-type reactions wherein a proton is injected or rejected from the lattice reversibly on discharge and on charge, respectively,

$$\beta\text{-Ni(OH)}_2 \quad \overset{\text{Transformation}}{\underset{\text{in KOH}}{\rightleftarrows}} \quad \alpha\text{-Ni(OH)}_2$$

| reduction (discharge) $\uparrow\downarrow$ | oxidation (charge) | | oxidation (charge) | reduction (discharge) $\downarrow\uparrow$ |

$$\beta\text{-NiOOH} \quad \overset{\text{overcharge}}{\underset{\text{in KOH}}{\rightarrow}} \quad \gamma\text{-NiOOH}$$

The oxidation (charge) voltage for the α- and β-materials is more positive than the discharge voltage by 60 mV and 100 mV, respectively. The β-Ni(OH)$_2$ is the usual electrode material. It is connected on charge to β-NiOOH with about the same molar volume. On overcharge the γ structure can form. This form also incorporates water and potassium (and lithium) into the structure. Its molar volume is about 1.5 times the β form. This is thought to be responsible in large part for the volume expansion (swelling) which occurs on charging the battery. The α form then results on discharge of the γ form. Its molar volume is about 1.8 times the β form, and the electrode can swell further on discharge. On discharge stand in concentrated electrolyte, the α form converts to the β form. Cobalt additions (2 to 5%) improve the charge acceptance (reversibility) of the nickel electrode.

20.3 CONVENTIONAL NICKEL-IRON BATTERIES

20.3.1 Construction

The construction of the nickel-iron cell is shown in Fig. 20.2. The active materials are filled in nickel-plated perforated steel tubes or pockets. The tubes are fastened into plates of desired dimensions and assembled into cells by interleaving the positive and negative plates. The container is fabricated from nickel-plated sheet steel. The cells may be assembled into batteries in molded nylon cases or mounted into wooden trays. The steel cases may be coated with plastic or rubber for insulation or spaced by insulating buttons.

The manufacturing process has remained relatively unchanged for over 50 y. The processes are designed to produce materials of highest purity and with special particle characteristics for good electrochemical performance.

Negative Electrode   To produce the anode active material, pure iron is dissolved in sulfuric acid. The FeSO$_4$ is recrystallized, dried, and roasted (815 to 915°C) to Fe$_2$O$_3$. The material is washed free of sulfate, dried, and partially reduced in hydrogen (from

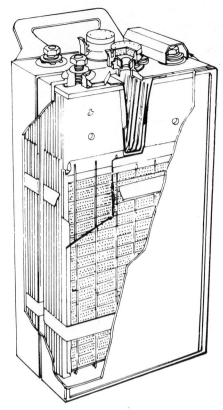

FIG. 20.2 Cross section of a typical nickel-iron battery. (*Courtesy of NIFE Inc.*)

iron and nickel dissolution processes). The resulting material (Fe_3O_4 and Fe) is partially oxidized, dried, ground, and blended. Small amounts of additives, e.g., sulfur, FeS, HgO, are blended in to increase battery life by acting as depassivators, reducing gas evolution, or improving conductivity.

To make the anode current collector, steel strips or ribbon are perforated and nickel-plated. After drying and annealing, the strip is formed into a pocket, about 13 mm wide and 7.6 cm long. One end is left open and filled with the iron active material. A machine automatically introduces the active material and tamps it into the pockets. After filling, the negative pockets are crimped and pressed into openings in a nickel-plated steel frame.

SAFT and Junger have developed pressed- and sintered-iron electrode structures which incorporate a copper matrix to yield better high rate capability. Iron and copper powders in special forms are mixed and pressed into a wire mesh. The final electrode is about 50% iron, 50% copper. An alternate procedure sinters the pressed electrodes in a neutral or reducing atmosphere at 600 to 650°C.

Positive Electrode The positive active material consists of nickel hydroxide in alternate layers with nickel flake. High-purity nickel powder or shot is dissolved in sulfuric acid. The hydrogen evolved is used in making the iron active material. The acidity of the resulting solution is adjusted to pH 3 to 4 and filtered to remove ferric iron and other insoluble materials. If needed, the solution may be further purified to remove traces of ferrous iron and copper. Cobalt sulfate may be added in the proportion of 1.5% to improve nickel electrode performance. The nickel sulfate solution is sprayed into hot 25 to 50% NaOH solution. The resulting slurry is filtered, washed, dried, crushed, and screened to yield particles which pass 20- but not 200-mesh screens.

Special nickel flake (1.6×0.01 mm) is produced by electrodepositing alternate layers of nickel and copper on stainless steel. The electroplate is stripped and cut into squares. The copper is dissolved out in hot sulfuric acid, and the resulting nickel flakes washed free of copper and dried at low temperature to prevent nickel oxide formation. With a new process[8] the flakes of proper shape and size can be produced as a single layer, eliminating the need for deposition of alternate copper layers. As in the negative electrode, the positive-electrode process starts with perforated steel ribbon which is nickel-plated and annealed. The ribbon is wound into tubes with an interlocked seam. Two types, right- and left-wound tubes, are produced, typically of 6.3-mm diameter.

The tubes are filled with alternate layers of nickel hydroxide and nickel flakes. Each layer is tamped (144 kg/cm^2) to ensure good contact. There are 32 layers of flake per centimeter. To prevent the seam from opening during the rigors of charge and discharge, rings are placed around the tubes at uniform intervals of about 1 cm. The tubes are enclosed, and the pinched ends are locked into the nickel-plated steel grid frame. The "rights" and "lefts" are alternated so that any tendency to distort on the part of one tube is counteracted by the next one. The positive electrode can also be made in the pocket plate construction as described above under "Negative electrode."

Cell Assembly   The configuration and size of the tubes and pockets determine the capacity for each plate. The plates are then assembled into electrodes to meet the capacity requirements of each cell.

Each plate group is assembled by bolting a terminal pole and a selected number of plates, depending on capacity, to a steel rod which passes through the grid at the top of the plates. Groups of positive and negative plates are intermeshed to form the element. A cell usually contains one or more negative than positive plates. The cells are made positive-limiting for best cycle life.

The positive and negative plates are separated by hard rubber or plastic pins called "hair pins" or "hook pins," which fit into spaces formed by the tubular positive and flat negative electrodes.

Electrolyte   The electrolyte is a 25 to 30% KOH solution with 50 g/L of LiOH added. The composition of the replacement electrolyte to compensate for losses due to gassing is about 23% caustic with about 25 g/L LiOH. Occasionally the electrolyte is replaced completely to rejuvenate the cell performance. The renewal electrolyte is about 30% KOH with 15 g/L LiOH.

Lithium additions to the electrolyte are important but poorly understood. Lithium improves cell capacity and prevents capacity loss on cycling. Lithium seems to facilitate nickel electrode kinetics. It expands the working plateau on charge and delays oxygen evolution. Some evidence exists for formation of Ni^{4+}, which improves electrode capacity. Lithium also decreases carbonate content in the electrolyte since Li_2CO_3 is not very soluble. Lithium increases the tendency for swelling of the positive active material and increases the resistivity of the cell electrolyte.

Shortly after initiation of charge, hydrogen evolution begins on the iron electrode. The considerable hydrogen evolution on charge presumably helps counteract iron passivation in alkaline solution. Mercury additions also have a similar effect, but only in the early formation cycles.

20.3.2 Performance Characteristics of Nickel-Iron Battery

Voltage   A typical discharge-charge curve of a commercial iron/nickel oxide cell is shown in Fig. 20.3. The cell's open-circuit voltage is 1.4 V; its nominal voltage is 1.2 V. On charge, at rates most commonly used, the maximum voltage is 1.7 to 1.8 V.

Capacity   The capacity of the nickel-iron battery is limited by the capacity of the positive electrode and, hence, determined by the length and number of positive tubes in each plate. The diameter of the tubes generally is held constant by each manufacturer. The 5-h discharge rate is commonly used as the reference for rating its capacity.

The conventional nickel-iron battery has moderate power and energy density and

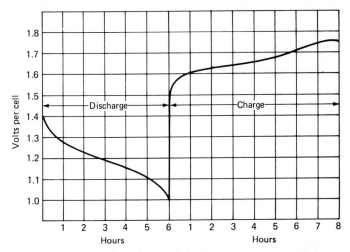

FIG. 20.3   Typical voltage characteristics during a constant-rate discharge and recharge. (*From Ref. 9.*)

is designed primarily for moderate to low discharge rates. It is not recommended for high-rate applications such as engine-starting. The high internal resistance of the cell significantly lowers the terminal voltage when high rates are required. The relationship between capacity and rate of discharge is shown in Fig. 20.4.

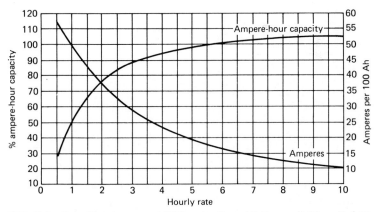

FIG. 20.4   Capacity-rate curves at 25°C, end voltage: 1.0 V per cell. (*From Ref. 9.*)

If a battery is discharged at a high rate and then at a lower rate, the sum of the capacities delivered at the high rate and low rate nearly equals the capacity that would have been obtained at the single discharge rate. This is illustrated in Fig. 20.5.

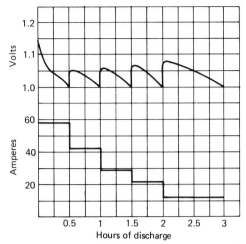

FIG. 20.5 Effect of decreasing rate on cell voltage of a nickel-iron cell.

Discharge Characteristics The nickel-iron battery may be discharged at any current rate it will deliver, but the discharge should not be continued beyond the point where the cell nears exhaustion. It is best adapted to low or moderate rates of discharge (1- to 8-h rate). Figure 20.6 shows the discharge curves at different rates of discharge at 25°C.

Effect of Temperature Figure 20.7 shows the effects of temperature on the discharge. The capacity at 25°C is normally taken as the standard reference value. The decrease in performance is generally attributed to passivity of the iron electrode, decreased solubility of the reaction intermediate, and, at low temperature, increased resistivity

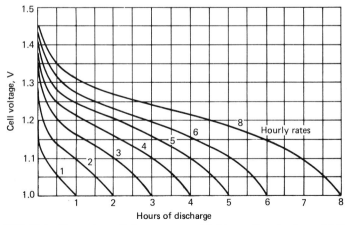

FIG. 20.6 Time-voltage discharge curves, nickel-iron cell, end voltage: 1.0 V per cell. (*From Ref. 9.*)

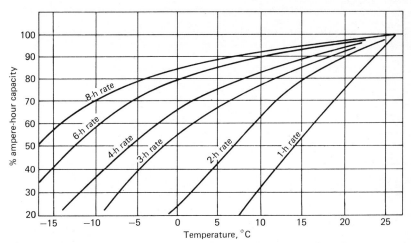

FIG. 20.7 Effect of temperature on capacity at various rates. (*From Ref. 9.*)

and viscosity of the electrolyte along with slower nickel electrode kinetics contribute to the fall-off capacity. Care must be exercised to keep the temperature from exceeding about 50°C as the self-discharge of the nickel positive electrode is accelerated. Also, the increased solubility of iron at high temperature can adversely affect operation of the nickel electrode by incorporation of soluble iron into the nickel hydroxide crystal lattice. The battery is seldom used below −15°C.

Hours of Service The hours of service on discharge that a typical nickel-iron cell, normalized to unit weight (kilogram) and volume (liter), will deliver at various discharge rates and temperatures is summarized in Fig. 20.8.

Self-Discharge The self-discharge rate, charge retention, or stand characteristic of the nickel-iron battery is poor. At 25°C, a cell will lose 15% of its capacity in the first 10 days and 20 to 40% in a month. At lower temperatures, the self-discharge rate is lower. For example, at 0°C the losses are less than one-half of those experienced at 25°C.

Internal Resistance To a rough approximation, the internal resistance R_i can be estimated for tubular Ni-Fe from the equation:

$$R_i \times C = 0.4$$

where R_i = internal resistance, Ω
C = cell capacity, Ah
For example, R_i = 0.004 Ω for a 100-Ah cell. The value of R_i remains constant through the first half of the discharge, then increases about 50% during the latter half of the discharge.

Life The main advantages of the tubular-type nickel-iron battery are its extremely long life and rugged construction. Battery life varies with the type of service but ranges from 8 y for heavy duty to 25 y or more for standby or float service. With moderate care, 2000 cycles can be expected; with good care, for example by limiting temperatures to below 35°C, cycle life of 3000 to 4000 cycles has been achieved.

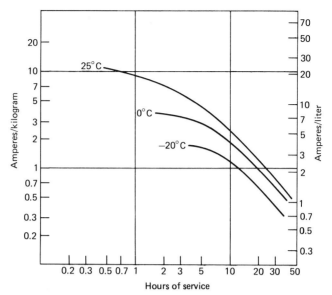

FIG. 20.8 Hours of service of the nickel-iron cell at various discharge rates and temperatures, end voltage: 1.0 V per cell.

The battery is less damaged by repeated deep discharge than any other battery system. In practice, an operator will drive a battery-operated vehicle until it stalls, at which point the battery voltage is a fraction of a volt per cell (some cells may be in reverse). This has a minimal effect on the nickel-iron battery in comparison with other systems.

Charging Charging of the batteries can be accomplished by a variety of schemes. As long as the charging current does not produce either excessive gassing (spray out of the vent cap) or temperature rise (above 45°C), any current can be used. Excessive gassing will require more frequent addition of water. If the cell voltage is limited to 1.7 V per cell, these conditions should not be a consideration. Typical charging curves are given in Fig. 20.9. The ampere-hour input should return 25 to 40% excess of the previous discharge to ensure complete charging. The suggested charge rate is normally between 15 and 20 A per 100 Ah of cell capacity. This rate would return the capacity in the 6- to 8-h time frame. The effect of temperature on charging is shown in Fig. 20.10.

Constant current and modified constant potential (taper) shown in Fig. 20.11 are common recharging techniques. The charging circuit should contain a current-limiting device to avoid thermal runaway on charge. Recharging each night after use (cycle-charging) is the normal procedure. The cells can be trickle-charged to maintain them at full capacity for emergency use. A trickle charge rate of 0.004 to 0.006 A/Ah of battery capacity overcomes the internal self-discharge and maintains the cell at full charge. Following an emergency discharge, a separate recharge is needed. For applications such as railroad signals, charging at a continuous average current may be most economic. Here a modest drain is required when no trains are passing but quite a heavy drain when a train passes, yet the total ampere-hours over a period of 24 h remains

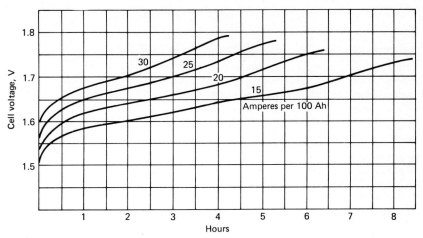

FIG. 20.9   Typical charging voltage for a nickel-iron cell at various rates. (*From Ref. 9.*)

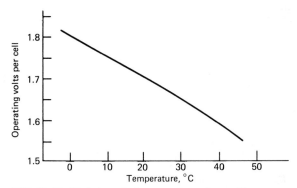

FIG. 20.10   Variation of relay operating voltage with temperature. (*From Ref. 9.*)

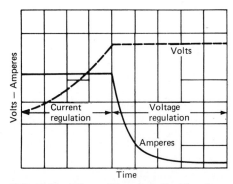

FIG. 20.11   Effects of "regulators" with voltage and current regulation. (*From Ref. 9.*)

fairly constant. For this situation, a constant current equal to that required to maintain the battery can be used.

State of Charge   An Edison test fork can be used to estimate the state of charge. The device connects a resistor across the cell terminals and measures the voltage on a calibrated voltmeter. The cell voltage on heavy discharge is roughly proportional to the state of charge and is indicated on the meter.

20.3.3 Sizes of Nickel-Iron Cells and Batteries

Nickel-iron batteries have been available in sizes ranging from about 5 to 1250 Ah. In recent years they have become less popular, giving way to the lead-acid and nickel-cadmium batteries, and are no longer manufactured by many of the original manufacturers. In Table 20.3 the physical and electrical characteristics of typical nickel-iron batteries are listed.

20.3.4 Special Handling and Use of Nickel-Iron Batteries

The battery should be operated in a well-ventilated ambient to prevent the accumulation of hydrogen. Under certain circumstances hydrogen can be ignited by a spark to cause an explosion with a resulting fire. In multicell batteries the usual precautions in dealing with high voltages should be taken.

If the battery is to be out of service for more than a month, it should be stored in the discharged condition. It should be discharged and short-circuited, then left in that condition for the storage period. Filling caps must be kept closed. If this procedure is not followed, several cycles are required to restore the capacity upon reactivation.

Constant-potential charging is not recommended for conventional nickel-iron batteries. It may lead to a thermal runaway condition which results in dangerous conditions and can severely damage the battery. When the battery nears full charge, the gassing reactions produce heat and the temperature rises, lowering the internal resistance and the cell emf. Accordingly, the charge current increases under constant-potential charge. This increased current further increases the temperature, and a vicious cycle is started. A modified constant-potential charging with current-limiting is, however, acceptable.

20.4   NEW DEVELOPMENTS

20.4.1 Advanced Nickel-Iron Batteries

The desire to use the attractive features of the nickel-iron couple, e.g., ruggedness and long life, in applications requiring high-rate performance and low manufacturing costs has led to the development of advanced nickel-iron batteries with performance characteristics suitable for electric automobiles and other mobile traction applications. The capability of these batteries permits an electric vehicle a range of at least 150 km between charges, acceleration rapid enough to merge into highway traffic, and a cycling life equivalent to 10 or more years of on-the-road service. The advanced battery utilizes sintered-fiber metal (steel wool) plaques, impregnated with active material, for both the

TABLE 20.3 Typical Nickel-Iron Batteries

Nominal capacity, Ah	169	225	280	337	395	450	560	675
Nominal current, A: 5-h discharge ⎫ 7-h discharge ⎭	34	45	56	67	79	90	112	135
Cell weight, filled, kg	8.8	10.8	12.9	15.3	17.4	19.5	24.3	28.6
Installed weight, kg	9.8	12.0	14.3	16.9	19.3	21.7	26.5	31.2
Electrolyte (1.17 kg/L), kg	1.8	2.2	2.6	3.0	3.4	3.8	4.9	5.9
Cell dimensions, [a] mm:								
Length	52	66	82	96	111	125	156	186
Width	130	130	130	130	130	130	135	135
Height	534	534	534	534	534	534	534	534
Battery dimensions, [b] mm:								
Length:								
2 cells	. . .	. . .	. . .	265	295	321	343	343
3 cells	. . .	. . .	. . .	376	421	460		
4 cells	284	367	431	487				
5 cells	346	448	545					
6 cells	408	546						
Width	161	161	161	161	161	161	197	228
Height	568	573	582	582	582	582	590	590

[a] Drawing of cell showing dimensions used in Table 20.3.
[b] Drawing of multicell battery showing dimensions used in Table 20.3. The tolerances are 5 mm, 3 mm, and 3 mm for dimensions L, B, and H, respectively.
SOURCE: Varta Batteries AG, Hannover, Germany.

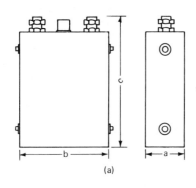

(a)

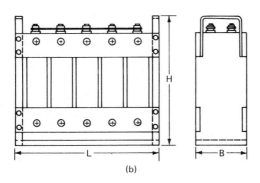

(b)

positive and negative electrodes. Nonwoven polypropylene sheets are used as separators between electrodes. The techniques for making plaques, impregnation and activation, stacking, and assembly are all amenable to high-volume production methods similar to those used in lead-acid battery manufacture.

The battery system design incorporates an electrolyte management system to minimize the maintenance problems associated with its widespread deployment in the public sector. This system, shown schematically in Fig. 20.12, provides for semiautomatic watering of the cells by utilizing a single-point watering port. Flow of electrolyte through the cells during the charge cycle enables heat removal and effective management of gas evolved during the charge. Uniform specific gravity for all cells is ensured and specific gravity maintenance is easily achieved by use of the system.

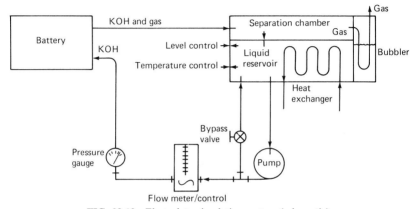

FIG. 20.12 Electrolyte circulation system (schematic).

Both the positive and negative plates of the Westinghouse nickel-iron battery use sintered–steel wool as the substrate. Two methods of active nickel impregnation are under development and are being used in demonstration batteries. An electroprecipitation process (EPP) has been used since the mid-1960s and has demonstrated good performance, ruggedness, and long cycle life. The EPP process electrochemically deposits nickel hydroxide into the porous steel wool substrate. Efficient use of the nickel material is achieved, with active material utilization of 0.14 Ah/g of total electrode. An alternate nickel electrode manufacturing process is being developed which entails the preparation of a nickel hydroxide paste that is then loaded into the steel wool substrate by roll pasting methods. Pasted nickel electrodes have demonstrated performance equivalent to EPP electrodes (0.14 Ah/g of total electrode) while holding promise for a less expensive manufacturing process. The iron electrodes are produced by a pasting process. Iron oxide, Fe_2O_3, is paste-loaded into the steel wool electrode substrate and then furnace-reduced in a hydrogen atmosphere. These electrodes have demonstrated 0.22 Ah/g of total electrode, or better, at C/3 discharge rates.

The performance characteristics of the advanced nickel-iron batteries, as typified by the Westinghouse system, are summarized in Table 20.4. Figure 20.13 shows a typical discharge curve for a 90-cell electric vehicle battery at the C/3 rate. The battery capacity as a function of discharge rate is shown in Fig. 20.14. Cell power and voltage characteristics, as a function of discharge rate and state of discharge, are shown in Fig. 20.15. The variation in battery capability with temperature, based on tests on five-cell

TABLE 20.4 Advanced Nickel-Iron Battery Performance Characteristics[a]

	Demonstrated as of Dec. 1981	Objectives, 1984–86
Capacity,[b] Ah	230	250
Energy density,[b] Wh/kg	55	65
Energy density,[b] Wh/L	110	130
Specific power,[c] W/kg	100	140
Cycle life[d]	> 900[e]	> 1200
Urban range,[e] km		
With regenerative braking	154	190
Without regenerative braking	125	160
Projected production cost,[f] $/kWh	110–180	70

[a]Based on the Westinghouse nickel-iron battery.
[b]At the C/3 rate.
[c]30-s average at 50% state of charge.
[d]Cycle to 80% depth of discharge; life to 75% of rated energy.
[e]Test continuing.
[f]Simulated test at NBTL (National Battery Test Laboratory, Argonne, Ill.) for GE/Chrysler car (ETV-1) on SAE J 227a, D cycle.

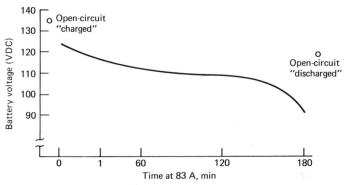

FIG. 20.13 Battery voltage at C/3 (83-A) discharge rate.

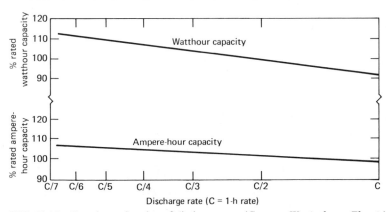

FIG. 20.14 Capacity as function of discharge rate. (*Courtesy Westinghouse Electric*)

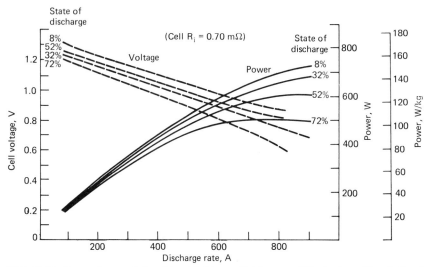

FIG. 20.15   Power characteristics of the nickel-iron cell. (*Courtesy of Westinghouse Electric Corp.*)

modules, is shown in Fig. 20.16.[10] The Eagle-Picher Co. also has developed a nickel-iron battery using sintered-nickel electrodes similar to those described in Part 3, Chap. 17. The iron electrode is similar to the Swedish National Development Corp. iron electrode discussed in Sec. 20.4.2. The performance of this battery is similar to those given in Figs. 20.13 to 20.16.[11]

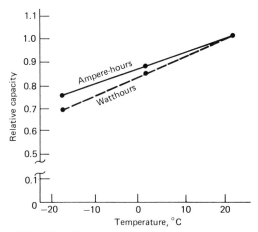

FIG. 20.16   Effect of temperature on capacity of nickel-iron cell (C/3 rate).

20.4.2 Iron/Air Batteries

Rechargeable metal/air batteries have an advantage over conventional systems as only one reactant (the anode material) need be contained within the battery. The electrically rechargeable iron/air cell has a lower specific energy than the mechanically rechargeable cells (see Part 4, Chap. 30) but has the advantage in potentially lower life cycle costs. Unlike zinc, the iron electrodes do not suffer a severe redistribution of active materials or gross shape upon prolonged electrical cycling. The iron/air cell is another candidate as a motive power source, especially for electric vehicles. The cell reactions are:

$$O_2 + 2Fe + 2H_2O \underset{\text{charge}}{\overset{\text{discharge}}{\rightleftarrows}} 2Fe(OH)_2 \qquad \text{(first plateau)}$$

$$3Fe(OH)_2 + \tfrac{1}{2}O_2 \underset{\text{charge}}{\overset{\text{discharge}}{\rightleftarrows}} Fe_3O_4 + 3H_2O \qquad \text{(second plateau)}$$

The iron electrode kinetics are covered in Sec. 20.2. The oxygen electrode reactions follow the kinetic path with peroxide as an intermediate. The oxygen electrode reactions in simple form are

$$O_2 + 2H_2O + 2e \rightarrow H_2O_2 + 2OH^-$$

$$H_2O_2 + 2e \rightarrow 2OH^-$$

The single most important limiting factor in this battery system is the stability of the air electrode, which loses its ability to function reversibly as it undergoes repeated charges and discharges. The oxygen and peroxide evolved on charge and discharge attack the substrate, alter the activity of the catalyst, and delaminate the wetproofing film. Separate air (oxygen) electrodes and circuits can be employed in the charge and discharge modes; however, considerations of system weight and volume favor the use of a bifunctional electrode, i.e., a single electrode capable of sustaining either oxygen reduction or evolution. These electrodes must be stable over the potential range of both reactions, a fact which poses constraint on materials stability.

Several designs are under development; three are described below.

The Swedish National Development Corp. iron/air cell uses the sintered-iron mesh anode construction.[5,12,13] A pore-forming material may be included to control the development of the optimum electrode structure. The resulting pressed matrix is sintered in H_2 at 650°C. The pore-forming material can be leached out after sintering. The active material utilization approaches 65%. The air electrode is a porous-nickel double-layer structure (0.6 mm thick) composed of sintered nickel of coarse and fine porosities. The coarse layer in the electrolyte side is catalyzed with silver and impregnated with hydrophobic agents. The electrodes are welded into a polymer frame and formed into cells, as shown in Fig. 20.17. There are two electrodes for each iron electrode. Air is forced past the electrode at about 2 times the stoichiometric requirement during operation. A schematic and photo of a 30-kWh battery are shown in Figs. 20.18 and 20.19, respectively. Electrolyte circulation is used to control heat balance and remove gases generated during operation. Carbon dioxide is scrubbed from the incoming air using NaOH. The air is then humidified in a moisture exchange. Overall the auxiliary systems require less than 10% of the system output.

Typical charge-discharge curves for an average cell in the iron/air battery are shown in Fig. 20.20. The marked difference in charge and discharge voltage accounts largely for the low overall system efficiency. Figure 20.21 shows the power-producing charac-

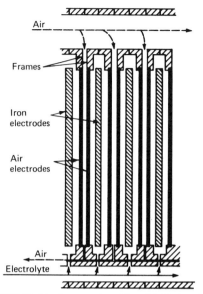

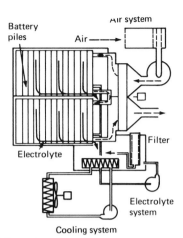

FIG. 20.18 Schematic cross section of Swedish National Development Corp. iron/air battery including auxiliary system. (*From Anderson and Ojefors, 1979.*)

FIG. 20.17 Cross section of Swedish National Development Corp. iron/air battery pile. (*From Anderson and Ojefors, 1979.*)

FIG. 20.19 Swedish National Development Corp. 30-kWh iron/air battery system. (*Courtesy of Swedish National Development Corp.*)

teristics. The system is capable of over 1000 cycles, limited by the gradual deterioration of the air electrode.

The Westinghouse iron/air battery uses a construction similar to that described for the Swedish National Development Corp. system.[14,15,16] The sintered-iron electrode is somewhat similar to that described before. The iron electrode for this iron/air battery has a high active iron content and lower cycle life compared with the electrode described previously. Particles of iron powder are sintered to form a structure without the steel fiber substrate. Electrodes with this construction have demonstrated up to 0.44 Ah/g. The air electrode is bifunctional and uses a Teflon-bonded carbon-based structure with complex silver catalysts (silver content is less than 1 mg/cm) supported on a silver-plated nickel screen. The Westinghouse system uses a horizontal flow concept to improve performance and control gas and heat. Good life has been demonstrated for over 500 cycles, toward a goal of 1000 cycles, with an air electrode of potentially very low cost. A summary of the projected characteristics of a 40-kWh battery is given in Table 20.5.

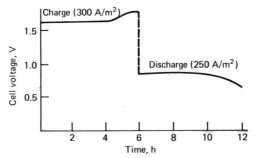

FIG. 20.20 Charge-discharge voltages for a cell in a Swedish National Development Corp. iron/air battery. (*From Bryant, Liu, and Buzzelli, 1978.*)

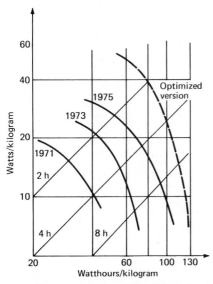

FIG. 20.21 Performance of Swedish National Development Corp. iron/air batteries. (*From Anderson and Ojefors, 1979.*)

TABLE 20.5 Characteristics of Iron/ Air Electric Vehicle Battery

Electric vehicle characteristics
Weight: 900 kg curb weight Range: 240 km
Battery characteristics
Capacity: 40 kWh Power: 10 kW continuous power Weight: 365 kg Volume: 0.25 m³ Cost: $40/kWh

The Siemens cell is similar except that the air electrode is fabricated with two layers: a hydrophilic layer of porous nickel on the electrolyte side for oxygen evolution and a hydrophobic layer [carbon black bonded with Teflon (PTFE) and catalyzed with silver] on the air side for oxygen reduction. The dual porosity helps to shield the silver catalyst from oxidation. As many as 200 cycles were achieved.[17]

20.4.3 Silver-Iron Battery

The silver-iron battery has been limited in use because of its high cost. Its theoretical energy density is essentially equal to that of the more popular silver-zinc system. However, the silver-iron battery has good cycle life compared with the silver-zinc and provides a battery of higher reliability, longer life, and more durability where high specific energy content is essential.[18-21] Figure 20.22 shows a 4.5-kWh battery designed for telecommunication use.

FIG. 20.22   4.5-kWh telecommunications iron/silver oxide battery. (*Courtesy of Westinghouse Electric Corp.*)

The battery reaction is

$$Fe + 2AgO + H_2O \underset{charge}{\overset{discharge}{\rightleftarrows}} Fe(OH)_2 + Ag_2O \qquad \text{(first plateau)}$$

$$Fe + Ag_2O + H_2O \underset{charge}{\overset{discharge}{\rightleftarrows}} Fe(OH)_2 + 2Ag \qquad \text{(second plateau)}$$

$$3Fe(OH)_2 + Ag_2O \underset{charge}{\overset{discharge}{\rightleftarrows}} Fe_3O_4 + 3H_2O + 2Ag \qquad \text{(third plateau)}$$

The overall reaction is

$$3Fe + 4AgO \underset{charge}{\overset{discharge}{\rightleftarrows}} Fe_3O_4 + 4Ag \qquad \text{voltage} = 1.26 \text{ V}$$

The iron electrodes are produced as described above for developmental nickel-iron cells. Six silver electrodes and seven iron electrodes are used to form a 140-Ah unit cell, with iron electrodes as the outside plates.

Typical charge-discharge curves are shown in Fig. 20.23. The electrolyte is KOH of 1.31 specific gravity with 15 g/L LiOH added. Multilayer microporous and non-woven felt polypropylene separators are used in the form of a bag surrounding the iron electrode. The cells are encased in molded styrene containers. The cycle life at 100% depth of discharge and 100% overcharge is shown in Fig. 20.24. The cells can withstand several complete reversals without appreciable adverse effect on cell capacity. The cells give 95% of their 25°C capacity at 0°C. The battery has been used in telecommunication systems and has been proposed as a small rechargeable battery for hearing-aid applications.

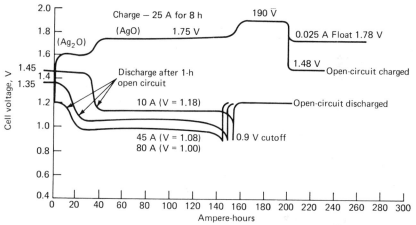

FIG. 20.23   Charge-discharge characteristics of a nominal 140-Ah iron/silver oxide cell. (*From Brown, 1977.*)

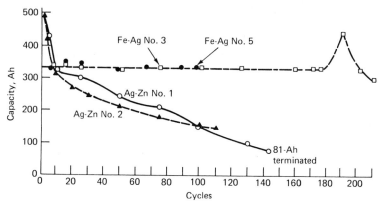

FIG. 20.24   Cyclic life performance on zinc/silver oxide and iron/silver oxide prototype cells. (*From Brown, 1977.*)

REFERENCES

1. S. U. Falk and A. J. Salkind, *Alkaline Storage Batteries,* Wiley, New York, 1969.
2. A. J. Salkind, C. J. Venuto, and S. U. Falk, "The Reaction at the Iron Alkaline Electrode," *J. Electrochem. Soc.* **111**:493 (1964).
3. R. Bonnaterre, R. Doisneau, M. C. Petit, and J. P. Stervinou, in J. H. Thompson (ed.), *Power Sources,* vol. 7, Academic, London, 1979, p. 249.
4. L. Ojefors, "SEM Studies of Discharge Products from Alkaline Iron Electrodes," *J. Electrochem. Soc.* **123**:1691 (1976).
5. B. Anderson and L. Ojefors, in J. H. Thompson (ed.), *Power Sources,* vol. 7, Academic, London, 1979, p. 329.
6. H. Bode, K. Dehmelt, and J. Witte, "Zur Kenntnis Der Nickeloxidelektrode. I. Uber Das Nickel (II) Hydroxidhydrat," *Electrochem. Acta* **11**:1079 (1966).
7. D. Tuomi, "The Forming Process in Nickel Positive Electrodes," *J. Electrochem Soc.* **123**: 1691 (1976).
8. INCO ElectroEnergy Corp. (formerly ESB, Inc.), Philadelphia, Pa.
9. "Nickel Iron Industrial Storage Batteries," Exide Industrial Marketing Division of ESB, Inc., 1966.
10. F. E. Hill, R. Rosey, and R. E. Vaill, "Performance Characteristics of Iron Nickel Batteries," *Proc. 28th Power Sources Symp.,* Electrochemical Society, Pennington, N.J., 1978, p. 149.
10a. R. Rosey and B. E. Tabor, "Westinghouse Nickel-Iron Battery Design and Performance," EV Expo 80, EVC #8030, May 1980.
10b. W. Feduska and R. Rosey, "An Advanced Technology Iron-Nickel Battery for Electric Vehicle Propulsion," *Proc. 15th IECEC,* Seattle, August 1980, p. 1192.
11. R. Hudson and E. Broglio, "Development of the Nickel-Iron Battery System for Electric Vehicle Propulsion," *Proc. 29th Power Sources Conf.,* Electrochemical Society, Pennington, N.J., 1980.
12. L. Carlsson and L. Ojefors, "Bifunctional Air Electrode for Metal-Air Batteries," *J. Electrochem. Soc.,* **127**:525 (1980).
13. L. Ojefors and L. Carlson, "An Iron-Air Vehicle Battery," *J. Power Sources,* **2**:287 (1977/78).
14. W. A. Bryant, C. T. Liu, and E. S. Buzzelli, "Iron-Air Battery Characteristics," *Proc. 28th Power Sources Symp.* Electrochemical Society, Pennington, N.J. 1978, p. 152.
15. E. S. Buzzelli, U.S. Patent 4,168,349, Sept. 18, 1979.
16. E. S. Buzzelli, C. T. Liu, and W. A. Bryant, "Iron Air Batteries for Electric Vehicles," *Proc. 13th IECEC,* San Diego, August 1978, p. 745.
17. H. Cnoblock, D. Groppel, D. Kuhl, W. Nippe, and G. Siemsen, in D. H. Collins (ed.), *Power Sources,* vol. 5, Academic, London, 1975, p. 261.
18. O. Lindstrom, in D. H. Collins (ed.), *Power Sources,* vol. 5, Academic, London, 1975, p. 283.
19. The Silver Institute Letter, vol. 7, no. 3, 1977.
20. J. T. Brown, Westinghouse R&D report 77-5E6-SILEL-R1, Oct. 4, 1977, paper presented at Electrochemical Society Meeting, Las Vegas (extended abstract).
21. E. Buzzelli, "Silver-Iron Battery Performance Characteristics," *Proc. 28th Power Sources Symp.,* Electrochemical Society, Pennington, N.J., 1978, p. 160.

21

Silver Oxide Batteries

by
Stephen F. Schiffer

21.1 GENERAL CHARACTERISTICS

The rechargeable silver oxide batteries are noted for their high energy and power density. The high cost of the silver electrode, however, has limited their use to applications where high energy density is a prime requisite, such as lightweight radio and electronics equipment, submarines and torpedoes, and space applications.

The characteristics of the silver oxide secondary batteries are summarized in Table 21.1.

The first recorded use of a "silver battery" was by Volta with his now-historic silver-zinc pile battery, which he introduced to the world in 1800.[1] This battery dominated the scene in the early nineteenth century, and, during the next 100 y, many experiments were made with cells containing silver and zinc electrodes. All these cells, however, were of the primary (nonrechargeable) type.

The first person to report on a workable secondary silver battery was Jungner in the late 1880s.[2] Although he experimented in the early stages with iron/silver oxide and copper/silver oxide batteries (which reportedly delivered as much as 40 Wh/kg), he settled on the cadmium silver oxide battery for his experiments with electric car propulsion. The short cycle life and high cost of these batteries, however, made them commercially unattractive. During the next 40 y other scientists experimented with various electrode formulations and separators, but without much practical success. It was the French professor Henri André who provided the key to the practical rechargeable zinc/silver oxide (silver-zinc) cell in his paper published in 1941.[3,4] He described the use of a semipermeable membrane, cellophane, as a separator which would retard the migration of the soluble silver oxide to the negative plate and also impede the formation of zinc "trees" from the negative to the positive plate, the two major causes of cell short circuits.

In the 1950s, interest was revived in the silver-cadmium cell using the then newly available silver-zinc and nickel-cadmium technology. This system provided improved

TABLE 21.1   Advantages and Disadvantages of Silver Oxide Secondary Cells

Advantages	Disadvantages
Silver-zinc (zinc/silver oxide)	
Highest energy per unit weight and volume	High cost
High discharge rate capability	Low cycle life
Moderate charge rate capability	Decreased performance at low temperatures
Flat discharge voltage curve	Sensitivity to overcharge
Low self-discharge	
Silver-cadmium (cadmium/silver oxide)	
High energy per unit weight and volume (approx. 60% of silver-zinc)	High cost
Flat discharge voltage curve	Decreased performance at low temperatures
Nonmagnetic construction	
Silver-iron (iron/silver oxide)	
High energy and power capability	High cost
Good capacity maintenance	Water and gas management requirements
Overcharge capability	Not yet proven in field operation

cycle life over silver-zinc. These batteries were first commercialized by Yardney International Corporation. More recently, Westinghouse has reported on the commercial application of a silver-iron battery (see Part 3, Chap. 20) in which they have sought to "eliminate the zinc plate problems with a trouble-free iron plate, ease the separator materials and life problem and shift the deep discharge capacity stability to that limited by the silver plate".[5] The goal now, as for the past two centuries, is to provide the high energy content and power capability of the silver electrode in an improved-life, lower-cost, commercially viable secondary battery.

Zinc/silver oxide batteries provide the highest energy per unit weight and volume of any commercially available secondary batteries. They can operate efficiently at extremely high discharge rates, and they exhibit good charge acceptance at moderate rates and low self-discharge. The disadvantages are low cycle life (ranging from 10 to 200 deep cycles, depending on design and use), decreased performance at low temperatures, sensitivity to overcharge, and high cost. Rates as high as 20 times the nominal capacity (20C rate) can be obtained from specially designed silver-zinc batteries because of their low internal impedance; these high rates, however, must often be limited in time duration because of a potentially damaging temperature rise within the cells.

Cadmium/silver oxide batteries have been viewed as a compromise between the high energy density but short life of the silver-zinc system and the long cycle life but low energy density of the nickel-cadmium system. Their energy density is roughly 2 to 3 times higher than that of nickel-cadmium, nickel-iron, or lead-acid batteries, with relatively long cycle life, especially during shallow cycling. Charge retention is excellent. In addition, the ability to fabricate the cells without use of magnetic materials has made them the battery of choice for several scientific satellite programs. The major disadvantage of the silver-cadmium system is cost; the cost per unit energy is even higher than that for silver-zinc. Additionally, their low-temperature discharge characteristics and their high-rate properties are not as good as those of silver-zinc.

Iron/silver oxide batteries may provide high energy and power capability, with long service life under deep discharge use. They are capable of withstanding overcharge and overdischarge without damage and could provide good capacity maintenance with cycling. Disadvantages are, once again, cost and also the need for gas and water management in overcharge applications. Sufficient field data have not been published for these batteries to date to permit complete characterization of their properties.

All three systems also offer the advantages of long dry shelf life and of providing a flat discharge voltage during the major portion of their discharge. This latter characteristic is related to the fact that as the silver oxide is reduced to metallic silver during discharge, the conductivity of the silver electrode increases and serves to counteract polarization effects.

21.2 CHEMISTRY

21.2.1 Cell Reactions

The overall electrochemical cell reactions for the silver-zinc, silver-cadmium, and silver-iron systems, all of which use aqueous solutions of potassium hydroxide (KOH) for electrolyte, can be summarized as follows:

$$AgO + Zn + H_2O \underset{\text{charge}}{\overset{\text{discharge}}{\rightleftarrows}} Zn(OH)_2 + Ag$$

$$AgO + Cd + H_2O \underset{\text{charge}}{\overset{\text{discharge}}{\rightleftarrows}} Cd(OH)_2 + Ag$$

$$4AgO + 3Fe + 4H_2O \underset{\text{charge}}{\overset{\text{discharge}}{\rightleftarrows}} Fe_3O_4 \cdot 4H_2O + 4Ag$$

These are simplified equations; there is still no general agreement on the detailed mechanisms of these reactions or on the exact form of all the reaction products, some of which are known to be unstable.

21.2.2 Positive-Electrode Reactions

The charge and discharge processes of the silver electrode in alkaline systems are of special interest because they are characterized by two discrete steps which manifest themselves as two plateaus in the charge and discharge curves. The reaction occurring at the silver electrode at the higher voltage plateau is shown as

$$2\,AgO + H_2O + 2e \underset{\text{charge}}{\overset{\text{discharge}}{\rightleftarrows}} Ag_2O + 2OH^-$$

and at the lower voltage plateau is shown as

$$Ag_2O + H_2O + 2e \underset{\text{charge}}{\overset{\text{discharge}}{\rightleftarrows}} 2Ag + 2OH^-$$

As shown, these reactions are reversible.

21.3 CELL CONSTRUCTION AND COMPONENTS

Secondary silver cells have been produced in prismatic, spirally wound cylindrical, and button shape configurations. The most common shape is the prismatic. The construction of a typical prismatic cell is shown in Fig. 21.1. This cell contains flat plates which are wrapped with multiple layers of separator to provide mechanical separation and

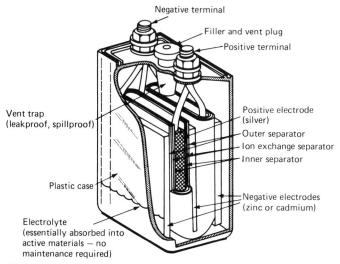

Negative terminal

Filler and vent plug

Positive terminal

Vent trap
(leakproof, spillproof)

Positive electrode
(silver)

Outer separator

Ion exchange separator

Inner separator

Plastic case

Negative electrodes
(zinc or cadmium)

Electrolyte
(essentially absorbed into
active materials — no
maintenance required)

FIG. 21.1   Zinc/silver oxide secondary battery—cutaway view.

inhibit migration of the silver to the zinc plate and the growth of zinc dendrites toward the positive plate. The plate groups are intermeshed and the pack is placed in a tightly fitting case (Fig. 21.2). Because of the relatively short shelf life of the activated silver cells, they are usually supplied by the manufacturers in the dry-charged or dry-unformed condition, with filling kits and instructions. The cells are filled with elec-

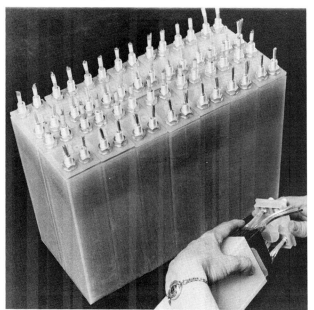

FIG. 21.2   Cell stack being assembled into cell case, model HR85, 100-Ah silver-zinc battery. (*Courtesy of Yardney Electric Corp.*)

trolyte and activated just prior to use. They may also be supplied in the filled and ready-to-use condition if required by the user.

The mechanical strength of these cells is generally excellent. The electrodes are generally strong and are fitted tightly into the containers. The cell containers are made of high-impact plastics. Various designs of these cells, when properly packaged, have met the high-shock and acceleration requirements of missiles and torpedoes with no degradation.

21.3.1 Silver Electrodes

The most common fabrication technique for silver electrodes is by sintering silver powder onto a supporting silver grid. The electrodes are manufactured either in molds (as individual or as master plates which are later cut to size) or by continuous rolling techniques; they are then sintered in a furnace at approximately 700°C.

Alternate techniques include the dry processing and pressing, or slurry pasting, with a binder of AgO or Ag_2O onto a grid. If pasted, the plates are often sintered, during which the silver oxide is converted into metallic silver and the organic additives are burned off. The grid may be a woven or expanded metal form of silver, silver-plated copper, or, in some cases, nickel.

After being cut to size and having wires or tabs welded onto an appropriately coined (compressed) area to carry current to the cell terminals, the electrodes are either electroformed (charged in tanks against inert counterelectrodes) before assembly into cells or assembled into the cells in the metallic state and later charged in the cell.

Grid material, density, and thickness; electrical lead type and size; and final electrode size, thickness, and density are all design variables which depend on the intended application for the cells.

21.3.2 Zinc Electrodes

Zinc electrodes are most widely made by dry-pressing, by a slurry or paste method, or by electrodeposition. In the dry-pressing method, a mixture of zinc oxide, binder, and additives is compressed around a metal grid; this is normally done in a mold. The grid usually has the current-carrying leads prewelded in place. As the unformed powder electrodes have little strength, one component of the separator system, the negative interseparator, is usually assembled around the electrode as part of the fabricating operation. Rolling techniques have also been developed to permit continuous fabrication of dry-powder electrodes.[6]

In the paste or slurry method, the mixture of zinc oxide, binder, and additives is combined with water and applied continuously to a carrier paper or directly to an appropriate metal grid. Again, the negative interseparator is usually integral to provide needed physical strength. After drying, multiple layers of these pasted slabs may be pressed together about a pretabbed grid to form the final electrode. These plates may be assembled unformed into the cell or may be electroformed in a tank against inert counterelectrodes.

Electrodeposited negative electrodes are manufactured by plating zinc, in tanks, onto metallic grids. The plates must then be dried and pressed or rolled to the desired thickness and density.

The zinc electrode is acknowledged as the life-limiting component in both the silver-zinc and nickel-zinc systems. Accordingly, much work has been done in the area of additives for these electrodes, both to reduce the hydrogen evolution and to improve cycle life. The common additive to reduce hydrogen evolution has traditionally been mercury (1 to 4% of the total mix), but this is being replaced, for personnel safety reasons, by small amounts or mixes of the oxide of lead, cadmium, indium, thallium, or gallium.[7-11] Many other (proprietary) additives have been introduced into the zinc electrode by various manufacturers in attempts to increase life.

Zinc electrodes also suffer capacity loss, which results from "shape change" or material migration from the surface of the zinc electrode. Several approaches have been taken to improve the stability of the zinc electrode: (1) an excess of zinc is used to compensate for losses during cycling, (2) oversized electrodes are used on the basis that shape change starts on the electrode edges where current densities are higher, and (3) binders, such as Teflon or potassium titanate, are used to hold the active materials together.

As is the case for the silver electrodes, the grid material, additives, and final electrode size, thickness, and density are all design variables which depend on the final application.

21.3.3 Cadmium Electrodes

Most silver-cadmium cells contain cadmium electrodes that are manufactured by pressed-powder or pasting techniques. Although other methods have been used, such as impregnating nickel plaque with cadmium salts as is done for nickel-cadmium cells, the most common method in silver-cadmium cells is to press or paste a mixture of cadmium oxide or cadmium hydroxide with a binder onto a silver or nickel grid. These processes are similar to those used for the pressed- and pasted-zinc electrodes described above.

21.3.4 Iron Electrodes

The iron electrodes used here are generally manufactured by powder-metallurgy techniques (see Part 3, Chap. 20).

21.3.5 Separators

The separators in the silver cells must meet the following major requirements:

1. Provide a physical barrier between the positive and negative electrodes
2. Have minimum resistance to the flow of electrolyte and ions
3. Prevent migration of particles and dissolved silver compounds between the positive and negative electrodes
4. Be stable in the electrolyte and cell operating environment

In general, secondary silver-zinc and silver-cadmium cells require a minimum of three different separators, as shown in Fig. 21.1. The "inner separator," or positive interseparator, serves both as an electrolyte reservoir and as a barrier to minimize oxidation of the main separator by the highly oxidative silver electrode. This separator is usually made of an inert fiber such as nylon or polypropylene.

The "outer separator," or negative interseparator, also serves as an electrolyte reservoir and can also, ideally, serve to stabilize the zinc electrode and retard zinc penetration of the main separator, thus minimizing dendrite growth. Much work has been done recently to develop improved inorganic negative electrode interseparators utilizing such materials as asbestos and potassium titanate. Improvements in cell life have been reported as a result of this work.[7,10–12]

The main "ion exchange" separator remains the key to the life of the secondary silver cell. It was André's[3] use of cellophane as a main separator that first made the secondary silver cells feasible. The cellulosics (cellophane, treated cellophane, and fibrous sausage casing) are usually employed in multiple layers as the main separators for these cells. Again, much work has been done in recent years to develop improved separators utilizing such materials as radiation-grafted polyethylene,[13] inorganic separators,[10–12,14] and other synthetic polymer membranes. Improved cell life has been reported through use of these new membranes either alone or in combination with the cellulosics. They have yet to be applied extensively to commercial silver cells, however, because of drawbacks usually involving high impedance or availability and cost.

21.3.6 Cell Cases

The cell cases must be chemically resistant to attack by the corrosive concentrated potassium hydroxide electrolyte and to oxidizing effects of the silver electrodes. They must also be strong enough to contain any internal pressure generated in the cells and to maintain structural integrity throughout the anticipated range of environmental conditions that will be experienced by the cells.

The majority of secondary silver cells are assembled in plastic cases. The plastic most commonly used is an acrilonitrile-styrene copolymer (SAN). This material is relatively transparent and can be easily sealed by solvent cement or epoxy. However, its relatively low softening temperature (80°C) precludes it for use in some applications. A wide variety of plastics have been used for cell cases. Table 21.2 lists some of these materials and gives their pertinent characteristics. Metal cases have been used for some sealed-cell and button cell applications; however, these can present problems in sealing and in electrically isolating the electrodes from the cases and are not often used.

21.3.7 Other Components

The electrolyte used in secondary silver cells is generally an aqueous solution (30 to 45% concentration) of potassium hydroxide (KOH). Terminals are typically made of steel or brass and are almost always silver-plated for maximum conductivity and corrosion resistance.

21.4 PERFORMANCE CHARACTERISTICS

21.4.1 Performance and Design Tradeoffs

The secondary silver cells provide high energy capability while utilizing minimum weight and volume. The advantages and disadvantages of each of the systems have been described earlier in this chapter. The performance of the cells for any specific application will depend on the internal design and history of the cells. It is rare that one can select an "off-the-shelf" cell that will meet all the requirements of a specific specialized application.

TABLE 21.2 Properties of Typical Plastic Cell Case and Cover Materials

Type	SAN	\	ABS	Nylon	Modified PPO	Poly-arylether	Poly-sulfone
Trade name	LAN 29	Cycolac X-27	Cycolac GSM — Lustran 440	Zytel 151	Noryl SE-1	Arylon T	P-1700
Characteristics							
Specific gravity	1.07	1.06	1.04 1.06	1.07	1.06	1.14	1.24
Transparency	Yes	No	No No	No	No	No	Yes
Tensile strength, 10^3 psi	11.9	7.3	6.3 6.5	8.8	9.6	7.5	10.2
Flexural strength, 10^3 psi	16.0	11.5	10.7 9.7	...	13.5	11.0	15.4
IZOD impact strength, ft·lb per inch notched	0.45	2.6	7.0 4.0	0.85	5.0	8.0	1.3
Hardness—Rockwell R	83	112	102 ...	114	119	117	120
Heat-deflection temp. at 264 psi, °F	193	244	212 202	180	265	300	345
Dielectric strength, V/mil	460	...	427 410	...	550	430	425
Thermal Conductivity, Btu/h/Ft²/°F/in	0.87	2.38	1.55 ...	1.5	...	2.07	1.8

Starting with the basic parameters, the cell design will consist of a series of compromises—to obtain the most favorable combination of voltage, electrical capacity, and cycle life characteristics within the allowable battery weight and volume.

Assuming, for example, a nominal 1.5 V per silver-zinc cell at low current densities (0.01 to 0.03 A/cm^2) and lower voltages at higher currents, the designer selects the number of cells for the application. The problem is increased if high current pulse loads are required and the battery must provide voltage above the minimum allowable at the high rate, while not exceeding the maximum allowable voltage at initial low rates. The size of the cell is then chosen by dividing the allowable volume by the calculated required number of cells.

The voltage, current, electrical capacity, and cycle-life requirements must then be reviewed in conjunction with the allowable weight and the environmental conditions that the battery must be able to withstand. Each of these will be a factor in determining the choice of separator material for the cell. The stability and number of layers of separator must be sufficient to provide the desired wet life under these conditions while having a resistance low enough to prevent undue voltage drops at the high current loads. Each of these requirements is also a factor in choosing the number of electrodes within the cell. As the number of electrodes (and thus the active electrode area) is increased, the current density during any discharge (amperes per square centimeter) is decreased—raising the output voltage. It should be noted that a cell design optimized for high discharge rates will, by nature of the design, have a reduced capacity under low discharge rates. This is a result of having many electrodes—each of which must be wrapped with the required number of layers of separators. Given a fixed internal volume, it follows that less space is available for active electrode material in such a high-rate cell.

The cell must also be designed to contain enough active electrode material (e.g., silver and zinc) to supply the required electrical capacity for the desired number of cycles. Theoretically, 2 g of silver and 1.2 g of zinc are required in the cell for each ampere-hour of electrical capacity desired. Since these values are the theoretical capability of the pure material, and since some of the active materials will go into solution with each charge-discharge cycle, the designer must work with higher values—in the order of magnitude of 4 g of silver and 3.5 g of zinc per nominal ampere-hour for multiple-cycle cells. Other design variables, such as silver powder particle size, will also ultimately affect cell performance.

Because of the above considerations, the performance curves shown below must be viewed in general terms as characteristic of the systems and not necessarily of specific cells for a specific application.

21.4.2 Discharge Performance for Zinc/Silver Oxide Cells

The open-circuit voltage of the zinc/silver oxide cell is 1.82 to 1.86 V. The discharge, as shown in Fig. 21.3, is characterized by two discrete steps, the first corresponding to the divalent oxide and the second corresponding to the monovalent oxide. The flat portion of the curves is referred to as "plateau voltage." This voltage is rate-dependent; at high rates the voltage steps may be obscured.

The performance of the cell at various discharge rates and temperatures can be seen

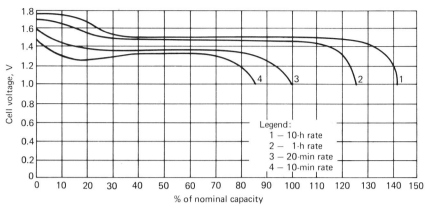

FIG. 21.3   Typical discharge curves of the silver-zinc cell at various rates, 20°C.

in Figs. 21.4 and 21.5, showing the effect on plateau voltage and capacity. The high rate capability of the zinc/silver oxide cell is due to the electrical conductivity of the silver grid and the conductivity of the positive electrode as it is discharged, as well as the thin multiplate design of the cell. The performance of the cell falls off with decreasing temperature, particularly below −20°C. Heating the cell with external heaters or by internal heat generated by high-rate discharge can improve the performance at low ambient temperatures.

The performance characteristics of the zinc/silver oxide cell are summarized in Figs. 21.6 and 21.7, which can be used to determine the capacity, service life, and voltage of a cell under a variety of discharge conditions. These figures present typical performance data. Performance differences can occur for each specific cell design and even for each cell, depending on the cycling history, state of charge, storage time, temperature, and other conditions of use.

Figures 21.3 to 21.7 are specifically for high-rate (HR) designs. For many applica-

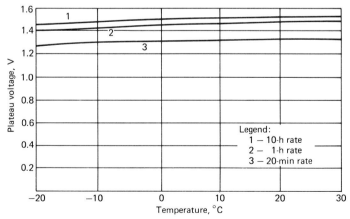

FIG. 21.4   Typical effect of temperature on plateau voltage for high-rate type silver-zinc cells (operated without heaters).

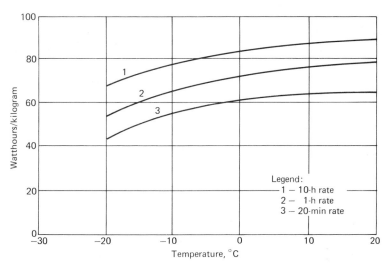

FIG. 21.5 Typical effect of temperature on energy per unit weight for silver-zinc cells (operated without heaters).

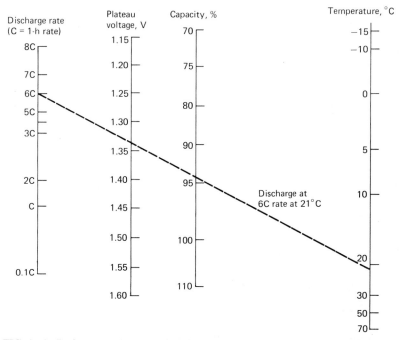

FIG. 21.6 Performance characteristics of silver-zinc cells under variable conditions. (To find the capacity and the plateau voltage of a silver-zinc cell, draw a straight line between the discharge rate and the ambient temperature at which the cell is stabilized.)

21-11

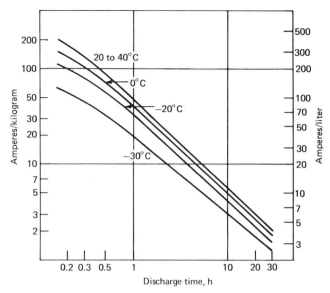

FIG. 21.7 Service life of the silver-zinc cell at various discharge rates and temperatures.

tions, tradeoffs can be made to provide longer life at the expense of somewhat lower energy density. Alternative low-rate (LR) designs contain additional layers of separator, meaning, of necessity, fewer electrodes with higher impedance and lower capacity within a given cell volume. Typically, the LR cell cannot be discharged at higher than the 1-h rate and will provide about 3 to 5% lower average voltage and capacity than its HR counterpart at the 1-h rate. However, the LR cells do provide substantial wet shelf life and cycle life advantages, as shown in Table 21.3.

21.4.3 Discharge Performance for Cadmium/ Silver Oxide Cells

The open-circuit voltage of the cadmium/silver oxide cell is 1.40 to 1.42 V. Typical discharge curves at 20°C are given in Fig. 21.8, with the two-step flat discharge typical of the silver oxide electrode. The discharge characteristics are similar to the zinc/silver oxide cell except for the lower operating voltage; ampere-hour capacities are about the same.

The capacity and discharge voltage of the cell are temperature-dependent, again similar to the zinc/silver oxide cell. The effect of temperature and discharge rate on voltage and capacity is shown in Figs. 21.9 and 21.10, respectively. The recommended operational temperature range is −25 to 70°C, with the optimum performance obtained between 10 to 55°C. With heating, the temperature range can be lowered to −60°C.

The performance characteristics of the cadmium/silver oxide cell are summarized in Figs. 21.11 and 21.12, which can be used to determine the capacity, service life, and voltage levels under a variety of discharge conditions.

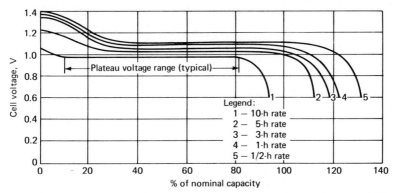

FIG. 21.8   Typical discharge curves of the silver-cadmium cell at various rates, 20°C.

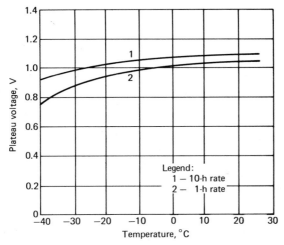

FIG. 21.9   Typical effect of temperature on plateau voltage for silver-cadmium cells (discharged without heaters).

21.4.4 Impedance

The impedance of the silver oxide cells is normally low but can vary considerably with many factors, including the separator system, current density, the state of charge, stand time, cell age, temperature, and, importantly, cell size. In a study of the effect of storage time on the impedance of silver/zinc cells, initial values of 5 to 11 mΩ for partially charged cells were reported,[16] with the values rising to as much as 3 Ω following 8 months storage at 21°C and 9 to 15 Ω following 8 months at 38°C. Cells stored in the fully discharged condition retained their low impedance (ranging from 2 to 10 mΩ) throughout the entire test period. The high-impedance cells returned to normal low values, however, within several seconds of the start of discharge.

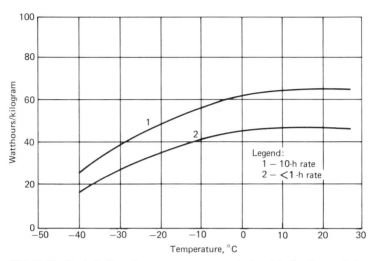

FIG. 21.10 Typical effect of temperature on energy/unit weight for silver-cadmium cells (discharged without heaters).

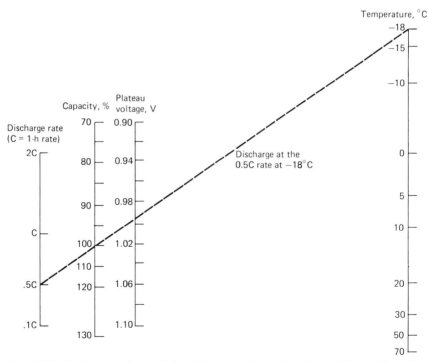

FIG. 21.11 Performance characteristics of silver-cadmium cells under variable conditions. (To find the capacity and the plateau voltage of a silver-cadmium cell, draw a straight line between the discharge rate and the ambient temperature at which the cell is stabilized.)

21-14

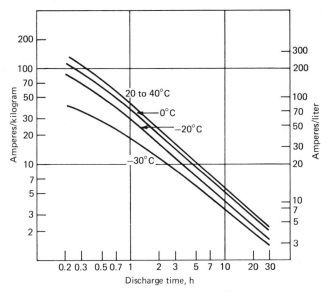

FIG. 21.12   Service life of the silver-cadmium cell at various discharge rates and temperatures.

21.4.5 Charge Retention

The charge retention of the activated and charged silver oxide cells is better than that of most secondary batteries, with retention of 85% of charge after 3 months storage at 20°C. As with other chemical reactions, the rate of loss of charge is dependent on storage temperature (see Fig. 21.13). Storage at -20 to 0°C is recommended for maximizing charge retention. In the dry and charged condition, properly sealed and stored cells will retain their charge for as long as 5 to 10 y.

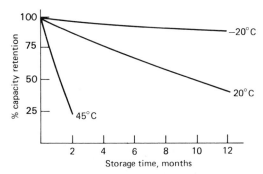

FIG. 21.13   Capacity retention for silver-zinc and silver-cadmium cells at various storage temperatures.

21.4.6 Cycle Life, Wet Life

The cycle life and wet life of these silver oxide cells will vary greatly with design and with operating and storage conditions. The life of the zinc/silver oxide cell is poor, which is one of its main disadvantages. This poor life is due to the solubility of the active materials, the properties of the separator, the growth of zinc dendrites, and the change of shape of the zinc electrode. The life of the cadmium/silver oxide cell is much better, but it is not as good as that of the nickel oxide batteries. The nominal wet life and cycle life ratings for the silver-zinc and the silver-cadmium cells are given in Table 21.3.

In a study to evaluate the capabilities of silver-cadmium batteries for satellite applications, an extensive test program was run on three-cell, 3-Ah silver-cadmium batteries at various depths of discharge.[15] These results are summarized in Table 21.4, showing the increase of cycle life with decreasing depth of discharge.

TABLE 21.3 Nominal Life Characteristics of Secondary Silver Cells*

	High-rate Ag-Zn	Low-rate Ag-Zn	Ag-Cd
Wet shelf life	6–12 months	1–3 y	2–3 y
Cycle life†	10–50 cycles	100–150 cycles	150–1000 cycles

*These characteristics are nominal and vary with operating conditions and design of individual models.
†Cycle life characteristics are for deep (80–100% of full capacity) discharge cycles. Cycle life improves considerably with partial discharges.

TABLE 21.4 Cycle Life vs. Depth of Discharge
3-V, 3-Ah Sealed Silver-Cadmium Batteries

Depth of discharge, %	Cycle at first cell failure
65	1800
50	3979
50	> 5400 (375 days)
35	> 5400 (375 days)
25	> 5400 (375 days)

SOURCE: Ref. 15.

21.5 CHARGING CHARACTERISTICS

21.5.1 Efficiency

The *ampere-hour efficiency* (ampere-hour output/ampere-hour input) of the silver-zinc and silver-cadmium systems under normal operating conditions is high—greater than 95% because practically no side reactions occur when charging at normal rates. The *watthour efficiency* (watthour output/watthour input) is about 70% under normal conditions because of the difference between the charge and discharge polarization potentials.

21.5.2 Zinc/Silver Oxide Cells

The manufacturers of these cells recommend constant-current charging at the 10- to 20-h rate for most applications. However, constant potential and pulsed charging techniques have also been applied.

A typical charge curve at constant current is shown in Fig. 21.14. The two level areas (plateaus) reflect the two levels of oxidation of the silver electrode: the first from silver to monovalent silver oxide (Ag_2O), which occurs at a potential of approximately 1.6 V; the second from the monovalent to the divalent silver oxide (AgO), which occurs at approximately 1.9 V.

Charging is normally terminated when the voltage during charge rises to 2.0 V. Above 2.1 V the cell begins to generate oxygen at the silver electrode and hydrogen at the zinc electrode, decomposing water from the electrolyte. Overcharge is also detrimental to cell life in that it promotes the growth of zinc dendrites and subsequent short circuits.

The importance of proper charging to the life of these cells cannot be overemphasized. Special provisions must be made for those applications which do not permit the use of a constant current with preset voltage cutoff or a similar controlled charging method.

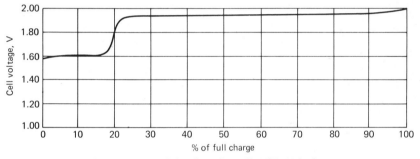

FIG. 21.14 Typical charge curve of the silver-zinc cell, 20°C, 10-h charge rate.

21.5.3 Cadmium/Silver Oxide Cells

Except for lower voltages on each of the plateaus (1.2 V on the lower level, 1.4 V on the upper level), the charging characteristics of the silver-cadmium cell are similar to those of silver-zinc. A typical charge curve is shown in Fig. 21.15.

As with silver-zinc, silver-cadmium cells are usually charged at constant current at the 10- to 20-h rates. The recommended cutoff voltage during charge is normally 1.6 V per cell. The silver-cadmium cell, however, is less sensitive to overcharge than the silver-zinc. Other charge methods can be and have been adapted to specific applications.

21.6 CELL TYPES AND SIZES

Tables 21.5 to 21.8 present excerpts from the catalogues of the two major silver cell manufacturers—Yardney Electric Corporation and Eagle-Picher Industries, Inc. They

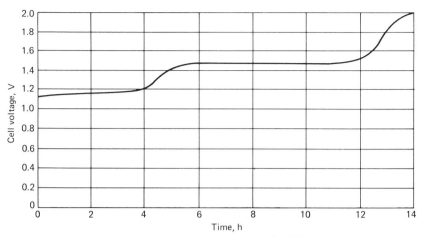

FIG. 21.15   Typical charge curve of the silver-cadmium cell, 20°C.

are intended as a guide only, since the design parameters can be varied to meet specific customer requirements. Many applications for the high-energy silver cells, in fact, require special designs, which often require new cell case and cover designs and tooling. These then become the "available" models for future applications.

In addition to the rectangular cells, there has been growing use of silver-zinc button cells in watches and calculators. Although the vast majority of applications of the silver-zinc button cell are for the nonrechargeable type, secondary cells have been manufactured. The characteristics of these cells are covered in Part 2, Chap. 9 and Part 3, Chap. 24 for the primary and secondary types, respectively.

21.7 SPECIAL FEATURES AND HANDLING

The silver cells are capable of providing extremely high currents if short-circuited. Accordingly, provisions must be made to insulate all tools used with the batteries and to protect the cells against grounding in their application.

The electrolyte is a caustic solution of potassium hydroxide. Precautions such as the use of gloves and safety glasses are required when handling the electrolyte. In most applications, addition of electrolyte or water is not required. However, the manufacturer's recommendations for periodic maintenance and electrolyte checks should be followed closely.

Proper ventilation of these as well as other vented cells, although not as much a problem here as with other battery systems, is required to avoid the accumulation of hazardous hydrogen, especially during charge. For larger installations, forced air or fans may also be required to prevent undesirable temperature buildups. When close voltage regulation is required at cold temperatures, thermostatically controlled heaters are often used with the batteries.

TABLE 21.5   Nominal Characteristics of Typical Vented Silver-Zinc Cells

Cell type	Capacity, Ah	Cell dimensions, mm (including terminals)			Cell weight, g (including electrolyte)
		Length	Width	Height	
		High-rate type			
HR-01	0.15	5.5	16.0	35.0	4.5
HR-05	1.40	13.7	27.4	39.6	22.7
HR-2	3.8	15.0	43.7	64.2	68.1
HR-5	8.0	20.0	52.8	73.9	130.6
HR-21	35.9	20.3	58.4	191.3	440
HR-85	132.0	46.0	71.4	239.8	1645
HR-100	122.8	70.6	87.4	132.2	1275
HR-140	190.0	72.4	82.5	183.4	2270
HR-190	237.7	39.4	152.6	165.4	2270
		Low-rate type			
LR-05	1.24	13.7	27.4	39.6	22.7
LR-1	1.6	13.7	27.4	51.3	31.2
LR-4	4.6	15.0	43.7	85.3	102.2
LR-12	14.0	20.3	47.8	101.6	173.1
LR-60	81.0	59.9	69.3	114.3	823.0
LR-90	136.0	54.9	82.9	179.3	1620.5
LR-190	203.2	39.4	152.6	162.6	2006
LR-660	766.0	79.7	178.8	178.8	6245

Cell (type)	Capacity, Ah	Cell dimensions, cm (including terminals)			Cell weight, kg (including electrolyte)
		Length	Width	Height	
		Special deep submersible types			
LR-670 (DS)	716	10.7	10.7	45.3	9.3
LR-700 (DS)	851	10.7	10.7	48.6	12.3
LR-750 (DS)	878	14.2	9.7	51.3	13.6
LR-850-15	956	12.1	11.6	47.9	14.3
LR-1000 (DS)	1072	13.7	13.7	51.3	18.5
LR-5000 (DS)	5576	27.3	27.3	77.2	100

SOURCE: Yardney Electric Corp., Pawcatuck, Conn.

Because of the sheer size and power of the batteries used, and because of critical personnel safety requirements (e.g., the Navy's NR-1 submersible has a 240-V, 850-Ah silver-zinc backup battery installed under the deck and venting into the operator's quarters), a whole new body of engineering technology has been developed for application of these batteries for underwater power. Some of the special features developed include the removal of all mercury, provision of fire walls inside cells, and provision for pressure compensation for those batteries that are external to the vessels' pressure hulls (see Ref. 8). Electronics systems have also been developed to permit continuous scanning of individual cell voltages to maximize battery life. Special applications such as these can be successfully developed only if the battery designers, manufacturers, and users work closely together during the design of the product.

TABLE 21.6 Nominal Characteristics of Typical Vented Silver-Zinc Cells

Dimensions, mm			High-rate cell						Low-rate cell						
				Rated capacity, Ah	Nominal capacity, Ah rates			Weight, g		Rated capacity, Ah	Nominal capacity, Ah rates			Weight, g
Length	Width	Height*	Cell type		15 min	30 min	60 min		Cell type		4 h	10 h	20 h	
10.9	26.9	40.6	SZHR	0.8	0.8	0.7	0.7	0.8	22.7	SZLR	0.8	0.8	0.8	0.8	0.8	22.7
12.5	30.7	47.0	SZHR	1.5	1.5	1.4	1.5	1.5	34.1	SZLR	1.5	1.5	1.5	1.5	1.5	34.1
14.2	35.1	53.3	SZHR	2.8	2.8	2.6	2.7	2.8	53.9	SZLR	3.0	3.0	3.0	3.0	3.0	53.9
16.3	40.1	60.7	SZHR	5.0	5.0	4.8	5.0	5.0	85.1	SZLR	5.3	5.3	5.3	5.3	5.3	85.1
18.0	44.7	67.1	SZHR	7.0	7.0	6.7	6.8	7.0	113.0	SZLR	8.0	8.0	8.0	8.0	8.0	107.8
20.1	49.5	74.9	SZHR	10.5	10.5	10.0	10.3	10.5	156.0	SZLR	11.5	11.5	11.4	11.5	11.5	147.0
22.6	55.9	84.3	SZHR	18.0	18.0	16.0	17.0	18.0	229.0	SZLR	20.0	20.0	19.0	20.0	20.0	224.0
25.4	62.7	94.7	SZHR	26.0	26.0	20.0	24.0	26.0	326.0	SZLR	30.0	30.0	28.0	30.0	30.0	312.0
28.7	71.4	107.4	SZHR	40.0	40.0	†	30.0	40.0	465.0	SZLR	45.0	45.0	42.0	45.0	45.0	442.0
32.5	80.8	121.4	SZHR	65.0	65.0	†	50.0	65.0	670.0	SZLR	70.0	70.0	65.0	70.0	70.0	641.0
37.3	92.7	139.4	SZHR	100.0	100.0	†	80.0	100.0	1000.0	SZLR	115.0	115.0	100.0	110.0	115.0	953.0
42.2	138.4	181.0	SZHR	290.0	290.0	†	†	290.0	2735.0	SZLR	320.0	320.0	†	290.0	320.0	2610.0

*Without terminal.
†Not applicable at this rate.
SOURCE: Eagle-Picher Industries, Inc., Joplin, Mo.

TABLE 21.7 Nominal Characteristics of Typical Vented Silver-Cadmium Cells

Cell type	Capacity, Ah	Avg. discharge voltage at 10-h rate, V	Cell dimensions (including terminals), mm			Cell weight, g (including electrolyte)
			Length	Width	Height	
YS-01	0.13	1.08	5.5	16.0	35.1	5.1
YS-05	1.9	1.10	13.7	27.4	39.3	21.3
YS-1	1.5	1.10	13.7	27.4	51.3	34.1
YS-3	4.2	1.10	15.0	43.7	72.7	90.8
YS-5	7.2	1.08	20.7	52.8	73.9	141.9
YS-10	14.0	1.10	18.8	58.9	122.0	261.1
YS-20	30.0	1.08	52.1	43.9	109	428.5
YS-40	54.0	1.10	25.2	82.5	179	746.4
YS-60	80.0	1.10	60.0	69.3	114	1206
YS-100	125.0	1.08	87.4	70.6	122	1504
YS-300	427.0	1.07	45.2	106.0	106	5195

SOURCE: Yardney Electric Corp., Pawcatuck, Conn.

21.8 APPLICATIONS

Because of the high cost of silver, the major applications of these batteries have historically been, and continue to be, governmental. However, these batteries have found many varied uses because of their high power and energy density where space and weight limitations are critical.

One of the original applications for the silver-zinc battery was in the underwater area for use in torpedoes.[17] Much of the original development work was sponsored by the U.S. Navy. Later, development expanded to other underwater applications including mines, buoys, special test vehicles, swimmer aids, and, presently, to the deep submergence and rescue vehicles (DSRV), such exploratory underwater vehicles as Trieste and NR-1, and various antisubmarine warfare (ASW) applications. The MK 66 torpedo battery is illustrated in Fig. 21.16. The discharge rate for this nominal 245-V battery is 580 A; operating time is 4 min.

Space applications include use in various satellites and space probes, powering the life-support equipment used by the Apollo astronauts, power for the Lunar Rover and Lunar Drill, power for space shuttle payload launching and controls, and power for many missile functions including guidance and control, telemetering, and destruct. Aircraft applications span the range from tethered balloons to helicopters to target drones to aircraft and missiles.

Ground applications include communications equipment, portable TV cameras, portable lights, camera drives, medical equipment, vehicle motive power, and similar ones requiring a high energy density rechargeable battery.

The user should keep in mind that no one type of battery is suitable for all applications. Optimum performance of a battery in an application can usually be achieved only by meeting the critical needs of the application and subordinating the others. The best approach for battery selection is to work with the battery manufacturers during the early stages of equipment design, rather than asking the battery designers, as is often done, to "design a battery that meets all my requirements and fits into this remaining cavity in my equipment."

TABLE 21.8 Nominal Characteristics of Typical Vented Silver-Cadmium Cells

Dimensions, mm			High-rate cells						Low-rate cells					
				Rated capacity, Ah	Nominal capacity, Ah rates			Weight, g		Rated capacity, Ah	Nominal capacity, Ah rates			Weight, g
Length	Width	Height*	Cell type		15 min	30 min	60 min		Cell type		1 h	4 h	10 h	
10.9	26.9	40.6	SCHR 0.5	0.5	0.5	0.5	0.5	22.7	SCLR 0.7	0.7	0.7	0.7	0.7	22.7
12.5	30.7	47.0	SCHR 1.0	1.0	1.0	1.0	1.0	36.8	SCLR 1.1	1.1	1.1	1.1	1.1	36.8
14.2	35.1	53.3	SCHR 2.0	2.0	2.0	2.0	2.0	59.6	SCLR 2.2	2.2	2.2	2.2	2.2	59.6
16.3	40.1	60.7	SCHR 3.3	3.3	3.3	3.3	3.3	90.8	SCLR 3.5	3.5	3.4	3.5	3.5	90.8
18.0	44.7	67.1	SCHR 3.8	3.8	3.7	3.8	3.8	107.8	SCLR 5.0	5.0	4.9	5.0	5.0	107.8
20.1	49.5	74.9	SCHR 5.6	5.6	5.5	5.6	5.6	141.9	SCLR 8.0	8.0	7.8	8.0	8.0	158.9
22.6	55.9	84.3	SCHR 8.0	8.0	7.8	8.0	8.0	201.5	SCLR 12.0	12.0	10.0	12.0	12.0	235.5
25.4	62.7	94.7	SCHR 12.0	12.0	11.7	11.9	12.0	298.0	SCLR 18.0	18.0	16.0	18.0	18.0	346
28.7	71.4	107.4	SCHR 17.0	17.0	16.5	16.8	17.0	434	SCLR 26.0	26.0	22.0	26.0	26.0	482
32.5	80.8	121.4	SCHR 31.0	31.0	28.0	30.0	31.0	689	SCLR 45.0	45.0	35.0	40.0	45.0	780
37.3	92.7	139.4	SCHR 55.0	55.0	†	40.0	55.0	1044	SCLR 70.0	70.0	50.0	65.0	70.0	1135
42.2	138.4	181.0	SCHR …	…	…	…	…	…	SCLR 175.0	175.0	†	150.0	175.0	2655

*Without terminal.
†Not applicable at this rate.
SOURCE: Eagle-Picher Industries, Inc., Joplin, Mo.

FIG. 21.16 MK 66 torpedo battery. (*Courtesy of Yardney Electric Corp.*)

21.9 RECENT DEVELOPMENTS

The vast majority of recent efforts in secondary silver cells have been concentrated in the areas of negative electrode and separator improvements for the zinc/silver oxide cell because of its proportionally large market volume.[7] The major developments are summarized in Table 21.9. These approaches have had mixed success, but some of the advances have been incorporated successfully into cells for medium- and long-life applications.

TABLE 21.9 Summary of Recent Developments in Silver-Oxide Secondary Cells and Components

Development area	Advantages	Disadvantages
Zinc Electrode		
Increase zinc to silver weight ratio	Delays onset of capacity losses.	Gains not proportional to extra material used. Reduces energy density.
Oversized negatives	Reduce current density at plate edges where "shape change" starts.	Reduce energy density.
"Contoured" negatives	Reinforce areas of maximum erosion.	Reduce energy density. High cost.
Teflon binder	Reduces "shape change" and dendrite growth. Improves low-temperature performance.	High cost. Difficult to disperse evenly, may interfere with normal electrode reactions.
Potassium titanate fibers	Reduces "shape change" and dendrite growth. Reduces the probability of "hot" shorts.	Somewhat higher cost.

TABLE 21.9 Summary of Recent Developments in Silver-Oxide Secondary Cells and Components (*Continued*)

Development area	Advantages	Disadvantages
	Zinc Electrode	
Lead or lead/cadmium additives as mercury substitutes	Reduce hazard to health and equipment in cases of "hot" shorts. Imporve capacity maintenance	Do not suppress gassing as well as mercury.
	Main separator	
Modified polyethylene film	High resistance to attachment by silver oxides and by electrolyte. Increase energy density because they are thinner than standard materials.	High cost. High electrolytic resistance. Difficult to handle. Low resistance to zinc penetration. Lack of uniformity.
Inorganic separators	Resist temperature above 150°C. Resist attack by silver oxides and by electrolyte.	High electrolytic resistance. Bulky, difficult to handle. High cost.
Cellulosics with metallic groups in molecule	Improve resistance to attack by silver oxides and by electrolyte; extend cycle life.	Somewhat higher cost.
	Interseparators	
Positive: Asbestos	Prevents or reduces magnitude of shorts.	Bulky, reacts with silver oxides. May contaminate cell with iron.
Negative: Potassium titanate mat	Reduces zinc "shape change"; reduces incidence and magnitude of shorts.	Bulky; reduces energy density. Somewhat costly.
	Silver electrode	
Close control of particle size distribution	Improves control of cell voltage and capacity.	
	Hardware	
New plastics for cases and covers (e.g., modified PPO, polysulfone)	High temperature operation. Better mechanical properties.	Higher cost.
New adhesives	More efficient at high temperatures. Comply with OSHA requirements.	

REFERENCES

1. A. Volta, *Philos. Trans. R. Soc. London* **90**:403, (1800).

2. S. U. Falk, and A. J. Salkind, *Alkaline Storage Batteries*, Wiley, New York, 1967.

3. H. André, *Bull. Soc. Fr. Electriciens* (6th series) **1**:132 (1941).

4. H. André, U.S. Patent 2,317,711 (1943).

5. J. T. Brown, *Iron-Silver Battery—A New High Energy Density Power Source*, Westinghouse Report no. 77-5E6-SILEL-RI, 1977.

6. *Final Report, Design & Cost Study, Zinc/Nickel Oxide Battery for Electric Vehicle Propulsion*, Contract no. 31-109-38-3543, Yardney Electric Corp., October 1976.

7. R. Serenyi, *Recent Developments in Silver-Zinc Batteries*, Yardney Electric Corp., Internal Report #2449-79, October 1979.

8. G. W. Work and P. A. Karpinski, "Energy Systems for Underwater Use," *Marine Tech. Expo. Int. Conf.*, New Orleans, October 1979.

9. A. Himy, "Substitutes for Mercury in Alkaline Zinc Batteries," *Proc. 28th Annu. Power Sources Symp.*, pp. 167–169 (1978).

10. R. Serenyi and P. Karpinski, *Final Report on Silver-Zinc Battery Development*, Contract no. N00140-76-C-6726, Yardney Electric Corp., November 1978.

11. *Final Report, Medium Rate Rechargeable Silver-Zinc 850Ah Cell*, USN Contract no. N00140-76-C-6729, Eagle-Picher Industries, Inc., March 1978.

12. A. Charkey, "Long Life Zinc-Silver Oxide Cells," Proc. 26th Annu. Power Sources Symp., pp. 87–89 (1976).

13. V. D'Agostino, J. Lee, and G. Orban, "Grafted Membranes," in A. Fleischer and J. Lander (eds.), *Zinc-Silver Oxide Batteries*, chap. 19, pp. 271–281, Wiley, New York, 1971.

14. C. P. Donnel, *Evaluation of Inorganic/Organic Separators*, Contract no. NAS3-18530, Yardney Electric Corp., October 1976.

15. *Evaluation of Silver-Cadmium Batteries for Satellite Applications*, Test no. D2-90023, Boeing Co., February 1962.

16. H. A. Frank, W. L. Long, and A. A. Uchiyama, "Impedance of Silver Oxide-Zinc Cells," *J. Electrochem. Soc.*, **123**(1):1–9 January 1976.

17. A. Fleischer and J. Lander (eds.), *Zinc-Silver Oxide Batteries*, Wiley, New York, 1971.

Nickel-Hydrogen Batteries

by
James D. Dunlop

22.1 GENERAL CHARACTERISTICS

A sealed nickel-hydrogen (Ni-H$_2$) secondary battery is a hybrid combining battery and fuel cell technologies.[1] The nickel oxide positive electrode comes from the nickel-cadmium cell, and the hydrogen negative electrode from the hydrogen-oxygen fuel cell. Major advantages and disadvantages are listed in Table 22.1. Over 6000 deep-discharge cycles have been demonstrated in laboratory tests using cells with an energy density approaching 60 Wh/kg. A potential disadvantage is the relatively high initial cost. However, on a cost per cycle basis, Ni-H$_2$ batteries are competitive with other secondary systems.

The application of the nickel-hydrogen battery has mainly been directed toward aerospace applications. Recently, however, programs have been started for terrestrial applications, for example, a sealed 100-Ah battery for deep-discharge and with a 20-y life.[2]

TABLE 22.1 Major Advantages and Disadvantages of the Nickel-Hydrogen Battery

Advantages	Disadvantages
High energy density	High initial cost
Long cycle life, even on deep discharging	Self-discharge proportional to H$_2$ pressure
Cell can tolerate overcharge and reversal	
H$_2$ pressure gives an indication of state-of-charge	

22.2 CHEMISTRY

The electrochemical reactions of the nickel-hydrogen cell for normal operation, overcharge, and reversal are

Normal operation:

Nickel electrode $\quad NiOOH + H_2O + e \underset{charge}{\overset{discharge}{\rightleftarrows}} Ni(OH)_2 + OH^-$

Hydrogen electrode $\quad \frac{1}{2} H_2 + OH^- \underset{charge}{\overset{discharge}{\rightleftarrows}} H_2O + e$

Net reaction $\quad \frac{1}{2} H_2 + NiOOH \underset{charge}{\overset{discharge}{\rightleftarrows}} Ni(OH)_2$

Overcharge:

Nickel electrode $\quad 2 OH^- \rightarrow 2e + \frac{1}{2} O_2 + H_2O$

Hydrogen electrode $\quad 2 H_2O + 2e \rightarrow 2 OH^- + H_2$

Chemical recombination $\quad \frac{1}{2} O_2 + H_2 \rightarrow H_2O$
of O_2

Reversal:

Nickel electrode $\quad H_2O + e \rightarrow OH^- + \frac{1}{2} H_2$

Hydrogen electrode $\quad \frac{1}{2} H_2 + OH^- \rightarrow e + H_2O$

22.2.1 Normal Operation

Electrochemically, the half-cell reactions at the positive nickel oxide electrode are similar to those occurring in the nickel-cadmium system. At the negative electrode, hydrogen gas is oxidized to water during discharge and is reformed, during charge, from the water by electrolysis. The next reaction shows hydrogen reduction of nickelic hydroxide to nickelous hydroxide on discharge with no net change in KOH concentration or in the amount of water within the cell.

22.2.2 Overcharge

Electrolysis occurs on overcharge with oxygen generated at the positive electrode and with hydrogen at the negative electrode. Oxygen is chemically recombined with hydrogen to form water at the catalytic platinum electrode. Again, there is no change in the KOH concentration or the amount of water in the cell with continuous overcharge.

The oxygen recombination rate at the negative platinum electrode is very rapid, and so even at high rates of overcharge there is no significant buildup of oxygen in the hydrogen gas (oxygen partial pressure of less than 1% is typical).

22.2.3 Reversal

During cell reversal, hydrogen is generated at the positive electrode and consumed at the negative electrode at the same rate. Therefore, the cell can be continuously operated in the cell reversal mode without pressure buildup or net change in electrolyte concentration. This is a unique feature of the system.

22.2.4 Self-Discharge

The electrode stack is surrounded by hydrogen under pressure. A salient feature is that the hydrogen reacts electrochemically but not chemically to reduce the nickelic hydroxide. Actually, the nickelic hydroxide is reduced, but at such an extremely low rate that performance for any of the aerospace applications is not affected.

22.3 CELL DESIGN AND CONSTRUCTION

Sealed Ni-H$_2$ cells are designed to contain hydrogen gas under pressure within a cylindrical pressure shell.[3] When charged, the internal gas pressure is about 30 to 40 $\times$ 10^5 Pa.

22.3.1 Cell Components (COMSAT/INTELSAT)

Components of the Ni-H$_2$ cell are shown in Fig. 22.1 with the electrode stack assembly positioned in front of the pressure shells. The pressure shells are fabricated from Inconel 718 material, 20 mils thick. Plastic compression seals are used at both ends to insulate the positive and negative feedthrough terminals from the pressure vessel. Burst pressure of 2400 psi provides a 4:1 safety factor. The diameter of the pressure vessel is 8.89 cm, which was an engineering compromise between maximum energy density and number of modules in the electrode stack.

The components that compose the electrode stack are shown in Fig. 22.2. The two positive electrodes, positioned back to back, are electrochemically impregnated nickel plates.[4] The hydrogen electrode consists of Teflon-bonded platinum black supported with a fine-mesh nickel screen with Teflon backing. A plastic gas diffusion screen facilitates hydrogen diffusion to the back of this platinum electrode. Fuel cell grade

FIG. 22.1   Ni-H$_2$ cell components (COMSAT/INTELSAT).

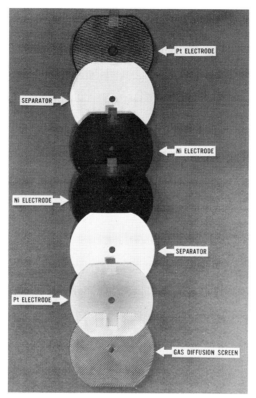

FIG. 22.2   Electrode stack components.

asbestos is used as the separator material. Busbars are used to connect the negative electrodes to one terminal and the positive electrodes to the other terminal.

This cylindrical cell has been used to fabricate three different size Ni-H$_2$ cells:

30-Ah INTELSAT V, 12 modules

35-Ah NTS-2, 15 modules

50-Ah laboratory test cell, 20 modules

The ampere-hour capacity in each case is a function of the number of modules used in the electrode stack, and the length of the pressure shell is varied to maintain the same operating pressure. Tables 22.2 and 22.3 show the weights and energy densities, respectively, for these three cells.[5]

22.3.2 "Pineapple Slice" Design

The cell design shown in Fig. 22.3 uses the same cylindrical concept but with the electrode stack in a "pineapple slice" configuration with a larger center hole.[6,7] The electrode tabs are brought out through this center hole.

The purpose of this design is to improve thermal control by reducing the thermal resistance between the electrode stack and the pressure vessel wall. Figure 22.4 com-

TABLE 22.2 Nickel-Hydrogen Cell Weights

Parameters	INTELSAT V, 30 Ah Wt, g	INTELSAT V, 30 Ah % of total	NTS-2, 35 Ah Wt, g	NTS-2, 35 Ah % of total	Laboratory test cell, 50 Ah Wt, g	Laboratory test cell, 50 Ah % of total
Positive electrode	282.4	31.7	348.6	33.9	480	40
Negative electrode	32.2	3.6	72.0	7.0	54.7	4.6
Separators	23.0	2.6	35.1	3.4	39.1	3.3
Screens	3.4	0.38	8.0	0.8	5.6	0.5
Electrolyte	110.0	12.4	159.0	15.5	187.0	15.6
End plates	35.1	3.9	34.0	3.3	35.1	2.9
Pressure vessel and miscellaneous hardware	379.9	42.7	371.7	36.1	397.3	33.0
Additional electrolyte allowance	24.0	2.7				
Total	890.2	100.0	1028.4	100.0	1198.8	100.0

Table 22.3 Nickel-Hydrogen Cell Energy Density

Parameters	INTELSAT V	NTS-2	50 Ah
Rated capacity, Ah	30	35	50
Cell weight, kg	0.890	1.028	1.198
Capacity, Ah			
20°C	31.91	38.5	52.8
10°C	34.80	42.96	57.6
0°C	35.31		
Weight pos., g	282.4	348.6	480.0
20°C, Ah/g	0.113	0.110	0.110
10°C, Ah/g	0.12	0.12	0.12
Average discharge voltage, V	1.25	1.25	1.25
Energy 10°C, Wh	43.5	53.7	72.0
Energy density, Wh/kg			
20°C	44.8	46.8	55.1
10°C	48.9	52.2	60.1

pares the hydrogen gap thermal resistance for the two designs. The difference is about 0.07 kWh, which is significant only at very high dissipation rates. At synchronous orbit, the temperature drop across the gap is about the same for both designs, less than 0.5°C difference.

Another cell design feature is to recirculate the electrolyte with a wick external to the stack to allow for return of any electrolyte or water vapor lost from the stack. This cell has a plasma-sprayed zirconium oxide wall wick in contact with knit zirconium oxide separators (Fig. 22.3). This recirculating design is used to solve the evaporation and condensation problem associated with low earth orbit, 80% DOD cycling.

22.3.3 Separator Material

Both bifunctional and asbestos fuel cell–type separator materials are being used in Ni-H$_2$ cells.[7,8]

(a)

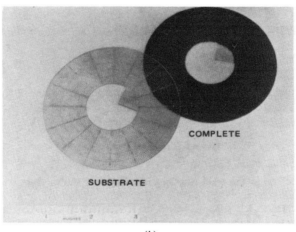

(b)

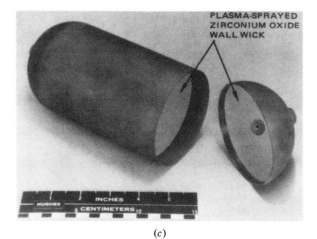

(c)

FIG. 22.3   Pineapple-slice Ni-H$_2$ cell (HAC/Air Force). (a) Stack assembly. (b) Negative electrode. (c) Pressure vessel cylinder and dome.

22-6

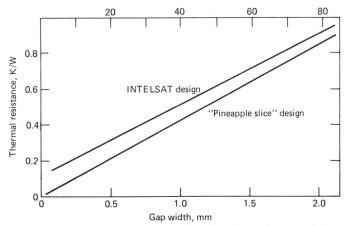

FIG. 22.4   Comparison of hydrogen gap thermal resistance for two cell designs.

Ni-H₂ Cell with Fuel Cell Separator   During charge and overcharge, oxygen that evolved at the nickel oxide electrodes is forced out between the back-to-back positive electrodes. This oxygen diffuses into the gas space between the electrode stack and the pressure vessel wall into the region of the gas diffusion screens on the backside of the negative electrodes, through the porous backing of the negative electrode, where it combines with hydrogen to form water.[5]

During development, three major concerns were inadequate electrolyte retention due to electrolyte entrainment loss when the oxygen is evolved, an unsafe buildup of oxygen gas partial pressure, and excessive heating at the periphery of the negative and positive electrodes. Extensive laboratory experiments were performed to investigate each area. It was shown that, for synchronous orbit application, there was no measurable electrolyte entrainment loss[9]; the partial pressure of oxygen was less than 1% of the hydrogen pressure, well below a hazardous level[8]; and peripheral heating did not occur because of $(OH)^-$ ion concentration gradient changes.[9] However, testing has shown that excessive overcharge at high rates does lead to electrolyte entrainment loss. With this type of operation, a wall wick is recommended for return of electrolyte to the cell stack.

Ni-H₂ Cell with Bifunctional Separator   During charge and overcharge, oxygen evolved at the nickel oxide electrode will permeate the separator and recombine with hydrogen at the platinum electrode to form water. The separator is extended to a wall wick to permit recirculation of the electrolyte. Potential advantages of this approach:

1. Electrolyte management (retention and recirculation) at high rates of charge and overcharge

2. No partial-pressure buildup of oxygen outside the electrode stack

Potential disadvantages are:

1. Localized rapid recombination of hydrogen at the platinum electrode resulting in excessive heating and possible catastrophic failure

2. Possible lack of stability of the bifunctional separator with long-term cyclic operation.

22.3.4 Positive Electrodes

The positive nickel oxide electrode composes approximately 35% of the total cell weight (by far the heaviest single component) and has the most influence on cell performance and cycle lifetime. It is the most difficult component to manufacture in terms of process variables, complexity, and process control. In recent years, several battery manufacturers have begun to produce electrochemically impregnated (EI) nickel oxide electrodes, primarily for aerospace applications. The advantages of these electrodes have been demonstrated through laboratory testing, as shown in Fig. 22.5. This figure compares the electrochemically impregnated electrode against the chemically impregnated electrode, impregnated to varying levels of active material (100% equals 2.1 g of active material per cubic centimeter of void). This Ni-H_2 battery system has the potential to exploit fully the performance capability (high-rate discharge) and cyclic life of these electrodes.

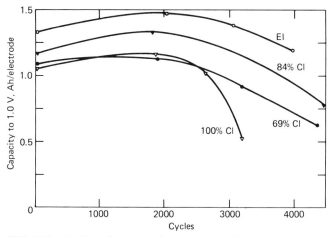

FIG. 22.5 Cyclic performance of electrochemically and chemically impregnated positive electrodes (60% DOD, C/2 rate for ½ h; 50% overcharge, C/2 rate for 1.8 h).

22.4 Ni-H_2 BATTERY DESIGN AND CONSTRUCTION

Three Ni-H_2 batteries are discussed: the NTS-2 battery launched in June 1977 and still performing well,[10-13] the INTELSAT V Ni-H_2 battery launched in 1983, and a larger 50-Ah battery. Figure 22.6 shows the NTS-2 battery and Fig. 22.7 the INTELSAT V battery.

22.4.1 Energy Density

A weight breakdown and the energy density for each of these three batteries are presented in Table 22.4. For the INTELSAT V design, the Ni-H_2 cells compose 80% of the total battery weight compared with 70% for the NTS-2 design. The NTS-2

FIG. 22.6 NTS-2 nickel-hydrogen battery.

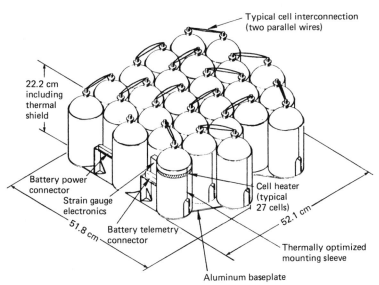

FIG. 22.7 INTELSAT V nickel-hydrogen battery [four temperature-sensing thermistors (not shown), each cell diode protected (not shown), pressure-sensing strain gauges on one cell (not shown)].

TABLE 22.4 Nickel-Hydrogen Battery Weight and Energy Density

Parameters	INTELSAT V Wt., kg	INTELSAT V %	NTS-2 Wt., kg	NTS-2 %	50 Ah Wt., kg	50 Ah %
Cell weight	0.890	...	1.0284	...	1.198	
Number of cells	27	...	14	...	27	
Total cell weight	24.03	80	14.3	70	32.35	80
Cell mounting shells	3.3	11	3.1	15.2	4.45	11
Base plates	0.73	2.4	1.8	8.8	0.97	2.4
Diode assembly	0.96	3.2	...	...	1.29	3.2
Mounting hardware	0.10	0.3	0.2	1.0	1.17	2.9
Connectors, etc.	0.23	0.8	1.0	4.9	6.32	0.8
Strain gauges						
Miscellaneous	0.77	2.6				
Total weight	30.12	100.00	20.50	100.00	40.44	100.00
Energy, Wh at 10°C	1174		750		1944	
Energy density, Wh/kg	39		36.5		48.1	

battery, which was first in orbit, was structurally overdesigned. The INTELSAT V battery uses the same design approach but with significant weight and volume reductions. The purpose of presenting the 50-Ah battery is to demonstrate the increase in energy density achievable by changing from the 30- to a 50-Ah size cell.

22.4.2 Energy per Unit Volume

The NTS-2 and INTELSAT V cells are designed to have a maximum pressure of 40 $\times$ 10^5 Pa. Table 22.5 shows cell and battery energy per unit volume. Changing the maximum pressure to 67 $\times$ 10^5 Pa increases the energy per unit volume by 34% as shown in Table 22.5. This represents the maximum energy per unit volume achievable in a cell with a diameter of 8.89 cm because almost all the hydrogen would be stored in cylindrical end caps, and the electrode stack would occupy the entire cylindrical section.

TABLE 22.5   Nickel-Hydrogen Cell and Battery Energy per Unit Volume

Parameter	INTELSAT V	NTS-2*	50 Ah	50-Ah high pressure
		Cell		
Capacity at 20°C, Ah	31.9	38.5	52.8	52.8
Maximum pressure, Pa	40×10^5	40×10^5	40×10^5	65×10^5
Length electrode stack, cm	4.2	5.08	7.0	7.0
Volume electrode stack, cm³	260	315	434	434
Volume to store hydrogen, cm³	426	514	686	409
Total volume required, cm³	686	829	1120	843
Actual cell volume, cm³	762	841	1120	841
Volume of domes, cm³	368	368	368	368
Diameter, cm	8.89	8.89	8.89	8.89
Length of cell, cm	15.2	16.5	21.3	16.5
Length with terminals, cm	21	24	29	23
Energy at 10°C, Wh	43.5	53.7	72	72
Energy per unit volume at 10°C, Wh/L	57	64	64	86
		Battery		
Number of cells	27	7*	27	27
Length, cm	51.8	48.26	51.8	51.8
Width, cm	52.1	24.13	52.1	52.1
Height, cm	22.2	25.4	29	24.2
Volume, L	59.9	29.6	78.3	75.3
Energy at 10°C, Wh	1174	376	1944	1944
Energy/volume, Wh/L	19.6	12.7	24.8	29.8

*NTS-2 battery system consists of two 7-cell batteries.

22.5 PERFORMANCE CHARACTERISTICS

22.5.1 Discharge

Electrochemically impregnated nickel oxide electrodes are used in Ni-H_2 cells because of their excellent cyclic performance capabilities.[14] The capacity of these electrodes increases as the temperature decreases (Fig. 22.8). Measured capacity of an electrochemically impregnated electrode is about 20% greater at 10 than at 20°C. The typical operating voltage is between 1.2 and 1.3 V per cell.

A typical discharge of an INTELSAT V cell, discharged at 200 A (12-min rate) is shown in Fig. 22.9. The cell temperature rises rapidly from -2 to over 60°C during the discharge. This temperature was measured on the dome of the pressure vessel; the interior temperature was significantly higher. The discharge profile is almost flat at 0.6 V; the potential drop is due to the terminal impedance of the cell of 3 mΩ.

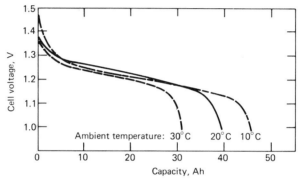

FIG. 22.8 Voltage performance on discharge vs. temperature (discharge: 21 A; charge: 3.5 A, 16 h).

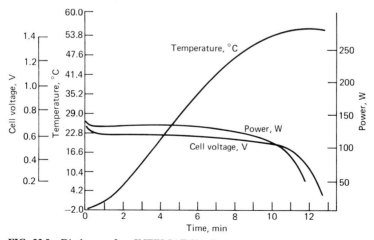

FIG. 22.9 Discharge of an INTELSAT V cell.

Figures 22.10 and 22.11 summarize the performance of aerospace type Ni-H$_2$ cells at various discharge rates and temperatures. The aerospace cell, not optimized for high-rate discharge, is capable of delivering about 60 Wh/kg and 60 Wh/L at discharge rates as high as the C rate. The temperature shown is the temperature at the start of the discharge. As previously discussed, there is a rapid rise of temperature within the cell during the discharge. Figure 22.11 shows the service life of the Ni-H$_2$ cell at various discharge loads on a weight and volume basis.

22.5.2 Charge

End-of-charge voltage as a function of temperature and rate of charge is shown in Fig. 22.12. These voltage values are typically about 19 mV higher for nickel-hydrogen cells than those observed for nickel-cadmium cells.

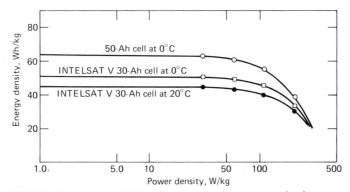

FIG. 22.10   Aerospace Ni-H$_2$ cell, energy density vs. power density.

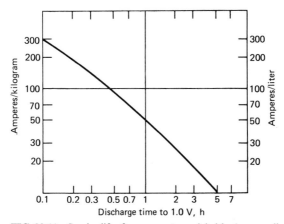

FIG. 22.11   Service life of aerospace-type nickel-hydrogen cell.

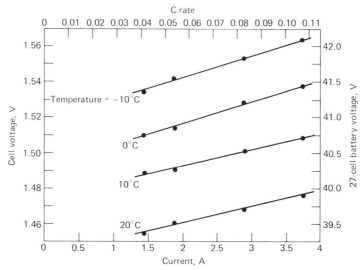

FIG. 22.12   End-of-charge voltage (35-Ah cells, NTS-2).

22.5.3 State of Charge

A salient feature of the Ni-H$_2$ cell is that the pressure is a direct indication of the state of charge of a cell. On charge the hydrogen pressure increases linearly until the nickel oxide electrode approaches the fully charged condition (Fig. 22.13). During overcharge, oxygen evolved at the positive electrode recombines with hydrogen at the negative electrode to form water, and the pressure stabilizes. On discharge the hydrogen pressure decreases linearly until the nickel oxide electrode is fully discharged. The hydrogen pressure of 3×10^5 Pa remaining within the cell is the precharge. If the cell is reversed by overdischarging, hydrogen generated at the positive electrode is consumed at the negative electrode, and again the pressure is constant.

22.5.4 Self-Discharge

Self-discharge characteristics of Ni-H$_2$ cells can be determined from the decrease in hydrogen pressure during open-circuit stand, recalling that hydrogen pressure is a direct indication of the state of charge. Figure 22.14 shows the cell capacity during open-circuit stand. Analysis of the self-discharge data shows that the rate of self-discharge is proportional to the hydrogen pressure.[9] The time constant for the self-discharge is approximately 500 h at room temperature.

22.5.5 Life Tests

Lightweight Ni-H$_2$ cells of the COMSAT/INTELSAT design are being life-tested in a real-time simulation of operation in synchronous orbit at 60% depth of discharge. Results for the first 5 y, 10 eclipse seasons, are shown in Fig. 22.15.

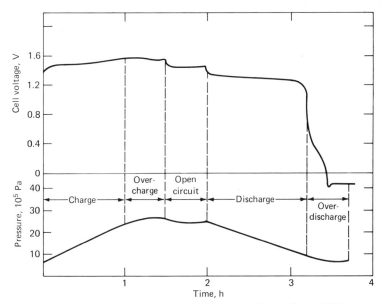

FIG. 22.13   Typical Ni-H$_2$ cell pressure and voltage characteristics, 23°C.

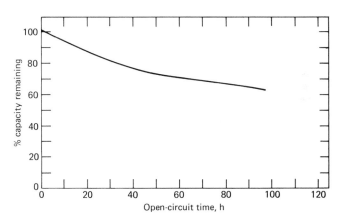

FIG. 22.14   Self-discharge at ambient temperatures (35-Ah cell, NTS-2 prototype).

22.5.6 Flight Tests

The NTS-2 spacecraft was launched in June 1977 with the first aerospace Ni-H$_2$ battery on board to provide power during the solar eclipse periods. Figure 22.16 shows the battery voltage and temperature characteristics during the longest eclipse day. The NTS-2 in-orbit performance has been highly satisfactory.

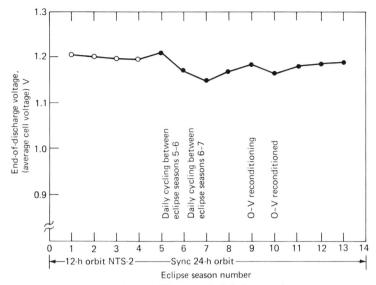

FIG. 22.15 Ni-H$_2$ cells (NTS-2) life test end-of-discharge voltage.

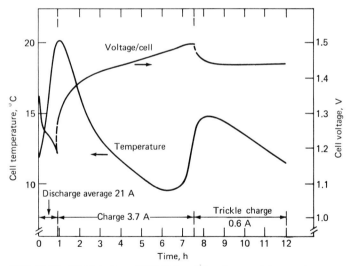

FIG. 22.16 Ni-H$_2$ battery (NTS-2) in-orbit voltage and temperature performance during longest eclipse day.

22-16

REFERENCES

1. J. Dunlop et al., Nickel-Hydrogen Cell, U.S. Patent 3,867,199, Feb. 18, 1975.

2a. J. F. Stockel et al., "A Nickel-Hydrogen Secondary Cell for Synchronous Orbit Applications," *Proc. 7th Intersoc. Energy Convers. Eng. Conf.,* San Diego, CA 1972, pp. 87–94.

2b. J. E. Clifford and E. W. Brooman, "Assessment of Nickel-Hydrogen Batteries for Terrestrial Solar Applications," SAND80-7191, Sandia National Laboratories, Albuquerque, N. M., Feb. 1981.

3. J. D. Dunlop, J. Stockel, and G. van Ommering, "Sealed Metal Oxide-Hydrogen Secondary Cells," *Power Sources,* vol. 5, 1974, Academic Press, London, pp. 315–329.

4. R. L. Beauchamp, Positive Electrode for Use in Nickel Cadmium Cells and the Method for Producing Same and Products Utilizing Same, U.S. Patent 3,653,967, April 4, 1972.

5. G. van Ommering and J. Stockel, "Characteristics of Nickel-Hydrogen Flight Cells," *Proc. 27th Power Sources Conf.,* Electrochemical Society, Pennington, N. J., 1976, pp. 124–128.

6. H. Rogers, "Failure Mechanisms in Nickel-Hydrogen Cells Final Report," APL-TR-77-90, Electrochemical Society, Pennington, N. J., U.S. Air Force, 1977.

7. H. Rogers, "Design of Long Life Nickel Hydrogen Cells," *Proc. 28th Power Sources Symp.,* Electrochemical Society, Pennington, N. J., 1978, pp. 142–144.

8. J. Giner and J. Dunlop, "The Sealed Nickel-Hydrogen Secondary Cell," *J. Electrochem. Soc.* Electrochemical Society, Pennington, N. J., **122**: 4–11 (1975).

9. G. L. Hollick, "Improvement and Cycle Testing of Ni-H_2 Cells," *Proc. 28th Power Sources Symp.,* Electrochemical Society, Pennington, N. J., 1978, pp. 139–141.

10. F. Betz, J. F. Stockel, and A. Gaudet, "Nickel-Hydrogen Storage Battery for Use on Navigation Technology Satellite-2," *11th Intersoc. Energy Convers. Eng. Conf.,* vol. 1, 1976, pp. 510–516.

11. J. D. Dunlop and J. F. Stockel, "Orbital Performance of NTS-2 Nickel-Hydrogen Battery," *COMSAT Tech. Rev.,* vol. 7, fall 1977, pp. 639–647.

12. J. F. Stockel, J. D. Dunlop, and F. Betz, "NTS-2 Nickel-Hydrogen Battery Performance," *AIAA 7th Commun. Satell. Syst. Conf.,* San Diego, Calif., 1978, pp. 66–71.

13. R. E. Patterson, W. Luft, and J. Dunlop, "Development of Spacecraft Power Systems Using Nickel-Hydrogen Battery Cells," *AIAA Commun. Satell. Syst. Conf.,* Montreal, Canada, 1976.

14. M. P. Bernhardt and D. W. Maurer, "Results of a Study on Rate of Thickening of Nickel Electrodes," *Proc. 29th Power Sources Conf.,* Electrochemical Society, Pennington, N. J., 1980.

Silver-Hydrogen Batteries

by
Allen Charkey and Martin Klein

23.1 GENERAL CHARACTERISTICS

The silver-hydrogen battery, like all metal-gas systems, is a hybrid couple, born from the use of a silver electrode of the type used in a silver-zinc secondary cell and an electrocatalytic gas electrode similar to that used in a fuel cell. Earliest development stems only from the early 1970s and was fostered by the need for a lightweight battery for satellites operating at high depths of discharge on synchronous earth orbits.[1]

Acceptance of this system has been slow because of the advanced state of development of the nickel-hydrogen battery and the high cost of the silver-hydrogen battery. It is expected, however, that increased power requirements and the need for greater energy storage in reduced volume on board spacecraft and satellites will undoubtedly generate increased interest in silver-hydrogen batteries. The battery also should find applications in special aerospace and military environments requiring high energy density and long cycle life. Among the potential applications are synchronous or low-orbit satellite power (25- to 50-Ah size batteries), space station backup power for solar arrays (100- 200-Ah size batteries), deep space probes, unmanned or manned Martian surface vehicles, and small submersibles or submarines. A summary of the characteristics of the battery is given in Table 23.1.

23.2 CHEMISTRY

On charge, discharge, and overcharge the following reactions occur at the electrodes:

Silver electrode:

$$\text{Ag} + \text{OH}^- \underset{\text{discharge}}{\overset{\text{charge}}{\rightleftharpoons}} \tfrac{1}{2}\,\text{Ag}_2\text{O} + \tfrac{1}{2}\,\text{H}_2\text{O} + \text{e}$$

$$\tfrac{1}{2}\,\text{Ag}_2\text{O} + \text{OH}^- \underset{\text{discharge}}{\overset{\text{charge}}{\rightleftharpoons}} \text{AgO} + \tfrac{1}{2}\,\text{H}_2\text{O} + \text{e}$$

TABLE 23.1 Characteristics of Silver-Hydrogen Batteries

Energy density	75 to 85 Wh/kg	Energy density exceeded only by silver-zinc secondary battery
Power density	80 to 90 W/kg	Moderate power capability
Cycle life	500 to 3000 cycles	Cycle life comparable to nickel-hydrogen cell
Operating temperature	0 to 50°C	Good low-temperature performance
Cost	$\sim$ \$2000/kWh (1982\$)	Extremely high because of cost of platinum and silver

Hydrogen electrode:

$$2H_2O + 2e \overset{charge}{\underset{discharge}{\rightleftarrows}} H_2 + 2OH^-$$

Overall reaction:

$$Ag + H_2O \overset{charge}{\underset{discharge}{\rightleftarrows}} H_2 + AgO \qquad E^0 = 1.398 \text{ V}$$

During overcharge oxygen is generated at the silver electrode according to

$$2OH^- \rightarrow H_2O + \tfrac{1}{2} O_2 + e$$

Oxygen is then electrochemically reduced on available catalytic sites on the hydrogen electrode

$$H_2 + \tfrac{1}{2} O_2 \rightarrow H_2O$$

The silver electrode operates effectively at three states of oxidation from 0 to $+2$ during deep cycling. The standard free energy E^0 of the silver-hydrogen couple is 1.398 V, and, on the basis of 100% utilization of active materials, the couple has a theoretical energy density of 535 Wh/kg. During actual operation, the cell operates at 1.05 to 1.15 V and delivers approximately 75 to 85 Wh/kg.

23.3 CELL COMPONENTS

23.3.1 Silver Electrode

Silver electrodes are fabricated by doctoring a fine silver powder (2 to 6 μ) onto an expanded silver current collector, compacting the strip to a density of between 4 and 5 g/cm^3, and then sintering the strip in a tunnel furnace at about 600 to 650°C. The sintering process bonds the metallic silver particles to each other and to the current collector, which provides the necessary mechanical integrity for good electrical conductivity. Silver electrodes are usually manufactured to a thickness between 0.75 and 1.5 mm, depending on cell capacity and volume considerations. In some instances, additives are included in the silver powder to improve utilization of silver and capacity maintenance or reduce polarization (see Fig. 23.1).[2, 3] The final operations in the electrode fabrication include die-cutting and current lead attachment.

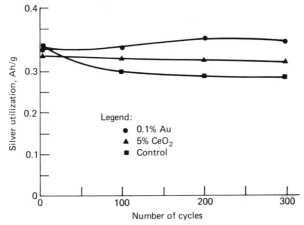

FIG. 23.1 Effect of additives on silver utilization in silver-hydrogen cells (deep discharge every 100th cycle).

23.3.2 Hydrogen Electrode

Hydrogen electrodes used in Ag-H_2 cells are similar to those developed for use in alkaline fuel cells and Ni-H_2 cells.[4] The electrodes contain Teflon-bonded platinum black or platinized carbon bonded to nickel collectors at loadings which range from 0.5 to 3 mg/cm². Microporous Teflon sheets about 0.05 mm thick are bonded to the opposite sides of the catalyzed electrodes. The hydrophobic Teflon backing serves to limit electrolyte entrainment which occurs when hydrogen is evolved from electrodes during charge.

23.3.3 Separator

Several types of separator systems have been studied in Ag-H_2 cells. These systems have varied with respect to both the absorber materials and the membrane materials employed. Electrolyte absorber materials evaluated have included nonwoven polyamide (Pellon), Teflon-bonded potassium titanate fibers, asbestos, and Viskon paper. Membrane materials evaluated have included cellophane, Permion 2291, inorganic separators (ERC), and inorganic-organic composites (NASA I.O.).

Among the absorber materials tested, asbestos and potassium titanate appear to offer the best combination of electrolyte retention and wicking properties. These properties are important since concentration changes which occur during charge dilute the electrolyte, and this dilution tends to expel water from the cell pack. During charge approximately 0.59 cm³ H_2O per ampere-hour is formed and must be wicked back into the cell pack to prevent dry-out during successive cycling.

Among the membrane materials tested, the inorganic separator developed at ERC and the NASA I.O. separator give the longest cycle life and act as good barriers against silver migration.[5, 6]

23.3.4 Electrolyte

Little work has been reported on the effect of electrolyte concentration on the performance and cycle life of silver-hydrogen cells. In practice, the cells employ an aqueous solution of potassium hydroxide between 25 and 30% by weight. Higher KOH concentrations favor increased Ag_2O solubility, increased active material utilization, and lower voltage. The reverse is true for lower KOH concentrations.

23.4 CELL CONSTRUCTION

23.4.1 Electrode-Separator Pack Arrangement

Cells are constructed by alternately stacking silver electrodes, absorbers, membranes, and hydrogen electrodes to form the required capacity (Fig. 23.2). Plastic gas diffusion screens are placed between adjacent hydrogen electrodes to allow gas access to the active catalytic sites. Plastic end plates and a centrally located tie rod support the active pack. Plastic edge masks are also used to protect all electrode edges from possible silver shorting.

Thermal analyses conducted on this type of pack design showed the need for

FIG. 23.2 Silver-hydrogen cell (developed for NASA). (*Courtesy of Energy Research Corp.*)

improved heat transfer from the electrodes through the gaseous medium to the walls of the cell case. To accomplish this, reticulated carbon foam spacers are used in place of the gas diffusion screens to increase the cell stack height to permit better heat dissipation to the cell housing.

23.4.2 Cell Hardware[5,6,7]

Cells employing lightweight hardware with capacities up to 50 Ah have been developed. The cell stack is contained within an Iconel type 625 or 718 cylinder 0.4 to 0.5 mm thick, which is fitted with two hemispherical end caps; all seams are electron-beam welded. Figure 23.2 shows a photograph of a silver-hydrogen cell developed for aerospace applications.[8] Immobility of the stack in the pressure vessel is accomplished by use of a ring which is welded to the can. The ring itself is fabricated with a cross-member which contains two holes. The stainless steel rods which hold the cell pack under compression are fed through these holes and fastened to the ring. The center of the cross-member is cut away to accommodate the connection between the positive leads and the terminal feedthroughs. The terminal feedthroughs for the cell are of a plastic compression type known as a Zeigler seal. Electrical contact to the cell is provided by a gold-plated copper-silver alloy rod which is 4.5 mm thick and is compressed within the plastic of the seal.

23.5 PERFORMANCE CHARACTERISTICS

Figure 23.3 shows the discharge characteristics of a 20-Ah Ag-H$_2$ cell at rates of discharge from C/4 to 1 C. The cell voltage at the C/4 rate is 1.12 V and decreases to 0.85 V at the 1C rate. Figure 23.4 shows the performance of two 20-Ah cells after 810 and 712 cycles. The cycling regime consisted of a modified synchronous orbit routine with 6.8 h of charge at 2.5 A and 1.2 h of discharge at 12.5 A. The depth of discharge corresponded to 75% based on a 20-Ah nominal cell capacity. Extended-life testing yielded cycle life greater than 1000 cycles before capacity fell below 15 Ah. Cycle life in excess of 3000 cycles at 27% depth of discharge on low-orbit (100 min) cycling has been reported.[9]

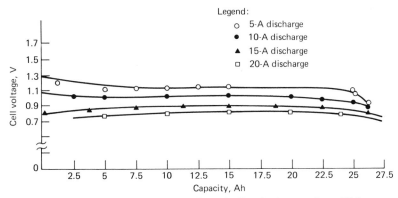

FIG. 23.3 Deep-discharge performance of 20-Ah silver-hydrogen cells at 20°C.

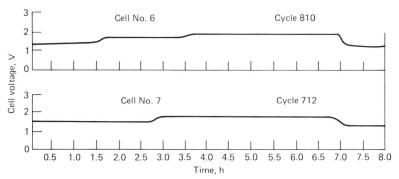

FIG. 23.4   Performance during cycling of 20-Ah silver-hydrogen cells (charge at 2.5 A for 6.8 h, discharge at 12.5 A for 1.2 h).

23.6 CHARGING AND CHARGE CONTROL

The constant-current method of charge has been found to be the most advantageous for recharging silver-hydrogen cells. As is generally true for all silver-type cells, no charging rate is considered too low for efficient recharge of the cell. At rates of charge greater than about C/8, the charge acceptance is considerably reduced. In practice, rates of charge from C/12 and lower yield almost 100% charge acceptance based on the previous discharge capacity. Voltage excursions encountered during charge range from about 1.55 V (Ag_2O plateau) to about 1.7 V at the onset of oxygen evolution. About 10 to 15% of the energy input at the beginning of charge is at a potential of about 1.3 V (Ag/Ag_2O plateau).

Two methods of charge control have been used successfully in terminating or controlling charge input. The first method utilizes a voltage cutoff based on the potential increase at the positive plate of about 100 mV at the onset of oxygen evolution. Unlike the $Ni-H_2$ cell, the ampere-hour charge efficiency of the silver-hydrogen cell approaches 100%, and no overcharge is necessary to obtain full capacity. The second method relies on a pressure switch to terminate charge which is set at the pressure achieved when the cell is fully charged. The pressure rise is directly proportional to the state of charge of the cell. Therefore, termination of charge at a given pressure is also a direct measure of the ampere-hours recharged.

REFERENCES

1. M. Klein, "Sealed Rechargeable Silver-Hydrogen Battery," in D.H. Collins (ed.), *Power Sources,* vol. 5, Academic, New York, 1975, pp. 347–359.

2. A. Charkey, "Silver-Hydrogen Cells," *Proc. 28th Power Sources Symp.* (1978), Atlantic City, N.J.

3. A. Charkey, "Long Life Silver-Zinc Cells," *Proc. 26th Power Sources Symp.* (1974), Atlantic City, N.J.

4. M. Klein and M. George, "Nickel-Hydrogen Secondary Batteries," *Proc. 26th Power Sources Symp.* (1974), Atlantic City, N.J.

5. G. L. Holleck, M. M. Turchan, F. S. Shuker, and D. J. DeBiccari, "Development of Silver-Hydrogen Cells," in J. Thompson (ed.), *Power Sources,* vol. 7, Academic, New York, 1979.

6. A. Charkey and D. J. DeBiccari, *Fabrication and Testing of Silver-Hydrogen Cells,* NASA, Final Report CR-159490, December 1978.

7. P. O. Offenhartz and G. L. Holleck, "On the Design of Silver-Hydrogen Electrochemical Cells," *J. Electrochem. Soc.* **127**:1213 (1980).

8. Energy Research Corporation, developed for NASA under Contracts NAS3-18543 and NAS3-10805.

9. P. Antoine and P. Fougere, "Development of Silver-Hydrogen Cells," in J. Thompson (ed.), *Power Sources,* vol. 7, Academic, New York, 1979.

Rechargeable "Primary-Type" Batteries

by
David Linden

24.1 GENERAL CHARACTERISTICS

Several of the conventional primary battery systems (e.g., the zinc/alkaline/manganese dioxide and the zinc/silver oxide systems) are also manufactured as rechargeable batteries. The rechargeable alkaline-MnO_2 cell is similar to the primary cell, using the cylindrical inside-out construction, and is available in several sizes. The main advantage of this rechargeable cell is its low initial cost and, in some applications, a lower operating cost than other secondary batteries. The zinc/silver oxide system is employed in a number of different types of primary, secondary, and reserve batteries because of its high energy and power density. The larger, more conventional zinc/silver oxide secondary batteries are covered in Part 3, Chap. 21. The zinc/silver oxide system also is used in rechargeable button-type cells similar to the primary cells (Part 2, Chap. 9) for application in watches, particularly those using solar cells as a prime energy source, and other small electronic devices.

24.2 RECHARGEABLE ZINC/ALKALINE/ MANGANESE DIOXIDE CELLS

24.2.1 General Characteristics

The rechargeable zinc/alkaline/manganese dioxide cells are an outgrowth of the primary cell. The cells use a zinc anode, a manganese dioxide cathode, and an electrolyte of potassium hydroxide. They are fabricated using an inside-out construction, and each

24-1

cell is hermetically sealed. The cell is manufactured in a fully charged state; capacity retention is similar to that of the primary cell. The major advantages and disadvantages of the rechargeable zinc/alkaline/MnO$_2$ battery are listed in Table 24.1.

TABLE 24.1 Major Advantages and Disadvantages of the Rechargeable Zinc/Alkaline/MnO$_2$ Battery

Advantages	Disadvantages
Low initial cost (and possible lower operating cost than other rechargeable batteries)	Useful capacity only one-third of primary cell
	Limited cycle life
Manufactured in a fully charged state	Available energy decreases rapidly with cycling
Good retention of capacity (compared to other rechargeable batteries)	
Completely sealed and maintenance-free	

The following features distinguish the rechargeable cell:

1. Improved cathode structure (binder and fiber additions)

2. Regenerated cellulose separators to avoid zinc shorts

3. Enlarged anode collector areas and electrolyte additives

A main problem is the expansion of the cathode after the discharge has progressed beyond MnO$_{1.6}$ (MnO$_{1.5}$ equals one electron change). In actual use the discharge is limited by voltage controls which prevent a discharge below, for example, 1.1 or 1.0 V, depending on the load and age of the battery. This is feasible with larger appliances (e.g., portable television sets) but not with simple consumer equipment or flashlights. A way to limit the cathodic discharge is to limit the zinc electrode capacity. Unfortunately this causes a generally poor rechargeability of the anode (in silver zinc/silver oxide or nickel zinc/nickel oxide batteries the zinc capacity is usually twice that of the cathode).

24.2.2 Performance Characteristics

The discharge characteristics of the rechargeable cells are similar to those of the primary cells, but with substantially lower energy density. The shape of the discharge curve changes slightly as the battery is repeatedly discharged and charged. The total voltage drop for a given energy withdrawal increases as the number of charge and discharge cycles increases. Coupled with this is the fact that the available energy per cell decreases with each cycle, even though the open-circuit voltage remains constant. This is shown graphically in Fig. 24.1 for a 15-V battery made from 10 G size cells discharged in 4-h cycles. The permissible discharge is about one-third of the primary cell capacity; about 50 cycles are available at that depth of discharge. Shallower cycling or decreasing the discharge current will increase cycle life.[1]

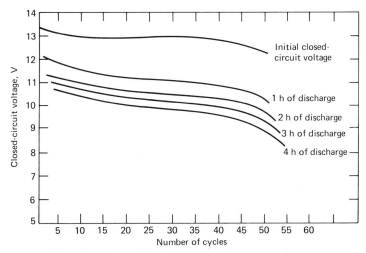

FIG. 24.1   Discharge characteristics of rechargeable zinc/alkaline/MnO$_2$ cells. Discharge cycle: 4 h through 9.6 Ω. Charge cycle: 16 h voltage-limited tapered-current charging (voltage limit = 17.5 V, filtered regulated power supply, 1% regulation with 4.0-Ω limiting resistor). (*From Eveready Battery Engineering Data, 1976.*)

24.2.3 Recharging Procedure

The cell should never be charged prior to the first discharge. D size cells can be charged with a 0.6-A (maximum) and G size cells with a 1.12-A (maximum) current. The charging characteristic is recommended to be linearly tapered up to the end voltage of 1.75 V per cell. Regulation of 1-V accuracy is desired, but it is possible to charge the cell completely between 1.7 and 1.725 V if more than 48 h are available. Overcharging damages the battery, and batteries should not be charged after 120% of the ampere-hours removed has been replaced and should not be float-charged for extended periods.[1]

24.2.4 Types of Cells and Batteries

The electrical characteristics of commercially available zinc/alkaline/manganese dioxide batteries are listed in Table 24.2.

24.3 RECHARGEABLE ZINC/SILVER OXIDE BUTTON CELLS

24.3.1 General Characteristics

The rechargeable zinc/silver oxide button cell is designed to have an extended life and, with proper usage, to deliver a total capacity significantly greater (5 to 8 times) than its rated capacity.

TABLE 24.2   Electrical Characteristics of Rechargeable Zinc/Alkaline/Manganese Dioxide Batteries

Voltage, V	Number and size of cells*	Rated capacity, Ah	Maximum recommended discharge current, A
4.5	3 D	2.5	0.625
6.0	4 G	5.0	1.25
7.5	5 D	2.5	0.625
13.5	9 G	5.0	1.25
15.0	10 G	5.0	1.25

*Cell Dimensions: O cell   32 mm diam, 61.5 mm high
               G cell   32 mm diam, 105 mm high
SOURCE: From *Battery Engineering Handbook,* Union Carbide Corp., New York, 1976.

Rechargeability   The rechargeability of the cell depends on many factors, such as the charging current, cutoff voltage of the charge, and depth of discharge to which the cell has previously been subjected. A maximum depth of discharge of 10 to 20% of the rated capacity is recommended. This shallow discharge is beneficial because recharging the limited portion of the capacity removed is easier. Restricting the depth to which the cell is discharged extends the number of cycles as well as the total capacity which can be obtained from the cell. The cells can be recharged after discharge depths greater than the 20% level; however, there will be some loss of rechargeability and capacity. Cells discharged to completion have also been successfully recharged, but the capacity delivered on subsequent discharge was low.

Shelf Life   The shelf life of the cell has been extended by means of special features of construction. Such cells have been stored at temperatures as high as 45°C for over 1 y and delivered 80% of initial capacity.

Discharge Characteristics   The voltage on discharge is stable throughout the discharge. The voltage will vary from about 1.58 V per cell on a C/2000 rate to 1.55 V on a C/100 rate (100,000- and 5000-Ω load, respectively, for a 28-mAh size cell).

Cell Impedance   The impedance of a fresh, undischarged 28-mAh cell, measured at 1000 Hz, is about 20 Ω. On discharge, the cell impedance decreases to about 15 Ω. On charge, the cell impedance increases slightly above these values but will decrease if left on stand or placed on discharge. There is a pronounced increase in impedance at the end of life if a cell is discharged to completion.

Cell Characteristics for Solar Panel Charging   The cell is designed to be charged in a two-cell series stack or as a single cell. There are certain limitations on charging with a solar panel array. The initial charge rate must be limited to a maximum of 1.5 mA at 1.6 V (28-mAh cell), and the charge current should decrease as the cell is recharged. The charge voltage must be limited to a maximum value of 1.9 V, and, at this point, the charging current should fall to zero. A maximum charge voltage of 1.85 V is preferable. The charging array should include a blocking diode. First, this will protect the cell from discharging through the solar cells during a low or no-light condition. Second, it acts to drop the charging voltage across the panel, which can be used to limit the maximum charge voltage. A minimum of 1.6 V per cell is necessary for the cell to accept the charge.

Service   The cell can give between 3 and 5 y of operation as long as the specified charge and discharge parameters are met.

24.3.2 Types of Cells

The characteristics of several of the rechargeable zinc/silver oxide cells that are available are listed in Table 24.3. These cells are designed for continuous low-rate discharge applications with a periodic high-rate pulse requirement.

Table 24.3   Rechargeable Zinc/Silver Oxide Button Cells

Type no.	Dimensions, mm		Volume, cm³	Weight, g	Nominal voltage, V	Rated capacity, mAh
	Diameter	Height				
Duracell 10SL19	7.75	3.58	0.164	0.8	1.55	28 (at 250 μA)
Maxell XR9527W	9.5	2.73	0.193	0.93	1.55	44

REFERENCE

1. *Eveready Battery Engineering Data,* Union Carbide Corp., New York, 1976.

25

Lithium Batteries (Ambient Temperature)

by
Paul A. Malachesky

25.1 GENERAL CHARACTERISTICS

A rechargeable battery, operating at ambient temperatures and using lithium as the anode material, offers the potential of high energy densities compared with conventional secondary batteries.

The early work, in the 1960s and 1970s, concentrated on metal halides, such as $CuCl_2$, metal oxides, and other soluble-cathode materials, such as bromine and sulfur.[1-3] While cells with these cathode materials had high energy densities, high rate capability, and good rechargeability, a major drawback for practical use was an unacceptably high self-discharge rate.

A more successful approach was found in the mid-1970s with certain insoluble-cathode materials such as several transition metal dichalcogenides which could reversibly intercalate lithium.[4-6] These intercalation reactions involve the interstitial introduction of a guest species, such as Li^+, into the host lattice with little or no structural modification of the host.

The first commercial cells, in button sizes, were introduced briefly in 1978 but were withdrawn from the market. Continued development of these cells, with improvement in their cycle life, should renew commercial interest and result in new markets and applications for these potentially attractive batteries. (See also Part 4 on advanced secondary battery systems using lithium anodes.)

25.2 CHEMISTRY

The key requirements for a solid-state cathode material which reacts with lithium to form a ternary-phase lithium intercalation compound, according to the following generalized reaction, are given in Table 25.1.

$$x\,\text{Li} + \text{MO}_z \underset{\text{charge}}{\overset{\text{discharge}}{\rightleftarrows}} \text{Li}_x\text{MO}_z$$

Table 25.2 presents the theoretical cell energy densities for the most attractive transition metal dichalcogenides when coupled with lithium. Of the materials in Table 25.2, TiS_2, in spite of its somewhat lower energy density, has been the most widely used because of its good rate capability, ease of synthesis (via gas phase reaction of TiCl_4 and H_2S), and potential low cost.

TABLE 25.1   Key Ternary Phase (Li_xMO_z) Cathode Requirements for Lithium Secondary Cell Operation

High free energy of reaction
Wide range of X with small ΔG changes with X
Little structural change on reaction
Reversibility of reaction
High diffusivity of Li
Good electronic conductivity and no solubility

SOURCE: Ref. 7.

TABLE 25.2   Theoretical Energy Densities of Lithium-Anode, Ternary-Phase Cathode Cells

Cathode material	Cell emf, V*	Energy density, Wh/kg
TiS_2	2.47	480
$\text{Cr}_x\text{V}_{1-x}\text{S}_2$	2.48	550
V_6O_{13}	2.80	800

*Initial value.

The excellent reversibility of these cathode systems has served only to highlight the technical problems associated with the lithium anode–electrolyte interface. While lithium primary cells, in general, owe their excellent shelf life to protective films formed by anode-electrolyte reactions, this electrolyte reactivity severely impedes the cycling characteristics of the lithium anode. This electrolyte-anode reactivity is accentuated by the high surface area of the lithium deposited during cell recharge. Anode-related cell failure modes are: (1) shorting by lithium dendrites and (2) electrical isolation of the electro-deposited lithium. This second failure mode can be quantified in terms of a figure-of-merit (FOM) limitation, where FOM is defined as the ratio of the cumulative cell discharge capacity to the theoretical amount of lithium originally incorporated into the cell. For example, a 100-mAh cathode limited-capacity cell might normally contain a 400-mAh theoretical lithium capacity. If the cell electrolyte is very reactive to lithium, cell failure would be expected to occur at or near an FOM of 1 or a cumulative capacity of ≈ 400 mAh (e.g., five 80% depth or ten 40% depth cycles). Such a cell is operating as a pseudo-primary cell, limited in available capacity by the amount of lithium in the cell.

There are a number of approaches to the rechargeable lithium anode which have included: (1) use of alloying substrates, (2) use of unreactive electrolytes—generally accompanied by rigorous electrolyte purification, and (3) deliberate modification of the Li-electrolyte interface by incorporation of film-forming additives. The use of the first approach has been successfully demonstrated in the LiAl/TiS_2 rechargeable cell system.[10] Lithium perchlorate in the cyclic ether solvent dioxolane is a very well-documented,[11] though demonstrated hazardous,[12] example of the second approach. While the third approach seems to be a contraindicated strategy, the formation of a "desirable" anode film has been demonstrated with $LiAsF_6$ (a reactive salt) in 2-methyltetrahydrofuran (2-MeTHF), an unreactive solvent.[13] This electrolyte system combines the second and third approaches and can give reasonable, albeit limited, cell-cycling performance.[14] Table 25.3 lists the properties of some electrolyte systems which have been used in secondary Li cells.

TABLE 25.3   Properties of Some Secondary Lithium Cell Electrolytes

Solvent system	Salt	Salt concentration, molal	Resistivity, $\Omega \cdot cm$
Dioxolane	$LiClO_4$	2.5	120
2-Methyltetrahydrofuran	$LiAsF_6$	1.0	330
Dioxolane	$Li(C_6H_5)_4B$	1.0	225
Dioxolane	$Li(CH_3)_4B$	2.0	94

25.3 CELL DESIGN AND PERFORMANCE CHARACTERISTICS—LiAl/TiS$_2$ CELLS

The only commercially available rechargeable lithium cells were the LiAl/TiS$_2$ series of button cells, which were withdrawn from the market in late 1979.[15] These LiAl/TiS$_2$ button cells were designed for microelectronic applications such as solar rechargeable watches and clocks and were characterized by limited deep-cycle performance at low discharge rates ($\approx C/100$).

Table 25.4 presents the basic characteristics of the two cell sizes which were developed. These cells were hermetically sealed with a glass-to-metal negative electrode feedthrough. The cell case (positive connection) was stainless steel and projection-welded as Fig. 25.1 illustrates. The lithium-aluminum anode (approximate composition LiAl) was formed in situ after cell closure. The TiS$_2$ cathodes were highly porous ($\approx 70\%$) to allow rapid electrolyte absorption during cell filling. The electrolyte used was $Li(C_6H_5)_4B$ in a dioxolane-1,2-dimethoxyethane mixed solvent. As such, the volumetric energy density was not optimized but sufficed for the potential applications. Still, at 100 Wh/L, the volumetric energy density is approximately twice that of small NiCd button cells.

Figure 25.2 shows typical 90-mAh cell voltage behavior during cycling. Due to the linear variation of $\triangle G$ (cell emf) with x in the discharge product, Li_xTiS_2, the discharge

TABLE 25.4   Characteristics of Rechargeable LiAl/TiS$_2$ Button Cells

Cell type	LTS-90	LTS-25
Diameter, cm	2.5	1.2
Height, cm	0.33	0.33
Voltage range, V	2.0–1.4	2.0–1.4
Cell capacity, mAh	90	25
Wh/L	110	110
Cell impedance, Ω	7	25
Self-discharge	$< 10\%/y$	$< 10\%/y$

SOURCE: Exxon Enterprises, Inc.

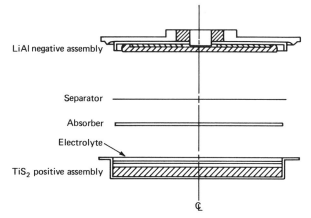

FIG. 25.1 Rechargeable LiAl/TiS$_2$ button cell assembly.

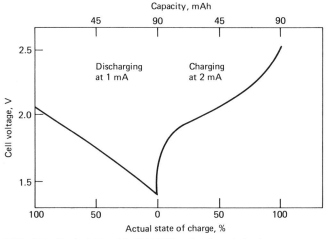

FIG. 25.2 Typical 90-mAh LiAl/TiS$_2$ cell voltage behavior.

curve is almost linear with state of charge at low discharge rates. During charge at the 50- to 100-h rate, alloy equilibration is slow and cell voltage behavior is more typical of a pure Li anode/TiS$_2$ cell.

Figure 25.3 shows the typical deep cycle performance of the 25-mAh cell, sized for use in a solar watch. Cell capacity, due to mechanical degradation of the alloy anode during such deep cycles, decreases to about 50% of its initial value in 8 to 10 cycles. This decrease in capacity is accompanied by an increase in cell impedance to 5 to 10 times its original value. In contrast, shallow-depth cycling at 5 to 10% DOD (more typical of actual watch applications), as shown in Table 25.5, gives cumulative cell capacities equivalent to greater than 10 y of electronic watch operation at average currents of 2 to 3 μA. The Li FOM limitation, common to pure Li anode cells, has not been observed with these LiAl anodes under these conditions which have extended to FOMs > 10.

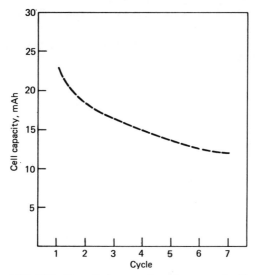

FIG. 25.3 Deep-discharge cycle performance of a 25-mAh LiAl/TiS$_2$ cell (discharged at 0.2 mA to 1.3 V, charged at 0.5 mA to 2.6 V).

TABLE 25.5 Extended Cycle Performance of a 25-mAh LiAl/TiS$_2$ Cell

Condition	Cell capacity, mAh	Cell impedance, Ω
Initial	. . .	24
25% discharged	6	22
Shallow cycling (5% DOD)	211	60
Deep discharge no. 1	15	34
Deep discharge no. 2	19	63
Deep discharge no. 3	17	68

Shelf life of these LiAl/TiS$_2$ cells is excellent, with no measurable capacity loss during accelerated storage testing for 3 months at 65°C. Microcalorimetric measurements at ambient temperature show heat output values of < 2 μW, indicating less than 10% capacity loss per year (assuming that this heat output is directly related to cell capacity loss) for the 90-mAh cell. (Preliminary capacity measurements after 1 y storage at ambient temperature show $\leq 5\%$ loss of cell capacity.)

25.4 ENGINEERING PROTOTYPE CELLS

Prototype units of secondary lithium cells have been fabricated[16] using TiS$_2$, Cr$_{0.5}$V$_{0.5}$S$_2$, and V$_6$O$_{13}$ cathode materials, and a LiAsF$_6$/2-MeTHF electrolyte.[14] Prismatic cells (approximately 5.8 × 2.0 × 5.6 cm) containing TiS$_2$ have delivered 4.5 Ah at the C/30 rate (175-mA drain) in the sixteenth cycle. This corresponds to a volumetric energy density of 145 Wh/L and a weight energy density of 85 Wh/kg. Cell failure, due to dendritic shorting, occurred after the eighteenth cycle. Spiral-wound D cells have also been fabricated[16] with 4-Ah initial capacity, also with a limited cycle life (15 to 20 cycles). Smaller cells have been reported to have a cycle life in excess of 100, with cathode capacity loss limiting cell performance.[14] This cathode capacity loss was attributed to degradation of cathode electrode structure, since decreasing the discharge rate by 50% restored cell capacity to its initial value.

Prototype Li/TiS$_2$ cells have also been fabricated[15] using LiAsF$_6$/2-MeTHF, in the 2.5-cm diameter button cell configuration described earlier, using pure Li anodes. Cell cycling at the C/20 rate is shown in Fig. 25.4. Initial capacities of 140 mAh were achieved corresponding to 215 Wh/L. Cell capacity declined rapidly after 20 cycles and was accompanied by a rise in cell impedance.

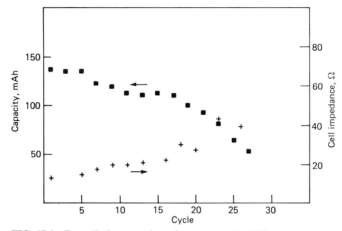

FIG. 25.4 Deep-discharge cycle performance at the C/20-rate for a pure lithium/1 M LiAsF$_6$/2-MeTHF/TiS$_2$ cell (2.5 cm diameter × 0.33 cm high) (7.5 = mA discharge, 2.0 = mA charge).

Larger Li/TiS$_2$ cells have been designed[15] with nominal cell capacity of 20 Ah (C/20 rate). In a prismatic cell configuration (13.2 × 8.4 × 2.7 cm), using a LiB(CH$_3$)$_4$/dioxolane electrolyte, a volumetric energy density of 110 Wh/L and a gravimetric energy density of 37 Wh/kg were achieved. Limited cycle life was obtained, due to electrolyte gassing as a result of LiB(CH$_3$)$_4$ decomposition via a slow heterogeneous reaction with TiS$_2$.

25.5 NEW DEVELOPMENTS

Now that attractive cathode couples have been successfully demonstrated, the major focus of future ambient-temperature rechargeable lithium cell research and development will be on the area of electrolytes capable of cycling without FOM limitations or dendritic plating problems. Other important needs, again electrolyte-related, are those of safety and tolerance of cell abuse (of particular importance for series-connected cells) under overcharge and overdischarge. A significant breakthrough in these areas would allow the development of a rechargeable lithium battery with 100+ deep cycle capability, which would offer the system designer an additional choice in the portable power source area now dominated by nickel-cadmium and lead-acid cells. Such secondary lithium batteries, operating at projected energy densities of 90 to 110 Wh/kg and 250 to 350 Wh/L at the 5 to 10-h rate, would be used in applications where they, in spite of higher costs, offer a clear advantage over aqueous batteries in terms of basic energy density operating characteristics and charge retention (shelf life). Using the present state-of-the-art technology, low-drain rate cells, in duty cycles involving shallow-depth cycling or float service regimes, appear well within the capabilities of this emerging lithium battery technology.

REFERENCES

1. G. Eichinger and J. O. Besenhard, "High Energy Density Lithium Cells Part II. Cathodes and Complete Cells," *J. Electroanal. Chem.* **72**:1 (1976).

2. J. Weininger and T. Rouse, "Improved Electrochemical Cells," French Patent 1,542,043 (1968).

3. J. R. Coleman and M. W. Bates, "The Sulfur Electrode," in D. H. Collins (ed.), *Power Sources,* vol. 2, Pergamon, New York, 1969, p. 299.

4. M. S. Whittingham, "Electrical Energy Storage and Intercalation Chemistry," *Science* **192**:1126 (1976).

5. M. S. Whittingham, "Chalcogenide Battery," U.S. Patent 4,009,052 (1977).

6. G. L. Holleck and J. P. Driscoll, "Transition Metal Sulfides as Cathodes for Secondary Lithium Batteries—II. Titanium Sulfides," *Electrochim. Acta* **22**:647 (1977).

7. M. S. Whittingham, "Chemistry of Intercalation Compounds: Metal Guests in Chalcogenide Hosts," *Prog. Solid State Chem.* **12**:1 (1978).

8. D. W. Murphy and P. A. Christian, "Solid State Electrodes for High Energy Batteries," *Science* **205**:651 (1979).

9. D. W. Murphy and F. A. Trumbore, "Metal Chalcogenides as Reversible Electrodes," *J. Crystal Growth* **39**:185 (1977).

10. B. M. L. Rao, R. W. Francis, and H. A. Christopher, "Lithium-Aluminum Electrode," *J. Electrochem. Soc.* **124**:1490 (1977).

11. B. M. L. Rao, G. H. Newman, R. W. Francis, and L. H. Gaines, "Titanium Disulfide Electrode," Extended Abstract no. 7, *Electrochem. Soc. Meeting,* Atlanta, October 1977.

12. G. H. Newman, R. W. Francis, L. H. Gaines, and B. M. L. Rao, "Hazard Investigations of LiClO₄/Dioxolane Electrolyte," *J. Electrochem. Soc.* **127**:2025 (1980).

13. V. R. Koch, "Aprotic Solvent Electrolytes and Batteries Using Same," U.S. Patent 4,118,550 (1978).

14. G. L. Holleck, K. M. Abraham, and S. B. Brummer, "Symp. Ambient Temperature Lithium Batteries," Electrochem. Soc. Meeting, Los Angeles, October 1979.

15. Exxon Enterprises Inc., Somerville, N.J.

16. EIC Corp., Newton, Mass.

PART 4 ADVANCED SECONDARY BATTERIES

Introduction

by
Warren L. Towle and David Linden

During the last decade, new applications have emerged as potentially important markets for batteries. These include batteries for electric vehicles, batteries for the storage of electrical energy to assist in handling peak loads for electric utilities, and batteries for the storage of electrical energy produced by solar or wind-driven devices to provide continuous power output from these intermittent-energy sources.

The performance characteristics required for these new applications far exceed the capabilities of conventional batteries and provide a strong incentive for significantly improving the performance of conventional battery systems and for the development of new electrochemical systems possessing advantages in energy density, long life, and low cost.

26.1 PERFORMANCE REQUIREMENTS FOR ADVANCED SECONDARY BATTERIES

The performance requirements for batteries for electric vehicles, electrical utility load leveling, and energy storage are given in Tables 26.1 to 26.3, respectively.[1] The dominant requisites for electric vehicle use are (1) high energy density to provide adequate driving range and (2) long cycle life to keep the cost and frequency of battery replacement low. For electric utility service the principal requisites are (1) low first cost and (2) very long cycle and calendar life. Long cycle and calendar life, low cost, and high efficiency are important for the solar photovoltaic applications, with high energy density being relatively unimportant.

26.1.1 Electric Vehicles

The major objectives of the electric vehicle program are to encourage the use of electric vehicles to alleviate dependence on oil and to improve the quality of air. Utility power, based on less critical prime energy sources, would be used for battery charging. The near-term program is directed toward commercialization of a compact electrical urban

TABLE 26.1 Proposed Battery Requirements for Electric Vehicle Applications

	Urban car	Van or truck	Bus
Gross vehicle wt., kg	1400 to 1500	1500 to 4000	13,000 to 14,000
Typical battery size, kWh	20 to 30	30 to 80	200
Battery voltage, V	←	90 to 440	→
Duty cycle	←	Variable	→
Specific energy,* Wh/kg	←	> 70	→
Wh/L	←	> 70	→
Specific power, W/kg			
Sustaining	←	> 20	→
15-s peak	←	> 100	→
Energy efficiency, %	←	> 60	→
Cycle life	←	> 1000	→
Initial battery cost, $/kWh	←	25–35	→
Environmental			
Ambient temperature, °C	←	−20 to 40	→
Vibration, g	←	> 5	→
Shock, g	←	> 30	→

*Battery rated at 3-h discharge, 80% depth of discharge.
SOURCE: Ref. 1.

TABLE 26.2 Proposed Battery Requirements for Electric Utility Applications

	Peaking duty	Intermediate duty
Typical installation size, MWh	50–200	50–200
Battery voltage, V	1000–2000	1000–2000
Daily duty cycle, h		
Discharge	5	10
Charge	5–7	5–7
Energy efficiency (round trip), %	> 70	> 80
Cycle life	> 1500 (10-y min)	> 2500 (10-y min)
Battery footprint, kWh/m²	> 1	> 1
Initial battery cost, $/kWh	25–30	> 20
Environmental impact	Minimal	Minimal

SOURCE: Ref. 1.

car having a range of about 160 km (80 km for a delivery van) in the late 1980s. Advanced battery systems, providing a range of 300 to 400 km, are projected for commercialization during the period from 1995 to 2000.

26.1.2 Energy Storage Applications

Battery energy storage facilitates the efficient use of cheap base-load energy generated from coal, nuclear sources, or other less critical prime energy sources and provides this energy to meet the peak power demand, which is presently being met with expensive gas- and oil-based turbines. The concept of load leveling is illustrated in Fig. 26.1.[1] Battery storage capacity, the specific duty cycle, and initial and operating costs are important considerations in optimizing the battery storage facility. Storage capacity sufficiently large to handle the weekly cycle is more attractive than one sized for the daily cycle because a larger portion of the available off-peak power can be used.

TABLE 26.3 Tentative Battery Requirements for Applications to Solar Photovoltaic Power Systems

Characteristic	Low power/remote site	Single residence	Multiple residence	Commercial/industrial	Central station
System					
Storage capacity, kWh	0.05–1000	5–30	10–2000	150–10⁴	10⁵–6 × 10⁶
Power, kW	0.001–2	2–10	5–500	50–5000	5 × 10⁴–10⁶
DC voltage, V	2–400	50–350	50–600	50–600	750–2500
Battery/cell					
Capacity, Ah	25–2000	20–500	50–10,000	100–10,000	2000–60,000
Maximum rate	$<C/50$	$C/3$–$C/5$	$C/5$–$C/7$	$C/5$–$C/9$	$C/2$–$C/6$
Duty cycle	Float, may be 1-y cycle	Daily cycle, rapid changes in rate, depth of discharge is design-dependent	Daily cycle, load diversity reduces rapid rate changes, depth of discharge is design-dependent		Daily cycle, possibly complete daily discharge
I-V curves	May seasonally lower array efficiency without regulator	Mostly adequate ($\approx$90% of maximum array efficiency without regulator) →			
Allowable cost, $/kWh	25–100	0–40	0–40	0–40	0–35
Efficiency		As high as practical without undue cost penalty →			
Life	4–10 y	~2500 cycles, 7–10 y (also as high as practical without undue cost penalty) →			
Energy density		Relatively unimportant		Minimum attainable without undue cost penalty →	
Maintenance	Must be minimal, several month interval	Mostly maintained by service contract		Maintained by site personnel →	
Safety	Access by skilled personnel	Limited user access			Access by skilled personnel
		Must be inherent in cell and system design →			

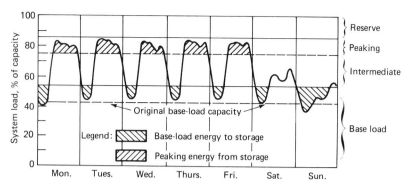

FIG. 26.1 Weekly load curve of an electric utility generation mix with energy storage. (*From Yao and Landgrebe, 1977.*)

However, the high initial cost of the battery tends to favor the daily storage cycle. The battery design is also influenced by the specific daily duty cycle. Typically, peaking duty is at the 2- to 8-h rate, and intermediate duty at the 10- to 14-h rate.

Battery storage also is required for solar, wind, and other such power systems where the prime energy source is not available on a continuous basis. The battery stores the energy when it is being generated for discharge during the period when the prime energy source is not available. The battery requirements for typical application of solar photovoltaic power sources are given in Table 26.3. In general, the duty cycle requirements are similar to those for application in electric utilities but are less predictable with regard to load and storage capacity and less stringent with regard to energy density. High energy efficiency and low self-discharge are particularly important because of the high cost of the generating equipment at present and the long storage periods required.

26.2 TYPES AND CHARACTERISTICS OF ADVANCED BATTERY SYSTEMS

A wide range of battery systems is being explored and developed to meet these new requirements with active support of such organizations as the U.S. Department of Energy (DOE), the Electric Power Research Institute (EPRI), and the battery industry. These activities can be categorized as follows:

1. Near-term programs directed to the improvement of the performance characteristics of conventional aqueous secondary batteries for deployment in the years 1985 to 1990

2. Advanced battery programs covering the development of new electrochemical systems offering the potential of higher energy and power output

Two test facilities have been established for the evaluation of these battery systems. Electric vehicle batteries are tested at the National Battery Test Laboratory of the Argonne National Laboratory at Argonne, Illinois. Energy storage systems for utility application are tested at the Battery Energy Storage Test (BEST) Facility, Hillsborough Township, New Jersey, where complete systems can be evaluated in a utility environment early in the development cycle.

The major electrochemical systems that have been considered as candidates for electric vehicle (EV) and energy storage applications are listed in Table 26.4. Table 26.5 lists some of the major commercial organizations that have done R&D work in these areas; some are no longer active. The characteristics of the more promising systems are summarized in Table 26.6.

TABLE 26.4 Candidate Systems for Electrical Vehicle and Energy Storage Applications

	Chapter and section reference
Conventional battery systems	
Lead-acid*	Part 3, Chap. 14
Nickel-iron*	Part 3, Chap. 20
Nickel-zinc	Part 3, Chap. 19
Nickel-hydrogen	Part 3, Chap. 22
Aqueous-type advanced battery systems	
Zinc/chlorine†	Part 4, Chap. 28
Zinc/bromine	Part 4, Chap. 29
Redox systems	Part 4, Chap. 27
Metal/air	Part 4, Chap. 30
Iron/air	Part 3, Chap. 20, Sec. 20.4.2
Zinc/air	Part 4, Chap. 30, Sec. 30.2
Lithium/air	Part 4, Chap. 30, Sec. 30.3
Aluminum/air	Part 4, Chap. 30, Sec. 30.4
Hydrogen/halogen	Part 4, Chap. 26, Sec. 26.4.1
High-temperature systems	
Lithium/chlorine	Ref. 2
Lithium/tellurium chloride	Ref. 3
Lithium/sulfur	Part 4, Chap. 31
Lithium-aluminum/iron sulfide†	Part 4, Chap. 31
Calcium/iron sulfide	Part 4, Chap. 26, Sec. 26.4.2
Sodium/sulfur†	Part 4, Chap. 32
Sodium/antimony trichloride	Part 4, Chap. 26, Sec. 26.4.2
Solid electrolyte batteries	Part 4, Chap. 26, Sec. 26.4.2
Ambient-temperature, nonaqueous systems	Part 4, Chap. 26, Sec. 26.4.3
	Part 3, Chap. 13
Polymer batteries	Part 4, Chap. 26, Sec. 26.44

*Major candidates, near term.
†Major candidates, advanced systems.

26.3 NEAR-TERM ADVANCED SECONDARY BATTERIES

The conventional aqueous secondary batteries, while limited in their performance characteristics, are the major candidates of the near-term program. Expected advances in energy density, power density, and life will permit these batteries to be used successfully but with some compromise in system objectives while awaiting the development of the advanced battery systems. The electric vehicle targets, for example, are 50 to 60 Wh/kg and 100 W/kg for the near-term electric commuter car with a 160-km range (80 km for an electric van) and an acceleration from 0 to 48 km/h in 8 s. The 800 cycles to an 80% DOD corresponds to a life of about 90,000 km. Table 26.7 summarizes the performance goals for the near-term battery systems.[4,5]

TABLE 26.5 Organizations Working on Advanced Secondary Battery Systems

Lead-Acid Batteries
  Johnson Controls, Inc., Globe Battery Div., Milwaukee, Wis.
  Exide Management & Technology Co., INCO, Philadelphia, Pa.
  Eltra Corp., C & D Batteries Div., Plymouth Meeting, Pa.

Nickel-Iron Batteries
  Westinghouse Electric Corp., Pittsburgh, Pa.
  Eagle-Picher Industries, Joplin, Mo.

Nickel-Zinc Batteries
  Energy Research Corp. Danbury, Conn.
  Yardney Electric Co., Pawcatuck, Conn.
  Exide Management and Technology Co., INCO, Philadelphia, Pa.

Nickel-Hydrogen Batteries
  Energy Research Corp., Danbury, Conn.
  Yardney Electric Co., Pawcatuck, Conn.
  Eagle-Picher Industries, Joplin, Mo.

Zinc/Chlorine Batteries
  Energy Development Associates, G & W Corp., Madison Heights, Mich.
Zinc/Bromine Batteries
  Energy Research Corp., Danbury, Conn.
  Exxon Research and Engineering Co., Linden, N.J.
  Gould, Inc., Rolling Meadows, Ill.
  Johnson Controls, Inc., Globe Battery Div., Milwaukee, Wisc.
Redox Batteries
  Lewis Research Center, NASA, Cleveland, Ohio.
  Lockheed Research Laboratory, Palo Alto, Calif.

Metal/Air Batteries
  Lawrence Livermore National Laboratory, Livermore, Calif.
  Westinghouse Electric Corp., Pittsburgh, Pa.
  Lockheed Research Laboratory, Palo Alto, Calif.
  Sanyo Electric Co., Osaka, Japan.
  Compagnie Generale d'Electricite, Marcoussis, France.
  Norwegian Defence Research Establishment, Kjeller, Norway.
  Swedish National Development Co., Akersberga, Sweden.

Lithium/Metal Sulfide Batteries
  Argonne National Laboratory, Argonne, Ill.
  Eagle-Picher Industries, Joplin, Mo.
  Gould, Inc., Rolling Meadows, Ill.
  Rockwell International, Canoga Park, Calif.

Sodium/Sulfur Batteries
  General Electric Co., Schenectady, N.Y.
  Ford Aerospace and Communication Corp., Newport Beach, Calif.
  Dow Chemical Co., Walnut Creek, Calif.

Solid Electrolyte Batteries
  Duracell International, Inc., Bethel, Conn.
Hydrogen/Halogen Batteries
  Brookhaven National Laboratory, Upton, N.Y.
  General Electric Co., Wilmington, Mass.
Ambient Temperature, Nonaqueous Batteries
  Duracell, Inc., Bethel, Conn.
  Exxon Research and Engineering Co., Linden, N.J.
  EIC Corp., Newton, Mass.
Polymer Batteries
  Allied Corp., C & D Batteries, Plymouth Meeting, Pa.

TABLE 26.6 Comparative Performance of Advanced Battery Systems*

Classification	System	Theoretical			Present performance†						
		Voltage, V	Capacity, Ah/kg	Energy density, Wh/kg	Operating temperature, °C	Working voltage, V	Energy density, Wh/kg	Energy density, Wh/L	Energy efficiency, %	Area impedance, Ω·cm²	Peak area power, W/cm²
Near-term	$Pb/H_2SO_4/PbO_2$	2.1	120	252	Ambient	2.0	35	70	80	2.7–6.5	0.17–0.4
	$Zn/KOH/NiOOH$	1.7	215	370	Ambient	1.6	60	120	70		
	$Fe/KOH/NiOOH$	1.5	224	340	Ambient	1.2	35	65	50		
Halogen	$Zn/ZnCl_2/Cl_2$	2.1	394	830	20–50	1.9	100	130	65	3.8	0.30
	$Zn/ZnBr_2/Br_2$	1.8	238	430	20–50	1.7	70	75	70	5.0	0.16
Air	$Fe/KOH/Air$	1.3	960‡	1250‡	50	1.0	60		40		
	$Zn/KOH/Air$	1.6	825‡	1320‡	50–60	1.2	110	100	40	1.7	0.5–0.6
	$Al/KOH/Air$	2.73	2980‡	8135‡	50–60	1.6	160		…		
	$Li/LiOH/Air$	3.4	3860‡	13,124‡	Ambient, 65	2.0					
Redox	$CrCl_2$/membrane/$FeCl_3$	1.2	108	130	Ambient	0.9	…	…	75	5.0	0.05
Lithium	$Li/LiCl/Cl_2$	3.4	638	2170	650	3.0					
	$Li/LiCl, KCl/S$	2.2	1165	2560	400–500	3.0					
	$LiAl/LiCl, KCl/FeS$	1.33	345	460	400–500	1.2	80	100	80	0.8–3	0.2–0.4
	$LiAl/LiCl, KCl/FeS_2$	1.7–1.33	418	650	400–500	1.5–1.2	100	200	80	0.8–3	0.2–0.4
Sodium	$LiAl/LiCl, KCl/TeCl_4$	3.1	360	1120	450	2.7					
	$Na/glass/S$	2.1	377	790	290–320	1.7		…	90	33	0.023
	$Na/ceramic/S$ β-Al_2O_3	2.1	377	790	300–350	1.7	140	280	80	2.8–3.5	0.31–0.38
	Na/β-$Al_2O_3/SbCl_3$	2.7	272	735	200–240	2.5	85	185	80	9.3	0.21

*In addition to the usual performance parameters, Table 26.6 presents comparative data on "peak area power" expressed in watts per square centimeter (W/cm²). This parameter represents the maximum rate at which power can be drawn per unit of electrode (i.e., separator) area. This parameter is of more significance for high-power applications, such as engine cranking or electric vehicle use. The peak area power can be calculated from the open-circuit voltage (OCV) and the "area impedance." The latter term represents the total cell impedance, assuming it to be distributed uniformly over the separator area. Most batteries are designed with positive and negative electrodes of approximately equal area, with a separator (when used) also of the same area. Most of the cell resistance is located in the separator itself or in the electrolyte, which for practical purposes also has the same area. Other resistances, not area-related (such as straps and posts), are normally very small. Thus the assumption in the large majority of cases is well justified. By making this pragmatic simplification, the area impedance itself becomes a very useful comparative parameter. It is calculated very simply, as the product of total cell impedance and separator area, and has dimensions ohms-square centimeter (Ω·cm²).

The peak area power can be calculated as

$$P_r = \left(\frac{OCV/2}{r}\right)^2 \times r = \frac{(OCV/2)^2}{r}$$

where r is the area impedance. The peak instantaneous delivery of power will be achieved when, by rapid increase in the flow of current, the terminal voltage has been reduced to 50% of the OCV.

†Based on present state-of-art cells.   ‡Based on metal anode only.

TABLE 26.7 Goals of Near-Term Battery Development Programs

	Lead-acid		Nickel-iron	Nickel-zinc
	ISOA*	Advanced		
Specific energy,† Wh/kg	40	50–60	70	70–85
Specific peak power,‡ W/kg	100	150	140	160
Cycle life, 80% DOD	800	1000	2000	800
Energy efficiency, %	> 60	> 60	> 60	> 60
Cost,§ $/kWh	50	40	70	70

*ISOA-improved state of the art.
†C/3 discharge rate.
‡30-s average at 50% state of charge.
§At production rate of 100,000 units (25 k/Wh)/y.
SOURCE: Ref. 4.

The *lead-acid battery* is attractive for both EV and power system applications because it is the most widely used battery system with an established manufacturing capability and is relatively inexpensive. Its main limitation has been its relatively low specific energy of less than 30 Wh/kg. Improved state-of-the-art batteries (ISOA) show promise of 40 to 45 Wh/kg and a lifetime of 500 to 800 cycles, with 60 Wh/kg and 1000 cycles a possibility by 1990 in a low-maintenance configuration.

The advantage of the *nickel-iron system* is its demonstrated ability to withstand abuse, ruggedness and long life with a moderate specific energy in the range of 50 to 55 Wh/kg. However, it has a relatively low specific energy, its performance at low temperatures is poor, and the battery has a high self-discharge rate. Another limitation is its high initial cost, although life cycle costs may be more favorable because of the long cycle life of this battery system and the possible increase in specific energy.

The *nickel-zinc system* has the advantage of high specific energy and good power output capability but suffers from poor lifetime and cycle life due to the difficulty with the zinc electrode. Lifetimes are now limited to 200 to 300 cycles, but design modifications show promise of significant improvements, particularly if the battery is not subjected to frequent deep cycling, making it a possible candidate for EV applications. Effort on this system has been deemphasized because of the poor cycle life.

The *nickel-hydrogen system* has been used primarily in space applications, demonstrating high reliability, long cycle life (in excess of 1000 cycles), a moderate specific energy (50 to 60 Wh/kg), and the ability to accept deep discharge. High initial cost due to the expensive catalyzed hydrogen electrode, and the nickel positive electrode, low volumetric energy density, high self-discharge rate, and the need to store hydrogen in the interior of the cell are potential barriers to the successful deployment of this battery system.

26.4 ADVANCED SECONDARY BATTERIES— GENERAL CHARACTERISTICS

A number of new battery systems are being investigated as potential candidates for energy storage and electric vehicles with the objective of attaining higher energy and power density (100 to 150 Wh/kg at > 100 W/kg) and longer life at acceptable cost levels by the years 1990 to 2000. Most of these systems are still in the relatively early development stage, with significant advances required in performance characteristics, reliability, and life as well as scaling-up to practical sizes for the intended applications. The systems require auxiliaries and controls, some operate at very high temperatures, and most use chemicals that will require special handling and battery design to avoid safety problems. Nevertheless, progress has been steady though slower than originally anticipated, and the eventual use of these systems will depend on advances achieved, as well as the performance of competing systems. Table 26.8 lists the performance goals for these advanced battery systems.

26.4.1 Aqueous Electrolyte Systems

These new battery systems include aqueous electrolyte systems, which have the advantage of operating close to the ambient temperature but which require more complex system design, often with a circulating electrolyte to meet the performance objectives.

Of these systems, the *zinc/chlorine battery* is a leading contender for both utility energy storage and electric vehicles, demonstrating good gravimetric energy density and life. The system does require refrigeration for the storage of the positive active material (chlorine hydrate) and other auxiliaries and may be limited in volumetric energy density to 100 Wh/L. It appears to be best for applications requiring larger-size batteries.

The *zinc/bromine* battery technology is similar to that of zinc/chlorine, requiring external halogen storage and a circulating electrolyte. Its advantage is storage in an organic phase rather than refrigeration. The system is being considered for utility and electric vehicle use, but, again, problems of bulkiness and safety, due to the use of halogen-active material, may limit applications.

The advantage of the *redox system* is that the reactants are stored in containers outside the cell and are introduced, as required, during use. This decoupling facilitates

TABLE 26.8 Goals for the Advanced Secondary Batteries

	Zn/Cl_2	Li/FeS	Li/FeS_2	Na/S (ceramic)	Na/S (glass)
Specific energy, Wh/kg	110	100	130–180	120–140	125–150
Volumetric energy density, Wh/L	100	150	200–250	> 150	> 150
Peak power, W/kg	120	120–160	190–360	150	200
Lifetime deep discharge, cycles	1000	800	1000	1000	1000

design, permits deep discharging, and contributes to longevity. Further, the capacity of the system is determined by the amount of reactants that can be stored or supplied and not by the size of the electrochemical converter. The redox system is being considered only for energy storage or standby power applications because of design considerations that are not amenable to EV use.

Another aqueous ambient temperature system in early development is the *hydrogen/halogen* system. This system is similar to the redox, storing the active materials external to the cell. The system consists of the reaction cell, hydrogen and halogen storage tanks, reaction product storage tanks, pumps, and controls. The active materials react on inert electrodes which are separated by an ion-selective membrane. At low current densities, the operating voltage is 1.3 V for H_2/Cl_2 and 1.0 V for H_2/Br_2. Possible applications include ground power sources for military and special use, standby power systems, and as an energy storage device. Rapid recharging or continuous operation is possible by supplying reactants; regeneration of reactants can be performed in the cell or in another electrochemical device.[6,7]

Secondary *metal/air batteries* have been investigated in electrically and mechanically rechargeable configurations. The mechanically rechargeable aluminum/air battery is emerging as a promising battery system for EV applications, because of its high energy density (long vehicle range) and rapid refueling (replacement of spent or discharged aluminum anodes), with claimed potential performance approaching that of the internal combustion engine. Major barriers are the need for a long-lived air electrode, achieving high anode efficiency, and establishing facilities for production anode and providing field servicing. The iron/air battery has an advantage over other metal-air systems for an electrically rechargeable EV battery. The iron electrode potentially is a rugged electrode with a long life, and the battery system has a high specific energy using relatively low cost materials. The electrolyte circulation system will limit volumetric energy density, and battery lifetime is dependent on the performance of the bifunctional air electrode. Performance objectives are similar to the other ambient temperature systems: about 100 Wh/kg and 90 to 120 Wh/L at 90 W/kg. The zinc/air battery has often been proposed for electric vehicle and other applications because of the attractive energy densities of the zinc electrode and the availability of zinc metal. The battery has been investigated in many configurations, as both an electrically and mechanically rechargeable system, and prototype units have been built. Electrochemical problems, usually with the difficulty of successfully recharging zinc and poor life, have led to effort in favor of other secondary systems.

26.4.2 High-Temperature Systems

The high-temperature systems, operating in the range of 300 to 400°C, have the advantages of high energy density and high specific power compared with the conventional ambient temperature systems. These battery systems use alkali metals, such as lithium and sodium, which have a high-voltage and electrochemical equivalence, as the negative active material. This precludes the use of aqueous electrolytes because of the reactivity of the alkali metals with water and necessitates the use of aprotic organic solvents, molten salt, or solid electrolytes. Although the use of molten salt or solid electrolytes requires operation at high temperatures, a benefit is the resulting high ionic conductivity which is needed for the required high power density. The high operating temperatures, however, do tend to increase the corrosiveness of the active materials and cell components, and this impacts on the life of the battery. Thermal insulation is also used to maintain operating temperatures during standby periods, and special design and operating procedures will be required to ensure safe operating conditions. The main candidates among the high-temperature battery systems are the sodium/sulfur and lithium/metal sulfide systems.

The *lithium/iron sulfide battery* system is one of the major high-temperature systems being considered for EV and energy storage applications. It potentially has a high energy density (120 Wh/kg) and a high power density (150 W/kg), and could be the most compact battery system with volumetric energy densities of about 200 Wh/L. The barriers to commercialization include the need for a low-cost durable separator and for improved cycle life and reliability.

The *calcium/iron sulfide* system is being developed as an alternative to the relatively expensive lithium anode with an inexpensive, more abundant material, in a structure similar to the lithium/iron sulfide cell. The battery shows promise of high energy densities and longer lifetimes than the lithium cell, but development is still in the early stages and major improvements in specific energy as well as power capability are required.[8]

The *sodium/sulfur battery system* is the second high-temperature system being given major development emphasis. Component materials are abundant and inexpensive. Two configurations are being investigated, one using a ceramic electrolyte (β-alumina) which is ionically conductive at 350°C, the second using thin, hollow fibers made from a sodium ion–conducting glass and operating at about 300 to 350°C. A major barrier is the difficulty in developing a satisfactory β-alumina electrolyte with required strength, durability, and conductivity. Packaging problems, low volumetric energy density, and safety considerations may limit interest in this battery for EV applications, but the system remains an important candidate for energy storage applications. The hollow-fiber glass electrolyte configuration is more promising for EV applications because of its low intrinsic cost and high energy density (125 to 150 Wh/kg, 175 Wh/L) with good power capability. The design, using thousands of the tiny fibers, presents a reliability problem due to the possible catastrophic effect of single-fiber failures. Attainment of acceptable lifetime in scaled-up cells remains a major objective.

The *sodium/antimony trichloride battery* is similar to the sodium/sulfur system, using the β-alumina electrolyte and antimony trichloride in place of the sulfur. The battery operates at a temperature of 220°C, with a voltage in the order of 2.3 V per cell at the 5-h rate. However, problems associated with corrosion and seal failures, the higher cell resistance due to the lower operating temperature, and high production cost estimates led to the termination of this program in favor of more energetic, lower-cost, advanced systems.[9,10]

The development of solid electrolytes, such as $LiI(Al_2O_3)$, capable of ion-conduction in the solid state, has enabled the development of *solid-state batteries* (see Part 2, Chap. 12). Operating these systems at high temperatures (300 to 400°C) results in a manyfold increase in ion conductivity over that at 20°C; thus the development of solid batteries for vehicular and utility applications can be considered. The cell components are an Li-Si alloy for the negative and a transition metal chalcogenide or a mixture of these active materials as the positive. Projections indicate that rechargeable cells having an energy density of 200 Wh/kg can be developed. On a battery basis, 60 Wh/kg and 150 Wh/L are feasible.[11]

26.4.3 Ambient Temperature, Nonaqueous Systems

Nonaqueous electrolyte systems, using lithium as an anode material and operating at ambient temperatures, have also been considered for EV applications. These include the lithium-intercalating batteries using organic electrolytes (covered in Part 3, Chap. 25) which have a high specific energy in excess of 100 Wh/kg, but have been used, to date, mainly in small button and cylindrical cell configurations. Improvements in power density, cycle life, and upgrading to larger size cells are needed for consideration for EV applications.

A rechargeable all-inorganic Li/SO_2 battery, which operates at ambient temperatures and is capable of delivering energy densities in excess of 250 Wh/kg, is being developed for possible EV application although work, to date, also has been limited to small cylindrical cells. The system consists of a lithium negative electrode and a porous carbon positive electrode. Liquid SO_2, containing an inorganic salt, is used as the electrolyte and as the active cathode material. During discharge, lithium dithionite $(Li_2S_2O_4)$, an insoluble solid, is formed within the porous carbon electrode. The solid $Li_2S_2O_4$ is converted back to liquid SO_2 and Li^+ ions during the recharge cycle. The reversibility of the lithium electrode and the carbon/SO_2 electrode in the inorganic electrolyte has been established.[12,13]

26.4.4 Polymer Batteries

A new concept for rechargeable batteries is the use of conducting polymers as battery electrodes.[14] The technology currently uses two organic polymers, polyacetylene and polyphenylene, which can be made conductive by doping with suitable cations and anions. The polymer battery can consist of a thin "sandwich" array of polymer strips, alternately oxidized and reduced, each separated from the other by a polypropylene separator, and the entire assembly immersed in an organic electrolyte. In the charged state, the electrolyte's anions serve as counterions for the oxidized or positively charged polymer (the cathode). Similarly, the electrolyte's cations (e.g., Li^+) serve as the counterions for the reduced or negatively charged polymer (the anode). During discharge, electrons flow from the anode to the cathode through the external circuit. As the two polymer strips return to their neutral (undoped) state, the counterions diffuse back into the solution. The cell can then be recharged. One complete charge/discharge cycle thus produces no net chemical change in the electrode material; the electrodes potentially could maintain their integrity over a large number of cycles.

The reaction mechanism is given as

$$xy\,Li + [CH^y(ClO_4)_y]_x \underset{\text{charge}}{\overset{\text{discharge}}{\rightleftarrows}} (CH)_x + xy\,LiClO_4$$

The shelf life and coulombic efficiency of the all-polymer battery are still low due to the instability of the reduced polymer in the organic electrolyte solvent. This problem can be alleviated by replacing the metal-doped polymer anode with metallic electrodes, such as lithium or lithium/aluminum. A prototype lithium/poly-p-phenylene battery had an operating voltage of 4.4 V, a charge/discharge efficiency of about 90%, and a theoretical energy density of 320 Wh/kg. The anode consisted of lithium dissolved in aluminum foil. The cathode was a partial hexafluorophosphate salt of poly-p-phenylene. The electrolyte was lithium hexafluorophosphate in propylene carbonate.

While a marketable polymer battery is at least 5 to 10 y away, the technology is very promising. Polymer batteries are projected to deliver as much as 100 Wh/kg at about 4 V, with power densities as high as 100 W/kg. The batteries should require little maintenance, and they can be hermetically sealed and fabricated in thin sections in a variety of shapes adaptable to many applications such as small portable batteries, for engine starting, and for electric vehicles.

REFERENCES

1. N. R. Yao and A. R. Landgrebe, "New Battery Technologies," *Battery Counc. Int. Conv.,* Washington, 1977.

2. T. G. Bradley and R. A. Sharma, "Rechargeable Lithium-Chlorine Cell with Mixed Ionic Molten Salt Electrolyte," *Proc. 26th Power Sources Symp.* Electrochem. Soc., Pennington, N.J., 1974.

3. J. C. Schaefer, "Production and Engineering Methods for Carb-Tek Batteries in Fork Lift Trucks," Government Accession no. AD-A018106.

4. N. P. Yao, C. C. Christianson, and T. S. Lee, "Improved Lead-Acid Batteries—the Promising Candidate for Near-Term Electric Vehicles," *Proc. 29th Power Sources Symp.,* Electrochem. Soc., Pennington, N.J., 1980.

4a. N. P. Yao, C. C. Christianson, R. C. Elliott, and J. F. Miller, "Status of Nickel/Zinc and Nickel/Iron Battery Technology for Electric Vehicle Applications," *Proc. 29th Power Sources Symp., Electrochem. Soc.,* Pennington, N.J. 1980.

5. "Development of Aqueous Batteries for Electric Vehicles—Summary Report October 1980-September 1981," Report ANL/OEPM-82-5, Argonne National Laboratory, Argonne, Ill., May 1982.

6. O. deNora, "Hydrogen/Halogen Energy Storage System: Safety, Performance and Cost Assessment," Brookhaven National Laboratory report 51070, New York, February 1979.

7. E. N. Balko and J. F. McElroy, "High Energy Density Hydrogen-Halogen Fuel Cell for Advanced Military Applications," *Proc. 29th Power Sources Symp.,* Electrochem. Soc., Pennington, N.J., 1980.

8. D. L. Barney et al., "Calcium/Metal Sulfide Battery Development Program," Argonne National Laboratory report ANL-81-14, Argonne, Ill., March 1981.

9. A. M. Creitzberg, "Sodium-Antimony Trichloride Battery Development Program for Load Leveling," EPRI Project 109-3 (EM-751), interim report, April 1978.

10. A. M. Creitzberg, "Final Status Report on Development of Na/$\beta\beta$"/$SbCl_3$ Battery for Load Leveling Applications," presented at *DOE/EPRI Sodium-Sulfur Workshop,* Schenectady, N.Y., May 14–16, 1979.

11. H. C. Kuo et al., "A New Solid State Secondary Battery System," *Proc. 29th Power Sources Symp.,* Electrochem. Soc., Pennington, N.J., 1980.

12. W. Bowden and A. N. Dey, "Initial Study of an Ambient Temperature Rechargeable Lithium-Sulfur Dioxide Battery," Report no. LBL-13264, Lawrence Berkeley Laboratory, Berkeley, Calif., June 1981.

13. A. N. Dey, "All Inorganic Ambient Temperature Rechargeable Lithium Battery," *Proc. Fifth*

U.S. Dept. of Energy Battery and Electrochem. Contractors' Conf., U.S. Department of Energy, Washington, D.C., December 1982.

14. *Chem. Eng. News,* American Chemical Society, Washington, Jan. 26, 1981; Oct. 12, 1981; April 19, 1982.

14a. P. J. Nigrey et al., "Lightweight Rechargeable Storage Batteries Using Polyacetylene, $(CH)_x$, as the Cathode-Active Material," *J. Electrochem. Soc.* **128**, August 1981.

BIBLIOGRAPHY

O'Connell, L. G., et al.: *Energy Storage Systems for Automobile Propulsion: Final Report,* Lawrence Livermore National Laboratory UCRL-53053-80, Livermore, Calif., December 1980.

Handbook for Battery Energy Storage Photovoltaic Power Systems, Bechtel National, Inc., San Francisco, report SAN-2192-T1 for U.S. Department of Energy, November 1979.

Proc. Fourth U.S. Dept. of Energy Battery and Electrochem. Contractors' Conf., U.S. Department of Energy, Washington, D.C., June 1981.

Proc. Fifth U.S. Dept. of Energy Battery and Electrochem. Contractors' Conf., U.S. Department of Energy, Washington, D.C., December 1982.

Rand, D. A. J.: "Battery Systems for Electric Vehicles—a State-of-the-Art Review," *J. Power Sources* **4**, Elsevier Sequoia S.A., Lausanne, Switzerland, 1979.

27

Aqueous Redox Flow Cells

by
Warren L. Towle and Ronald A. Rizzo

27.1 GENERAL CHARACTERISTICS

In most electrochemical batteries the active electrode materials are solids, in both the charged and discharged conditions. A liquid electrolyte provides a pathway for ionic flow as required to complete the circuit during charge or discharge. The total capacity of a cell in such a system may be limited by either the anode or cathode active mass, or conceivably by the electrolyte if it participates in the reaction.

In an aqueous redox flow cell, both anode and cathode reactions take place in solution, on the surface of inert electrodes serving as current collectors. The reactants flowing across these electrodes are supplied from containers outside the cell. The reactant solutions themselves serve as the electrolyte, being prevented from mixing together by a highly specialized, semipermeable membrane. A typical system is shown schematically in Fig. 27.1. In such a system the electrochemical reactor unit is decoupled from the storage unit, yielding advantages in ease of design and in costs of both operation and maintenance of the system. Another advantage is the fact that both the reactants and reaction products are soluble and thus possess no physical form to be maintained. As a result, there are no inherent life-limiting factors associated with electrode morphology changes. Mild reaction conditions contribute to equipment longevity. The absence of highly reactive or toxic substances generally minimizes safety and environmental problems.

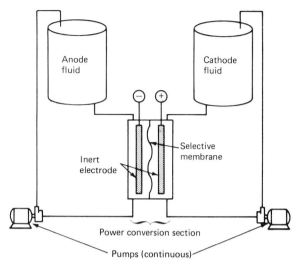

FIG. 27.1 Two-tank electrically rechargeable redox flow cell. (*From Thaller, Ref. 1.*)

27.2 DESCRIPTION OF ELECTROCHEMICAL SYSTEM

The term "redox" is obtained from a contraction of the words "reduction" and "oxidation." Although reduction and oxidation occur in all battery systems, the term "redox battery" is used for those electrochemical systems where the oxidation and reduction of two chemical species take place on inert electrodes and these active materials are stored externally from the cell.

In early work, a system using $FeCl_3$ as the oxidizing agent (cathode) and $TiCl_3$ as the reducing agent (anode) was studied.[1] Better performance has since been obtained by use of $CrCl_2$ as the reducing agent. Other redox systems have also been proposed, such as the zinc/alkaline/sodium ferricyanide ($Na_3Fe(CN)_6 \cdot H_2O$) couple.[1a]

For the chromium-iron battery, both solutions are 1 M in concentration and contain 2 M HCl as a supporting electrolyte. The electrode reactions[2] are

At anode
$$Cr^{2+} \underset{\text{charge}}{\overset{\text{discharge}}{\rightleftarrows}} Cr^{3+} + e$$

At cathode
$$Fe^{3+} + e \underset{\text{charge}}{\overset{\text{discharge}}{\rightleftarrows}} Fe^{2+}$$

Cross-mixing of the reactants would result in a permanent loss in energy storage capacity for the system because of the resulting dilution of the active materials. Migration of other ions (H^+ and/or Cl^-) for charge balancing, however, must be permitted. Thus selective membranes are required.

Large-area carbon felts are used as inert current collectors. Charge and discharge reactions of the Fe^{2+} and Fe^{3+} ions are highly reversible. For the Cr reactions a catalyst is needed. A very light coating of gold[2,3] overlaid with a light coating of lead has proved very effective.

A small amount of hydrogen, which is evolved on charge, can be recombined in a "rebalancing cell." A separate cell with a fuel-cell type H_2 anode and an iron chloride solution cathode is used. Since it operates only as hydrogen becomes available, it is self-regulating. Air oxidation of either Fe^{2+} to Fe^{3+} or Cr^{2+} to Cr^{3+} can also readily be corrected by addition of H_2 from an external supply, using the rebalance cell for the reaction.

A characteristic of systems with soluble reactants is that as their concentration is depleted, the voltage drops, following the Nernst equation. For the chromium-iron system, this would be

$$E = 1.075 - 0.059 \log \frac{Cr^{3+}}{Cr^{2+}} \times \frac{Fe^{2+}}{Fe^{3+}}$$

A plot[2] of this equation as applied to 1 M solutions with the 50% discharge point fixed experimentally, is shown in Fig. 27.2. An idle cell in the cell stack can readily be used to provide an OCV reading and thus provide an indication of state of charge of the system.

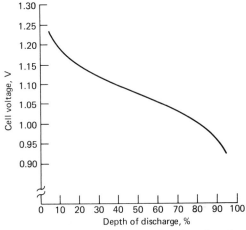

FIG. 27.2   Open-circuit voltage of an iron-chromium redox cell as a function of depth of discharge. (*From Thaller, Ref. 2.*)

Another characteristic of redox systems offers an easy means of counteracting the falling voltage. Additional cells ("trim cells") can be added to the stack.[3] They can sit idle (i.e., on open circuit) at the beginning of discharge and be switched in progressively as the voltage falls. Overcharge or overdischarge is relatively harmless to the system.

27.3 CONSTRUCTION

Construction of a redox flow cell is remarkably simple. Bipolar cells are used. Each bipolar cell, as shown in Fig. 27.3, provides entrance and exit ports for the reactant solutions, a chamber through which each flows, carbon felt for current collection, the separating membrane, and a conductive plate to separate it from the next cell in the stack. With care, the total structure can be reduced to an overall thickness of 0.33 cm; thus a 100-cell stack will be only 33 cm long.

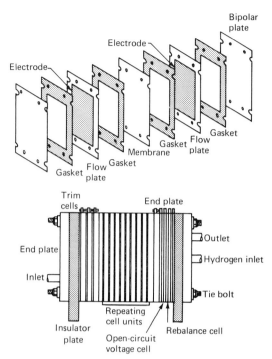

FIG. 27.3 Components of a single redox cell and full-function stack. (*From Thaller, Ref. 2.*)

In addition to the cell stack, with supply lines and pumps, the system contains storage tanks of appropriate size. The cell stack itself, in addition to the working cells, also includes the trim and the rebalance cells. Finally, an extra cell similar to the working cells is included. It is left on open circuit at all times to allow monitoring of the state of charge of the system from a reading of the OCV. A diagram of the full system is shown in Fig. 27.4.

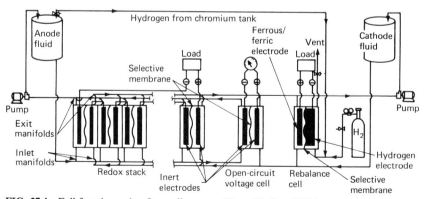

FIG. 27.4 Full-function redox flow cell system. (*From Thaller, 1979.*)

27.4 PERFORMANCE CHARACTERISTICS

The performance of a redox flow cell differs from that of solid systems in at least one important respect: it has an additional variable, namely, the flow rate of the reactants. If the flow rate were zero, the discharge curves would look rather like those of a solid system of low capacity. With a finite flow rate, the capacity, as far as the cell is concerned, is infinite, being fixed only by the capacity of the supply tanks. With a high flow rate the diffusional resistance is low and the composition change is minimal, and so polarization is also minimal. At progressively lower flow rates polarization increases. The flow rate at which complete exhaustion of reactants occurs as the fluid passes through the cell is called the "stoichiometric rate." A convenient way of expressing these relationships is in a "performance map"[3] as shown in Fig. 27.5.

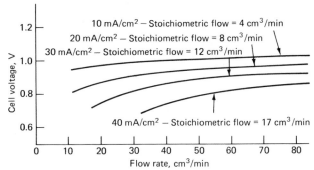

FIG. 27.5 Performance map of a redox cell, 50% DOD, 325-cm^2 cell. (*From Thaller, Ref. 3.*)

By holding the flow rate at a moderately high level, its effect as a variable is minimized. In this way conventional polarization curves can be drawn for each of several different state-of-charge (SOC) levels. Figure 27.6 shows cell voltage on charge and discharge at an SOC level of 50%. The curved lines on this plot are power density reference lines. For other SOC levels, the "wedge" is shifted up or down to meet the appropriate OCV value (i.e., zero current density), as defined in Fig. 27.2. With adequate flow rate of reactants, the principal cause of polarization in this system is the *IR* drop through the membrane. The membrane is the performance-limiting component. The major problem is that the resistivity of the membrane rises by a factor of around 2.5 during the first 100 h of exposure to the redox environment. This appears to be due to an interaction with a chloro complex of iron in the ferric state.[3] Impedance ascribable to other cell components is negligible by comparison, largely because of the bipolar construction.

Coulometric efficiency in this system is customarily around 98%. Hydrogen generation on charge could become high but is readily controlled in a constant-voltage charge control system. Pumping consumes around 2 to 3% and shunt currents 0.5 to 3% of input energy.

Because the bulk of the reactant materials are in tanks remote from the electrochemical cells, spontaneous self-discharge cannot take place in a manner similar to many self-contained systems. Liberation of H_2 by reaction of H^+ with Cr^{2+} occurs too slowly to be observed under most circumstances. (Air absorption could cause signifi-

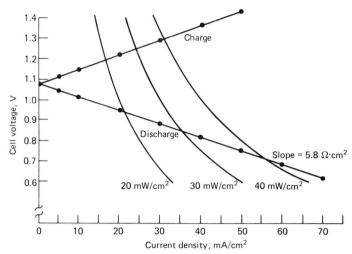

FIG. 27.6 Current-voltage characteristics of single redox cell. (Cell size, 325 cm²; flow rate, 60 cc/min; DOD, 50%.) (*Courtesy of NASA, Lewis Research Center.*)

cant capacity loss without care in design and operation.) Within the cell, ion crossover through the membrane will take place very slowly and at known rates. Shunt currents represent the only significant cause of energy loss during shutdown. Protection could be achieved by a stack design in which the cells and manifolds could be drained during shutdown. Even with such protection, only the fluids trapped within the cell will be discharged.

The ultimate cycle life limitations, if any, of such a system have yet to be determined. Single cells have been cycled up to 3000 times under flow conditions with no observed effects other than the reversible one of H_2 rebalance requirement. Under static conditions cycling has been continued to 20,000 cycles.

Operation at temperatures in the neighborhood of 65°C has been shown to have certain advantages with regard to the chromium solution chemistry.[6] In particular, the equilibrium between the two predominant chromic ion species is shifted to favor the one that is electrochemically active. However, at these temperatures, the membranes presently available are no longer capable of preventing rapid cross-mixing of the reactant species. This has led to the "mixed-reactant" mode of operation, in which both reactant systems, when fully discharged, are identical. For example, the composition of each could be 1.0 M $CrCl_3$ and 1.0 M $FeCl_2$ and 1.0 N HCl. When operating in this mode, it is possible to use membranes that are very "open" and which, therefore, have a low resistivity. The performance of single cells in this mode of operation is shown in Fig. 27.7.

Charging of this system is relatively uncomplicated as long as excessive evolution of H_2 is avoided at the Cr electrode. The OCV gives an easy measure of state of charge. The upper limit on the voltage per cell at which the current is supplied should be fixed at a level which will avoid "blinding" (i.e., interference with current flow) of the electrode by H_2 evolution as full charge is approached.

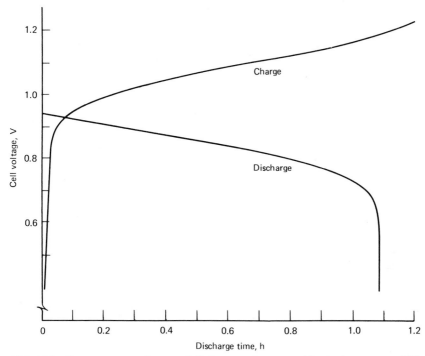

FIG. 27.7 Constant-current charge and discharge performance with mixed reactant at 66°C. [60 mA/cm²; mixed reactants (fully discharged: 1.1 M FeCl$_2$, 0.9 M CrCl$_3$, 2.0 N HCl); cation exchange membrane; 0 to 87% SOC; gold-lead chromium electrode catalyst; energy efficiency, 72.3%; coulombic efficiency, 92.2%; voltage efficiency, 78.4%; hydrogen evolution, 0.9% of charge capacity; cell resistivity, 1.76 Ω·cm²; membrane resistivity, 0.36 Ω·cm²] (*Courtesy of NASA Lewis Research Center.*)

27.5 APPLICATIONS AND SYSTEM DESIGN

The near-term application of the redox system is for small village power systems where solar arrays or wind turbines would supply the power but where storage is required for nighttime or periods of low sun or wind levels. This application is characterized by low power requirements and long durations of dependence on storage devices. Such demands are ideal for redox systems since long storage time requires only larger reactant fluid and storage tank volumes. A longer-term application is in electric utility service where the redox system would serve in a load-leveling capacity.[4,5]

Initial development work on the redox system was done on cells 16.8 cm × 19.4 cm. The design features of a larger 1-kW system are given in Table 27.1. The test data and practical operating experience obtained with this system should provide the design basis for scale-up to the planned 10-kW system.[4] This 10-kW system, designed for remote village solar energy storage, is visualized as a cell stack about 60 cm × 60 cm × 120 cm in overall dimensions with a 10-kW output over a 50-h period (500 kWh).

Preliminary cost projections (manufacturer's selling price) range from $74 to $116 per kilowatthour and $63 to $110 per kilowatthour for the 500-kWh system and the

TABLE 27.1   Redox System Design Parameters

Gross power	1260 W
Nominal net power	1000 W
Voltage	120 ± 5%, VDC
Number of stacks	4
Number of cells per stack	39
Trim packages	10 packages, 6 cells each
Depth of discharge range (utilization)	80–20% (0.60)
Reactant volume (each)	700 L
Reactant energy density	14.5 Wh/L
Cell active area	320 cm²
Nominal current density	30 mA/cm²
Reactants	1 M/L FeCl₃, 2 N HCl
	1 M/L CrCl₂, 2 N HCl
Reactants flow rates (nominal)	100–150 cm³/min per cell
Parasitic losses	Pumps: 200 W
	Shunt currents: 125 W
Number of rebalance cells	5
Number of charge-indicator cells	1

electric utility 100-MWh design, respectively (in terms of 1980 dollars).[7] Operation in the mixed-reactant mode will result in an increased cost for the reactants. However, when projected to high levels of production, the increase is estimated to be about 50% of the cost of the unmixed reactants on a dollar per kilowatthour basis.

REFERENCES

1. L. H. Thaller, "Electrically Rechargeable Redox Flow Cells," NASA TMX-71540, August 1974.

1a. R. P. Hollandsworth et al., "Development of an Alkaline Zinc/Ferricyanide Battery," *Proc. 5th U.S. Dept. of Energy Battery and Electrochemical Contractors' Conf.*, December 1982.

2. L. H. Thaller, "Redox Flow Cell Energy Storage Systems," DOE/NASA/1002-79/3, NASA TM-79143.

3. L. H. Thaller, "Recent Advances in Redox Flow Cell Storage Systems," DOE/NASA/1002-79/4, NASA TM 79186, August 1979.

4. Margaret A. Reid and Lawrence H. Thaller, "Improvement and Scale-up of the NASA Redox Storage System," NASA report TM-81632, August 1980.

5. Norman H. Hagedorn and Lawrence H. Thaller, "Redox Storage Systems for Solar Applications," *Power Sources,* vol. 8, Academic, New York, 1981.

6. Norman H. Hagedorn, "The NASA Redox Project, Status Summary," *Proc. 5th U.S. Dept. of Energy Battery and Electrochemical Contractors' Conf.,* December, 1982.

7. Kenneth Michaels and Gene Hall, "Cost Projections for Redox Energy Storage Systems," report no. DOE/NASA/0126-1, February 1980.

Zinc/Chlorine Batteries

by
Warren L. Towle and Ronald A. Rizzo

28.1 GENERAL CHARACTERISTICS

The zinc/chlorine battery is an aqueous flowing-electrolyte system being developed both for stationary energy storage (SES) for electric utilities and for electric vehicles.[1] Starting with an aqueous solution of zinc chloride, elemental zinc and chlorine are generated electrolytically during the charge cycle on inert (graphite) electrodes. Storage of the zinc takes place on the face of the negative electrode. The chlorine gas, formed on the positive electrode, is transported to a separate chamber and converted to solid chlorine hydrate by chilling, then stored in a chamber separate from the electrochemical cells. On discharge the hydrate is slowly decomposed by warming, and the chlorine is fed in solution to the cells. The system operates at temperatures and pressures close to ambient.

28.2 DESCRIPTION OF ELECTROCHEMICAL SYSTEM

The electrochemical reactions at the zinc and chlorine electrodes and the formation of solid chlorine hydrate are highly reversible

$$ZnCl_2(aq) \underset{\text{discharge}}{\overset{\text{charge}}{\rightleftarrows}} Zn(s) + Cl_2(g)$$

$$Cl_2(g) + xH_2O \underset{\text{discharge}}{\overset{\text{charge}}{\rightleftarrows}} Cl_2 \cdot xH_2O(s)$$

The open-circuit voltage at standard conditions is 2.12 V. The value of x, depending on conditions, lies in the range of 5.7 to 8. The chlorine hydrate may be formed directly from the battery electrolyte by chilling it in a chamber remote from the cells. Alternatively the chlorine may be desorbed from the electrolyte and reabsorbed in chilled water. A schematic representation of the charge-discharge sequence is shown in Fig. 28.1. In the

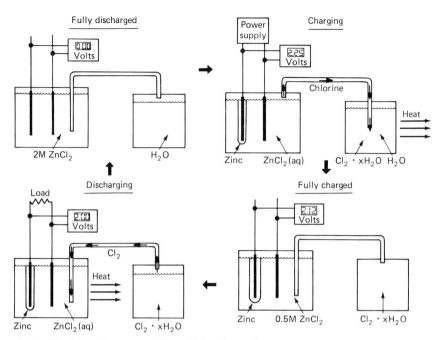

FIG. 28.1 Schematic representation of the charge-discharge sequence for a zinc/chlorine battery.

charge mode, as the electrolysis proceeds, zinc is plated onto one inert graphite plate and chlorine is evolved at the other. The chlorine passes from the gas space above the electrolyte into the "store," where it reacts with water cooled to less than 10°C. The formation of solid chlorine hydrate is exothermic, approximately -18 kcal/mole. Therefore the store must be chilled constantly during charge, requiring external refrigeration. The zinc chloride concentration drops from an initial value of 2 to 3 M down to a final level of 0.5 M at full charge. The electrolyte usually contains about 1.25 M KCl to improve conductivity, and its pH is adjusted with HCl to 0.1 at room temperature in the completely discharged state. On charge the pH rises about 0.5 unit.

Provisions for the necessary flows of chlorine, electrolyte, water, and heat are shown diagrammatically in Figs. 28.2 and 28.3. Pump P1 delivers electrolyte through line E to pockets between pairs of porous graphite chlorine electrodes. The electrolyte passes through the porous graphite, flows up between the zinc and chlorine electrodes, and eventually spills back to the sump. Chlorine gas is pumped by P2 through line C. Chilled water from the store enters P2 via line W, forming chlorine hydrate, which enters the store through line H. Refrigerated glycol chills the water in the store by heat exchange between lines W and R.

In discharge, the valve in line D opens. The electrolyte, warmed to 20 to 50°C by thermodynamic, voltaic, and chemical inefficiencies in the cells, warms the store by heat exchange. This decomposes the hydrate, raising chlorine pressure in the store, causing the valve in line G to open and chlorine to dissolve in the electrolyte going to the cells. (The electrolyte flows in the same direction on both charge and discharge.) Closing the

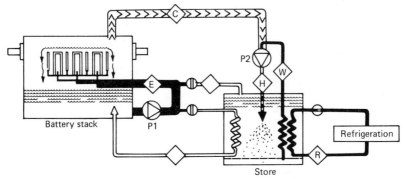

FIG. 28.2   Schematic of a zinc/chlorine battery system on charge. (*Courtesy of Energy Development Associates.*)

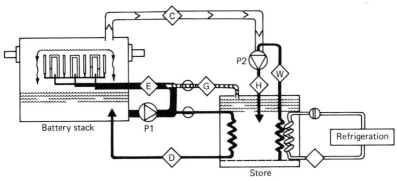

FIG. 28.3   Schematic of a zinc/chlorine battery system on discharge. (*Courtesy of Energy Development Associates.*)

electrical circuit then permits electrochemical discharge at the zinc and chlorine electrodes.

For load-leveling applications the most important performance goal is an energy efficiency of $\geq 65\%$ for the battery system as a whole. In view of the energy losses due to pumping, refrigeration, and voltage polarization, this requires a coulombic efficiency of about 85%. The most important problem is, therefore, the coulombic loss due to the chemical reaction of dissolved chlorine with the zinc electrode. This is greater on charge than on discharge, because the chlorine concentration is higher. For this reason the cell stack is operated under a partial vacuum (approximately 60 kPa) to reduce chlorine solubility in the electrolyte. The pump P2 maintains a positive pressure of approximately 30 kPa in the gas space of the store.

In spite of the need for high coulombic efficiency, and in contrast to the zinc/bromine cell construction, current practice with zinc/chlorine is to use no separators. The low solubility of chlorine of course limits the potential for coulombic loss, which is to be balanced against any added voltage drop caused by the separators.

Since a small amount of hydrogen is evolved from the zinc surface, a fluorescent lamp is used to promote the gas phase reaction of $H_2 + Cl_2$. The resulting HCl is reabsorbed in the electrolyte, thus helping to stabilize its pH. Hydrogen evolution is promoted by metallic impurities such as Sb, Co, Ge, Ni, Sn, Fe, and Cu and by graphite degradation products.[2]

Electrochemical kinetics at the zinc electrode are highly reversible, requiring no catalyst or other treatment. However, activation of the porous graphite chlorine electrode is required. This is done electrochemically after cell assembly is completed. Current is passed through the cell in a charging mode for several hours with a dilute $ZnCl_2$ electrolyte flowing through the cell.

During normal cell operation CO_2 gas and a smaller amount of O_2 are formed at the chlorine electrode. Methods for their removal have been developed using a chlorine-chlorine cell for electrochemical separation.[3] Zinc dendrite formation is controlled[2] by use of "zinc-plate leveling agents."

For electric vehicles, battery weight and volume (energy and power density) are more important than efficiency. The major difference between the two battery designs lies in the use of battery electrolyte for hydrate formation in EV batteries, whereas water is used as the store fluid in load-leveling batteries. This minimizes battery weight for the EV application, while the use of water results in a higher coefficient of performance for heat rejection during charge in load-leveling batteries. It should be noted, however, that the electric vehicle battery requires one less heat exchanger as hydrate decomposition can be accomplished by passing warm stack electrolyte at approximately 40°C directly into the store. This is in contrast to the load-leveling batteries in which warm electrolyte is pumped through a titanium tube, located in the store, in order to effect hydrate decomposition.

28.3 CONSTRUCTION

The basic cell construction of the zinc/chlorine battery is shown in Fig. 28.4. The cell assembly is an interesting variation on the classic bipolar construction. The "intercell

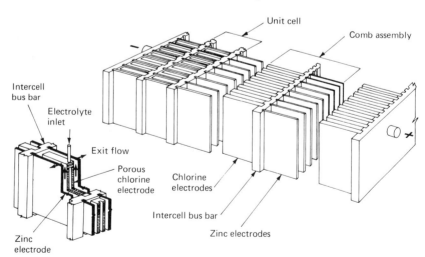

FIG. 28.4 Comb-type bipolar arrangement of electrodes in zinc/chlorine battery. (*Courtesy of Energy Development Associates.*)

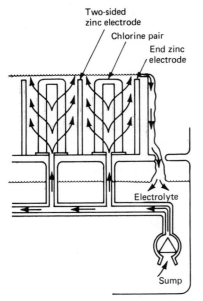

Two-sided zinc electrode

Chlorine pair

End zinc electrode

Electrolyte

Sump

FIG. 28.5 Schematic of electrodes showing electrolyte flow path. (*Courtesy of Energy Development Associates.*)

bus bar" may be regarded as the bipolar plate, in that it divides one cell from another, and (being made of graphite) carries the current from one cell to the next. The electrode plates themselves, then, are simply projections from each face of the bipolar plate, to increase its area. Zinc is used for the negative electrode; flat porous graphite plates, in plastic frames to provide a cavity into which the electrolyte is fed, are used for the chlorine electrode. The details of the electrolyte flow through and past the electrodes are shown in Fig. 28.5.

The interelectrode separations are 0.20 cm and 0.33 cm for the electric vehicle and load-leveling batteries, respectively. This difference results from the need to maximize power density for the EV batteries and efficiency for the load-leveling batteries.

28.4 PERFORMANCE CHARACTERISTICS

The OCV of the zinc/chlorine battery is 2.12 V, with a working voltage of 1.9 to 2.0 V at the 4- to 5-h rate and 1.5 V at the 1-h rate.[1]

A *typical discharge* curve for SES application is shown in Fig. 28.6. The 7 h of charging at constant-voltage and -current density are followed by slightly over 5 h of constant-current discharge which shows a slightly falling voltage. At this point most of the zinc is gone, and the voltage drops to the cutoff point of 1.75 V. However, there

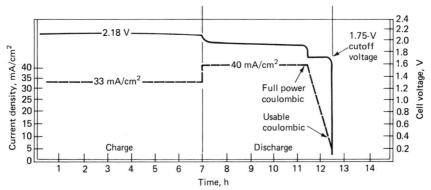

FIG. 28.6 Complete cycle regime for zinc/chlorine load-leveling battery. (*Couresy of Energy Development Associates.*)

is still enough zinc to maintain this voltage at progressively falling current densities. When it is no longer feasible to maintain 1.75 V per cell, the system is considered discharged. The small amount of zinc remaining at the end of each discharge (1.7 to 4%) is "unusable coulombic" loss. It is discharged by short-circuiting, in order not to impair the geometrical uniformity of subsequent deposits. Thus the duration of discharge for a given electrode spacing depends on the uniformity of the zinc deposit.

The effect of *current density* on cell voltage is shown in Fig. 28.7 for constant temperature and flow rate. As in the redox flow cell, the *rate of flow* of electrolyte influences the voltage characteristics of the zinc/chlorine battery. The *effect of temperature* is illustrated in Fig. 28.8. Charge voltage increases with decreasing temperature, and discharge voltage will increase with increasing electrolyte temperature.

There are two *self-discharge* mechanisms. The first is the slow evolution of hydrogen

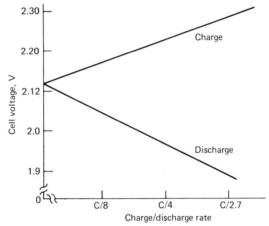

FIG. 28.7 Typical voltage profile for a zinc/chlorine battery (electrolyte temperature = 50°C).

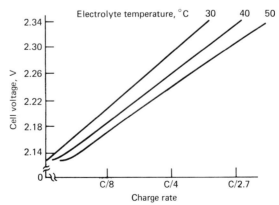

FIG. 28.8 Effect of temperature on charge voltage of a zinc/chlorine battery.

from the zinc electrode mentioned above. The second is the reaction of dissolved chlorine with the zinc electrode. During charged stand with pumps off, chlorine in the electrolyte migrates to the zinc electrode and reacts. After approximately 1 h the OCV will have dropped almost to zero. Further loss will occur only as heat from other parts of the battery or from the surroundings enters the insulated case and decomposes chlorine hydrate.

The *power* performance of the zinc/chlorine cell is plotted in Fig. 28.9.[4] From the indicated value of 295 mW/cm² for maximum power density at 1.05 V, an effective area resistivity of 3.75 Ω·cm² is calculated for the cell.

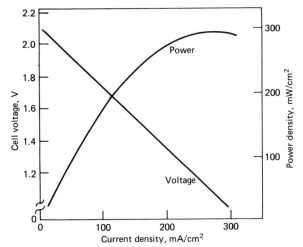

FIG. 28.9 Zinc/chlorine cell discharge characteristics. (*From Chi, Carr, and Symons, 1979.*)

Charge control is simple since the cell accepts a constant-current charge at essentially constant voltage. Limiting the charge coulometrically (i.e., by ampere-hours supplied) will avoid a drop-off in coulometric efficiency due to zinc dendrite formation associated with heavy zinc deposits. *Cycle life* goals are over 1000 cycles for the electric vehicle system and 2500 cycles and 10 y for the SES or load-leveling battery. The graphite electrode is the life-limiting component.

The efficiency data for full charge-discharge on a load-leveling battery are summarized in Fig. 28.10.[5]

The only significant *hazardous material* in this system is chlorine. Since it exists in quantity only in hydrate form, the measure of risk in case of accident is the rate at which heat can be supplied for its decomposition. A 50-kWh module will contain a maximum of around 100 lb of chlorine. As a worst case this could result in a chlorine concentration of 50 ppm or more for a maximum of 50 ft from the module.[2]

28.5 APPLICATIONS AND SYSTEM DESIGN

The zinc/chlorine batteries, for both the utility SES and electric vehicle applications, are in the 40- to 50-kWh size range, and were originally based on an integrated module

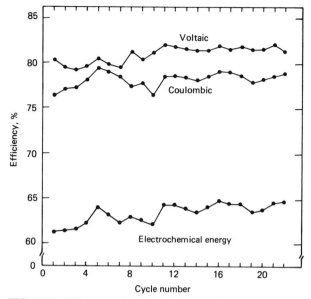

FIG. 28.10 Efficiency data for full charge-discharge cycles on the zinc/chlorine load-leveling battery. (*Courtesy of Energy Development Associates.*)

design. However, more recent studies indicated that significant cost advantages would result from packaging in commercially available structures.[5-7]

28.5.1 Load-Leveling Battery

The load-leveling battery consists of two separate units connected by appropriate lines for gas and electrolyte (as shown schematically in Fig. 28.11) and are mounted in a module rack. The lower cylinder houses the battery stack and sump and the upper unit houses the battery store. The dimensions of the module are 2.6 m in length, 1.2 m in width, and 2.1 m in height. In normal operation, the module will charge at 200 A for 6.7 h and discharge at 200 A for 5.4 h. The average charge and discharge voltages are 55 and 46 V, respectively.

A cutaway of the stack module is shown in Fig. 28.12. Two 24-cell units are located side by side in the upper portion of the cylindrical case. The electrolyte sump occupies the lower portion of the stack vessel. Reinforced end caps are sealed to the flanged cylindrical case using O-ring seals. A plug-in electrolyte pump is mounted on one of the end caps. The discharge gas-injection and the decomposition-electrolyte connections to the outlet of the pump are located on the same end cap. The charge gas-transfer and discharge decomposition return lines penetrate the opposite end cap as do the sump-cooling heat exchanger lines. Two sets of terminals also penetrate the case through these end caps.

A cutaway view of the store module is shown in Fig. 28.13. A domed end cap is used to seal the flanged cylinder. A plug-in hydrate-former pump is mounted on the end cap. All other penetrations—Freon coolant, discharge gas injection, and discharge decomposition lines—are through the walls of the cylinder.

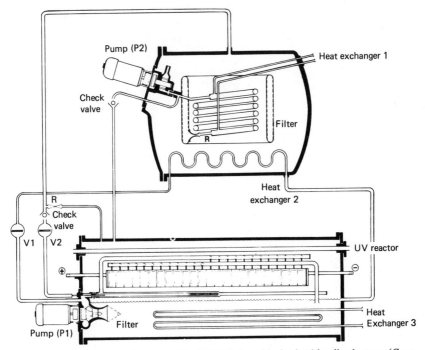

FIG. 28.11 Operational schematic of the 50-kWh zinc/chlorine load-leveling battery. (*Courtesy of Energy Development Associates.*)

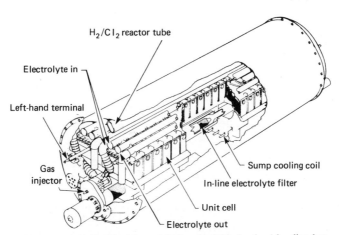

FIG. 28.12 Details of stack module for zinc/chlorine load-leveling battery. (*Courtesy of Energy Development Associates.*)

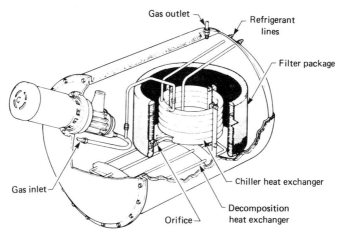

FIG. 28.13 Cutaway view of store module for zinc/chlorine load-leveling battery. (*Courtesy of Energy Development Associates.*)

28.5.2 Electric Vehicle Battery

The prototype zinc/chlorine electric vehicle battery consists of 60 cells arranged into two 30-cell assemblies as shown in Fig. 28.14. The low profile of the package (230 cm $\times$ 56 cm $\times$ 19 cm) allows mounting underneath the floor of a vehicle resulting in maximum space utilization and a low vehicle center of gravity. The weight of the battery, including the microprocessor battery controller and pump power supplies, is 636 kg and the overall volume is 464 L. The corresponding energy densities are 64 Wh/kg and 88 Wh/L; the power density is 59 W/kg.

The performance characteristics of the EV battery are shown in Fig. 28.15. The load of the battery is ramped from 0 to 468 A in 6 s and removed for 5 s before ramping up again for the next cycle. The upper trace is the load voltage taken at the load terminals, the lower trace is the instantaneous current, and the middle trace is the

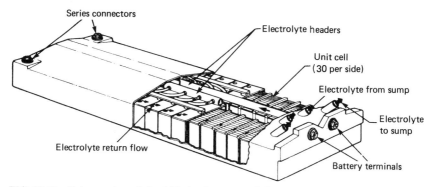

FIG. 28.14 Cutaway view of zinc/chlorine battery stack for electric vehicle battery. (*Courtesy of Energy Development Associates.*)

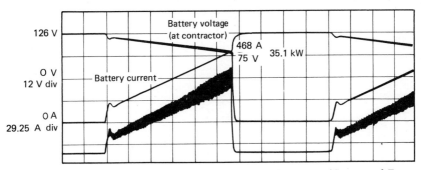

FIG. 28.15   Power response of zinc/chlorine electric vehicle battery. (*Courtesy of Energy Development Associates.*)

smoothed average of the instantaneous current. This prototype battery was subjected to over 200 charge-discharge cycles.

REFERENCES

1. P. Carr and P. C. Symons, "Zinc-Chlorine Batteries," *Symp. Adv. Batteries,* Chem. Soc. Congr., Honolulu, April 1979.

2. Energy Development Associates, *Development of the Zinc-Chlorine Battery for Utility Applications,* EPRI report EM 1051, April 1979.

3. Energy Development Associates, "Development of the Zinc-Chloride Battery for Utility Applications," EPRI report EM-1417, May 1980.

4. C. H. Chi, P. Carr, and P. C. Symons, "Modeling of Zinc-Chloride Batteries for Electric Vehicles," *14th IECEC,* pp. 692–697, 1979.

5. P. Carr et al., "Recent Advances in Zinc-Chloride Battery Technology," *Proc. 30th Power Sources Symp.,* Electrochem. Soc., Pennington, N.J., June 1982.

6. R. D. Clubb, C. J. Warde, and P. Carr, "45 kWh Zinc-Chloride Load-Leveling Battery Module: Recent Performance Improvements," *Proc. 29th Power Sources Conf.,* Electrochem. Soc., Pennington, N.J., June 1980.

7. C.M. Blevins, "Life Testing of 1.7 kWh Zinc-Chloride Battery System," *J. Power Sources,* vol. 7, Elsevier Sequoia S.A., Lausanne, Switzerland, 1981/82.

Zinc/Bromine Batteries

by
Warren L. Towle and Ronald A. Rizzo

29.1 GENERAL CHARACTERISTICS

The zinc/bromine battery is another promising aqueous secondary system for energy storage and electric vehicles because of its inherent chemical simplicity, good electrochemical reversibility at the electrodes, ambient temperature operation, good energy density, and low-cost materials. These characteristics attracted early attention to this battery,[1,2] but two inherent technical problems impeded its development. The first of these is the well-known propensity of the zinc electrode to form dendrites, with resultant danger of shorting. The second is poor coulombic efficiency. This results from the high solubility of bromine in aqueous zinc bromide, permitting corrosive reaction with the zinc anode as diffusion occurs.

The recent interest in improved battery systems for energy storage and electric vehicles has renewed development of this battery. The zinc/bromine system, as it has now evolved, may be regarded as a close relative of the redox flow cell, using zinc bromide as both anolyte and catholyte. In the classic redox flow cell the reactants and reaction products are stored outside the electrochemical cell, giving the cell itself limitless capacity. Cell area (i.e., electrode area) need be provided in any given application only as necessary to meet its required electric power delivery. In the zinc/bromine system, however, the zinc metal at the anode is alternately dissolved and redeposited on the electrode face, thus limiting cell capacity. Because of the tendency to form dendrites, and because this tendency becomes much more pronounced as the zinc thickness builds up, the practical limit to cell capacity appears to lie in the range of 2000 to 3000 Ah/m².

29.2 DESCRIPTION OF ELECTROCHEMICAL SYSTEM

The zinc/bromine system consists of a flow cell through which an aqueous zinc bromide stream from each of two separate tanks is circulated, one for the anode and one for the cathode. Means are also provided for contacting the catholyte with a complexing agent to extract the bulk of the free bromine from the aqueous base.

When aqueous zinc bromide is electrolyzed, metallic zinc is deposited at the negative. At the positive, the bromine, as liberated, dissolves in the zinc bromide, in which it is highly soluble.

The reactions at the electrodes in a zinc/bromine cell (dissolved Br_2 associates with Br^- ion resulting in Br_3^-) are[3]:

		$E^0(25°C)$	$E^0(50°C)$
Negative	$Zn(s) \underset{charge}{\overset{discharge}{\rightleftarrows}} Zn^{2+}(aq) + 2e$	−0.763 V	−0.760 V
Positive	$Br_3^- (aq) + 2e \underset{charge}{\overset{discharge}{\rightleftarrows}} 3Br^-(aq)$	1.085 V	1.056 V
Overall	$Zn(s) + Br_3^- (aq) \underset{charge}{\overset{discharge}{\rightleftarrows}} Zn^{2+} (aq) + 3Br^-(aq)$	1.848 V	1.816 V

The growth of high-quality (i.e., smooth, nonporous) zinc deposits at high loadings appears to depend on the following requirements:[3]

1. Keep the pH of the electrolyte above 1.0 (to minimize gassing) and below 3.5 (to avoid porous deposits)

2. Maintain uniform flow distribution across the face of the electrode and from cell to cell in a series (electrical) configuration

3. Keep electrolyte flow rates up, current densities down

4. Avoid foreign organic impurities in the electrolyte, especially dewetting agents

5. Limit the deposit of zinc to a maximum of 2000 to 3000 Ah/m²

6. Periodically strip the zinc completely to renew the smoothness of the electrode surface

Use of the organic additions common to industrial zinc plating has been intentionally avoided. The concern was that (1) their instability would lower coulombic efficiency and (2) the special provisions necessary to maintain them at the proper level would further complicate electrolyte control.

Limiting the corrosive attack on the zinc anode by dissolved bromine has been addressed in at least two ways. Coulombic efficiencies as high as 98% have been achieved by using a cation exchange membrane (NAFION) as a barrier to bromine diffusion.[4] The more customary approach has been to extract the bromine into a separate phase, usually a liquid. Materials which form loose chemical associations with bromine, such as quaternary ammonium bromides, are often chosen. (A preferred quaternary is N-methyl, N-ethyl morpholinium bromide.[5])This polybromide phase, referred to as a "bromine oil," is circulated in a separate loop outside the cell.[3] The polybromide phase may be emulsified in the catholyte and circulated with it through the cathode chambers of the cells. This ensures availability of enough bromine to forestall concentration polarization and permits the use of a nonporous electrode plate such as required for bipolar construction. Further control of bromine diffusion is achieved by the separator.

For the negative electrode, both carbon and titanium are within a reasonable cost, give good zinc adherence, and have a hydrogen overvoltage high enough to function satisfactorily. With the selection of the bipolar cell design for the energy storage system, dense carbon sheets can be used for the electrodes.[6] An alternative material, referred to as "carbon-plastic," is a proprietary composition of carbon blended in polypropylene.

The conductivity of $ZnBr_2$ electrolyte solutions is lower than aqueous H_2SO_4 or NaOH; the maximum at 25°C being 0.12 $\Omega^{-1}\cdot cm^{-1}$ at 2.1 molal. The acceptable operating range is from 1 to 4 molal. As a supporting electrolyte, KCl has been used, which at 4 molal raises the conductivity to around double this value. KCl also reduces solution viscosity, thus reducing pumping power. Unfortunately, it increases the tendency to form zinc dendrites, cutting the maximum satisfactory deposit to around 2000 Ah/m^2, and also increases the parasitic current losses in bipolar cells. Zinc chloride has also been considered as an additive, to enable more complete utilization of the more expensive $ZnBr_2$ without the deterioration in quality of zinc deposit that occurs when Zn^{2+} concentration drops below about 2 molal.

Large systems are expected to run at around 50°C because of the need for heat rejection from the system. There is an improvement in voltage efficiency at this temperature because of the improved electrolyte conductivity. This tends to be offset by a decrease in coulombic efficiency because of increased bromine diffusion. Thus operation at 50°C closely approximates that at 20°C. In the operation of the cell, the unavoidable evolution of H_2 causes a pH management problem, even though not enough to affect coulombic efficiency significantly. Recombination of H_2 with the dissolved bromine in the catholyte using ultraviolet light has been employed with some success.

29.3 CONSTRUCTION

The schematic of the zinc/bromine system is shown in Fig. 29.1 for the system being developed for utility load-leveling and in Fig. 29.2 for the electric vehicle (EV) system. The overall system consists of the cells, a pumping system, and storage tank for the anolyte, a pumping and storage system for the catholyte, plus a storage system for free bromine with means for transferring it to and from the cathode. In a practical unit a

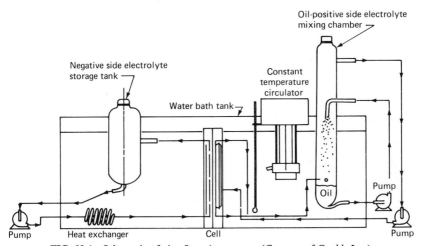

FIG. 29.1 Schematic of zinc/bromine system. (*Courtesy of Gould, Inc.*)

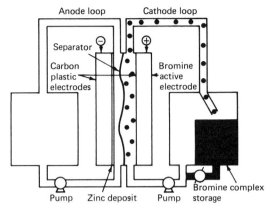

FIG. 29.2 Schematic of zinc/bromine cell. (*Courtesy of Exxon Research and Engineering Co.*)

heat exchanger may be included for temperature control. Because of the difference in intended uses, there are significant differences in construction between the system for utility load leveling and the one for electric vehicles. For utility use, a 5- to 10-h discharge period is desired, at relatively uniform power. Also, a required lifetime of 10 or more years places a severe requirement on components' durability. For EVs much higher power peaks are encountered, shifting the design emphasis toward minimizing electrical resistance.

The load-leveling system was originally based on a monopolar cell construction with titanium electrodes.[6] This was changed to a bipolar design, using dense carbon plate electrodes to eliminate corrosion problems that occurred at the end of discharge with the titanium electrode. In addition, the bipolar design has a potentially major cost advantage. The submodule, shown in Fig. 29.3, is the building block of the module and consists of a filter press stack of bipolar carbon electrodes mounted in a polypropylene electrode frame. The frame also provides the function of directing the electrolyte flow to and from the electrode surfaces. A microporous separator, such as Daramic, is located between each pair of electrodes to limit the rate of self-discharge; spacing ribs are added on each side to hold the separator off the electrode surfaces. The polypropylene end plates are used to compress the cell to provide the plumbing connections for electrolyte flow and to collect the electrical current from the two end electrodes.

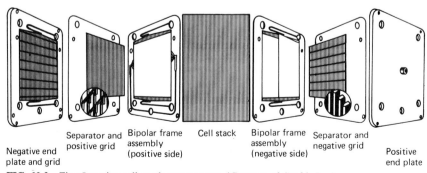

FIG. 29.3 Zinc/bromine cell stack components. (*Courtesy of Gould, Inc.*)

In the 80-kWh module, ten 8-kWh submodules are connected electrically in a series-parallel arrangement. Since the submodules require electrolyte circulation on both the negative and positive sides, two plumbing circuits are provided. The flow loop on the negative side (Fig. 29.1) contains the submodule, an electrolyte storage tank (and within it an immersed coil heat exchanger), and a circulation pump. The positive electrolyte is pumped concurrently with the liquid polybromide through a static mixing device where the bromine dissolves in the electrolyte. This bromine solution then flows to the submodules and next, via two outlets, to the tanks. In the tank, the bulk of polybromide returns via gravity to its sump, from which it is recirculated to the static mixer via the polybromide pump. During charge, the electrolyte, freed of most of the polybromide, is pumped through the mixer via the positive electrolyte pump, reversing the bromination process.

The electric vehicle design is based on a bipolar construction. Work on this design has progressed through electrodes of 6 dm² in size in a 52-cell, 80-V, 3½-kWh unit. The schematic of the cell module is shown in Fig. 29.4. Six of these modules have been assembled in parallel to produce a 20-kWh prototype.

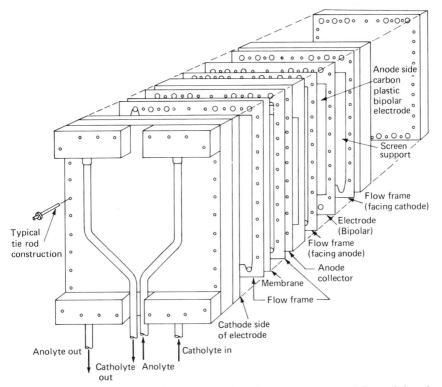

FIG. 29.4 Module for 20-kWh zinc/bromine battery. (Prototype battery module consisting of three 52-cell battery stacks. Two modules required per 20-kWh battery. Each cell consists of one electrode, two flow frames, and one membrane.) (*Courtesy of Exxon Research and Engineering Co.*)

29.4 PERFORMANCE CHARACTERISTICS

A *charge-discharge curve* for the zinc/bromine cell operating at about 20°C is shown in Fig. 29.5 and for a 12-V, 500-Wh bipolar battery in Fig. 29.6. The zinc loading on the anode is approximately 600 Ah/m²; the coulombic efficiency is about 84% and the energy efficiency between 70 and 73%. A typical *polarization* curve is shown in Fig. 29.7. Each cell in this 1.3-kWh load-leveling battery has an electrode area of 1000 cm², with current densities in the range of 20 to 50 mA/cm². Typical polarization and power curves for the EV cell are shown in Figs. 29.8 and 29.9. *Area resistivity* values are from 3.5 and 10 Ω·cm².

No significant coulombic loss was observed over an 8-h period on *open-circuit stand* with the electrolyte pumps off. With the pumps on, a 10% loss was recorded.[3]

Cycle life of these cells, like those of a redox flow cell, is less meaningful than in a solid-solid system. While the zinc is deposited as a solid, it is periodically stripped off and completely renewed. Thus the service life of a battery will depend on the long-term stability of such components as the electrodes and the plastic parts, such as

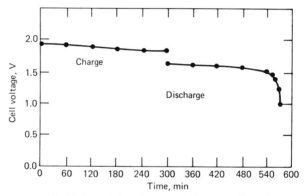

FIG. 29.5   Voltage during charge and discharge of Zn/Br₂ cell.

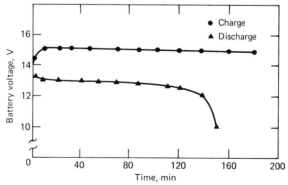

FIG. 29.6   Charge-discharge profile, 12-V, 500-Wh bipolar battery (charge-discharge current 11.6 A).

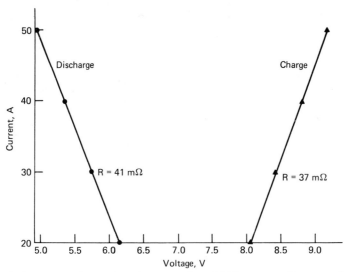

FIG. 29.7 Typical polarization curve for 1.3-kWh battery. (*Courtesy of Gould, Inc.*)

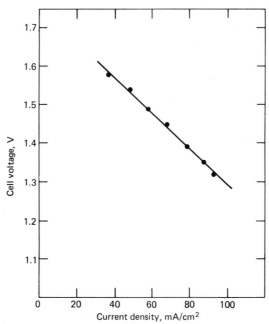

FIG. 29.8 Polarization curve for 52-cell (80-V) 3-1/2-kWh bipolar battery. (*Courtesy of Exxon Research and Engineering Co.*)

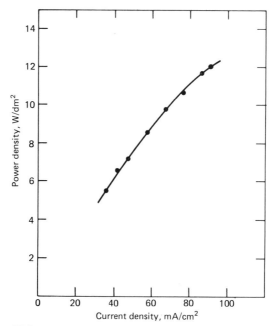

FIG. 29.9   Power curve for 52-cell (80-V) 3-1/2-kWh bipolar battery (approximate weight = 306 kg.). (*Courtesy of Exxon Research and Engineering Co.*)

the cell frames and piping. Two critical factors are the separator life and the long-term stability of the bromine complexing agent. To date, no problems in these areas have been noted; cycle testing of the cell components has shown no life-limiting decay mechanisms. A cell submodule has accrued over 200 continuous cycles in an automatic cycling routine with energy efficiencies in the range of 55 to 65%. The EV system (3½-kWh, 80-V module) has been cycled about 100 times with greater than 85% coulombic efficiency and 75% energy efficiency, neglecting auxiliaries. These losses will amount to less than 10% in the full-scale system.

Charge control in the zinc/bromine systems appears very simple, requiring only a constant-current supply with a timed cutoff.

29.5 APPLICATIONS AND SYSTEM DESIGN

The zinc/bromine system, intended for electric utility load leveling, is planned to be made in modular form. A 400-kWh module is conceived as a truck-transportable building block for a full-size 100-MWh load-leveling battery. The Mod 0 design specifies exterior dimensions of 2.44 m by 1.83 m by 6.10 m. The design characteristics for this system are shown in Table 29.1[3] The Mod 1 design is significantly more economical of space, being 1.83 m by 1.53 m by 4.89 m in size.

The projected 20-kWh, 80-V, battery for electric vehicle applications is shown in Fig. 29.10. The battery is based on a 600-cm² electrode. The system uses 6 substacks

TABLE 29.1 Mod 0 Zinc/Bromine Battery Module Design
Characteristics

	5-h rate	10-h rate
Capacity, MWh	100	105
Power, average, MW	20	10.6
Charge voltage, V		
Cell	2.0	1.95
Module	20	19.5
String	1,000	975
Discharge voltage, V		
Cell	1.6	1.7
Module	16	17
String	800	850
Charge current density, A/m^2	490	245
Charge current, string A	5,560	2,780
Total battery charge current, A	27,800	13,900
Discharge current density, A/m^2	440	220
Discharge current, string, A	5,000	2,500
Total battery discharge current, A	25,000	12,500
Voltaic efficiency, %	80	87
Coulombic efficiency, %	90	90
Energy efficiency, %	72	78
Energy efficiency, all system losses except inverter-converter, %	70	76
Operating temperature, °C	50	50

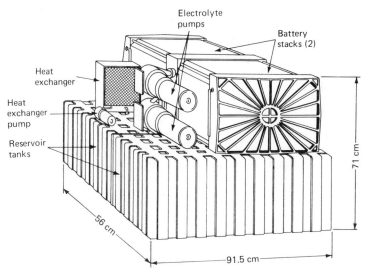

FIG. 29.10 Advanced 20-kWh zinc/bromine battery (120 V). (*Courtesy of Exxon Research and Engineering Co.*)

of 52 bipolar cells which are connected in parallel. Electrolyte is contained in reservoirs beneath the electrochemical module. Complexed bromine is surrounded by anolyte as an additional safeguard. Specific energy for the complete package, based on a weight of 300 kg and a volume of 265 L, is about 65 Wh/kg and 75 Wh/L. A prototype unit has been cycled more than 100 times. The problem of maldistribution of zinc in this type of bipolar module is avoided by the injection of a "protective current" in the common manifold to null the shunt-driving potentials and produce a uniform current distribution within the series battery.[7]

The selling price of a 100-MWh utility battery is estimated to range from $77 to $109/kWh for the Mod O and from $47 to $64/kWh for the Mod 1, depending on thickness of zinc deposit. These costs are based on a manufacturing facility capable of producing 25 such batteries per year. Further cost reductions, as the technology matures, should permit a selling price of $20 to $40/kWh in terms of 1980 dollars.[3] The EV battery is estimated to sell at less than $45/kWh in 1980 dollars in volumes of 100,000 units per year. The type of circulating system used also offers the possibility of less than full battery replacement in the event of premature failure which would substantially lower replacement costs.

REFERENCES

1. C. S. Bradley, U.S. Patent 312,802 (1885).

2. H. H. Dow, *Trans. Am. Electrochem. Soc.* **1**:127 (1902).

3. R. A. Putt et al., *Assessment of Technical and Economic Feasibility of Zinc/Bromine Batteries for Load-Leveling,* EPRI EM-1059, May 1979.

4. F. G. Will, Zinc/Bromine Workshop, Memphis, 1978.

5. Exxon Research and Engineering Co., Linden, N.J.

6. Ronald A. Putt et al., *Development of Zinc-Bromine Batteries for Utility Energy Storage,* Report no. DOE/ET/29345, May 1980.

7. R. J. Bellows et al., "Advances in Zinc Bromine Batteries for Motive Power," *EV Expo. 1980 Conf.,* St. Louis, May 20–22, 1980.

BIBLIOGRAPHY

Bellows, Richard J.: "Recent Progress on Exxon's Zinc-Bromine Battery Technology," *Proc. 5th U.S. Dept. of Energy Battery and Electrochemical Contractors' Conf.,* December 1982.

Will, F. G., and H. S. Spacil: "Performance Analysis of Zinc-Bromine Batteries in Vehicle and Utility Applications," *J. Power Sources* **5**:173, 1980.

Metal/Air Batteries

by
Ernest L. Littauer and John F. Cooper

30.1 GENERAL CHARACTERISTICS

The electrochemical coupling of a reactive anode to an air electrode provides a battery with an inexhaustible cathode reactant and, in some cases, a very high theoretical system specific energy. The capacity limit of such systems is determined by the ampere-hour capacity of the anode and the technique for handling and storage of the reaction product. These considerations have long intrigued electrochemists, and many attempts have been made over the years to develop practical systems.[1-3] Zinc has received by far the most attention because it is the most electropositive metal which can be electroplated from aqueous solution. Problems of dendrite formation, nonuniform zinc dissolution and deposition, limited solubility of the reaction product, and unsatisfactory air electrode performance have prevented, until recently, the demonstration of the battery as an electric vehicle power source.

Primary (see Part 2, Chap. 10), secondary, and mechanically rechargeable metal/air battery concepts have been explored and developed to prototype status. In principle, mechanical recharging (the discharged metal electrode is replaced) permits short recharge time and allows mass production of battery fuel at facilities dedicated to that function. This approach makes use of simple, "unifunctional" air electrodes, which need operate only in a discharge mode. Conventional electrical recharging of metal/air batteries requires either a third electrode (to sustain oxygen evolution on charge) or a "bifunctional" electrode. The latter is a single electrode capable of both oxygen reduction and evolution.

Recent advances have provided potentially inexpensive oxygen electrode structures and catalysts; anode alloys and electrolyte additives which reduce parasitic corrosion and increase cell potential; and novel battery designs involving flowing electrolyte technology. A result has been a resurgence of interest in metal/air systems.

The metal/air batteries being developed today use neutral or alkaline electrolytes. The oxygen reduction half-cell reaction may be written:

$$O_2 + 2H_2O + 4e = 4OH^- \qquad E^0 = + 0.401 \text{ V}$$

The theoretical cell voltages, the equivalent weights of the metals, and the theoretical specific energies obtained when this oxygen electrode is coupled with various metal anodes are given in Table 30.1. Polarization effects at both electrodes degrade these voltages to those shown in the table at practical operating discharge rates.

TABLE 30.1 Characteristics of Metal/Air Cells

Metal anode	Electrochemical equivalent of metal, Ah/g	Theoretical cell voltage,* V	Valence change	Theoretical specific energy (of metal), Wh/g	Practical operating voltage, V
Li	3.86	3.4	1	13	2.4
Ca	1.34	3.4	2	4.6	2.0
Mg	2.20	3.1	2	6.8	1.4
Al	2.98	2.7	3	8.1	1.6
Zn	0.82	1.6	2	1.3	1.2
Fe	0.96	1.3	2	1.2	1.0

*Cell voltage with oxygen cathode.

The zinc/air battery has been developed to electric vehicle prototype status. Large lithium/air and aluminum/air mechanically rechargeable batteries have been demonstrated in the laboratory. Magnesium/air,[4] calcium/air,[5] lead/air,[6] and cadmium/air[6] batteries have been studied, but anode polarization or instability, parasitic corrosion, nonuniform dissolution, and other problems have prevented successful development. The lithium/air, aluminum/air, and some zinc/air batteries are recharged by mechanical replacement of their anodes. Iron/air batteries have also received considerable attention. This battery is designed as an electrically rechargeable system. It potentially has very low life cycle costs but has lower specific energy and power density than the mechanically rechargeable systems. (The iron/air battery is covered in Part 3, Chap. 20, Sec. 20.4.)

30.2 ZINC/AIR BATTERIES

Much work has been done in recent years on the zinc/air system as a candidate for electric vehicle propulsion. It has failed, however, to attract strong support because of its low energy efficiency (about 40%), low operating voltage, and related chemical and operational difficulties. The overall cell reaction may be represented as

$$Zn + \tfrac{1}{2}O_2 + H_2O + 2OH^- = Zn(OH)_4{}^{2-} \qquad E^0 = 1.62 \text{ V}$$

Common electrochemical problems include anode passivation at high discharge rates, zinc dendrite formation on charge (which can be controlled by electrolyte circulation), and polarization and stability of the air electrodes. Chemical problems include carbonate precipitation from atmospheric carbon dioxide (within both the air electrode and electrolyte) and limited solubility of the zinc reaction products in the alkaline (KOH) electrolyte. The development of an efficient high-rate bifunctional air electrode remains a formidable challenge. An additional oxygen-evolving electrode is provided for recharging secondary zinc/air batteries.

30.2.1 Circulating Electrolyte Systems

Two significantly different approaches have been taken in the development of large-capacity secondary zinc/air batteries. One approach, exemplified by systems pioneered by Sanyo, makes use of forced circulation of air and electrolyte.[7,8] This system is perhaps the most advanced of metal/air batteries developed to date. A 124-V, 560-Ah traction battery has been tested by Sanyo in large vehicles. Also 15-V, 560-Ah modules are available for various stationary applications. Individual cells within these systems are rated at 560 Ah and 1 V; nominal current density is 80 mA/cm^2 with a maximum of 130 mA/cm^2.

TABLE 30.2 Specifications of 15- and 124-V Sanyo Zinc/Air Battery Systems

	15-V battery system	124-V battery system
Battery voltage, V	15 V (0.2 C)	124 V (0.2 C)
Capacity, Ah	560	540, 130-A max.
Specific energy, Wh/kg	116.5	109
Specific energy, Wh/L	99	c. 100
Weight, kg	70	565
Cycle life	200–300	200–300
Dimension (L × W × H), mm	560 × 330 × 384	1550 × 1050 × 334

SOURCE: Sanyo Electric Co., Ltd., Osaka, Japan.

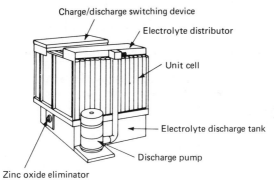

FIG. 30.1 Zinc/air 15-V module. (*Courtesy of Sanyo Electric Co.*)

Table 30.2 gives the specifications of the 15- and 124-V systems, which are shown in Figs. 30.1 and 30.2, respectively; Fig. 30.3 shows the current-voltage characteristic of the cell and a typical discharge curve at an 80-A discharge rate.

30.2.2 Circulating Zinc Slurry Cells

A different approach to the development of an electrically rechargeable zinc/air system for vehicle propulsion has been developed by Compagnie Générale d'Electricité in France[9–11] and Sony in Japan.[12] Here a slurry containing zinc particles flows through tubular (Compagnie Générale d'Electricité) or plane-parallel cells (Sony). Zinc particle regeneration occurs in a special component external to the discharge cell modules.

The Sony effort, which culminated in a 3-kW prototype, was terminated in 1974.

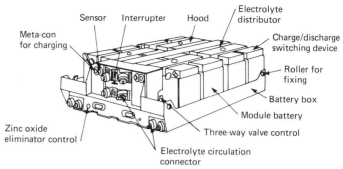

FIG. 30.2 Zinc/air 124-V battery system. (*Courtesy of Sanyo Electric Co.*)

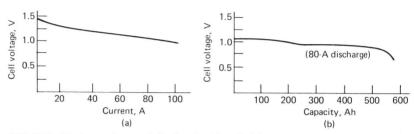

FIG. 30.3 Discharge characteristics for zinc/air cell: (*a*) current voltage characteristic and (*b*) discharging characteristic. (*Courtesy of Sanyo Electric Co.*)

The Compagnie Générale d'Electricité battery system is shown schematically in Fig. 30.4 and the tubular cell and battery are depicted in Figs. 30.5 and 30.6, respectively. The zinc powder–electrolyte (12N KOH) slurry is pumped through the cell during discharge, and the zinc reacts to form zincate, which dissolves in the electrolyte. During charge, which is done on board in an electrolysis unit, zinc is deposited in the cell and removed in particulate form in the electrolyte flow. The system can also be recharged by replacing the spent slurry and regenerating it externally.

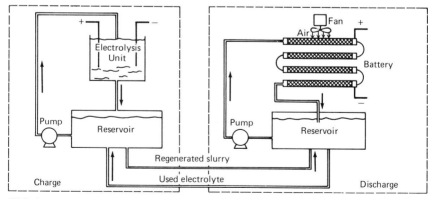

FIG. 30.4 Schematic diagram of the separate charge and discharge modules of the Compagnie Générale d'Electricité circulating zinc/air battery. (*From Ref. 10.*)

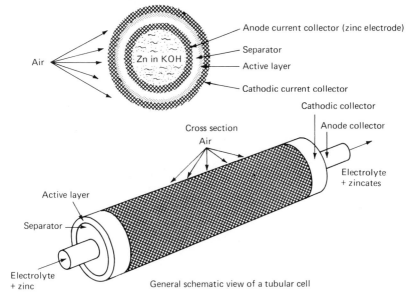

FIG. 30.5   Tubular cell design of the Compagnie Générale d'Electricité circulating zinc/air battery. (*From Ref. 10.*)

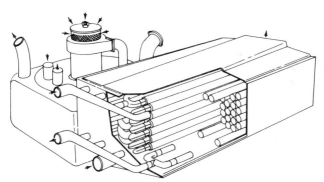

FIG. 30.6   Overall view of system concept—Compagnie Générale d'Electricité zinc/air battery. (*From Ref. 10.*)

Typical cell polarization data are shown in Fig. 30.7. An overall energy efficiency of 40% has been obtained (discharge voltage 1.23 V, 85% coulombic efficiency; recharge coulombic efficiency 90% at 2.4 V). Up to 1000 3-h cycles have been reported on individual cells, and up to 600 cycles on small-cell assemblies. A traction battery based on this technology is expected to have an energy density of 115 Wh/kg at 80 W/kg. The Compagnie Générale d'Electricité development culminated in a demonstration of a 12- to 14-kW system, but it has not been made commercially available.

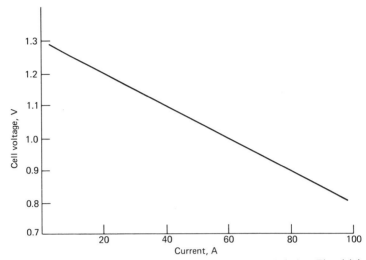

FIG. 30.7 Current-voltage relationship for Compagnie Générale d'Electricité zinc/air cell (50°C). (*From Ref. 10.*)

30.2.3 Mechanically Rechargeable Batteries

Mechanically rechargeable or refuelable batteries are designed with replaceable metal anodes; the discharged or "spent" anode is replaced and the battery is thus "recharged." Mechanically rechargeable zinc/air batteries were seriously considered for powering portable military electronic equipment in the 1970s because of their high energy density and ease of recharging.[13] They were never deployed, however, due to their short activated life, poor intermittent performance, and development of new high-performance primary batteries, which were superior in energy density, rate capability, and ease of handling in the field. In the 1970s, a mechanically rechargeable 35-kWh zinc/air battery was demonstrated by General Motors in a 1350-kg test bed vehicle, but the development was not continued to commercialization.[14]

Sanyo manufactures a series of 14-pack batteries ranging from 20 to 200 Ah at voltages from 1 to 12 V.[8] They are designed as water-activated reserve batteries but are also mechanically refuelable and have provisions for replacement of the zinc electrode cassettes. Specifications of four-pack batteries are given in Table 30.3, and typical discharge characteristics are shown in Fig. 30.8.

TABLE 30.3 Specifications of Sanyo Zinc/Air Pack Batteries

Sanyo designation	OCV	Working voltage, V	Capacity, Ah	Discharge current,* A	Dimensions, mm L W H	Wt, g†
6 AZ-20	8.4	6.0–7.2	20	1	165 × 83 × 90	1,320
12 AZ-20	16.8	12.0–14.4	20	1	298 × 83 × 90	2,460
12 AZ-100	16.8	12.0–14.4	100	5	455 × 180 × 182	10,530
12 AZ-200H	16.8	10.2–12.0	200	35	475 × 230 × 240	18,640

*Continuous maximum.
†Excludes H_2O, includes case.
SOURCE: Sanyo Electric Co., Ltd., Osaka, Japan.

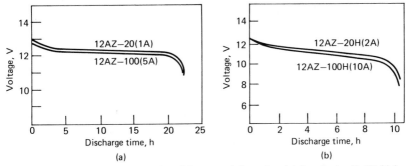

FIG. 30.8 Discharge characteristics of Sanyo pack batteries. (*a*) Standard cell; (*b*) high-rate cell.

30.3 LITHIUM/AIR BATTERIES[15–20]

The lithium/air battery is an attractive candidate for electric vehicle propulsion because lithium has the highest theoretical voltage and electrochemical equivalency (3860 Ah per kilogram of lithium) of any metal anode considered for a practical battery system.

The cell discharge reaction is

$$2\text{Li} + \text{H}_2\text{O} + \tfrac{1}{2}\text{O}_2 \rightarrow 2\text{LiOH} \qquad E^0 = 3.35 \text{ V}$$

Lithium metal, atmospheric oxygen, and water are consumed during the discharge, and excess LiOH is generated. The cell can operate at high coulombic efficiencies because of the formation of a protective film on the metal that retards rapid corrosion after formation. On open-circuit and low-drain discharge, the self-discharge of the lithium metal is rapid, due to the parasitic corrosion reaction

$$\text{Li} + \text{H}_2\text{O} \rightarrow \text{LiOH} + \tfrac{1}{2}\text{H}_2$$

This reaction degrades the anode coulombic efficiency and must be controlled if the full potential of the lithium anode is to be realized. This self-discharge also necessitates the removal of the electrolyte during stand.

The concentration of the lithium hydroxide governs the anode current efficiency as it controls the rate of corrosion and the cell power density. Therefore, the concentration must be controlled. This can be done by precipitation or dilution. In applications requiring maximum specific energies, carbon dioxide may be used to control the concentration through precipitation of solid lithium carbonate with the resultant production of water

$$\text{CO}_2 + 2\text{LiOH} \rightarrow \text{Li}_2\text{CO}_3 + \text{H}_2\text{O}$$

The theoretical open-circuit voltage of the lithium/air cell is 3.35 V, but, in practice, this value is not achieved because the lithium anode and air cathode exhibit mixed potentials. Figure 30.9 shows that the actual OCV is near 2.85 V. It is also evident that the major voltage loss is at the air cathode and that the cell cannot be discharged efficiently at voltages much above 2.2 V unless a substantial reduction in cathode polarization is achieved.

During discharge, the lithium dissolves at a rate determined by the electrolyte concentration, temperature, and flow rate. In addition to controlling the parasitic corrosion reaction, another limitation is that the lithium anode will not discharge

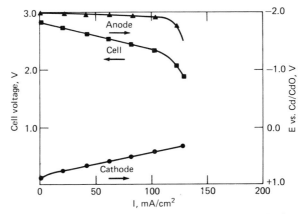

FIG. 30.9 Typical individual electrode and cell polarization curves—lithium/air cell.

efficiently at temperatures above 40°C, which hinders the oxygen reduction kinetics and makes waste heat rejection and thermal control difficult. A lithium-alloy anode, with lower parasitic corrosion, may be required for the vehicular battery.

The elements of the lithium/air cell are illustrated in Fig. 30.10 and the system schematic in Fig. 30.11. The cell consists of a consumable lithium anode, an insulated wire screen electrode separator-flow channel, and an air cathode. Cells are connected in series to achieve the desired battery voltage. The cells are activated by pumping LiOH electrolyte through the separators. A cell stack compression mechanism is

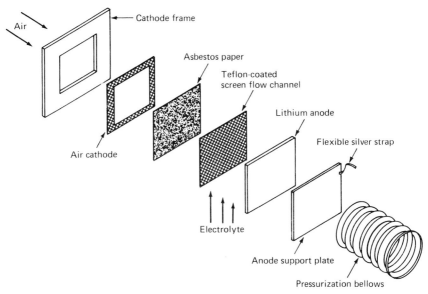

FIG. 30.10 Elements of lithium/air cell.

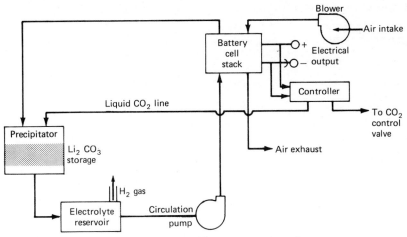

FIG. 30.11   Lithium/air automotive propulsion system—schematic.

required in order to keep the anode pressed against the separator during the discharge. The battery uses a flowing electrolyte and generates waste heat and hydrogen as by-products during the discharge. The LiOH reaction product is converted to Li_2CO_3, which is recovered for recycling to lithium anodes. The battery is deactivated by draining the electrolyte into a holding-tank reservoir.

A 1.2-kW lithium/air battery, made of six monopolar modules, has been constructed using Teflon-bonded, carbon-catalyzed air electrodes. The performance of this battery is shown in Fig. 30.12.[15,16] The air electrodes suffer some polarization loss when operated near room temperature, which is characteristic of the lithium discharge. Significant hydraulic and pneumatic subtlety is involved in the mechanization of the battery because the electrodes must be closely juxtaposed and the electrolyte flow over

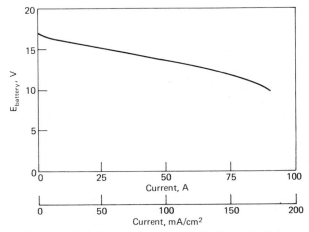

FIG. 30.12   Polarization curve of a six-cell lithium/air battery.

the lithium must be turbulent. A typical discharge is shown in Fig. 30.13. The automatic diluent controller (ADC) was set so the terminal voltage of the battery remained constant at about 12.5 V. Then, as the load was adjusted, diluent water was admitted or the LiOH concentration allowed to increase according to the power demands. The current line reflects these changes.

A lithium/air battery for a vehicle with a 400-km range will weigh about 250 kg, similar to the design objectives for the aluminum/air battery, and projected to deliver about 300 Wh/kg. The principal advantage of the lithium battery is its higher cell voltage, which translates into higher power. However, in view of the availability and safety advantages of aluminum, it is most likely that future effort on mechanically rechargeable batteries will concentrate on the aluminum anode.

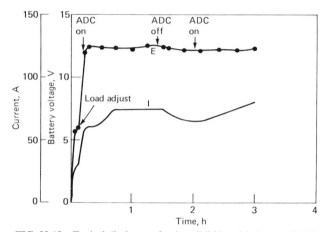

FIG. 30.13 Typical discharge of a six-cell lithium/air battery (ADC represents automatic diluent controller).

30.4 ALUMINUM/AIR BATTERIES

Aluminum has long attracted attention as a potential battery anode because of its high ampere-hour capacity (2980 Ah per kilogram of Al) and attractive voltage. While its voltage is about a volt below that of lithium, the complications of exposure to water, nitride formation, and strict temperature control necessary with lithium are circumvented with aluminum. Aluminum is also much more abundant and less expensive, and it presents a much lower safety risk than lithium.

The discharge reaction for the aluminum/air cell is

$$Al + 3/2H_2O + \tfrac{3}{4}O_2 \rightarrow Al(OH)_3 \qquad E^0 = 2.73 \text{ V}$$

The aluminum/air cells fall into two general categories according to their electrolyte: neutral (usually saline) electrolytes and concentrated alkaline electrolytes. The neutral electrolytes are attractive because of the extremely low open-circuit corrosion rates and the reduced hazards of these solutions compared with concentrated caustic.

The air electrodes show considerable concentration polarization; the oxygen reduction reaction produces hydroxide ion which accumulates within the air electrode's porous structure, raising the local pH. The formation of an aluminum hydroxide gel within the electrode is a serious limiting factor because this ties up a considerable quantity of water (up to 7.5 ml per gram of aluminum). Concentrated caustic electrolytes prevent the formation of thick, resistive surface films, and the metal dissolves with low polarization even in highly supersaturated caustic-aluminate electrolytes, forming a metastable solution. Gel formation may be avoided in alkaline solutions which are seeded with particles of aluminum trihydroxide. The loss of the protection of the surface film tends to enhance the water-reduction reaction, resulting in hydrogen evolution

$$Al + 3H_2O \rightarrow Al(OH)_3 + 3/2H_2$$

This effect may be greatly reduced by the addition of corrosion inhibitors (tin or zinc) to the electrolyte or by surface amalgamation.

Alloying agents are used in both types of aluminum/air batteries, in the 0.01 to 0.5% range, to modify the rates of both water-reduction and aluminum-dissolution reactions. Small quantities of gallium, thallium, or indium, added to the metal, depolarize the aluminum-dissolution reaction. Phosphorus, tin, and, to a lesser extent, zinc strongly reduce the hydrogen-evolution reaction. The effects of low-level alloying agents on the dissolution reaction have been variously attributed to the resultant expansion or contraction of the metal lattice[21] or to the displacement of either oxygen or aluminum ions from the sublattices of the oxide surface film.[22]

30.4.1 Aluminum/Air Cells in Neutral Electrolytes

Aluminum/air cells using neutral electrolytes have been developed for marine applications as well as stationary power sources. A 90-Ah, six-cell module, with a total weight of 0.65 kg and an electrolyte volume of 0.2 L, delivered a specific energy of 150 to 200 Wh/kg. The cells could be mechanically recharged 20 to 40 times by renewing the electrolyte and, less often, by replacement of the anodes. Cells open to the seawater were designed, with energy densities based on the weight of aluminum electrodes, of over 2 kWh per kilogram of aluminum.[23]

The properties of aluminum-alloy electrodes for use in seawater or other neutral electrolytes have been explored, including the use of gallium-, indium-, and thallium-doped anodes and low (0.1%) quantities of phosphorus and manganese to offset the rates of water reduction in alloys containing gallium and iron, respectively.[24-26] Polarization curves showing a potential plateau and hydrogen evolution characteristics are shown in Figs. 30.14 and 30.15, respectively; the aluminum/air test cell is shown in Fig. 30.16. Designs of a rapidly refuelable aluminum/air battery, using wedge-shaped electrodes in a novel continuous-feed configuration, were developed for electric vehicle use with energy densities estimated in the 170 to 190 Wh/kg range.[27] Subsequently, the designs have been refined to overcome the problem of gel formation on the anode. A pumping arrangement has been incorporated into the reservoir which provides oscillating electrolyte flow across the anode. This system has been highly successful in overcoming passivation in neutral solutions. Further work to explore the advantages of the wedge-shaped cell has resulted in design concepts similar to those shown in Figure 30.24. A multicell unit with oscillating electrolyte, wedge-shaped anodes and front-current collectors, providing about 300 W, is planned.

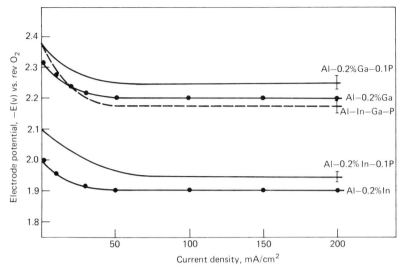

FIG. 30.14 Polarization curves for aluminum anodes for saline electrolyte aluminum/air batteries. Composition given in weight %.

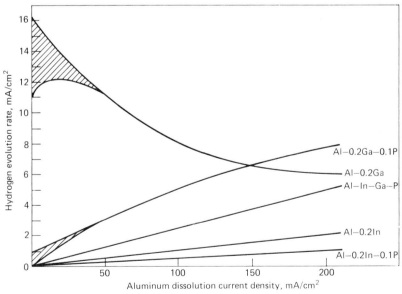

FIG. 30.15 Dependence of parasitic corrosion rate on aluminum dissolution rate (for alloys described in Fig. 30.14).

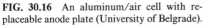

FIG. 30.16 An aluminum/air cell with re-placeable anode plate (University of Belgrade).

FIG. 30.17 Aluminum/air battery system (Norwegian Research Defense Establishment).

A 40-W mechanically refuelable aluminum/air battery illustrated in Fig. 30.17 was developed for use in stationary power generators for military applications. The system makes use of flowing electrolyte (a mixture of KCl and KF at pH = 11), ambient temperatures, and natural convection of air. The reaction product, a solid, highly hydrated alumina, is formed in the electrolyte and separated from the electrolyte streams with the use of hydrocyclones and stored in an internal barrel. The system uses 10 series-connected cells; each cell consists of the single aluminum sheet anode (230 cm^2) with two air electrodes made of uncatalyzed, Teflon-bonded carbon.

Limitations posed by the large volumes of reaction product (7.5 cm^3 per gram of aluminum), and the occasional anode passivation, are major problems. Air cathodes have been discharged for intervals as long as 700 h. The current efficiency for aluminum dissolution was reported as 100%, with the hydrogen evolution essentially eliminated, even at open circuit. Relevant data are shown in Table 30.4, and electrical characteristics are shown in Fig. 30.18.[28] A 36-cell unit with a total anode area of 0.94 m^2, with forced electrolyte, air circulation, and filters for reaction separation, is now under construction.

30.4.2 Aluminum/Air Cells in Alkaline Electrolyte

Aluminum/air and aluminum/hydrogen peroxide cells were pioneered in the early 1960s.[29–31] The use of aluminum/air batteries in electric vehicles was described conceptually, including electrolyte exchange (for reaction product removal) and extended ranges between anode replacement.

Aluminum/oxygen cells and batteries of very high power (over 15 kW/m^2) have been developed. These operate at elevated temperatures (80°C) in concentrated KOH electrolyte and use high-purity (99.99+ %) aluminum anodes and oxygen cathodes.

TABLE 30.4 Characteristics of the NRDE Aluminum/Air Battery

Characteristic	Unit
Nominal net power, W	40
Peak power, W	50
Nominal voltage, V	12
Net weight, kg	5.3
Volume, L	13
Auxiliary pumping power, W	1.2
Coulombic efficiency, %	100 approx.
Energy efficiency, %	40–50
Energy density, practical, kWh per kilogram of aluminum	3.4

SOURCE: Reference 28.

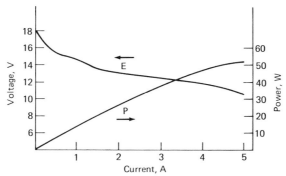

FIG. 30.18 Dependence of voltage and power of a 10-cell aluminum/air stack on current (each cell has an area of 230 cm³).

Two cell designs were tested: one with plane-parallel sheet electrodes and one employing 5-cm-long cylinder anodes which dissolved from one end. The cylinders were continuously advanced to maintain a constant interelectrode gap. After 38 h of accumulated dissolution, the active surface remained flat and smooth.[32]

A program to develop mechanically refuelable aluminum/air batteries for electric vehicle propulsion is now underway under the sponsorship of the U.S. Department of Energy with corporate contributors.

The overall system concept for the aluminum/air battery and its use in vehicles has been described.[33,34,35] The battery is mechanically recharged by addition or replacement of aluminum anode plates and by the addition of water. The electrolyte is typically an aqueous $4M$ NaOH + $1M$ Al(OH)$_3$ solution with $0.04 - 0.10M$ sodium stannate [Na$_2$Sn(OH)$_6$] added as a corrosion inhibitor. The cell reactions are:

Anode $\quad$ Al + 4OH$^-$ → Al(OH)$_4^-$ + 3e $\quad$ $E^0 = -2.35$

Cathode $\quad$ 3/4O$_2$ + 3/2H$_2$O + 3e → 3OH$^-$ $\quad$ $E^0 = 0.40$

The concentration of aluminate increases during discharge to 3 times the saturation level. In order to maintain a stable electrolyte composition during discharge and prevent uncontrolled precipitation, the aluminum ion is allowed to precipitate slowly as aluminum trihydroxide (hydrargillite) by seeding with $Al(OH)_3$:

$$Al(OH)_4^- \rightarrow Al(OH)_3 + OH^-$$

This precipitate is washed onboard the vehicle to form an alkali-free powder which is then withdrawn periodically.

This system is shown schematically in Fig. 30.19. Cells operate near 60°C with forced convection of electrolyte and air. Heat rejection is achieved by a combination of evaporative cooling and a radiator. Air flow is typically 4 times the stoichiometric requirement. The removal of carbon dioxide from the incoming airstream is required to prevent the gradual degradation of the air electrodes, and a moisture exchanger may be used to offset water losses. The crystallizer provides the electrolyte concentration control through the precipitation of aluminum trihydroxide from a supersaturated solution of caustic aluminate. Finally, tanks are provided for the electrolyte, water, and hydrate storage.[36,37]

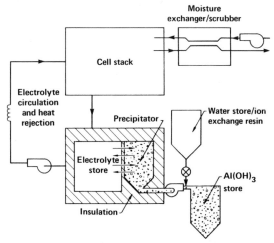

FIG. 30.19 Aluminum/air power cell system. The design provides for forced convection of air and electrolyte, heat rejection, electrolyte concentration control via $Al(OH)_3$ precipitation, and storage for reactants and products.

The essential structure of the aluminum/air cell is shown in Fig. 30.20. The cell consists of the aluminum anode plate and a cathode module with grooves for electrolyte flow. The aluminum plate dissolves during discharge (at about 0.2 mm/h) and oxygen is reduced at the cathode (a porous carbon-Teflon composite doped with an appropriate catalyst). Constant cell geometry is achieved by advancing the anode against the surface of the fixed air cathode module.

The electrical characteristics of a 25-cm² aluminum/air cell operating at 60°C are shown in Fig. 30.21. Peak energy yield was 4.1 kWh/kilogram of aluminum. Parasitic corrosion losses occur at a rate of 10 to 20 mA/cm² and is nearly independent of the net current withdrawn from the cell.

Aluminum/air cells have been scaled to 0.1-m² anodes and discharged with crystal-

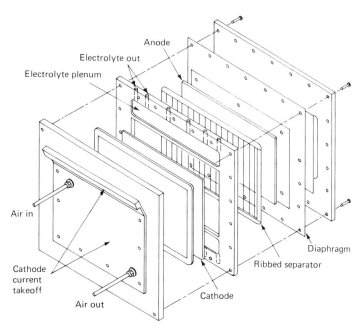

FIG. 30.20 Large-scale aluminum/air test cell.

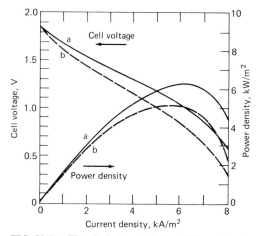

FIG. 30.21 Electrical characteristics of 25-cm² aluminum/air cell (temperature: 60°C; electrolyte composition: $4M$ NaOH, $1M$ Al(OH)$_3$, $0.06M$ Na$_2$Sn(OH)$_6$; interelectrode spacing: (a) 1.5 mm, (b) 3.2 mm).

lizers to determine operating parameters.[38] Four different approaches to achieving mechanical refuelability have been experimentally investigated in test cells, bicells, or cell stacks.[39,40]

1. Moving anode, dry-side current collection cells (Fig. 30.22). Aluminum anode plates are soldered to copper-clad circuit board and placed between air-cathode modules. The modules feature an air-electrode suspended on a flexible diaphragm which advances toward the anode during discharge.[41]

FIG. 30.22   Aluminum/air six-cell stack. The refuelable cell provides for replacement of anode sheets affixed to copper-clad support sheets (dry-side current collection). The cell includes internal manifolding of the electrolyte and pneumatically advances the cathode against the anode module. (*Courtesy of Lawrence Livermore National Laboratory.*)

2. Moving anode, solution-side current collection (SSCC). Anode current collection is effected by a permanent network of conducting metal teeth placed in the gap between the air electrode (rigid) and an anode slab which is held in place and advanced by an air bladder[42] (Fig.30.23).

3. Moving cathode (SSCC). An anode slab is fixed between two air electrode modules, which are advanced pneumatically.[42]

4. Wedge-shaped cells (SSCC). A wedge-shaped anode is consumed between two rigid air electrodes held at an acute angle to form a dimensionally stable cell.[27,40] Anode current collection is achieved by parallel tracks of metal, which also define the interelectrode separation as in earlier designs. This cell is designed for continuous,

FIG. 30.23 Refuelable 1000-cm² bicell (moving anode design). Moving anode design using solution-side anode current collection. Recharging is by replacement of partially consumed anode. (*Courtesy of Lockheed Missiles and Space Company.*)

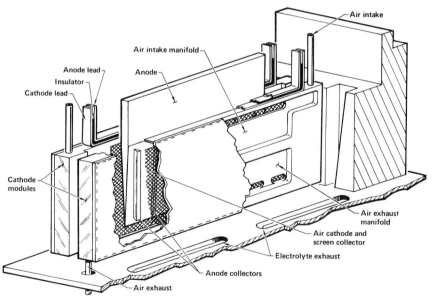

FIG. 30.24 Refuelable wedge-shaped 600 aluminum/air cell. The cell provides for continuous feed of aluminum slabs into a dimensionally stable cell: full anode utilization and partial anode recharge. Anode current collection is by parallel tracks of copper in the interelectrode gap. (*Courtesy of Lawrence Livermore National Laboratory.*)

30-18

gravity feed of rectangular slab anodes into the cell, partial recharge capability, and full utilization of anode material (Fig 30.24).

The estimated weight and volume of a 60-cell, 70-kWh, 40-kW (peak) aluminum/air battery power unit for electric vehicle use are 200 to 250 kg and 350 to 420 dm^3. This unit will require the addition of 25 L of tap water and the removal of 50 kg of hydrargillite every 400 km, a procedure taking from 6 to 9 min. Every 1600 km, the 60 aluminum plates, weighing 64 kg, will be replaced at a service station, with 15 min being required for this refueling. These estimates are based on a vehicle with a gross weight of 1320 kg. Air electrodes have been cycled over 1000 times over a sequence of current levels and a standby period designed to simulate the operating and shutdown modes of the vehicle.[43] The goal is a cost and total energy consumption equivalent to automobiles of similar performance running on liquid fuels synthesized from coal and selling for $0.65 to $0.75 per liter (1981 dollars).[44,45]

REFERENCES

1. D. A. J. Rand, "Battery Systems for Electric Vehicles: State of Art Review," *J. Power Sources* **4**(2):101 (1979).

2. K. D. Beccu, "The Characteristics of Metal-Air Systems," in *Electrocatalysis to Fuel Cells,* University of Washington Press, Seattle, 1972.

3. K. F. Blurton and A. F. Sammells, "Metal/Air Batteries: Their Status and Potential—a Review," *J. Power Sources* **4**:263, 1979.

4. W. N. Carson and C. E. Kent, "The Magnesium-Air Cell," in D. H. Collins (ed.), *Power Sources,* 1966, p. 119.

5. J. F. Cooper and P. K. Hosmer, "The Behavior of the Calcium Electrode in Aqueous Electrolyte," *Fall Meeting Electrochem. Soc.,* October 1977, Atlanta.

6. M. Klein, "Metal-Gas Rechargeable Batteries," *Proc. Seventh Intersoc. Energy Convers. Eng. Conf.,* 1972, p. 79.

7. H. Ikeda, N. Furukawa, and M. Ide, "Metal-Air Batteries," paper presented at *J. Congr. Am. Chem. Soc. Jap.,* Honolulu, April 1979.

8. *Research and Development of an Electrolyte Circulating System Zinc-Air Storage Battery,* Sanyo Electric Co., Ltd., Osaka, Japan, 1979.

9. A. J. Appleby, J. Jacquelin, and J. P. Pompon, "Current Status of the CGE Circulating Zinc-Air Vehicle Battery," Paper no. 316, *Spring Meeting Electrochem. Soc.,* Philadelphia, May 1977.

10. A. J. Appleby and M. Jacquier, "The CGE Circulating Zinc/Air Battery: A Practical Vehicle Power Source," *J. Power Sources* **1**:17 (1976).

11. A. J. Appleby, J. Jacquelin, and J. P. Pompon, SAE Paper 770381, *Trans. Automot. Eng. Conf.,* Detroit, 1977.

12. H. Baba, "A New Zinc-Air Fuel Battery System," Society of Automotive Engineers, Warrendale, Pa., Paper 710327, *Automot. Eng. Conf.,* Detroit, 1971, p. 930.

13. A. L. Almerini and S. J. Bartosh, "Simulated Field Tests on Zinc-Air Batteries," *Proc. 26th Power Sources Symp.,* 1974.

14. R. Witherspoon, E. Zeitner, and H. Schulte, *Proc. Sixth Intersoc. Energy Convers. Eng. Conf.,* Boston, August 1971, Soc. Automot. Eng., Warrendale, Pa., p. 96.

15. H. F. Bauman and G. B. Adams, "Lithium-Water-Air Battery for Automotive Propulsion," Final Report for period Oct. 28, 1976 to Sept. 28, 1977, Lockheed Palo Alto Research Laboratory, COO/1262-1, October 1977.

16. W. R. Momyer and E. L. Littauer, "Development of a Lithium-Water-Air Primary Battery," *Proc. Fifteenth Intersoc. Energy Convers. Eng. Conf.,* Seattle, August 1980, p. 1480.

17. J. F. Cooper, P. K. Hosmer, and R. V. Homsy, "The Anodic Behavior of Lithium in Aqueous Lithium Hydroxide Solutions," *J. Electrochem. Soc.* **125**:1 (1978).

18. E. L. Littauer and K. C. Tsai, "Anodic Behavior of Lithium in Aqueous Electrolytes, ii. Mechanical Passivation," "Corrosion of Lithium in Aqueous Electrolytes," "Anodic Behavior of Lithium in Aqueous Electrolytes, iii. Influence of Flow Velocity, Contact Pressure and Concentration," *J. Electrochem. Soc.* **123**(7):964 (1976); **124**(6):850 (1977); **125**(6):845 (1978).

19. E. L. Littauer and K. C. Tsai, "Observation of the Diffusion Coefficient of the Perhydroxyl Ion (HO_2^-) in Lithium Hydroxide Solutions," *Electrochim. Acta* **24**:681 (1979).

20. W. R. Momyer and J. L. Morris, "Reactive Metal-Air Batteries for Automotive Propulsion," Report no. LMSC-D-683375, Lockheed Missiles & Space Co., Dec. 20, 1979.

21. D. S. Kier, M. J. Pryor, and P. R. Sperry, "The Influence of Ternary Alloying Addition on the Galvanic Behavior of Al-Tin Alloys," *J. Electrochem. Soc.* **116**(3):319 (1969).

22. T. Valand and G. Nilsson, "The Influence of F Ions on the Electrochemical Reaction on Oxide-Covered Al," *Corros. Sci.* **17**:449, 1977.

23. G. Hoffman, M. Ritschel, and W. Vielstich, "Neutral Electrolyte Aluminum Air Cell," *Proc. First Int. Workshop Reactive Metal-Air Batteries,* University of Bonn, West Germany, July 9–11, 1979, published by Lockheed Missiles and Space Company, Palo Alto, Calif.

24. A. R. Despic, "Problem Areas of a Neutral Electrolyte Al/Cell," *Proc. First Int. Workshop React. Metal-Air Batteries,* University of Bonn, West Germany, July 9–11, 1979, published by Lockheed Missiles and Space Company, Palo Alto, Calif.

25. D. M. Drazic, "Features of an Al/Air Cell with Aqueous Sodium Chloride Electrolyte," *Proc. First Int. Workshop React. Metal-Air Batteries,* University of Bonn, West Germany, July 9–11, 1979, published by Lockheed Missiles and Space Company, Palo Alto, Calif.

26. D. M. Drazic et al., "Neutral Electrolyte Aluminum-Air Battery," *Eleventh Power Sources Symp.,* Brighton, England, 1978.

27. A. R. Despic and P. D. Milanovic, "Aluminum-Air Battery for Electric Vehicles," *Rec. Trav. Inst. Sciences Techniques Academie Serbe Sciences Arts* **12**(1):1–18 (1979).

28. T. Valand, O. Mollstad, and G. Nilsson, "Al-Air Cells—Potential Small Electric Generators for Field Use," *Power Sources Symp.,* Brighton, England, October 1980.

29. S. Zaromb, "The Use and Behavior of Aluminum Anodes in Alkaline Primary Batteries," *J. Electrochem. Society* **109**:1125 (1962).

29a. S. Zaromb, "Feasibility of Electrolyte Regeneration in Al Batteries," *J. Electrochem. Soc.* **109**:1191 (1962).

30. L. Bocstie, D. Trevethan, and S. Zaromb, "Control of Al Corrosion in Caustic Solutions," *J. Electrochem. Soc.* **110**:267(1963).

31. S. Zaromb, "Aluminum Fuel Cell for Electric Vehicles," *Proc. Symp. Power Systems Electric Vehicles,* National Center for Air Pollution Control, Cincinnati (PB 177 706), 1967.

32. J. Ruch "High Power Density Metal-Air Cells," Paper no. 42, *Fall Meeting Electrochem. Soc.,* Atlanta, October 1977.

33. J. F. Cooper and E. L. Littauer, "Mechanically-Refuelable Metal-Air Batteries for Automotive Propulsion," *Proc. Thirteenth Intersoc. Energy Convers. Eng. Conf.,* SAE, San Diego, CA, August 1978; Lawrence Livermore Laboratory, UCRL-81178, August 1978.

34. J. F. Cooper, R. V. Homsy, and J. H. Landrum, "The Aluminum-Air Battery for Electric Vehicle Propulsion," in *Proc. Fifteenth Intersoc. Energy Convers. Eng. Conf.,* Seattle, August 18–22, 1980, Lawrence Livermore National Laboratory, Livermore, Calif., UCRL-84443, June 1980.

35. J. D. Salisbury and E. Behrin, "An Analysis of Aluminum-Air Battery Propulsion Systems for Passenger Vehicles," *Proc. Fifteenth Intersoc. Energy Convers. Eng. Conf.,* Seattle, August 18–22, 1980, Lawrence Livermore National Laboratory, Livermore, Calif., UCRL-83824, May 1980.

36. R. V. Homsy, "Aluminum-Air Power Cell System Design: Mass and Enthalpy Balance," *Fall Meeting Electrochem. Soc.,* Hollywood, Fla., October 1980, Lawrence Livermore National Laboratory, Livermore, Calif., UCRL-84444, June 1980.

37. J. F. Cooper and R. V. Homsy, "Development of the Aluminum-Air Battery for Electric Vehicle Applications," *Proc. Fourth International Conference and Demonstration,* Electric Vehicle Development Group, London, September 1981; Lawrence Livermore National Laboratory, Livermore, Calif., UCRL-86560, August 1981.

38. R. V. Homsy, "The 1000-cm^2 Aluminum-Air Power Cell," Lawrence Livermore National Laboratory, Livermore, Calif., UCRL-86971, November 1981.

39. "Metal-Air Battery Research and Development Summary Report," E. Behrin and J. Cooper, eds., Lawrence Livermore National Laboratory, Livermore, Calif., UCID-19440, May 1982.

40. J. F. Cooper, "Preliminary Design and Analysis of Aluminum-Air Cells Providing for Continuous Feed and Full Utilization of Anodes," Lawrence Livermore National Laboratory, Livermore, Calif. UCID-19178, August 16, 1981.

40a. B. J. McKinley, "Aluminum-Air Process Development," *Proc. Fifth U.S. DOE Battery and Electrochemical Contractors' Conference,* December 1982.

41. K. K. Burr, J. F. Cooper, R. V. Homsy, and B. J. McKinley, "Testing of Refuelable Aluminum-Air Multicell Batteries," Paper No. INDE 68, Spring Meeting of American Chemical Society, April 1982.

42. "Aluminum-Air Battery Cell Hardware Development," Lockheed Missiles and Space Company, Report on Subcontract, January 1 - April 30, 1982, P.O. 5513309, prepared for the Aluminum Air Battery R&D Program, Lawrence Livermore National Laboratory, Report UCRL-15479, April 30, 1982.

43. I. Malkin, "Air Depolarized Cathodes," Final Report, Diamond Shamrock, Painesville, Ohio, Lawrence Livermore National Laboratory, Livermore, Calif., UCRL-15447, December 14, 1981.

43a. L. Gestaut, "Development of Air Cathodes for the Aluminum-Air Battery," *Proc. Fifth U.S. DOE Battery and Electrochemical Contractors' Conference,* December 1982.

44. J. F. Cooper, "Estimates of the Cost and Energy Consumption of Aluminum-Air Electric Vehicles," Paper 118, Fall Meeting of the Electrochemical Society, Hollywood, Fla., October 1980, Lawrence Livermore National Laboratory, Livermore, Calif., UCRL-84445, June 1980; Update August 1981; UCRL-84445 Rev. 1.

45. C. J. McMinn and J. A. Branscomb, "Production of Anodes for Aluminum-Air Power Cells Directly from Hall Cell Metal," Reynolds Aluminum, Reduction Research Division, Sheffield, Ala., Lawrence Livermore National Laboratory, Livermore, Calif., UCRL-15354, February 12, 1981.

31

Lithium/ Iron Sulfide Batteries

by
Warren L. Towle and Ronald A. Rizzo

31.1 GENERAL CHARACTERISTICS

The lithium/iron sulfide battery operates between 400 and 500°C and uses a molten-salt electrolyte, a mixture of LiCl and KCl. High power densities, close to 200 W/kg, and other performance advantages can be realized at this high operating temperature, but the high temperature does increase problems with material stability and corrosion.

Originally, metallic lithium was used as the anode material and elemental sulfur as the positive. Because of difficulties in keeping the molten lithium in place, particularly on recharging, shorting was a problem. Alloys of lithium with aluminum or silicon were then found which showed good reversibility to lithium penetration and which functioned at voltages only slightly less negative than that of liquid lithium.

Problems also arose with the sulfur cathode both because of its high vapor pressure at the operating temperature and because of its corrosiveness. The use of metal sulfides solved this problem. Both FeS and FeS_2 are now used. FeS, because it is less corrosive, allows a less expensive construction and is the more advanced technology; FeS_2 develops higher specific energy levels and, if successfully developed, would be less costly than the FeS cells.

Applications in electric vehicles (EV) and stationary energy storage (SES) are envisioned for this battery system.

31.2 DESCRIPTION OF ELECTROCHEMICAL SYSTEM

The Li/FeS_x cells that are currently under development consist of an Li-Al or Li-Si negative electrode, an FeS or FeS_2 positive electrode, a separator to provide electrical isolation of the electrodes, and molten LiCl-KCl electrolyte. The melting point of the electrolyte (352°C at the eutectic composition of 58.2 mol % LiCl) requires a battery operating temperature in the range of about 400 to 500°C. LiCl-rich (67 vs. 58 mol % LiCl) or all-lithium cation (LiF-LiCl-LiBr) electrolytes are now being used in the FeS cell. The optimum electrolyte for the FeS$_2$ cell has not yet been established.

The overall electrochemical reaction for the Li-Al/FeS cell involves the transfer of two electrons

$$2\text{Li-Al} + \text{FeS} \underset{\text{charge}}{\overset{\text{discharge}}{\rightleftarrows}} \text{Li}_2\text{S} + \text{Fe} + 2\text{Al}$$

The theoretical specific energy for this reaction is about 460 Wh/kg, and the discharge voltage curve has a single voltage plateau at about 1.3 V. The reaction is actually much more complex than shown as intermediate compounds are formed through a reaction with the KCl in the electrolyte.

The overall reaction for the Li-Al/FeS$_2$ cell is shown in two steps:

$$2\text{Li-Al} + \text{FeS}_2 \underset{\text{charge}}{\overset{\text{discharge}}{\rightleftarrows}} \text{Li}_2\text{S} + \text{FeS} + 2\text{Al}$$

$$2\text{Li-Al} + \text{FeS} \underset{\text{charge}}{\overset{\text{discharge}}{\rightleftarrows}} \text{Li}_2\text{S} + \text{Fe} + 2\text{Al}$$

The total theoretical specific energy for these reactions is approximately 650 Wh/kg. This discharge has two voltage plateaus, the first at about 1.7 V and the second at 1.3 V. The Li-Al/FeS$_2$ cells are often designed to operate only on the upper voltage plateau; these are referred to as "upper-plateau" cells. The Li-Al/FeS$_2$ reactions also involve several complex intermediate phases (generally compounds of lithium, iron, and sulfur).[1,2]

31.3 CONSTRUCTION

Most of the early cells that were fabricated used a prismatic, "bicell" design (see Fig. 31.1) with a central positive electrode and two facing negative electrodes. Later, multiplate cells (see Fig. 31.2), which have two or more positive electrodes and facing negative electrodes, were developed. Porous separator sheets, usually of boron nitride cloth or felt or thin layers of MgO or AlN powder, located between the electrodes serve as electronic insulators. Cloths or screens, currently in the form of sheet metal that has been perforated mechanically or by photoetching, are used to prevent the escape of particulate material from the electrodes into the separator. In the positive electrodes, metallic current collectors (steel for FeS, molybdenum for FeS$_2$) are used. The negative electrodes are usually grounded to the housing. The terminal of the positive electrode extends through the top of the cell can via an electrically insulating feedthrough. The top of the cell can is also provided with a tube, which is closed off later by a weld, to permit the addition of molten electrolyte to the cell after assembly. In many of the cells, "picture-frame" structures around the perimeter of the electrodes are used to hold the electrode components together as a unit.

The lithium/metal sulfide cell can be assembled in a charged, uncharged, or partially charged state. To assemble an Li-Al/FeS or Li-Al/FeS$_2$ cell in the charged state, the negative electrodes are normally cold- or hot-pressed from Li-Al powder (usually 46 to 50 atom % lithium), which may or may not be mixed with some of the LiCl-KCl electrolyte powder. The positive electrodes are formed similarly by cold- or hot-pressing FeS or FeS$_2$ powder with or without added electrolyte powder. In the

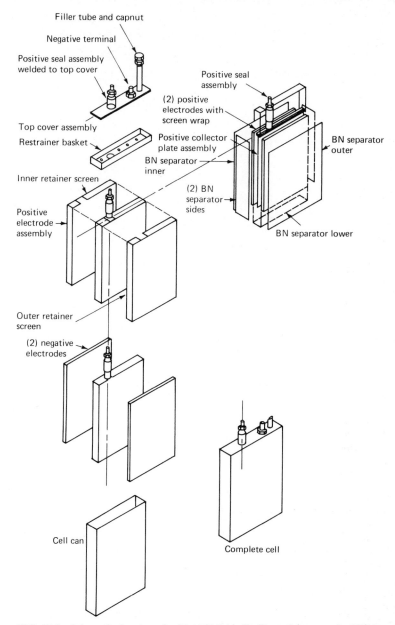

Filler tube and capnut

Negative terminal

Positive seal assembly welded to top cover

Positive seal assembly

(2) positive electrodes with screen wrap

Top cover assembly

Restrainer basket

Positive collector plate assembly

BN separator inner

BN separator outer

Inner retainer screen

(2) BN separator sides

Positive electrode assembly

BN separator lower

Outer retainer screen

(2) negative electrodes

Cell can

Complete cell

FIG. 31.1 Schematic drawing of a Li-Al/FeS bicell. (*From Barney et al., 1980.*)

case of the uncharged cells, the electrode plaque is pressed from a mixture of Li_2S and iron powder in the appropriate proportions. The negative electrode in this case is an aluminum structure (e.g., pressed wire, porous metal, or solid plate), which is

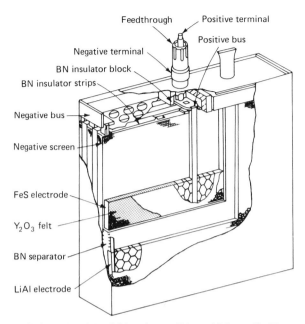

FIG. 31.2   Mark IA lithium/iron sulfide multiplate cell. (*From Kolba et al., 1980.*)

converted to the Li-Al alloy electrochemically when the cell is charged. Partially charged cells can be fabricated from mixtures of the above materials in intermediate ratios.[1]

31.4 PERFORMANCE CHARACTERISTICS

31.4.1 Voltage

The open-circuit voltage (OCV) for the Li-Al/FeS cell is 1.34 V and for the Li-Al/FeS$_2$ cell is 1.74 V and is a function of both temperature and state of discharge.[3,4] The operating voltage for the former is in the range of 1.1 to 1.3 V; the operating voltage for the Li-Al/FeS$_2$ system is in the range of 1.5 to 1.7 V while on the "upper plateau" and drops to 1.1 to 1.3 V during the second part of the discharge.

31.4.2 Discharge Characteristics

A typical set of discharge curves for the multiplate Li-Al/FeS cells is given in Fig. 31.3 covering discharge periods from the 2.5- to 9-h rate. Similar curves for the "upper plateau" discharge of the FeS$_2$ cell are shown in Fig. 31.4. As is the case with most batteries, the capacity of the lithium/metal sulfide cell decreases with increasing discharge rate.

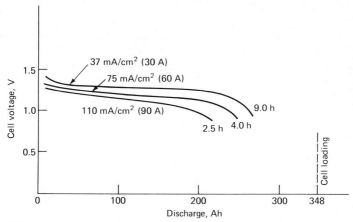

FIG. 31.3 Discharge curves for the Li-Al/FeS cell. (*From Miller et al., 1979.*)

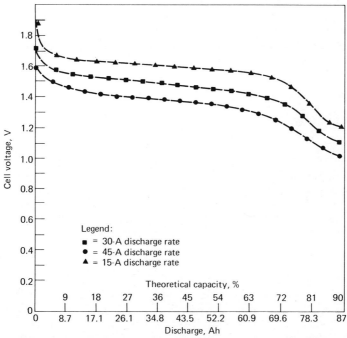

FIG. 31.4 Discharge curves for the Li-Al/FeS$_2$ cell upper plateau. (*From Miller et al., 1979.*)

31.4.3 Effect of Temperature

The effect of temperature on the performance of the cell is shown in Fig. 31.5. Capacity increases with higher operating temperatures; however, higher temperatures will also increase corrosion rates and shorten life.

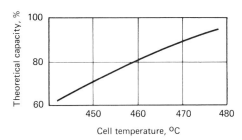

FIG. 31.5   Effect of temperature on cell capacity (multiplate Li-Al/FeS cell; LiCl·KCl electrolyte, 52 w/o LiCl; 200-Ah capacity).

31.4.4 Self-Discharge

The rate of self-discharge at the operating temperature of the lithium/iron sulfide cell is remarkably low. Trickle-charge tests on a 300-Ah multiplate FeS cell at 450°C showed that only 0.15 A was the level of current needed to maintain the cell in a fully charged condition.

31.4.5 Power and Energy Characteristics

The specific power and energy characteristics of several Li-Al/FeS cells are shown in Table 31.1. The specific power of these cells was improved over earlier designs through changes in the current-collector configuration and cell terminal/feedthrough components. The specific energy was improved through the use of additives to the Li-Al electrode and by a reduction in the current collector weight. The higher specific power was thus maintained with cells having specific energies above 100 Wh/kg.[4,5,6]

TABLE 31.1   Specific Energy and Specific Power of Li-Al/FeS Cells*

Specific energy, Wh/kg	Specific power, W/kg		Number of cycles tested
	95% SOC	50% SOC	
77	233	192	94
88	142	129	100
92	178	130	> 180
101	180	132	75
98	208	161	106

*Experimental cells (1981-82 design).
SOURCE: Refs. 4 and 5.

31.4.6 Cycle life

The end of life of the Li-Al/FeS cell has been defined as a 20% loss of capacity or a decrease in the coulombic efficiency to less than 95%. FeS cells have demonstrated the capability of up to 1000 cycles; the FeS_2 cells, because of the more corrosive conditions and less advanced technology, have not shown equivalent performance. Figure 31.6 is a plot of the cycle life performance of a group of Li-Al/FeS cells. The mean time to failure for these cells was 345 cycles and the average capacity-loss rate was 0.02% per cycle.[5,6,7]

31.4.7 Efficiency

The coulombic efficiency of the Li-Al/FeS cell is > 99%. It remains essentially at this value throughout its life. The energy efficiency of the cell usually is in the range 80 to 85% for the 4-h discharge rate, and it also remains essentially constant throughout the useful life of the cell.

31.4.8 Charging

Two characteristics of the Li alloy/FeS_x system require special attention in charging. The first is that voltages sufficiently negative to cause the formation of liquid lithium must be avoided. The second is that provision must be made in any series arrangement of cells to bypass current around any cells which reach full charge before their neighbors.

The equilibrium voltage of the Li-Al alloy is about 300 mV less negative than liquid

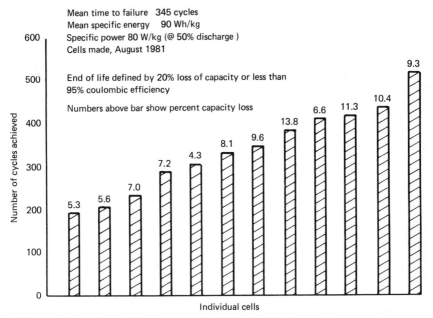

FIG. 31.6   Cycle life test, full-scale Li-Al/FeS EV cells, 395-Ah capacity.

lithium. Thus with careful voltage control the formation of liquid Li can be avoided. For FeS cells, maximum charging voltage is usually 1.6 V, and for FeS_2 it is 2.16 V.

The need for a bypass around individual cells arises from the fact that inevitably there will be small differences in coulombic efficiency from cell to cell. Thus an external bypass is needed since there is no innocuous side reaction which can take place within the cell to allow passage of current, once full charge is reached. (In the familiar lead-acid system, for example, electrolysis of water can take place in individual cells which reach full charge before their neighbors.) Overcharge of individual cells could result in the formation of liquid lithium at the negative and the development of more corrosive conditions and possibly other undesirable effects at the positive.

Typical charge and discharge curves for the Li-Al/FeS cell, using a LiCl-KCl electrolyte with 53.8% LiCl, are shown in Fig. 31.7a to c. The cell represented has a theoretical capacity of 360 Ah. The input current during charge was limited to 40 A. The voltage was allowed to rise to 1.53 V (Fig. 31.7a). Charging was terminated when current flow dropped to 10 A (Fig. 31.7b). The corresponding discharge at 70 A is shown in Fig. 31.7c.

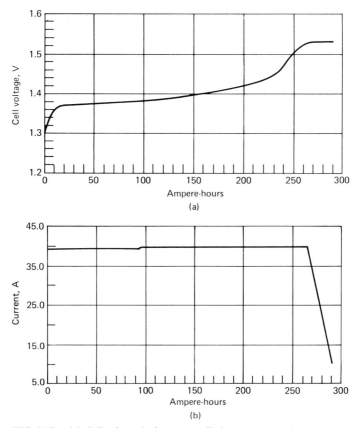

FIG. 31.7 (a) Cell voltage during current-limited constant-voltage charge. (b) Constant input during constant-voltage charge (current limit of 40 A, cutoff at 10 A).

Typical charge and discharge curves for the Li-Al/FeS$_2$ cell are shown in Fig. 31.8. These data also show the stable capacity exhibited by the cell during cycling.[4,6]

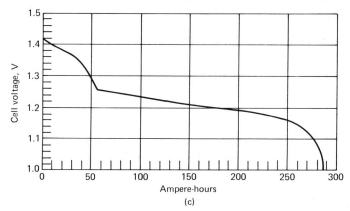

FIG. 31.7 (*Continued*)   (*c*) Constant-current discharge after constant-voltage charge (70-A discharge current).

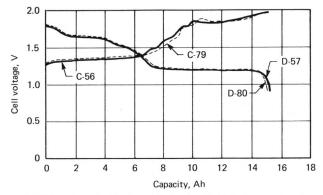

FIG. 31.8   Charge and discharge curves, Li-Al/FeS$_2$ experimental 25.2-Ah cell; 425°C, 3-h discharge (D), 10-h charge (C). (*From Proc. 5th U.S. Dept. of Energy Battery and Electrochemical Contractors' Conf., 1982.*)

31.5 APPLICATIONS AND SYSTEM DESIGN

The two principal areas of application of the lithium/metal sulfide batteries are the electric vehicle and stationary energy storage (SES) systems. These can also include hybrid electric vehicles and electrical storage systems for solar and wind-driven generators. It is evident, because of the high operating temperature and the resulting need for insulation to ensure heat economy, that small-size applications would not be feasible.

The performance and lifetime goals for the lithium/iron sulfide battery for electric vehicle use are summarized in Table 31.2. Depending on the type of vehicle, 20- to 60-kWh batteries are expected to be used, most likely based on a 60-cell (72-V) modular

TABLE 31.2 Program Goals for the Lithium/Iron Sulfide Electric-Vehicle Battery

	1983	1986	Post-1990
System	Li-alloy/FeS	Li-alloy/FeS	Li or Li-alloy/disulfide
Specific energy,[a] Wh/kg			
Cell	100	100–125	160–225
Battery	—	80–100	130–180
Energy density, Wh/L			
Cell	240	300	400–500
Battery	—	150	200–250
Peak specific power,[b] W/kg			
Cell	125	150–250	240–450
Battery	—	120–160	190–360
Battery heat loss,[c] W	—	75–150	50–100
Lifetime,[d] deep discharges			
Cell	500	1000[e]	1200
Battery	—	800[e]	1000
Equivalent kilometers		240,000	300,000

[a]Calculated at the 4-h discharge, 8-h charge rate.
[b]Peak power sustainable for 20 s at 0 to 50% state of discharge.
[c]Heat loss of battery through the insulated case; under some operating conditions, additional heat removal may be required.
[d]End of lifetime is defined as a 20% loss of the initial peak capacity or a decrease of coulombic efficiency to 95%.
[e]Ultimate lifetime of Li-Al/FeS vehicle battery projected to be > 1000 cycles.
SOURCE: Ref. 6.

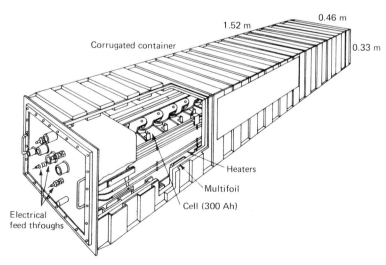

FIG. 31.9 Lithium/iron sulfide module battery, 20 kWh. (*From Nelson et al., 1978.*)

TABLE 31.3 Prototype Li-Al/FeS cell

Positive electrode size, cm × cm	18.4 × 17.5
Theoretical capacity, Ah	394.8
Thickness, cm	0.33
FeS weight, g	647.2
Active material loading density, Ah/cm^3	1.4 (1.36)*
Number of parallel electrodes	3
Electrode utilization, %	91.5
Negative electrodes	
Theoretical capacity, Ah	511.8
Thickness, cm	0.64
48 at % Li-Al weight, g	741.7
Active material loading density, Ah/cm^3	0.93 (0.83)*
Number of parallel electrodes	4
Negative to positive capacity ratio	1.3
Electrode utilization, %	70.6
Electrolyte weight (54 wt % LiCl-KCl), g	1350
Current collector	
Particle retainer thickness, cm	0.013
Total current collector weight, g	1376
Open area of particle retainer	0.26
Terminal diameter, cm	0.95
Cell resistance, mΩ	0.86
Cell housing	
Thickness, cm	0.046
Weight, g	590†
Total cell weight, kg	4.793
Specific energy (C/4), Wh/kg	93.9
Specific power (50% DOD), W/kg	102
Power to energy ratio	1.06

*Value in parentheses includes volume of retainer frame at edge of electrode.
†Top of cell housing and feedthrough shell weigh 185 g.
SOURCE: Ref. 4.

TABLE 31.4 Goals for Lithium/Metal Sulfide SES Batteries

Goal	Demonstration
Battery performance:	
Energy output, kWh	100,000
Peak power, kW	25,000
Sustained power, kW	10,000
Cycle life	3000
Discharge time, h	5–10
Charge time, h	10
Minimum energy efficiency (dc out/dc in)	75

SOURCE: Ref. 1, Argonne National Laboratory, Argonne, Ill.

unit of 20 kWh using 350-Ah cells. A 40-kWh EV battery, consisting of two 20-kWh modules, is shown schematically in Fig. 31.9. Design details of a prototype cell are listed in Table 31.3. The energy density is expected to more than double from 240 Wh/L with the early Li-Al/FeS cells to over 500 Wh/L with the higher-performance Li-Al/FeS$_2$ cells.

The goals for the SES batteries are given in Table 31.4. An artist's concept of the 100-MWh SES installation is shown in Fig. 31.10. Several designs have been considered for the submodule and module. The original concept considered that the largest practical cell was 2.5 kWh (17 × 28 × 23 cm). The 43.2-V submodule, the smallest replaceable unit, consisted of 96 cells with a design weight of 5440 kg (Fig. 31.11). Thirty-six of these submodules composed the truckable module with an output voltage of 1555 V (end-of-charge) and a nominal capacity of 8.64 MWh. The dimensions of the module were designed to be height, 7.0 m; width, 2.9 m; length, 15.2 m.[8] The high estimated cost of the battery hardware led to other system designs, including one based on large 30-kWh cells and others based on cells 1.25 to 2.5 kWh in capacity.[8,9]

Cost estimates for the EV and SES cells are summarized in Fig. 31.12. These estimates are for the ultimate cost by a matured manufacturing technology.[10]

REFERENCES

1. P. A. Nelson et al., *Development of Lithium/Metal Sulfide Batteries at Argonne National Laboratory: Summary Report for 1978,* Argonne Report ANL-79-64, Argonne National Laboratory, Argonne, Ill., July 1979.

2. D. L. Barney et al., *High Performance Batteries for Electric Vehicle Propulsion and Stationary Energy Storage,* Argonne Report ANL-79-94, Argonne National Laboratory Argonne, Ill., March 1980.

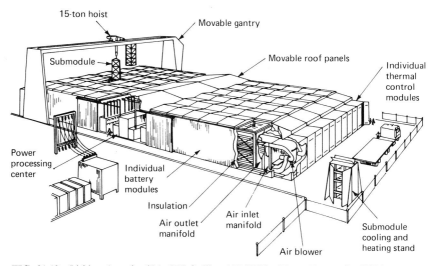

FIG. 31.10   Lithium/metal sulfide SES facility, 100 MWh. (*From Zivi et al., 1980.*)

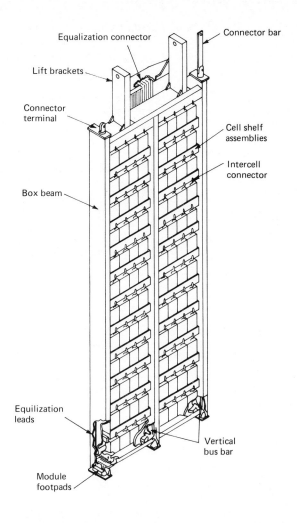

Height	5.84 m
Width	1.8 m
Thickness	0.275 m
Weight	5442 kg
Capacity	240 kWh (nominal maximum)
Voltage	43.2 V dc (end of charge (EOC))
Cell quantity	96
Equalization circuits	24

FIG. 31.11 Submodule design and characteristics. (*From Zivi et al., 1980.*)

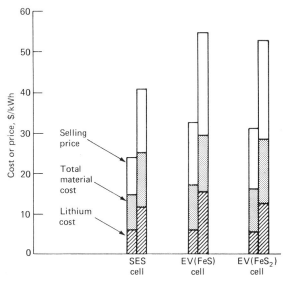

FIG. 31.12 Material and manufacturing cost range for lithium/metal sulfide cells in terms of 1979 dollars. (SOURCE: Argonne National Laboratory. *From Chilenskas et al., 1979.*)

3. P. A. Nelson et al., *High Performance Batteries for Electric Vehicle Propulsion and Stationary Energy Storage,* Argonne Report ANL-78-94, Argonne National Laboratory, Argonne, Ill., November 1978.

4. *Proc. Lithium/Metal Sulfide Battery Workshop,* Argonne National Laboratory, Argonne, Ill., August 1982.

5. *Proc. 5th U.S. Dept. of Energy Battery and Electrochemical Contractors' Conf.,* U.S. Department of Energy, Washington, D.C., December 1982.

6. *Annu. Tech. Rept.,* Chemical Engineering Division, Argonne National Laboratory, Argonne, Ill., "Report No. ANL 82-23," 1981 and 1982 Annual Report for Review Committee.

7. V. M. Kolba et al., *Failure Analysis of Mark IA Lithium/Iron Sulfide Battery,* Argonne Report ANL-80-44, Argonne National Laboratory, Argonne, Ill., October 1980.

8. S. M. Zivi, et al., *Reference Design of 100 MW-h Lithium/Iron Sulfide Battery System for Utility Load Leveling,* Argonne Report ANL-80-19, Argonne National Laboratory, Argonne, Ill., March 1980.

9. S. M. Zivi et al., *Conceptual Designs for Utility Load-Leveling Battery with Li/FeS Cells,* Argonne Report ANL-80-20, Argonne National Laboratory, Argonne, Ill., July 1980.

10. A. A. Chilenskas et al., *A Preliminary Estimate of Manufacturing Cost for Lithium/Metal Sulfide Cells for Stationary and Mobile Applications,* Argonne Report ANL-79-59, Argonne National Laboratory, Argonne, Ill., fall 1979.

32

Sodium/Sulfur Batteries

by
**Ronald A. Rizzo, Warren L. Towle,
and Marjorie L. McClanahan**

32.1 GENERAL CHARACTERISTICS

The concept of a secondary sodium/sulfur battery was first announced in 1967.[1] The system uses a molten sodium anode, a molten sulfur cathode, and a solid material which functions as both a sodium ion–conducting electrolyte and a separator.

Conventional battery systems, such as the lead-acid battery, rely on solid electrodes and ion-conducting liquid electrolytes for their reaction medium. The sodium/sulfur concept, however, is based on liquid electrodes and an ion-conducting solid electrolyte. The system must be operated at temperatures above 285°C to maintain the sodium, sulfur, and the reaction products in a liquid state and to obtain adequate electrolyte conductivity. Typical operating temperatures are 300 to 350°C. These temperatures present special problems in the areas of corrosion and thermal cycling. Special handling procedures also are required during cell fabrication because of the highly reactive nature of the sodium and sulfur cell components with respect to oxygen and water.

The sodium/sulfur concept is being developed primarily for electric utility load-leveling applications and for electric vehicles. It is attractive for these applications due to its potential low cost through its use of abundant, noncritical materials and promising energy density and performance characteristics. Several design concepts for the cell have been developed. One uses a ceramic tube of sodium beta″-alumina (β''-$A1_2O_3$) for the solid electrolyte material. An alternate candidate material which is being considered for the solid electrolyte is a sodium zirconium phosphosilicate ($Na_3Zr_2PSi_2O_{12}$) called NASICON.[2] A second concept uses bundles of hollow glass fibers for this purpose.

32.2 DESCRIPTION OF ELECTROCHEMICAL SYSTEM

The individual half-cell and total cell reactions are

Negative electrode $\qquad$ $2Na \underset{charge}{\overset{discharge}{\rightleftarrows}} 2Na^+ + 2\,e$

Positive electrode $\qquad$ $xS + 2e \underset{charge}{\overset{discharge}{\rightleftarrows}} S_x^{2-}$

Overall reaction $\qquad$ $2Na + xS \underset{charge}{\overset{discharge}{\rightleftarrows}} Na_2S_x$

During discharge, sodium is oxidized at the sodium/solid electrolyte interface. The sodium ions then migrate through the electrolyte to the cathode compartment, where their initial reaction product is sodium pentasulfide, Na_2S_5. The pentasulfide is immiscible with sulfur, and a two-phase mixture forms. Upon further discharge, first the free sulfur phase is consumed. Then the Na_2S_5 is progressively converted in a single phase to sodium trisulfide, Na_2S_3. If discharge is continued, a two-phase mixture again forms as solid Na_2S_2 separates out of the liquid. Because of the severe drop-off in performance which this causes, cells are not designed to be discharged into this region.

Figure 32.1 shows the equilibrium potential for the Na/S system as a function of the Na:S molar ratio.[3] The fully charged state is represented by pure sulfur as liquid L_1. The sulfur–sodium pentasulfide mixture is represented by $L_1 + L_2$. This discharge plateau (henceforth referred to as the two-phase region) has an open-circuit voltage of 2.078 V down to the composition $Na_2S_{5.19}$. The mixture becomes single phase between $Na_2S_{5.19}$ and $Na_2S_{2.98}$. The open-circuit voltage falls during this transition to 1.782 V. The system has a theoretical specific energy density of 760 Wh/kg when discharged

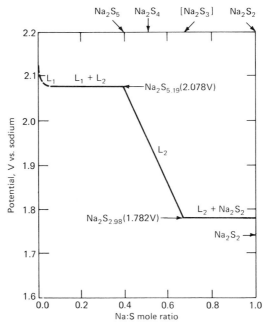

FIG. 32.1 Equilibrium potential vs. molar ratio Na:S. (*From Topouzian.*)

to this point. The precise voltages and compositions at the phase boundaries have a slight temperature dependence in the range of 300 to 400°C. The cells operate at 100% coulombic efficiency with no self-discharge. Shelf life, therefore, is indefinite.

Discharge of the cathode to a composition richer in sodium than Na_2S_3 (with an OCV of 1.78 V) risks the precipitation of Na_2S_2 and would be considered overdischarge. Since structural damage can result, care must be exerted to limit discharge. On charge, the higher the degree of recharge, the greater will be the proportion of the nonconductive liquid sulfur phase in the cathode chamber, with the inevitable increase in internal resistance of the cell. This effect on terminal voltage can be used to signal the end of charge. Carried too far, overcharge can result in electrolyte cracking. An internal safety device is very effective for controlling the highly exothermic reaction which occurs as the sodium and polysulfides mix in the event of electrolyte failure. This device restricts the flow of sodium and thereby prevents burn-through of the cell container. The situation in which a burn-through of the cell container wall is observed is called a "breaching failure." It can occur on uncontrolled charging for reasons mentioned above. Electrolyte tube cracking and breaching failures also have been observed in poorly designed cells near the end of discharge when all the cell void volume has been used up.

32.3 HOLLOW-FIBER SODIUM/SULFUR BATTERY

A schematic of the Na/S cell developed by Dow Chemical Co. is shown in Fig. 32.2.[4] It consists of a multitude of thin-walled, hollow glass fibers interacting with a common sodium reserve. The fibers are 10 to 15 μm thick with outside diameters of 70 or 80 μm. The 10-μm-thick (70-μm-OD) fibers have a packing density about 3000/cm². The large surface-area-to-volume ratio characteristic of this type of construction results in low current density and reduced internal cell resistance. To reduce cathode resistance and cathode polarization, a metal foil, which is interspersed among the fibers serves as the cathode current collector. The cell design is unique; small cells, or single-fiber cells for that matter, have virtually identical electrical behavior as a volumetric fraction of larger cells. This should simplify scale-up to practical-size units.[4]

32.3.1 Construction

This system has a basic cell design consisting of a molten sodium anode separated from a molten sulfur cathode by bundles of hollow fibers. The hollow fibers which serve as electrode separator and sodium ion–conducting electrolyte are made from a sodium borosilicate glass having the composition $Na_2O \cdot B_2O_3 \cdot 0.16NaCl \cdot 0.2SiO_2$. NaCl is added to improve the ionic conductivity of the material. The glass has a specific resistivity of 2.4×10^4 $\Omega \cdot$cm at 300°C. However, the net internal cell resistance is greatly reduced as a result of the thin-walled nature of the fibers. The fibers are sealed at the bottom and opened at the top where they penetrate the tube sheet material. The tube sheet performs a dual function in assembling the bundles of fibers as an integral unit and in helping the fibers to separate the anode and cathode compartments. A low-melting sodium borate glass (4 to 8% Na_2O) is used to make the tube sheet. The glass is applied around the open end of the fibers as a paste. The paste solvent is evaporated and the tube sheet–fiber assembly is fired as a solid disk. The preparation and application of

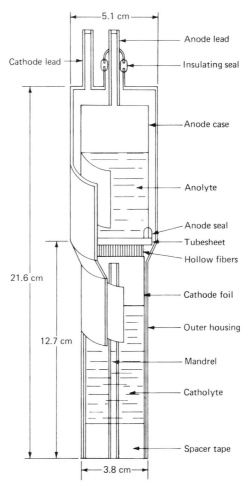

FIG. 32.2 Hollow-fiber Na/S cell, 40 Ah, developed by Dow Chemical Co. (*From Tsang, Anand, and Levine, 1978.*)

both the tube sheet material and the glass fibers are detailed along with crucial aspects of this system.[4,5]

The large electrolyte-anode surface area is complemented by a large cathode current collector area of coated metal foil. Aluminum foil coated with either graphite or a 10-nm-thick layer of molybdenum has been used to prevent formation of nonconductive corrosion products such as Al_2S_3. The foil is interspersed among the fibers with an electrical connection to the external stainless steel cell case. The fiber-fiber spacing and the fiber-foil spacing are determined on the basis of the charge-discharge rates required from the cell and the desired cell capacity. Typical fiber-fiber spacing is 200 to 500 μm, and fiber-foil spacing is about 20 μm.

The system normally is operated at 300°C, although 290 to 320°C is an acceptable temperature range. The electrolyte ionic resistance is very sensitive to temperature. At

300°C, 90% of the internal cell resistance is due to the ionic resistance of the glass fibers. This decreases by 50% for each 25°C rise in temperature. Cell performance is therefore very temperature-sensitive. The basic case design and insulation requirements to control temperature are dictated by battery size. Because of their large surface-area-to-volume ratio, small batteries (sizes less than 40 kWh at 4-h rate) require evacuated double-walled cases with inner radiation shields and some external insulation. Larger sizes, however, require only ordinary insulation a few inches thick to control heat loss.

32.3.2 Performance Characteristics

Cells have been tested in sizes up to 40-Ah capacity having the configuration depicted in Fig. 32.2. The ultimate design goal involves scale-up to a 400-Ah size. This cell is the prototype building block to be used in eventual construction of advanced batteries. Approximately 13,000 fibers are used in the 40-Ah size.[6]

The basic cell is designed to operate at 90% energy efficiency between a sulfide composition of $Na_2S_{3.5}$ and Na_2S_{20}. The *open-circuit cell voltage* varies from 1.886 V at the end of discharge to 2.078 V at full charge. During cycling, the open-circuit voltage differs slightly from the theoretical equilibrium value because of concentration gradients. This is shown in Fig. 32.3 for a typical cell cycled at the 4-h rate.[6] The corresponding change in internal cell resistance during cycling is shown in Fig. 32.4. IR losses account for a 5% voltage drop under load and on charge. From Fig. 32.4 it can be seen that the average resistance on discharge is about 0.011 Ω for the 40-Ah cell. The area resistivity for a 40-Ah cell, based on an electrolyte area of 3000 cm², is 33 $\Omega \cdot$cm². This is about an order of magnitude greater than for comparable Na/S systems which use a ceramic electrolyte tube of β''-Al_2O_3 and not unexpected due to the higher specific resistance of the hollow glass fiber electrolyte.

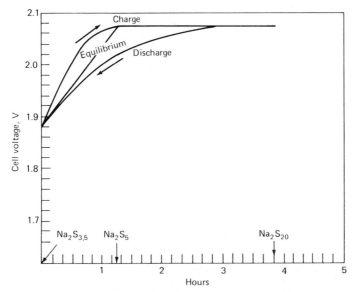

FIG. 32.3 Comparison of open-circuit voltage with equilibrium voltage (Na/S cell cycled at 4-h rate). (*From Anand, 1979.*)

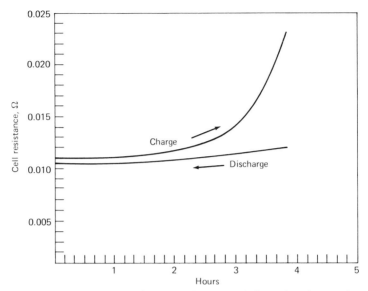

FIG. 32.4   Change in cell resistance on charge and discharge (40-Ah cell resistance). (*From Anand, 1979.*)

The *maximum power curves* for a 6-Ah cell are shown in Fig. 32.5 for two states of charge.[7] The dashed line represents a power curve obtained after the cell had been discharged less than 0.1 Ah, i.e., essentially a fully charged cell. The solid line represents a power curve obtained after the cell had discharged about 80% of its capacity. The power that can be drawn from a fully charged cell is lower because the internal cell resistance is higher due to the greater proportion of the high-resistance sulfur phase

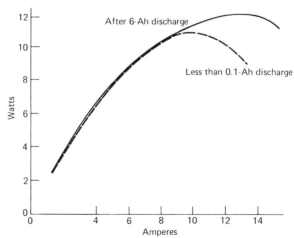

FIG. 32.5   Maximum power curves for 6-Ah Na/S cell. (*From Levine, 1980.*)

which is present. As discharge proceeds, the sulfur reacts and the internal cell resistance drops. The maximum power that can be drawn from a cell rises about 10% by the end of discharge. On the basis of the electrolyte area the data depicted on the graph correspond to an average maximum power of 0.023 W/cm^2.

The effects of *high-rate discharge* on the internal cell resistance and cell voltage are shown in Fig. 32.6 for a 6-Ah cell.[7] The initial open-circuit voltage is 2.08 V. When discharged at 17 A (cell designed for 1-A operation), the terminal voltage declines to 0.4 V. The cell temperature rises because of I^2R heating. This lowers the internal cell resistance and causes the terminal voltage to rise because of reduced IR loss. High discharge rates or overdischarge can result in the formation of supercooled polysulfide in the glass fibers. The supercooled polysulfide breaks the fibers as it solidifies. This problem is avoided by keeping the discharge cutoff above 1.79 V and operating at moderate rates.

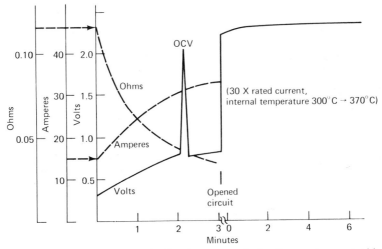

FIG. 32.6 Variation in internal cell resistance, terminal voltage, and current with time, high-rate discharge, 6-Ah Na/S cell. (*From Levine, 1980.*)

It is not possible to use open-circuit voltage as an indicator of the state of *charge* since it remains constant at 2.078 V between Na_2S_5 and the full-charge point, Na_2S_{20}. As this point is approached, however, cell resistance increases because of the increasing proportion of the nonconductive sulfur phase. The charging is terminated once the cell resistance equals twice the discharge endpoint resistance (see Fig. 32.4). Capacity mismatch in a string of cells in series can create a potential problem on charge. The lowest-capacity cell will reach the high-resistance region first. This can result in different heating rates from cell to cell. The vertical assembly of the sodium-filled glass fibers in conjunction with the aluminum cathode current collector helps to limit the problem by transferring heat to the outer cell case.

Cycle life tests have been conducted on prototype 0.5- and 6-Ah cells. The cells were cycled between 80 and 95% DOD at the 2- to 10-h rate. These cells had lifetimes in excess of 200 cycles and delivered consistent performance over this repeated cycling. The longest lifetime reported at 100% DOD was in excess of 3700 cycles; at the 1-h rate to 7% DOD, it was in excess of 40,000 cycles.[4]

32.4 CERAMIC TUBE SODIUM/SULFUR BATTERY—DESIGN 1

The sodium/sulfur battery is being developed by Ford Aerospace & Communications Corporation for electric vehicles and for stationary energy storage applications such as electric utility load leveling. The goals of this program are to develop an electric vehicle battery and to develop technology for eventual large batteries for utility applications. As with all sodium/sulfur batteries, the system operates at 300 to 350°C and requires thermal insulation. Lifetimes exceeding 5 y for electric-vehicle and 10 y for load-leveling applications are required for economic viability.

32.4.1 Design

A schematic diagram of the Ford cell is shown in Figure 32.7. The basic cell design was developed around a tubular electrolyte geometry. The anode compartment contains molten sodium which is separated from the molten sulfur cathode by a β''-alumina electrolyte tube. The β''-alumina electrolyte conducts sodium ions, but is an electronic insulator. The anode and cathode containers are sealed separately to an α-Al_2O_3 electronic insulator. The anode and cathode containers are typically fabricated from stainless steel, and the cathode compartment interior is protected by a corrosion-resistant material, such as chromium electroplate to prevent attack by molten sodium polysulfides. The cathode current collector consists of a felt and mat of fibrous graphite and carbon into which the sulfur cathode is impregnated.

All cells include a protection tube which is inserted into the electrolyte tube. It is made from stainless steel and extends from inside the sodium reservoir down to the base of the electrolyte tube. It incorporates a sodium metering device which serves to control

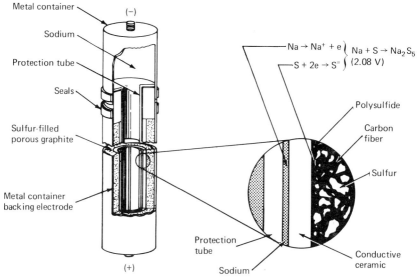

FIG. 32.7   Schematic of ceramic tube sodium/sulfur cell. (*Ford Aerospace & Communications Corp.*)

the rate of the exothermic reaction between sodium and sulfur in the event of electrolyte fracture, limiting the temperature excursions during failure to approximately 100°C above the normal operating temperature.

Seals play an important role in the cell design because of high temperatures and corrosive liquids present in the system. The cell depicted in Fig. 32.7 has two types of seals: α-Al_2O_3 to β''-alumina, and α-Al_2O_3 to the metal containers. The ceramic components are joined using a glass which must be stable in molten sodium, sulfur, and sodium polysulfides at operating temperatures of 300 to 350°C. The glass must remain electrically insulating and have a thermal expansion coefficient matching the materials it joins. The hermetic seals which join the metal containers to the α-Al_2O_3 insulator are accomplished using soft metal gaskets and radial-compression rings of high-strength steel.

Cell designs are strongly influenced by the power and energy requirements of the intended applications. A high power-to-energy ratio is required for the electric vehicle cell. Load-leveling cells, however, are designed more for energy storage than power capability, and high-energy cells are designed for extended discharge periods, such as utility applications with weekend charging or support of renewable energy sources such as solar and wind. Higher energy is achieved by adding more reactants to the cell, keeping the electrolyte area constant.

Higher power capability is achieved in electric vehicle cells by increasing the area of the electrolyte surface relative to the volume of reactants. This is accomplished by using electrolyte tubes with smaller diameter (about half the diameter of those used in load-leveling cells) and by narrowing the annular gap containing the cathode. These smaller, lighter cells have a high-power capability but store less energy than larger diameter cells which have the capacity necessary for the load-leveling and high-energy applications.

The open-circuit voltage and the operating voltage of a typical cell during charge and discharge are shown in Fig. 32.8. The OCV of a fully charged cell is approximately 2.08 V. Cells are normally discharged to an operating voltage of about 1.6 V. The voltage is relatively constant during the first 60% of the capacity and then decreases linearly during the remainder of the discharge. Abrupt voltage changes at the end of the discharge and end of charge provide effective signals for charge/discharge control circuitry.

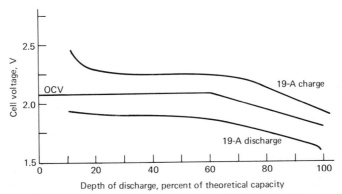

FIG. 32.8 Typical charge and discharge curves, Na/S cell. (*Ford Aerospace & Communications Corp.*)

32.4.2 Stationary Energy Storage

A development prototype of the load-leveling cell has been designated as the Mark II, illustrated in Fig. 32.9. This cell was designed for a daily duty cycle of 5-h discharge and 7-h charge. Mean performance for a group of these cells is summarized in Table 32.1. The overall cell impedance Z was calculated using the formula:

$$Z = (\overline{V}_C - \overline{V}_D) / (I_C + I_D)$$

where C and D indicate the charge and discharge portions of the cycle, the $\overline{V}$ values are average cell voltages over the period indicated, and I is the current.

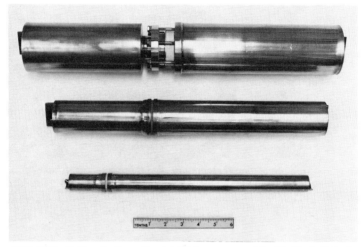

FIG. 32.9 Sodium/sulfur cells. (*a*) High-energy model. (*b*) Mark II load-leveling cell. (*c*) Electric vehicle cell. (*Ford Aerospace & Communications Corp.*)

TABLE 32.1 Performance of Mark II Cells

Parameter	Mean* value (23 cells)
Energy, Wh	286
Efficiency, 100 (Wh_o/Wh_i), %	81
Capacity, Ah	159
Utilization of sulfur, % of theoretical capacity to form Na_2S_3	82.7
Impedance, mΩ	9.9
Energy density	
Wh/kg	143
Wh/L	286

*Discharged at 24 A to 1.6 V, and charged at 18 A to 2.5 V; 340°C.
SOURCE: Ford Aerospace & Communications Corp.

A 100-kWh battery, constructed from 512 Mark II cells and composed of four series-connected 25-kWh modules, is shown in Fig. 32.10. Each module consisted of two series-connected submodules each with 64 parallel-connected cells; this 4-V module was rated for 1500 A.[9,10]. The performance of the 100-kWh battery under two test conditions is shown in Fig. 32.11. The battery delivered 102 kWh when discharged at its rated current of 1500 A.

FIG. 32.10   25-kWh module of 100-kWh load-leveling battery. (*Ford Aerospace & Communications Corp.*)

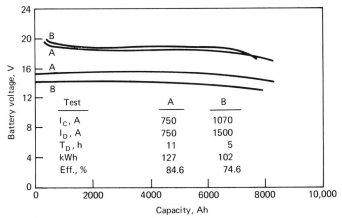

Test	A	B
I_C, A	750	1070
I_D, A	750	1500
T_D, h	11	5
kWh	127	102
Eff., %	84.6	74.6

FIG. 32.11   Charge-discharge performance of 100-kWh battery. (*Ford Aerospace & Communications Corp.*)

The mean time before failure for Mark II cells constructed early in 1980 was about 600 cycles, where failure is defined as: (1) a decline in capacity of 20% from the rated value; (2) impedance rise to 14.4 mΩ (a value representing 75% efficiency in a preliminary design for a large load-leveling battery); or (3) nonfaradaic cell behavior due to ceramic failure. Figure 32.12 compares charge and discharge curves for one of the cells early in the test (cycle 99) with similar curves approximately 1000 cycles later. The small voltage changes observed are attributed to a slight change in overall cell impedance. Weibull analysis of data from tests of the 100-kWh battery shows the characteristic life of the cells to be about 1500 days (equivalent to more than 1000 cycles when operated at this cycle rate).

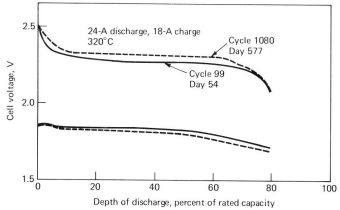

FIG. 32.12 Performance stability of Mark II cell. (*Ford Aerospace & Communications Corp.*)

A high-energy cell, rated to deliver 460 Ah over an extended discharge period, is also shown in Figure 32.9.[11] These cells utilize the Mark II size electrolytes and are designed to be discharged at the same current level as the Mark II cell, but over longer periods. The cells have containers of larger diameter to accommodate more sodium and sulfur. These cells have impedances and reactant utilization similar to those of Mark II cells.

32.4.3 Mobile Applications

Smaller cells with high power-to-energy ratios are being developed for electric vehicle (EV) applications.[11] Designs for horizontal operation are being developed to provide improved packaging flexibility. The data in Figure 32.13 are for horizontally operated cells weighing 0.5 kg and rated to deliver 35.2 Ah (66 Wh) with an efficiency (watthours, output/watthours, input) of 85% at rates of 3-h charge and 3-h discharge. An EV cell is also shown in Fig. 32.9.

Figure 32.14 shows data for 35.2-Ah EV cells of two designs discharged at currents up to 62.5 A. Corresponding test conditions are shown in Table 32.2. Even at the very high discharge rates, the cells operated at efficiencies of greater than 70%. The apparent slight improvement in the capacity and energy at the higher rates is due

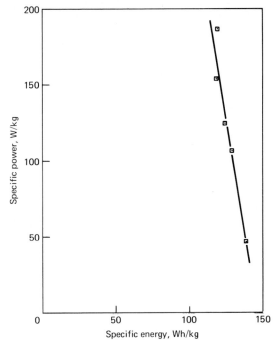

FIG. 32.13 Performance of electric vehicle cell. (*Ford Aerospace & Communications Corp.*)

to inexact adjustment of the end-of-discharge voltage limit and to slight internal heating above the 350°C nominal test temperature. Tests also were performed in which cells were operated with charge and discharge currents of equal magnitude. At equal rates as high as C/1.4, the cells exhibited an efficiency (Wh_o/Wh_i) of approximately 80%.

The EV cells are capable of very high peak power under pulse-load conditions. The response of experimental EV cells to short duration loads at various states of charge is shown in Fig. 32.15. The power values were sustained for > 15 s at cell voltages above 1.0 V. The cells can deliver a large pulse (in terms of ampere-hours) without substantially altering the average concentration gradients. Therefore, the response of these cells to pulse loads reflects primarily ohmic polarization, resulting in peak power capability exceeding the values for maximum sustained power. This figure also shows the power values normalized to electrolyte area, indicating the ability of the electrolyte to operate under conditions of high pulse power per unit area.

Figure 32.16 shows the rapid recharge of the 32.5-Ah cells. The rated vehicle range is defined as 80% of the rated cell capacity. Tests in which cells were charged at various high constant currents showed that capacity equal to the full rated vehicle range can be returned to the cells in approximately 30 min (or half of vehicle range in 15 min). No apparent degradation in cell performance due to rapid charging has been observed, but the effects on cell life are as yet unknown.

A small utility vehicle was used as a test bed for a sodium-sulfur battery.[11] Fifty-two

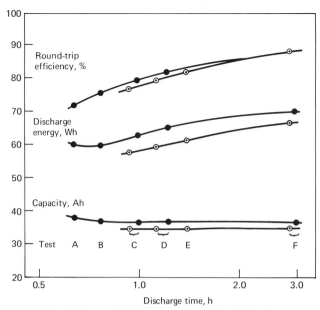

FIG. 32.14   High-rate discharge of electric vehicle cell. (*Ford Aerospace & Communications Corp.*)

TABLE 32.2   Test Conditions for High-Rate Discharge of 32.5-Ah EV Cells

Test	A	B	C	D	E	F
Current, A						
Charge	12.5	12.5	12.5	12.5	12.5	12.5
Discharge	62.5	50.0	37.5	31.2	25.0	12.5
Voltage limits, V						
Charge	2.5	2.5	2.5	2.5	2.5	2.5
Discharge	1.3	1.4	1.5	1.5	1.6	1.7

Mark II load-leveling cells were configured into a 14.4-kWh battery by series-connecting 13 submodules of four parallel-connected cells each. The cells were surrounded by an aluminum housing and 5-cm-thick conventional insulation which was covered with a polypropylene jacket. The use of conventional insulation and cells not optimized for the application resulted in a battery weight of 205 kg (compared to 115 kg for the lead-acid batteries originally supplied with the vehicle). Projections for the sodium-sulfur battery showed more than a fivefold improvement in range over that projected for the lead-acid battery, as shown in Table 32.3. Vehicle range tests were done with each battery. In each case, the total vehicle weight was 1135 kg. The sodium-sulfur battery was kept hot by utilizing 2.4 kWh of its own energy. Despite the thermal loss, the range was 117 km (10.1 h driving time) as shown in Figure 32.17. The vehicle range with the lead-acid batteries was 22 km (2.0 h driving time). Based on this preliminary test, it appears that sodium-sulfur batteries will be able to provide a full 8-h shift capability for industrial vehicles.

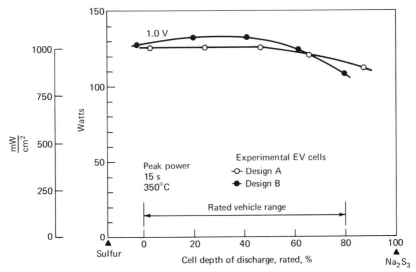

FIG. 32.15   Pulsed response of electric vehicle cells. (*Ford Aerospace & Communications Corp.*)

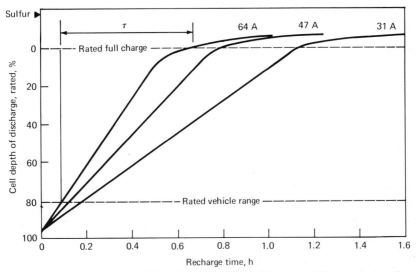

FIG. 32.16   High-rate recharge of electric vehicle cells. (*Ford Aerospace & Communications Corp.*)

32.5 CERAMIC TUBE SODIUM/SULFUR BATTERY—DESIGN 2

The second ceramic tube sodium/sulfur battery is being developed at General Electric Co. solely for electric utility load-leveling application. The ultimate goal is a 100-MWh battery facility. This system has been scaled up from laboratory-size cells with 16 Ah theoretical capacity to intermediate-size cells rated at 240 Ah.

TABLE 32.3 Projections for Utility Vehicle
Batteries

	Lead-acid	Sodium/sulfur*
Capacity, Ah	115 (1.5 h)	600 (8 h)
Energy, Wh	2800	14,400
Weight, kg	115	205
Cost, $	250	600 (goal)
Life, cy	100–500	2500 (goal)
Range, km	25 (16 km/h)	125 (16 km/h)

*Constructed with cells designed for load leveling; not optimum
for this application.

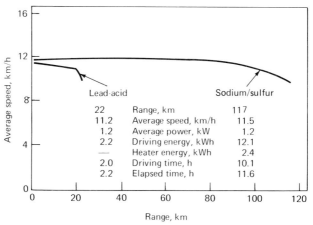

FIG. 32.17 Utility vehicle range performance. (*Ford Aerospace &
Communications Corp.*)

Although the original work centered on cells made with β-Al_2O_3 electrolyte tubes,
cost consideration redirected project efforts to concentrate on cells using β''-Al_2O_3 for
the electrolyte. Cost is lower primarily because fewer cells are required in a given system
due to a twofold decrease in resistivity associated with the β''-Al_2O_3 electrolyte. The
size of the commercial cell design also will be increased to further reduce the number
of cells used.

32.5.1 Construction

The physical layout of the General Electric cell, shown in Fig. 32.18, is similar to that
of the Ford unit. The solid-electrolyte tube contains the sodium and is connected to
an overhead anode reservoir. The cathode which surrounds the solid electrolyte is
impregnated into a carbon current collector. The sodium insert indicated at the top of
the cell is a steel tube with a 0.04-cm-diameter hole in the bottom. The tube acts as
a sodium flow restrictor which prevents rapid mixing of the reactants from taking place
in the event of tube failure. The cell case is made of a special chromized coated mild
steel for corrosion inhibition.

The solid electrolyte is attached to the α-Al_2O_3 insulating header using a special

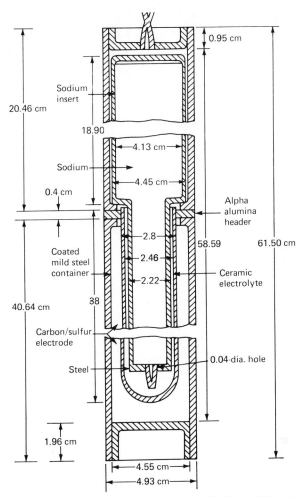

FIG. 32.18 Schematic of 168-Ah Na/S cell. (*General Electric Co.; from Mitoff and Bast, 1979.*)

sealing glass. The α-Al₂O₃ header electrically insulates the anode and cathode compartments from each other. The seal between the header and the two compartments must be hermetic to prevent contamination by atmospheric oxygen and water. The design uses a technique called "thermo-compression bonding" (TCB) to accomplish the α-Al₂O₃ ceramic-to-metal cell case seal. Aluminum, chromium-coated mild steel, and stainless steel were found to provide acceptable seals. Uncoated mild steel did not.

Aluminum and low-carbon steel appear to be suitable container materials for the sodium anode reservoir. However, the sulfur cathode compartment is subject to the highly corrosive polysulfide environment. The most corrosion-resistant materials tested were specially coated chromized mild steel and molybdenum. These are acceptable either as solid tubing or a layer on low-carbon steel tubing.

The design of the sulfur electrode current collector was shown to affect the ability of the cell to be charged into the two-phase region, thus governing complete utilization of the reactants. Improved rechargeability in the two-phase region is attributed not to enhanced free flow of reactants, but rather to a gradient in electronic resistance within the sulfur electrode. A dual carbon mat arrangement is used with a high-resistance graphite (≈ 2500 Ω·cm at 350°C) immediately adjacent to the ceramic electrolyte tube and a low-resistance graphite (2 Ω·cm) filling the rest of the electrode compartment. By using a graphite layer of higher electronic resistance toward the axial center, formation of the sulfur phase is displaced preferentially toward the outer portions of the cell, where it causes less electrical resistance.

The battery system requires thermal management. Once the battery is brought to operating temperature, inefficiencies in the electrochemical process will, in fact, necessitate cooling. Several designs were considered, including a gas-cooled module using forced nitrogen and a liquid-cooled module which used an organic-based coolant. These designs were abandoned because they were expensive and used a low-flammability liquid. In a recent design, General Electric uses a module that is cooled by natural, i.e., free, convection air circulation.[14]

32.5.2 Performance Characteristics

The open-circuit voltage and voltage under load as a function of state of charge for a 16-Ah laboratory-size cell are shown in Figs. 32.19 and 32.20. The open-circuit voltages were recorded by placing the cell off load for 6 min every hour. The charge and discharge curves for voltage under load can be approximated by three straight lines each. The open-circuit voltage curves are approximated by two straight lines each. The cells are designed to operate with an energy efficiency around 75%. The first 50% of the discharge corresponds to the formation of Na_2S_5 during the initial discharge phase; the subsequent decrease results from the formation of other polysulfide compositions. The reverse is true during the charge.

The cells operate with an ampere-hour efficiency close to 100%. The round-trip energy efficiency is a function of the discharge rate and is governed by the difference between the charge and discharge voltages. In load-leveling applications, the energy efficiency is about 75%.

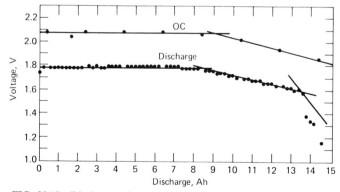

FIG. 32.19 Discharge voltage and open-circuit voltage as a function of ampere-hours. (*From Mitoff and Bast, 1979.*)

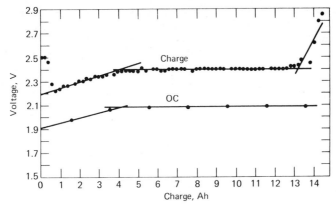

FIG. 32.20 Charge voltage and open-circuit voltage as a function of ampere-hours. (*From Mitoff and Bast, 1979.*)

The performance of cells cycled at a constant current of 2 A between 1.3 and 2.8 V is shown in Fig. 32.21. The capacity decline coincides with an increase in internal cell resistance and possible sulfur electrode degradation.

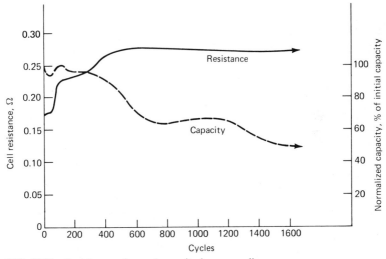

FIG. 32.21 Resistance aging and capacity loss on cycling.

The major contributors to the total internal resistance of the 16-Ah cell are shown in Fig. 32.22 as a function of the amount of capacity discharged and charged. There are four sources of cell resistance: the sodium anode, the ceramic electrolyte, the sulfur electrode, and the metal cell case. The sodium contribution is negligible. In this instance the cell case is coated with chromium, and the resistance between the coated case wall and the graphite current collector is very small. Hence, the total cell resistance is composed essentially of the sulfur electrode contribution and the electrolyte tube

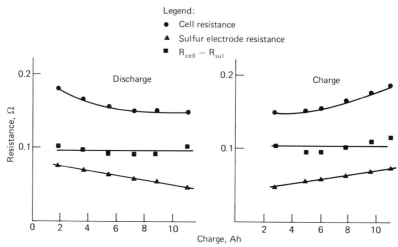

FIG. 32.22 Characteristic cell resistance. (*From Mitoff and Bast, 1979.*)

contribution. Because the electrolyte tube resistance is essentially steady during a particular cycle, the change in the shape of the total cell resistance curve is controlled by the shape of the sulfur electrode resistance curve for a cycle. The total cell resistance of 0.15 to 0.18 Ω as shown corresponds to an area resistivity of 2.8 to 3.5 $\Omega \cdot cm^2$ based on the geometric surface of the electrolyte tube.

The design of the sulfur electrode exerts a significant influence on the utilization of cathode reactants, i.e., the extent to which a cell can be cycled into the two-phase region. Modeling has resulted in the molded sulfur electrode with a graduated resistance profile which permits cycling into the two-phase region.[12, 13] Cells are cycled at constant current between preset voltage limits. Charging is terminated when the applied cell voltage reaches 3 V.

32.6 APPLICATIONS AND SYSTEM DESIGN

The sodium/sulfur battery is being developed for both mobile and stationary energy storage. Design considerations for the two types of applications are quite different. Electric vehicle batteries are constrained by weight and volume considerations and yet must deliver high power and energy to provide the desired vehicle acceleration and range at reasonable cost. For the stationary energy storage applications, power density and energy density are relatively less important, and designs are influenced more by goals of long life with low maintenance, high reliability, and low life-cycle costs.

32.6.1 System Designs for the Hollow-Fiber Na/S Battery

A hollow-fiber sodium/sulfur cell of 40-Ah capacity has been selected as the intermediate size for scale-up for the utility load-leveling applications and is also being considered as the unit cell for the electric vehicle application. A 400-Ah size cell is being considered as the building block for the proposed 10-MWh load-leveling battery shown in Fig. 32.23. It contains 20 submodules connected in parallel. Each submodule contains 400 of the 400-Ah cells connected in series. The maximum charge voltage is 1000 V. The cost of the hollow-fiber Na/S cell (bare cell cost) is estimated at $35–40 per kilowatt-hour (1980 $).

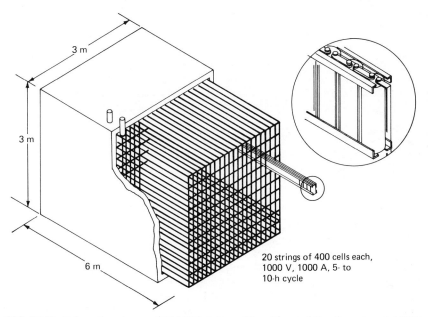

3 m

3 m

6 m

20 strings of 400 cells each, 1000 V, 1000 A, 5- to 10-h cycle

FIG. 32.23   Schematic cutaway of 10-MWh battery. (*Dow Chemical Co.; from Anand, 1979.*)

32.6.2 System Designs for the Ceramic Tube Na/S Battery—Design 1

Concepts have been developed for stationary battery systems for utility applications and for use in stand-alone solar or wind electrical power systems. Several design concepts for a vehicle battery have been considered, with testing of an experimental battery in a passenger vehicle planned for the mid-1980s.

Stationary Energy Storage Batteries   A conceptual design was completed for a 20-MW, 5-h discharge system for use in a daily discharge/charge cycle, 5 days/week.[15] The system design includes five 20-MWh unit batteries and a central control/power conversion building. The conceptual design for the 20-MWh unit battery is shown in Fig. 32.24. The illustration shows the modular units of cells stacked within the structure. One module is shown removed for servicing. The module includes a cooling air circulation manifold, a thermally insulated access cover, and a module controller enclosure. The battery unit is about 5.5 m wide by 6.2 m high by 24 m long and is thermally insulated and weatherproofed. The module controllers provide thermal control in the form of cooling air during battery discharge, electric heater power after extended idle periods, and charge equalization circuits to trim the individual modules for maximum system efficiency.

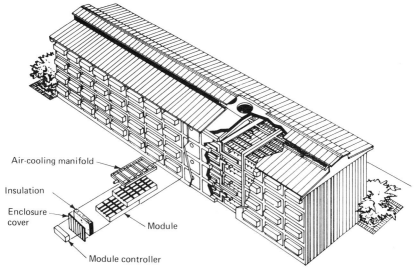

FIG. 32.24 20-MWh load-leveling battery. (*Ford Aerospace & Communications Corp.*)

Figure 32.25 shows the details of a 200-kWh module (2600 A at 16.5 V) with cells arranged in two vertically stacked trays. This module includes 960 cells arranged in 10 submodules. The 96 cells in each submodule are connected in parallel to interconnecting bus bars with fuse-link straps to disconnect any short-circuited cell from the battery circuit. The submodules are connected in series. The 20-MWh battery consists of a string of 112 modules and is designed for an average discharge voltage of 1850 V and an end-of-charge voltage of 2600 V. Cycle life is projected at 2500 cycles over

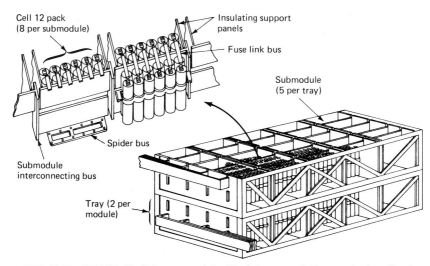

FIG. 32.25 200-kWh Na/S battery module. (*Ford Aerospace & Communications Corp.*)

10 y on a 5-h discharge/7-h charge cycle. The battery is designed to provide an average efficiency over life of 75%.

The cost for the 100-MWh load-leveling battery has been estimated at about $90 per kilowatthour (1980 $).[16]

Two additional conceptual designs have been completed for systems utilizing high-energy sodium-sulfur cells, optimized for discharge times in the range of 30 to 70 h. One design was for a small, stand-alone solar or wind electrical power system, and the second for a 10-MW utility load-leveling system for 10 h discharge per day with weekend charge recovery.

The design for the solar application is a 1.0-MWh$_e$(rated electrical energy capacity) battery, intended for use in small ($\approx$15-kW), stand-alone electrical power systems. The design incorporates about 1400 sodium-sulfur cells of a high-energy configuration. The cells are connected in parallel (18 cells/module), and the modules (78 modules/battery) are connected in series to provide a minimum discharge voltage of 120 V. Cell redundancy is such that only one module repair is expected per battery over its 10-y life, assuming a cell failure rate of 0.5%/y.[17] The battery structure is a simple steel frame mounted on nine steel piers. Six rows of modules are mounted within the frame using high-temperature mineral board to isolate the module rows. This battery core measures 2.5 m long by 2.1 m wide by 1.9 m high, and weighs about 6.3 m tons. This size permits complete assembly of the core, including cells, within a production plant. The core may then be transported by truck to the battery site. The battery enclosure is a free-standing steel structure with 46 cm of glass fiber insulation between the core and the ground, on all sides, and in the roof. The normal operating mode of the battery in a stand-alone solar power system is to provide a constant power source for overnight system use, with a full recharge the following day. The battery is sized to provide power, if necessary, for two continuous cloudy days. Following this 100% discharge duty cycle, the battery can recover its full charge in 2 to 5 days, depending upon the season and the system's generating capacity. Overall battery efficiency, including thermal losses, is expected to exceed 70% throughout the 10-y life.

The second design is based on a high-energy cell[11] for a utility battery to be used in a weekly storage cycle. The battery is rated for a 100-MWh daily discharge capacity, operated at 10 MW for 10 h/day. The battery charging schedule includes four 7-h intervals during weeknights and 42 h over the weekend. The weekly battery state-of-charge profile is shown in Figure 32.26. The arrangement of cells into submodules, and submodules into strings is similar to that of the previously described design for stationary energy storage. Each cell is connected through a fuse designed to blow at about 4 times the rated current, thus disconnecting failed cells from the submodules. Approximately 10% extra cells are provided to compensate for random cell failures, thereby reducing battery maintenance requirements. The bus bars are sized to limit their total loss to 3% of the rated discharge energy. The electrical efficiency of the battery will decline linearly over its 10-y life, and, including auxiliary equipment losses, is estimated to average 74%. The additional cell cost related to the added capacity is estimated to be from $2 to $4 per kilowatthour, with the dominant cost increment related to sodium and sulfur.

Electric Vehicle Batteries   Design concepts for a higher power battery specifically for electric vehicle applications have been evaluated. A typical concept is for a battery consisting of six 60-cell modules connected in series to provide a 120-V, 36-kWh nominal system.[11] Each module consists of six parallel strings of 10 cells in series. The

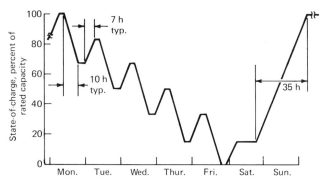

FIG. 32.26 State-of-charge profile for utility load-leveling battery with weekend charging. (*Ford Aerospace & Communications Corp.*)

configuration provides the possibility of reconnection into a 240-V battery, if desired. Each module is self-contained and provides insulated support for each cell. By orienting horizontal cells head to toe, stacking patterns can be selected so that series bus bars extend only to the adjacent cell. Hexagonal patterns provide improvement in density, despite wasted volume at irregular edges. Module designs are based on heat exchange from cell ends, with electrical feedthroughs and leads for monitoring and sensing extending downward through the insulated base platform to which the module frame is anchored. This design allows close module positioning and permits simple installation/removal. The battery enclosure consists of an insulated base platform and a vacuum "superinsulation" cover which fits around the battery assembly. The overall size of the battery plus enclosure is about 48 cm × 94 cm × 48 cm. The platform attaches to the vehicle frame and provides support to the modules. Thick (7.6-cm) standard insulation is penetrated by the structural supports, by the electrical feedthroughs, and by the air ducts required for battery cooling.

The platform design is critical in that it has a strong impact on the total volume and thermal losses of the battery. The "superinsulating" enclosure design follows the developments of load-bearing vacuum insulation.[18] Thin, high-quality insulation is essential to maintain the energy density benefits of sodium-sulfur batteries sized for EV applications.

The tentative specifications for two configurations (120 V and 240 V) of the battery are listed in Table 32.4. Iterations of battery design and vehicle projections, coupled with refined packaging constraints, are continuing. The batteries described in Table 32.4 are based on 1979–80 projections of the EV cell characteristics. Based on the actual characteristics of 1982 cells, projections of battery power capability are much improved, permitting a ratio of peak pulse power (80% DOD) to available energy (C/3) of 1.5 or greater.

32.6.3 System Design for the Ceramic Tube Na/S Battery—Design 2

The design of the 100-MWh battery system for utility load-leveling application, using the ceramic tube Na/S cell (Design 2), is shown schematically in Fig. 32.27. This system has been designed around a module which utilizes air in natural convection as

TABLE 32.4 EV Battery Electrical Performance and Preliminary Weight Estimates

	Configuration A	Configuration B
Electrical Performance		
Nominal voltage, V dc	120	240
Rated capacity at C/3, Ah	342	171
Available energy, kWh at C/3	36	36
Average impedance, mΩ	106	424
Rated discharge efficiency, %	90	90
Peak pulse power (80% DOD), kW	34.4	34.4
Preliminary Weight Estimates		
Cells (360), kg	181.5	
Thermal enclosure, kg	31.8	
Balance of system, kg	58.9	
Total, kg	272.2	
Performance Summary		
Specific energy, Wh/kg	132	
Energy density, Wh/L	165	
Specific peak power, W/kg	126	
Peak power density, W/L	158	
Round-trip electrical efficiency, %	85	
Overall efficiency, %	80	

NOTE: Data based on 1979–80 projection. A ratio of 1.5 or greater is projected for peak pulse power/available energy based on actual 1982 test data.
SOURCE: Ford Aerospace & Communications Corp.

the coolant. The 100-MWh battery will consist of 1280 modules with a total of 345,600 1.7-V, 300-Wh cells using the β''- Al_2O_3 tubes. The battery system will occupy about ½ acre and provide electricity for about 5000 homes for 5 h. The specifications and design objectives for the cell and battery system are summarized in Table 32.5.[19]

The projected cost of the 100-MWh battery system, based on 1980 $, is $95 to 115 per kilowatthour.[12]

REFERENCES

1. J. T. Kummer and N. Weber, "A Sodium-Sulfur Secondary Battery," *Automot. Eng. Cong.*, Detroit, Jan. 1–13, 1967.

2. J. B. Goodenough, H. Y.-P. Hong, and J. A. Kafalas, *Mat. Res. Bull.* **11**:203 (1976).

3. A. Topouzian, "Research on Electrodes and Electrolyte for the Ford Sodium-Sulfur Battery," Annual Report for June 30, 1976 to Oct. 31, 1977, Contract NSF-C805.

4. F. Y. Tsang, J. Anand, and C. A. Levine, "Current Status of the Hollow Fiber Sodium-Sulfur Cell," *Proc. 28th Power Sources Symp.*, June 12–15, 1978, Electrochemical Society, Pennington, N.J.

5. C. A. Levine, "Sodium-Sulfur Battery System," ERDA, Report EY-66-C-02-2565, 1976.

6. J. N. Anand, "Dow Sodium-Sulfur Battery for Energy Storage," *Proc. 19th Intersoc. Energy Conv. Eng. Conf.*, Aug. 5–10, 1979, Boston, ACS.

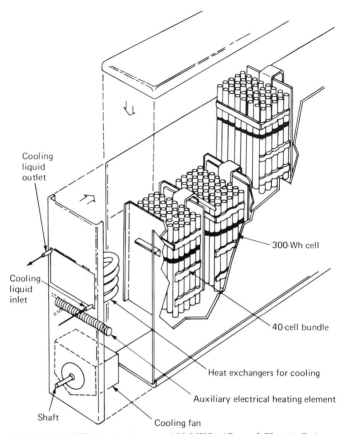

Cooling liquid outlet

Cooling liquid inlet

300-Wh cell

40-cell bundle

Heat exchangers for cooling

Auxiliary electrical heating element

Shaft

Cooling fan

FIG. 32.27   Utility battery system, 100 MWh. (*General Electric Co.*)

7. C. A. Levine, Dow Chemical Co., Walnut Creek, Calif., 1980.

8. R. A. Harlow and M. L. McClanahan, "Status of the DOE/Ford Sodium-Sulfur Battery Program," *Proc. 29th Power Sources Conf.,* Electrochemical Society, Pennington, N.J., 1980.

9. D. W. Bridges and R. W. Minck, "Evaluation of Small Sodium-Sulfur Batteries for Load-Leveling," *16th Intersoc. Energy Conv. Eng. Conf.,* Aug. 9–14, 1981, Atlanta, Paper No. 819392.

10. D. W. Bridges and H. J. Haskins, "Operation of a 100-kWh Na-S Battery," *17th Intersoc. Energy Conv. Eng. Conf.,* Aug. 8–13, 1982, Los Angeles.

11. "Sodium-Sulfur Battery Development," Final Report for Phase IV and Phase VA, for work performed under Dept. of Energy Contract No. AM02-79CH10012 by Ford Aerospace & Communications Corp., Oct., 1982.

12. S. P. Mitoff and J. A. Bast, "Development of Advanced Batteries for Utility Application," EPRI Project 128-4 and 128-5, Final Report, May, 1979.

13. M. W. Breiter and B. Dunn, "Structure of the Sulfur Electrode in Sodium-Sulfur Cells and Comparison with Model Predictions," *Proc. 28th Power Sources Symp.,* June 12–15, 1978, Electrochemical Society, Pennington, N.J.

TABLE 32.5   Specifications for 100-MWh Utility Battery System (General Electric Co.)

	Charge	Discharge
Energy storage capacity	100 MWh	
Nominal discharge power	20 MW	
Duty cycle, h	7	5
Current, avg. A	19,200	27,000
Voltage, V	994 (max.)	680 (min.)
Power, MW	17.5	20.4
Energy, MWh	130	100
Efficiency, %	75	
Lifetime	10 y, 2500 cycles	
Operating temperature, °C	300 to 350	
Weight, kg	1.9×10^6	
Dimensions, m	39 L $\times$ 31 W $\times$ 6 H	
Energy density	53 Wh/kg, 13.5 Wh/L	
Number of modules	1280	
Number of batteries	345,600	
Battery characteristics:		
Theoretical capacity, Ah	240	
Nominal capacity, Ah	168	
Wh	300	
Dimensions, cm		
Length	61	
Height	5.1	

SOURCE: Asher and Bast, 1980.

14. J. A. Asher and J. A. Bast, "GE Advanced Battery for Utility Application," General Electric Co. in-house publication.

15. "Sodium-Sulfur Battery Development," Phase IV Interim Report for work performed under Dept. of Energy Contract No. AM02-79CH10012 by Ford Aerospace & Communications Corp., Newport Beach, Calif., Report No. DOE/CH/10012-T2, Oct., 1980.

16. "Sodium-Sulfur Battery Development Economic Analysis," Phase IV topical status report for work performed under Dept. of Energy Contract No. AM02-79CH10012 by Ford Aerospace & Communications Corp., Newport Beach, Calif., Report No. DOE/CH/10012-T1, Sept., 1980.

17. H. J. Haskins and A. G. Domaszewicz, "Small Sodium-Sulfur Battery for Solar and Wind Energy Systems," *16th Intersoc. Energy Conv. Eng. Conf.,* Aug. 9–14, 1981, Atlanta, Paper No. 819393.

18. "Insulation and Enclosure Development for High Temperature Electric Vehicle Batteries," Final Report for work performed under Dept. of Energy Contract No. DE-AC02-80ET25426, by the Linde Division, Union Carbide Corp., Tonawanda, N. Y., May 1, 1982.

19. J. A. Asher and J. A. Bast, "Advanced Battery Development at General Electric," *Proc. 29th Power Sources Conf.,* 1980, Electrochemical Society, Pennington, N.J.

PART 5 RESERVE AND SPECIAL BATTERIES

33

Introduction

by
David Linden

33.1 CLASSIFICATION OF RESERVE BATTERIES

Batteries, which use highly active component materials to obtain the required high energy, high power, and/or low-temperature performance, are often designed in a reserve construction to withstand deterioration in storage and to eliminate self-discharge prior to use. These batteries are used primarily to deliver high power for relatively short periods of time after activation, in such applications as radiosondes, missiles, torpedoes, and other weapon systems. The reserve design also is used for batteries required to meet extremely long or environmentally severe storage requirements.

In the reserve structure, one of the key components of the cell is separated from the remainder of the cell until activation. In this inert condition, chemical reaction between the cell components (self-discharge) is prevented, and the battery is capable of long-term storage. The electrolyte is the component that is usually isolated, although in some water-activated batteries the electrolyte solute is contained in the cell and only water is added.

The reserve batteries can be classified by the type of activating medium or mechanism that is involved in the activation:

Water-activated batteries: Activation by fresh- or seawater.

Electrolyte-activated batteries: Activation by the complete electrolyte or with the electrolyte solvent (the electrolyte solute is contained in or formed in the cell).

Gas-activated batteries: Activation by introducing a gas into the cell. The gas can be either the active cathode material or part of the electrolyte.

Heat-activated batteries: A solid salt electrolyte is heated to the molten condition, becomes ionically conductive, thus activating the cell.

Activation of the reserve battery is accomplished by adding the missing component just prior to use. In the simplest designs, this is done by manually pouring or adding the electrolyte into the cell or placing the battery in the electrolyte (as in the case of seawater-activated batteries). In more sophisticated applications, the electrolyte storage and the activation mechanism are contained within the overall battery structure and the electrolyte is automatically brought to the active electrochemical components by remotely activating the activation mechanism. The trigger for activation can be a mechanical or electrical impulse, the shock and spin accompanying the firing of a shell or missile, etc. Activation can be completed very rapidly if required, usually in less than 1 s. The penalty for automatic activation is a substantial reduction in energy density of the battery due to the volume and weight of the activating mechanism. It is, therefore, not general practice to rate these batteries in terms of specific energy or watthours per unit weight or volume.

The gas-activated batteries are a class of reserve batteries which are activated by introduction of a gas into the battery system. There are two types of gas-activated batteries: those in which the gas serves as the cathodic active material and those in which the gas serves to form the electrolyte. The gas-activated batteries were attractive because they offered the potential of a simple and positive means of activation. In addition, because the gas is nonconductive, it can be distributed through a multicell assembly without the danger of shorting out the battery through the distribution system.

The thermal or heat-activated battery is another class of reserve battery. It employs a salt electrolyte, which is solid and, hence, nonconductive at the normal storage temperatures when the battery must be inactive. The battery is activated by heating it to a temperature sufficiently high to melt the electrolyte, thus making it ionically conductive and permitting the flow of current. The heat source and activating mechanism, which can be set off by electrical or mechanical means, can be built into the battery in a compact configuration to give very rapid activation. In the inactive stage, the thermal battery can be stored for periods of 10 y or more.

33.2 CHARACTERISTICS OF RESERVE BATTERIES

Reserve batteries have been designed using a number of different electrochemical systems to take advantage of the long unactivated shelf life achieved by this type of battery design, including conventional systems such as the Leclanche zinc-carbon and zinc-alkaline cells. Relatively few of these have achieved wide usage because of the lower capacity of the reserve structure (compared with a standard battery of the same system), poorer shelf life after activation, higher cost, and generally acceptable shelf life of active primary batteries for most applications. For the special applications that prompted their development, however, the reserve structure offers the needed advantageous characteristics.

The reserve batteries are usually designed for specific applications, each design optimized to meet the requirements of the application. A summary of the major types of reserve batteries, their major characteristics and advantages, disadvantages, and key areas of application is given in Table 33.1.

Reserve batteries employing the conventional electrochemical systems, such as the Leclanche zinc-carbon system, date back to the 1930 to 1940 period. This structure, in which the electrolyte is kept in a separate vial and introduced into the cell at the time of use, was employed as a means of extending the shelf life of these batteries, which was very poor at that time. Later, similar structures were developed using the zinc-alkaline systems. Because of the improvement of the shelf life of the primary batteries in more recent years and the higher cost and lower capacity of the reserve structure, batteries of this type never became popular.

A reserve battery that is used widely is the water-activated type. This battery was developed in the 1940s for applications such as weather balloons, radiosondes, sono-buoys, and electric torpedoes requiring a low-temperature, high-rate, or high-capacity capability. These batteries use an active electrochemical system, generally magnesium as the anode and a metal halide for the cathode, which is capable of providing the electrical performance and the reserve structure to meet the shelf life requirement. The battery is activated by introduction of water or an aqueous electrolyte. The batteries are used at moderate to high discharge rates, for periods up to 24 h after activation. More recently, lithium has been considered as an anode in an aqueous reserve battery for high-rate applications in a marine environment.

Another important reserve battery uses the zinc/silver oxide system which is noted for its high rate capability and high energy density. For missile and other high-rate applications, the cell is designed with thin plates and large surface area electrodes which increase the high rate and low-temperature capability of the battery and give a flatter discharge profile. This construction, however, reduces the activated shelf life of the cell, necessitating the use of a reserve battery design. The cells can be filled and activated manually, but for missile applications the zinc/silver oxide battery is used in an auto-matically activated design. This use requires a long period in a state of readiness (and storage), necessitating the reserve structure, a means for rapid activation, and an efficient high-rate discharge at approximately the 2- to 20-min rate. Activation is accomplished within a second by electrically firing a gas squib which forces the stored electrolyte into the cells. Shelf life of the unactivated battery is 10 y or more at 25°C storage.

The spin-activated design provides another means of activating reserve batteries using liquid electrolytes, taking advantage of the forces available during the firing of an artillery projectile. The electrolyte is stored in a vial in the center of the battery. The shock of the firing breaks or opens the vial, and the electrolyte is distributed into the annular-shaped cells by the centrifugal force of the spinning of the projectile.

Nonaqueous electrolyte systems are also used in reserve batteries to take advantage of their lower freezing points and better performance at low temperatures. The liquid ammonia battery, using liquid ammonia as the electrolyte solvent, has been employed as the power source for fuzes which require a shelf life in excess of 10 y. The battery is operable at cold as well as normal temperatures with little change in cell voltage and energy output. The battery typically uses a magnesium anode, a meta-dinitrobenzene-carbon cathode, and an electrolyte salt system based on ammonium and potassium thiocyanate. Activation is accomplished by igniting a gas generator which forces the electrolyte into the cells. Depending on the application, the battery can be designed for efficient discharge for several minutes or up to 50 or more hours of service.

More recently, the lithium anode electrochemical system is being developed in reserve configurations to take advantage of its high energy density and good low-temperature performance. These cells use either an organic electrolyte or a nonaqueous inorganic electrolyte because of the reactivity of lithium in aqueous electrolytes. Even though the active lithium primary cells are noted for their excellent storability, the reserve structure is used to provide a capability of essentially no capacity loss even after storage periods in the inactive state of 10 y or more. The performance characteristics

TABLE 33.1 Characteristics of Reserve Batteries

System	Conventional systems		Water-activated batteries
General characteristics	Conventional cylindrical cells in reserve design (electrolyte separated in cell during storage).		Battery activated by adding or placing battery in water.
Advantages	Low-cost, available materials; reserve structure extends shelf life.		High energy density; moderate to high rate capability; good low-temperature performance after activation; simple designs; easy activation.
Disadvantages/ limitations	Lower capacity than conventional active cells; low to moderate discharge rates.		Rapid self-discharge after activation; AgCl system is expensive.
Chemistry:			
Anode	Zn	Zn	Mg
Cathode	MnO_2	HgO	AgCl, CuCl, MnO_2, $PbCl_2$, and others
Electrolyte	Salt	KOH	H_2O, seawater, aqueous solutions
Nominal cell voltage, V	1.5		1.5–1.6
Performance characteristics: Operating temperature, °C	0 to 50		−60 to 65 (after activation) Performance almost independent of ambient temperature after activation.
Energy density, Wh/kg Wh/L	30 60	(at moderate rates)	AgCl 100 Others 45–80 200 50–100
Status	Zinc-carbon system obsolete; some use of alkaline systems		In production
Major applications	Mines		Marine applications (torpedoes, sonobuoys); air-sea rescue; emergency lights; weather balloons
Representative manufacturers	Duracell, Inc.		Magnavox Co. Eagle-Picher Industries

Lithium/water batteries	Zinc/Silver Oxide Batteries	
	Manually activated	Automatically activated
Primary reserve system, depending on controlled reaction of Li with H_2O.	Battery activated by adding KOH electrolyte just prior to use.	Electrolyte separately stored in battery; built-in device to automatically activate from remote or local position.
High energy density.	Highest capacity of practical aqueous systems for high-rate use.	High capacity, no maintenance; automatic activation. Excellent unactivated shelf life.
Need to control Li reaction with H_2O; complex system and controls.	Manual activation is inconvenient and undesirable for field use; low temperature performance is poor.	Activation device reduces energy density; costly, but warranted for special applications.
Li H_2O, H_2O_2,O_2, AgO	Zn AgO, Ag_2O	Zn AgO, Ag_2O
H_2O, LiOH	KOH	KOH
2.2	1.6	1.6
0 to 30	0 to 60	0 to 60 ($<0°C$ operation with heaters)
160 (at 20-h 135 rate)	60–80 (at high 100–160 rates)	20–50 (at high 100–200 rates)
In development	In production	In production
Marine applications (torpedoes, sonobuoys, submersibles)	Special applications requiring high-rate, high-capacity batteries	Missile applications
Lockheed Corp.	Eagle-Picher Industries Yardney Electric Corp.	Eagle-Picher Industries Yardney Electric Corp.

TABLE 33.1 Characteristics of Reserve Batteries *(Continued)*

System	Spin-activated batteries	Lithium-nonaqueous batteries	Liquid ammonia batteries
General characteristics	Electrolyte separately stored in battery; activated by shock and spin of projectiles.	Battery activated by introducing liquid electrolyte (contained in ampul) into battery system.	Battery activated by introducing liquid NH_3 into battery system (NH_3 can be stored in ampul in battery).
Advantages	Excellent unactivated shelf life; convenient, reliable, rapid "built-in" activation.	High energy density; wide operating temperature range; flat discharge profile; excellent unactivated storage; high to low discharge rate capability.	Wide operating temperature range; high and moderate rate applications.
Disadvantages/ limitations	Activation device reduces energy density.	Reserve structure has lower energy density than active primary systems.	

Chemistry:

Anode	Pb	Zn	Li	Li	Li	Li	Mg	
Cathode	PbO_2	AgO	$SOCl_2$	V_2O_5	SO_2	$SOCl_2$	m·DNB	
Electrolyte	HBF_4	KOH	$SOCl_2$	Organic	Organic	$SOCl_2$	NH_4SCN, $KSCN(NH_3)$	
Nominal cell voltage, V	2.0	1.6	3.5	3.3	3.0	3.5	2.2	

System	Spin-activated batteries	Lithium-nonaqueous batteries	Liquid ammonia batteries
Performance characteristics: Operating temperature, °C	−40 to 60 (For HBF_4 system, other systems may require heating for low-temperature operation.)	−55 to 70	−55 to 70
Energy density, Wh/kg	—	50–150 (depending on	45 (at high 60 (at low
Wh/L	—	100–300 battery system)	100 rates) 130 rates)
Status	In production	In production	In production
Major applications	Artillery and spin stabilized projectiles—fuzing control or arming	Mine fuzing	Mine fuzing, missiles
Representative manufacturers	Globe Union, Inc. Honeywell, Inc.	Honeywell, Inc.	Honeywell, Inc. Tadiran Israel E I Phillips US FABV

Gas-Activated Batteries		Thermal batteries		
Chlorine depolarized	Ammonia-vapor-activated (AVA)			
Battery activated by introducing chlorine gas to act as the depolarizer.	Battery activated by introducing ammonia gas to form the electrolyte with salt already in the battery.	Battery activated by heating to a temperature sufficient to melt solid electrolyte, making it conductive.		
Potential for high-rate, high-capacity, good low-temperature performance; simple activation even at $-20°C$.	Potential for good low-temperature performance; simple activation; excellent unactivated shelf life; high and moderate rate applications.	Performance independent of ambient temperature; rapid activation; excellent unactivated shelf life.		
Short shelf life even in unactivated condition.	Activation slow and nonuniform.	Short lifetime; activation device reduces energy density. New designs, using Li anode, however, has higher energy density and longer lifetime.		
Zn Cl_2	Zn PbO_2	Ca, $CaCrO_4$	Mg V_2O_5	Li FeS_2
Salt $(CaCl_2, ZnCl_2)$	$NH_4SCN(NH_3)$	LiCl/KCl	LiCl/KCl	LiCl/KCl
1.5	1.9	2.22–2.6	2.2–2.7	1.6–2.1
-20 to 50	-55 to 75	-55 to 75		
40 60	25 50	10 $\}$ (for Ca up to 30 $\}$ batteries)		25 $\}$ (for Li 80 $\}$ batteries)
Development effort terminated	Effort redirected to liquid ammonia batteries	In production		
		Military ordnance (projectiles, rockets, missiles, fuzing)		
Aerojet-General Corp.	Honeywell, Inc. Eastman Kodak Co.	Catalyst Research Corp. Eagle-Picher Industries SAFT		

of the reserve battery, once activated, are similar to those of the active lithium batteries, but with a penalty of 50% or more in energy density due to the need for the activation device and the electrolyte reservoir.

The gas-activated batteries were attractive because their activation was potentially simpler and more positive than liquid or heat activation. The ammonia vapor–activated (AVA) battery was representative of a system in which the gas served to form the electrolyte. (Solids such as ammonium thiocyanate will absorb ammonia rapidly to form electrolyte solutions of high conductivity.) In practice, ammonia vapor activation was found to be slow and nonuniform, and the development of the ammonia battery was directed to liquid ammonia activation.[1] The chlorine-depolarized zinc/chlorine battery was representative of the gas depolarizer systems. This battery used a zinc anode, a salt electrolyte, and chlorine, which was introduced into the cell at the time of use, as the active cathode material. The battery was designed for very high rate discharge ranging from 1 to 5 min, but its poor shelf life while inactivated limited further development and use.[2]

The thermal battery has been used extensively in fuzes, mines, missiles, and nuclear weapons which require an extremely reliable battery that has a very long shelf life, can withstand stress environments such as shock and spin, and has the ability to develop full voltage rapidly, regardless of temperature. The life of the battery after activation is short—the majority of applications are high rate and require only 1 to 10 min of use—and is primarily dependent on the time the electrolyte can be maintained above its melting point. The energy density of the thermal battery is low; in this characteristic it does not compare favorably with other batteries except at the extremely high discharge rates. New designs, using lithium anode, have resulted in a significant increase in the energy density as well as an increase in the discharge time to 30 to 60 min.

REFERENCES

1. John M. Freund and William C. Spindler, "Low Temperature Nonaqueous Cells," in G. W. Heise and N. C. Cahoon (eds.), *The Primary Battery,* vol. I, Wiley, New York, 1971, chap. 10.

2. E. M. Langworthy, C. L. Randolph, and H. E. Lawson, "Chlorine-Depolarized Battery," *Proc. 11th Annu. Battery Res. Dev. Conf.,* Electrochemical Society, Pennington, N.J., 1957.

34

Water- Activated Batteries

by
Ralph F. Koontz

34.1 GENERAL CHARACTERISTICS

The water-activated battery is a popular reserve battery and was first developed in the 1940s to meet a need for a high energy density, long shelf life battery, with good low-temperature performance, for military applications.

The battery is constructed dry, stored in the dry condition, and activated at the time of use by the addition of water or an aqueous electrolyte. Most of the water-activated batteries use magnesium as the anode material; several cathode materials have been used successfully in different types of designs and applications.

The magnesium/silver chloride seawater-activated battery was developed by Bell Telephone Laboratories as the power source for electric torpedoes.[1] This work, coupled with that of the General Electric Company, resulted in the development of small, high energy density batteries readily adaptable for use as the power source for sonobuoys, electric torpedoes, weather balloons, air-sea rescue equipment, pyrotechnic devices, marine markers, and emergency lights.

The magnesium/cuprous chloride system became commercially available in 1949.[2,3] Compared with the magnesium/silver chloride battery, this system has lower energy density, lower rate capability, and less resistance to storage at high humidities, but significantly lower cost. Although the magnesium/cuprous chloride system can be used for the same purposes as the magnesium/silver chloride battery, its primary use is in airborne meteorological equipment where the use of the more expensive silver chloride system is not warranted. The cuprous chloride system does not have the physical or electrical characteristics required for use as the power source for electric torpedoes.

Because of the high cost of silver and its irrecoverable loss during use and the

34-1

disadvantages of the magnesium/cuprous chloride system, nonsilver water-activated batteries were developed recently primarily as the power source for antisubmarine warfare (ASW) equipment.

The systems which have been developed and used successfully are the magnesium/lead chloride,[4] magnesium/cuprous iodide-sulfur,[5-7] magnesium/cuprous thiocyanate-sulfur,[8] and magnesium/manganese dioxide utilizing an aqueous magnesium perchlorate electrolyte.[9-11] None of these systems can compete with the magnesium/silver chloride system in almost every attribute except cost. The advantages and disadvantages of these systems are given in Table 34.1.

TABLE 34.1 Comparison of Silver and Nonsilver Cathode Batteries

Advantages	Disadvantages
Silver chloride cathodes	
Reliability	High raw material costs
Simplicity	Not rechargeable
Ease of fabrication	High rate of self-discharge after activation
100% active material	
Wide voltage range	
High cell voltage	
High current densities	
High energy density	
Good response to pulse loading	
Instantaneous activation	
Long unactivated shelf life	
No maintenance	
Nonsilver cathodes	
Abundant domestic supply	Not 100% active material
Low raw material cost	Requires supporting conductive grid
Instantaneous activation	Operates at low current densities
Reliability	Low energy density
Long unactivated shelf life	Not rechargeable
No maintenance	High rate of self-discharge after activation

34.2 CHEMISTRY

The principal overall and current-producing reactions for the water-activated batteries currently available are as follows:

1. Magnesium/silver chloride

Anode $\quad Mg - 2e \rightarrow Mg^{2+}$

Cathode $\quad 2AgCl + 2e \rightarrow 2Ag + 2Cl^-$

Overall $\quad Mg + 2AgCl \rightarrow MgCl_2 + 2Ag$

2. Magnesium/cuprous chloride

Anode $\quad Mg - 2e \rightarrow Mg^{2+}$

Cathode $\quad 2CuCl + 2e \rightarrow 2Cu + 2Cl^-$

Overall $\quad Mg + 2CuCl \rightarrow MgCl_2 + 2Cu$

3. Magnesium/lead chloride
   Anode $Mg - 2e \rightarrow Mg^{2+}$
   Cathode $PbCl_2 + 2e \rightarrow Pb + 2Cl^-$
   Overall $Mg + PbCl_2 \rightarrow MgCl_2 + Pb$

4. Magnesium/cuprous iodide, sulfur
   Anode $Mg - 2e \rightarrow Mg^{2+}$
   Cathode $Cu_2I_2 + 2e \rightarrow 2Cu + 2I^-$
   Overall $Mg + Cu_2I_2 \rightarrow MgI_2 + 2Cu$

5. Magnesium/cuprous thiocyanate, sulfur
   Anode $Mg - 2e \rightarrow Mg^{2+}$
   Cathode $2CuSCN + 2e \rightarrow 2Cu + 2SCN^-$
   Overall $Mg + 2CuSCN \rightarrow Mg(SCN)_2 + 2Cu$

6. Magnesium/manganese dioxide
   Anode $Mg - 2e \rightarrow Mg^{2+}$
   Cathode $2MnO_2 + H_2O + 2e \rightarrow Mn_2O_3 + 2OH^-$
   Overall $Mg + 2MnO_2 + H_2O \rightarrow Mn_2O_3 + Mg(OH)_2$

A side reaction also occurs between the magnesium anode and the aqueous electrolyte resulting in the formation of magnesium hydroxide, hydrogen gas, and heat.

$$Mg + 2H_2O \rightarrow Mg(OH)_2 + H_2$$

In immersion-type batteries the hydrogen evolved creates a pumping action which helps purge the insoluble magnesium hydroxide from the battery. Magnesium hydroxide remaining within a cell can fill the space between the electrodes which can become devoid of electrolyte, prevent ionic flow, and cause premature cell and battery failure.

The heat evolved improves the performance of immersion-type batteries; it enables dunk-type batteries to operate at low ambient temperatures and forced-flow batteries to operate at high current densities.

Those cathodes containing sulfur exhibit a higher potential vs. magnesium than cathodes possessing only the prime depolarizer. During discharge the sulfur probably reacts with the highly active copper formed when the prime depolarizer is reduced producing a copper sulfide, thus accounting for the fact that no copper is observed at end of discharge. This reaction may also prevent copper from plating out on the magnesium, thus deterring premature voltage drop. In those cases where the battery is allowed to discharge past the point where all prime depolarizer is gone and magnesium is present, hydrogen sulfide can be produced. Hydrogen sulfide can also result if the cell is shorted.

34.3 TYPES OF WATER-ACTIVATED BATTERIES

Water-activated batteries are manufactured in the following basic types:

1. *Immersion* batteries are designed to be activated by immersion in the electrolyte and have been constructed in sizes to produce from 1.0 V to several hundred volts at currents up to 50 A. Discharge times can vary from a few seconds to several days. A typical immersion-type water-activated battery is shown in Fig. 34.1.

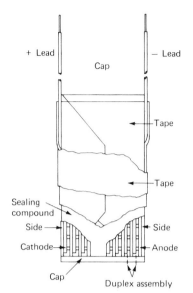
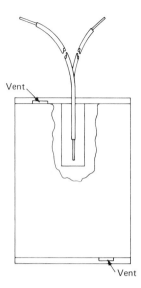

FIG. 34.1 Seawater battery, immersion type.

2. *Forced-flow* batteries are designed for use as the power source for electric torpedoes. The name is derived from the fact that seawater is forced through the battery as the torpedo is driven through the water. Because of the heat generated during discharge and electrolyte recirculation, these systems can perform at current densities up to 500 mA/cm² of cathode surface area. Batteries containing from 118 to 460 cells which will produce from 25 to 460 kW of power are currently in use. Discharge times are about 10 to 15 min. A diagrammatic representation of a torpedo battery and a torpedo battery with recirculation voltage control is shown in Fig. 34.2.[12]

3. *Dunk-type* batteries are designed with an absorbent separator between the electrodes and are activated by pouring the electrolyte into the battery, where it is absorbed by the separator. Batteries of this type have been designed to produce from 1.5 to 130 V at currents up to about 10 A. Lengths of discharge vary from about 0.5 to 15 h. Figure 34.3 is a diagrammatic representation of a magnesium/cuprous chloride battery used in radiosonde applications. A pile-type construction is used. A sheet of magnesium is separated from the cuprous chloride cathode by a porous separator which also serves to retain the electrolyte. The cathode is a pasted type made by applying a paste of powdered cuprous chloride and a liquid binder onto a copper grid or screen. The assembly is taped together to form the battery. The batteries are also made in spiral or jelly-roll design.

34.4 CONSTRUCTION

Water-activated cells consist of an anode, a cathode, a separator, terminations, and some form of encasement. A battery consists of a multiplicity of cells connected in series or series-parallel. Such an assembly requires a method to connect the cells in the desired

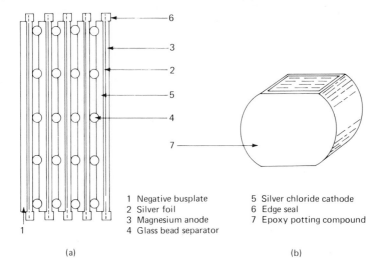

1 Negative busplate   5 Silver chloride cathode
2 Silver foil   6 Edge seal
3 Magnesium anode   7 Epoxy potting compound
4 Glass bead separator

(a)   (b)

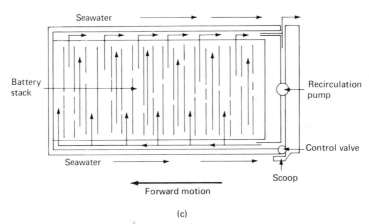

(c)

FIG. 34.2   Diagrammatic representation of torpedo battery construction: (a) cell construction; (b) battery configuration; (c) recirculation voltage control.

configuration plus a method to control leakage currents. The voltage of a cell depends primarily upon the electrochemical system involved. To increase voltage, a number of cells must be connected in series. The capacity of a cell in ampere-hours is primarily dependent upon the quantity of active material in the electrodes. The ability of a cell to produce a given current at a usable voltage depends upon the area of the electrode. To decrease current density so as to increase load voltage, the electrode area must be increased. Power output depends upon the temperature and salinity of the electrolyte. Power output can be increased by increasing the temperature and/or the salinity of the electrolyte.

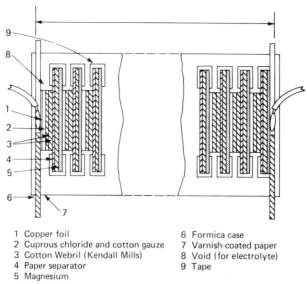

1 Copper foil
2 Cuprous chloride and cotton gauze
3 Cotton Webril (Kendall Mills)
4 Paper separator
5 Magnesium

6 Formica case
7 Varnish-coated paper
8 Void (for electrolyte)
9 Tape

FIG. 34.3   Diagrammatic representation of magnesium/cuprous chloride dunk-type battery. (*Courtesy of Eagle-Picher Industries, Inc.*)

The basic components of a single cell, a duplex assembly for connecting cells in series, and a finished battery are illustrated in Figs. 34.4, 34.5, and 34.1, respectively.[13–16] The illustrations represent a battery designed for use by immersion in the electrolyte as contrasted to a dunk-type (radiosonde) battery, which is activated by pouring the electrolyte into the battery, or a forced-flow electric torpedo battery. The construction principles with slight variations are similar in all cases.

34.4.1 Components

A more-detailed description of the various cell and battery components and construction elements is as follows.

Anode (Negative Plate)   The anode is made from sheet magnesium. Magnesium AZ61A is preferred because it tends to sludge and polarize less. In some cases AZ31B alloy is used; however, this alloy gives slightly lower voltage, polarizes at high current densities, and sludges more. In recent years magnesium alloys AP-65 and MTA-75 have been developed and evaluated. These are high-voltage alloys giving load voltages from 0.1 to 0.3 V higher than AZ61A. MTA-75 is a higher-voltage alloy than AP-65. These alloys sludge more; however, under some forced-flow discharge conditions, the sludging problem may be controlled. These alloys are not used extensively in the United States; however, they are used in the United Kingdom and Europe in electric torpedo batteries. Composition ranges of these alloys are shown in Table 34.2.[12]

Zinc can be used as the anode in low-current, low-power, long-life water-activated batteries. Zinc has the advantage of not sludging, but the disadvantage of being an extremely low power density system. Zinc/silver chloride seawater batteries have been

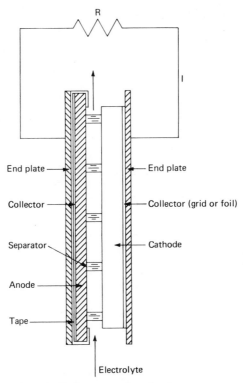

FIG. 34.4   Basic water-activated cell.

used as the power source for repeaters for submarine telephone cables (for example, 5 mA at 0.9 to 1.1 V for 1 y).

Attempts have been made to use aluminum as the anode metal; however, the presence of the insulating oxide film on the metal surface has not as yet been successfully overcome.[17-19]

Cathode (Positive Plate)   The cathode consists of a depolarizer and a current collector. These depolarizers are powders and are nonconductive. In order for the depolarizer to function, a form of carbon is added to impart conductivity; a binder is added for cohesion and a metal grid is used as a current collector, a base for the cathode, to facilitate intercell connections and battery terminations. Possible cathode formulations are shown in Table 34.3.[1,3-5,8]

Silver chloride is a special case. Silver chloride can be melted, cast into ingots, and rolled into sheet stock in thicknesses from about 0.08 mm up. Since this material is malleable and ductile, it can be used in almost any configuration. Silver chloride is nonconductive and is made conductive by superficially reducing the surface to silver by immersion in a photographic developing solution. No base grid need be used with silver chloride.

Nonsilver cathodes are usually prismatic in shape and are flat. Silver chloride cathodes are used flat and corrugated in many configurations.

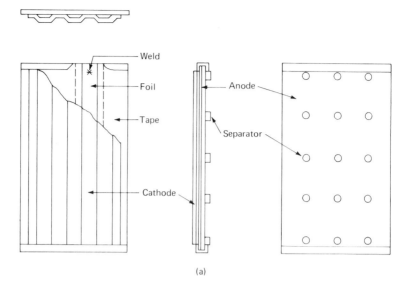

(a)

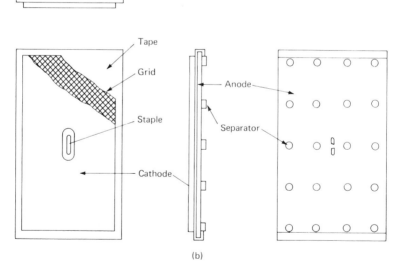

(b)

FIG. 34.5   (*a*) Silver and (*b*) nonsilver duplex electrode assemblies.

Separators   Separators are nonconductive spacers placed between the electrodes of immersion- and forced-flow-type batteries to form a space for free ingress of electrolyte and egress of corrosion products. Separators in the form of disks, rods, glass beads, or woven fabrics may be used.[13,14]

Dunk-type batteries utilize a nonwoven, absorbent, nonconductive material for the dual purpose of separating the electrodes and absorbing the electrolyte.

TABLES 34.2   Composition Range for Battery Plate Alloys

Element	AZ31 % Min.	AZ31 % Max.	AZ61 % Min.	AZ61 % Max.	AP65 % Min.	AP65 % Max.	MELMAG 75 % Min.	MELMAG 75 % Max.
Al	2.5	3.5	5.8	7.2	6.0	6.7	4.6	5.6
Zn	0.6	1.4	0.4	1.5	0.4	1.5	. . .	0.3
Pb	. . .	. . .	. . .	. . .	4.4	5.0	. . .	. . .
Tl	. . .	. . .	. . .	. . .	. . .	. . .	6.6	7.6
Mn	0.15	0.7	0.15	0.25	0.15	0.30	. . .	0.25
Si	. . .	0.1	. . .	0.05	. . .	0.3	. . .	0.3
Ca	. . .	0.04	. . .	0.3	. . .	0.3	0.3	. . .
Cu	. . .	0.05	0.05	0.05	0.05	. . .	. . .	. . .
Ni	. . .	0.005	. . .	0.005	. . .	0.005	. . .	0.005
Fe	. . .	0.006	. . .	0.006	. . .	0.010	. . .	0.006

TABLE 34.3   Cathode Compositions

	Silver chloride[1]	Cuprous iodide[5,6]	Cuprous thiocyanate[8]	Lead chloride[4]	Cuprous chloride
Depolarizer, %/w	100	73	75–80	80.7–82.5	95–100
Sulfur, %/w	. . .	20	10–12	. . .	. . .
Additive, %/w	. . .	. . .	0–4	2.3–4.4	. . .
Carbon, %/w	. . .	7	7–10	9.6–9.8	. . .
Binder, %/w	. . .	. . .	0–2	1.5–1.6	0–5
Wax, %/w	. . .	. . .	. . .	3.8	. . .

Intercell Connections   In a series-arranged battery of pile construction, the anode of one cell is connected to the cathode of the adjacent cell. To accomplish this without producing a shorted cell, an insulating tape or film is placed between the electrodes on nonsilver batteries. For silver batteries silver foil alone or in conjunction with an insulating tape is used.

For nonsilver cells the connection is made by stapling the electrodes together through the insulator.[15] For silver cells, the silver chloride, surface-reduced to silver, is heat-sealed to silver foil which has been previously welded to the anode. Where large surface areas are involved, contact between the silver and silver foil can be made by pressure alone.

Terminations   For silver chloride cathodes the lead is soldered directly to silver foil which has been heat-sealed to one surface of the silver chloride.

Leads are soldered directly to the collector grid of nonsilver cathodes or soldered to a piece of copper foil which has been stapled to the collector grid.

The anode connection is made by soldering the lead to silver foil which has been welded to the anode, or by welding directly to the anode.

Encasement   The battery encasement must effectively rigidize the battery and provide openings at opposite ends to allow free ingress and egress of electrolyte and corrosion products.

The periphery of the battery must be sealed in such a manner that the cells contact the external electrolyte only at the openings provided at the top and bottom of the battery. The encasement can be accomplished by using premolded pieces, caulking compounds, epoxy resins, an insulating sheet, or hot-melt resins.[13–16] For single cells these precautions are not necessary.

34.4.2 Leakage Current

All the cells in the immersion- and forced-flow-type batteries operate in a common electrolyte. Since the electrolyte is conductive and continuous from cell to cell, conductive paths exist from each point in a battery to every other point. Current will flow through these conductive paths to points of different potential. This current is referred to as "leakage current" and is in addition to current flowing through the load. Electrodes must be designed to compensate for these leakage currents.

Leakage currents for a small number of cells can be reduced by increasing the resistance path from a cell to the common electrolyte or that of the common electrolyte between adjacent cells. Leakage currents for a large number of cells can be reduced by increasing the resistance of the common electrolyte external to the individual cells.

By construction the conducting paths from cell to cell are made as long as possible. In many instances the negative or positive of the battery is connected to an external metal surface. Leakage currents flow from the battery to this surface. These leakage currents are controlled by placing a cap containing a slot over the battery openings. If one terminal is connected to an external conductive surface, the slot in the cap is opened to the electrolyte only on that side of the battery. Where neither terminal is connected to an external conductive surface, either end of the cap may be opened; however, only those on one side of the battery should be opened.

The resistance (ohms) of the slot in the cap may be calculated using the formula

$$R = p\frac{l}{a}$$

where R = resistance, Ω
l = length of slot, cm
a = cross-sectional area of slot, cm^2
p = resistance of electrolyte for temperature and salinity in which battery is operating, $\Omega \cdot$cm

For dunk-type batteries the electrolyte continuity from cell to cell is broken when the electrolyte is absorbed in the separator; the excess is poured off the battery or spun away from the cells by some external force applied to the battery.

34.4.3 Electrolyte

Seawater-activated batteries are designed to operate in an infinite electrolyte, namely, the oceans of the world. However, for design, development, and quality control purposes, it is not practical to use ocean water. Thus it is common practice throughout the industry to use a simulated ocean water conforming to ASTM D1141-52 diluted to the required salinity and adjusted to the required pH. A commercial product composed of a blend of all the ingredients required simplifies the manufacture of simulated ocean water test solutions.

Dunk-type batteries activated by pouring the electrolyte into the battery where it is absorbed by the separator can utilize water or seawater when the temperature is above freezing. At lower temperatures special electrolytes can be used. The use of a conducting aqueous electrolyte will result in faster voltage buildup. However, the introduction of salts in the electrolyte will increase the rate of self-discharge.

34.5 PERFORMANCE CHARACTERISTICS

34.5.1 General

A summary of the performance characteristics of the major water-activated batteries currently available is given in Table 34.4. A 10-day humidity cycle is given in Table 34.5.

Voltage vs. Current Density   Figures 34.6 and 34.7 are representative voltage vs. current density curves for several of the water-activated battery systems at 35 and 0°C, respectively, using a simulated ocean water electrolyte.

Discharge Curves   Discharge curves of the magnesium/silver chloride, magnesium/ cuprous thiocyanate-sulfur, magnesium/cuprous iodide-sulfur, and magnesium/lead chloride electrochemical systems, discharged continuously through various resistances in simulated ocean water and high and low temperatures and salinities, are shown in

TABLE 34.4   Performance Characteristics—Water-Activated Batteries

	Silver chloride	Lead chloride	Cuprous iodide	Cuprous thiocyanate	Cuprous chloride[a]
Cathode					
Anode			Magnesium		
Electrolyte	Tapwater, seawater, or other conductive aqueous solutions				
OCV, V	1.6–1.7	1.1–1.2	1.5–1.6	1.5–1.6	1.5–1.6
V per cell at 5 mA/cm^2[b]	1.42–1.52	0.90–1.06	1.33–1.49	1.24–1.43	1.2–1.4
Activation:					
+35°C[c]	<1	<1	<1	<1	
RT[d]	. . .	. . .	. . .	. . .	1–10
0°C[e]	45–90	45–90	45–90	45–90	
Internal resistance, Ω[f]	0.1–2	1–4	1–4	1–4	2
Ah/g cath. theor.[g]	0.187	0.193	0.141	0.220	0.271
Usable cap. % of theor.			60–75		
Wh/kg	100–150	50–80	50–80	50–80	50–80
Wh/L	180–300	50–120	50–120	50–120	20–200
Operating temp., °C[h]			−60 to +65		
Storage:					
Ambient			Indefinite if stored in dry condition		
−60 to + 70°C			5 y		
Per MIL-T-5422E			See Table 34.5		
−50 to +40°C at 90% RH			90 days		
Shock, g			18–400		
Vibration, Hz			5–500		
Availability			Batteries are designed for specific applications		

[a]All but cuprous chloride are immersion type. Cuprous chloride is dunk type.
[b]See voltage vs. current density curves.
[c]Battery preconditioned at +55°C, then immersed in simulated ocean water of 3.6% by wt.
[d]Electrolyte at room temperature poured into battery and absorbed by separator.
[e]Battery preconditioned at −20°C, then immersed in simulated ocean water of 1.5% by wt.
[f]Depends upon battery design.
[g]100% active material.
[h]Following activation at room temperature.

TABLE 34.5 10-Day Humidity Cycle, MIL-T-5422E

In ready-for-use condition, out of packing, the battery shall be capable of withstanding the following humidity cycle:

Step 1. Place battery in appropriate environmental chamber capable of maintaining the following conditions:
   a. Air velocity shall not exceed 45.7 m/s.
   b. Moisture by steam or evaporated tap water possessing a pH of 6.5 to 7.5 at 25°C.
   c. Temperature change shall not exceed 1°C/s.
   d. Altitude change shall not exceed 12.7 mm Hg/s.
Step 2. Increase temperature of chamber to +50°C during 2-h period, maintaining relative humidity greater than 95%.
Step 3. Maintain 50°C $\pm$2°C at a relative humidity of 95 +5 −0% for 6 h.
Step 4. During next 16 h drop temperature gradually to 38°C, keeping relative humidity as high as possible and not allowing it to drop below 85%.
Step 5. Repeat Steps 2, 3, and 4 for a total of 10 cycles (240 h).

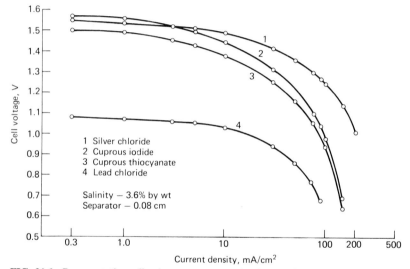

FIG. 34.6 Representative cell voltages vs. current density at 35°C.

Figs. 34.8 to 34.15. These data show the advantageous performance of the silver chloride system.

Service Life The capacities per unit of weight vs. the average power output of these same electrochemical systems, at high and low temperatures and salinities, are shown in Figs. 34.16 and 34.17, respectively.

34.5.2 Immersion-Type Batteries

The performance of these same systems, designed as immersion-type batteries to meet the physical, electrical, and environmental specifications listed in Table 34.6, are shown in Figs. 34.18 to 34.20. The performance characteristics are also summarized in Table 34.7.

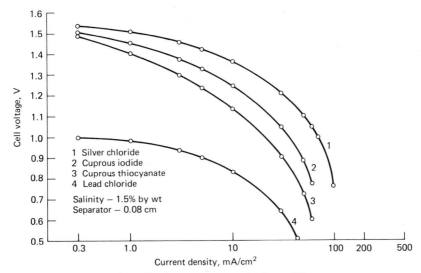

FIG. 34.7 Representative cell voltages vs. current density at 0°C.

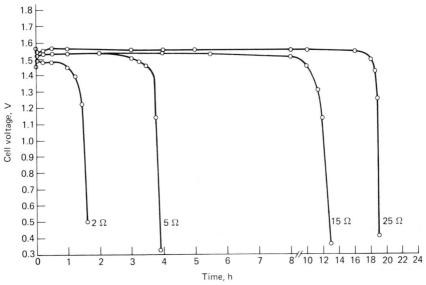

FIG. 34.8 Magnesium/silver chloride seawater-activated cell discharged continuously at 35°C in simulated ocean water, 3.6% salinity.

34.5.3 Forced-Flow Batteries

With the development of the recirculation system in which the inflow of fresh electrolyte can be controlled, thereby maintaining the temperature and conductivity of the electrolyte, the performance of electric torpedo batteries has been improved markedly.

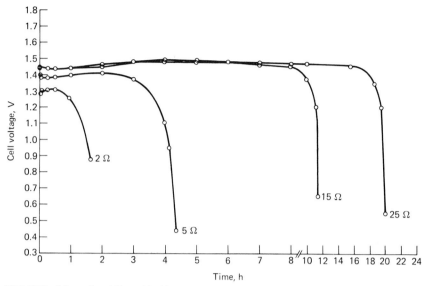

FIG. 34.9   Magnesium/silver chloride seawater-activated cell discharged continuously at 0°C in simulated ocean water, 1.5% salinity.

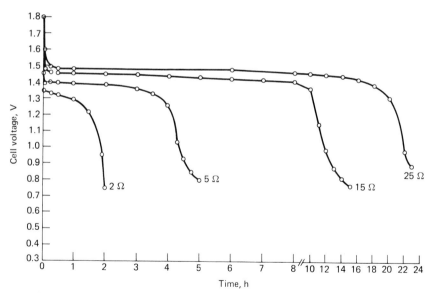

FIG. 34.10   Magnesium/cuprous thiocyanate seawater-activated cell discharged continuously at 35°C in simulated ocean water, 3.6% salinity.

With recirculation and flow control, a recirculation pump (see Fig. 34.2) and voltage-sensing mechanism are added to the battery system. By this method the temperature of the battery and the conductivity of the seawater electrolyte increase. Since battery

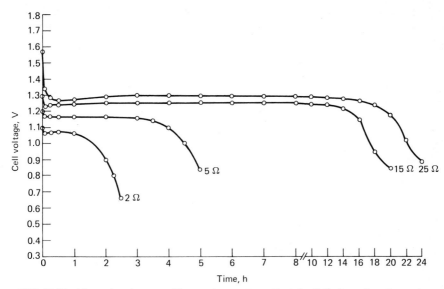

FIG. 34.11 Magnesium/cuprous thiocyanate seawater-activated cell discharged continuously at 0°C in simulated ocean water, 1.5% salinity.

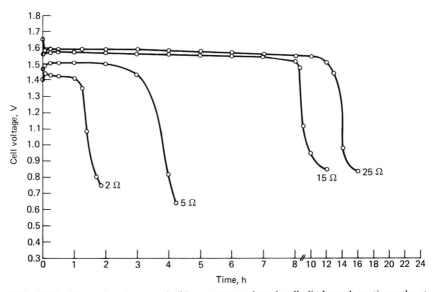

FIG. 34.12 Magnesium/cuprous iodide seawater-activated cell discharged continuously at 35°C in simulated ocean water, 3.6% salinity.

voltage increases directly with temperature and conductivity, it is possible to control the output of the battery by controlling the intake of electrolyte by means of the voltage-sensing mechanism.

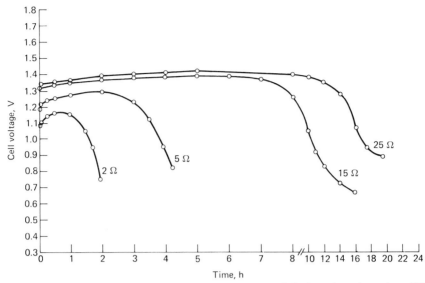

FIG. 34.13 Magnesium/cuprous iodide seawater-activated cell discharged continuously at 0°C in simulated ocean water, 1.5% salinity.

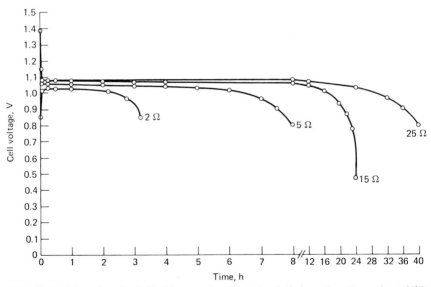

FIG. 34.14 Magnesium/lead chloride seawater-activated cell discharged continuously at 35°C in simulated ocean water, 3.6% salinity.

The performance of one type of torpedo battery with and without recirculation voltage control is shown in Fig. 34.21.[20] The blocked-in area represents the limits within which an electric torpedo battery with recirculation and flow control will perform when

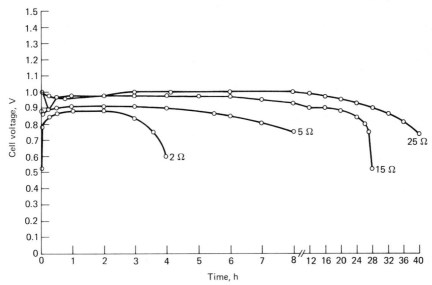

FIG. 34.15 Magnesium/lead chloride seawater-activated cell discharged continuously at 0°C in simulated ocean water, 1.5% salinity.

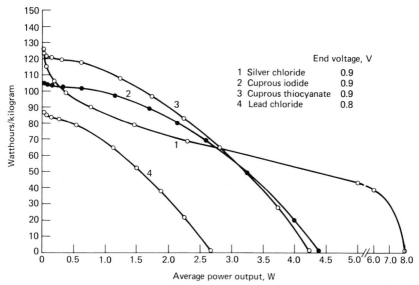

FIG. 34.16 Capacity vs. power output, seawater-activated cells discharged continuously at 35°C in simulated ocean water, 3.6% salinity.

discharged under any of the conditions shown by the three individual curves. All voltages pertinent to the start and finish of the battery are shown by the three individual curves.

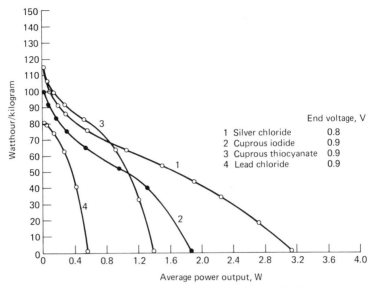

FIG. 34.17   Capacity vs. power output, seawater-activated cells discharged continuously at 0°C in simulated ocean water, 1.5% salinity.

TABLE 34.6   Performance Specification for Seawater-Activated Battery

Load	$80 \pm 2 \ \Omega$	
Life	9 h	
Voltage	15.0 V min. from 90 s to 9 h	
	19.0 V max.	
Activation*	60 s to 13.5 V	
	90 s to 15.0 V	
Battery Size:	Silver	Nonsilver
Height, cm	7.7 Max.	10.6 Max.
Width, cm	5.7 Max.	7.6 Max.
Thickness, cm	4.2 Max.	5.7 Max.
Weight, g	255 ± 14	482 ± 85
Environmental:		
Storage	From -60 to $+70°C$ for 5 y†	
	90 days at -50 to $+40°C$ at a RH of 90% (see Table 34.5)	
	10 days per MIL-T-5422E (see Table 34.5)	
Vibration, Hz	5–500	
Electrolyte:		
Low temp.	Ocean water of 1.5% salinity by weight at $0° \pm 1°C$	
High temp.	Ocean water of 3.6% salinity by weight at $+34° \pm 1°C$	

*Battery preconditioned at $-20°C$ prior to immersion in ocean water of 1.5% salinity by weight at $0° \pm 1°C$.
†In equipment packed in sealed plastic container with appropriate desiccant.

34.5.4 Dunk-Type Batteries

Magnesium/Cuprous Chloride Batteries   The magnesium/cuprous chloride battery is widely used in applications requiring low-temperature performance, such as radiosondes, having replaced the more expensive magnesium/silver chloride system in ap-

TABLE 34.7 Performance Summary of Seawater-Activated Batteries

	Silver chloride	Cuprous iodide	Cuprous thiocyanate	Lead chloride
No. of cells	11	12	13	16
Battery dimensions:				
Height, cm	7.5	9.8	10.2	10.5
Width, cm	5.5	7.6	7.4	7.5
Thickness, cm	3.9	4.4	5.7	4.5
Weight, g	252	516	478	458
Activation:				
Low temp.:				
To 13.5 V, s	< 15	< 15	< 15	< 15
To 15.0 V, s	60	60	60	15
High temp.:				
To 15.0 V, s	< 1	< 1	< 1	< 1
Life:				
High temp., h	9.67	9.4	9.3	9.5
Low temp., h	9.80	10.3	10.3	10.7
Load resistance (per cell), Ω*	7.27	6.67	6.15	5.0
Cut-off voltage (per cell), V*	1.364	1.25	1.154	0.9375
Average current, A	0.206	0.220	0.236	0.219
Average volts per cell, V*	1.497	1.463	1.378	1.048
Wh/L	204	110	90	100
Wh/kg	130	70	79	75

*As each battery system contains a different number of cells, cell load resistances and cell voltages are different for each battery.

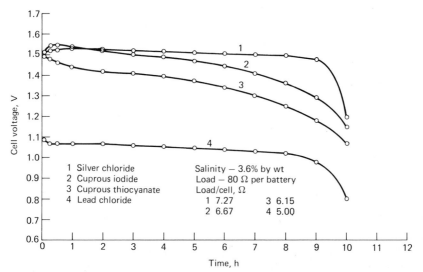

FIG. 34.18 Discharge curves, seawater-activated batteries, 35°C.

plications where weight and volume are not critical. Figure 34.22 illustrates a typical magnesium/cuprous chloride battery. The pile-type construction shown in Fig. 34.3 is used.

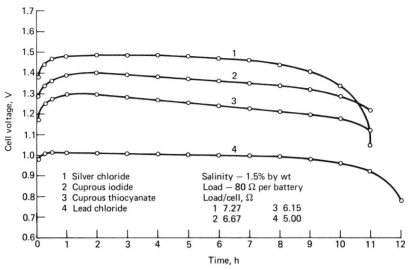

FIG. 34.19 Discharge curves, seawater-activated batteries, 0°C.

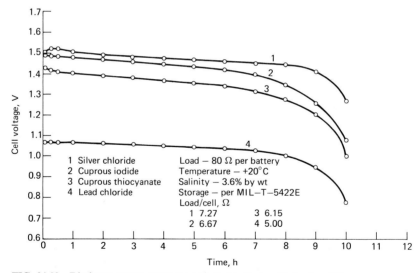

FIG. 34.20 Discharge curves, seawater-activated batteries, 10-day humidity.

The battery is activated by filling it with water, and full voltage is reached within 1 to 10 min. The battery is best suited for discharge at about the 1- to 3-h rate at temperatures from +60 to −50°C after activation at room temperature. Overheating and dry-out will occur on high current drains, and self-discharge limits the life activation. For best service, these batteries should be put into use soon after activation. The heat developed during discharge can be used to advantage in cells which are operated

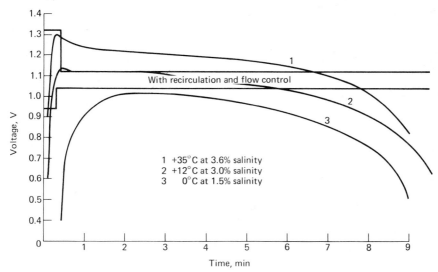

FIG. 34.21   Discharge curves, torpedo battery—effect of recirculation and flow control.

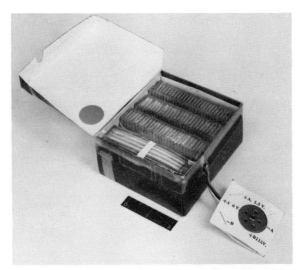

FIG. 34.22   Magnesium/cuprous chloride radiosonde battery. (Size: 10.2 × 11.7 × 1.9 cm. Weight: 450 g. Rated capacity: A_1 section, 1.5 V, 0.3 Ah; A_2 section, 6.0 V, 0.4 Ah; B section, 115 V, 0.08 Ah.)

at low temperatures; hence, the energy output varies little with decreasing temperature. Figure 34.23 shows the discharge curve for this battery at various temperatures. Figure 34.24 gives some typical discharge curves for this type of battery with a similar design at various discharge loads.[21]

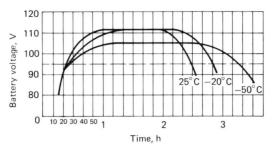

FIG. 34.23 Discharge curves of magnesium/cuprous chloride radiosonde battery, 115-V section; discharge load: 3050 Ω.

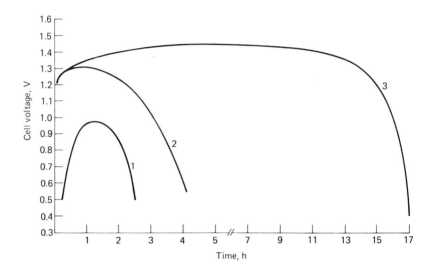

Cell no.	Load, Ω	Dimensions, cm			
		Volume	Length	Height	Thickness
1	2.5	10.2	8.2	2.5	0.5
2	8.0	2.5	2.2	3.8	0.3
3	125	1.3	2.0	2.0	0.3

FIG. 34.24 Discharge curves, magnesium/cuprous chloride water-activated batteries, 20°C (electrolyte: tapwater).

Magnesium/Manganese Dioxide Battery   This reserve battery consists of a magnesium anode and a manganese dioxide cathode.[9-11,22] It is activated by pouring an aqueous magnesium perchlorate electrolyte into the cells of the battery, where it is absorbed by the separators. Electrolyte absorption occurs within a few seconds at 0°C or above, but 3 min or more are required at −40°C due to the viscosity of the electrolyte.

The battery can deliver between 80 and 100 Wh/kg over the temperature range of −40 to +45°C at the 10- to 20-h discharge rate. Over 75% of the battery's fresh capacity is available after 7 days activated stand at 20°C and 4 days storage at 45°C.

Typical discharge curves are shown in Fig. 34.25[23] for a five-cell 10-Ah battery, weighing about 1 kg and being 655 cm³ in size.

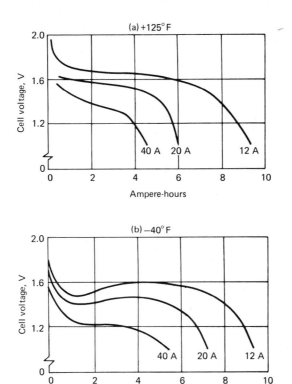

FIG. 34.25 Typical discharge curves of magnesium/ manganese dioxide cell, 10-Ah size.[23]

34.6 NEW DEVELOPMENTS

Many materials have been evaluated for use as cathode depolarizers for water-activated batteries. In every case alloys of magnesium, aluminum, or zinc have been used as anodes. Prime candidates as cathode depolarizers have been manganese dioxide,[9,22,24,25] steel wool,[26] activated iron,[27] lead dioxide,[23] cupric halides,[7] cupric oxalate,[29] cuprous oxalate,[30] cuprous formate,[30] cuprous tartrate,[30] air,[31] mercuric oxide,[22] cupric oxide,[22] and mercurous chloride.[6] To this time none has become commercially important; however, with the demise of silver chloride as the premier cathode due to cost, all old systems will be reevaluated and new systems investigated.

34.7 BATTERY APPLICATION

Water-activated batteries can be viable candidates as the power source for many types of equipment. The choice of which battery to use becomes one of economics. By proper design all will perform similarly. Where high current densities are required and cost is secondary, the magnesium/silver chloride system is best. All can be used as immersion or dunk-type batteries; however, all but the magnesium/cuprous chloride system will withstand long storage times at high temperatures and high humidities. At the present state of the art only the magnesium/silver chloride system is suitable for use in forced-flow batteries.

Since almost all available water-activated batteries have been designed for a specific application, there is no readily available standard line of these cells or batteries.

REFERENCES

1. National Defense Research Committee, *Final Report on Seawater Batteries,* Bell Telephone Laboratories, Inc., New York, 1945.

2. L. Pucher, "Cuprous Chloride-Magnesium Reserve Battery," *J. Electrochem. Soc.* **99**:203C (1952).

3. B. N. Adams, "Batteries," U.S. Patent 2,322,210, 1943.

4. H. N. Honer, F. P. Malaspina, and W. J. Martini, "Lead Chloride Electrode for Seawater Batteries," U.S. Patent 3,943,004, 1976.

5. H. N. Honer, "Deferred Action Battery," U.S. Patent 3,205,696, 1965.

6. N. Margalit, "Cathodes for Seawater Activated Cells," *J. Electrochem. Soc.* **122**:1005 (1975).

7. J. Root, "Method of Producing Semi-Conductive Electronegative Element of a Battery," U.S. Patent 3,450,570, 1969.

8. R. F. Koontz and L. E. Klein, "Deferred Action Battery Having an Improved Depolarizer," U.S. Patent 4,192,913, 1980.

9. E. P. Cupp, "Magnesium Perchlorate Batteries for Low Temperature Operation," *Proc. 23rd Annu. Power Sources Conf.,* Electrochemical Society, Pennington, N.J., 1969, p. 90.

10. C. K. Morehouse, R. Glicksman, and G. S. Lozier, "Cells Electric," *Encyclopedia of Chemical Technology,* 2d Suppl., September, 1960, pp. 126–161.

11. N. T. Wilburn, "Magnesium Perchlorate Reserve Battery," *Proc. 21st Annu. Power Sources Conf.,* Electrochemical Society, Pennington, N.J., 1967, p. 113.

12. J. F. King and W. Unsworth, "Magnesium in Seawater Batteries," *Int. Magnesium Assoc.,* Spokane, Wash., 1978.

13. M. E. Wilkie and T. H. Loverude, "Reserve Electric Battery with Combined Electrode and Separator Member," U.S. Patent 3,061,659, 1962.

14. K. R. Jones, L. J. Burant, and D. R. Wolter, "Deferred Action Battery," U.S. Patent 3,451,855, 1969.

15. H. N. Honer, "Seawater Battery," U.S. Patent 3,966,497, 1976.

16. H. N. Honer, "Multicell Seawater Battery," U.S. Patent 3,953,238, 1976.

17. J. J. Stokes, Jr. and D. Belitskus, in N. C. Cahoon and G. Heise (eds.), in *Primary Batteries,* vol. 2, Wiley, New York, 1976, chap. 3.

18. D. Belitskus, "Performance of Aluminum-Manganese Dioxide Dry Cells," *J. Electrochem. Soc.* **119**:295, 1972.

19. J. J. Stokes, Jr., "Primary Cell Anode," U.S. Patent 2,796,456, 1957; *Electrochem. Technol.* **6:**36, 1968.

20. J. F. Donahue and D. S. Pierce, "A Discussion of Silver Chloride Seawater Batteries," winter meeting *Am. Inst. Elect. Eng.,* New York, 1963.

21. Commercial Literature, Battery Engineering Publication, ESB-Ray-O-Vac, Madison, Wis.

22. H. R. Knapp and A. L. Almerini, "Perchlorate Reserve Batteries," *Proc. 17th Annu. Power Sources Conf.,* Electrochemical Society, Pennington, N.J., 1963, p. 125.

23. Commercial literature, Eagle-Picher Industries, Joplin, Mo.

24. J. B. Okerman, "Seawater Battery Having Magnesium or Zinc Anode and Manganese Dioxide Cathode," U.S. Patent 3,433,678, 1969.

25. C. P. Wales, "MnO₂ Cathodes for Seawater Cells," *J. Electrochem. Soc.* **124:**809, 1977.

26. C. L. Opitz, "Salt Water Galvanic Cell with Steel Wool Cathode," U.S. Patent 3,401,063, 1968.

27. J. C. Duddy, "Seawater Reserve Battery Having Magnesium Anode and Lead Dioxide-Graphite Fabric Cathode," U.S. Patent 3,481,790, 1969.

28. P. R. Juckniess and R. D. Blue, "Galvanic Cell Employing Iron Cathode and Method of Producing Galvanic Cathode Having Activated Iron Surface," U.S. Patent 3,477,876, 1969.

29. C. P. Wales, "Cupric Oxalate Cathodes for Seawater Cells," *J. Electrochem. Soc.* **126:**351, 1979.

30. R. F. Koontz, "Deferred Action Battery Having an Improved Depolarizer," U. S. Patent 4,007,316, 1977.

31. M. Katoh, K. Schimizer, and S. Katoh, *Denki Kagaku, J. Electrochem. Soc., Japan,* **38:**753, 1970.

35

Lithium/Water Batteries

by
Ernest L. Littauer

35.1 GENERAL CHARACTERISTICS

The high energy density of lithium (3.86 Ah/g) can be used in a reserve battery in a reaction with water to give an efficient, high power output system. In a marine environment, the battery can be especially attractive because the weight of the water need not be taken into consideration.

Generally, the combination of lithium with water may be considered hazardous because of the high heat of reaction. However, in the presence of hydroxyl (OH$^-$) ion at concentrations greater than $1.5M$, a protective film is formed which exists in a dynamic steady state. The film is pseudoinsulating, which permits the cathode to be pressed against it without causing a short circuit, thus reducing IR losses and concentration polarization and achieving high current outputs. The electrochemical cell reaction competes with a parasitic corrosion reaction evolving hydrogen at the cathode which is wasteful and requires unique gas disposal considerations for safety.

In situations where H_2 gas must not be evolved, such as in torpedo propulsion, the coupling of lithium to a reactive cathode (e.g., AgO) provides for the highest energy density and power density of any reserve system. This battery is entirely safe in storage because it is activated only after launch by admittance of seawater. The coupling of lithium to a dissolved cathode reactant (e.g., H_2O_2) gives high energy density, little gas evolution, and far greater voltage (power) per unit cell than with water. Because H_2O_2 is inexpensive in comparison with lithium, systems utilizing it have appreciable economic advantage.

The major difficulty in use of lithium in aqueous solution is the fact that it will not discharge efficiently at current densities less than about 0.2 to 0.4 kA/m^2 and the battery cannot be placed on open circuit with electrolyte present within it. Also, under certain conditions the lithium will passivate. This feature can be used to advantage, however, to temporarily terminate reaction for standby with an electrolyte-filled battery.

The aqueous lithium systems are in principle quite simple, but in practice the electrolyte management subsystem and internal cell features involve sophisticated design and a level of complexity characteristic of fuel cells, requiring a reservoir, pump, heat exchanger, and controller, and using a flowing electrolyte. The electrodes are held in close juxtaposition, and frequently they are pressed together. A film on the lithium prevents short-circuiting but permits high flux rates. The rate of discharge is inversely proportional to the concentration of electrolyte, and the output of the cell can be uniquely controlled by adjusting the molarity *(M)* of the LiOH produced at the anode. The batteries' potential is influenced more by the characteristics of the cathode than the lithium anode.

35.2 CHEMISTRY

35.2.1 Reaction of Lithium with Water

The dissolution of lithium in aqueous solutions appears either as a charge-transfer reaction

$$Li + OH^- \rightarrow LiOH + e \qquad E^0 = -2.95 \text{ V}$$

subsequently, by dissociation

$$LiOH \rightleftharpoons Li^+ + OH^-$$

or a corrosion reaction

$$Li + H_2O \rightarrow LiOH + \tfrac{1}{2} H_2 \qquad \triangle H^0 = -53.3 \text{ kcal/mole}$$

The energy density of the Li/H_2O reaction is 8450 Wh/kg of lithium.

35.2.2 Nature of Anodic Film

In strongly alkaline aqueous electrolyte at open-circuit voltage (OCV) or when polarized, Li is covered with a thin, conducting, coherent, probably nonstoichiometric, LiOH film adjacent to the metal, overlaid by 5×10^{-2} cm thick porous anhydrous LiOH, intermixed with $LiOH \cdot H_2O$. The active-electrode surface area is less (5 to 40%) than its projected area. The film thickness changes only slightly with M LiOH, Li polarization, and electrolyte flow velocity. In LiOH solutions greater than $4M$, the anode can be passivated if it becomes polarized excessively.[1,2] The passivation is due to blocking of the pores in the film. Activity can be reestablished by diluting the solution or reducing the load. This feature provides a means to temporarily terminate the reaction in electrolyte-filled cells.

35.2.3 Corrosion Reaction

At OCV, Li corrodes very rapidly. The rate is influenced by temperature and the concentration of LiOH, as indicated in Fig. 35.1. The corrosion activation energy (E_{act}) is 15.5 kcal/mole. An empirical equation[3] to calculate the OCV corrosion rate is

$$i_{corr} \, \frac{kA}{m^2} = [a - 10 \, (M \text{ LiOH})] \exp \frac{b - E_{act}}{RT}$$

where a = 54.1
$ b$ = 25.04
$ R$ = gas constant
$ T$ = temperature, K

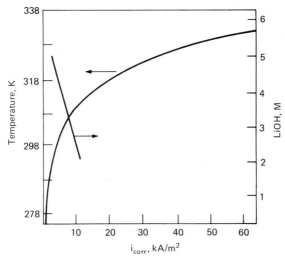

FIG. 35.1 Lithium corrosion rate as a function of temperature and LiOH solution concentration.

35.2.4 Electrochemical Efficiency

Figure 35.2 shows typical anode constant load polarization curves; the current density at invariant potential N is influenced greatly by M LiOH. Flow rate shifts N, but to a far smaller extent (see also Table 35.1). Operation of the cell at any point more anodic

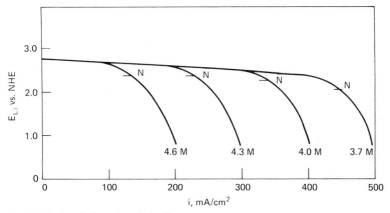

FIG. 35.2 Typical anode polarization curves.

TABLE 35.1   Influence of Electrolyte Concentration, Temperature, and Flow Rate on the Normal Operating Current Density (CD) of Lithium*

Influence of LiOH concentration At 25°C, 20 cm/h		Influence of temperature M LiOH = 4.5		Influence of flow rate, T = 25°C M LiOH = 4.5	
M LiOH	CD, kA/m² ±10%	Temp. of LiOH, °C	CD, kA/m2 ±10%	Flow, cm/s	CD, kA/m² ±20%
3	7.5	30	2.0	20	1.0
3.5	5.0	35	3.3	30	1.2
4.0	3.0	40	4.6	40	1.4
4.5	1.0	45	5.9	50	1.6
4.8	0.5	50	7.2		
5.0	0.2				

*$E_{\text{Li/H}_2\text{O}}$ = 1.1 ± 0.1 V.
$E_{\text{Li/H}_2\text{O}_2}$ = 2.4 ± 0.2 V.
$E_{\text{Li/O}_2}$ = 2.2 ± 0.1 V.
$E_{\text{Li/AgO}}$ = 2.5 ± 0.2 V.

than N decreases faradic efficiency η_F. Efficiency does not increase significantly at potentials more cathodic.

The calculated OCV depends on the half-cell reaction specified

$$\text{Li} = \text{Li}^+ + \text{e} \qquad E^0 = -3.045 \text{ V}$$

$$\text{Li} + \text{OH}^- = \text{LiOH} + \text{e} \qquad E^0 = -2.95 \text{ V}$$

The observed Li OCV is usually -2.7 to -2.8 V, and N is approximately -2.4 V. Accordingly, the Li voltaic efficiency $\eta_{v\text{Li}}$ is approximately 80%. In the Li/H$_2$O cell, the cathode polarizes by approximately 0.5 V and the cell voltaic efficiency η_{V_c} reduces to 50%. With an H$_2$O$_2$ depolarized cathode, cell potential is 2.6 V. The calculated OCV for Li/H$_2$O$_2$ is 3.92 V, then η_{V_c} = 60%. In Li/O$_2$ cells the η_{V_c} is about 65%. With all Li/H$_2$O cells, η_F is > 80%.

35.2.5 Cathode Reactions

For Li/H$_2$O, the cathode reaction is

$$\text{H}_2\text{O} + \text{e} = \text{OH}^- + \tfrac{1}{2}\text{ H}_2 \qquad E^0 = -0.828 \text{ V}$$

and occurs at an iron cathode with minimal polarization. The overall Li/H$_2$O cell voltage is 2.12 V. With H$_2$O$_2$, the situation is complex.[4] H$_2$O$_2$ converts to the perhydroxyl ion HO$_2^-$, which may be reduced directly at a catalytic surface, such as silver or palladium

$$\text{HO}_2^- + \text{H}_2\text{O} + 2\text{e} = 3\text{ OH}^-$$

or it may decompose to give oxygen

$$2\text{ HO}_2^- = 2\text{ OH}^- + \text{O}_2\uparrow$$

which may also be reduced electrochemically at the cathode

$$\text{O}_2 + \text{H}_2\text{O} + 2\text{e} = \text{HO}_2^- + \text{OH}^- \qquad E \text{ observed } = +0.1 \text{ to } +0.2 \text{ V (NHE)}$$

When O_2 is the reactant at an electrode

$$O_2 + 2H_2O + 4e = 4 OH^- \qquad E^0 = + 0.401 \text{ V}$$

and $\qquad O_2 + H_2O + 2e = HO_2^- + OH^- \qquad E^0 = + 0.076 \text{ V}$

The Li/O_2 cell typically has an OCV of 2.9 to 3.0 V, and a mixed reaction therefore occurs at the cathode.

AgO reduces according to

$$Ag_2O + H_2O + 2e = 2 Ag + 2 OH^- \qquad E^0 = + 0.345 \text{ V}$$

or $\qquad 2 AgO + H_2O + 2e = Ag_2O + 2 OH^- \qquad E^0 = + 0.607 \text{ V}$

The observed Li/AgO OCV is 3.1 V, and at N the cell potential is 2.3 to 2.5 V. Specially formulated silver oxide primary cathodes are not rate-limiting, but they are capacity-limiting components in the battery.

35.3 GENERAL SYSTEM DESCRIPTION

In general, the Li/H₂O power system is tailored to satisfy a specific requirement. Regardless of the application, however, the major elements of the system are basically the same functionally, although they will vary in size, configuration, and placement, depending on electrical performance and physical, environmental, and operational requirements. The power system block diagram is shown in Fig. 35.3. It comprises three subsystems: the power subsystem, the electrolyte management subsystem, and the control subsystem.

35.3.1 Structure of Anode

The lithium is shaped in a dry room and pressure-bonded to its substrate-current collector (typically iron). The edges of the lithium are protected with a stop-off lacquer.

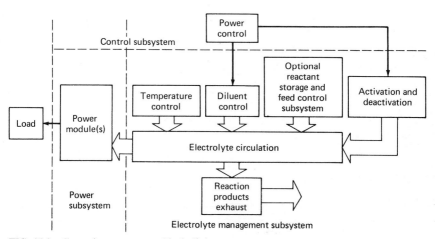

FIG. 35.3 General power system block diagram.

35.3.2 Structure of Cathode

The structure is dependent on the specific couple. For Li/H_2O and Li/H_2O_2 it comprises an iron backplate with vertical flow channels overlaid by a mesh screen. The active screen is Fe or Fe plated with Ag or Pd.

For Li/O_2, the structure is a Teflon-bonded, catalyzed graphite electrode supported on a metal screen and attached to an air cavity. Electrical connection is made at the perimeter. To provide an interelectrode flow channel, a plastic screen is inserted between the anode and cathode. Likewise, the Li/AgO cell comprises a silver oxide cathode bonded to a conducting substrate with an interelectrode flow screen.

35.3.3 Features of Cell Modules

Most lithium-aqueous systems utilize bipolar stacks. The backing plates serve to control shunt currents and to balance electrolyte flow. The cell module is also equipped with shunt loss inhibitors and electrolyte flow manifolds and channels. A compression system is used to push the electrodes together as the lithium is consumed. Figure 35.4 gives design details of a typical power module.

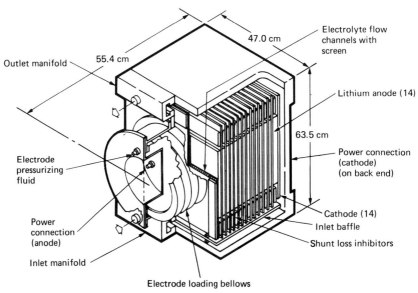

FIG. 35.4 Features of a typical power module.

35.3.4 Auxiliary Systems

The lithium/water systems are organized as in Fig. 35.3. They contain a power module subsystem and an electrolyte management subsystem. This subsystem contains as a minimum the following components

Reservoir

Gas-liquid separator

Circulation pump

Heat exchanger

Controller

Li/H_2O_2 batteries also have an H_2O_2 tank, an H_2-O_2 reactor, and sometimes an Li^+ precipitator.[5,6]

The reservoir frequently doubles as a gas-liquid separator, and additional separation is accomplished dynamically. The pump provides a flow rate across the electrodes of 20 to 50 cm/s depending on the system rating. Heat exchangers may be of the flat-plate or tube-and-shell type. Heat dissipation capacity equals the power rating of the battery. The controller monitors the cell stack voltage and adds diluent (fresh water or seawater) and/or H_2O_2 when it drops below a specified point. Alternatively, in terrestrial applications where diluent is not available, the controller adjusts the addition of an Li^+ precipitating agent, such as CO_2 or H_3PO_4. The H_2O_2 concentration in LiOH is variable from 0.2 to 1.5 M depending on the power rating of the battery.

Additional controller functions include temperature maintenance, start-up, and shutdown. A microprocessor and supporting integrated circuits are used for this purpose.

35.3.5 General Performance Characteristics

Table 35.1 provides data obtained from various lithium-aqueous systems and shows the influence of electrolyte composition, flow rate, and temperature. The cell performance is dominated by the discharge characteristics of the cathode.

The current voltage curves for all lithium aqueous electrolyte systems are, of course, a composite of the polarization characteristics of the lithium anode (Fig. 35.2) and of the cathode in question. The behavior of the Li is dominated by electrolyte concentration and temperature, and the efficiency of the discharge process is then controlled by the concentration. In the case of a single-cell Li/H_2O system (Fig. 35.5), the i-E curve for cathodic evolution of H_2 on an iron cathode is characterized by initially significant and subsequently little polarization. The cell discharge features are shown by the dotted curves at three LiOH concentrations. The Li polarization and the observed current efficiency (ϵ_{OBS}) are shown by the solid lines. It is apparent how important it is to discharge beyond the linear polarization region if good efficiency is to be obtained.[1,7]

35.4 CHARACTERISTICS OF SPECIFIC SYSTEMS

35.4.1 Lithium/Water

The lithium/water battery is suited for marine applications such as sonobuoys, torpedoes,[8] and submersibles.[9] Seawater which is consumed in the electrochemical reaction is drawn from the surroundings. The free-flooding system is essentially insensitive to the ocean ambient pressure. Typical of the Li/H_2O prototype batteries is a

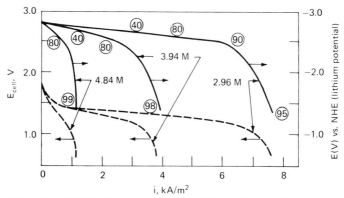

FIG. 35.5 Li/H$_2$O cell polarization (dotted line) and anode characteristics at 298 K in three LiOH concentrations (E$_{OBS}$%, shown in ○, represents Li faradic efficiency).

0.45-kW, 4-kWh marine battery.[10] Figure 35.6 shows a block diagram of the system; the specifications are given in Table 35.2. Figure 35.7 is a photograph of the entire system, with heat exchanger, cell stack, and reservoir. The battery is designed to deliver power over the range of 10 to 450 W. Figure 35.8 shows typical polarization levels of

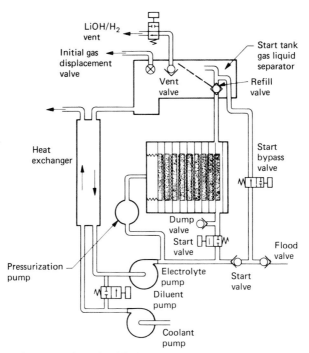

FIG. 35.6 Schematic of lithium/water system.

FIG. 35.7   Lithium/water battery, 0.45 kW, 4 kWh. (*Courtesy of Lockheed Missiles and Space Company.*)

the cell stack at various electrolyte concentrations and indicates the efficiency of lithium utilization at three power levels. At the 0.45- and 0.2-kW power levels, 90% current efficiency was obtained. At the lower 10-W level, efficiency is degraded because the direct corrosion reaction is not entirely suppressed.

35.4.2 Lithium/Hydrogen Peroxide System

A large-capacity lithium/hydrogen peroxide reserve system has been developed for a mission of extended duration in which minimal gas evolution is desirable. This 600-kWh system contains five 14-cell, 34-V, 5.5-kW modules each for a 24-h discharge capacity. Figure 35.9 shows the system functional diagram, and Fig. 35.4 shows the 14-cell power module. Typical performance of this module is given in Fig. 35.10.[5,6] At the nominal operating voltage of 34 V, the lithium efficiency was about 90% and hydrogen peroxide utilization about 70% after more than 20 h of operation. The catalyst selection is critical in the Li/H_2O_2 system for optimized cell voltage and peroxide utilization. Both silver and palladium are acceptable for this purpose.

TABLE 35.2   Characteristics of 12.5-V, 0.45-kW Li/H_2O Marine Battery

Number of cells	10	Weight	25 kg
Voltage per cell	1.25 V	Volume	0.03 m³
Power, max.	0.45 kW	Current	
Min.	0.01 kW	density	2.8 kA/m² max.
Discharge life	20 h		0.4 kA/m² min.
Delivered energy	4 kWh	Weight of	
Temperature	0–30°C	lithium	157 g per anode

35.4.3 Lithium/Air System

The lithium/air battery is one of the candidates for electric vehicles (see Part 4, Chap. 30, Sec. 30.3).

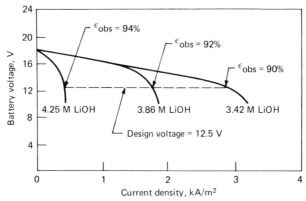

FIG. 35.8 Battery polarization curves for a 12.5-V Li/H_2O system.

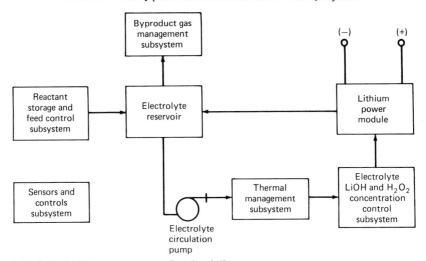

FIG. 35.9 Li/H_2O_2 power system functional diagram.

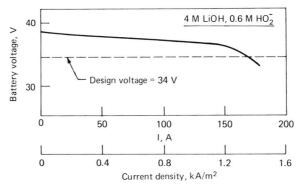

FIG. 35.10 Polarization curves of 5.5-kW, 14-cell, Li/H_2O_2 battery.

35.4.4 Lithium/Silver Oxide System

The lithium/silver oxide battery is being developed for torpedo propulsion.[11] This battery provides a very high discharge rate for short durations. At current densities greater than 10 kA/m², each cell operates at above 2.3 V. Its configuration is similar to that of the magnesium/silver chloride seawater torpedo battery. Figure 35.11 shows the typical design features of a > 100-kW system. Energy densities above 400 Wh/kg are anticipated for this application. The battery operates well above ambient temperature to minimize cathode polarization.

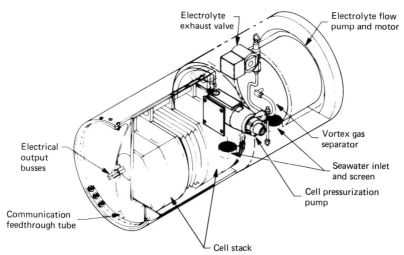

FIG. 35.11   Lithium/silver oxide torpedo battery system.

35.5 SPECIAL FEATURES AND HANDLING

35.5.1 Mechanical Recharging

The batteries can be rapidly recharged by replacing cell stacks. Module design allows for removal of an end plate for ready access to the stack. Refurbished electrodes are prepared by bonding fresh Li onto them. With the Li/AgO battery, both active materials are replaced. For Li/O₂, techniques are being explored to install Li plates without having to dismantle the cell.

35.5.2 Safety

Lithium should never be exposed to dilute, nonflowing electrolyte. Automatic purging systems must be incorporated in batteries in the event of excessive temperature or hydrogen evolution. Unlike most other batteries, short-circuiting is not destructive. When stored, because electrolyte is not present the battery is inert and quite safe. In the Li/H_2O_2 systems, it is necessary to combine H_2 (from the anode) and O_2 (from the cathode). Special reactors have been developed for this purpose.

35.5.3 Shelf Life

Cell stacks are normally stored under inert gas and therefore have indefinite shelf life. After activation and deactivation, some moisture remains in the system, and this in conjunction with CO_2 and N_2 from the air will, in time, form Li carbonate and nitrides. This aging can compromise start-up after a few days' storage in a deactivated state.

REFERENCES

1. E. L. Littauer and K. C. Tsai, "Anodic Behavior of Lithium in Aqueous Electrolytes—iii. Influence of Flow Velocity, Contact Pressure and Concentration," *J. Electrochem. Soc.* **125**(6):845, 1978.

2. E. L. Littauer and K. C. Tsai, "Anodic Behavior of Lithium in Aqueous Electrolytes—ii. Mechanical Passivation," *J. Electrochem. Soc.* **123**(7):964, 1976.

3. E. L. Littauer and K. C. Tsai, "Corrosion of Lithium in Alkaline Solutions," *J. Electrochem. Soc.* **124**(6):850, 1977.

4. E. L. Littauer and K. C. Tsai, "Observations of the Diffusion Coefficient of the Perhydroxyl Ion (HO^-_2) in Lithium Hydroxide Solutions," *Electrochim. Acta* **24**:681, 1979.

5. E. L. Littauer, W. R. Momyer, and E. S. Schaller, "Development and Evaluation of a 600-kWh Lithium/Hydrogen Peroxide Reserve Power System," *Proc. 13th Intersoc. Energy Convers. Eng. Conf.,* pp. 750–754, 1978.

6. Lockheed Missiles & Space Co., *Lithium Power Cell Feasibility Demonstration for Minuteman Survivable Power Applications,* LMSC-D556214, 1977, under USAF Contract F42600-76-C-1214, Ogden Air Logistics Center, Odgen, Utah.

7. E. L. Littauer, W. R. Momyer, and K. C. Tsai, "Current Efficiency in the Lithium/Water Battery," *J. Power Sources* **2**:163 (1977/1978).

8. E. L. Littauer and K. C. Tsai, "Electrochemical Characteristics of the Lithium/Aqueous Electrolyte/Passive Cathode Battery System," *Proc. 26th Power Sources Conf.,* p. 570, Electrochemical Society, Pennington, N. J. 1974.

9. H. J. Halberstadt, E. L. Littauer, and E. S. Schaller, "Physical and Economic Characteristics of the Lithium/Water Marine Battery," *Proc. 10th Intersoc. Energy Convers. Eng. Conf.,* p. 1120, 1975.

10. D. D. Kemp, E. L. Littauer, W. R. Momyer, and J. J. Redlien, "Design and Performance Features of a 0.45-kW, 4-kWh Lithium/Water Marine Battery," *Proc. 11th Intersoc. Energy Convers. Eng. Conf.,* p. 462, 1976.

11. W. R. Momyer and E. L. Littauer, *Lithium-Silver Oxide Seawater Battery for Torpedo Application,* LMSC-D636668, 1979, under Naval Underwater Systems Center Contract no. N00140-78-C-6435.

36

Zinc/Silver Oxide Reserve Batteries

by
James M. Dines and Elliott M. Morse

36.1 GENERAL CHARACTERISTICS

An important reserve battery, particularly for missile and aerospace applications, is the zinc/silver oxide electrochemical system, which is noted for its high rate capability and high energy density. The cell is designed with thin plates and large surface area electrodes which augment its high rate and low-temperature capability and provide a flat discharge characteristic. This design, however, reduces the activated or wet shelf life of the battery, necessitating the use of a reserve-battery design to meet storage requirements.

The zinc/silver oxide electrochemical system was the metallic couple with which Volta demonstrated the possibility of using dissimilar metals in a "pile-type" multicell construction to obtain a substantial electric potential. The system existed somewhat as a laboratory device until Professor André designed a practical secondary cell in the early World War II era. Subsequent to World War II, the U.S. military became interested in a dry-charged, primary version for use in airborne electronics and missiles because of its very high energy output per unit weight and volume and high rate capability. The ultimate result of this interest was the development of lightweight batteries for the aerospace industry, both military and civilian. The entire manned space program was keyed to zinc/silver oxide reserve batteries as the power sources for the various flight vehicles.

Zinc/silver oxide reserve batteries are divided into two classes, the manually activated and the remotely activated. In general, the manually activated types are used for space systems and accessible terrestrial applications and are usually packaged in more conventional configurations. The remote or automatically activated types are used principally for weapon and missile systems. This use requires a long period of readiness (in storage), a means for rapid remote activation, and an efficient discharge at high discharge rates, typically the 2- to 20-min range. The performance of manually activated types ranges from about 60 to 220 Wh/kg and 120 to 550 Wh/L; for remotely activated types, the energy density is reduced because of the self-contained activating device and ranges from about 11 to 88 Wh/kg and 24 to 320 Wh/L.

36.2 CHEMISTRY

The electrochemical reactions associated with the discharge of a zinc/silver oxide battery as a primary system are generally considered to proceed as follows. The cathode or positive electrode is silver oxide and may be either Ag_2O (monovalent), AgO (divalent), or any composition between the two. The anode or negative electrode is metallic zinc, and the electrolyte is an aqueous solution of potassium hydroxide. The chemical reactions and the associated voltages at standard conditions are:

$$Zn + 2AgO + H_2O \rightarrow Zn(OH)_2 + Ag_2O \qquad E^0 = 1.815 \text{ V}$$

$$Zn + Ag_2O + H_2O \rightarrow Zn(OH)_2 + 2Ag \qquad E^0 = 1.589 \text{ V}$$

The total cell reaction with 31% KOH electrolyte at 25°C is:

$$Zn + AgO + H_2O \rightarrow Zn(OH)_2 + Ag \qquad \text{Cell voltage} = 1.852 \text{ V}$$

36.3 CONSTRUCTION

A typical assembly of the manually activated reserve zinc/silver oxide cell is shown in Fig. 36.1. These batteries are designed to be manually filled with electrolyte just before use. The conventional cell design is a prismatic container with positive and negative terminals and a combination fill/vent cap. Batteries are formed by connecting single cells in series and packaging in a unit container. Batteries used in the space programs utilized thin-gauge stainless-steel, titanium, or magnesium containers to minimize weight.

36.3.1 Cell Components

The components of a reserve zinc/silver oxide cell consist of the positive plates, the negative plates, and the separators. The components are assembled such that each negative plate is protected from direct electronic contact with the adjacent positive plate by a separator. The cell components are assembled and packaged in a container; the plates are prepared in a dry and charged condition.

Positive Plates The positive plates are prepared by applying silver or silver oxide powder to a metallic grid. Copper, nickel, and silver have all been used for grid material, with silver the most prevalent for reasons of electrochemical stability and conductivity. After the silver powder is pressed or sintered to the grid, the plates are electroformed in an alkaline solution, then washed thoroughly and air-dried at a moderate temperature (usually 20 to 50°C). The nominally divalent oxide thus formed is relatively stable

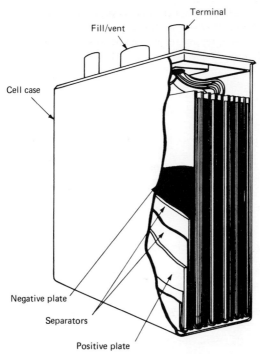

FIG. 36.1 Typical construction of a primary reserve zinc/silver oxide cell. (*Courtesy of Eagle-Picher Industries, Inc., Joplin, Mo.*)

at room ambient temperatures but tends to lose oxygen and degrade to the monovalent state with increasing temperatures and time. Continuous exposure to high temperatures (70°C) causes reduction to the monovalent oxide in a few months.

Negative Plates The negative plates may be prepared by pasting or pressing zinc powder or zinc oxide onto a grid or electroplating zinc from an alkaline bath to form a very active spongy zinc deposit.

Both positive and negative electrodes may vary in thickness from 0.12 mm as a practical minimum to 2.5 mm maximum for positives and 2 mm maximum for negatives. The extremely thin plates are utilized for very short life, high discharge rate, automatically activated batteries; the thick plates are employed in manually activated batteries designed for continuous discharge over several months at very low currents.

Separator Materials Typical separator materials used in zinc/silver oxide cells include regenerated cellulose films (cellophane, Visking), nonwoven synthetic fiber mats of nylon, Dacron, and polypropylene, and nonwoven rayon fiber mats. The synthetic fiber mats are frequently placed adjacent to the positives to protect the cellophane from the highly oxidizing influence of that material. The cellophane, a semipermeable film, prevents buildup of particles between plates (while allowing ion transfer), thus preventing interplate short circuits. The rayon mat absorbs the electrolyte solution and distributes it over the plate surfaces. Cells intended for automatic activation normally are not designed with the film separators for they require too long for complete wetting. The open-mat separators provide sufficient protection from interplate shorting for several hours.

Separator materials are necessary to the cell operation because they prevent short circuits, but they also impede current flow causing an IR drop within the cell. Very high discharge rate cells must have very low internal impedance, hence, a minimum of separator material; as a result, this type cell is restricted to very short wet-life applications. The semipermeable film is the separator which contributes most to IR drop and also to protection against shorts. Long-life cells may contain five or six layers of cellophane. They are therefore better suited to medium or low discharge rates.

Electrolyte   The electrolyte used for reserve zinc/silver oxide cells is a water solution of potassium hydroxide. High and medium discharge rate cells use a 31% by weight electrolyte solution because this composition has minimum resistance. Low-rate cells may use a 40% solution.

36.3.2 High- and Low-Rate Designs

A battery intended to be discharged at a 5- to 60-min rate is considered a high-rate design. These cells are designed primarily to deliver high current and require a large plate surface area. They contain many very thin plates. The separators also have as low impedance as possible, i.e., one or two layers of cellophane versus five or six layers for low-rate cells. Thirty-one percent potassium hydroxide electrolyte has the highest conductivity and is therefore employed in high-rate cells.

Low-rate batteries are in a class intended for discharge at rates ranging from 10 to 1000 h with the emphasis on high energy density. The plates are thick (2 mm), and relatively high impedance separator wraps are used. A higher concentration of electrolyte (40%), which permits a greater ampere-hour capacity, can also be used. This design configuration also gives a substantial improvement in the activated or wet stand capability of the cell.

FIG. 36.2   Zinc/silver oxide primary reserve battery designed for automatic activation. (*Courtesy of Eagle-Picher Industries, Inc., Joplin, Mo.*)

36.3.3 Automatically Activated Types

The automatically activated type battery is a class of reserve battery intended for quick preparation for use after an undetermined period subsequent to installation. The very high energy output of the primary zinc/silver oxide system and the use of an integrally designed system for injecting the electrolyte into the cells combine to provide an effective and efficient power source for weapons and other systems requiring a long-term ready state. Figure 36.2 is a photograph of a typical automatically activated battery for a missile application.

Four kinds of activation systems have been utilized in this type battery for transferring electrolyte from a reservoir to the cells. All the systems depend on gas pressure to move the electrolyte, and the most conventional source of gas is pyrotechnic. The "gas generator" is a small cartridge which contains an ignitable propellant material and an electrically fired ignitor or "match." Figure 36.3 shows the four types of activators.

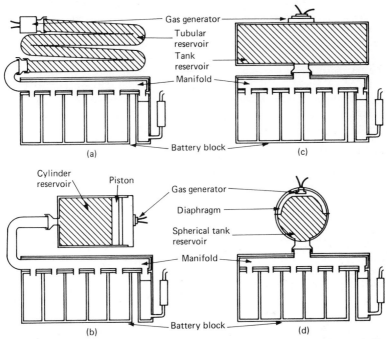

FIG. 36.3 Schematic drawings of four types of activation systems used in automatically activated batteries: (*a*) tubular reservoir; (*b*) piston activator; (*c*) tank activator; (*d*) tank-diaphragm activator. (*Courtesy of Eagle-Picher Industries, Inc., Joplin, Mo.*)

FIG. 36.4 Assembly of an automatically activated zinc/silver oxide primary battery with a tubular reservoir.

The tubular reservoir *(A)* can assume many forms. It is usually coiled around the battery as shown in Fig. 36.4 (an assembly of a battery with a tubular reservoir), but it can also be formed with 180° bends into a flat shape, or it can be configured to fit into available nonstandard volumes into which a missile battery is often mounted. The tubular reservoir is fitted with foil diaphragms at each end. For activation, the gas generator located at one end can be electrically ignited; the gas causes the diaphragms to break, and the electrolyte is forced into a manifold which distributes it to the cells of the battery. The piston activator *(B)* operates by pushing the electrolyte out of a cylinder reservoir when a gas generator is fired behind it. The tank activator *(C)* contains the electrolyte in a variable-geometry tank with a gas generator located at the top. When the gas enters at the top, the electrolyte is forced out through an aperture at the bottom. The system is position-sensitive and will operate properly only when in an upright position, relative to the components. The tank-diaphragm activator *(D)* uses a sphere or spheroid tank with a diaphragm attached internally at the major circumference. When the gas generator is fired, the diaphragm moves to the opposite side, forcing the electrolyte out through an aperture in the reservoir side of the tank.

Of the four systems, the tubular system is the most versatile, but in simple battery shapes may be heavier. The piston and diaphragm systems have moving parts and thus can be less reliable; they are also less adaptable to special shapes. The tank is efficient but position-sensitive.

The operating sequence of an automatically activated battery involves (1) application of ignition current, (2) gas generator burning and associated gas production, (3) movement of electrolyte out of the reservoir into the distribution manifold, and (4) filling of the cells with electrolyte. In a typical operation the total sequence involves about one second. In many applications the electrical load is wired directly to the battery, and so the battery activates under load. Figure 36.5 shows the rise time of the voltage under load *(A),* and rise time under no-load condition *(B)* for a battery not used until 6 h later. The delayed-use battery has film separators, and the slower wetting is reflected by the longer rise time.

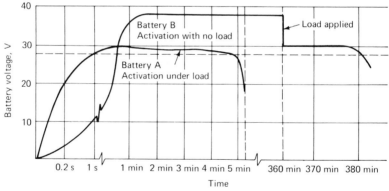

FIG. 36.5   Voltage rise time for automatically activated zinc/silver oxide batteries, 25°C. (*Courtesy of Eagle-Picher Industries, Inc., Joplin, Mo.*)

Automatically activated batteries suffer a weight and volume penalty compared with manually activated types, but the design permits the use of a high performance battery when there is not time available to activate manually or the unit is inaccessible. In many applications both conditions exist. The volume penalty is usually about 2 times, and the weight penalty about 1.6 times the basic battery. Most automatically activated battery designs utilize an integral electrical heater. The heater maintains the electrolyte at about 40°C or at a temperature which will raise a cold battery to 40°C when activation occurs. The use of heaters permits the design of batteries which can meet close voltage tolerances, thus improving the capability of the weapons' electrical and electronic systems.

36.4 PERFORMANCE CHARACTERISTICS

Zinc/silver oxide reserve batteries as a class are somewhat unique in that they are almost entirely committed to specific applications. These applications require the flat voltage profile and the high energy density available from this system, and they often demand a special design for each requirement. If a low-temperature environment is involved, battery heaters are used. If the discharge requires a wide range of current with only a small voltage variation, many very thin plates are used. A very high capacity requirement at low rates requires the use of thick plates and more concentrated electrolyte. There is no standard design or size because there is no typical application. The applications always demand the maximum from the battery design in capacity and voltage regulation at a minimum weight and volume.

36.4.1 Voltage

The open circuit of the zinc/silver oxide cell will range from 1.6 to 1.85 V per cell. The nominal load voltage is 1.5 V, and typical end voltages are 1.4 V for low-rate cells, 1.2 V for high-rate cells. At high rates of discharge, e.g., a 5- to 10-min discharge rate, the output voltage would be about 1.3 to 1.4 V per cell, whereas the 2-h-rate discharge voltage would be slightly above 1.5 V. Figure 36.6 shows a family of discharge curves at four different current densities. The voltage level is inversely related to the current density (calculated from the area of the active plate surface). Thus, based on 100 cm² of positive plate surface, a 10-A discharge rate would be 0.1 A/cm² current density. If the discharge rate is doubled to 0.2 A/cm², the voltage level would decrease, and if the rate is lowered to 0.05 A/cm², the voltage level would increase.

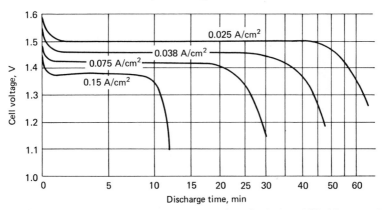

FIG. 36.6 Effect of changing current density on the cell voltage at 25°C. (*Courtesy of Eagle-Picher Industries, Inc., Joplin, Mo.*)

In cell design, the ampere-hour capacity of the cell is determined by the amount of silver oxide active material present (zinc active material is provided in excess because of the cost relationship to silver), but the voltage is determined by the current density. In a fixed volume, higher discharge rates can be obtained without lowering the cell voltage by using thinner plates (thus providing more plates per cell element and lowering the current density), but with a reduction of cell capacity. The lower the current density at which a cell can operate, the better the voltage regulation with changing rates of discharge.

36.4.2 Discharge Curves

A set of discharge curves for high-rate cells is shown in Fig. 36.7 and for low-rate cells in Fig. 36.8. The design for these two types of cells is quite different with the principal difference being the thickness of the plates. The thin plates used in high-rate cells provide more surface area for lower current density, thus better voltage control and also more efficient utilization of the active material. At lower rates of discharge, as in Fig.

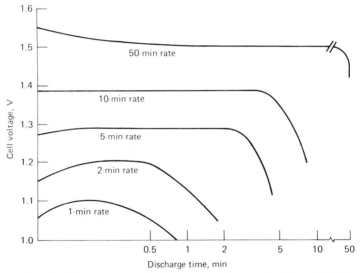

FIG. 36.7 Discharge curves for high-rate zinc/silver oxide cells at 25°C. (*Courtesy of Eagle-Picher Industries, Inc., Jopin, Mo.*)

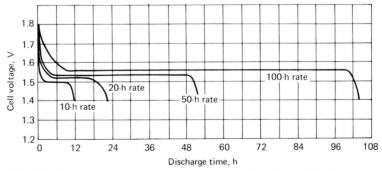

FIG. 36.8 Discharge curves for low-rate zinc/silver oxide cells at 25°C. (*Courtesy of Eagle-Picher Industries, Inc., Joplin, Mo.*)

36.8, the voltage level is higher and active material utilization is also excellent, both because of lower current density. It will be noted that the low-rate discharge curves are above 1.6 V for a period of time. This is the effect of the divalent oxide which affects voltage only at low rates. Most of the divalent capacity is obtained at high rates, but its voltage is subdued by the higher current density imposed.

36.4.3 Effect of Temperature

The family of curves shown in Fig. 36.9 illustrates the performance obtained from a high-rate cell when discharged over a range of temperatures. It should be understood that the change in voltage levels caused by temperature is closely related to the changes caused by current density. Thus the adverse effect of cold temperature can be improved by reducing the current density of the cell, and the voltage and capacity of cells discharged at high current densities can be improved by increasing their operating temperature. Figure 36.10 shows a family of curves for low-rate cells discharged at various temperatures. The two sets of curves show that the zinc/silver oxide system is significantly affected at temperatures below 0°C and thus is not recommended for applications in this environment without heaters.

36.4.4 Impedance

Figure 36.11 shows the dynamic internal resistance (DIR) of a high-rate cell at various stages of discharge and temperature. These curves show a declining ratio, $\triangle V/\triangle A$, until the end of discharge, at which time the dynamic resistance rises rapidly. The declining impedance is caused by an improvement in the positive-plate conductance and a temperature rise during the discharge. This feature can vary considerably, depending on cell design, ambient temperature of the discharge, and the point in time after the change of discharge rate when the voltage change is observed.

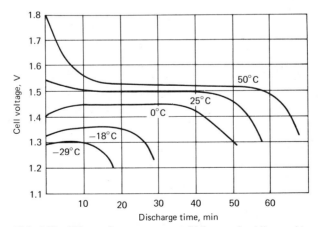

FIG. 36.9 Effects of temperature on high-rate zinc/silver oxide primary cells, discharged at the 1-h rate. (*Courtesy of Eagle-Picher Industries, Inc., Joplin, Mo.*)

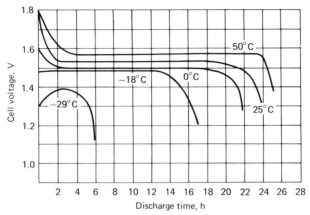

FIG. 36.10 Effects of temperature on low-rate zinc/silver oxide primary cells, discharged at the 24-h rate. (*Courtesy of Eagle-Picher Industries, Inc., Joplin, Mo.*)

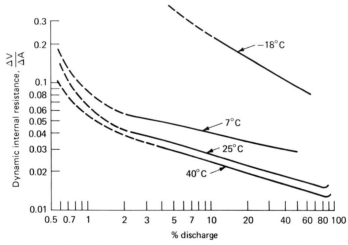

FIG. 36.11 Dynamic internal resistance of zinc/silver oxide primary cells. (*Courtesy of Eagle-Picher Industries, Inc., Joplin, Mo.*)

36.4.5 Service

The performance of zinc/silver oxide batteries in amperes per unit weight and volume versus service time is given in Fig. 36.12. It can be noted, again, that this battery system is particularly sensitive to temperatures below 0°C. These data are applicable, within reasonable accuracy, for both high- and low-rate designs.

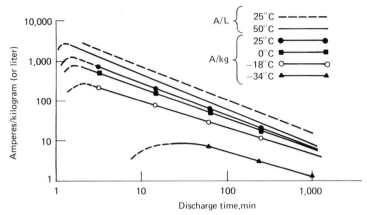

FIG. 36.12   Service life of zinc/silver oxide primary cells. (*Courtesy of Eagle-Picher Industries, Inc., Joplin, Mo.*)

36.4.6 Shelf Life

The dry shelf life of the zinc/silver oxide battery is shown in Fig. 36.13, which gives storage data at 25, 50, and 74°C for periods up to 2 y. The losses shown are based on the assumption that the positive active material is divalent silver oxide, which slowly degrades to monovalent oxide at temperatures above about 20°C. Degradation of the negative plate is minimal. It is expected that the monovalent oxide level would be reached in about 30 months when the storage temperature is 50°C. Experience has shown that batteries stored at average ambient temperatures of 25°C or lower retain capacity at or above the monovalent oxide level for a period of 10 y or longer.

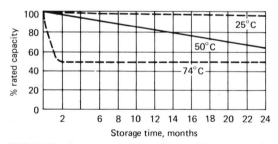

FIG. 36.13   Dry storage of zinc/silver oxide primary cells. (*Courtesy of Eagle-Picher Industries, Inc., Joplin, Mo.*)

The wet shelf life of the zinc/silver oxide battery varies considerably with design and method of manufacture. Figure 36.14 provides a guide to the expected performance of most designs. The wet shelf life degradation is caused principally by loss of negative-plate capacity (dissolution of the sponge zinc in the electrolyte) or development of shorts through the cellulosic separators.

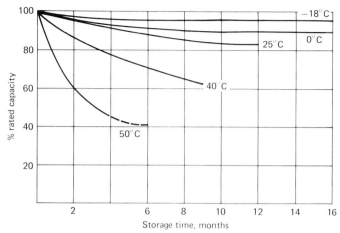

FIG. 36.14 Wet (activated) storage of zinc/silver oxide primary cells. (*Courtesy of Eagle-Picher Industries, Inc., Joplin, Mo.*)

36.5 CELLS AND BATTERY TYPES AND SIZES

Single-cell units of the reserve zinc/silver oxide type are available in sizes from about 1 Ah as a minimum up to about 400 Ah. Table 36.1 provides the specifications for a series of high-rate cells ranging in capacity from 1 to 68 Ah and a series of low-rate cells ranging in capacity from almost 2 to 410 Ah.

Table 36.2 lists a number of automatically activated batteries which have been designed to meet various specific applications. Most of these batteries are high-rate, short wet-life batteries. The weight and volume of this type are more a function of the load requirements and the space envelope provided than the voltage and capacity.

36.6 SPECIAL FEATURES AND HANDLING

Both manually and automatically activated zinc/silver oxide batteries were developed to meet highly stringent requirements with regard to performance and reliability. The time and temperature of storage prior to use are of importance, and records should be maintained to ensure use within allowable limits. Special care must be exercised to ensure that the proper amount of the specified type of electrolyte is added to each cell of a manual-type battery and that after activation the unit is discharged within the shelf life limitation at the proper temperature. Some battery containers have pressure-relief valves and/or heaters, and these must be carefully maintained and monitored.

Automatically activated batteries require special preinstallation checking of gas generator ignitor circuits, heater circuits, and vent fittings. For long-term installations there should be monitoring of the ambient temperature to prevent degradation caused by exposure to high temperatures. Periodic checks should be made to ensure that the ignitor circuits are intact because some circuits are sensitive to electromagnetic fields. After activation, if the battery is not discharged within the specified time, it must be replaced.

TABLE 36.1 Zinc/Silver Oxide Manually Activated Cells

| | | High-rate cells—15-min rate | | | | | | Low-rate cells—20-h rate | | | | | | |
| | | | Energy density | | | | | | Energy density | | | Physical dimensions, cm | | |
Cell type*	Capacity, Ah	Wh/kg	Wh/L	Wt, g		Cell type*	Capacity, Ah	Wh/kg	Wh/L	Wt, g	L	W	H
SZH 1.0	1.0	57	104	25		SZL 1.7	1.7	84	171	30	1.09	2.69	5.16
SZH 1.6	1.6	66	110	35		SZL 2.8	2.8	88	201	50	1.25	3.07	5.72
SZH 2.4	2.4	66	116	55		SZL 4.5	4.5	92	220	75	1.42	3.50	6.32
SZH 4.0	4.0	66	128	90		SZL 7.5	7.5	97	250	120	1.63	4.00	7.09
SZH 5.0	5.0	66	128	110		SZL 11.2	11.2	99	275	170	1.80	4.47	7.72
SZH 7.0	7.0	66	134	160		SZL 16.8	16.8	106	305	240	2.00	4.95	8.48
SZH 10.9	10.9	66	134	250		SZL 25.0	25.0	115	318	330	2.26	5.59	9.53
SZH 16.0	16.0	66	140	370		SZL 43.2	43.2	125	397	520	2.54	6.27	10.39
SZH 24.0	24.0	66	146	540		SZL 65.0	65.0	136	410	680	2.87	7.14	11.99
SZH 38.0	38.0	73	159	780		SZL 105.0	105.0	158	458	1010	3.25	8.08	13.39
SZH 68.0	68.0	80	196	1290		SZL 160.0	160.0	187	470	1330	3.73	9.27	15.09
						SZL 410.0	410.0	210	560	3000	4.22	13.84	19.35

*Eagle-Picher Industries, Joplin, Mo.

TABLE 36.2 Zinc/Silver Oxide Automatically Activated Batteries

Part no.*	Application	Wt, kg	Volume, L	Voltage, V	Current, A	Capacity, Ah	Energy density Wh/kg	Wh/L
EPI 4331	AIM-7	1.0	0.18	26	10.0	0.8	20	108
EPI 4500	Patriot	3.6	0.76	51	18.0	1.5	21	101
EPI 4108	Atlas	4.2	1.00	28	4.5	5.0	33	140
YEC 15148	Trident	5.0	1.20	28	6.0	12.0	65	284
YEC 15700	SAM-D	7.3	1.90	56	8.0	5.4	41	159
EPI 4445	Torpedo	9.3	1.90	28	30.0	20.0	60	294
YEC 15066	Trident	14.5	3.80	30, 31	15, 23	4, 10	30	112
EPI 4470	Harpoon	8.9	2.00	28	30.0	16.0	50	223
EPI 4367	SRAM	11.8	2.70	28	10.0	8.0	43	186

*EPI = Eagle-Picher Industries, Joplin Mo.; YEC = Yardney Electric Corp., Pawcatuck, Conn.

The proper electrical performance of these types of batteries is most confidently ensured by operating them at temperatures at or slightly above room temperature. Temperatures below 15°C can adversely affect the voltage regulation of high-rate batteries, and below 0°C there is also considerable loss of capacity for both types.

36.7 COST

The cost of high performance primary zinc/silver oxide batteries is dependent upon the specifications to which they are built and the quantity involved. Manual-type batteries may cost anywhere from $5 to $15 per watthour; remote-activated types will cost about $15 to $20 per watthour. When the price of silver is high, material cost becomes one of the chief disadvantages of these batteries. There are many applications, however, in which no other current technology can meet the high energy density of the zinc/silver oxide primary system.

BIBLIOGRAPHY

Bauer, Paul: *Batteries for Space Power Systems,* U.S. Government Printing Office, Washington, 1968.

Cahoon, N.C., and G. W. Heise: *The Primary Battery,* Wiley, New York, 1969.

Chubb, M. F., and J. M. Dines: Electric Battery, U.S. Patent 3,022,364.

Fleischer, Arthur, and John J. Lander: *Zinc Silver Oxide Batteries,* Wiley, New York, 1971.

Hollman, E. G., et al.: Silver Peroxide Battery and Method of Making, U.S. Patent 2,727,083.

Jasinski, Raymond: *High Energy Batteries,* Plenum, New York, 1967.

37

Spin-
Activated
Batteries

by
Asaf A. Benderly

37.1 GENERAL CHARACTERISTICS

Various military, and a few civilian, applications with long shelf life requirements must turn to reserve batteries for their electrical power. This is particularly true when the system requires that the power supply be integrally packaged with the electronics and not replaced throughout the storage life of the system. Typical of such applications are fuzing, control, and arming systems for artillery and other spin-stabilized projectiles.

High-spin forces, such as those encountered in artillery projectiles, may constitute a difficult environment for many battery designs. However, special designs for liquid reserve batteries have evolved that take advantage of spin to bring about their activation.

A typical spin-dependent reserve battery is illustrated in Figs. 37.1 and 37.2. The electrode stack consists of electrodes and cell spacers of an annular configuration packaged dry and therefore capable of long-term storage. A metal ampule, inserted in the center hole of the stack, houses the electrolyte. Upon firing of the gun, the ampule opens; the electrolyte is released and is then distributed into the annular-shaped cells centrifugally, thereby causing the battery to become active.

The electrode stack may be arranged in two ways, one to favor a high-voltage output and the other to allow for high-current output. The former is generally accomplished by using bipolar electrodes, i.e., electrodes wherein anodic and cathodic materials are applied respectively to the opposite sides of a metal substrate. Such bipolar electrode plates are stacked in a pile or series configuration, making automatic contact from one cell to the next. The voltage output of such a stack is the sum of each of the cells. In the second type of array, electrode plates coated with anodic material on both sides of

FIG. 37.1 Cross section of lead/fluoboric acid/lead dioxide multicell reserve battery showing the "dashpot" cutter for the copper ampule. (*Courtesy of Harry Diamond Laboratories, Department of the Army.*)

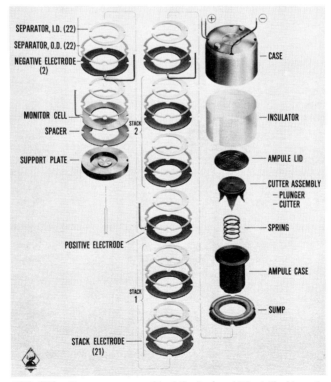

SEPARATOR, I.D. (22)
SEPARATOR, O.D. (22)
NEGATIVE ELECTRODE (2)
MONITOR CELL
SPACER
STACK 2
SUPPORT PLATE
POSITIVE ELECTRODE
STACK 1
STACK ELECTRODE (21)

CASE
INSULATOR
AMPULE LID
CUTTER ASSEMBLY
 – PLUNGER
 – CUTTER
SPRING
AMPULE CASE
SUMP

FIG. 37.2 Component parts of lead/fluoboric acid/lead dioxide multicell reserve battery, PS 416 power supply. (*Courtesy of Harry Diamond Laboratories, Department of the Army.*)

the substrate are stacked alternately with those coated with cathodic material. All anodic plates are electrically connected through tabs, and all cathodic plates are similarly connected, the two electrical connections constituting the effective terminals of the battery. This type of parallel stack is, in effect, a single electro-chemical cell with considerable electrode area (see Fig. 37.3). Where required by the application, multiples of series stacks can be connected in parallel, thereby yielding both high-voltage and high-current outputs.

FIG. 37.3 Electrode stack and case of lead/fluoboric acid/lead dioxide parallel-construction (single cell) battery. (*Courtesy of Harry Diamond Laboratories, Department of the Army.*)

37.2 CHEMISTRY

The chemistry most commonly employed in spin-activated, liquid reserve batteries is that of the lead/fluoboric acid/lead dioxide cell represented by the following simplified reaction:

$$Pb + PbO_2 + 4HBF_4 \rightarrow 2Pb(BF_4)_2 + 2H_2O$$

To a lesser extent, the zinc/potassium hydroxide/silver oxide system has been employed in spin-activated, reserve batteries. (More frequently, this reserve system is used in nonspin applications, such as missiles, where the electrolyte is driven into place by a gas generator or other pumping method. See Part 5, Chap. 36.) The chemistry of the silver oxide–zinc couple can be represented by either of two reactions, depending on the oxidation state of the silver oxide

$$2AgO + Zn \rightarrow Ag_2O + ZnO$$

$$Ag_2O + Zn \rightarrow 2Ag + ZnO$$

Very recently, development work has been initiated on spin-activated reserve batteries employing lithium anodes. The most promising system is that in which thionyl chloride serves in the dual role of electrolyte carrier and active cathodic depolarizer (see Part 5, Chap. 39). The generally accepted cell reaction for this system is

$$4Li + 2\,SOCl_2 \rightarrow 4\,LiCl + S + SO_2$$

37.3 DESIGN CONSIDERATIONS

37.3.1 Volume Optimization

The electrolyte capacity of the ampule must be matched to the composite volume of all the cells in the battery. A parallel construction battery is reasonably tolerant of electrolyte flooding or starvation since it is a single cell. A series configuration, however, can tolerate no flooding since that condition produces intercell short circuits in the electrolyte fill channel or manifold. The opposite condition, i.e., insufficient electrolyte, may leave one or more cells empty and therefore fail to provide continuity throughout the cell stack.

Since the expansion and contraction of the liquid electrolyte is more affected by temperature extremes than is the volume of the cells, a match must be made at low temperature so that the cells will be reasonably full. This, however, leads to an excess of electrolyte at higher temperatures, and this excess must be provided for in the design of the battery by the use of a "sump." In some short-life batteries, a match is established at high temperature with the recognition that cells will be less than filled at lower temperatures. To ensure that some electrolyte enters each cell (so that continuity can be maintained), a leveling hole may be provided from cell to cell. Though kept very small to reduce the effect of inevitable intercell short circuits, this hole does dissipate some of the capacity of the battery.

37.3.2 Activation Time

This time is defined as that from initiation of the battery to the point at which it delivers and sustains a requisite level of voltage across a specified electrical load. For a spin-activated liquid reserve battery, this time would include the times for ampule opening, electrolyte distribution, clearing of electrolyte short circuits in the filling manifold, depassivation of electrodes, and elimination of any form of polarization. Activation time is protracted to a maximum at low temperature where increased viscosity of electrolyte and decreased ion mobility are most significant.

The application normally establishes the maximum allowable activation time, and reserve batteries are frequently designed to reach 75 or 80% of their peak voltage within this required time. Typical of an application requiring a very short activation time, perhaps less than 100 ms, would be a time fuze for an artillery projectile. Battery power is required to start the timer. Hence, a stretch-out or uncertainty of time to reach timer voltage could readily result in a serious timing error with corresponding ineffectiveness of gunfire. In some cases, safety can be adversely affected by timing error. In less critical situations, 0.5 to 1 s is allowed for activation.

37.3.3 Cell Sealing

Since the individual cells of a spin-activated liquid reserve battery are generally annular in shape and are filled by centrifugal force, it is necessary that the periphery of the cell be sealed to keep electrolyte from leaking out. This may be accomplished by molding a plastic around the outside of the electrode-spacer stack and relying on the plastic to bond to the outside edges of the stack components. A method widely in vogue employs fish paper (a dense, impervious paper) coated with a polyethylene that melts at a relatively low temperature, similar to that used on milk cartons. Cell spacers are punched from the coated paper and placed between the electrodes. The stack is then clamped together and heated in an oven at a temperature sufficient to fuse the polyethylene, which then acts as an adhesive and sealer between the electrodes.

37.3.4 Ampules

Early designs of liquid reserve batteries primarily relied on the use of glass ampules to house the electrolyte, and, in fact, some modern batteries still use such ampules. These ampules are generally smashed by the acceleration force of gunfire or by the explosive output of a primer or squib. Although these forces are ample, there is also a tendency for rough handling or a drop on a hard surface to cause inadvertent glass ampule breakage. This would then result in destruction of the battery by premature leakage of electrolyte into the cells.

A major advance in battery ruggedness resulted from the design of metal, usually copper, ampules with internal cutting mechanisms. One version employs a cutter that is activated by a combination of spin and acceleration (see Fig. 37.4), both provided by the act of gunfire. Another relies on a dashpot mechanism (Figs. 37.1 and 37.2) which requires a sustained acceleration (several milliseconds experienced in gunfire) but will not function when subjected to the much shorter (a portion of a millisecond) shock pulse resulting from drop on a hard surface. The use of these "intelligent" ampules, ones that are capable of discriminating between the forces of gunfire and those of rough handling, have resulted in a substantial improvement in battery reliability.

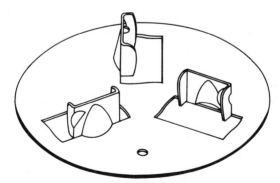

FIG. 37.4   Three-bladed cutter for copper ampule requiring spin and acceleration for activation. (*Courtesy of Harry Diamond Laboratories, Department of the Army.*)

37.4 PERFORMANCE CHARACTERISTICS

37.4.1 General

It is clear that spin-activated liquid reserve batteries suffer in space efficiency due to the need to provide double volume for electrolyte, first in an ampule and later in the cells themselves. Space is also consumed by the ampule-opening mechanism and the cell-sealing material. Finally, cell area is sometimes dissipated by the eccentricity of spin of the vehicle which houses the battery. Consequently, such batteries are generally limited to short-time applications, e.g., not exceeding 3 min. It is therefore not the practice to rate such batteries in terms of watthours per pound or per unit volume.

Like most other batteries, the performance of liquid reserve batteries is affected by temperature. Military applications frequently demand battery operations at all temperatures between -40 and $60°C$ with storage limits of -55 to $70°C$. These requirements are routinely met by the lead/lead dioxide/fluoboric acid systems and, with some difficulty at the low-temperature end, by the lithium/thionyl chloride and zinc/potassium hydroxide/silver oxide systems. Provision is occasionally made to warm the electrolyte prior to the activation of the latter system.

Since the voltage sustained by a liquid reserve battery at low temperature and under heavy electrical loading is much lower than that which it delivers at high temperatures, a serious problem of voltage regulation frequently results. In some situations, the ratio of high- to low-temperature voltage may be as much as 2:1. This problem may be avoided by the use of thermal batteries (Part 5, Chap. 40) which provide their own pyrotechnically induced, operating temperature irrespective of the ambient temperature. Until recently, thermal batteries were extremely ineffective at high spin rates but progress has been made in this field and thermal batteries capable of withstanding spin rates of 300 rps should soon be available.

Shelf life of liquid reserve batteries is highly dependent on storage temperature, high temperature being the more deleterious. Zinc/silver oxide cells are probably the most vulnerable of the generally used systems, due to the reduction of silver oxide and the passivation of zinc. Ten-year storage life is probably the best that can be expected unless the battery is substantially overdesigned. Lead/lead dioxide/fluoboric acid batteries do degrade with time, in both the loss of capacity and the lengthening of activation time. However, if objectionable organic materials are avoided in battery construction, 20 to 25 y of shelf life may be derived from them provided the battery is designed with some safety factor. Lithium/thionyl chloride reserve systems are too new to have a reliable storage history; however, a long storage capability is projected for a properly (dry) built and sealed battery.

Since spin-activated batteries are normally expected to be used in environments where guns are used, they must be built to withstand the forces of gunfire. With the development of the ampules and construction methods described above, such batteries can withstand acceleration to the 20,000- to 25,000-g level, and in small sizes, intended for small caliber (20- to 40-mm) projectiles, g levels 2 to 5 times that high.

The physical and electrical characteristics of several typical spin-activated reserve batteries are presented in Table 37.1.

TABLE 37.1   Typical Spin-Activated Reserve Batteries

Reference	Electrochemical system	Height, cm	Diameter, cm	Weight, g	Nominal voltage, V	Nominal capacity, Wh
Fig. 37.1	Pb/HBF$_4$/PbO$_2$	4.1	5.7	280	35	0.5
Fig. 37.3	Pb/HBF$_4$/PbO$_2$	2.5	3.8	75	1.5	0.05
Fig. 37.6	Li/SOCl$_2$	2.5	10.9	350	30	30
Fig. 37.8	Zn/KOH/AgO	1.3	5.1	80	1.4	0.65

37.4.2 Lead/Fluoboric Acid/Lead Dioxide Battery

Discharge curves for a typical lead/fluoboric acid/lead dioxide liquid reserve battery employed to power the proximity fuze of an artillery shell are given in Fig. 37.5. The slight rise in the low-temperature curve is due to its gradual rise in temperature in a room-temperature spinner. Similarly, the high-temperature curve is falling faster than it would in a true isothermal situation.

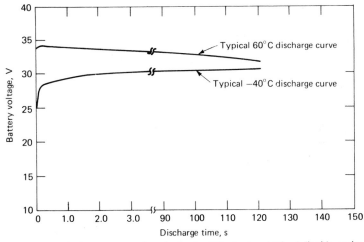

FIG. 37.5   Discharge curves of a spinning lead/fluoboric acid/lead dioxide, series configuration reserve battery (current density: 100 mA/cm²). (*Courtesy of Harry Diamond Laboratories, Department of the Army.*)

37.4.3 Lithium/Thionyl Chloride Battery

Until recently, spin-activated batteries were expected to function for short periods of time and only under sustained spin (necessary to keep the electrolyte within the cells). New applications have arisen that require a battery capable of withstanding artillery fire and spin for a short time followed by some substantial operating time in a nonspin mode. Such applications include artillery delivery of mines or communication jammers intended to function after impact with the ground or projectiles and submunitions that are operative while being slowed down by parachute.

The lithium-based liquid reserve battery holds promise of fulfilling this difficult combination of requirements. A typical cell, as illustrated in Fig. 37.6, incorporates an absorbing separator between the electrodes and a long, high-resistance electrolyte filling path. After cell filling under spin, the absorbing material causes electrolyte retraction away from the manifold and retention of the electrolyte within the cell after cessation of spin. These design features, coupled with the long wet-stand capability of the lithium/thionyl chloride system, have paved the way for reserve batteries in applications that previously had to depend on the use of active batteries with relatively shorter storage capability. The discharge curves for such a multicell, liquid reserve battery are given in Fig. 37.7 (also see Part 5, Chap. 39).

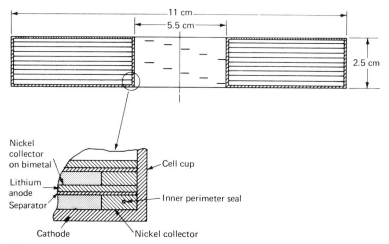

FIG. 37.6   Lithium anode/carbon cathode cell stack for thionyl chloride liquid reserve battery. (*Courtesy of Honeywell Power Sources Center.*)

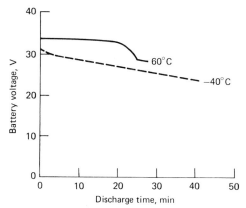

FIG. 37.7   Discharge curve of a lithium/thionyl chloride, series configuration reserve battery (current density: 50 mA/cm²). (*Courtesy of Harry Diamond Laboratories, Department of the Army.*)

37.4.4 Zinc/Silver Oxide Battery

A zinc/silver oxide, reserve, single-cell battery has been developed to power an artillery-delivered, scatterable mine. This cell is illustrated in Fig. 37.8. The low voltage of the cell is raised to the operating voltage by use of a built-in dc/dc converter. Although zinc/silver oxide batteries are considered marginal at low temperatures, they can be activated and will function at −40°C, as shown in Fig. 37.9, if the current density is low enough.

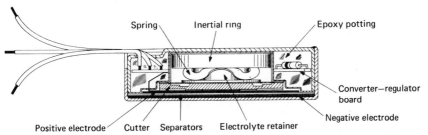

FIG. 37.8 Zinc/silver oxide reserve cell with integral dc/dc converter. (*Courtesy of Harry Diamond Laboratories, Department of the Army.*)

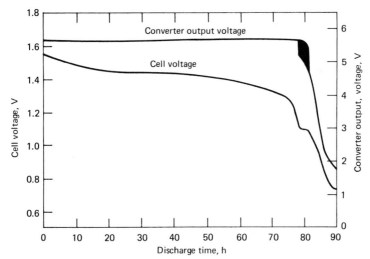

FIG. 37.9 Discharge curve at $-40°C$ of the zinc/silver oxide reserve cell (current density: 1 mA/cm²) and converter output voltage (6800-Ω load across converter output). (*Courtesy of Harry Diamond Laboratories, Department of the Army.*)

BIBLIOGRAPHY

Benderly, Asaf A.: "Power for Ordnance Fuzing," *National Defense* March-April 1974.

Biggar, Allan M.: "Reserve Battery Requiring Two Simultaneous Forces for Activation," *Proc. 24th Annu. Power Sources Symp.,* Electrochemical Society, Pennington, N.J., 1970.

———— R. C. Proestel, and W. H. Steuernagel: "A 48-Hour Reserve Power Supply for a Scatterable Mine," *Proc. 26th Annu. Power Sources Symp.* Electrochemical Society, Pennington, N.J., 1974.

Doddapaneni, H., D. L. Chua, and J. Nelson: "Development of a Spin Activated, High Rate, Li/SOCl₂ Bipolar Reserve Battery," *Proc. 30th Power Sources Symp.,* Electrochemical Society, Pennington, N.J., 1982.

Turrill, Fred G., and Walter C. Kirchberger: "A One-Dollar Power Supply for Proximity Fuzes," *Proc. 24th Annu. Power Sources Symp.,* Electrochemical Society, Pennington, N.J., 1970.

Ammonia
Batteries

by
Peter N. Lensi

38.1 GENERAL CHARACTERISTICS

Ammonia batteries include all those in which the electrolyte solvent is ammonia. The use of ammonia as a battery electrolyte was initiated in 1947 as an outgrowth of a process for electrowinning alkali metals from ore using liquid ammonia solutions as the electrolyte.[1] As active ammonia batteries were found to have poor wet-stand life, development concentrated on reserve battery configurations in which the dry salts of the electrolyte were placed between the electrodes and the battery activated by introducing ammonia vapor to the electrode compartment.[2] Ammonia-activated (AVA) batteries were developed for safing, arming, and fuzing applications with the potential advantage of good low-temperature performance. An undesirable characteristic of the ammonia vapor battery was that the activation was slow (several minutes) and often nonuniform; development was then directed toward liquid ammonia reserve batteries.[3-6]

The general characteristics of liquid ammonia batteries are:

1. Wet cell life is limited, due to corrosiveness of the ammonia electrolyte; therefore, practical battery designs are of the reserve type.

2. The shelf life for nonactivated reserve ammonia batteries is in excess of 10 y and is unaffected by normal storage temperature (-70 to $74°C$).

3. The low viscosity of liquid ammonia (approx. 0.35 cp at $-55°C$) enhances activation times (10- to 300-ms activation time).

4. The high conductivity of ammonia electrolytes over a wide temperature range results in operation at temperatures between -55 and $74°C$, with little change in voltage or capacity.

5. Hermetic seals as well as heavy case strength are required to contain the high pressure of ammonia.

6. Open-circuit voltages vary from 1.1 to 3.0 V per cell, depending on the electrochemical system that is used.

7. Nominal current densities are dependent on the electrochemical system selected; e.g., Mg/PbO_2 can be rated at 50 mA/cm^2 while Mg/m-DNB may be rated at 20 mA/cm^2.

8. Energy densities for short-term (minutes) discharge are up to 0.10 Wh/cm^3 and 45 Wh/kg and are up to 0.13 Wh/cm^3 and 60 Wh/kg for long-term (days) discharge.

38.2 CHEMISTRY

Many electrochemical couples can operate in ammonia-solvent electrolytes. Of these, battery development has centered on the electrochemical couples listed in Table 38.1. The only ammonia battery system that has advanced to major production uses the Mg/m-DNB couple.

TABLE 38.1 Open-Circuit Voltages for Ammonia Electrolyte Electrochemical Couples

System	OCV, V
$Mg/NH_4SCN/m$-DNB	2.2
$Mg/NH_4SCN/HgSO_4$	2.2
$Mg/NH_4SCN/PbO_2$	3.0
$Zn/NH_4SCN/PbO_2$	1.9
$Pb/NH_4SCN/PbO_2$	1.1
$Mg/KSCN/S$	2.3

In the Mg/m-DNB batteries, the anode is magnesium (Mg), the active ingredient in the cathode is meta-dinitrobenzene (m-DNB), and the electrolyte salt system is based on ammonium thiocyanate (NH_4SCN) and potassium thiocyanate (KSCN). The salts, in a dry condition, are stored in the electrode stack. KSCN is a neutral salt and is stored in the separator in direct contact with the magnesium anode and the m-DNB cathode. Placement of the KSCN directly in the separator provides shorter battery rise time than does placement elsewhere. NH_4SCN is stored in the cathode because it is reactive with magnesium. The cathodes are made by mixing paper pulp, carbon, m-DNB, and the NH_4SCN together and pressing into cathode pads. The NH_4SCN salt plays several roles; primarily, it ensures ionic conductivity; secondly, it provides the acid environment which enhances activity at the anode; and thirdly, by solvation with ammonia it reduces the vapor pressure of ammonia. When the ammonia is introduced into the battery cell, it combines with the catholyte and anolyte to form the electrolyte

$$NH_4SCN \xrightarrow{NH_3} NH_4^+ + SCN^-$$

$$KSCN \xrightarrow{NH_3} K^+ + SCN^-$$

At the anode, the magnesium is oxidized

$$Mg \xrightarrow{NH_3} Mg^{2+} + 2e$$

At the cathode the m-DNB undergoes reduction

$$\text{(ring with -NO}_2\text{, -NO}_2\text{)} + 8\,NH_4^+ + 8e \rightarrow \text{(ring with -NHOH, -NHOH)} + 6\,NH_3 + 2NH_4OH$$

The complete cell reaction is summarized as

$$4Mg + 8\,NH_4SCN + \text{(ring with -NO}_2\text{, -NO}_2\text{)} \rightarrow \text{(ring with -NHOH, -NHOH)}$$
$$+ 4Mg(SCN)_2 + 6NH_3 + 2\,NH_4OH$$

In the high-rate discharge ammonia systems, using magnesium anodes and heavy metal sulfates, typical reactions are as follows:
When the ammonia enters the cell

$$KSCN \overset{NH_3}{\rightarrow} K^+ + SCN^-$$

At the anode

$$Mg \overset{NH_3}{\rightarrow} Mg^{2+} + 2e$$

At the cathode

$$HgSO_4 + 2K^+ + 2e \rightarrow Hg + K_2SO_4$$

A limiting factor in the magnesium/heavy metal sulfate system is that relatively insoluble products are formed at the anode (magnesium amide) and at the cathode (ammonium sulfate). The precipitation of these two compounds tends to restrict further reaction at anode and cathode.

38.3 CONSTRUCTION

The ammonia battery cell consists of an anode, a cathode, and an electrolyte, plus a separator which prevents direct contact between electrodes. The salts of the electrolyte are deposited on the separator and on, or in, the cathode at the time of construction and become conductive when ammonia enters the cell and dissolves them. As the active cathode ingredient itself is relatively nonconductive and the cathode base material a poor conductor, a metallic plate conductor is used as a positive (cathode) terminal. In production models, magnesium is used as the anode, filter paper serves as the separator, and paper pulp or glass fiber plus carbon forms the binder and conductor for the active-cathode material. Potassium thiocyanate (KSCN) as the anolyte and ammonium thiocyanate (NH₄SCN) as the catholyte compose the electrolyte solution. The two most popular active cathode materials have been m-dinitrobenzene (m-DNB) and mercuric sulfate (HgSO₄). Carbon is added in the formation of the cathode pads to increase conductivity to the metallic cathode electron collector plate.

The flat-plate configuration is the only cell design which has been used for volume production of ammonia batteries (Fig. 38.1). The components are flat disk-shaped plates, usually with holes in the center (similar to flattened doughnuts). In some designs these flat plates are divided into segments. The disks are stacked with the cathode collector plate and the cathode on the bottom, then the separator, and then the anode, in repeating units, to form the cell stack.

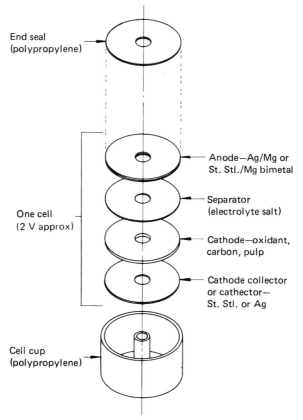

End seal
(polypropylene)

Anode—Ag/Mg or
St. Stl./Mg bimetal

Separator
(electrolyte salt)

One cell
(2 V approx)

Cathode—oxidant,
carbon, pulp

Cathode collector
or cathector—
St. Stl. or Ag

Cell cup
(polypropylene)

FIG. 38.1 Components of liquid ammonia cell. (*Courtesy of Honeywell, Inc., Power Systems Division.*)

The cell stack is positioned in a plastic cell cup to prevent shorting to the steel housing of the battery. The plastic cup is designed to interlock with the metallic electron collector plates to form a tight electrolyte seal between cells in order to prevent or reduce intercell electrolyte leakage. The channels used to transport the liquid ammonia to the individual cells are kept to a minimum diameter. Liquid ammonia is relatively nonconductive, but the intercell diffusion of the conductive electrolyte, formed by the liquid ammonia and the salts at the anode-cathode interface, is inhibited, to some degree, by the narrow transport channels.

A typical liquid ammonia reserve battery (Honeywell model G2514) is shown in an exploded view in Fig. 38.2 and in a cutaway view in Fig. 38.3. Activation of the battery is by an external force, including gun firing setback, shown in Fig. 38.3. When the glass ampul is broken, the ammonia passes through ports in the bulkhead, combines with salts in the cells to form electrolyte, and activates the battery. The vapor pressure of ammonia is sufficient to give adequate distribution to the cells for loads up to 40 mA at room temperature, but in order to induce cell operation at higher loads or at low temperature, the liquid ammonia distribution must be aided by mechanical force, such as spinning or setback.

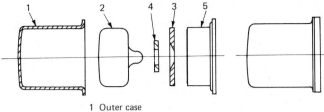

1 Outer case
2 Electrolyte glass ampul
3 Bulkhead
4 Cushion gasket
5 Cover/electrode stack assembly

FIG. 38.2 Exploded view of liquid ammonia reserve battery, Honeywell model G2514. (*Courtesy of Honeywell, Inc., Power Systems Division.*)

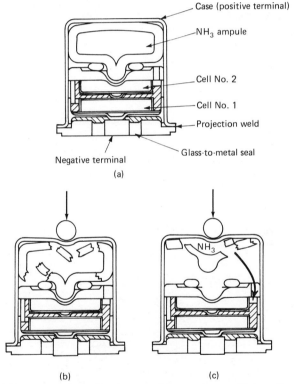

FIG. 38.3 Activation of liquid ammonia reserve battery, Honeywell model G2514: (*a*) inactive cell; (*b*) ampul broken by external force; (*c*) ammonia activates battery stack. (*Courtesy of Honeywell, Inc., Power Systems Division.*)

The other major ammonia battery is built with an internal activation system, as shown in Fig. 38.4 (Honeywell model G2492). When the pyrotechnic gas generator is initiated by the percussion primer, gas pressure causes the ammonia reservoir to collapse, driving the lance, which is attached to the reservoir wall, to penetrate the

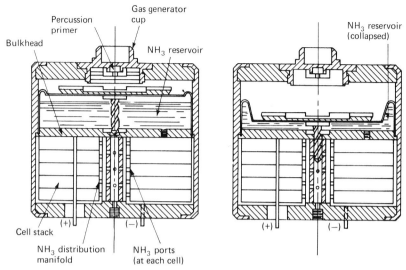

FIG. 38.4 Pyrotechnic gas generator activation of liquid ammonia reserve battery, Honeywell model G2492. (*Courtesy of Honeywell, Inc., Power Systems Division.*)

bulkhead. Ammonia is forced into the cells, dissolves the electrolyte salts, and activates the cell stack. Gas generator–activated batteries are generally slower in activation than setback-activated or spin-activated batteries (approximately 300 vs. 100 vs. 10 ms).

Another liquid ammonia battery, in limited production, is shown in Fig. 38.5 (Honeywell model EX-90). While similar to the model G2492 in that it has a gas generator to pressurize the reservoir compartment for activation, the electrode stack

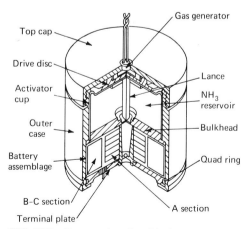

FIG. 38.5 Cross section of multivoltage, multistack liquid ammonia reserve battery, Honeywell model EX-90. (*Courtesy of Honeywell, Inc., Power Systems Division.*)

is more complex. It uses a center stack composed of 15 flat disk-shaped cells, each cell connected in parallel, to form a 2-V section, and an outer ring divided into six segments of bipolar cells. One segment with a 16-cell stack is for the 28- to 30-V section, and the remaining five bipolar segments are connected in series for a total series string of 98 cells to produce the 150- to 180-V section.

38.4 PERFORMANCE CHARACTERISTICS

The two-cell liquid ammonia reserve battery, Honeywell model G2514, was developed to fill the need for a power source that combined long storage life with an operating life of up to several weeks for use in low-power ordnance devices. The characteristics of this battery are listed in Table 38.2. The capacity of the battery, as a function of temperature, is shown in Figs. 38.6 and 38.7 for two different spin orientations, but the

TABLE 38.2   Honeywell G2514 Battery Characteristics

Size	1.3 cm diam. $\times$ 1.47 cm height
Weight	6 g
Case	Stainless steel
Terminal seal	Glass-to-metal
Closure	Welded
Activation	External impact
Activation time	Approx. 200 ms
Open-circuit voltage	4.25 V
Short-circuit voltage	Approx. 200 mA
Capacity	14 mAh (at 10-h rate)
Operating temperature	-54 to $+74°C$
Storage	> 10 y

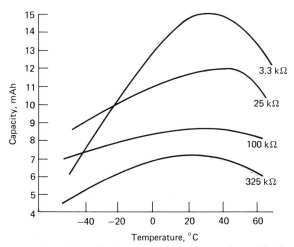

FIG. 38.6   Effect of temperature on battery capacity, axial spin activation, Honeywell model G2514. (*Courtesy of Honeywell, Inc., Power Systems Division.*)

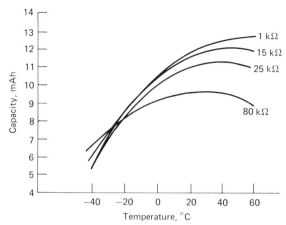

FIG. 38.7 Effect of temperature on battery capacity, transverse spin activation, Honeywell model G2514. (*Courtesy of Honeywell, Inc., Power Systems Division.*)

battery was designed and optimized for axial spin initiation. Typical discharge curves for this battery at several discharge temperatures are shown in Fig. 38.8. These data show the similarity of voltage and capacity over a very wide range of operating temperatures; the low-temperature capacity is within 60 to 80% of the high-temperature capacity.

The characteristics of the five-cell gas generator–activated liquid ammonia (Honeywell model G2492) fuze battery are listed in Table 38.3. This power source has a nominal operating output of 9 V and supplies a current of 12 mA above 7 V for 70 h, well above the required 36 h, and over 100 h at a 2-mA load, over the operating range of −55 to 70°C. Typical discharge curves for this battery are shown in Fig. 38.9. Activation is completed in less than 500 ms, as shown in Fig. 38.10.

The Honeyweil model EX-90 multivoltage liquid ammonia fuze battery also uses the magnesium/meta-dinitrobenzene system. The typical performance of the battery for the three sections at −55 and 74°C are shown in Fig. 38.11. Voltage regulation is within the specified $\pm 5\%$. The batteries were also subjected to typical missile environment test for shock, vibration, and acceleration without evidence of damage or effect on discharge characteristics.

38.5 BATTERY TYPES AND CHARACTERISTICS

The major attraction of the liquid ammonia reserve battery is its long shelf life (in excess of 10 y) combined with its ability to operate over a wide temperature range (−55 to 75°C) after activation. The battery can be expected to deliver a given voltage reliably at any temperature within that operating range because of a low voltage-temperature coefficient, and it can reach the operating voltage within an extremely short activation period (milliseconds). The system is flexible to electrode stack design, utilizing bobbin, spiral-wound, flat-plate concepts as well as bipolar a:.d series or parallel-cell arrangements, and can be configured to meet a variety of capacities, voltages, and load require-

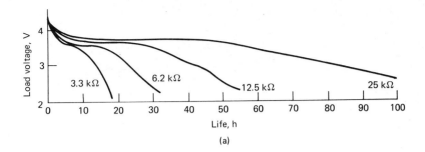

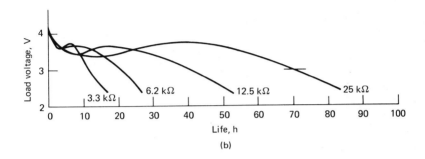

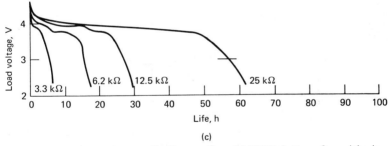

FIG. 38.8 Discharge voltage profile, Honeywell model G2514, battery after axial spin activation: (*a*) 20°C; (*b*) 70°C; (*c*) −55°C. (*Courtesy of Honeywell, Inc., Power Systems Division.*)

TABLE 38.3 Honeywell Model G2492 Battery Characteristics

Size	6.6 cm diam. × 5.1 cm height
Weight	600 g
Activation	Gas generator, percussion primer–initiated
Rise time	9 V within 500 ms
Voltage	9 V
Capacity	12 mA for 36 h (minimum)
Operating temperature	−55 to 70°C
Shelf life	> 10 y

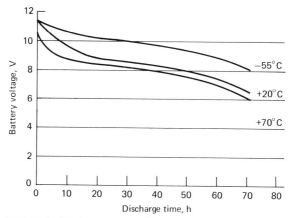

FIG. 38.9 Discharge curves, Honeywell model G2492 (load: 833 Ω). (*Courtesy of Honeywell, Inc., Power Systems Division.*)

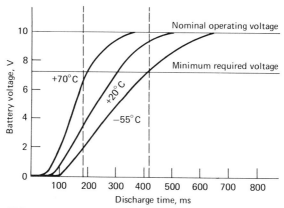

FIG. 38.10 Voltage rise in gas generator–activated battery, Honeywell model G2492. (*Courtesy of Honeywell, Inc., Power Systems Division.*)

ments. No costly, rare, or highly dangerous materials are used. The battery is thus well-suited for ordnance and selected space applications.

The liquid ammonia reserve batteries that have advanced to major production use the magnesium/meta-dinitrobenzene couple and the flat-plate configuration. The two batteries described in Sec. 38.6.4 (Honeywell models G2514 and G2492) have been produced in large quantities approximating 2 million units for use in low-power ordnance devices; Honeywell model EX-90 is in limited production. Few others have gone beyond the development stage. The only manufacturer of these batteries is Honeywell, Inc., Power Sources Center, Horsham, Pa. and St. Louis Park, Minn.

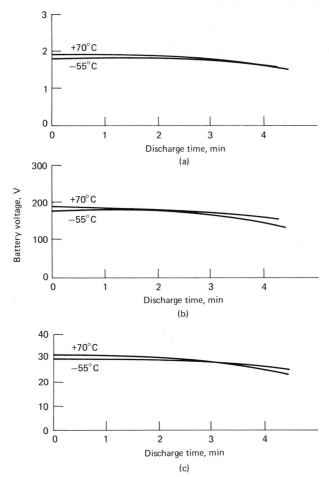

FIG. 38.11 Discharge curves, Honeywell model EX-90: (*a*) 15 cells in parallel, load = 2.7 Ω, current drain = 667 mA; (*b*) 98 cells in series, load = 7500 Ω, current drain = 24 mA; (*c*) 16 cells in series, load = 150 Ω, current drain = 20 mA. (*Courtesy of Honeywell, Inc., Power Systems Division.*)

REFERENCES

1. L. J. Minnick and C. Presgrave (to G. and W. H. Corson, Inc.), U.S. Patent 2,615,838, Oct. 28, 1952.

2. H. S. Gleason and L. J. Minnick, "Ammonia Vapor Activated Batteries," *Proc. 11th Annu. Battery Res. Dev. Conf.,* Electrochemical Society, Pennington, N.J., 1957.

3. D. J. Doan and L. R. Wood, "Liquid NH₃ Battery Systems," *Proc. 16th Annu. Power Sources Conf.,* Electrochemical Society, Pennington, N.J., 1962.

4. L. J. Minnick, "Ammonia Batteries," *Proc. 17th Annu. Power Sources Conf.,* Electrochemical Society, Pennington, N.J., 1963.

5. H. R. Smith and B. C. Tierney, "Ammonia Batteries," *Proc. 19th Power Sources Conf.,* Electrochemical Society, Pennington, N.J., 1965.

6. John M. Freund and William C. Spindler, "Low Temperature Nonaqueous Cells," in G. W. Heise and N. C. Cahoon (eds.), *The Primary Battery,* vol. I, Wiley, New York, 1971, chap. 10.

39

Ambient-Temperature Lithium Anode Reserve Batteries

by
David L. Chua, William J. Eppley, and Robert J. Horning

39.1 GENERAL CHARACTERISTICS

The use of lithium metal as an anode in reserve batteries provides a significant energy advantage over the traditional reserve batteries because of the high potential and low equivalent weight (3.86 Ah/g) of lithium. A lithium reserve cell can operate at a voltage close to twice that of the conventional aqueous types. Due to the reactivity of lithium in aqueous electrolytes, with the exception of the special lithium-water and lithium-air cells (Part 5, Sec. 37.3 and Part 4, Sec. 30.3), lithium cells must use a nonaqueous electrolyte with which lithium is nonreactive. Lithium anode cells operating at ambient temperatures use organic or inorganic liquid electrolytes (Part 2, Chap. 11) or solid electrolytes (Part 2, Chap. 12). High-temperature cells may use molten electrolytes (Part 4, Chap. 31 and Part 5, Chap. 40).

The various ambient-temperature active (nonreserve) lithium cells are covered in Part 2, Chap. 11. Of these systems, the ones demonstrating the higher energy densities and rate capabilities are Li/SO_2, Li/V_2O_5, and $Li/SOCl_2$. The discharge characteristics of these cells are shown in Fig. 39.1. These are the electrochemical systems that are employed in the reserve-type configurations.

In the reserve construction, the electrolyte is separated from the electrode active materials until the battery is used and is stored in a reservoir. This reserve design feature provides a capability of essentially undiminished output even after storage periods, in the inactive state, of over 10 y. The reserve feature results in an energy density penalty

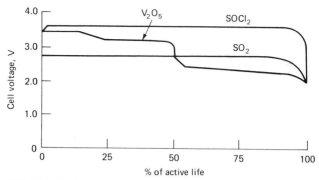

FIG. 39.1   Performance comparison of lithium anode primary systems at 20°C [thionyl chloride (SOCl₂) = 3.6 V; vanadium pentoxide (V₂O₅) = 3.4 V; sulfur dioxide (SO₂) = 2.9 V].

of as much as 50% compared with the active lithium primary cells. Key contributors to this penalty are the activation device and the electrolyte reservoir.

In the selection of a lithium anode electrochemical system for packaging into a reserve battery, besides such important considerations as physical properties of the electrolyte solution and performance as a function of the discharge conditions, factors such as the stability of the electrolyte and the compatibility of the electrolyte with the materials of construction of the electrolyte reservoir are of specific importance.

39.2 CHEMISTRY

39.2.1 Lithium/Vanadium Pentoxide (Li/V₂O₅) Cell

The basic cell structure of this system consists of a lithium anode, a nonwoven polypropylene separator, and a cathode that is usually composed of 90% V_2O_5 and 10% graphite, on a weight basis. When it is used in a reserve battery, the prevalent electrolyte solution is $2M$ LiAsF₆ + $0.4M$ LiBF₄ in methyl formate (MF) because of its excellent stability during long-term storage.

As shown in Fig. 39.1, the Li/V₂O₅ system has a two-plateau discharge characteristic. A net cell reaction, involving the incorporation of Li in V₂O₅, has been postulated to account for the first plateau:

$$Li + V_2O_5 \rightarrow LiV_2O_5$$

The initial voltage level ranges from 3.4 to 3.3 V, decreases to 3.2 to 3.3 V for approximately 50% of the active life of the first discharge plateau, at which point the range again decreases to a level of 3.1 to 3.2 V which is maintained for the balance of the first plateau of discharge. After completion of the first plateau, the Li/V₂O₅ system undergoes a rapid change in voltage to the second discharge plateau around a voltage range of 2.3 and 2.4 V. This step involves the formation of reduced forms of V₂O₅, although specific mechanisms remain unclear.[1] This second plateau is relatively more sensitive to temperature and discharge rate, and it is for this reason that most Li/V₂O₅ cells (active and reserve) are designed to operate at only the first discharge plateau level.[2]

The long-term storage capability of the Li/V_2O_5 reserve cells is heavily dependent on the stability of the electrolyte solution. $LiAsF_6$:MF electrolyte is unstable due to the decomposition reactions involving the hydrolysis of methyl formate followed by the dehydration of the hydrolysis product(s).[3] These reactions result in a premature fracture of the glass ampul used as the electrolyte reservoir. The stability of the $LiAsF_6$:MF electrolyte solution was achieved by making the solution either neutral or alkaline. In practice, this is accomplished by using a double-salt electrolyte ($LiAsF_6 + LiBF_4$:MF) and by immersing lithium metal in the glass ampul.

39.2.2 Lithium/Thionyl Chloride (Li/SOCl₂)

The basic cell structure that is generally used for this system consists of a lithium anode, a nonwoven glass separator, and a Teflon-bonded carbon cathode which serves only as the reaction site medium. One unique feature of this chemistry is the fact that thionyl chloride ($SOCl_2$) serves two functions—as the solvent of the commonly used $LiAlCl_4$:$SOCl_2$ electrolyte solution and as the active cathode material (see Part 2, Sec. 11.6).

Figure 39.1 shows the marked advantage in discharge performance of the $Li/SOCl_2$ system. The popularly accepted net cell reaction for this system is:

$$4 \ Li + 2 \ SOCl_2 \rightarrow 4 \ LiCl + S + SO_2$$

Most of the sulfur dioxide formed during discharge is dissolved in the electrolyte [possibly by partial chemical bond(s)], and practically no gas pressure is generated.[4] Depending on the discharge rate and temperature, the $Li/SOCl_2$ system normally exhibits a working voltage range between 3.0 and 3.6 with a flat discharge characteristic. These excellent discharge characteristics—high voltage and flat discharge—are best attained in a reserve cell, especially if the designed discharge current density is high. In an active primary $Li/SOCl_2$ cell, the lithium anode is coated with a passive LiCl film. Under sustained storage coupled with high-temperature storage, the passivating film will limit the current-handling capability of the anode as well as increasing the time required to reach operating voltage.[4]

The conventional $LiAlCl_4$:$SOCl_2$ electrolyte solution has been proved to have excellent stability. Electrolyte glass ampuls exposed to $+74°C$ did not show any sign of apparent degradation up to at least 7 y of storage. Because of this and the overall performance superiority offered by this system, the $Li/SOCl_2$ reserve cell is a serious contender for the next generation of high-energy reserve cells.

Recently, the use of excess $AlCl_3$ Lewis acid in the conventional $LiAlCl_4$:$SOCl_2$ electrolyte solution was shown to improve the rate capability of the $Li/SOCl_2$ system.[6,7] It should be noted, however, that the inherent stability of this high-rate electrolyte has yet to be established so as to ensure its application in reserve cells.

39.2.3 Lithium/Sulfur Dioxide (Li/SO₂)

The Li/SO_2 system uses a basic cell structure consisting of a lithium anode, a separator, and a Teflonated carbon cathode, similar to one used in the $Li/SOCl_2$ system, which serves as the reaction site medium. The electrolyte solution commonly employed contains a mixture of lithium bromide (LiBr), acetonitrile (AN), and sulfur dioxide (SO_2), which also serves as the active cathode material.

One serious problem in using the $LiBr:AN:SO_2$ electrolyte solution for reserve cells is its instability during storage. Although this electrolyte solution is commonly employed in active primary cells, it is unsuitable for reserve battery applications because it decomposes to form highly reactive and solid products when stored in the absence of cell components. Replacing LiBr with lithium hexafluoroarsenate ($LiAsF_6$) results in an electrolyte solution with good stability. The functional performance of the $LiAsF_6$ electrolyte is equivalent or superior to that of the LiBr solution for low to moderate rates.[8-10]

The Li/SO_2 reserve battery, using the stable $LiAsF_6:AN:SO_2$ electrolyte solution, follows the same net cell reaction

$$2\ Li + 2\ SO_2 \rightarrow Li_2S_2O_4$$

as in the active primary cell. It should be noted, however, that $LiAsF_6:AN:SO_2$ electrolyte solution is limited to moderate- or lower-rate application due to poorer electrolyte conductivity. Because of this, the emphasis of using the Li/SO_2 system for reserve applications has been shifted to the higher-performance $Li/SOCl_2$ system.

39.3 CONSTRUCTION

39.3.1 General Considerations

Lithium anode reserve batteries are basically composed of three major components:

Activation and electrolyte delivery system

Electrolyte reservoir

Cell and/or battery unit

However, the actual design can vary widely depending on the application. The design can vary from a simple, small, single cell with an ampul, manually activated, to a very large, complex, multicell battery with an automatic electric initiation mechanism to transfer the electrolyte from a bellows chamber to a high-voltage multicell battery stack. Both the electrodes and hardware components are essentially the same as the primary active units but with allowances made for the electrolyte storage and electrolyte delivery into the cells at the time of activation. In addition, the electrochemical and hardware components must be constructed of a rugged maintenance-free design to survive severe environmental and performance requirements as most are used in military or special applications. For example, Table 39.1 lists typical requirements of lithium reserve batteries and illustrates the reason for many of their unique construction and design features.

Some common construction features are used in the design of lithium reserve batteries. The outer case is generally made of a 300-series stainless steel since it offers the corrosion resistance against both the internal system and the external environment during its long-term use. Various welding techniques such as laser, tungsten inert gas (TIG), resistance, and election beam can be applied to the 300-series stainless steel. Thus the outer case provides a true 20-y reliable storage life, capable of maintaining the hermeticity required for reserve lithium batteries. The electrical terminals used are generally glass-to-metal types, which also provide the hermeticity required for long-term storage.

TABLE 39.1 Typical Requirements of Lithium Anode Reserve Batteries

---

Operating temperature range of -55 to $70°C$
10 to 20 y unactivated storage life
Hermetically sealed
High energy density
High reliability
Low electrical noise
Flat discharge voltage profile
Rapid voltage rise after initiation
Mechanical environmental capability:
    Acceleration shocks up to 20,000 g
    High spin up to 20,000 rpm
    Vibration
Operating life from several seconds up to 1 y

39.3.2 Types of Lithium Anode Reserve Batteries

Three basic lithium reserve battery types are being manufactured at the present time; they are:

1. Single-cell battery with electrolyte stored in a glass ampul

2. Multiple single cells using bellows for the electrolyte storage reservoir

3. Multicells of bipolar construction with either a glass ampul or a metal reservoir for electrolyte storage

Ampul Type Single-cell reserve types using an ampul as the electrolyte storage reservoir are the most reliable of the reserve designs due to their simple construction and lack of intercell leakage problems associated with multicell ampul types. One group of these cells is sized to the standard ANSI specifications, and the other group consists of those cells built for special-purpose applications which are not sized to the ANSI specifications. Both groups, however, are very similar in construction.

Figure 39.2 shows the cross section of a reserve lithium anode cell in an A size configuration of about 1 Ah, using the Li/SOCl₂ system.[7] The cell consists of concentrically arranged components; a li-

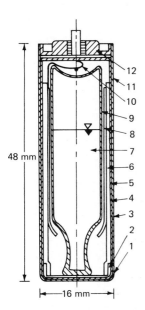

48 mm

16 mm

1 Insulator	7 Electrolyte
2 Bottom separator	8 Current collector
3 Cell can	9 Glass ampul
4 Lithium anode	10 Positive terminal tab
5 Separator	11 Top spacer
6 Carbon cathode	12 Cell cover

FIG. 39.2 Cross section of Li/SOCl₂ A size reserve cell. (*Courtesy of Tadiran Industries, Ltd.*)

thium anode is swagged against the inner wall of a stainless-steel cylindrical can. A nonwoven glass separator is located adjacent to the anode. The Teflon-bonded carbon cathode is inserted against the separator. A cylindrical nickel current collector provides the electrical contact to the positive terminal and houses the hermetically sealed glass ampul. The ampul is held firmly in place by upper and lower insulating supports which protect the ampul from premature breakage while permitting transmission of a direct force at the bottom of the case to shatter the ampul at the time of activation. The unit is hermetically sealed to ensure long shelf life in the unactivated condition. Activation is achieved by applying a sharp directed force at the bottom of the cell case to shatter the glass ampul. The electrolyte, a solution of lithium tetrachloroaluminate with an excess of aluminum chloride in thionyl chloride, is absorbed by the porous cathode and the glass separator, thereby activating the battery.

Another design has been developed for mine and fuze applications, using both the Li/V_2O_5 and $Li/SOCl_2$ systems, in the capacity range of 100 to 500 mAh.[11] The cross sections of these two cells are shown in Figs. 39.3 (Li/V_2O_5) and 39.4 ($Li/SOCl_2$). Both cells are similar with respect to the external hardware and the internal arrangement of the components. The case and header assembly are projection-welded together at the case flange. The header serves as the cover for the cells and incorporates a glass-to-metal seal for the center terminal pin made of nickel/iron alloy. The terminal pin has negative polarity (both cell designs), and the balance of the header and case surface have positive polarity. The hermetically sealed hardware in conjunction with the reserve feature of the design makes it possible to achieve storage in excess of 20 y.

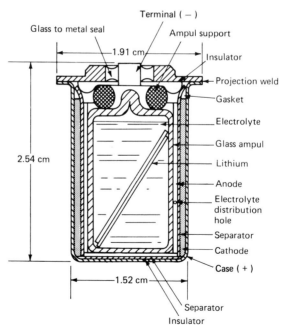

FIG. 39.3 Cross section of Li/V_2O_5 reserve cell, Honeywell model G2659.

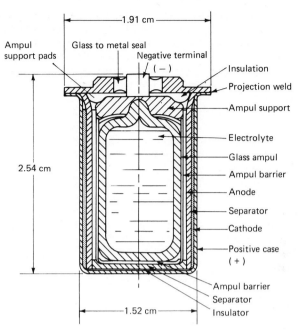

FIG. 39.4 Cross section of Li/SOCl$_2$ reserve cell, Honeywell model G2659B1.

The internal arrangement of the components consists of annularly located electrodes about a central glass ampul used as the electrolyte solution reservoir. In addition, there are various insulating components in the upper and lower portions of the cell used to prevent internal shorting.

Several features account for most of the design differences between these two cells. In the Li/SOCl$_2$ reserve cell, the glass ampul also contains the cathode oxidant, SOCl$_2$, while the cathode oxidant of the Li/V$_2$O$_5$ reserve cell is contained in the cathode structure. Directly adjacent to the Li/SOCl$_2$ cell case is the Teflonated carbon, while in the case of the Li/V$_2$O$_5$ cell, the cathode is molded from a dry mixture of V$_2$O$_5$ and graphite. The Teflonated-carbon cathode for the reduction of SOCl$_2$ is made in sheet form and is attached to a metal grid rolled to shape and inserted against the inside wall of the case. Another difference is the way the electrical connection is made for the two cathodes. The V$_2$O$_5$ connection is made by the direct-pressure contact of the molded cathode, whereas with the SOCl$_2$ system the cathode lead is welded to the case at the time the cover is welded. The lithium anode structure consists of pure lithium metal which is pressed onto an expanded metal grid anode of 316L stainless steel. One end of a flat 316L stainless-steel lead is spot welded to the pin of the glass-to-metal seal. Rolled into a cylinder, the anode is inserted into the cell next to the separator. Both cells are provided with ampul support in order to survive the shock environment specified. In the Li/SOCl$_2$ system, Teflon and glass have been found to be chemically stable for use as insulators, separators, and supports. The Li/V$_2$O$_5$ system allows more flexibility because many rubbers and plastics can be used.

Multicell Single Activator Design   For those applications where higher than single cell voltages are required, a battery is constructed of two or more cells, depending, of course, on the voltage needed. Typical voltages are 12 and 28 V and, for lithium anode cells with a 2.7- to 3.3-V operating voltage, this would require anywhere from 4 to 10 cells for each battery. This family of batteries is unique with respect to the method of cell activation and the containment of electrolyte in multiple cells initiated from a single self-contained reservoir of electrolyte. Batteries of this design are used in preference to the bipolar type to achieve higher cell capacities and to allow discharge times up to 1 y or more through the tight control of intercell leakage. The leakage currents are controlled and limited to usually less than several percent of the discharge current. This feature, however, limits these batteries from being miniaturized which is possible with many bipolar designs.

The most advanced design using this concept is a battery using the Li/SO_2 system illustrated in Fig. 39.5. The battery is cylindrical and contains three main components: (1) the electrolyte storage reservoir section, (2) the electrolyte manifold and activation system, and (3) the reserve cell compartment. About one-half of the internal battery volume contains the electrolyte reservoir. The reservoir section consists primarily of a collapsible bellows in which the electrolyte solution, a solution of SO_2, CH_3CN, and electrolyte salt, is stored. Surrounding the bellows, between it and the outer battery case, is volume that holds a specific amount of Freon gas/liquid. The Freon gas is selected such that its vapor pressure always exceeds that of the electrolyte, thereby providing the driving force for eventual liquid transfer into the cell chamber section once the battery has been activated.

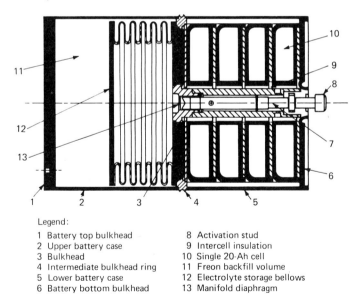

Legend:
1 Battery top bulkhead
2 Upper battery case
3 Bulkhead
4 Intermediate bulkhead ring
5 Lower battery case
6 Battery bottom bulkhead
7 Activation manifold
8 Activation stud
9 Intercell insulation
10 Single 20-Ah cell
11 Freon backfill volume
12 Electrolyte storage bellows
13 Manifold diaphragm

FIG. 39.5   Cross section of a 20-Ah Li/SO_2 multicell battery.

In the remaining half of the battery volume is the centrally located electrolyte manifold and activation system housed in a 1.588-cm-diameter tubular structure plus the series stack of four torroidally shaped cells that surround the manifold/activation system.

The manifold and cells are separated from the reservoir by an intermediate bulkhead. In the bulkhead, there is a centrally positioned diaphragm of thin-enough section to be pierced by the cutter contained within the manifold. In fabrication, the diaphragm is assembled as part of the tubular manifold which, in turn, is welded as a subassembly to the intermediate bulkhead. Figure 39.6 is a more detailed cross-sectional view of the electrolyte manifold and activation system with the major components identified.

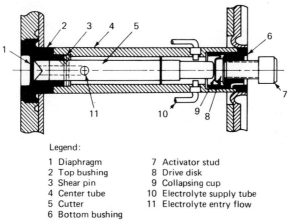

Legend:

1 Diaphragm	7 Activator stud
2 Top bushing	8 Drive disk
3 Shear pin	9 Collapsing cup
4 Center tube	10 Electrolyte supply tube
5 Cutter	11 Electrolyte entry flow
6 Bottom bushing	

FIG. 39.6   Electrolyte manifold and activation system cross section.

The activating mechanism consists of a cutter that is manually moved into the diaphragm, cutting it and thereby allowing electrolyte to flow. The movement of the cutter is accomplished by the turning of an external screw that is accessible in the bottom base of the battery. The cutter section and the screw mechanism are isolated from one another by a small collapsible metal cup that is hermetically sealed between the two sections. This prevents external electrolyte leakage. The manifold section is a series of small nonconductive plastic tubes connected to one end of the central cylinder and to each of the individual cells at the other end. The long length and small cross-sectional area of the tubes minimize intercell leakage losses during the period of time that electrolyte is present in the manifold structure.

In this application, four individual cells are required to meet the voltage requirement. (The number of cells is, of course, adjustable with minor modification to meet a wide range of voltage needs.) Each cell contains flat circular anodes and cathodes that are separately wired in parallel to achieve the individual cell capacity and plate area needed for a given set of requirements. To fabricate, the components, with intervening separators, are alternately stacked around the cell center tube, after which the parallel connections are made. The cells are individually welded about the inner tube and outer perimeter to form hermetic units ready for series stacking within the battery. Connections from the cells are made to external terminals which are located in the bottom bulkhead of the battery.

Figure 39.7 is a photograph of the major battery components prior to assembly. The components shown are fabricated primarily from 321 stainless steel, and the construc-

FIG. 39.7 Pictorial view of a 20-Ah Li/SO$_2$ multicell battery.

tion is accomplished with a series of TIG welds. The hardware shown is designed specifically for use with the lithium/sulfur dioxide electrochemical system; however, it is adaptable, with minor modifications, to other liquid and solid oxidant systems. The battery can also be adapted to electrical rather than manual activation.

Multicell Bipolar Construction with Single Activator Reservoir Lithium anode reserve batteries using bipolar construction, are relatively new and the trend is to adapt the lithium reserve cells to bipolar design such as the one shown in Fig. 39.8 for a 28-V ammonia battery. Advantages of the bipolar construction—one component is used as both the anode collector of one cell and the cathode collector of the next cell in the stack—are several:

Very high energy and power density for high-voltage batteries

Rugged construction to withstand spin and setback forces from artillery firing

Flexibility to adjust voltages in the cell stack

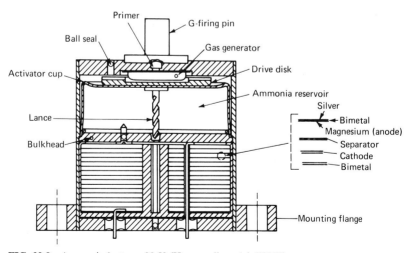

FIG. 39.8 Ammonia battery, 28 V (Honeywell model G3023).

Adaptability to varying energy and power requirements

Enhancement of voltage rise after activation

Figure 39.8 shows all the major components of a bipolar multicell reserve battery and illustrates the rigid construction used. There are four key components for this type of battery: initiator, activation energy, electrolyte reservoir, and bipolar cell stack.

Activation of the reserve battery is accomplished by supplying an electrical pulse to the battery by firing an electric squib or actuator or by some mechanical means. This type of reserve battery has been used chiefly in artillery shells for electronic fuze power supplies and in missiles for the electronic power supply. Therefore, the electric pulse can be supplied prior to firing or at the time of launch. However, for artillery fuze power supplies, the battery is usually activated by the launch acceleration (set back) and/or the spin forces. The acceleration force of the artillery shell releases a firing pin which strikes and fires a primer. The primer can ignite a gas generator or directly release a stored gas by breaking open a metal diaphragm.

Once the battery has been initiated as described above, the gas pressure (from a gas generator or stored Freon or CO_2, etc.) forces the electrolyte into each of the cells through a manifold (electrolyte distribution network). For the battery shown in Fig. 39.8, the gas pressure collapses the metal reservoir cup and at the same time drives a lance through the diaphragm in the bulkhead, releasing the electrolyte to the cell stack.

The electrolyte reservoir is generally made using a collapsible cup, a bellows, or a wound tubing design. These serve to hold the electrolyte during the long inactive storage period and act as the delivery mechanism during activation. Each reservoir has some type of diaphragm which is broken with high pressure or mechanical means to allow the electrolyte to enter the cell stack part of the battery hardware.

The bipolar cell stack with the electrolyte distribution manifold in the center comprises the battery section. When electrolyte enters the center manifold, it is distributed to each cell through holes or passageways in the plastic housing encompassing the battery. The design of the manifold is the key to controlling intercell leakage. For bipolar batteries, life requirements are relatively short (seconds to several hours); therefore, the manifolding is relatively simple. But when longer life is needed, the parasitic leakage currents are controlled by the length and area of the leakage path.

39.4 PERFORMANCE CHARACTERISTICS

39.4.1 Ampul-Type Batteries

Voltage characteristics at the "time of activation" are unique and an important feature of reserve cells and batteries. This is especially true for military applications, where reserve cells must normally be designed to meet operational voltage in less than 1 s and in many cases even less than ½ s. For nonmilitary use, activation times to operating voltage level are less critical. However, for a given reserve battery design and the electrochemical couple used, the activation time for any reserve cell is dependent on the discharge, load, and temperature.

In general, the voltage rise times for both $Li/SOCl_2$ and Li/V_2O_5 have similar characteristics. Figure 39.9 shows the rise time characteristics for the $Li/SOCl_2$ cell (illustrated in Fig. 39.4) at five temperatures at a current density of 0.1 mA/cm^2 (approx

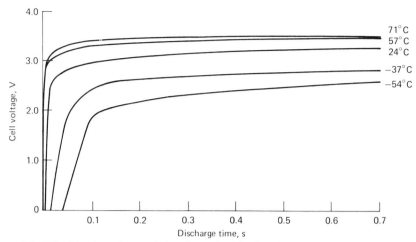

FIG. 39.9   Rise time characteristics after activation of a Li/SOCl₂ reserve cell (Honeywell model G2659B1) (load = 4.35 kΩ).

C/500 rate). Rise times are typically below 20 ms at ambient (24°C) and higher temperatures but increase up to 500 ms at the lower temperatures. The ability of these small cells to activate rapidly is primarily due to the cell design, which allows the electrolyte to penetrate and wick into the porous electrodes and separator at the instant of ampul breakage.

The voltage levels of both the Li/V₂O₅ and Li/SOCl₂ systems (batteries illustrated in Figs. 39.3 and 39.4) under steady-state discharge conditions are shown in Fig. 39.10. These two systems are very close in voltage at the lower temperatures, ranging from 3.3 to 3.0 V at current densities less than 1 mA/cm². At higher temperatures, ambient and above up to 74°C, the Li/SOCl₂ cell operates above 3.5 V, whereas the V₂O₅ cells

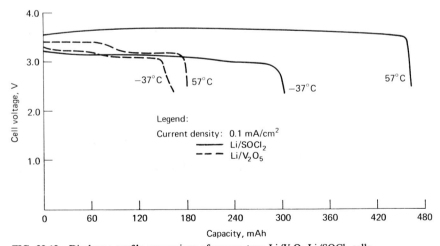

FIG. 39.10   Discharge profile comparison of reserve-type Li/V₂O₅ Li/SOCl₂ cells.

normally operate between 3.2 and 3.4 V. The higher voltage and increased capacity account for the significant increase in energy density of the $SOCl_2$ over the V_2O_5 cell. As shown in Fig. 39.10, the V_2O_5 system has very little change in capacity over the wide temperature range but is still much lower in capacity than the $SOCl_2$ cell when discharged at the same rate, 0.1 mA/cm^2. Although the capacity and voltage of the Li/$SOCl_2$ cell are lower at cold temperatures, its ouput is still higher than most other systems and its voltage profile is characterized by a flat single-step plateau. The high-temperature curve is also extremely flat and typically discharges above 3.6 V at a current density of 0.1 mA/cm^2. Ambient discharge voltage characteristics are similar to those at high temperature except for a slightly lower load voltage when discharged at the same rate, averaging 3.5 V. Table 39.2 compares the output parameters of the two systems in identical hardware and shows the superior performance of the Li/$SOCl_2$ cell. The similarity in voltage and the fact that the same hardware is used for both systems permits a one-for-one replacement.

TABLE 39.2 Performance Comparison Between Li/$SOCl_2$ and Li/V_2O_5 Systems

System	Temper- ature, °C	Cell voltage, V	Capacity, mAh	Cell volume, cm^3	Cell weight, g	Energy density Wh/kg	Energy density Wh/dm^3
Li/V_2O_5*	−37	3.15	160	5.1	10	50.4	98.8
	57	3.30	180	5.1	10	59.4	116.5
Li/$SOCl_2$†	−37	3.05	300	5.1	10.5	87.1	179.4
	57	3.60	450	5.1	10.5	154.3	317.6

*Honeywell's model G2659.
†Honeywell's model G2659B1.

Figure 39.11 shows the effect of inactive storage of up to 12 months at 71°C on the Li/$SOCl_2$ cell performance over the temperature range of −54 to 71°C. No significant effect on performance was found as a result of the storage. The slightly lower voltages during discharge or the voltage delays when the load is first applied on active (non-reserve) cells were not present with the reserve batteries. Figure 39.11 also gives a summary of the performance of fresh cells at various discharge loads and temperatures.

Figure 39.12 shows the discharge curves of the A size reserve Li/$SOCl_2$ cell (illustrated in Fig. 39.2). The current drain capability of a reserve system significantly exceeds that of the corresponding active primary cell. Figure 39.12 shows the discharge characteristics at 1.25 kΩ (about 3 mA) or 0.15 mA/cm^2. Currents higher than 1.5 A (current density of 100 mA/cm^2) can be obtained at voltages higher than 2.0 V for several minutes at −10°C. The energy density of the cells, to a cutoff voltage of 2.0 V, as a function of the discharge current at various temperatures is shown in Fig. 39.13. The performance at higher temperatures is close to that for 25°C.[7]

39.4.2 Multicell Battery Designs

Performance characteristics of the Li/SO_2 multicell single-activation design battery are shown in Fig. 39.14. The activation and discharge profiles for a 12-V, 100-Ah battery using an $LiAsF_6$:AN:SO_2 electrolyte are illustrated. Because the battery is manually activated, the slower voltage rise time is expected because the cutting of the diaphragm in the center bulkhead requires several turns on the activation bolt. The battery could

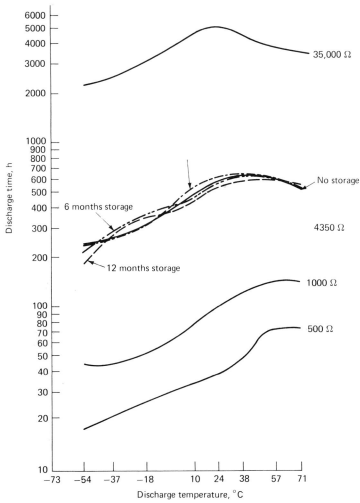

FIG. 39.11 Effect of discharge rate and inactive storage on the Li/SOCl₂ reserve cell (Honeywell model G2659B-1).

easily be activated with an electric or mechanical input to a piston actuator or squib to cut the diaphragm to improve the voltage rise time. Although the operational life of the battery can be very short, the data illustrate its capability for a long-term discharge at low discharge rates if a stable electrolyte is used.

Typical discharge curves for a high-rate reserve Li/SOCl₂ battery are shown in Figure 39.15. The battery uses a bipolar construction and a noncirculating electrolyte system. Operating temperatures range from −40 to 70°C, with electrolyte heaters required for operation below 0°C. A typical battery, hermetically sealed in stainless-steel case, is illustrated in Fig. 39.16.

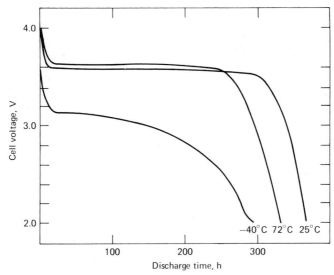

FIG. 39.12 Typical discharge curves of the Li/SOCl₂ reserve cell (Tadiran model TL-5160).

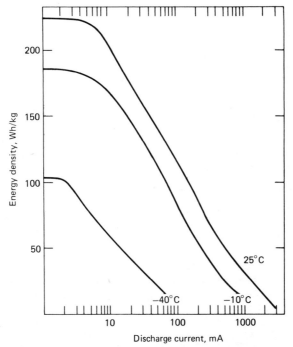

FIG. 39.13 Energy density as a function of the discharge current and temperature (Li/SOCl₂ reserve cell, Tadiran model TL-5160).

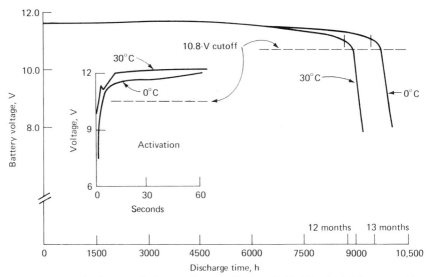

FIG. 39.14 Activation and discharge voltage profile for a 12-V, 100-Ah Li/SO₂ battery (electrolyte: LiAsF₆:AN:SO₂).

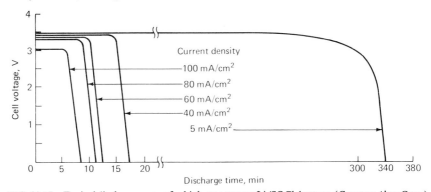

FIG. 39.15 Typical discharge curves for high-rate reserve Li/SOCl₂ battery. (*Courtesy Altus Corp.*)

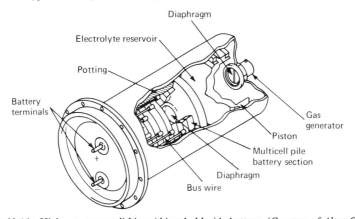

FIG. 39.16 High-rate reserve lithium/thionyl chloride battery. (*Courtesy of Altus Corp.*)

TABLE 39.3 Commercially Available Lithium Reserve Batteries

Company	Chemistry	Voltage, V	Capacity, mAh	Model number	Dimensions, mm		
					Height	Diameter	Weight, g
Honeywell, Inc.,	Li/V$_2$O$_5$	3.30	100	G2666	21.3	12.7	5.1
Horsham, Pa.	Li/SOCl$_2$	3.50	250	G2666B1	21.3	12.7	5.1
Tadiran Israel Electronics Industries, Ltd., Rehovot, Israel	Li/SOCl$_2$	3.50	1250	TL-5160	48	16.0	16.5
Philips USFA BV,	Li/V$_2$O$_5$	3.30	250	UA 6172	26	17	9
Eindhoven,		3.30	500	UA 6171	46	17	16
The Netherlands		6.60	500	UA 6170	64(L)× 39(W)× 18(H)		60

39.5 BATTERY TYPES AND CHARACTERISTICS

Table 39.3 lists the characteristics of typical lithium reserve batteries that are commercially available.

REFERENCES

1. A. N. Dey, "Lithium Anode Film and Organic and Inorganic Electrolyte Batteries," *Thin Solid Films,* vol. 43, Elsevier Sequoia, S.A., Lausanne, Switzerland, 1977, p. 131.

2. R. J. Horning, "Small Lithium/Vanadium Pentoxide Reserve Cells," *Tenth Intersoc. Energy Convers. Eng. Conf.,* 1975.

3. W. B. Ebner and C. R. Walk, "Stability of LiAsF$_6$-Methyl Formate Electrolyte Solutions," *27th Annu. Proc. Power Sources Conf.,* June, 1976.

4. B. Ravid, *A Reserve-Type Lithium-Thionyl Chloride Battery,* Tadiran Israel Electronics Industries, Ltd., 1979.

5. R. J. Horning and M. J. Faust, *High Energy Density Reserve Cell Storage Evaluation,* Contractual Effort by Honeywell with ARRADCOM, Dover, N.J.

6. M. J. Domenicomi and F. G. Murphy, High Discharge Rate Reserve Cell and Electrolyte, U.S. Patent 4,150,198, April 17, 1979.

7. M. Baboi, U. Meishar, and B. Ravid, "Modified Li/SOCl$_2$ Reserve Cells with Improved Performance," *29th Power Sources Conf.,* June 1980.

8. P. M. Shah, "A Stable Electrolyte for Li/SO$_2$ Reserve Cells," *Proc. 27th Power Sources Symp.,* 1976.

9. P. M. Shah and W. J. Eppley, "Stability of the LiAsF$_6$:AN:SO$_2$ Electrolyte," *Proc. 28th Power Sources Symp.,* 1978.

10. R. J. Horning and K. F. Garoutte, "Li/SO$_2$ Multicell Reserve Structure," *Proc. 27th Power Sources Symp.,* 1976.

11. W. J. Eppley and R. J. Horning, "Lithium/Thionyl Chloride Reserve Cell Development," *Proc. 28th Power Sources Symp.,* 1978.

Thermal Batteries

by
Frederick Tepper and David Yalom

40.1 GENERAL CHARACTERISTICS

Thermal (heat-activated) batteries employ inorganic salt electrolytes that are nonconductive solids at ambient temperatures and an integral pyrotechnic mixture scaled to supply sufficient heat to melt the electrolyte. The battery is activated by an electrical or mechanical signal to a built-in squib or primer, which ignites the pyrotechnic, melting the electrolyte which becomes conductive and permitting the battery to deliver high power for relatively short durations. The shelf (preactivation) life of thermal batteries is greater than 10 y, and batteries of more recent manufacture are expected to last at least 20 y. The operational (postactivation) life ranges from seconds for high-power "pulse" batteries to more than 1 h for suitably insulated designs. Thermal batteries are primary, reserve batteries and, once expended, are not rechargeable.

Thermal cells originated in the 1940s and were developed mainly for weapons applications.[1,2] More than 10 million have been produced since Catalyst Research Corporation manufactured the first models in 1947.[3] Their characteristics are ideally suited for military ordnance, and they are widely used as power sources in projectiles, rockets, bombs, mines, missiles, decoys, jammers, and torpedoes. Figure 40.1 is a photograph of some of the thermal batteries that have been manufactured.

The advantages of thermal batteries are:

1. Very long shelf life (longer than 10 y) in a "ready ammunition" state without degradation in performance

2. "Instant" activation—fast-start designs can be activated and provide useful power in tens of milliseconds

3. High peak power exceeding 10 W/cm^2

4. Very high demonstrated reliability and ruggedness following long-term storage at extremes of ambient temperatures, and severe stockpile-to-target environmental

FIG. 40.1   Typical thermal batteries. (*Courtesy of Catalyst Research Corp.*)

forces such as shock, vibration, acceleration and/or spin, concurrent with operation at temperature extremes

5. No maintenance or servicing; can be permanently installed in equipment

However, the design considerations required to achieve these characteristics can result in reduction in energy density and higher manufacturing costs compared with other battery systems.

40.2 DESCRIPTION OF ELECTROCHEMICAL SYSTEMS

Thermal battery electrochemical cells consist of an alkali or alkaline earth metal anode, a fusible salt electrolyte, and a metallic salt depolarizer. The pyrotechnic heat source is usually inserted between cells in a series cell stack for rapid activation. The term "thermal battery" does not refer to a single electrochemical system but to a family of batteries using different electrochemical systems. The active materials, the electrolyte, and other battery components, as well as the design, can be varied according to the performance that is desired. Table 40.1 lists some of the popular types in use.

40.2.1 Anode Materials

Until recently, most thermal battery designs employed a calcium metal anode, with the calcium generally attached to an iron, stainless steel, or nickel current collector. A typical anode sheet called "bimetal" is manufactured by vapor deposition of calcium onto 0.1-mm sheets. Calcium thicknesses range between 0.03 and 0.25 mm. In other calcium anode designs, calcium sheet is pressed onto perforated "cheese-grater" surfaced metal sheets.

TABLE 40.1 Types of Thermal Batteries

Electrochemical system: anode/electrolyte/cathode	Operating cell voltage, V	Characteristics and/or applications
$Ca/LiCl\text{-}KCl/WO_3$	2.4–2.6	Used principally for fuse applications where a low level of electrical noise is essential and where dynamic environments are not severe
$Ca/LiCl\text{-}KCl/CaCrO_4$	2.2–2.6	Used in applications requiring short-term operation in severe dynamic environments
$Mg/LiCl\text{-}KCl/V_2O_5$	2.2–2.7	Used in applications requiring short-term operation in severe dynamic environments
$Li(M)/LiCl\text{-}KCl/FeS_2$	1.6–2.1	Overall advantages: low electrical noise, can operate in severe dynamic environments, long service life (up to 1 h)

Magnesium metal is another anode metal that has been used extensively,[4] most often with a vanadium pentoxide depolarizer. Recent and substantial improvements in the state of the art have occurred through the use of lithium metal, or lithium alloys such as lithium-aluminum, lithium-boron, or lithium-silicon, as the anode. (The lithium anode is discussed in Sec. 40.7.)

40.2.2 Electrolyte

Most thermal battery designs use molten lithium chloride-potassium chloride eutectic as the electrolyte (mp = 352°C). Halide melts are preferred over lower-melting oxygen-containing salts since the latter are susceptible to gas generation by thermal degradation or side reactions. The lithium bromide-potassium bromide eutectic melts lower than LiCl-KCl and has been used but is more hygroscopic and expensive. The use of sodium chloride-aluminum chloride eutectics in conjunction with lithium anodes is covered in Sec. 40.7.

40.2.3 Cathode Materials

A wide variety of cathode materials (depolarizers) have been used for thermal batteries, including calcium chromate ($CaCrO_4$), potassium chromate ($K_2Cr_2O_7$ and K_2CrO_4), metal oxides (V_2O_5, WO_3), and sulfides (CuS, FeS_2). The criteria for suitable depolarizers include high potential vs. a suitable anode, compatibility with halide melts, and thermal stability to approximately 600°C. Calcium chromate has been most often used with calcium anodes because of its high potential (at 500°C $\cong$ 2.7 V) and its thermal stability (> 600°C).

40.2.4 Pyrotechnic Heat Sources

The two principal pyrotechnic heat sources used in thermal batteries are a paper-type composition of zirconium and barium chromate powders supported on inorganic fibers[5] and a pressed tablet made up of iron powder and potassium perchlorate, commonly known as a heat pellet.[6]

The Zr-$BaCrO_4$ "heat paper" is manufactured from pyrotechnic-grade zirconium and $BaCrO_4$, both with particle size of about 1 to 10 μm. Inorganics such as ceramic and asbestos fibers are added. The mix is formed into paper as individual sheets by use of a mold or continuously by use of a paper-making process. The resultant sheets are cut into parts and dried. Once dry, they must be handled very carefully since they are very susceptible to ignition by static discharge. Heat paper has a burning rate of about 10 to 15 cm/s and a heat content of about 1675 J/g. Heat paper combusts to an inorganic ash with high electrical resistivity. Consequently, its use necessitates addition of nickel or iron electrode collectors and intercell connectors to conduct current around each cell's heat paper pad (see Fig. 40.3*b* to *e*).

Fine iron powder (1 to 10 μm) and potassium perchlorate are blended dry and pressed to form heat pellets. The iron content ranges from 80 to 88% by weight and is considerably in excess of stoichiometry. Excess iron provides the combustible pellet with sufficient electronic conductivity, eliminating the need for intercell connectors. The heat content of Fe-$KClO_4$ pellets ranges from 920 for 88% iron to 1420 J/g for 80% iron. Burning rates of pellets are generally slower than those of heat paper, and the energy required to ignite them is greater. The heat pellet has higher activation energy and is, therefore, less susceptible to inadvertent ignition during battery manufacture. However, the battery must be designed so that there is good contact of the heat pellet with the first fire or ignition source.

After combustion, the heat pellet is an electronic conductor, simplifying intercell connection and battery design. The heat pellet burns at a lower rate than heat paper, and so heat pellet batteries generally start about 0.2 s slower than heat paper batteries under load. Upon combustion, however, the heat pellet "ash" retains its original shape, and since it has a higher enthalpy than heat paper ash, it serves as a heat reservoir, retaining considerable heat, reaching lower peak temperatures, and releasing its heat to the cell as the electrolyte starts to cool. The heat paper combusts to shapeless refractory oxides causing slumping of the battery stack during ignition and less resistance to environmental effects.

40.2.5 Methods of Activation

Thermal batteries are initiated by either mechanical action using a percussion-type primer (M-42) or an electrical pulse to an integral electric match (squib). For most military applications, safety considerations require the squib to be nonignitable under a load of 1 W, 1 A. (Thermal battery users are cautioned against zealous no-fire testing of squibs, since excessive no-fire testing will deactivate the squib.) In heat pellet batteries, an intermediate heat paper firing train is often added to carry the ignition from the primer or squib to each pellet.

40.2.6 Insulation Materials

Thermal batteries are designed to maintain hermeticity throughout service life, even though internal temperatures approach 600°C. Thermal insulation used to minimize peak surface temperatures must be anhydrous, and, if organic, must have high thermal stability. Dehydrated asbestos, ceramic fibers, and Kapton (polyimide) have been used.

Insulative layers are tightly bound around the periphery of the battery stack, and several asbestos or ceramic fiber disks are placed on each end to ensure a tight pack in the metal can. Special end reservoir pellets are often used.[7] Long-life (> 10 min) thermal batteries often use a very low thermally conductive product called "Min-K" (Johns-Manville Co.), which is manufactured from titania and silica.

40.3 CHEMISTRY

The overall reaction for the calcium/calcium chromate system is dependent on the discharge parameters. A postulated cell reaction is

$$3 \text{ Ca} + 2 \text{ CaCrO}_4 + 6 \text{ LiCl} \rightarrow 3 \text{ CaCl}_2 + \text{Cr}_2\text{O}_3 \cdot 2\text{CaO} + 3\text{Li}_2\text{O}$$

although the exact composition of the mixed oxide has not been determined. It is possible that lithium formed by reaction of calcium with LiCl electrolyte enters into the electrochemical reaction.

Side reactions occur between calcium and LiCl-KCl that can limit the full utilization of coulombic capacity between calcium and suitable depolarizers. The reaction

$$\text{Ca} + 2\text{LiCl} \rightleftarrows 2\text{Li} + \text{CaCl}_2$$

occurs spontaneously in a thermal battery with the result that an alloy (Li_2Ca) is formed which is liquid at thermal battery operational temperatures. This melt, known as an "alloy," can be responsible for internal shorts, resulting in intermittent cell shorts (noise) and cell misfunction during the discharge. This alloy permits high anode current density but limits the use of the couple on open circuit or light loads because an excess causes the cell shorting. The rate of alloy formation can be controlled by deactivating or passivating the cell to slow down the Ca + LiCl reaction. Techniques include controlling by current density,[3] acetic acid treatment of calcium,[8] addition of passivating agents and excess binder to the electrolyte, and reduction of electrolyte pellet density.

Calcium chloride, formed by the reaction of calcium with lithium chloride, reacts further with potassium chloride to form the double salt $\text{CaCl}_2 \cdot \text{KCl}$. This salt has a melting point of 575°C and has been identified as a precipitate in calcium anode thermal cells. It has been suggested[2] that it can co-exist with molten chloride electrolytes up to about 485°C.

The self-discharge reaction of calcium with calcium chromate is highly exothermic, forming complex chromium III oxides. Above about 600°C the self-discharge reaction accelerates, probably due to the markedly increasing solubility of the chromate in the chloride electrolyte. This acceleration increases the rate of formation of calcium-lithium alloy. The resulting thermal runaway is characterized by short battery lives, overheating and cell step-outs, shorts, and noise characteristics of excess alloy.

The Ca/CaCrO_4 couple can be separated by a discrete electrolyte (plus binder) layer or can be a homogeneous pelletizing mix of depolarizer, electrolyte, and binder (DEB).[9] The calcium/DEB battery developed in the 1960s has been used considerably in military fuzing.[2,10]

Much of the effort associated with thermal battery design involves determining the amount of pyrotechnic heat to give acceptable performance over the necessary temperature range. Many of the required temperature extremes range from -55 to $+75$°C for initial ambient. This 130°C difference is well within the operational limits of 352°C, the melting point of LiCl-KCl eutectic, and 600°C, the approximate temperature of thermal runaway. However, as discussed earlier for $Ca/CaCrO_4$, a temperature of at least 485°C is necessary for minimal performance at moderate drain ($> 50 \ mA/cm^2$) due to freeze out of double salt (KCl, $CaCl_2$) below that temperature. Thus, it is seen that the working temperature range is reduced to 115°C, whereas military extremes exceed this by as much as 15°C.

Figure 40.2 shows the performance and physical characteristics of a 28 $\pm$ 4-V Ca-DEB pellet-heat pellet battery over the temperature range -54 to $+71$°C. The life to a 24-V limit is shown as a function of temperature at a 1.5-A drain. The heat balance is adjusted for optimum performance at about 15°C, with the cold performance limited to freeze out on the one hand, and hot performance limited by self-discharge on the other. It is apparent that the thermal balance can be shifted to give optimum performance at something other than room temperature.

40.4 CELL CONSTRUCTION

Most thermal battery applications require steady-state power under moderate dynamic environments. The cup-type and pellet-type electrolyte batteries and variants of these technologies satisfied most of the power needs. Special applications have resulted in special designs—high-voltage thermals,[11] spin thermals,[12] and pulse thermals. Figure 40.3 shows six typical cell configurations.

40.4.1 Cup Cell

Figure 40.3a is a closed-cell $Ca/LiCl-KCl/WO_3$ design which was employed in the earliest thermal batteries produced. The cup cell features a common anode (calcium metal affixed to both sides of a metal substrate) and two depolarizer pads, thereby doubling the cell area. The LiCl-KCl electrolyte is impregnated onto glass tape. The sides of the cell are crimped over a metal disk to seal the cell. The heat pad is

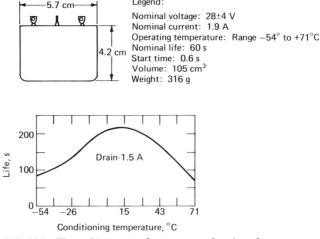

Legend:

Nominal voltage: 28±4 V
Nominal current: 1.9 A
Operating temperature: Range $-54°$ to $+71°$C
Nominal life: 60 s
Start time: 0.6 s
Volume: 105 cm^3
Weight: 316 g

FIG. 40.2 Thermal battery performance as a function of temperature.

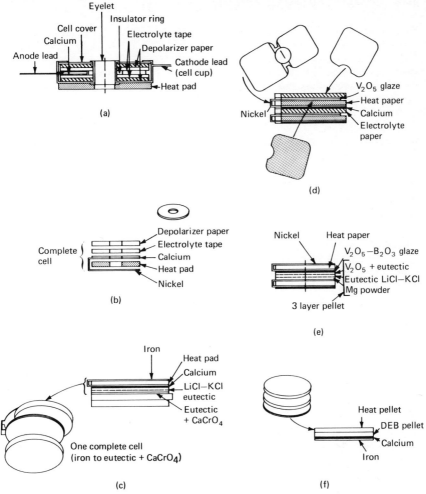

FIG. 40.3 Thermal battery—cell configurations: (a) closed-cell WO₃; (b) open-cell WO₃; (c) open-pellet cell; (d) glazed V₂O₅ depolarizer; (e) magnesium-V₂O₅ pellet cell; (f) heat pellet-DEB pellet cell.

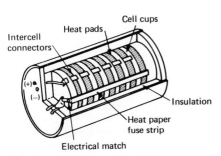

FIG. 40.4 Assembly of thermal battery—cup design.

fabricated from heat paper and is laid between each cell. Calcium chromate depolarizers were also used in this design. Figure 40.4 illustrates the construction of the thermal battery.[2] An electric match and fuse strip are used for activation. The entire battery is hermetically sealed in a steel can, and all electrical connections are brought through the header plate.

40.4.2 Open Cell

Figure 40.3*b* shows an open-cell version of the design in Fig. 40.3*a* which eliminates the cup. A dumbbell-shaped nickel part, coated with calcium on one side, is folded around heat paper disks and serves as the connector between cells. Elimination of the cup is made possible by reducing the quantity of electrolyte and by restricting the battery to low or moderate dynamics. This design is useful for light to moderate loads (< 50 mA/cm^2) for durations of 1 to 1-½ min. Potassium dichromate depolarizer is also used in this cell and allows very fast start-up, with high current density discharges for fractions of seconds to about 5 s.

40.4.3 Electrolyte Pellet

With the development of pellet-electrolyte cells, sufficient electrolyte and depolarizer could be included in a cell to provide a coulombic capacity equivalent to the cup design without the necessity of a complex cell configuration. Binders added to the electrolyte tend to retain the shape of the pellet during thermal battery operation. One such configuration is shown in Fig. 40.3*c*. The two-layer pellet consists of an electrolyte layer (LiCl-KCl) with a binder such as kaolin or microsize silica and a second layer composed of CaCrO$_4$, plus some electrolyte. Calcium metal is used as the anode and Zr-BaCrO$_4$ heat paper is again the heat source, with dumbbell-shaped parts serving as intercell connectors.

40.4.4 Glazed Depolarizers

The glazed-depolarizer cell is shown in Fig. 40.3*d*.[13] One side of one face of the nickel strip is coated with a glaze of vanadium pentoxide using boric oxide as a flux. The other side of the same nickel face is coated with calcium, and the composite was folded around heat paper. An asbestos mat served as a matrix for electrolyte. The cell had a square configuration and a thickness of only 1.0 mm. The thin cell characteristics and high voltage (open-circuit voltage is approximately 3.2 V) allow a significant voltage per unit height of cell stack.

40.4.5 Multilayer Pellet

Figure 40.3*e* shows the Mg/LiCl-KCl/V$_2$O$_5$ pellet cell developed[3] in the mid-1950s. It was the first design in which the active cell components were combined in a single three-layer pellet. The tablet was formed by compacting, in sequence, an anode layer of magnesium powder, a layer of LiCl-KCl electrolyte and kaolin binder, and a layer of V$_2$O$_5$ and electrolyte, and then compressing the three layers into a single pellet.

40.4.6 Heat Pellet—DEB Pellet Cell

The DEB cell design, in which no distinct electrolyte layer was necessary, was conceived in 1959. Instead, LiCl-KCl, CaCrO$_4$, and kaolin binder were blended and formed into a homogeneous pellet.[12] Since it was homogeneous, it could not be misassembled during construction as could happen with multilayer pellets. Shortly thereafter, a similar pellet was conceived[2,9] in which the binder was Cab-O-Sil, a microporous silica. Using a new heat source, Fe-KClO$_4$ heat pellets and the homogeneous DEB pellet, an advanced cell design was created (Fig. 40.3f) with the advantages of ease of assembly, low cost, high dynamic stability, and suitability for longer life than had been possible with heat paper, two-layer pellet systems.

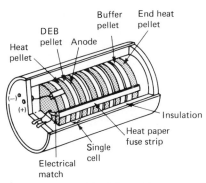

FIG. 40.5   Assembly of thermal battery—pellet construction.

A cross section of a thermal battery employing this homogeneous DEB pellet with the heat pellet is shown in Fig. 40.5.[2] The anode is sheet calcium or calcium metal vapor deposited upon a metal substrate. The heat pellet is composed of powdered iron and potassium perchlorate. This pellet contains excess iron which causes the residue to be electronically conducted after burning, eliminating the need for intercell connectors. The buffer pellets act as heat reservoirs to maintain the battery temperature for a longer time.

40.5 PERFORMANCE CHARACTERISTICS

Thermal batteries are specially designed to meet the specific requirements of each application. Batteries with output voltages from 2 to over 100 V, power levels up to 125 to 200 W, and service life from 1 to 800 s (and more recently for discharge periods up to 1 h) are in use. Sizes have ranged from 8 to 3800 cm^3 and weights from 20 g to 12 kg. Because of their specific design considerations, thermal batteries do not lend themselves to general-performance characterization as do other battery systems. Table 40.2 summarizes the size, weight, and average performance characteristics (under the loads to which they were designed) of a number of thermal batteries that represent the cell designs described. The data presume performance under the full military temperature extremes (-54 to $+74°C$) to the specified end of life voltage. Improved performance generally results over narrower temperature ranges.

40.5.1 Discharge

Typical discharge curves for a pulse and power-type thermal battery are given in Fig. 40.6. Calcium anode cells are limited in active life to a few minutes. The lithium anode batteries can deliver higher energy densities, with service life up to 60 min.

40.5.2 Current Density/Life

Table 40.3 shows the performance of the different cell designs, comparing them on the basis of attainable current density for 10- and 100-s rates. These values are composites using data derived from testing production batteries, such as those described in Table 40.2. They represent the capability for a production battery to reliably meet a specified

TABLE 40.2   Characteristics of Various Thermal Batteries

Cell type	Volume, cm³	Weight, g	Nominal voltage, V	Current, A	Peak power, W	Average life, s	Energy density Wh/kg	Energy density Wh/L
Cup/WO₃	450	850	7	5.8	41	70	2.3	4.3
			126	0.47	60	70		
Open cell/tape/WO₃	100	385	50	0.36	15	70–100	1.3	5.0
Open cell/tape/ dichromate	44	148	18	26	462	1.2	1.0	3.5
Open cell/tape/ dichromate	1	5.5	10	5.0	50	0.15	0.4	2.1
High-voltage array	81	225	203	0.02	4	45	0.2	0.72
			1.7	0.32	0.5	45		
Two-layer CaCrO₄/heat paper	123	310	42	2.9	125	25	2.8	6.8
Ca/V₂O₅/heat paper	16	52	31	0.12	3.8	14	0.3	1.0
Heat-DEB	2.9	7.5	20	0.045	0.9	25	0.8	2.1
Heat-DEB	105	307	28	1.2 or	34	150	4.6	13.4
			28	2.5	75	60		
Heat-DEB	592	1560	28	2.8	78	800	11.1	29
Activated heat-DEB	83	250	45	6.0	385	25	7.5	23
LAN-FeS₂	570	1485	28	5.0	135	1200	31	80

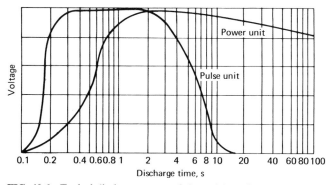

FIG. 40.6   Typical discharge curves of thermal batteries.

life (defined as the time required for the voltage to degrade to 75% of the peak at the given rate) over the full military temperature spectrum of -54 to $+74°C$. Composite data for several different batteries are used since heat content, processing, and minor design variations result in optimum performance as a function of load.

As indicated earlier, the DEB pellet cell can be optimized for current density. An increase in the capability of a heat pellet-DEB pellet battery is shown at the 10-s rate (515 to 930 mA/cm²) where the DEB has been specially treated to effect high pellet density. Such treatments yield excessive alloying under light loads, and so use at the 100-s rate is not desirable. The current-carrying capability of the dichromate cell is only 54 mA/cm² at 10 s, but it is significantly better than other designs below about 2 s. Current densities of at least 6 A/cm² are attainable for lives less than about 1 s.

The LAN-FeS₂ data are based on the performance of the prototype Li/FeS₂ thermal battery (Catalyst Research Corp. P/N 406620) and are taken from Fig. 40.11. This

TABLE 40.3 Attainable Current Density of Cell Designs

Cell structure	Current density, mA/cm^2	
	10-s rate	100-s rate
Cup cell	620	35
Open cell/tape/WO$_3$	100	32
Open cell/dichromate	54	
Two-layer/heat paper	790	46
Heat pellet/DEB pellet	515	122
Heat pellet/DEB pellet	930	
Glazed V$_2$O$_5$	50	
Mg/V$_2$O$_5$ pellet	1200	61
Li/FeS$_2$	> 2500	580

battery employs lithium anodes (LAN) and FeS$_2$ cathode material. The system shows almost a tenfold increase in attainable current density in the 10- to 100-s rate range.

40.5.3 Voltage

The designers of military devices are often concerned with the voltage density rather than the current density of a thermal battery. The "high voltage array" battery, listed in Table 40.2, is one solution to such a problem.[10] The open-cell WO$_3$ battery performance shows relatively low current-carrying capability. However, as shown in Fig. 40.7, this design has a relatively thin cell (1.5 mm) allowing a greater number of cells per unit length of stack than cup or pellet cells.

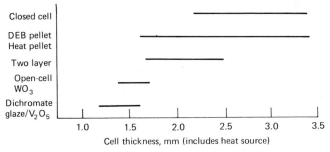

FIG. 40.7 Thickness of thermal battery cells (including heat source).

40.5.4 Activation Time

The time required for a battery to reach working voltage is often a primary criterion for battery acceptance. Figure 40.8 shows the range of activation time for different cell designs. It is notable that heat paper batteries generally start faster, with the dichromate cell the fastest, then the heat pellet, which burns at a lower rate.

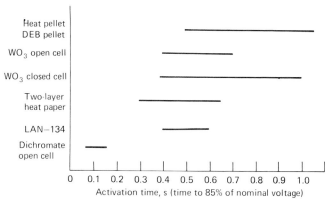

FIG. 40.8 Activation time (time to 85% of nominal voltage) of different thermal battery cell designs.

Techniques exist for improving the start characteristics of DEB pellet and Li/FeS$_2$ batteries, including the use of fast-burning heat pellets. Burning speeds as high as 1 m/s, equivalent to that of heat paper, have been obtained with pellets.

40.5.5 Surface Temperature

Batteries using lithium chloride-potassium chloride eutectic all have internal working temperatures in excess of about 450°C. Battery surface temperatures are therefore most affected by the amount of insulation used internally and its thermal conductivity. Temperatures as high as 400°C have been measured with some designs, while those using efficient insulation have peak temperatures as low as 135°C.

40.6 TESTING AND LIFE SURVEILLANCE

The safety and reliability of thermal batteries have been a matter of continuing study since they were first developed. To identify defective units, most designs are 100% leak-tested in hot water and checked for polarity and stack-to-case shorts (and squib resistance where appropriate) at manufacture. When safety and/or reliability consider-ations warrant, helium leak detection methods are used and the batteries may be radiographed. Almost all thermal batteries are fabricated in homogeneous groups or lots, and a sample from each lot is discharged to demonstrate compliance with the performance requirements. Usually the samples are discharged at maximum loads and temperature extremes, often with concurrently imposed environmental forces. By use of such testing programs, reliability values greater than 99% and safety values greater than 99.9% have been demonstrated innumerable times in the last three decades. Tests show that certain thermal batteries are in compliance with the confines of NAVSEA INST 9310.1A on lithium battery safety.[14] (However, it is recommended that safety shields be used when thermal batteries are discharged.)

In addition, many surveillance studies of thermal batteries have been carried out, and no evidence of degradation found after 15 y shelf life for the Ca/CaCrO$_4$ and Mg/V$_2$O$_5$ systems. This has been correlated with microcalorimetric studies of Ca/Ca-

CrO_4 and Li/FeS_2 test batteries, where the coulombic capacity decrement is less than 0.5%/y at 37°C.

40.7 LITHIUM ANODE THERMAL BATTERIES

The use of a lithium anode in thermal batteries, as in other battery systems, provides the potential for significantly improved performance. First reported in 1974,[16] thermal batteries based on lithium metal anodes and transition metal sulfide cathodes are being developed in a variety of configurations.[15-36]

Since lithium metal is molten at thermal battery discharge temperatures, it is retained on high surface area metals, by immersion of the metal matrix in molten lithium to form anodes.[29] Often this structure is contained within a metal cup to prevent leakage during cell operation.[27,28] Another method is the fabrication of lithium alloy anodes, such as Li-B, Li-Al, and Li-Si, which are solid at battery discharge temperatures and thus offer the possibility of simpler construction. However, the lithium alloys are more difficult to fabricate than the metal matrix anodes and do not achieve this same peak current density. Most of the lithium anode batteries use the LiCl-KCl electrolyte and an iron disulfide (FeS_2) cathode.

The lithium/iron disulfide electrochemistry is more straightforward than that of the calcium chromate cell. The overall reaction is

$$2Li + FeS_2 \rightarrow Li_2S + FeS$$

Iron disulfide begins to thermally decompose at about 550°C into sulfur and iron sulfide, but good cathode efficiencies have been obtained up to 600°C. Above 600°C, the rate of decomposition increases, but experience with this system shows that thermal runaway is not as much a problem as with calcium/calcium chromate.

Another advantage to the lithium/iron disulfide system is the absence of high-melting-salt phases such as $CaCl_2 \cdot KCl$. The cell can thus operate close to 352°C. While calcium/calcium chromate may be used with homogeneous electrolyte-depolarizer blends, iron disulfide must be separated from the anode by a distinct electrolyte layer. Otherwise, the FeS_2, which is a fairly good conductor, will be electronically shorted to the anode.

Most experience has been with molten lithium anode batteries where the lithium is mechanically retained. Over 100,000 batteries have been built with this design. Figure 40.9 shows one version of a lithium/iron disulfide thermal battery cell. The design uses a heat pellet, but heat paper versions are also feasible. Figure 40.10 is an illustration of a battery which uses a lithium-silicon alloy anode and does not require any asbestos disk within the cup.

Figure 40.11 shows the average service life of a 14-cell Li/FeS_2 thermal battery designed for power applications discharged at various constant-current loads to 24 V over a temperature range of − 40 to 71°C. This battery is 5.1 cm OD by 4.4 cm high and uses a modification of the cell shown in Fig. 40.9. For comparison, the performance of a similar-sized $Ca/CaCrO_4$ thermal battery is also plotted. The Li/FeS_2 battery has a significantly higher capacity at the high power levels; at the lighter loads, the performance of both cells is limited by the cooling of the cell below the operating temperature.

Figure 40.12 shows the discharge curves comparing an $LiAl/FeS_2$ cell with an Mg/FeS_2 cell as well as with a typical $Ca/CaCrO_4$ cell.[35] Both the magnesium and

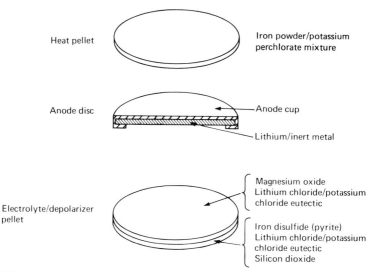

Heat pellet — Iron powder/potassium perchlorate mixture

Anode disc — Anode cup
— Lithium/inert metal

Electrolyte/depolarizer pellet —
{ Magnesium oxide
Lithium chloride/potassium chloride eutectic }
{ Iron disulfide (pyrite)
Lithium chloride/potassium chloride eutectic
Silicon dioxide }

FIG. 40.9   Typical LAN (Li/FeS$_2$) cell construction.

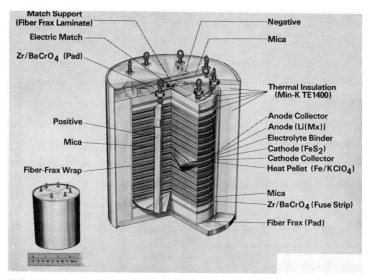

Match Support (Fiber Frax Laminate)
Electric Match
Zr/BaCrO$_4$ (Pad)
Negative
Mica
Positive
Mica
Fiber-Frax Wrap
Thermal Insulation (Min-K TE1400)
Anode Collector
Anode (Li(Mx))
Electrolyte Binder
Cathode (FeS$_2$)
Cathode Collector
Heat Pellet (Fe/KClO$_4$)
Mica
Zr/BaCrO$_4$ (Fuse Strip)
Fiber Frax (Pad)

FIG. 40.10   Long-life Li/FeS$_2$ thermal battery. (*Courtesy of Sandia Laboratories.*)

lithium anode cells give longer performance than the calcium cell; the main difference between magnesium and lithium is the higher voltage of the lithium cell. Figure 40.13 shows the activated lifetime and rise time for a 60-min, 28-V, 0.5-A thermal battery with a volume of 400 cm^3 in the Li(Si)/LiCl·KCl/FeS$_2$ electrochemical system. The rise time decreases with increasing temperature; the activated life, however, maximizes in the range of 25 to 50°C.[36]

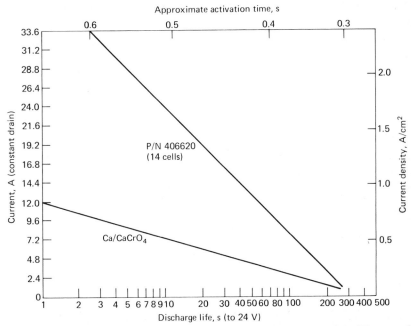

FIG. 40.11 Average performance of Li/FeS₂ thermal battery (P/N 406620). (*Courtesy of Catalyst Research Corp.*)

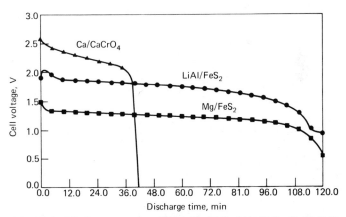

FIG. 40.12 Discharge curves at 520°C, Mg/FeS₂, LiAl/FeS₂, Ca/CaCrO₄ thermal battery cells. (*From Bush and Nissen, 1978.*)

Table 40.4 presents data on two types of Li/FeS₂ thermal batteries and illustrates the advantage of the lithium anode systems. While the liquid lithium anode battery shows better performance than the lithium alloy battery, particularly its rate capability, some of the advantage is due to size and design differences which caused the smaller LiSi battery to cool more rapidly.[31]

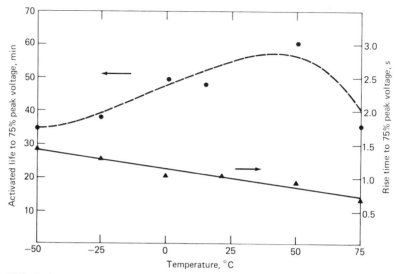

FIG. 40.13   Activated life and rise time, Li(Si)/FeS$_2$ thermal battery. (*From Quinn and Baldwin, 1980.*)

TABLE 40.4   Comparison of Li/FeS$_2$ Thermal Batteries

Duration of discharge, min	Li$_{(liq)}$/LiF·LiCl·LiBr/FeS$_2$		Li(Si)/LiCl·KCl/FeS$_2$	
	Power density, W/cm^3	Energy density of cell, Wh/L	Power density, W/cm^3	Energy density of cell, Wh/L
3.5	5.9	340		
5.0			4.3	360
5.5	5.2	480		
10.0			2.6	430
12	2.6	520		
21	1.9	520		
24	. . .	. . .	0.8	390
41	. . .	. . .	0.5	370

The performance data for the Li/FeS$_2$ thermal battery cover a range from high power (½-min rate) to long life (40- to 60-min rate). Although the data are still based on prototype batteries, they already show a magnitude improvement over the conventional thermal battery systems. The advantages are so pronounced that the Li/FeS$_2$ thermal battery, it can be confidently predicted, will be the dominant system in the 1980s.

REFERENCES

1. O. G. Bennett et al. (for Catalyst Research Corp.), U.S. Patent 3,575,714, Apr. 20, 1971.

2. B. H. Van Domelen and R. D. Wehrle, "A Review of Thermal Battery Technology," *Intersoc. Energy Convers. Conf.,* 1974.

3. F. Tepper, "A Survey of Thermal Battery Designs and their Performance Characteristics," *Intersoc. Energy Convers. Conf.,* 1974.

4. N. C. Nielsen, "Three Layer Pelletized Cells for Thermal Batteries," *Proc. 23rd Annu. Power Sources Conf.,* 1969.

5. W. H. Collins (for Catalyst Research Corp.), U.S. Patent 4,053,337, Oct. 11, 1977.

6. W. H. Collins (for Catalyst Research Corp.), U.K. Patent 1,482,738, Aug. 10, 1977.

7. D. M. Bush (for U.S. Atomic Energy Commission), U.S. Patent 3,677,822, July 18, 1972.

8. R. P. Clarke and K. R. Grothaus, "An Improved Calcium Anode for Thermal Batteries," *J. Electrochem. Soc.* **118**:1680 (1971).

9. D. M. Bush et al. (for U.S. Atomic Energy Commission), U.S. Patent 3,898,101, Aug. 3, 1975.

10. D. M. Bush, "Advancements in Pellet Type Thermal Batteries," *Proc. 25th Annu. Power Sources Conf.,* 1972.

11. G. C. Bowser et al. (for Catalyst Research Corp.), U.S. Patent 4,034,143, July 5, 1977.

12. G. C. Bowser, et al. (for Catalyst Research Corp.), U.S. Patent 4,044,192, Aug. 23, 1977.

13. R. Mead, "A Low Cost Thermal Battery," *Proc. 23rd Annu. Power Sources Conf.,* 1969.

14. "Lithium Batteries, Responsibilities, Policies and Guidelines for," NAVSEA Inst. 9310.1A.

15. C. S. Winchester, "The LAN/FeS₂ Thermal Battery System," *Power Sources* **13** (1982).

16. B. A. Askew and R. Holland, "A High Rate Primary Lithium-Sulfur Battery," *Power Sources* **4** (1972).

17. D. Birt, C. Feltham, G. Hazzard, and L. Pearce, "The Electrochemical Characteristics of Iron Sulfide in Immobilized Salt Electrolytes," *Power Sources* **7** (1978).

18. M. D. Baird, A. J. Clark, C. R. Feltham, and L. H. Pearce, "Recent Advances in High Temperature Primary Lithium Batteries," *Power Sources* **7** (1978).

19. G. C. Bowser, D. E. Harney, and F. Tepper, "A High Energy Density Molten Anode Thermal Battery," *Power Sources* **6** (1976).

20. A. A. Schneider et al. (for Catalyst Research Corp.), U.S. Patent 4,119,796, Oct. 10, 1978.

21. R. K. Quinn et al., "Development of a Lithium Alloy/Iron Disulfide 60-Minute Primary Thermal Battery," SAND79-0814, Sandia Laboratories, Albuquerque, N.M., April 1979.

22. J. A. DeGruson, "Improved Thermal Battery Performance," AFAPL-TR-79-2042, Eagle-Picher Industries, Joplin, Mo., June, 1979.

23. P. V. Dand et al., "Studies on the Ca-CaCrO₄ and Li-Al-FeS₂ Systems for Thermal Battery Applications," *Power Sources* **7** (1978).

24. D. A. Ryan et al., "A Low Temperature Thermal Battery," *Proc. 28th Annual Power Sources Symp., Atlantic City,* June 1978, p. 90.

25. B. F. Larrick et al., "Lithium-Boron Alloy—A New Battery Anode Material," ibid.

26. R. Szwarc, "Study of Li-B Alloys in LiCl-CKl Eutectic Thermal Cells Utilizing Chromate and Iron Disulfide Depolarizer," Gepp-TM-426, General Electric Co., Neut. Dev. Dept., April 1979.

27. G. C. Bowser et al. (for Catalyst Research Corp.), U.S. Patent 3,891,460, June 24, 1975.

28. G. C. Bowser et al. (for Catalyst Research Corp.), U.S. Patent 3,930,888, Jan. 6, 1976.

29. D. E. Harney (for Catalyst Research Corp.), U.S. Patent 4,221,849, Sept. 9, 1980.

30. R. T. M. Fraser et al., "High-Rate Lithium-Iron Disulfide Batteries," *Proc. 29th Annu. Power Sources Symp.,* Electrochem. Soc., Pennington, N.J., 1978.

31. A. Attewell and A. J. Clark, "A Review of Recent Developments in Thermal Batteries," *Power Sources* **12** (1980).

32. A. J. Clark and W. R. Young, "Thermal Battery for Military Applications," 2d ERA Battery Seminar, October 3, 1979.

33. J. Q. Searey et al., "Improvements in Li(Si)/FeS$_2$ Thermal Battery Technology," SAND82-0565, Sandia National Laboratories, Albuquerque, N. M., June 1982.

34. R. Szwarc et al., "Discharge Characteristics of Lithium-Boron Alloy Anode in Molten Salt Thermal Cells," *J. Elect. Soc.* **129** (1168) (1982).

35. D. M. Bush and D. A. Nissen, "Thermal Cells and Batteries Using the Mg/FeS$_2$ and LiAl/FeS$_2$ Systems," *Proc. 28th Power Sources Symp.,* Electrochem. Soc., Pennington, N.J., 1978.

36. R. K. Quinn and A. R. Baldwin, "Performance Data for a Lithium-Silicon/Iron Disulfide Long-Life Primary Thermal Battery," *Proc. 29th Power Sources Symp.,* Electrochem. Soc., Pennington, N. J., 1980.

PART 6 FUEL CELLS

41

General Characteristics

by
Arnold P. Fickett

41.1 DESCRIPTION OF THE FUEL CELL

A fuel cell is an electrochemical device that continuously converts the chemical energy of a fuel (and oxidant) to electrical energy. The essential difference between a fuel cell and a battery is the continuous nature of the energy supply. The fuel and the oxidant, which is usually oxygen, are supplied continuously to a fuel cell from an external source. In a battery, the fuel and oxidant are contained within; when the contained reactants have been consumed, the battery must be replaced or recharged.

The fuel cell uses liquid or gaseous fuels, such as hydrogen, hydrazine, hydrocarbons, and coal gas. The oxidant in a fuel cell is gaseous oxygen (or air).

A practical fuel cell power plant, depicted in Fig. 41.1, consists of at least three basic subsystems:

1. A power section which consists of one or more fuel cell stacks—each stack containing many individual fuel cells usually connected in series to produce a stack output ranging from a few to several hundred volts (direct current). This section converts processed fuel and the oxidant into dc power.

2. A fuel subsystem that manages the fuel supply to the power section. This subsystem can range from simple flow controls to a complex fuel-processing facility. This subsystem processes fuel to the type required for use in the fuel cell (power section).

3. A power conditioner that converts the output from the power section to the type of power and quality required by the application. This subsystem could range from a simple voltage control to a sophisticated device that would convert the dc power to an ac power output.

In addition, a fuel cell power plant, depending on size, type, and sophistication, may require an oxidant subsystem as well as thermal and fluid management subsystems.

Although fuel cells were invented nearly 150 y ago, the first practical application

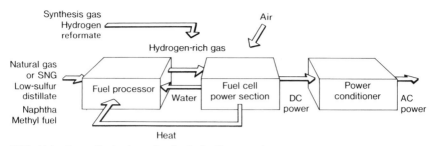

FIG. 41.1   Generalized schematic of a fuel cell power plant.

was in the 1960s as a spacecraft power source for Gemini and Apollo missions. Storage batteries were adequate for short space missions, but fuel cells were necessary to provide power (and energy) for the extended missions. Future fuel cell applications will include remote power generation (space, undersea), military, commercial electric power generation (by both electric and gas utility industries), and possibly, in the longer range, hybrid electric vehicles.

41.2 PROPERTIES OF FUEL CELL POWER PLANTS

Fuel cells are classified as power generators because they can operate continuously, or for as long as fuel and oxidant are supplied, and are rated by the kilowatt of power output.

The major characteristics of fuel cells are summarized in Table 41.1. Fuel cells produce power by an electrochemical rather than a thermal cycle and are not subject to the Carnot cycle limitation of thermal machines, thus offering the potential for highly efficient conversion of chemical to electrical energy. Furthermore, the efficiency is essentially independent of size; small power plants operate nearly as efficiently as large ones. Fuel cell power plants are quiet and clean, the byproducts being water, carbon

TABLE 41.1   Advantages of Fuel Cell Power Plants

Very efficient:
    In small as well as in large sizes
    At part power

Modular:
    Can be sited where needed

A good neighbor:
    Quiet
    Low emissions
    Operates without external water

Waste heat recoverable

Fuel flexibility

Durable and reliable

Attractive for diverse applications

dioxide, and nitrogen. They can be considered for applications where noxious emissions or noise would be objectionable and where water is unavailable, and they can be sited where used rather than in a remote location. Fuel cell power plants offer considerable flexibility and can be configured to use a wide variety of fuels and produce a wide range of dc or ac power outputs. Like batteries, they are inherently modular down to the level of the individual cell. Fuel cells can be configured in virtually any size from milliwatts to megawatts. This modularity should also lead to economies resulting from mass production techniques.

These properties of fuel cell power plants lead to their consideration for a variety of applications. Fuel cell power plants operating on hydrogen and oxygen offer high energy density; that is, a relatively small weight and volume of total system can produce large energy outputs. Thus, fuel cells are preferred power generators in remote applications where system weight and volume are important parameters, e.g., space, undersea. Fuel cell power plants operating on logistic fuels and air offer the potential for environmentally acceptable, highly efficient, and low-cost power generation. Thus, fuel cells are being seriously considered for applications where these attributes are important, e.g., military and commercial electric power generation and, possibly, vehicles.

41.3 OPERATION OF THE FUEL CELL

A simple fuel cell is illustrated in Fig. 41.2. Two catalyzed carbon electrodes are immersed in an electrolyte (acid in this illustration) and separated by a gas barrier. The fuel, in this case hydrogen, is bubbled across the surface of one electrode while the oxidant, in this case oxygen from ambient air, is bubbled across the other electrode. When the electrodes are electrically connected through an external load, the following events occur:

1. The hydrogen dissociates on the catalytic surface of the fuel electrode, forming hydrogen ions and electrons.
2. The hydrogen ions migrate through the electrolyte (and a gas barrier) to the catalytic surface of the oxygen electrode.
3. Simultaneously, the electrons move through the external circuit to the same catalytic surface.
4. The oxygen, hydrogen ions, and electrons combine on the oxygen electrode's catalytic surface to form water.

The reaction mechanisms of this fuel cell, in acid and alkaline electrolytes, are shown in Table 41.2. The major differences, electrochemically, are that the ionic conductor in the acid electrolyte is the hydrogen ion (or, more correctly, the hydronium ion, H_3O^+) and the OH^- or hydroxyl ion in the alkaline electrolyte. Further, in the acid electrolyte the product, water, is produced at the cathode and in the alkaline electrolyte fuel cell at the anode.

The net reaction is that of hydrogen and oxygen producing water and electrical energy. As in the case of batteries, the reaction of one electrochemical equivalent of fuel will theoretically produce 26.8 Ah of dc electricity at a voltage that is a function of the free energy of fuel-oxidant reactions. At ambient conditions, this potential is ideally 1.23 V dc for a hydrogen/oxygen fuel cell.

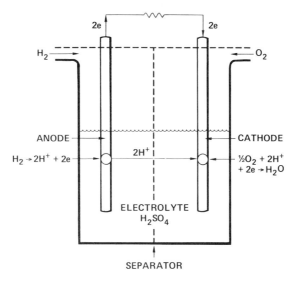

COMPONENTS

CATHODE: Oxygen Electrode
ANODE: Fuel Electrode
ELECTROLYTE: H_2SO_4

OVERALL REACTION

$H_2 + \frac{1}{2}O_2 \rightarrow H_2O$

(a)

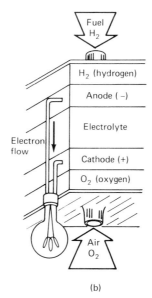

(b)

FIG. 41.2 Operation of the fuel cell. (*a*) Reaction mechanisms; (*b*) general construction features.

TABLE 41.2 Reaction Mechanisms of the H_2/O_2 Fuel Cell

	Acid electrolyte	Alkaline electrolyte
Anode	$H_2 \rightarrow 2H^+ + 2e$	$H_2 + 2OH^- \rightarrow 2H_2O + 2e$
Cathode	$1/2O_2 + 2H^+ + 2e \rightarrow H_2O$	$1/2O_2 + 2e + H_2O \rightarrow 2OH^-$
Overall	$H_2 + 1/2O_2 \rightarrow H_2O$	$H_2 + 1/2O_2 \rightarrow H_2O$

41.4 MAJOR COMPONENTS OF THE FUEL CELL

The important components of the individual fuel cell are:

1. The *anode* (fuel electrode) must provide a common interface for the fuel and electrolyte, catalyze the fuel oxidation reaction, and conduct electrons from the reaction site to the external circuit (or to a current collector that, in turn, conducts the electrons to the external circuit).

2. The *cathode* (oxygen electrode) must provide a common interface for the oxygen and the electrolyte, catalyze the oxygen reduction reaction, and conduct electrons from the external circuit to the oxygen electrode reaction site.

3. The *electrolyte* must transport one of the ionic species involved in the fuel and oxygen electrode reactions while preventing the conduction of electrons (electron conduction in the electrolyte causes a short circuit). In addition, in practical cells, the role of gas separation is usually provided by the electrolyte system. This is often accomplished by retaining the electrolyte in the pores of a matrix (or inert blotter). The capillary forces of the electrolyte within the pores allow the matrix to separate the gases, even under some pressure differential.

Other components may also be necessary to seal the cell, to provide for gas compartments, separate one cell from the next in a fuel cell stack, etc. Figure 41.3 depicts such a stack. This particular stack is a bipolar configuration. In this design, cells are placed electrically in series by the electronically conducting separator located between the anode and cathode of adjacent cells. The merit of such a configura-

FIG. 41.3 250-kW fuel cell stack. (*Courtesy of United Technologies Corp.*)

tion is that no external connections are required to electrically connect individual cells. Because of its simplicity, the bipolar approach is used in most fuel cell power plants.

A primary constraint imposed upon fuel cell components is that of material compatibility. The components must be stable in their respective environments for thousands (or tens of thousands) of hours. Thus, the pacing challenge to the fuel cell technologist becomes that of developing the materials and components required for high-performance fuel cells in a hostile environment for extended periods of time.

41.5 GENERAL PERFORMANCE CHARACTERISTICS

The performance of a fuel cell is represented by the current density vs. voltage (or "polarization") curve (Fig. 41.4). Whereas ideally a single H_2/O_2 fuel cell could produce 1.23 V dc at ambient conditions, in practice, fuel cells produce useful voltage outputs that are somewhat less than the ideal and decrease with increasing load (current density). The losses or reductions in voltage from the ideal are referred to as "polarization," as illustrated in Fig. 41.4.

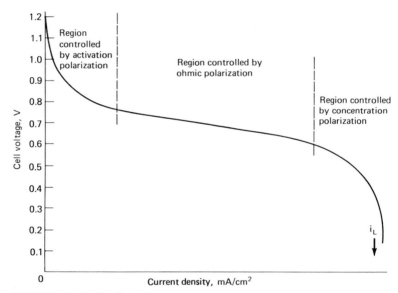

FIG. 41.4   Fuel cell polarization curve.

These losses include:

1. Activation polarization represents energy losses that are associated with the electrode reactions. Most chemical reactions involve an energy barrier that must be overcome for the reactions to proceed. For electrochemical reactions, the activation energy lost in overcoming this barrier takes the form

$$\eta_{\text{act}} = a + b \ln i$$

where η_{act} = activation polarization, mV
  a, b = constants
    i = current density, mA/cm^2
Activation polarization is associated with each electrode independently and

$$\eta_{act(cell)} = \eta_{act(anode)} + \eta_{act(cathode)}$$

2. Ohmic polarization represents the summation of all the ohmic losses within the cell, including electronic impedances through electrodes, contacts, and current collectors and ionic impedance through the electrolyte. These losses follow Ohm's law

$$\eta_{ohm} = iR$$

where η_{ohm} = ohmic polarization, mV
    i = current density, mA/cm^2
    R = total cell impedance, $\Omega \cdot$cm^2

3. Concentration polarization represents the energy losses associated with mass transport effects. For instance, the performance of an electrode reaction may be inhibited by the inability for reactants to diffuse to or products to diffuse away from the reaction site. In fact, at some current, the limiting current density i_L, a situation will be reached wherein the current will be completely limited by the diffusion processes (see Fig. 41.4). Concentration polarization can be represented by

$$\eta_{conc} = \frac{RT}{nF} \ln \left(1 - \frac{i}{i_L}\right)$$

where η_{conc} = concentration polarization, mV
    R = gas constant
    T = temperature, K
    n = number of electrons
    F = Faraday's constant
    i = current density, mA/cm^2
    i_L = limiting current density, mA/cm^2
Concentration polarization occurs independently at either electrode. Thus, for the total cell

$$\eta_{conc(cell)} = \eta_{conc(anode)} + \eta_{conc(cathode)}$$

The net result of these polarizations is that practical fuel cells produce between 0.5 and 0.9 V dc at currents of 100 to 400 mA/cm^2 of cell area. Fuel cell performance can be increased by increasing cell temperature and reactant partial pressure. For any fuel cell, the trade-off always exists between achieving higher performance by operating at higher temperature or pressure and confronting the materials and hardware problems imposed at the more severe conditions.

41.6 CLASSIFICATION OF FUEL CELL SYSTEMS

Fuel cell systems can take a number of different configurations, depending on the combination of type of fuel and oxidant, whether the fueling is direct or indirect, the type of electrolyte, the temperature of operation, etc. Some of these parameters are listed in Table 41.3. In actual practice, the number of combinations are limited for the following reasons.

TABLE 41.3 Fuel Cell Parameters

Fuel		Oxidant	Temperature	Electrolyte
Direct	Indirect			
Hydrogen	Hydride	Oxygen	Low (120°C)	Aqueous Acid
Hydrazine	Ammonia	Oxygen (air)	Intermediate	Sulfuric
Ammonia	Hydrocarbon	Hydrogen peroxide	(120–260°C)	Phosphoric
Hydrocarbon	Methanol		High (260–750°C)	Solid polymer
Methanol	Ethanol		Very high (750°C)	electrolyte (SPE)
Coal gas	Coal			Aqueous alkaline
Coal				Molten alkaline
				Aqueous carbonate
				Molten Carbonate
				Solid oxide

A simple classification approach can be achieved by:

1. Considering direct- or indirect-fueled systems: In a direct fuel cell, the fuel used is readily oxidized electrochemically in the fuel cell and is fed directly. In the indirect system, the fuel is first converted in a fuel-processing subsystem to an easily oxidized hydrogen-rich gas, which is then fed into the fuel cell.

2. Considering the requirements of achieving reasonable fuel electrode reaction rates or of integrating the fuel-processing subsystem within the power plant: This sets the fuel cell operating temperature for specific fuels as follows:

   Direct: hydrogen, hydrazine, methanol; indirect: ammonia or hydrazine—low temperature (<120°C)

   Indirect: hydrocarbon, methanol, ethanol, coal (120–260°C)

   Direct: coal gas, ammonia—intermediate temperature (120–260°C)

   Direct: coal—high temperature (260–750°C)

3. Considering the logical relationships between application and fuel type:

   Remote (space, undersea): direct hydrogen, hydrazine

   Military: logistic fuel such as liquid hydrocarbon or methanol

4. Considering pure oxygen and hydrogen peroxide as special oxidant cases. Air will normally be the source of the oxidant except in special situations such as in space or underseas where air is not available and weight or volume are critical.

5. Coupling the electrolyte systems with temperature based upon the electrolyte's thermal capability. This results in combinations of

   Low temperature: sulfuric acid, solid polymer electrolyte, aqueous alkaline

   Intermediate temperature: phosphoric acid, aqueous alkaline

   High temperature: molten carbonate, molten alkaline

   Very high temperature: solid oxide

6. Considering the reaction of alkaline electrolytes with carbon oxides from the fuel or air. This essentially precludes the use of alkaline electrolytes in high-power applications involving carbonaceous fuels or air.

Combining the above considerations results in the relatively few practical fuel cell combinations shown in Table 41.4. Furthermore, systems based on sulfuric acid, aqueous carbonate, and molten alkaline electrolytes have not found a niche in a practical fuel cell application. The sulfuric acid electrolyte has marginal stability; aqueous carbonate systems tend to be too low in performance for all but a very few specialty applications; and the molten alkaline system seems to have been dropped by developers due to its complexity.

41.7 DESCRIPTIONS OF FUEL CELL SYSTEMS

Table 41.4 lists two types of acid fuel cells (solid polymer electrolyte and phosphoric acid), aqueous alkaline fuel cells, molten carbonate fuel cells, and solid oxide fuel cells.

41.7.1 Acid Fuel Cell

The *acid fuel cell* has been described in Fig. 41.2. Acid fuel cells are characterized by:

Ionic conduction is provided by hydrogen ions [or by hydronium ions (H_3O^+)].

Platinum or platinum alloys (in very small quantity) are the active electrocatalysts.

Carbon (graphite) is an acceptable material of construction for current collectors, gas separators, etc., and is commonly used.

Solid Polymer Electrolyte System   The solid polymer electrolyte (SPE) system uses an ion exchange membrane as the electrolyte. The advantages of the SPE fuel cell are (1) the electrolyte, being a solid, cannot change, move about, or vaporize from the system. (2) The only liquid in the fuel cell is water, minimizing corrosion. The disadvantages are (1) the SPE must be hydrated (water-saturated) to perform; consequently, operation must be under conditions where the by-product water does not vaporize into the reaction air stream faster than it is produced; this constrains cell operation to under 60°C at ambient pressure and about 120°C at elevated pressures. (2) The SPE freezes at about 0°C and undergoes a freeze-drying phenomenon. This constrains applications to those where low-temperature capability is not a requirement.

Due to their inability to operate much above 120°C, SPE fuel cells are best suited for use with hydrogen-rich gases that contain little or no carbon monoxide. Carbon monoxide inhibits the fuel cell anode reaction, the degree of inhibition decreasing with increasing temperature. Consequently, SPE fuel cells have found their important applications in the space program operating on pure hydrogen or in military applications operating on hydrogen obtained by the decomposition of a hydride. Approaches are being evaluated that would allow future SPE cells to operate with fuel gases containing carbon monoxide. These techniques involve improving the anode electrocatalysts and externally prehumidifying the air to allow the operating temperature to be increased.

TABLE 41.4 Classification of Practical Fuel Cells

Application	Fuel	Oxidant	Electrolyte	Temperature
Remote:				
Space	Direct H_2	Liquid O_2	Aqueous alkaline Solid polymer	Low, intermediate Low
Undersea	Direct H_2 Direct hydrazine	Liquid O_2 Hydrogen peroxide	Aqueous alkaline	Low
Military:				
Low power, 100 W	Indirect hydride	Air	Aqueous alkaline Solid polymer	Low
High power, 500 W	Indirect hydrocarbon Indirect methanol	Air	Phosphoric acid	Intermediate
Commercial power:				
Dispersed (or on-site)	Indirect hydrocarbon Indirect methanol, ethanol Direct coal gas	Air	Phosphoric acid Molten carbonate	Intermediate High
Central station	Indirect coal Direct coal gas	Air	Phosphoric acid Molten carbonate Solid oxide	Intermediate High Very high
Vehicle	Hydrogen Hydride Indirect methanol Indirect hydrocarbon	Air	Phosphoric acid	Intermediate

Phosphoric Acid Electrolyte System   The phosphoric acid electrolyte system operates at 150 to 220°C. At lower temperatures, phosphoric acid is a poor ionic conductor. At higher temperatures, material stability (carbon and platinum) becomes limiting. The advantages of phosphoric acid fuel cells are (1) the electrolyte is very stable, (2) the phosphoric acid can be highly concentrated ($\sim$100%) where the water vapor pressure is very low and steady-state water removal by the reactant gases will always equal product water rate, and (3) at 150 to 220°C, the anode performance is very good even on fuels containing up to 5% carbon monoxide. The disadvantage of phosphoric acid fuel cells is that the cathode performance is sluggish. In fact, the major technology thrusts in phosphoric acid are toward improvement of the cathode. The phosphoric acid fuel cell is a preferred system for use with fuels containing carbon oxides.

41.7.2 Alkaline Fuel Cells

Although early alkaline fuel cells operated at relatively high temperature ($\approx$250°C) with concentrated (85 wt %) potassium hydroxide, systems developed more recently operate at much lower temperatures ($<$120°C) using less concentrated (35 to 50 wt %) potassium hydroxide. The lower temperature enables the use of matrices to retain the electrolyte and increases the life of other components.

Alkaline fuel cells are characterized by:

Ionic conduction is provided by hydroxyl (OH^-) ions.

A wide range of electrocatalysts can be used including nickel, silver, metal oxides, spinels, and noble metals—although the truly high performance systems use at least small amounts of noble metal.

Construction materials include carbon, nickel, and stainless steel.

The advantages of alkaline fuel cells are (1) cathode performance is much better than for acid fuel cells and (2) materials of construction tend to be low in cost. The primary disadvantage is that the electrolyte reacts with carbon oxides to form potassium carbonate. This severely limits the cells' performance. Thus, alkaline fuel cells have only limited application where carbonaceous fuels or air are used as reactants. The important applications (space and underseas) involve pure hydrogen and oxygen.

41.7.3 Molten Carbonate Fuel Cells

Molten carbonate fuel cells use an alkali metal (Li, K, Na) carbonate as the electrolyte. Since these salts can function as electrolytes only when in the liquid phase, the cells operate at 600 to 700°C, which is above the melting points of the respective carbonates. Molten carbonate cells are characterized by:

Ionic conduction is by the carbonate ion; thus the carbonate ion must be involved in the two electrode reactions

$$1/2O_2 + CO_2 + 2e \rightarrow CO_3^{2-} \qquad \text{(cathode)}$$
$$\underline{CO_3^{2-} + H_2 \rightarrow CO_2 + H_2O + 2e \quad \text{(anode)}}$$
$$1/2O_2 + H_2 \rightarrow H_2O \qquad \qquad \text{(net)}$$

A consequence of this is that CO_2 must be recycled from the anode to the cathode. At 600 to 700°C, electrode reactions proceed without highly specific catalysts. Nickel and nickel oxide work quite well; noble metals are not used.

Construction materials include nickel and ceramics.

The advantages of molten carbonate fuel cells are (1) cell performance is good, activation polarization is small, and (2) at 600 to 700°C, any carbon monoxide in the fuel converts to hydrogen on the anode via the water gas shift reaction

$$CO + H_2O \rightarrow CO_2 + H_2$$

(as a result, fuel gases high in carbon monoxide are readily used), and (3) waste heat from the fuel cell can be available at a relatively high temperature ($> 500°C$), enabling its use in bottoming or industrial heating cycles.

Disadvantages are (1) the high temperature imposes severe constraints on materials suitable for long lifetimes and (2) a source of carbon dioxide is required to complete the cathode reaction (this is provided by recycling CO_2 from the anode exhaust to the cathode inlet). As a result, molten carbonate fuel cells are best suited for applications that integrate the fuel cell with a carbonaceous fuel processor, i.e., a reformer or coal gasifier.

41.7.4 Solid Oxide Fuel Cells

As the name implies, solid oxide fuel cells employ a solid, nonporous metal oxide electrolyte which allows ionic conductivity by the migration of oxygen ions through the lattice of the crystal. Stabilized zirconia is commonly used as the electrolyte. The cells operate at 900 to 1000°C. Whereas practical cells of the technologies previously discussed are normally packaged into "filter press" or "plate and frame" stack assemblies (Fig. 41.3), solid oxide fuel cells are configured into tubular cell stacks.

Characteristics of the solid oxide fuel cell include:

Ionic conduction is provided by oxide ions.

The cathode employs metal oxides, such as praseodymium oxide or indium oxide; the anode uses nickel or nickel cermet.

Because of the high temperature, materials of construction will likely be confined to ceramics or metal oxides.

Solid oxide fuel cells offer advantages similar to those of molten carbonate cells, that is, good performance on fuels containing hydrogen or hydrogen and carbon monoxide, the elimination of noble-metal catalysts, and the availability of high-grade reject heat. In addition, they do not suffer the constraint of molten carbonate cells which require a carbon dioxide recycle to the cathode. The primary disadvantages are the very high temperature of operation and the severe material constraints imposed by the $\sim 1000°C$ temperature.

BIBLIOGRAPHY

A. P. Fickett: "Fuel Cell Power Plants," *Scientific American* (December 1978).

"Fuel Cells, A Bibliography," *Report No. TID-3359,* U.S. Energy Research and Development

Administration, Washington, D.C., June 1977.

K. V. Kordesch: "25 Years of Fuel Cell Development (1951–1976)," *J. Electrochem. Soc.* **125**(3) (March 1978).

H. A. Liebhafsky and E. J. Cairns: *Fuel Cells and Fuel Batteries,* Wiley, New York, 1968.

Low-Power Fuel-Cell Systems

by
David Linden

42.1 GENERAL CHARACTERISTICS

The advantageous characteristics of fuel cells led to the development of a number of different systems ranging in size from the portable units, 5 W or smaller to kilowatt power levels (where ease of operation, low maintenance, and silence are important), to large stationary plants delivering megawatts of power [where the high efficiency over the range from full to partial load and reduced pollution are significant (see Part 6, Chap. 43)].

The lower-power fuel cells were designed mainly for military or special applications such as the space program. For the space applications and for forward-area military use, the fuel cell offers high energy densities that exceed the performance of batteries when operated over long periods of time. Figure 3.20 in Part 1 compares the performance of typical primary and secondary batteries with fuel cell systems (at two levels of fuel energy density) and shows the weight advantage of the fuel cell for long-term operation.

The fuel-cell systems built between 1960 and 1970 for the Gemini and Apollo space programs and in 1980 for the Space Shuttle Orbiter are still among the most successful demonstrated to date. They used hydrogen as the fuel and oxygen as the oxidizer. The Gemini fuel cell, designed as a 1-kW system, used an ion exchange membrane separator-electrolyte, known as a solid polymer electrolyte (SPE). The larger 1.5-kW Apollo and 7-kW Space Shuttle Orbiter units were based on a potassium hydroxide electrolyte system.

For the ground applications, air-breathing, rather than pure oxygen systems, is used.

In the smaller-sized units (up to about 250-W) hydrogen is supplied in a pressure vessel or, more commonly, generated from a metal hydride. In the larger-sized systems, including both electric vehicle and power unit applications, hydrogen is generated by steam reforming or thermocatalytic cracking of fuels, such as methanol or hydrocarbons.

Direct-type fuel cells, using gaseous or liquid fuels that are more readily oxidized in the fuel cell than the fossil fuels, were also considered for the smaller fuel cells, particularly for military application. No system was ever built beyond the development stage, and work on this approach has been deemphasized.

42.2 SPACECRAFT FUEL-CELL SYSTEMS

42.2.1 Gemini System

The Gemini fuel-cell power system [1,2] is based on the SPE fuel-cell technology, with hydrogen as the fuel and oxygen as the oxidizer. The basic building block is the individual fuel cell (Fig. 42.1). The assembly consists of the ion-exchange membrane–electrode assembly with associated components for gas distribution, current collection, heat removal, and water management. Hydrogen enters the fuel cell on one side of the ion-exchange membrane and is ionized by the platinum catalyst, and the hydrogen ions are transferred through the membrane, as hydronium ions, to react with the oxygen. Each side of the membrane is covered by a titanium screen, which, in turn, is coated with the platinum catalyst. One side of the membrane is bonded at its edges to a titanium sheet to form the cavity for the hydrogen gas. The hydrogen is introduced into the cavity by a tube that feeds through the edge seal. A similar tube passes through the edge seal to facilitate purging of the hydrogen cavity. Two loops of coolant tubing are placed against the other face of the titanium sheets, and wicks are put between each pass of the tubing. The wicks remove water produced at the cathode; they are necessary

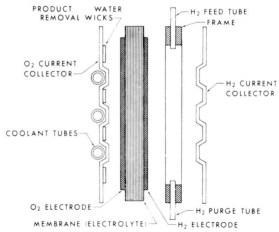

FIG. 42.1 Gemini basic fuel cell assembly. (*From Cohen, 1966.*)

in a zero-gravity environment to prevent water from accumulating in undesirable locations. A total of 32 of these assemblies are bolted together between end plates to form the stack assembly (Fig. 42.2). Each stack contains its own hydrogen and coolant manifolding and water-oxygen separator. Three stacks are installed in a cylindrical container to form the fuel-cell section. Two sections were used in each Gemini system. Each weighed about 30 kg, was 63.5 cm long and 31.7 cm in diameter, and produced 1 kW at 26.5 to 23.3 V; reactant consumption was 400 g/kWh and water production about 0.5 L/kWh.

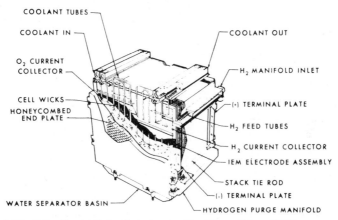

FIG. 42.2   Gemini fuel cell stack assembly. (*From Cohen, 1966.*)

The performance of the fuel cell installed in the Gemini V spacecraft is plotted in Fig. 42.3. The dropoff in performance shown for the first day was probably due to the water imbalance caused by prelaunch low standby-current usage. During the Gemini VII mission, decay rates of 0.003 and 0.005 V/h at 10 and 24 A, respectively, were observed during the 14-day mission.

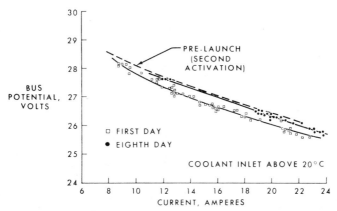

FIG. 42.3   Gemini V mission—fuel cell section 1 performance. (*From Cohen, 1966.*)

42.2.2 Apollo System

The Apollo fuel-cell power system[3,4] utilized the Bacon hydrogen/oxygen fuel-cell technology, operated at 200°C and at a cell pressure near 3.5 kg/cm². Concentrated potassium hydroxide (45%) was used as the electrolyte; the reactant gases were supplied to the cells at a pressure of 0.7 kg/cm² above the cell pressure. The liquid-gas reactant sites were maintained by the use of two levels of porosity in the sintered-nickel electrodes in combination with a pressure differential which was maintained between the reactant compartments or cavities and the electrolyte. The by-product water was formed at the hydrogen electrode. Heat removal and water vapor control were accomplished by means of a closed-loop hydrogen recirculation system. Figure 42.4 is a schematic of the fuel-cell power supply system, designated as PC3A-2.

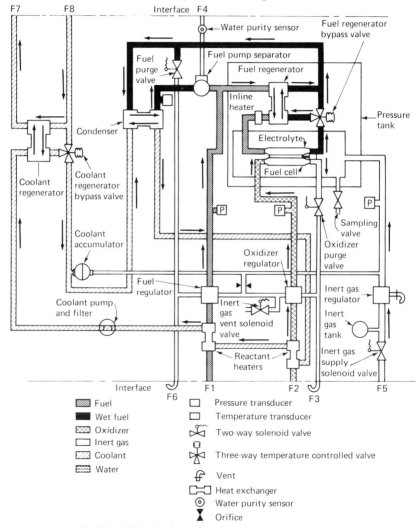

FIG. 42.4   Apollo fuel-cell electrical power supply. (*From Ferguson, 1969.*)

A fuel-cell assembly is shown in Fig. 42.5; 31 cells were located in the lower cylindrical portion of the assembly. The accessories, which included the hydrogen recirculation loop, the glycol cooling pump and accumulator, and the reactant controls, were located in the unpressurized section above the cell stack. Each of these assemblies or modules weighed about 110 kg and was approximately 57 cm in diameter and 112 cm high.

FIG. 42.5   Apollo fuel-cell assembly. (*From Morrill, 1965.*)

Three of the PC3A-2 units were mounted in the service compartments adjacent to the command module. The hydrogen and oxygen reactants were supercritically stored in cryogenic Dewar flasks within the same sector.

The average mission power level required from each fuel-cell assembly was about 600 W. Each assembly was capable of producing 1420 W, within the normal spacecraft voltage requirements of 27 to 31 V. Thus, in the event of the loss of two of the three fuel-cell assemblies, the third assembly could still provide sufficient power to return the spacecraft safely from any point in the lunar mission. Total reactant consumption at 1420 W was a maximum of 0.55 kg/h; product water was delivered from the power plant.

The power plant was designed to operate for a period in excess of 400 h. The moon flight, Apollo 8, lasted for 440 h; the fuel-cell system produced 292 kWh of electricity and 100 L of water. The accrued individual fuel-cell time in space was 1995 h; the total energy produced was 1325 kWh with 450 kg of water. No further units were produced after the Apollo program ended.

42.2.3 Space Shuttle Orbiter System

The Space Shuttle Orbiter fuel-cell power system also is based on the Bacon hydrogen-oxygen technology, but incorporates significant advances in design. The Space Shuttle fuel cells are 20 kg lighter and deliver 6 to 8 times as much power as those of the Apollo system. The Space Shuttle system is made up of three fuel-cell power plants and normally supplies 14 kW of power, with peak loads of 36 kW. The fuel-cell power plant is illustrated in Fig. 42.6. Each power plant is 35 cm high, 38 cm wide, and 101 cm long and weighs 91 kg.

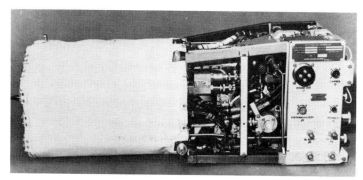

FIG. 42.6 Space Shuttle Orbiter fuel-cell power plant. (*Courtesy of United Technologies Corp.*)

Each fuel-cell power section contains two parallel stacks of 32 cells in series. Each is capable of supplying 12 kW at peak (27.5 V and 436 A), 7 kW average power; at 2 kW, each power plant provides 32.5 V and 61.5 A. Cryogenically stored hydrogen and oxygen are delivered to each fuel. On a 3-day mission, the fuel cells utilize about 450 kg of H_2 and O_2; about 600 L of water are produced. The power plants will be serviced between flights and reused until each has accumulated 2000 h of on-line service.

42.3 PORTABLE FUEL CELLS (1 to 100 W)

Fuel cells, in the power range of up to about 200 W, can be an attractive alternative to batteries for long-term operation, with potential reduction in the weight (Fig. 3.20, Part 1) and cost. The size limitation of these portable fuel-cell systems is generally too small to permit the use of elaborate fuel conditioning. Hence, easily handled and readily oxidized fuels (e.g., liquid or gaseous fuels such as hydrazine, methanol, ammonia) and fuels that provide hydrogen through a simple physical or chemical reaction have been used in most fuel-cell systems of this size and type.

While the fuel-cell systems in this size and power range potentially have advantageous characteristics, system complexities and the development of new battery systems with higher energy densities have limited the interest and successful use of the low-power fuel cell.

42.3.1 Direct Fuel-Cell Systems

Direct-type fuel cells, in which the fuel can be introduced into the fuel cell without requiring conversion to hydrogen, were considered for small fuel-cell systems because they eliminated the need for a fuel-conditioning unit, thus saving important space and weight. Methanol (CH_3OH) and hydrazine (N_2H_4) were the main liquid fuels used. Methanol[5] is directly oxidizable, but removal of the carbonate, one of the reaction products of the dissolved methanol fuel cell in alkaline electrolytes, from the electrolyte is extremely difficult. Efforts then shifted to hydrazine.[5-8] Hydrazine decomposes easily into hydrogen and nitrogen at the electrode surface; in fact, the voltage observed is that of hydrogen.

Major effort was directed toward a silent power source for forward-area military use. A 60-W, 24-V hydrazine/air fuel cell was developed in a configuration similar to the one used later for the metal hydride cell (Fig. 42.9). The schematic for this system is given in Fig. 42.7. The fuel cell used a 35% potassium hydroxide electrolyte and a 64% hydrazine monohydrate fuel and operated between 55 and 70°C with a fuel utilization of 600 Wh/kg. A larger 300-W, 24-V power source was also developed for forward area use. This system weighed 20 kg, with electrolyte and 4 L of fuel, and had a volume of 35 dm³. The fuel was sufficient for 12 h of operation at 300 W. Field tests[8] confirmed the successful electrochemical functioning of the cell, but mechanical deficiencies caused early failure of the system.

Ammonia (NH_3) was also used in direct fuel cells because the direct oxidation of ammonia to nitrogen and water occurs on noble-metal catalysts.[5] No successful and

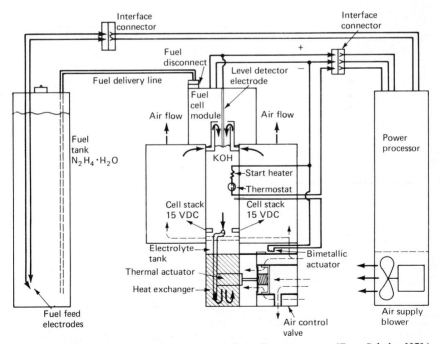

FIG. 42.7   Schematic of manpack hydrazine/air fuel-cell power source. (*From Salathe, 1970.*)

practical direct fuel cells for this type of application were developed; most of the effort ended early in the 1970s.

42.3.2 Metal Hydride Fuel Cells

The majority of current portable fuel-cell developments use a metal hydride as the source of hydrogen fuel. Metal hydrides are attractive because they can store large amounts of hydrogen more conveniently and with a higher energy density (total equivalents of hydrogen per total weight of hydrogen source and container) than hydrogen in a pressurized or liquefied form.

One type of metal hydride produces hydrogen by the reaction with water, e.g., calcium hydride (CaH_2)

$$CaH_2 + 2H_2O \rightarrow Ca(OH)_2 + H_2$$

A second type of metal hydride, a reversible hydride, is based on the principle that certain metals or alloys (e.g., iron titanium, lanthanum nickel, and various other rare-earth metal and nickel alloys) have the ability to take up large amounts of hydrogen gas within their crystal structure. A reduction in pressure or an increase in temperature releases the hydrogen. These hydrides can deliver hydrogen at about 500 Wh/kg.

Remote Fuel-Cell Power Source This high energy density, refuelable metal hydride fuel cell was developed for military use[9] for remote, unattended equipment operating at the 1- to 50-W level for long periods of unattended operation. The power source, shown in schematic style in Fig. 42.8, consists of a small solid-polymer electrolyte fuel-cell stack and a miniature hydrogen generator operating on calcium hydride and water vapor. No auxiliary components are required, and the fuel cell is capable of being throttled from full to no-load. Total volume of the system is less than 750 cm³. Water vapor from the reservoir flows into the water chamber adjacent to the porous hydrophobic membrane. Water vapor diffuses through the membrane, where hydrogen production is self-regulated according to the demand of the fuel cell. At no-load, hydrogen

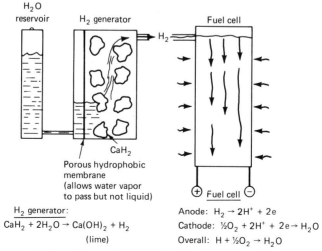

H_2 generator:
$$CaH_2 + 2H_2O \rightarrow Ca(OH)_2 + H_2$$
(lime)

Anode: $H_2 \rightarrow 2H^+ + 2e$
Cathode: $\frac{1}{2}O_2 + 2H^+ + 2e \rightarrow H_2O$
Overall: $H + \frac{1}{2}O_2 \rightarrow H_2O$

FIG. 42.8 Schematic of a remote fuel-cell power source.

FIG. 42.9   Hydrogen/air fuel-cell system, 60 W.

is not consumed, and the pressure within the reaction chamber increases, forcing water into the reservoir and out of the water chamber and reducing hydrogen production. As hydrogen is consumed by the fuel cell, the water level will self-adjust to generate hydrogen at the demand rate. The fuel-cell power source is refuelable at 1000 Wh/kg of hydride.

Metal Hydride-Air Fuel-Cell System, 30- and 60-W   Small portable fuel-cell systems are being developed for military use as power supplies for radio sets and other electronic equipment and as battery chargers. A 60-W, 28-V hydrogen/air fuel cell is illustrated in Fig. 42.9.[10] A similar package was also used for a 30-W fuel cell. The system consists of three sections: the hydrogen generator, the fuel-cell stack, and the power conditioner (Fig. 42.10). The overall unit weighs 7 kg and has a volume of 10 dm³.

Hydrogen is supplied by a Kipp generator using the reaction between sodium aluminum hydride ($NaAlH_4$) and water

$$NaAlH_4 + H_2O \text{ (excess)} \rightarrow Al(OH)_3 + NaOH + 4H_2$$

The hydride, in the form of a solid pellet, delivers in excess of 2000 Wh/kg. Hydrogen for 4 h of operation is supplied to the fuel cell with a single 120-g charge. The Kipp generator delivers hydrogen to the fuel cell on demand; when no hydrogen demand

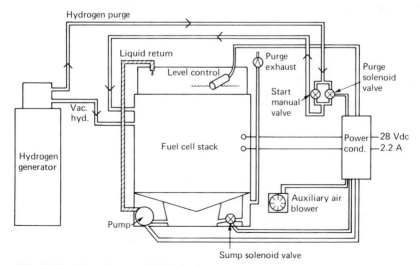

FIG. 42.10   Schematic of a 60-W fuel-cell system.

exists, the pressure builds up in the generator, forcing water away from the fuel pellet and stopping the reaction.

The fuel-cell stack is an assembly of 11 dual-element cells, each dual element producing 1.7 V; the stack is designed for an 80-W output. The cells use KOH electrolyte and are filled by a common reservoir tank located above the stack. Electrolyte is collected in a bottom sump tank and recirculated. Air is circulated between cathodes by natural convection. The free electrolyte maintains a gas diffusion interface with the highly hydrophobic Teflon-bonded platinum-black cathodes bonded to a nickel screen. A porous Teflon backing is bonded on the gas side of both the anode and cathode to prevent electrolyte weeping.

The power conditioner converts the 19-V dc from the fuel-cell stack into 28-V regulated direct current and also contains the logic and control circuits.

The fuel-cell stack operates at a temperature of 50°C. The performance of the 22-cell stack is shown in Fig. 42.11. The stack is designed to operate at a high voltage efficiency of approximately 70%; overall system efficiency is about 55%. Five hundred hours of life have been demonstrated, but major problems existed with carbonate formation in the electrolyte due to CO_2 absorption and with heat and mass balance.

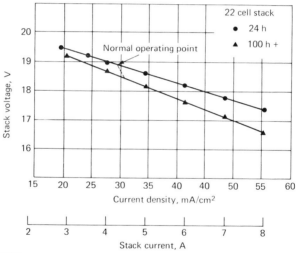

FIG. 42.11 Performance of the 60-W fuel-cell stack.

A later design[11] used a reversible metal hydride, lanthanum pentanickel hydride, as the source of hydrogen

$$LaNi_5 + 3H_2 \rightleftarrows LaNi_5H_6$$

This change, replacing the exothermic sodium aluminum hydride generator with the reversible metal hydride source which absorbs heat with the release of hydrogen, could reduce the total heat output of the system by 65%. The system was found to operate satisfactorily at 20°C, but higher ambient temperatures still caused problems because of the much higher operating temperature of the fuel-cell stack.

SPE Fuel Cell   There is renewed interest in the SPE fuel cell for applications in which mobility is important and power requirements are low.[12,13] The advantages of the SPE cell are ease of product water removal, simple construction, stable electrode-electrolyte interface, and favorable life characteristics.

The SPE cell is being considered for power levels from a few watts to 500 W, operating in the −40 to 50°C range with no restrictions on humidity. Special designs, including insulation for low temperatures and methods for waste heat disposal, will probably be required to achieve this performance. The cell operates on hydrogen and ambient air. Preferred fuel sources for hydrogen generation are magnesium and aluminum, which are reacted with saltwater. Bottled hydrogen, hydride-stored hydrogen, or hydrides may also be used. With magnesium, it is expected to obtain energy densities (fuel consumption) of about 220 Wh/kg on a wet basis and 1300 Wh/kg on a dry basis. For longer missions, where a larger system weight can be tolerated, reformed methanol combined with carbon monoxide absorption, is being considered.

A 30-W SPE hydrogen/air stack is illustrated in Fig. 42.12. In a complete device, the fuel cell is integrated with the hydrogen source. The complete unit weighs about 7 kg and has a volume of about 15 L. A flow diagram for this unit is shown in Fig. 42.13. In this case, CaH_2, contained in thin aluminum tubes, is used for hydrogen generation. Discharge characteristics are given in Fig. 42.14; the generator was charged with 280 g of CaH_2. Current-voltage characteristics for this system are plotted in Fig. 42.15. Field tests have demonstrated operation in excess of 3 months.

FIG. 42.12   SPE fuel-cell stack, 30 W. (*Courtesy of Engelhard Industries.*)

42.4 INDIRECT METHANOL FUEL-CELL POWER PLANTS

42.4.1 General Characteristics

A family of fuel cells, with output ratings of 500 W to 5 kW, is being developed for forward-area military use to provide a silent, lightweight source of electrical energy. Originally, conventional hydrocarbon fuels were considered for these power plants, but the changing availability in recent years of traditional petroleum-based fuels prompted a consideration of methanol as an alternative fuel. Methanol is attractive for fuel-cell use; it is more convenient to store and handle than either hydrogen or ammonia, and

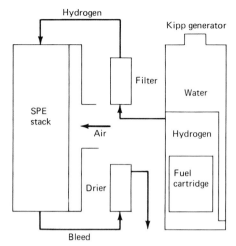

FIG. 42.13   Flow schematic for the SPE fuel-cell power source.

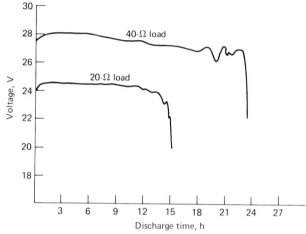

FIG. 42.14   Discharge characteristics of the 30-W SPE fuel-cell (33-cell SPE stack, hydrogen generator with two fuel cartridges containing 280 g CaH$_2$; temperature approximately 22°C). (*Courtesy of Engelhard Industries.*)

it can be more readily reformed to hydrogen than the long-chain hydrocarbons. Methanol can not be used successfully in alkaline electrolyte fuel cells because the carbon oxides that form during reforming can not be tolerated. The development of the phosphoric acid electrolyte fuel cell, which could tolerate these carbon oxides, reactivated interest in the methanol fuel cell, even though the acid electrolyte presented more problems with corrosion and requires the use of noble-metal catalysts.

Three sizes of fuel-cell power plants are being developed. The characteristics of these units are summarized in Table 42.1. Major development emphasis is on the 1.5-kW size.

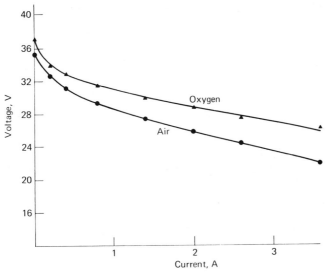

FIG. 42.15 Current-voltage characteristics of the 30-W, 35-cell SPE prototype system (25°C). (*Courtesy of Engelhard Industries.*)

TABLE 42.1   Characteristics of Methanol Fuel-Cell Power Plants

Power rating, kW	Size (volume), dm³	Dry weight, kg	Fuel consumption, g/kWh
1.5	200	70	1000
3.0	340	135	800
5.0	510	225	800

SOURCE: Reference 16.

The methanol fuel-cell power plant is based on a low-temperature steam reformer
and a phosphoric acid fuel-cell stack. The power plant consists of three subassemblies
as shown in the schematic in Fig. 42.16: the fuel conditioner converting methanol to
hydrogen, the fuel-cell assembly converting the hydrogen to electric power (direct
current), and the power conditioner which converts the fuel-cell output power to
regulated direct current or to an ac output.

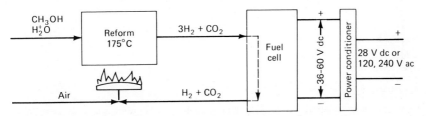

FIG. 42.16 Simplified schematic of a 15-kW indirect methanol air fuel-cell system.

42.4.2 Methanol fuel-cell power plant, 1.5 kW

The 1.5-kW methanol fuel-cell power plant is illustrated in Fig. 42.17, and the schematic of the system, including water recovery, is shown in Fig. 42.18.[15]

In the fuel-conditioning subsystem, an aqueous methanol feed (58% by weight of methanol) is vaporized and superheated to a temperature of 160°C and passed through

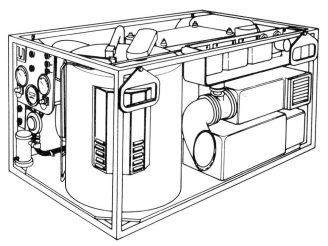

FIG. 42.17   Methanol fuel-cell power plant, 1.5 kW. (*From Meyer, 1978.*)

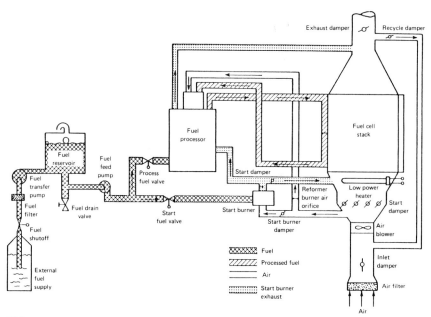

FIG. 42.18   Schematic of a 1.5-kW methanol fuel-cell power plant (with water recovery subsystem). (*From Stedman and Fanciullo, 1982.*)

the catalyst bed where the water and methanol react at a temperature of 250–300°C. Hydrogen and carbon oxides (approximately 75% H_2, 25% CO_2, 2% CO) are formed with residuals of water and methanol. The gaseous products are cooled with ambient air to about 50°C and routed to the fuel cell. The fuel-cell subassembly consists of 80 phosphoric acid electrolyte cells and the ancillary equipment for thermal control. The fuel-cell stack is built up of four repeating cell elements: graphite intercell plates, grooved to permit gas flow; the two graphite electrodes, with platinum-black catalyst bonded to one face; and a porous phenolic resin mat between the electrodes to retain the electrolyte and serve as a gasket to prevent leakage. The cells are clamped between honeycomb end plates to ensure good electrical contact and set the gasketing function of the mat. Air flows in a single pass across the short dimension of the cell and fuel in a double pass across the long dimension. The fuel cell operates at about 170°C.

The power plant can have either a dc or ac output, depending on the power conditioner that is used. The inverter converts the 36- to 60-V dc to either 120- or 240-V ac. The dc/dc converter takes the same input voltage and delivers an adjustable output voltage in a range of 26- to 36-V dc. The initial and 6000-h stack performance is shown in Fig. 42.19. Fuel consumption was less than 650 g/kWh, considerably less than the design objective.

This indirect methanol fuel-cell system uses few moving parts and operates at relatively low temperatures; it should have a long and reliable life, with minimum maintenance (MTBF: 750 h; overhaul: 6000 h) and operate over the specified temperature range (−50 to +50°C) and environmental conditions.[16]

42.4.3 Fuel Cells for Vehicular Applications

A similar fuel-cell system, sized at about 20 kW, is also being considered for electric vehicle use in a hybrid fuel-cell/secondary battery power system.[17] The most advanced approach uses methanol for the fuel, a reformer and the phosphoric acid fuel cell. Also under consideration are fuel-cell systems based on the solid polymer electrolyte (SPE)

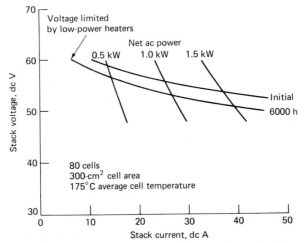

FIG. 42.19 Methanol fuel-cell power plant (1.5 kW): fuel-cell stack performance characteristics. (*From Stedman and Fanciullo, 1982.*)

technology and the use of superacid electrolytes, such as trifluoromethanesulfonic acid (TFMSA). The aim of this program is to use the fuel cell's high efficiency, low pollution (both air and noise), and the ability to use nonpetroleum fuels in a vehicle having characteristics competitive with internal combustion engines. Figure 42.20 shows an illustration of a proposed near-term unit; Table 42.2 summarizes the characteristics of the fuel-cell power plant and the goals of the program.

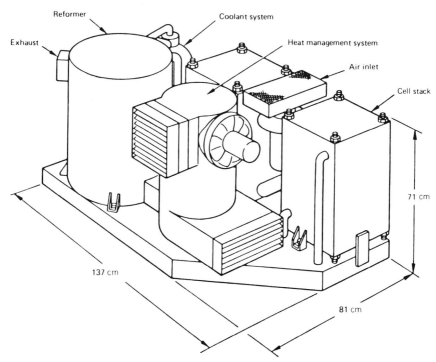

FIG. 42.20   Near-term vehicle fuel-cell power plant. (*Courtesy of United Technologies Corp.*)

TABLE 42.2   Characteristics of Fuel-Cell System for Vehicular Applications

	Target	Near-term	Advanced
Rated power, kW	20	20	20
Peak power, kW	67	30	60
Weight, kg	310	350	250
Volume, m^3	0.7	0.7	0.35
Operating temperature range, °C	120 to 180	120 to 180	120 to 180
Fuel-cell system efficiency, %	40	45	55
Volts per cell	0.6	0.7	0.85
Production cost, $/kW (1982 $)	minimum	250-300	150-250
Overhaul period, h	5000	5000	5000

SOURCE: Reference 17.

REFERENCES

1. John H. Russell, "Gemini Fuel Cell System," *Proc. 19th Power Sources Conf.*, 1965.
2. R. Cohen, "Gemini Fuel Cell System," *Proc. 20th Power Sources Conf.*, 1966.
3. C. C. Morrill, "Apollo Fuel Cell System," *Proc. 19th Power Sources Conf.*, 1965.
4. Richard B. Ferguson, "Apollo Fuel Cell Power System," *Proc. 23d Power Sources Conf.*, 1969.
5. K. V. Kordesch, "25 Years of Fuel Cell Development (1951–1976)," *J. Electrochem. Soc.* **125**(3) (March 1978).
6. Leonard J. Rogers, "Hydrazine-Air (60/240 Watt) Manpack Fuel Cell," *Proc. 23d Power Sources Conf.*, Electrochem. Soc., Pennington, N.J., 1969.
7. R. E. Salathe, "Evolution of Replaceable Hydrazine Module as a Basic Building Block," *Proc. 24th Power Sources Conf.*, Electrochem. Soc., Pennington, N.J., 1970.
8. F. G. Perkins, "Experience with Hydrazine Fuel Cells in SEA," *Proc. 24th Power Sources Conf.*, Electrochem. Soc., Pennington, N.J., 1970.
9. U.S. Army Mobility Equipment R & D Command, Ft. Belvoir, Va.
10. Michael Onischak and Bernard S. Baker, "Metal Hydride-Air Battery," ECOM Report 72-0317-F, Ft. Monmouth, N.J., March 1975.
11. Michael George and Joseph Scozzafava, "Reversible Metal Hydride-Air Fuel Cell," ECOM Report 77-2644-F, Ft. Monmouth, N.J., June 1978.
11a. J. F. Nachman et al., "Development of Lightweight Hydrides," Report no. DOE/CS/52059-1 for Department of Energy, Washington, D.C., July 1982.
12. O. J. Adlhart, "An Assessment of the Air Breathing, Hydrogen Fueled SPE Cell," *Proc. 28th Power Sources Symp.*, Electrochem. Soc., Pennington, N.J., 1978.
13. O. J. Adlhart, "Environmental Testing of SPE Fuel Cell Assemblies," *Proc. 29th Power Sources Symp.*, Electrochem. Soc., Pennington, N.J., 1980.
14. S. S. Kurpit, "1.5 and 3 kW Indirect Methanol-Air Fuel Cell Power Plants," *Proc. 10th Intersoc. Energy Convers. Conf. (IECEC)*, IEEE Catalog no. 75CHO 983-7, 1975.
15. Alfred P. Meyer, "1.5 kW Methanol Fuel Cell Power Plant System Design Study," *Proc. 28th Power Sources Symp.*, Electrochem. Soc., Pennington, N.J., 1978.
15a. J. K. Stedman and S. Fanciullo, "1.5 kW Fuel Cell Engineering Development Program," *Proc. 30th Power Sources Symp.*, Electrochem. Soc., Pennington, N.J., 1982.
15b. S. Abens et al., "3 and 5 kW Methanol Power Plant Program," *Proc. 30th Power Sources Symp.*, Electrochem. Soc., Pennington, N.J., 1982.
16. "Specification of Purchase Description for Fuel Cell Power Units," U.S. Army Mobility R & D Command, Ft. Belvoir, VA 22060.
17. *Fifth U.S. Dept. of Energy Battery and Electrochemical Contractors' Conf.*, U.S. Department of Energy, Washington, D.C., December 1982.

Utility
Fuel Cell
Power Plants

by
Arnold P. Fickett

43.1 GENERAL CHARACTERISTICS

Interest in the fuel cell as a utility power plant results from its efficiency, its environmental acceptability, and its modular configuration. In terms of efficiency the key point is that the fuel cell, not being a heat engine, is not limited by the Carnot cycle, which describes the limits on the efficiency of such engines. The fuel cell therefore offers the potential for higher conversion efficiencies than thermal generators can achieve. Since the efficiency of the fuel cell is determined chiefly by the voltage of the single cell, the efficiency of a fuel cell power plant is (to a degree) independent of the size of the plant. Moreover, in contrast to conventional generators, the fuel cell is nearly constant in efficiency over the range from 25 to 100% of its rated power output. Hence, fuel cells providing power for utility load-following applications offer the potential for conserving fuel. And the only emissions other than air and water are those from the fuel processor.

If a power plant made up of fuel cells is to be useful as a utility generator, it must use fuel that is economically available to the utility, and it must produce alternating current compatible with the utility's transmission system. Therefore, as shown in Fig. 41.1, a fuel cell power plant must include not only a fuel cell but also a fuel processor and a power conditioner. The fuel processor converts a utility fuel, such as natural gas, into a hydrogen-rich gas, and the power conditioner converts direct current into alternating current.

Fuel cell power plants that operate on fuels derived from petroleum or coal will produce sulfur dioxide and nitrous oxide emissions of less than 0.05 and 10 g per 10^9 J, respectively. These emissions and others will be at least 10 times lower than federal standards. The sulfur emissions are low because fuel cell power plants do not tolerate sulfur compounds well, and so such compounds are removed by a special subsystem of the fuel processor. The other emissions are low because of the inherent cleanness of the fuel cell. A fuel cell plant is expected to operate quietly and to require no makeup water at ambient temperatures of 35°C and below. As a result of its environmental attributes, a fuel cell plant can be located in the area it is to serve. The utility company can thus defer investments in new lines for the transmission and distribution of power and can reduce the losses inherent in transmission lines by dispersing fuel cell plants in this way.

Since the fuel cell is essentially modular in configuration, the components of a fuel cell power plant, even a large multimegawatt station, can be made in a factory by mass production techniques instead of having to be built at the site. Construction costs will therefore be lower. In addition, the availability of small modular generators enables a utility to add relatively small blocks of capacity matching the growth of demand in an economical way. The existence of small, reliable modular plants could also increase the inherent reliability of a utility system and allow a reduction in the reserve-generating capacity the utility maintains.

As a result of these characteristics, which are not available in conventional power generators, the fuel cell could serve utilities in several ways, as shown in Fig. 43.1 and summarized in Table 43.1. Relatively small fuel cell power plants (with a capacity ranging from 40 to 300 kW) could be set up in commercial and residential buildings. Such a plant would use natural gas as a fuel. The plant would provide both electric and thermal energy (the latter from the waste heat of the fuel cells), consuming the same amount of fuel ordinarily required for the thermal demand alone. Overall efficiencies approaching 90% have been projected for fuel cell power plants of this type. An advantage of the fuel cell is that the waste heat can be used without altering the power production characteristics. This is not the case with Carnot-type machines.

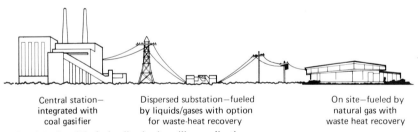

| Central station—
integrated with
coal gasifier | Dispersed substation—fueled
by liquids/gases with option
for waste-heat recovery | On site—fueled by
natural gas with
waste heat recovery |

FIG. 43.1   Possible fuel cell roles in utility applications.

Larger plants, ranging in capacity from 5 to 25 MW, could be dispersed throughout an electric utility system to perform load-following duty efficiently. The conservation potential of such fuel cells can be seen in Fig. 43.2, which compares the heat rate (the number of joules of fuel required to produce one kilowatt hour of electricity) with that of other types of power plants. (The break at 25% is caused by switching from a low-flow turbocompressor to a high-flow turbocompressor. Turbocompressors are used to pressurize the reactant air; below 25% load, a small turbocompressor is used to minimize parasitic energy losses.) Fuel cells installed for load-following duty would enable other generators to operate at their most efficient rate. In addition, if fuel cells were equipped to recover waste heat, they would represent a highly efficient end use of fossil-fuel resources. Efficiencies of more than 80% (based on fuel consumption) are attainable.

TABLE 43.1 Utility Fuel Cell Power Plant Programs

Role	Size	Fuel	Projected cost, 1983 $	Efficiency, %	Electrolyte
On-site power plants	40–300 kW	Pipeline gas	$800/kW	39–42 or 90 (with reject-heat recovery)	Phosphoric acid
Dispersed (substation) power plants	5–25 MW	Petroleum- or coal-derived gas or liquid	$600/kW	41–47 or 80 (with reject-heat recovery)	Phosphoric acid
Central station power plants	150–600 MW	Coal	$1500/kW*	45–50	Molten carbonate or phosphoric acid

*Includes cost of coal gasifier.

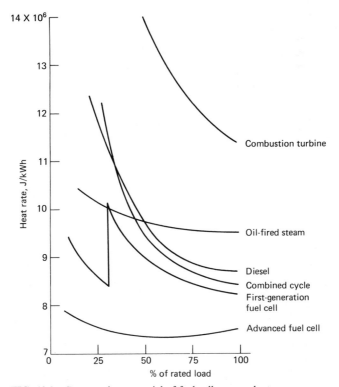

FIG. 43.2 Conservation potential of fuel cell power plants.

In the future, fuel cells could be integrated with coal gasifiers to provide large, central-station, base-load power plants that utilize coal directly. The capacity of such plants would range from 150 to 1000 MW. A plant of this kind is projected to be more than 45% efficient, on the basis of the heating value of the coal consumed.

43.2 UTILITY FUEL CELL POWER PLANT PROGRAMS

Three major programs aimed at hastening the development of the utility fuel cell as a commercial technology are in progress. These include the on-site program, the FCG-1 program, and the molten-carbonate fuel cell program. In addition to these three major programs, the high-temperature solid-oxide fuel cell has received intermittent support as a utility fuel cell.

43.2.1 On-Site Program

The on-site program is an extension of a project named TARGET (Team to Advance Research for Gas Energy Transformation). The project was originally sponsored jointly by the United Technologies Corporation and a consortium of gas and gas and electric utilities. Starting in 1967, the group supported the development of phosphoric acid fuel cells for on-site residential and commercial applications.

The on-site power plant concept utilizes gas (natural or synthetic) as the fuel, converting it into electricity and into thermal energy for heating and cooling. The efficiency is quite high, since virtually all the waste heat is directed toward the heating and cooling part of the operation. The industry has spent about $100 million on the concept so far. In 1972 and 1973 more than 60 plants of 12.5-kW capacity were tested under field conditions; the tests led to the development and demonstration of a 40-kW plant in 1975. Now, in an extension of the program with the support of the Department of Energy and the Gas Research Institute, the objective is the testing of 50 plants of 40-kW capacity and the attainment of commercial availability of the technology by the end of the decade.

Recently, two other development teams, Westinghouse/Energy Research Corporation and Engelhard Corporation, initiated programs supported by the Department of Energy aimed at the on-site fuel cell application. The Westinghouse/Energy Research Corporation effort, now discontinued, was directed at a 120-kW power plant fueled by natural gas, whereas Engelhard is developing a 5-kW methanol-fueled system.

43.2.2 FCG-1 Program

The second program originated with the electric utility industry in 1971, when a group of utility companies joined the Edison Electric Institute and United Technologies in an assessment of the potential benefit of fuel cells to the industry. The venture led in 1972 to the FCG-1 (for fuel cell generator-1) program, an effort sponsored by United Technologies and nine utility companies to develop a 26-MW phosphoric acid power plant for commercial service by 1980. The FCG-1 program developed and demonstrated (in 1976 and 1977) a 1-MW pilot plant. The demonstration showed that a power plant fueled by naphtha could provide large amounts of energy to a utility system while satisfactorily meeting the utilities' operational requirements for heat rate, emissions, and load-following capability.

In 1976, seeking to expedite the commercial application of fuel cells, the Electric Power Research Institute and the Energy Research and Development Administration (now the Department of Energy) became involved in the FCG-1 program. One result was a project to design, build, and test a 4.5-MW module of the FCG-1 power plant in a utility system. The Consolidated Edison Company of New York was chosen as the host utility for the demonstration.

A related demonstration activity involves an agreement between Tokyo Electric Power Company and United Technologies Corp. that will result in the installation and test of a similar 4.5-MW power plant in Tokyo.

The FCG-1 program has recently been expanded to include a 3- to 4-y effort that will result in the design and development of a commercial FCG-1 prototype configuration. In addition, the Westinghouse/Energy Research Corporation team has recently announced plans to extend their on-site fuel cells efforts to the larger dispersed multimegawatt range.

43.2.3 Molten-Carbonate Fuel Cell Program

The third major program grew out of efforts undertaken in 1971 by the electric and gas utility industries to advance fuel cell technology in order to expand its applicability to a wider range of fuels, including coal, by the direct integration of the fuel cell to a coal gasifier. The focus of this program is the molten-carbonate fuel cell power plant expected to be available for commercial service in about a decade. General Electric, the Institute of Gas Technology, Oak Ridge National Laboratory, Energy Research Corporation, and United Technologies Corp. are involved in the program. Argonne National Laboratory is managing the Department of Energy effort as well as conducting supporting research and development efforts.

43.2.4 Solid-Oxide Fuel Cells

Serious interest in the high-temperature solid-oxide (HTSO) fuel cell developed in the early 1960s[1,2] as a result of several anticipated advantages offered by this technology including:

Low inherent cost

High power density

Fuel flexibility

Lack of a liquid phase

Very high power plant efficiency (if integrated with coal gasification systems)

However, by 1970 much of the enthusiasm toward HTSO fuel cells had waned, and by 1972 the major programs had been terminated. The loss of interest was principally due to problems of material stability at 1000°C; problems with cathode and intercell-connection materials were especially severe. Therefore, many of the advantages cited for this technology could not be determined in prototype systems. Specifically, the cost advantage became questionable as more exotic materials were employed to achieve high-temperature stability, and the fuel capability as well as the efficiency (considered from a total system point of view) turned out to be only slightly better than those of the phosphoric acid or molten carbonate fuel cells.

More recently, interest in the HTSO fuel cell has been renewed and concentrated in two areas: first, between 1973 and 1976, HTSO electrolytes were investigated that could operate at lower temperatures (700 to 850°C),[3-5] and second, since

1976 HTSO fuel cells using a zirconia electrolyte operating at 1000°C have been explored.[6]

43.3 PHOSPHORIC ACID FUEL CELLS

Despite the increasing number of developers and the seeming wide range of applications (kilowatts to megawatts), the basic phosphoric acid fuel cell (PAFC) technology that is being developed is virtually the same in all cases.

43.3.1 Cell Structure

The basic cell structure is shown in Fig. 43.3. It consists of:

1. *A carbon or graphite separator-current collector plate* that separates hydrogen from the air of the adjacent cell (in a multicell stack) and also provides the electrical series connection between cells. This plate must be impermeable to hydrogen and oxygen, a good electronic conductor, and stable to both fuel and air environments in the presence of concentrated, 200°C phosphoric acid.

2. *Anode current collector ribs* that conduct the electrons from the anode to the separator plate. The ribbed configuration provides gas passages for hydrogen distribution to the anode. These carbon or graphite ribs are made either a part of the separator current collector plate or of the anode. In either event, they must have good electronic conductivity and be stable to the fuel environment. If made a part of the separator plate, they would be fabricated of dense impermeable material, whereas if part of the anode, they could be porous. The latter configuration would appear to have advantages over the ribbed separator plate; one is that ribs can be formed in a continuous process, and another is that the porous ribs can be used to store phosphoric acid, increasing endurance.

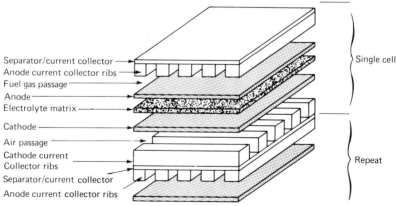

Separator/current collector
Anode current collector ribs
Fuel gas passage
Anode
Electrolyte matrix
Cathode
Air passage
Cathode current
Collector ribs
Separator/current collector
Anode current collector ribs

Single cell

Repeat

FIG. 43.3 Basic phosphoric acid electrolyte fuel cell.

3. *An anode* that consists of a porous graphitic substrate with the surface adjacent to the electrolyte treated with a platinum or platinum alloy catalyst. Practical anodes will contain only a fraction of a milligram of noble metal for a square centimeter of cell area (0.25 mg/cm^2 is a typical state-of-the-art loading). Teflon is usually employed as a catalyst binder.

4. *An electrolyte matrix* that retains the concentrated phosphoric acid. The state-of-the-art matrices are fabricated from silicon carbide powder and a Teflon binder and range between 0.01 and 0.03 cm in thickness. The electrolyte matrix has minimum ionic resistance, while separating the fuel and oxidant gas streams. To accomplish this separation in a practical power plant, the matrix has pores or capillaries which are sufficiently small that, when filled with electrolyte, they can withstand pressure differentials of 6 $\times$ 10^3 Pa or more without gas "blow-through."

5. *A cathode* that is similar to the anode but uses a modified noble-metal catalyst and an increased catalyst loading (usually 0.5 mg/cm^2) to enhance the oxygen reduction kinetics. The cathode also is made hydrophobic by increasing the Teflon content to prevent flooding of the electrode and provide for adequate oxygen diffusion to the active sites.

6. *Cathode current collector ribs* that are also virtually identical to the anode ribs. The only differences are in the height and width of the ribs which control the mass transport characteristics in the gas passage and the contact resistance between components. In a stack, the cathode ribs are perpendicular to the anode ribs to enable the simple location of external manifolds.

These single cells are stacked in series (Fig. 43.3) to produce the desired output power and voltage. For instance, the 4.5-MW power plant being installed in New York contains 20 stacks, each having nearly 500 cells of 3400 cm^2 area each. Fuel and air supply and exhaust manifolds are then connected along the respective sides of the stacks. Both cell seals and manifold seals are critical to proper operation. The cell seals are located between the separator plate and the electrolyte matrix to prevent overboard leakage of fuel or air. The manifold seals are located between the respective sides of the stack and the manifolds. The cell sealing is usually accomplished by a "wet seal" —that is, the surface tension of the phosphoric acid wetting the component surfaces is sufficient to provide sealing. The manifold seal is accomplished with a Teflon type of caulking.

To mitigate sealing difficulties during operation at high pressures, the stacks are housed in a pressure container. This allows the container to be pressurized to operating pressure and minimizes the differential pressure across the seals.

43.3.2 Performance Characteristics

Figure 43.4 indicates typical performance for a PAFC at the start of test. Important PAFC performance characteristics include:

Present PAFC performance is almost entirely determined by the cathode. That is, concentration polarization, anode activation polarization, and ohmic losses are small compared with the activation losses exhibited at the cathode. For instance, almost 300 mV is lost at the cathode at very low current densities.

Cathode current density (at fixed potential) is proportional to oxygen partial pressure. Thus, increasing the system pressure can have a dramatic effect on performance.

Cathode current density (at fixed potential and pressure) has been shown[7] to be essentially proportional to catalyst loading (for loadings of 1 mg/cm^2 and less). Beyond this loading, gains in current density diminish with further increases in catalyst.

Utility PAFC power plants will derive hydrogen by the conversion of a wide variety of primary fuels such as methane (natural gas), petroleum products (such as naphtha), coal liquids (such as methanol), or coal gases. The conversion process will result in a hydrogen-rich fuel gas containing CO_2 and CO. The CO_2 is inert and acts as a diluent; the CO, however, is a potential fuel cell poison. The low anode overvoltage portrayed in Fig. 43.4 can be achieved only under relatively low CO concentrations.[8,9] Figure 43.5 indicates the areas of CO concentration and temperature where the anode performance shown in Fig. 43.4 is attainable.

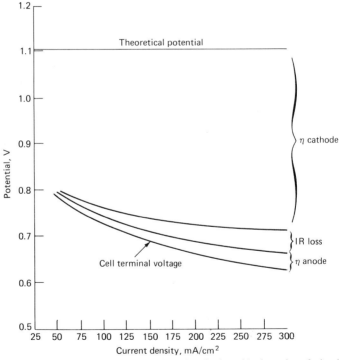

FIG. 43.4 Typical performance of a phosphoric acid electrolyte fuel cell (190°C, 3 × 10^5 Pa, air, H$_2$ fuel, platinum loading: 0.75 mg/cm^2).

Increasing cell temperature for practical power plant conditions has little effect on performance except with respect to CO tolerance, as discussed above. For a given pressure, an increase in temperature increases the phosphoric acid concentration. Thus, any gain due to the effect of temperature on kinetics is offset by the reduced intrinsic

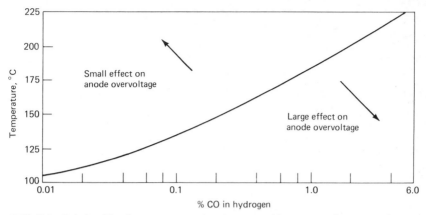

FIG. 43.5   Relationship of temperature and carbon monoxide concentrations to anode overvoltage.

activity of the more concentrated acid. However, for a practical power plant, it is not possible to decouple pressure and temperature if the waste heat from the fuel cell power section is used to raise steam for the fuel processor. The fuel cell temperature must be high enough to produce steam at a pressure sufficient for injection into the fuel processor whose pressure is greater than fuel cell pressure. This results in certain pressure-temperature combinations at elevated pressure, i.e., 3×10^5 Pa at 190°C, 6×10^5 Pa at 210°C, 8×10^5 Pa at 220°C. These combinations allow sufficient temperature to overcome thermal losses and heat exchanger pinch points and still produce steam at the required pressure.

Given the above, the major opportunities for improving PAFC performance include:

Developing catalysts to improve the cathode kinetics: to this end, recent results[10-12] have shown significant performance increase with a Pt-V intermetallic catalyst (Fig. 43.6). At a cathode potential of 0.7 V dc, the current density can be approximately doubled.

Increasing the oxygen partial pressure by operating the system at higher overall pressure: at fixed-cathode potential, the current density will be roughly proportional to oxygen partial pressure or approximately proportional to system total pressure (if water partial pressure is neglected). This will require increasing cell temperature because of the pressure-temperature coupling described above.

Figure 43.7 projects initial PAFC cell performances that might be expected with conventional and improved catalysts at several pressure-temperature conditions.

43.3.3 Endurance

Figure 43.8 shows typical PAFC performance decay rates as a function of operating time at a constant-current density. The interesting feature of this figure is the linear relationship of voltage decay with log (time) following the initial thousand hours of operation. This decay results from the tendency of the platinum catalyst to lose surface

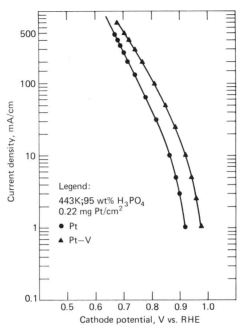

FIG. 43.6 Performance of the Pt-V intermetallic catalyst vs. platinum.[12]

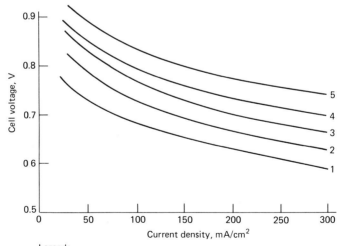

Legend:

1 Atmospheric pressure — 190°C, 0.5 mg/cm^2 Pt on carbon
2 3 X 10^5 Pa — 190°C, 0.5 mg/cm^2 Pt on carbon
3 6 X 10^5 Pa — 210°C, 0.5 mg/cm^2 Pt on carbon
4 8 X 10^5 Pa — 220°C, 0.5 mg/cm^2 Pt on carbon
5 8 X 10^5 Pa — 220°C, 0.5 mg/cm^2 Pt —V intermetallic on carbon

FIG. 43.7 Phosphoric acid fuel cell—performance for a range of pressures and an improved cathode catalyst (at start of test).

area and hence intrinsic activity as a function of operating time. This is principally due to three mechanisms.

1. The carbon substrate slowly oxidizes at the fuel cell cathode, resulting in both a reduction of surface area and a small spalling of platinum.

2. The platinum crystallites migrate on the carbon surface and aggregate to form larger crystallites (smaller surface area) with time.

3. Platinum oxide tends to be very slightly soluble in the acid and undergoes dissolution-reprecipitation resulting in a slow growth in crystallite size with time.

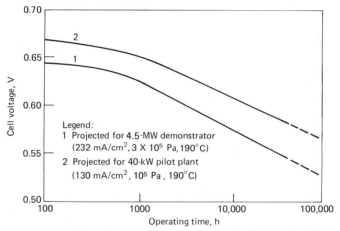

FIG. 43.8   Typical performance decay for phosphoric acid fuel cells. (*Curve 1 from Ref. 13; curve 2 from Ref. 13a.*)

The net result of these three mechanisms is a loss of cathode catalyst surface area that is a function of log (time). These mechanisms are becoming understood, as are the parameters that control the decay rates.[14,15] Cathode potential, temperature, acid concentration (water content), and oxygen partial pressure are all important parameters. However, even more important are the characteristics of the carbon (graphite) surface and the platinum alloy composition. The carbon oxidation as well as the platinum migration and dissolution can be markedly reduced by careful attention to the nature of the materials. Thus, the decay rates shown in Fig. 43.8 will decrease with a change in operating parameters as well as with subtle material changes. It is important that the end-of-life performance can be modeled and projected from relatively short-term cell data; there are now sufficient data to verify that such projections as those in Fig. 43.8 do, indeed, portray the performance characteristics of the better cells and cell stacks.

There are other potential decay and failure modes that can cause cells to lose performance and fail. These other mechanisms which would result in deviations from Fig. 43.8 include:

Resistance increase and/or gas crossover due to insufficient electrolyte

Concentration polarization at the cathode due to inadequate hydrophobicity (Teflon) with resulting loss of cathode reaction area

Loss of catalyst activity due to poisoning of the cathode or anode

Resistance increase due to loss of contact pressure

External gas leakage

While all these problems can and do occur, they can be avoided through proper attention to design and fabrication; they are not considered to be inherent in the PAFC technology.

Figure 43.9 shows the performance of a 24-cell stack that has operated for over 12,000 h. This stack verifies the ability to achieve or better the endurance characteristics depicted in Fig. 43.8.

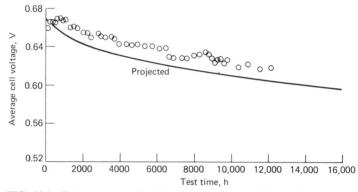

FIG. 43.9 Endurance test of a 24-cell stack (test conditions: 130 mA/cm², 190°C, 10⁵ Pa pressure). (*Courtesy of United Technologies Corp.*)

43.3.4 Phosphoric Acid Power Plants

General Description   The characteristics targeted by the utility industry for commercial first-generation phosphoric acid power plants under development by United Technologies Corp. are summarized in Table 43.2. The similarity in the 40-kW on-site and the 10-MW dispersed power plants is apparent. The significant differences between the two systems (other than size) are the operational pressure-temperature and current densities. The 40-kW unit operates at ambient pressure, 190°C, 200°C, and between 100 and 200 mA/cm²; the 10-MW unit will operate at 6.5 to 8×10^5 Pa, 200°C, and between 200 and 300 mA/cm². A system schematic for the pressurized power plant is shown in Fig. 43.10. This specific system under development by United Technologies Corp. is described below.

Fuel entering the power plant is pumped to system pressure, vaporized, and mixed with a H_2-rich recycle stream. This mixture passes through a hydrodesulfurizer, which converts the fuel's sulfur compounds into hydrogen sulfide (H_2S). The H_2S is then removed by absorption on a zinc oxide bed. The desulfurized fuel is then combined with steam and enters the reformer, where the mixture is catalytically converted into a

TABLE 43.2 First-Generation Fuel Cell Program Targets

Characteristic	On-site	Dispersed
Modular size	~ 40 kW	~ 10 MW
Stack temperature/pressure	190°C, 1 × 10⁵Pa	210°C, 6 × 10⁵Pa
Electrical efficiency		
(based on HHV)	39%	41%
Total efficiency		
(including heat recovery)	85%	85%
Fuel capability	All pipeline	Pipeline natural gas
	quality fuels	Peak shaving gas
		Synthesis gas
		Distillate (coal or
		petroleum derived)
		Methanol
Design life	20 y	20 y
Noise	< 60 dbA at 3 m	< 55 dbA at 30 m
Operation	Automatic	Automatic
Water requirements	None	None
Output	3φ, 120/208 V ac	3φ, 4–69 kV ac
Footprint	5 m²	~ 1000 m²
Available heat:		
at 80°F	3.75 × J/kWh × 10⁶	3.3 × 10⁶ J/kWh
at 150°F		0.4 × 10⁶ J/kWh

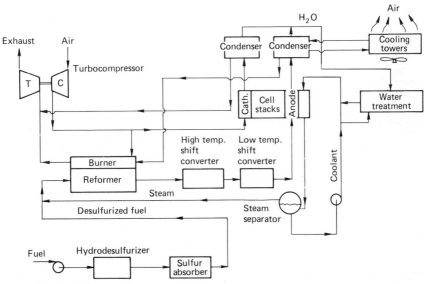

FIG. 43.10 Simplified system schematic of dispersed power plant. (*Courtesy of United Technologies Corp.*)

hydrogen-rich gas. The H_2 content of the product gas is further increased by two stages of shift conversion ($H_2O + CO \rightarrow CO_2 + H_2$). The gas is then cooled and passes into the power section (cell stacks), where hydrogen and oxygen from the process gas and air streams are electrochemically combined, producing dc electricity and by-product water. Both the air and fuel streams entering the fuel cell are pressurized. The depleted fuel gas stream leaves the power section and passes through a water-recovery condenser. After the condenser, it enters the reformer burner, where it is combusted to provide thermal energy for the steam-reforming reaction. Process air leaving the power section contains by-product water, which is recovered by a condenser. The dried, pressurized air is mixed with the hot, high-pressure reformer burner exhaust. This combined stream is then expanded through a power recovery turbine which, in turn, drives the process air compressor.

Water recovered in the condensers is recycled to the thermal management subsystem. The primary function is to control the power section temperature by circulating water (steam) through the cell stacks. Waste heat generated in the process of producing power is removed by evaporating a portion of the circulating water. Steam is separated for use in the reformer. The remaining water is collected, purified, and circulated through the cell stacks. Direct current power from the fuel cells is converted to utility quality alternating current by a self-commutated inverter and conventional transformer.

The 40-kW on-site unit employs a similar system concept. The major difference is the operation at ambient pressure, eliminating the need for the turbo-compressor.

On-Site Power Plants   A prototype 40-kW on-site system design is shown in Fig. 43.11. This design is a modification of that fabricated and tested as a pilot 40-kW fuel cell power plant. The pilot power plant exceeded 18,000 h of operation, more than 8000 h with one power section and with a continuous run of more than 3000 h.

The prototype subsystem design and engineering development are complete, and testing has been initiated to verify all operating characteristics. A coordinated effort is presently being implemented by the Department of Energy, Gas Research Institute, United Technologies Corp., and several utilities that will place up to fifty 40-kW prototypes into the field for evaluation. The primary objective of this operational feasibility program is to establish the utilities', manufacturers', and sponsors' acceptance of and commitment to on-site fuel cell systems.

Parallel development efforts were initiated at Westinghouse and Engelhard. The former was aimed at developing a commercially viable on-site, integrated energy system (OS/IES). The activity was focused on a PAFC cell and stack development and verification, component design, and system conceptualization. The concept (Fig. 43.12) was centered on a nominal 120-kW system, although sizes of 100 to 1000 kW were considered, and differed from that employed by United Technologies Corp. in the means of heat removal. In the United Technologies Corp. systems (typified by Fig. 43.10), heat removal is by the circulation of a two-phase water-steam coolant through stack coolant passages located between blocks of four to six cells in the cell stack. Westinghouse/Energy Research Corp. use a patented distributed gas (DIGAS) cooling concept that allows a portion of the reaction's air stream to pass through coolant plates, similarly located between blocks of cells. The steam necessary for integration in the fuel-processing subsystem is raised external to the stack. The DIGAS concept should result in a simpler, more reliable, stack design but will likely cause the fuel cell stack to operate at a slightly higher temperature for a given system pressure due to the

FIG. 43.11   40-kW on-site fuel cell system. (*Courtesy of United Technologies Corp.*)

thermal pinch characteristics of the external steam generator. In 1980, Westinghouse terminated work on this application in order to concentrate on larger dispersed power plants for electric utilities. The Engelhard efforts are focused on smaller power plants (5 to 12 kW) for industrial or traction power, although they could be applicable to on-site utility use. These activities are also concentrating on cell and stack development.

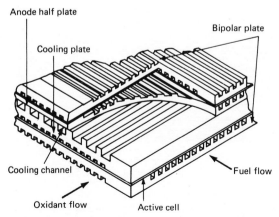

FIG. 43.12   DIGAS stack concept. (*Courtesy of Westinghouse Electric Corp.*)

Dispersed Power Plants   Dispersed power plant programs are under way to:

Demonstrate the sitability and operational characteristics of a 4.5-MW (ac) fuel cell demonstration module in New York City

Demonstrate the characteristics of a similar 4.5-MW fuel cell in Tokyo, Japan

Develop a ~10-MW commercial prototype design which could be available for utility evaluation by the mid-1980's

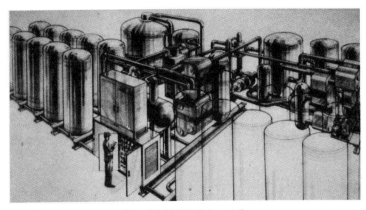

FIG. 43.13   Artist's drawing of 4.8-MW (dc) module.

FIG. 43.14   4.5-MW site (New York, N.Y.). (*Courtesy of United Technologies Corp.*)

An artist's drawing of the 4.8-MW dc (4.5-MW ac) module (fuel processing and power sections) is shown in Fig. 43.13; the power plant is currently undergoing installation at the downtown New York City site as shown in Fig. 43.14. The purposes of this demonstration are similar to those of the 40-kW field test, discussed previously, with a focus on establishing the power plant's sitability and operating characteristics.

A parallel effort is being implemented that will result in the design, development, and component-level verification of a commercial prototype power plant (Table 43.2) by 1983. The commercial prototype will be nominally rated at 10 MW and will contain improvements on the 4.5-MW design including:

Longer-lived (40,000 h) fuel cell stacks rather than the 4.5-MW demonstrator's life of <5000 h

Improved performance—41% efficiency rather than the 4.5-MW demonstrator's 37% efficiency

Higher-pressure operation—essentially doubling the throughput and the power rating

This effort will lead to a prospectus for commercial prototype power plants with the expectation that the industry will subsequently deploy up to 500 MW of such units as part of a commercial feasibility program, leading to commercial service between 1985 and 1990. One possible configuration of three units to provide a 30-MW power plant is depicted in Fig. 43.15.

In 1979, Westinghouse announced an interest in and commitment to a dispersed PAFC power plant program. They have conceptualized a 7.5-MW power plant based upon their air-cooled DIGAS stack (Figs. 43.12 and 43.16) and have undertaken a comprehensive program to lead to the fabrication and test of a 7.5-MW demonstration on utility system by 1987. This power plant would use the PAFC technology developed for their OS/IES power plant.

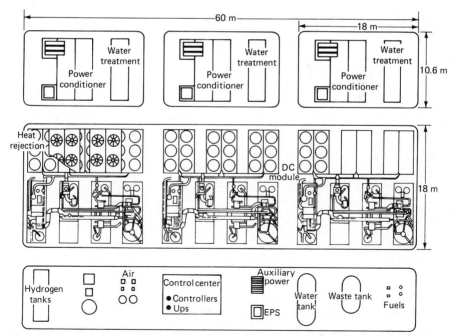

FIG. 43.15   30-MW power plant—layout of three-power-unit power plant.

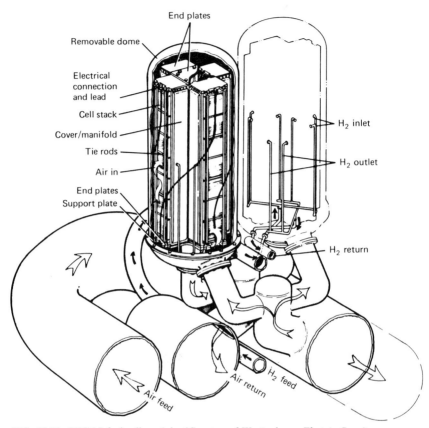

FIG. 43.16 DIGAS fuel cell module. (*Courtesy of Westinghouse Electric Corp.*)

Central-Station Power Plant Phosphoric acid fuel cell power plants are also being considered for use in central stations. These power plants would utilize integrated coal gasifiers to supply the hydrogen-rich fuel and would provide base load power. Such a power plant,[16] based upon 1977 PAFC technology, projects an efficiency for coal to ac power of 34%. By use of current PAFC technology, this efficiency would approach 40%. Thus, such a power plant is not an unreasonable alternative.

43.4 MOLTEN-CARBONATE FUEL CELLS

Interest in the molten-carbonate fuel cell (MCFC) technology stems from its high efficiency, excellent environmental character, potential for competitive costs, and modularity. In contrast to the PAFC, MCFC development efforts are presently focused on large, multimegawatt ($\sim$600 MW) central-station power plants integrated with a coal gasifier. This focus is, in part, due to their high operating temperature and the problems inherent in starting (heating) and stopping (cooling) MCFC power plants.

43.4.1 Cell Structure

The MCFC structure is geometrically very similar to that of the PAFC, as shown in Fig. 43.3. The materials that are used are, however, very different from those used in the PAFC. As reference to Fig. 43.3 shows, the MCFC consists of:

1. *A separator and current-collector plate* that separates the fuel gas from the air of the adjacent cell in a multicell stack and also provides the electrical connection between cells. Like its PAFC counterpart, it must be impermeable to hydrogen and oxygen, a good electronic conductor, and stable to fuel and air environments in the presence of 650°C carbonate salts. No single material has been found that is completely satisfactory for this duty; those satisfactory in the fuel environment are undesirable in the oxidizing environment, and vice versa.

2. *An anode current collector* that conducts the electrons from the anode to the separator plate. This current collector must also provide passage for fuel flow. In some configurations, this function is provided by ribbing or folding the separator plate. In other configurations, the current collector ribs are formed into the porous electrodes, as discussed under PAFCs. The current collector must have good electronic conductivity and be stable to the fuel environment. Nickel is a satisfactory material, although copper is a possibility if it can be treated to reduce sintering.

3. *An anode that* consists of a porous nickel treated with a refractory oxide to reduce sintering. At the 650°C temperature, no other catalyst is required.

4. *An electrolyte system* comprising a mixture of lithium-potassium carbonate and inert powder (presently a lithium aluminate). This mixture forms a paste when molten and freezes to form a "tile" when cooled. This electrolyte system presents major challenges to MCFC development. It must have minimum ionic resistance, while separating the fuel and oxidant gases at pressure differentials in excess of 0.07×10^5 Pa. In addition, it must be electronically insulating. A historical problem with MCFCs is the tendency for the electrolyte system to crack because of stresses induced during the cooling of the hardware below the carbonate freezing point. If these cracks do not "heal" upon subsequent melting, the fuel and oxidant gases mix, causing premature failure.

5. *A cathode* that is similar to the anode except that it uses nickel oxide (doped with lithium to impart electronic conductivity). Since hydrophobic agents such as the Teflon used in PAFCs are not stable at 650°C, cathode flooding is prevented by careful attention to the pore (capillary) sizes in the anode, electrolyte system, and cathode. The cathode pore size must be larger than the other two to prevent flooding. (The electrolyte system must, of course, have the smallest pores to retain the molten carbonate.) Alternative cathode materials will be needed as nickel oxide is not stable for long periods at higher operating pressures which will likely be required for good performance.

6. *A cathode current collector* that has similar requirements and configurational options as the anode current collector. Since nickel is thermodynamically unstable, material options include lithium-doped nickel oxide and corrosion-resistant stainless steel.

As with the PAFC, individual cells are stacked in series to result in a cell stack of the required power and voltage output. Cell and manifold sealing are accomplished by "wet seals" and the use of an inert caulking (again similar to PAFC with a change in materials), respectively. And when operating at pressure, the stacks will be contained in a pressure vessel to minimize the pressure differential across the cell and stack seals.

43.4.2 Performance Characteristics

Figure 43.17 illustrates a typical MCFC performance at the start of test. The performance of MCFCs tends to be better than that of PAFCs primarily due to the improved behavior of the air cathode. Performance considerations include:

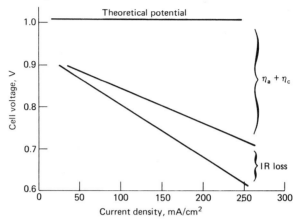

FIG. 43.17   Typical molten carbonate fuel cell performance, 650°C, 1×10^5 Pa, air and reformate fuel.

The performance losses in practical MCFCs are distributed among the various polarizations with η_{ohm} and η_{act} (cathode) being the largest contributors.

Since *IR* losses due to the electrolyte and contact resistance can be severe, attention must be given to developing thin electrolyte structures (< 0.05 cm) and maintaining good pressure contacts. Total cell *IR* loss should be maintained below 40 mV at 100 mA/cm².

MCFC involves the reactions of CO_2 at the cathode, $O_2 + 1/2\ CO_2 + 2e \rightarrow CO^{2-}$, and its release at the anode, $CO^{2-} + H_2 \rightarrow CO_2 + H_2O + 2e$. The kinetics of the cathode are extremely complex and not well understood but likely involve peroxides or superoxides as well as carbonates. Thus, although thermodynamics would suggest that the ratio of CO_2 to O_2 should be 2:1, the kinetics may optimize at a different ratio.

Current density will increase with increased operating pressure. The precise theoretical relationship is not well understood; however, empirical data for the pressure effect are shown in Fig. 43.18. System considerations have generally caused MCFC cost and efficiency to optimize at 6 to 10×10^5 Pa.

Any CO in the fuel gas will undergo the water-gas shift reaction ($CO + H_2O \rightarrow CO_2 + H_2$) at the 650°C cell temperature. The equilibrium constant for this reaction is a function of temperature, causing the theoretical potential to decrease with increasing

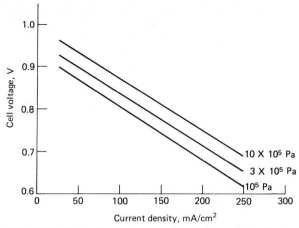

FIG. 43.18   Effect of pressure on molten-carbonate fuel cell performance at 650°C, air and reformate fuel.

temperature. However, all other performance aspects improve with increasing temperature. The net effect is improved cell voltage with increased temperature (Fig. 43.19). The relatively small effect of temperature above 650°C provides little incentive for operating above this temperature and encountering the additional corrosion problems that are incurred at the higher temperature.

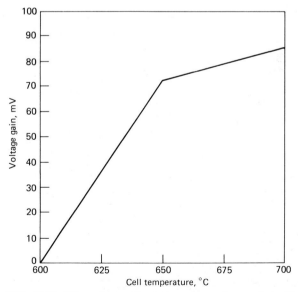

FIG. 43.19   Effect of temperature on molten-carbonate fuel cell performance, 1×10^5 Pa, 100 mA/cm². (*From Ref. 16a.*)

In addition to the water-gas shift reaction, other reactions can occur at the anode; the methanation reaction, CO (or CO_2) + $3H_2 \rightarrow CH_4 + H_2O$, and carbonization, $2CO \rightarrow C + CO_2$, may take place, removing available hydrogen from the anode stream and reducing performance. Thus, the fuel composition and flow rate (utilization) can have significant impacts on performance as well as other system characteristics. Figure 43.20 illustrates the effect of fuel utilization on performance at a fixed inlet composition.[15] Figure 43.21 illustrates the effect of fuel composition on performance at zero utilization.[15] Both figures are for ambient pressure operation. At elevated pressure, the precise relationships are not yet understood due to the complex interactions of the various chemical and electrochemical reactions.

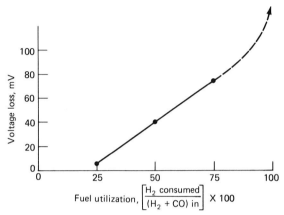

FIG. 43.20 Effect of fuel utilization on MCFC performance. (*From Ref. 16a.*)

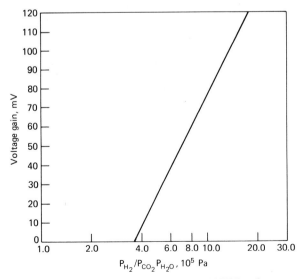

FIG. 43.21 Effect of fuel partial pressure on MCFC performance at 1×10^5 Pa total pressure, 650°C, 160 mA/cm². (*From Ref. 16a.*)

Because CO reacts via the water-gas shift at the MCFC anode, it is a desirable component of the fuel gas (virtually indistinguishable from H_2 in its effect on the fuel cell
rather than a poison as in the PAFC case). The most obvious MCFC poison is H_2S or
COS. Concentrations as low as 1 ppm can have a deleterious effect. Thus, MCFC
systems must exhaustively scrub sulfur-containing species before they enter the cell. It
turns out that this does not inflict a severe cost or efficiency penalty upon the power
plant. To reduce the sulfur content of a typical coal gas to < 1 ppm may add $40/kW
to the capital cost and reduce the efficiency by 1%.[17]

State-of-the-art MCFC performance is good, and future gains will be derived from
reduced *IR*, slightly improved anode and cathode transport characteristics, increased
operating pressure, and careful attention to the total system to optimize fuel and
oxidant composition and utilization.

43.4.3 Endurance

Figure 43.22 shows the best life test data for small single cells. The decrease in performance with time tends to be associated with a corresponding *IR* increase.

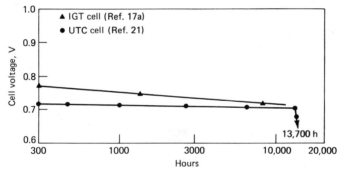

FIG. 43.22 Performance vs. time for longest-lived MCFC (data in both
cells based on simulated reformate fuel).

Loss of electrolyte is one of the major modes of MCFC performance decay. This
loss can occur either through the reaction of the electrolyte with other cell components
or through evaporation into the fuel or air exhaust streams. The United Technologies
Corp. cell shown in Fig. 43.22 had electrolyte added throughout the test. The Institute
of Gas Technology cell was operated without electrolyte replenishment; this may
account for the slightly higher decay rate. As electrolyte is lost from the cell to the point
that unfilled pores or cracks exist in the electrolyte system, reactant gases will cross
over (mix), resulting in cell overheating and failure. Practical MCFCs will require
adequate electrolyte inventory within the cell hardware for 40,000 h. This problem will
be significantly reduced at higher operating pressure as the evaporative losses will likely
be inversely proportional to pressure.

In addition to electrolyte loss, performance decay can be caused by the corrosion of
the various cell components. Even stainless steel may be expected to corrode within
MCFCs. Thus, long-term material corrosion can significantly increase *IR*. Another
potential decay mechanism is the tendency of the lithium aluminate filler in the elec-

trolyte system to undergo phase changes as a function of temperature and time. The phase changes result in volume (pore size) changes leading to a loss in the system's ability to retain electrolyte. Alternative filler materials have been (and are being) investigated. However, low-cost materials that are stable in this system are difficult to find.

The most serious endurance problem for state-of-the-art MCFCs is not long-term decay but rather a precipitous failure caused by reactant mixing through open (unfilled) cracks in the electrolyte system. This problem usually results from a thermal cycle, that is, a cell cool-down to below the electrolyte's melting point. Thermal stresses that develop during a shutdown (cool-down) cause the electrolyte system to crack. If this crack cannot heal itself upon the next heat-up (melting), then the reactants will mix and the cell will fail.[18, 19] Major research and development efforts are addressing both material and configurational solutions.

The longest endurance demonstrated by a multicell stack is 2000 h.[18] This 20-unit, 930-cm^2 cell stack also completed five thermal cycles without failure, confirming at least limited success in resolving this problem.

43.4.4 Molten Carbonate Fuel Cell Power Plants

The thrust of the MCFC program is the development of a central-station power plant, comprising a coal gasifier, gas clean-up system, MCFC (topping cycle), and gas or steam turbine (bottoming cycle). An analysis of this power plant concept was completed under the Energy Conversion Alternatives Study.[20] That study compared all coal-fueled electric power generating alternatives and concluded that the MCFC power plant offered the most attractive potential, considering its overall efficiency, cost, and environmental characteristics. As a result, the Department of Energy initiated a major MCFC development program in 1977 to expedite the availability of a coal-fueled MCFC power plant.

A typical system energy flow for a central-station MCFC power plant is shown in Fig. 43.23. Design requirements and goals for a similar power plant advanced by General Electric to produce 675 MW are described in Table 43.3. Key features are listed in Table 43.4. This particular system uses an oxygen-blown Texaco gasifier fed with a coal/water slurry. The coal gas is produced at 1370°C and 42×10^5 Pa at a rate of 2.18 kg gas per/kg coal. The gasifier effluent passes through a high-temperature steam generator dropping the temperature to 650°C and then through a regenerative heat exchanger train that cools the gas to 38°C, condensing the water. The cool gas proceeds through an NH_3 scrubber, COS converter, and sulfur clean-up (Selexol). The clean gas leaves the Selexol system at 25°C and is reheated (by the regenerative heat exchanger train) to 620°C (35×10^5 Pa). This is expanded through the turbine to the fuel cell pressure of 6.7×10^5 Pa (350°C). After reheat, the gas enters the cell stacks at a utilization rate of 0.85, that is, 85% of the H_2 and CO are converted to electricity with the excess exhausted from the stack. The fuel exhaust is catalytically burnt and mixed with air (compressed by energy from the clean gas "letdown" turbine) to provide a proper CO_2/O_2 ratio for the fuel cell cathode. This cathode inlet gas is heated to 538°C and then supplied to the fuel cell stacks at an oxygen utilization of 25%. The cathode exhaust gas powers a simple gas turbine with a pressure ratio of 6 and discharge temperature of 387°C that produces about 75 MW(e). In addition, a steam generator (operating from the various high-pressure steam flows) produces about 150 MW(e). A possible power plant configuration would comprise:

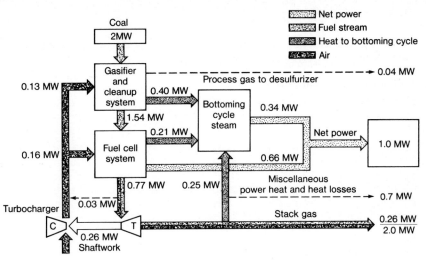

FIG. 43.23 Typical integrated fuel cell coal gasifier energy flow.

TABLE 43.3 Design Requirements and Goals for a Central Station MCFC Power Plant

Requirements	
Central station plant	
Power level	— 675 MW(e)
Fuel specification	Illinois no. 6 coal
Modular construction	
Environmental	Projected 1985 federal requirements
Site characteristics	"Middletown" except for cooling tower heat rejection

Goals	
Base load duty with daily load following capability	
Heat rate	6.8×10^6 J/kWh
Capital installed cost (1983 dollars)	$1500/kW(e)
Plant availability	85%
Life goals (75% capacity factor)	
Fuel cell stacks	6 y
Balance of plant	30 y
Startup/shutdown:	
Startup: Cold startup in 4 to 6 h	
Shutdown: 100% to zero load in 3 h	
Daily load following:	
Large-load change response time of 2 h	
Small-load change response rate up to 2%/min	
Abnormal conditions:	
Complete-load rejection (breakers opening)	
Partial-load rejection (from power system breakup)	
Sustained abnormal voltage or frequency operation	
Limit fault current to 1.1 per unit current (rms basis)	
Other:	
Independent volt-ampere reactive (VAR) control	

600 MCFC stacks (having five hundred, 1-m^2 cells each) for a total 450-MW output

15 coal gasifiers capable of handling 3×10^{11} J/h each

10 heat recovery steam generators

Five 15-MW gas turbines

One 150-MW steam turbine

Other alternatives have been considered including other gasifiers, clean-up systems, and bottoming cycle arrangements. While it is likely that other configurations will be developed which are improvements over the above, these improvements will probably not be dramatic because the overall power plant efficiency and resulting bus bar cost are relatively insensitive to the many variables.[22, 23] An artist's concept of a 675-MW central station MCFC power plant is shown in Fig. 43.24.

43.4.5 Other MCFC Power Plants

Although the ultimate application of the MCFC will likely be as a large coal-fueled, central-station power plant, other applications such as that of dispersed or on-site generators with and without reject heat recovery are also being considered. Because of the high operating temperature, the quality of the MCFCs waste heat could be compatible with a variety of industrial heating applications. Also, at 600°C, in-situ reforming of methane or methanol is possible; this could result in a very efficient, small power plant.

To date, however, these applications have been viewed as the substitution of MCFC power sections for PAFC power sections within a dispersed or on-site power plant. As the MCFC stack capability progresses, more sophisticated treatments of such systems will be justified. The major question will ultimately be that of trading off the characteristics of MCFCs and advanced PAFCs as they compete for the central station as well as the smaller fuel cell applications.

43.5 HIGH-TEMPERATURE SOLID-OXIDE FUEL CELLS

43.5.1 Cell Structure

HTSO fuel cells employ a "tubular" rather than a "filter press" stack configuration. This tubular stack, shown in Fig. 43.25, avoids the problem of sealing the edges of the individual cells that is necessary in filter press stacks. Such edge seals are considered impractical due to the lack of nonporous, insulating, gasket materials for use at 1000°C. The state-of-the-art HTSO fuel cell stacks[6] consist of five basic components: a porous support tube, a fuel electrode, a solid electrolyte, an air electrode, and an electronically conducting interconnection.

The porous tube provides mechanical support for the other components. It is nominally 1 to 2 cm in diameter with a wall thickness of 0.1 cm. The fuel electrode is typically a porous nickel-zirconia cermet that serves as both the electrocatalyst and current collector. The electrolyte is a dense film of yttria-stabilized zirconia. The air

TABLE 43.4   Key Features* of MCFC Power Plant

Rated power plant output	675 MW(e)
Individual cell voltage*	0.78 V dc
Power density*	125 MW/cm²
Cell area/power plant	3 × 10⁵ m²
Fuel cell output	∼365 MW(e)
Fuel cell pressure	
Bottoming cycle output	310 MW(e)
Overall power plant	50% (∼6800 Btu/kWh heat rate)
Efficiency (coal to ac power)	
Gasifier	Texaco (oxygen blown)
Clean-up	Selexol + ZnO

*At rated power.

FIG. 43.24   675-MW MCFC fuel cell power plant—artist's conception. (*Courtesy of United Technologies Corp.*)

electrode consists of a discontinuous catalyst layer coated with a porous, doped indium oxide current collector. The fuel electrode, electrolyte, and air electrode are each approximately 50 μm thick. Individual cells are series-connected by electronically shorting the fuel electrode of one cell to the air electrode of the adjacent cell. This interconnection must pass through the electrolyte via a gastight seal. The interconnection material has historically been a major concern because of the severe environment.

Practical considerations in developing the individual components are:

Porous support tubes are required to provide a mechanically strong structure. The support tube must allow access of the fuel gas to the fuel electrode. Thus, a proper trade-off between strength and porosity is important. It is also necessary that the thermal expansion characteristics of the support tube match those of the other components. These considerations coupled with cost have led to the selection of calcia-stabilized zirconia as the material presently in use in the Westinghouse program.[6] The Westinghouse requirements include 25 vol % open porosity, 3 × 10⁷ Pa tensile strength, and 10 × 10⁻⁶ m/m°C thermal expansion.

The fuel electrode must be electronically conductive, not crack or otherwise lose

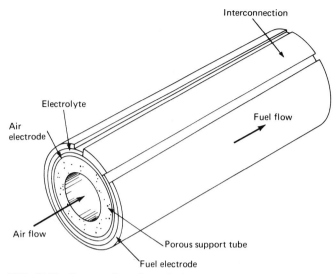

Interconnection

Electrolyte

Air electrode

Fuel flow

Air flow

Porous support tube

Fuel electrode

FIG. 43.25 Cross section of Westinghouse thin-film, high-temperature, solid-electrolyte fuel cell stack. (*Courtesy of Westinghouse Electric Corp.*)

conductivity after a thermal cycle, allow fuel gas to reach the electrolyte interface, and catalyze the fuel oxidation reaction. The use of a porous nickel-zirconia cermet ensures a thermal expansion match and reasonable conductivity. The zirconia also provides a bond to the porous support and mitigates the sintering of the nickel catalyst.

The solid electrolyte must be an absolute gas barrier, provide good ionic conductivity, but be an electronic insulator. The HTSO electrolyte must tolerate thermal cycling without loss of integrity. Figure 43.26 plots the ionic conductivity of several oxide conductors. As can be seen, the best conductivities are obtained with 10% yttria-stabilized zirconia (about 10^{-1} S/cm at 1000 °C) and 4% yttria/4% ytterbia–stabilized zirconia (about 2×10^{-1} S/cm at 1000°C). In addition, these electrolytes have negligible electronic conductivity and are stable in the fuel cell environment. Successful operation with cells containing yttria/ytterbia-doped zirconia has been demonstrated for 4 y.[24]

The air electrode must conduct electrons, withstand the oxidizing environment, allow oxygen to reach the electrolyte interface, adhere to the electrolyte thermal cycle without cracking, and catalyze the reduction of oxygen to oxide ions. Only electron-conducting metal oxides are able to satisfy these requirements. Candidate materials include tin-doped indium oxide and doped lanthanum manganite. To enhance performance, a catalyst such as praseodymium oxide can be incorporated into the electrode-electrolyte interface. A concern is the mechanical adhesion between the air electrode and the electrolyte. Spalling or flaking during thermal cycling has been a problem.

The interconnection between cells in a stack represents one of the major challenges in HTSO fuel cell development. The interconnection must have high electron conductivity

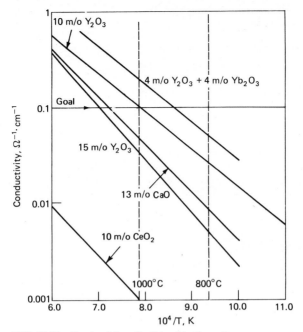

FIG. 43.26   Conductivity of solid-oxide electrolytes.

(and negligible ionic conductivity), gastightness, low volatility, a thermal expansion match with other components, and chemical compatibility with both oxidizing and reducing environments. Compatibility with an oxidizing environment dictates the use of an oxide (nothing else is stable in air in 1000°C); yet most oxides are thermodynamically unstable in a reducing atmosphere, and only a few offer electronic conductivity. Modified lanthanum chromites have been identified as suitable cell interconnection materials.

43.5.2 Single-Cell and Stack Performance Characteristics

Figure 43.27 depicts typical single-cell performance on hydrogen and air and hydrogen/carbon monoxide and air. Figure 43.28 shows typical stack performance on hydrogen and air as well as the voltage loss contribution of the various components.

Endurance capability of HTSO fuel cells has steadily improved. Single-cell lifetimes of 34,000 h have been achieved at a constant output of 120 mA/cm² and 0.7 V dc.[24]

More recently, over 3000 h of operation have been achieved on a 10-cell stack.[6] After 2000 h of operation at 400 mA/cm² on hydrogen-air, stack voltage had dropped 6%. This stack subsequently operated stably for an additional 3000 h at 150 mA/cm² on a simulated coal gas. Both tests involved 11 cool-down and reheat cycles, providing some confidence that cells and stacks can tolerate thermal cycling without performance deterioration.

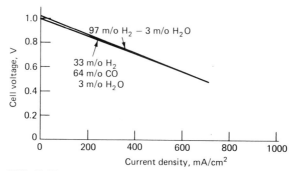

FIG. 43.27 Typical cell performance at 1000°C (both fuel and airflow unlimited). (*From Ref. 28.*)

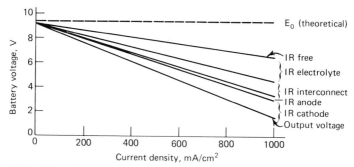

FIG. 43.28 Typical voltage losses for a 10-cell battery of HTSO cells at 1000°C with hydrogen and air. (*From Ref. 24.*)

43.5.3 Power Plant Considerations

Under a NASA contract, Westinghouse[25] analyzed power plant concepts involving the HTSO fuel cell integrated with a coal gasifier. The preferred power plant was similar to that previously shown (Fig. 43.23) for molten-carbonate fuel cells and utilized a steam bottoming cycle to convert waste heat from the gasifier and the stack to additional power. An overall efficiency of 40 to 44% was calculated. This efficiency assumed a fuel cell operating point of 400 mA/cm² at 0.66 V dc. It is likely that efficiencies approaching 48% would have been achieved if lower current densities (200 mA/cm²) had been used. That study also assumed a 10,000-h fuel cell life, and as a result the HTSO power plant was not economically competitive with molten-carbonate fuel cell power plants. If a 40,000-h lifetime assumption is made, the resulting economics would likely have been competitive.

43.5.4 Lower-Temperature Solid-Oxide Fuel Cells

A patent issued to Hitachi Ltd. in 1971[3] described the development of a solid-oxide electrolyte with sufficient conductivity at 700°C to be useful in practical fuel cells. Subsequently, there was considerable activity[4,5] aimed at developing fuel cells based upon solid-oxide materials that conduct at temperatures of 700 to 800°C. Operation at these lower temperatures would circumvent the materials and lifetime problems inherent at the 1000°C-temperature level. Ceria doped with gadolinia, yttria, and calcia are typical of oxides that exhibit high-oxide conductivity at 700 to 800°C.

Further investigations[4,26,27] determined that these solid oxides become mixed ionic-electronic conductors at low-oxygen partial pressures (on the fuel side). This, in effect, creates an electronic short across the electrolyte; even though seemingly small, this electronic short drastically reduces the achievable cell efficiency to less than 40% which, in turn, limits the total power plant efficiency to less than 35%.

43.6 PROGNOSIS

Individual phosphoric acid cells can today meet the environmental, efficiency, and endurance characteristics that are required for utility service. Realizing these capabilities at the stack and power plant levels will require further effort and experience. Projected costs for production quantities appear marginal when compared with costs for alternative power plants. PAFC technology is, however, progressing well; first-generation power plants will likely enjoy an efficiency of 41% (using clean hydrocarbon liquids or gases) rather than the 37% projected earlier. It is reasonable to believe that evolutionary advances will carry the PAFC into the marketplace in this decade. The first utility application will be as a dispersed or on-site generator fueled by a liquid or gas. However, PAFCs may well find applications in central-station power plants fueled by coal (with coal gasifiers). Even today's PAFC could operate as part of a coal-fueled, central-station power plant at a 34% efficiency (coal to ac power). With the current emphasis on cogeneration, it is also possible that the PAFC's unique environmental character and small size will lead to its early use in applications where its electric and thermal energy can be employed.[29]

MCFCs are perhaps 5 y from enjoying the endurance (at allowable cost) required for utility service. However, efficiency and other performance characteristics are now satisfactory. With appropriate support, MCFC power sections could be qualified and available for central-station power plant service in the early 1990s or as soon as significant numbers of coal-gasifier facilities are available. It is also likely that, toward the end of the decade, MCFCs will be competing with the advanced PAFCs, especially for the larger (> 1 MW) utility applications. At that time, both fuel cells will offer features that are considerably improved over those we can envisage today.

Recent HTSO cell and stack progress has been excellent, as evidenced by the successful endurance tests described above. However, this progress must be extended to longer lifetimes (up to 40,000 h) at the stack level, to multiple stack assemblies operating under real power plant conditions, and ultimately to power plants. This will involve tuning the materials and components to ensure that for extended operating periods and real reactant conditions, the support tube does not crack or become electronically conductive, the interconnections are stable and leak-tight, and the air electrode does not spall. In addition, fabrication techniques will need to be improved to ensure reproducibility when large numbers of cell stacks are produced. And finally, large numbers of cells, stacks, and modules will need to be assembled and tested to verify that the technology can indeed meet the endurance and performance requirements of utility power plants. Nevertheless, the prognosis is good that this can be accomplished in 5 to 10 y, and serious consideration can be given at that time to the incorporation of HTSO fuel cells in multimegawatt power plants.

REFERENCES

1. H. S. Spacil and C. S. Tedmon, Jr., Papers 356 and 357, *Fuel Cell Symp., Electrochem. Soc. Meeting,* Montreal, 1968.

2. "Final Report, Project Fuel Cell," Research and Development Report no. 57 (1970), Office of Coal Research, Department of the Interior, Washington.

3. Y. Maki, M. Matsuda, and T. Kudo (for Hitachi Ltd.), U.S. Patent 3,607,424.

4. "Advanced Fuel Cell Technology," EPRI Project 114-1 Final Report no. EM335, October 1976, Power Systems Div., United Technologies Corp.

5. H. Tuller and A. Nowick, "Doped Ceria as a Solid Oxide Electrolyte," *J. Electrochem. Soc.* **122**(2) 255–259 (February 1975).

6. "Thin Film Battery—Fuel Cell Power Generating System," Sixth Quarterly Report under Contract no. DE-AC-0379ET11305, July 1979, Department of Energy, Washington.

7. H. R. Kunz and G. A. Gruver, *J. Electrochem. Soc.,* **122**:1279 (1975).

8. A. P. Fickett, "Fuel Cell Electrocatalysts—Where Have We Failed?," *Electrochem. Soc. Spring Meeting,* Philadelphia, 1977.

9. P. Stonehart and J. Baris, "Preparation and Evaluation of Advanced Electrocatalysts for Phosphoric Acid Fuel Cells," First Quarterly Report NASA CR 159843, January–March 1980.

10. Noble Metal-Refractory Metal Alloys as Catalysts and Method for Making, U.S. Patent 4,186,110.

11. Noble Metal/Vanadium Alloy Catalyst and Method for Making, U.S. Patent 4,202,934.

12. P. N. Ross, Jr., "Oxygen Reduction on Supported Pt Alloys, and Intermetallic Compounds in Phosphoric Acid," EPRI Project 1200-5, Final Report no. EM1553, September 1980, Lawrence Berkeley Laboratories.

13. "Integral Cell Scale-up and Performance Verification," EPRI Project 842-4, Final Report no. EM1134, June 1979, Power Systems Div., United Technologies Corp.

13a. "Improvement of Fuel Cell Technology Base," Technical Progress Report no. 30, Contract no. DE AC 03 79 ET 11301, prepared for the U.S. Department of Energy by United Technologies Corp., September 1979.

14. "Stability of Acid Fuel Cell Cathode Materials," EPRI Project 1200-2, Interim Report no. EM1664, Stonehart Associates, January 1983.

15. "Fuel Cell Catalyst Sintering Studies," EPRI Project 583-1, Final Report no. EM833, Exxon Research and Engineering Co., July 1978.

16. "Assessment of Fuels for Power Generation by Electric Utilities Fuel Cells," EPRI Project 1042, Final Report no. EM695, vol. 2, Arthur D. Little, Inc., Cambridge, Mass., March 1978.

16a. T. G. Benjamin, E. H. Camara, and L. G. Marianowski, *Handbook of Fuel Cell Performance,* prepared for U.S. Department of Energy, Contract no. EC77CO3-1545, Institute of Gas Technology, May 1980.

17. "Sulfur Removal Processes for Advanced Fuel Cell Systems," EPRI Project 1041-5, Final Report EM1333, C. F. Braun, January 1980.

17a. "Fuel Cell Research on Second Generation Molten Carbonate System," prepared for the U.S. Department of Energy, Contract no. DE AC 037-78 ET 11276, Institute of Gas Technology, May 1980.

18. "Advanced Technology Fuel Cell Program," EPRI Project 114-2, Annual Report no. EM1328, Power Systems Div., United Technologies Corp., 1979.

19. J. Sim, R. Singh, and K. Kinoshita, *J. Electrochem. Soc.* **127**:1766 (1980).

20. "Integrated Coal Gasifier/Molten Carbonate Fuel Cell Power Plant Conceptual Design and Implementation Assessment," Energy Conversion Alternatives Study, Contract NAS 3-19586, Phase II Final Report no. FCR 0237, Power Systems Div., United Technologies Corp., November 1976.

21. "Advanced Technology Fuel Cell Program," EPRI Project 114-2, Annual Report no. EM956, Power Systems Div., United Technologies Corp., December 1978.

22. "Fuel Cell Power Plant Integrated Systems Evaluations," EPRI Project 1085-1, Interim Report no. EM1097, General Electric Co., June 1979.

23. "Fuel Cell Power Plant Integrated Systems Evaluation," EPRI Project 1085-1, Final Report, General Electric Co., to be published.

24. F. J. Rohr, "High Temperature Solid Oxide Fuel Cells—Present State and Problems of Development," Extended Abstracts, *Workshop High Temp. Solid Oxide Fuel Cells,* Brookhaven National Laboratory, Brookhaven, N.Y., May 1977.

25. C. J. Warde, R. J. Ruka, and A. O. Isenberg, "Energy Conversion Alternatives Study (ECAS)," Westinghouse Phase I Final Report, Vol. XII, "Fuel Cell Power Plants," NASA CR-134941, February 1976.

26. T. Kudo and H. Obayashi, "Ion-Electron Mixed Conduction in the Fluorite Type Ce_{1-x} Gd_x $O_{2-x/2}$," *J. Electrochem. Soc.* **123**:415.

27. P. N. Ross, Jr., and T. G. Benjamin, "Thermal Efficiency of Solid Electrolyte Fuel Cells with Mixed Conduction," *J. Power Sources,* **1**:311–321 (January 1977).

28. A. O. Isenberg, "Processing and Performance of High Temperature Solid Oxide Fuel Cells —State of the Art," abstracts, *1980 Natl. Fuel Cell Semin.* San Diego, Calif., 1980.

29. "An Analysis of the Application of Fuel Cells in Dual Energy Systems," EPRI Project 1135-1, Final Report EM 981 SY, Mathtech Inc., February 1979.

PART 7
APPENDICES

Definitions

Accumulator   See SECONDARY BATTERY.

Activated Stand Life   The period of time, at a specified temperature, that a cell can be stored in the charged condition before its capacity falls below a specified level.

Activation   The process of making a reserve cell functional, either by introducing an electrolyte, by immersing the cell into an electrolyte, or by other means (see Part 5).

Activation Polarization   Polarization resulting from the rate-determining step of the electrode reaction (see POLARIZATION).

Active Cell   A cell containing all components and in a charged state ready for discharge (as distinct from a RESERVE CELL).

Active Material   The material in the electrodes of a cell or battery which takes part in the electrochemical reactions of charge or discharge.

Ambient Temperature   The average temperature of the surroundings.

Ampere-Hour Capacity   The quantity of electricity measured in ampere-hours (Ah) which may be delivered by a cell or battery under specified conditions.

Ampere-Hour Efficiency   The ratio of the output of a secondary cell or battery, measured in ampere-hours, to the input required to restore the initial state of charge, under specified conditions (also coulombic efficiency).

Anion   Particles in the electrolyte carrying a negative charge and moving toward the anode during operation of the cell.

Anode   The electrode in an electrochemical cell where oxidation takes place. During discharge, the negative electrode of the cell is the anode. During charge, the situation reverses and the positive electrode of the cell is the anode.

Anolyte   The portion of the electrolyte in a galvanic cell adjacent to the anode; if a diaphragm is present, the electrolyte on the anode side of the diaphragm.

Aprotic Solvent   A nonaqueous solvent that does not contain a reactive proton although it may contain hydrogen in the molecule.

Available Capacity   The total capacity, Ah or Wh, that will be obtained from a cell or battery at defined discharge rates and other specified discharge or operating conditions.

Battery   Two or more electrochemical cells electrically interconnected in an appropriate series/parallel arrangement to provide the required operating voltage and current levels. Under common usage, the term "battery" is often also applied to a single cell.

Bipolar Plate   An electrode construction where positive and negative active materials are on opposite sides of an electronically conductive plate.

Bobbin   A cylindrical electrode (usually the positive) pressed from a mixture of the active material, a conductive material, such as carbon black, the electrolyte and/or binder with a centrally located conductive rod or other means for a current collector.

Boundary Layer   The volume of electrolyte solution immediately adjacent to the electrode surface in which concentration changes occur due to the effects of the electrode process.

C Rate   (also see HOURLY RATE) Discharge or charge current, in amperes, expressed in multiples of the rated capacity. For example, the C/20 discharge current for a battery rated at the 5-h discharge rate is

$$\frac{C_5 \ (Ah)}{20} = \text{current (A)}$$

As a cell's capacity is not the same at all discharge rates and usually increases with decreasing rate, a discharge at the $C_5/20$ rate will run longer than 20 h.

Capacity   The total number of ampere-hours or watthours that can be withdrawn from a fully charged cell or battery under specified conditions of discharge. (Also see AVAILABLE CAPACITY, RATED CAPACITY.)

Capacity Current   The fraction of the cell current consumed in charging the electrical double layer.

Capacity Retention   The fraction of the full capacity available from a battery under specified conditions of discharge after it has been stored for a period of time.

Cathode   The electrode in an electrochemical cell where reduction takes place. During discharge, the positive electrode of the cell is the cathode. During charge, the situation reverses, and the negative electrode of the cell is the cathode.

Catholyte   The portion of an electrolyte in a galvanic cell adjacent to a cathode; if a diaphragm is present, the electrolyte on the cathode side of the diaphragm.

Cation   Particle, in the electrolyte, carrying a positive charge and moving toward the cathode during operation of the cell.

Cell   The basic electrochemical unit used to generate or store electrical energy.

Charge   The conversion of electrical energy, provided in the form of a current from an external source, into chemical energy within a cell or battery.

Charge Rate   The current applied to a secondary cell or battery to restore its capacity. This rate is commonly expressed as a multiple of the rated capacity of the cell or battery. For example, the C/10 charge rate of a 500-Ah cell or battery is expressed as

$$\frac{C}{10} \text{ rate} = \frac{500 \ Ah}{10} = 50 \ A$$

Charge Retention   See CAPACITY RETENTION.

Chlorine Hydrate/Zinc Cell   See ZINC/CHLORINE CELL.

Closed-Circuit Voltage (CCV)   The potential or voltage of a cell or battery when it is discharging.

Concentration Polarization   Polarization caused by the depletion of ions in the electrolyte at the surface of the electrode. (See also POLARIZATION.)

Continuous Test   A test in which a cell or battery is discharged to a prescribed end-point voltage without interruption.

Creepage   The movement of electrolyte onto surfaces of electrodes or other components of a cell with which it is not normally in contact.

Current Collector   An inert member of high electrical conductivity used to conduct current from or to an electrode during discharge or charge.

Current Density   The current per unit active area of the surface of an electrode.

Cutoff Voltage The cell or battery voltage at which the discharge is terminated. The cutoff voltage is specified by the cell manufacturer and is generally a function of discharge rate.

Cycle The discharge and subsequent charge of a secondary battery such that it is restored to its original conditions.

Cycle Life The number of cycles under specified conditions which are available from a secondary battery before it fails to meet specified criteria as to performance.

Cycle Service A duty cycle characterized by frequent and usually deep discharge-charge sequences, such as motive power applications.

Deep Discharge Withdrawal of at least 80% of the rated capacity of a cell or battery.

Density The ratio of a mass of material to its own volume at a specified temperature.

Depolarization A reduction in the polarization of an electrode.

Depolarizer A substance or means used to prevent or decrease polarization. The term "depolarizer" is often used to describe the positive electrode of a primary cell.

Depth of Discharge (DOD) The ratio of the quantity of electricity (usually in ampere-hours) removed from a secondary cell or battery on discharge to its rated capacity.

Diaphragm A porous or permeable means for separating the positive and negative electrode compartments of a galvanic cell and preventing admixture of catholyte and anolyte.

Diffusion The movement of species under the influence of a concentration gradient.

Discharge The conversion of the chemical energy of a cell or battery into electrical energy and withdrawal of the electrical energy into a load.

Discharge Rate The rate, usually expressed in amperes, at which electrical current is taken from the cell or battery.

Double Layer The region in the vicinity of an electrode-electrolyte interface where the concentration of mobile ionic species has been changed to values differing from the bulk equilibrium value by the potential difference across the interface.

Double-Layer Capacitance The capacitance of the double layer.

Dry Cell A cell with immobilized electrolyte. The term "dry cell" is often used to describe the Leclanche cell.

Dry Charged Battery A battery in which the electrodes are in a charged state, ready to be activated by the addition of the electrolyte.

Duplex Electrode or Plate See BIPOLAR PLATE.

Duty Cycle The operating regime of a cell or battery including factors such as charge and discharge rates, depth of discharge, cycle length, and length of time in the standby mode.

Efficiency The ratio of the output of a secondary cell or battery on discharge to the input required to restore it to the initial state of charge under specified conditions. (See also AMPERE-HOUR EFFICIENCY, VOLTAGE EFFICIENCY, and WATTHOUR EFFICIENCY.)

Electrical Double Layer See DOUBLE LAYER.

Electrocapillarity The surface tension between liquid mercury and an electrolyte solution is modified by the potential difference across the interface. The effect is termed "electrocapillarity."

Electrochemical Couple The system of active materials within a cell that provides electrical energy storage through an electrochemical reaction.

Electrochemical Equivalent Weight of a substance that is deposited at an electrode when the quantity of electricity which is passed is one coulomb (see FARADAY).

Electrochemical Series A classification of the elements according to the values of the standard potentials of specified electrochemical reactions.

Electrode The site, area, or location at which electrochemical processes take place.

Electrolyte   The medium which provides the ion transport mechanism between the positive and negative electrodes of a cell.

Electrolytic Cell   A cell in which electrochemical reactions are caused by supplying electrical energy or which supplies electrical energy as a result of electrochemical reactions: if the first case only is applicable, the cell is an electrolysis cell; if the second case only, the cell is a galvanic cell.

Electromotive Force (emf)   The standard potential of a specified electrochemical action.

Electromotive Series   See ELECTROCHEMICAL SERIES.

Element   The negative and positive electrodes together with the separators of a single cell. It is used almost exclusively in describing lead-acid cells and batteries.

End Voltage   The prescribed voltage at which the discharge (or charge, if end-of-charge voltage) of a cell or battery may be considered complete (also cutoff voltage).

Energy Density   The ratio of the energy available from a cell or battery to its volume (Wh/L) or weight (Wh/kg).

Equalization   The process of restoring all cells in a battery to an equal state of charge.

Equivalent Circuit   An electrical circuit describing the impedance properties of a cell or battery.

Equilibrium Electrode Potential   The difference in potential between an electrode and an electrolyte when they are in equilibrium for the electrode reaction which determines the electrode potential.

Exchange Current   Under equilibrium conditions, the forward and backward currents of an electrochemical process are equal. This equilibrium current is defined as the exchange current.

Faraday   One gram equivalent weight of matter is chemically altered at each electrode of a cell for each 96,494 international coulombs, or one Faraday, of electricity passed through the electrolyte.

Fauré Plate   See PASTED PLATE.

Flash Current   See SHORT-CIRCUIT CURRENT.

Flat-Plate Cell   A cell fabricated with rectangular flat-plate electrodes.

Float Charge   A method of maintaining a cell or battery in a charged condition by continuous, long-term constant-voltage charging, at a level sufficient to balance self-discharge.

Flooded Cell   A cell design which incorporates an excess amount of electrolyte.

Forced Discharge   Discharging a cell or battery, with external battery or power source, below zero volts into voltage reversal.

Formation   Electrochemical processing of a battery plate or electrode between manufacture and first discharge which transforms the active materials into their usable form.

Fuel Cell   A cell in which the active materials are continuously supplied and the reaction products continuously removed (see Part 6).

Galvanic Cell   An electrolytic cell that converts chemical energy into electrical energy by electrochemical action.

Gassing   The evolution of gas from one or more of the electrodes in a cell. Gassing commonly results from local action (self-discharge) or from the electrolysis of water in the electrolyte during charging.

Grid   A framework for a plate or electrode which supports or retains the active materials and acts as a current collector.

Half-Cell   An electrode (either the anode or cathode) immersed in a suitable electrolyte.

Hourly Rate   A discharge rate, in amperes, of a cell or battery which will deliver the specified hours of service to a given end voltage (also see C RATE).

Hydrogen Electrode   An electrode of platinized platinum saturated by a stream of pure hydrogen.

Hydrogen Overvoltage   The activation overvoltage for hydrogen discharge.

Initial (Closed-Circuit) Voltage   The on-load voltage at the beginning of a discharge.

Inner Helmholtz Plane   The plane of closest approach of ions solution. It corresponds to the plane which contains the contact-adsorbed ions and the innermost layer of water molecules.

Intermittent Test   A test during which a cell or battery is subjected to alternate periods of discharge and rest according to a specified discharge regime.

Internal Impedance   The opposition or resistance of a cell or battery to an alternating current of a particular frequency.

Internal Resistance   The opposition or resistance to the flow of an electric current within a cell or battery; the sum of the ionic and electronic resistances of the cell components.

Local Action   Chemical reactions within a cell that convert the active materials to a discharged state without supplying energy through the battery terminals (self-discharge).

Luggin Capillary   The salt bridge from a reference electrode to a cell solution often has a restricted (capillary) junction between the bridge and the cell solution. The restriction, which is often situated close to the working electrode, is called a Luggin capillary.

Maintenance-Free Battery   A secondary battery which does not require periodic "topping up" to maintain electrolyte volume.

Mean Diffusion Current   In polarography, the periodic detachment of mercury drops from the dropping mercury electrode impart an oscillation to the measured current. The average value of this current is termed the mean diffusion current.

Mechanical Recharging   Restoring the capacity of a cell by replacing a spent or discharged electrode with a fresh one.

Memory Effect   A phenomenon in which a cell, operated in successive cycles to the same, but less than full, depth of discharge, temporarily loses the rest of its capacity at normal voltage levels (usually applies to a nickel-cadmium cell).

Midpoint Voltage   The voltage of a cell or battery midway in the discharge between the fully charged state and the end voltage.

Migration   The movement of charged species under the influence of a potential gradient.

Motive Power Battery   See TRACTION BATTERY.

Negative Electrode   The electrode acting as an anode when a cell or battery is discharging.

Negative-Limited   The operating characteristics (performance) of the cell is limited by the negative electrode.

Nominal Voltage   The characteristic operating voltage or rated voltage of a cell or battery (as distinct from WORKING VOLTAGE, MIDPOINT VOLTAGE, etc.).

Off-Load Voltage   See OPEN-CIRCUIT VOLTAGE.

Ohmic Overvoltage   Overvoltage caused by the ohmic drop at an electrode-electrolyte interface.

On-Load Voltage   The difference in potential between the terminals of a cell or battery when it is discharging.

Open-Circuit Voltage (OCV)   The difference in potential between the terminals of a cell or voltage when the circuit is open (no-load condition).

Outer Helmholtz Plane   The plane of closest approach of those ions which do not contact-adsorb but approach the electrode with a sheath of solvated water molecules surrounding them.

Overcharge   The forcing of current through a cell after all the active material has been converted to the charged state. In other words, charging continued after 100% state of charge is achieved.

Overvoltage The potential difference between the equilibrium potential of an electrode and that of the electrode under an imposed polarization current.

Oxygen Recombination The process by which oxygen generated at the positive plate during charge is reacted at the negative plate.

Paper-Lined Cell Construction of a cell where a layer of paper, wetted with electrolyte, acts as the separator.

Parallel Term used to describe the interconnection of cells or batteries in which all the like terminals are connected together.

Passivation The phenomenon by which a metal, although in conditions of thermodynamic instability, remains indefinitely unattacked because of certain surface conditions.

Paste-Lined Cell Leclanche cell constructed so that a layer of gelled paste acts as the separator.

Pasted Plate A plate, usually for a lead-acid battery, manufactured by coating a grid or support strip with active materials.

Planté Plate A plate for a lead-acid battery in which the active materials are formed directly from a lead substrate by electro-chemical processing.

Plate An assembly of active materials on a supporting framework grid, frame, or support strip.

Pocket Plate A plate for a secondary battery in which active materials are held in perforated metal pockets on a support strip.

Polarization The lowering of the potential of a cell or electrode from its equilibrium value caused by the passage of an electric current.

Positive Electrode The electrode acting as a cathode when a cell or battery is discharging.

Positive-Limited The operating characteristics (performance) of the cell is limited by the positive electrode.

Power Density The ratio of the power available from a battery to its weight (W/kg) or volume (W/L).

Primary Cell or Battery A cell or battery which is not intended to be recharged and is discarded when the cell or battery has delivered all its electrical energy.

Rate Constant At equilibrium, the forward and backward Faradic currents of an electrode process are equal and referred to as the exchange current. This exchange current can be defined in terms of a rate constant called the standard heterogeneous rate constant for the electrode process.

Rated Capacity The number of ampere-hours a cell or battery can deliver under specific conditions (rate of discharge, end voltage, temperature); usually the manufacturer's rating.

Recharge See CHARGE.

Recovery See RECUPERATION.

Recuperation The lowering of the polarization of a cell during rest periods.

Redox Cell A secondary cell in which two reactant fluids, separated by a membrane, form the active materials. (See Chap. 27 in Part 4.)

Reference Electrode A specially chosen electrode which has a poised, reproducible potential against which other electrode potentials may be referred.

Reserve Cell A cell which may be stored in an inactive state and made ready for use by adding electrolyte or, in the case of a thermal battery, melting a solidified electrolyte.

Reversal The changing of the normal polarity of a cell or battery.

Secondary Battery A galvanic battery which, after discharge, may be restored to the fully charged state by the passage of an electric current through the cell in the opposite direction to that of discharge.

Self-Discharge The loss of useful capacity of a cell or battery on storage due to internal chemical action (local action).

Separator An ionic permeable electronically nonconductive spacer which prevents electronic contact between electrodes of opposite polarity in the same cell.

Series The interconnection of cells in such a manner that the positive terminal of the first is connected to the negative terminal of the second, and so on.

Service Life The period of useful life of a primary cell or battery before a predetermined end-point voltage is reached.

Shape Change Change in shape of an electrode due to migration of active material during charge/discharge cycling.

Shelf Life The duration of storage under specified conditions at the end of which a cell or battery still retains the ability to give a specified performance.

Short-Circuit Current (SCC) The initial value of the current obtained from a cell or battery in a circuit of negligible resistance.

Sintered Electrode An electrode construction in which active materials are deposited in the interstices of a porous metal matrix made by sintering metal powder.

SLI Battery A battery designed to start internal combustion engines and to power the electrical systems in automobiles when the engine is not running (starting, lighting, ignition).

Specific Energy See ENERGY DENSITY.

Specific Gravity The specific gravity of a solution is the ratio of the weight of the solution to the weight of an equal volume of water at a specified temperature.

Specific Power See POWER DENSITY.

Spiral Wound The shape of the internal roll in a cylindrical cell. It is made by winding the electrodes and separator into a spiral-wound, jelly-roll, construction.

Standard Electrode Potential The equilibrium value of an electrode potential when all the constituents taking part in the electrode reaction are in the standard state.

Standby Battery A battery designed for emergency use in the event of a main power failure.

Starved Electrolyte Cell A cell containing little or no free fluid electrolyte. This enables gases to reach electrode surfaces during charging and facilitates gas recombination.

State-of-Charge (SOC) The available capacity in a cell or battery expressed as a percentage of rated capacity.

Stationary Battery A secondary battery designed for use in a fixed location.

Storage Battery See SECONDARY BATTERY.

Storage Life See SHELF LIFE.

Taper Charge A charge regime delivering moderately high rate charging current when the battery is at a low state of charge and tapering the charging current to lower rates as the battery is charged.

Thermal Runaway A condition whereby a cell or battery on charge or discharge will destroy itself through internal heat generation caused by high overcharge or overdischarge current or other abusive condition.

Traction Battery A secondary battery designed for the propulsion of electric vehicles or electrically operated mobile equipment operating in a deep-cycle regime.

Transfer Coefficient The transfer coefficient determines what fraction of the electrical energy of a system resulting from the displacement of the potential from the equilibrium value affects the rate of electrochemical transformation. (See Chap. 2 in Part 1.)

Transition Time   The time of an electrode process from the initiation of the process at constant current to the moment an abrupt change in potential occurs signifying that a new electrode process is controlling the electrode potential.

Transport Number   The fraction of the total cell current carried by the cation of an electrolyte solution is called the "cation transport number." Similarly, the fraction of the total current carried by the anion is referred to as the "anion transport number."

Trickle Charge   A charge at a low rate, balancing losses through local action and/or periodic discharge, to maintain a cell or battery in a fully charged condition.

Tubular Plate   A battery plate in which an assembly of perforated metal or polymer tubes holds the active materials.

Unactivated Shelf Life   The period of time, under specified conditions of temperature and environment, that an unactivated or reserve cell or battery can stand before deteriorating below a specified capacity.

Vent   A normally sealed mechanism which allows for the controlled escape of gases from within a cell.

Vented Cell   A cell design in which a vent mechanism operates to expel gases that are generated during the operation of the cell.

Voltage Delay   Time delay for a cell or battery to deliver the required operating voltage after it is placed under load.

Voltage Efficiency   The ratio of average voltage during discharge to average voltage during recharge under specified conditions of charge and discharge.

Watthour Capacity   The quantity of electrical energy measured in watthours which may be delivered by a cell or battery under specified conditions.

Watthour Efficiency   The ratio of the watthours delivered on discharge of a battery to the watthours needed to restore it to its original state under specified conditions of charge and discharge.

Wet Shelf Life   The period of time that a cell or battery can stand in the charged or activated condition before deteriorating below a specified capacity.

Working Voltage   The typical voltage or range of voltage of a cell or battery during discharge.

SOURCE: Dr. Douglas H. Spencer, Berec Group Ltd., and Dr. John Broadhead, Bell Laboratories, contributed to this Appendix.

Standard Reduction Potentials

TABLE B.1 Standard Reduction Potentials of Electrode Reactions at 25°C

Electrode reaction	E^0, V
$Li^+ + e \rightleftarrows Li$	-3.01
$Rb^+ + e \rightleftarrows Rb$	-2.98
$Cs^+ + e \rightleftarrows Cs$	-2.92
$K^+ + e \rightleftarrows K$	-2.92
$Ba^{2+} + 2e \rightleftarrows Ba$	-2.92
$Sr^{2+} + 2e \rightleftarrows Sr$	-2.89
$Ca^{2+} + 2e \rightleftarrows Ca$	-2.84
$Na^+ + e \rightleftarrows Na$	-2.71
$Mg(OH)_2 + 2e \rightleftarrows Mg + 2OH^-$	-2.69
$Mg^{2+} + 2e \rightleftarrows Mg$	-2.38
$Al(OH)_3 + 3e \rightleftarrows Al + 3OH^-$	-2.34
$Ti^{2+} + 2e \rightleftarrows Ti$	-1.75
$Be^{2+} + 2e \rightleftarrows Be$	-1.70
$Al^{3+} + 3e \rightleftarrows Al$	-1.66
$Zn(OH)_2 + 2e \rightleftarrows Zn + 2OH^-$	-1.25
$Mn^{2+} + 2e \rightleftarrows Mn$	-1.05
$Fe(OH)_2 + 2e \rightleftarrows Fe + 2OH^-$	-0.88
$2H_2O + 2e \rightleftarrows H_2 + 2OH^-$	-0.83
$Cd(OH)_2 + 2e \rightleftarrows Cd + 2OH^-$	-0.81
$Zn^{2+} + 2e \rightleftarrows Zn$	-0.76

TABLE B.1 Standard Reduction Potentials of
Electrode Reactions at 25°C *(Continued)*

Electrode reaction	E^0, V
$Ni(OH)_2 + 2e \rightarrow Ni + 2OH^-$	-0.72
$Ga^{3+} + 3e \rightleftarrows Ga$	-0.52
$Fe^{2+} + 2e \rightleftarrows Fe$	-0.44
$Cd^{2+} + 2e \rightleftarrows Cd$	-0.40
$PbSO_4 + 2e \rightleftarrows Pb + SO_4^-$	-0.36
$In^{3+} + 3e \rightleftarrows In$	-0.34
$Tl^+ + e \rightleftarrows Tl$	-0.34
$Co^{2+} + 2e \rightleftarrows Co$	-0.27
$Ni^{2+} + 2e \rightleftarrows Ni$	-0.23
$Sn^{2+} + 2e \rightleftarrows Sn$	-0.14
$Pb^{2+} + 2e \rightleftarrows Pb$	-0.13
$O_2 + H_2O + 2e \rightleftarrows HO_2^- + OH^-$	-0.08
$D^+ + e \rightleftarrows \frac{1}{2}D_2$	-0.003
$H^+ + e \rightleftarrows \frac{1}{2}H_2$	0.000
$HgO + H_2O + 2e \rightleftarrows Hg + 2 OH^-$	0.10
$CuCl + e \rightleftarrows Cu + Cl^-$	0.14
$AgCl + e \rightleftarrows Ag + Cl^-$	0.22
$Cu^{2+} + 2e \rightleftarrows Cu$	0.34
$Ag_2O + H_2O + 2e \rightleftarrows 2 Ag + 2OH^-$	0.35
$\frac{1}{2}O_2 + H_2O + 2e \rightleftarrows 2OH^-$	0.40
$Cu^+ + e \rightleftarrows Cu$	0.52
$I_2 + 2e \rightleftarrows 2I^-$	0.54
$2 AgO + H_2O + 2e \rightleftarrows Ag_2O + 2OH^-$	0.57
$Hg^{2+} + 2e \rightleftarrows 2Hg$	0.80
$Ag^+ + e \rightleftarrows Ag$	0.80
$Ir^{3+} + 3e \rightleftarrows Ir$	1.00
$Br_2 + 2e \rightleftarrows 2Br^-$	1.07
$O_2 + 4H^+ + 4e \rightleftarrows 2H_2O$	1.23
$Cl_2 + 2e \rightleftarrows 2Cl^-$	1.36
$PbO_2 + 4H^+ + 2e \rightleftarrows Pb^{2+} + 2H_2O$	1.46
$PbO_2 + SO_4 + 4H^- + 2e \rightleftarrows PbSO_4 + 2H_2O$	1.69
$F_2 + 2e \rightleftarrows 2F^-$	2.87

Properties of
Battery Materials

TABLE C.1 Electrochemical Equivalents of Battery Materials

Material	Symbol	Atomic no.	Atomic wt., g	Density, g/cm³	Valence change	Electrochemical Equivalents		
						Ah/g	g/Ah	Ah/cm³
Elements								
Aluminum	Al	13	26.98	2.699	3	2.98	0.335	8.05
Antimony	Sb	51	121.75	6.62	3	0.66	1.514	4.37
Arsenic	As	33	74.92	5.73	3	1.79	0.559	10.26
Barium	Ba	56	137.34	3.78	2	0.39	2.56	1.47
Beryllium	Be	4	9.01		2	5.94	0.168	
Bismuth	Bi	83	208.98	9.80	3	0.385	2.59	3.77
Boron	B	5	10.81	2.54	3	7.43	0.135	18.87
Bromine	Br	35	79.90		1	0.335	2.98	
Cadmium	Cd	48	112.40	8.65	2	0.477	2.10	4.15
Cesium	Cs	55	132.91	1.87	3	0.574	1.74	1.07
Calcium	Ca	20	40.08	1.54	2	1.34	0.748	2.06
Carbon (graphite)	C	6	12.01	2.25	4	8.93	0.112	20.09
Chlorine	Cl	17	35.45		1	0.756	1.32	
Chromium	Cr	24	52.00	6.92	3	1.55	0.647	10.72
Cobalt	Co	27	58.93	8.71	2	0.910	1.10	7.93
Copper	Cu	29	63.55	8.89	2	0.843	1.19	7.49
					1	0.422	2.37	3.75
Fluorine	F	9	19.00		1	1.41	0.709	
Gold	Au	79	197.00	19.3	1	0.136	7.36	2.62
Hydrogen	H	1	1.08		1	26.59	0.0376	
Indium	In	49	114.82	7.28	3	0.701	1.43	5.10
Iodine	I	53	126.90		1	0.211	4.73	
Iron	Fe	26	55.85	7.85	2	0.96	1.04	7.54
					3	1.44	0.694	11.30
Lead	Pb	82	207.2	11.34	2	0.259	3.87	2.94
Lithium	Li	3	6.94	0.534	1	3.86	0.259	2.06
Magnesium	Mg	12	24.31	1.74	2	2.20	0.454	3.83
Manganese	Mn	25	54.94	7.42	2	0.976	1.02	7.24
Mercury	Hg	80	200.59	13.60	2	0.267	3.74	3.63
Molybdenum	Mo	42	95.94	10.2	6	1.67	0.597	17.03
Nickel	Ni	28	58.71	8.6	2	0.913	1.09	7.85
Nitrogen	N	7	14.01		3	5.74	0.174	
Oxygen	O	8	16.00		2	3.35	0.298	
Platinum	Pt	78	195.09	21.37	4	0.549	1.82	11.73
Potassium	K	19	39.10	0.87	1	0.685	1.46	0.59
Silver	Ag	17	107.87	10.5	1	0.248	4.02	2.60
Sodium	Na	11	22.99	0.971	1	1.17	0.858	1.14
Sulfur	S	16	32.06	2.0	2	1.67	0.598	3.34
Tin	Sn	50	118.69	7.30	4	0.903	1.11	6.59
Vanadium	V	23	50.95	5.96	5	2.63	0.380	15.67
Zinc	Zn	30	65.38	7.1	2	0.820	1.22	5.82
Zirconium	Zr	40	91.22	6.44	4	1.18	0.851	7.60

TABLE C.1 (*Continued*)

Material	Symbol	Molecular wt., g	Density, g/cm³	Valence change	Electrochemical Equivalents		
					Ah/g	g/Ah	Ah/cm³
Compounds							
Bismuth trioxide	Bi₂O₃	466	8.5	6	0.35	2.86	2.97
Calcium chromate	CaCrO₄	156.1		2	0.34	2.91	
Carbon monofluoride	(CF)ₙ	(31)ₙ	2.7	1	0.86	1.16	2.32
Copper chloride	CuCl	99	3.5	1	0.27	3.69	0.95
Copper chloride	CuCl₂	134.5	3.1	2	0.40	2.50	1.22
Copper fluoride	CuF₂	101.6	2.9	2	0.53	1.87	1.52
Copper oxide	CuO	79.6	6.4	2	0.67	1.49	4.26
Copper sulfate	CuSO₄	159.6	3.6	2			
Copper sulfide	CuS	95.6	4.6	2	0.56	1.79	2.57
Iron sulfide	FeS	87.9	4.84	2	0.61	1.64	2.95
Iron sulfide	FeS₂	119.9	4.87	4	0.89	1.12	4.35
Lead bismuthate	Bi₂Pb₂O₅	912	9.0	10	0.29	3.41	2.64
Lead chloride	PbCl₂	278.1	5.8	2	0.19	5.18	1.12
Lead dioxide	PbO₂	239.2	9.3	2	0.22	4.45	2.11
Lead iodide	PbI₂	461	6.2	2	0.12	8.60	0.72
Lead oxide	Pb₃O₄	685	9.1	8	0.31	3.22	2.85
Lead sulfide	PbS	239.3	7.5	2	0.22	4.46	1.68
Manganese dioxide	MnO₂	86.9	5.0	1	0.31	3.22	1.54
Mercuric oxide	HgO	216.6	11.1	2	0.247	4.05	2.74
Molybdenum trioxide	MoO₃	143	4.5	1	0.19	5.26	0.84
Nickel oxide	NiOOH	91.7	7.4	1	0.29	3.42	2.16
Nickel sulfide	Ni₃S₂	240		4	0.47	2.12	
Silver chloride	AgCl	143.3	5.56	1	0.19	5.26	1.04
Silver chromate	Ag₂CrO₄	331.8	5.6	2	0.16	6.25	0.90
Silver oxide (monovalent)	Ag₂O	231.8	7.1	2	0.23	4.33	1.64
Silver oxide (divalent)	AgO	123.9	7.4	2	0.43	2.31	3.20
Sulfur dioxide	SO₂	64	1.37	1	0.419	2.39	
Sulfuryl chloride	SO₂Cl₂	135	1.66	2	0.397	2.52	
Thionyl chloride	SOCl₂	119	1.63	2	0.450	2.22	
Vanadium pentoxide	V₂O₅	181.9	3.6	1	0.15	6.66	0.53

TABLE C.2 Properties of Sulfuric Acid Solutions

| Specific gravity | | Temperature coefficient α | H_2SO_4 | | | Freezing point, °C | Electrochemical equivalent (per liter of acid), Ah |
At 15°C	At 25°C		Wt., %	Vol., %	Mol/L		
1.00	1.000	—	0	0	0	0	0
1.05	1.049	33	7.3	4.2	0.82	−3.3	22
1.10	1.097	48	14.3	8.5	1.65	−7.7	44
1.15	1.146	60	20.9	13.0	2.51	−15	67
1.20	1.196	68	27.2	17.7	3.39	−27	90
1.25	1.245	72	33.2	22.6	4.31	−52	115
1.30	1.295	75	39.1	27.6	5.26	−70	141
1.35	1.345	77	44.7	32.8	6.23	−49	167
1.40	1.395	79	50.0	38.0	7.21	−36	
1.45	1.445	82	55.0	43.3	8.2	−29	
1.50	1.495	85	59.7	48.7	9.2	−29	
1.55	1.545	89	64.2	54.1	10.3	−38	
1.60	1.594	92	68.6	59.7	11.4	−40	
1.65	1.644	95	72.9	65.3		−40	
1.70	1.694	100	77.1	71.2		−14	
1.75	1.743	107	81.5	77.5		+5	
1.80	1.193	110	86.7	84.8		+6	

NOTE: To calculate the specific gravity for any temperature, °C, SG (t) = SG (15°C) + $\alpha \times 10^{-5}$ $(15 - t)$

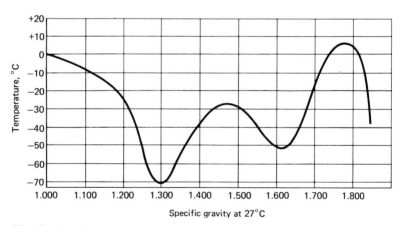

Fig. C.1 Freezing points of sulfuric acid solutions at various specific gravities.

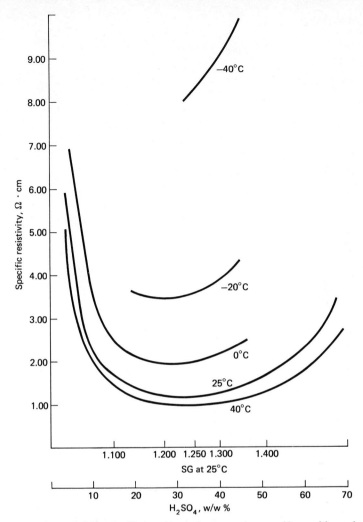

Fig. C.2 Specific resistivity of sulfuric acid solutions at various specific gravities and temperatures.

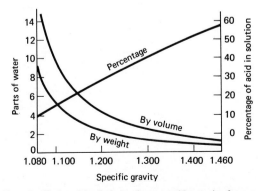

Fig. C.3 Preparation of sulfuric acid solutions of any specific gravity from concentrated sulfuric acid, 1.835 specific gravity. (*From G. W. Vinal, Storage Batteries, Wiley, New York, 1955, p. 129.*)

TABLE C.3   Properties of Potassium Hydroxide Solutions

Wt., %	KOH concentration		Density d_4^{20}	g/L at 20°C	Specific conductance at 20°C, $\Omega^{-1}cm^{-1}$	Freezing point, °C	Viscosity at 20°C, cP
	Molality	Molarity at 20°C					
4	0.7427	0.7372	1.034	41.36	0.15	−2	1.08
8	1.550	1.527	1.071	85.68	0.28	−6	1.18
12	2.431	2.373	1.109	133.1	0.38	−11	1.29
16	3.395	3.271	1.147	183.5	0.47	−17	1.44
20	4.456	4.229	1.186	237.2	0.54	−24	1.62
24	5.629	5.246	1.226	294.2	0.60	−35	1.86
28	6.932	6.326	1.267	354.8	0.62	−49	2.17
32	8.388	7.469	1.309	418.9	0.60	−61	2.57
36	10.03	8.678	1.352	486.7	0.56	−46	3.11
40	11.88	9.911	1.396	558.4	0.52	−35	3.88
44	14.00	11.31	1.442	634.5	—	−29	5.11
48	16.45	12.73	1.488	714.2	—	−4	6.73

Standard Symbols and Constants

TABLE D.1 SI Base Units

Quantity	Unit	Symbol
Length	meter	m
Mass	kilogram	kg
Time	second	s
Electric current	ampere	A
Thermodynamic temperature*	kelvin	K
Amount of substance	mole	mol
Luminous intensity	candela	cd

*Celsius temperature is, in general, expressed in degrees Celsius (symbol °C).

SOURCE: From Fink and Beaty; reproduced from IEEE Standard 268-1976, by permission.

TABLE D.2 SI Prefixes Expressing Decimal Factors

Factor	Prefix	Symbol	Factor	Prefix	Symbol
10^{18}	exa	E	10^{-1}	deci	d
10^{15}	peta	P	10^{-2}	centi	c
10^{12}	tera	T	10^{-3}	milli	m
10^{9}	giga	G	10^{-6}	micro	μ
10^{6}	mega	M	10^{-9}	nano	n
10^{3}	kilo	k	10^{-12}	pico	p
10^{2}	hecto	h	10^{-15}	femto	f
10^{1}	deka	da	10^{-18}	atto	a

SOURCE: From Fink and Beaty; adapted from IEEE Standard 268-1976, by permission.

TABLE D.3 Derived Units of the International System

Quantity	Name of unit	Unit symbol or abbreviation, where differing from basic form	Unit expressed in terms of basic or supplementary units*
Area	square meter		m^2
Volume	cubic meter		m^3
Frequency	hertz, cycle per second†	Hz	s^{-1}
Density	kilogram per cubic meter		kg/m^3
Velocity	meter per second		m/s
Angular velocity	radian per second		rad/s
Acceleration	meter per second squared		m/s^2
Angular acceleration	radian per second squared		rad/s^2
Volumetric flow rate	cubic meter per second		m^3/s
Force	newton	N	$kg \cdot m/s^2$
Surface tension	newton per meter, joule per square meter	N/m, J/m^2	kg/s^2
Pressure	newton per square meter, pascal†	N/m^2, Pa†	$kg/m \cdot s^2$
Viscosity, dynamic	newton-second per square meter, poiseuille†	$N\ s/m^2$, Pl†	$kg/m \cdot s$
Viscosity, kinematic	meter squared per second		m^2/s
Work, torque, energy, quantity of heat	joule, newton-meter, watt-second	J, $N \cdot m$, $W \cdot s$	$kg \cdot m^2/s^2$
Power, heat flux	watt, joule per second	W, J/s	$kg \cdot m^2/s^3$
Heat flux density	watt per square meter	W/m^2	kg/s^3
Volumetric heat release rate	watt per cubic meter	W/m^3	$kg/m \cdot s^3$
Heat transfer coefficient	watt per square meter degree	$W/m^2 \cdot deg$	$kg/s^3 \cdot deg$
Heat capacity (specific)	joule per kilogram degree	$J/kg \cdot deg$	$m^2/s^2 \cdot deg$
Capacity rate	watt per degree	W/deg	$kg \cdot m^2/s^3 \cdot deg$
Thermal conductivity	watt per meter degree	$W/m \cdot deg$, $\dfrac{Jm}{s \cdot m^2 \cdot deg}$	$kg \cdot m/s^3 \cdot deg$
Quantity of electricity	coulomb	C	$A \cdot s$
Electromotive force	volt	V,W/A	$kg \cdot m^2/A \cdot s^3$
Electric field strength	volt per meter	V/m	
Energy density (volumetric)	joule per cubic meter, watthours per liter	J/m^3, Wh/L	$kg/m \cdot s^2$
Specific energy or	joule per kilogram	J/kg	m^2/s^2
Energy density (gravimetric)	watthours per kilogram	Wh/kg	
Coulombic energy density (volumetric)	ampere-hours per liter	Ah/L	
(gravimetric)	ampere-hours per kilogram	Ah/kg	
Electric resistance	ohm	Ω, V/A	$kg \cdot m^2/A^2 \cdot s^3$
Electric conductivity	ampere per volt meter	$A/V \cdot m$	$A^2 s^3/kg \cdot m^3$
Electric capacitance	farad	F, A·s/V	$A^2 s^4/kg \cdot m^2$
Magnetic flux	weber	Wb, V·s	$kg \cdot m^2/A \cdot s^2$
Inductance	henry	H, V·s/A	$kg \cdot m^2/A^2 s^2$
Magnetic permeability	henry per meter	H/m	$kg \cdot m/A^2 s^2$
Magnetic flux density	tesla, weber per square meter	T, Wb/m^2	$kg/A \cdot s^2$
Magnetic field strength	ampere per meter		A/m
Magnetomotive force	ampere		A
Luminous flux	lumen	lm	cd sr
Luminance	candela per square meter		cd/m^2
Illumination	lux, lumen per square meter	lx, lm/m^2	$cd \cdot sr/m^2$

*Supplementary units are plane angle, radian (rad); solid angle, steradian (sr).
†Not used in all countries.
SOURCE: McGraw-Hill Metrication Manual.

TABLE D.4 Greek Alphabet

Greek letter	Greek name	English equivalent	Greek letter	Greek name	English equivalent
A α	Alpha	a	N ν	Nu	n
B β	Beta	b	Ξ ξ	Xi	x
Γ γ	Gamma	g	O o	Omicron	ŏ
Δ δ	Delta	d	Π π	Pi	p
E ϵ	Epsilon	ĕ	P ρ	Rho	r
Z ζ	Zeta	z	Σ σ	Sigma	s
H η	Eta	ē	T τ	Tau	t
Θ θ	Theta	th	Y υ	Upsilon	u
I ι	Iota	i	Φ ϕ	Phi	ph
K κ	Kappa	k	X χ	Chi	ch
Λ λ	Lambda	l	Ψ ψ	Psi	ps
M μ	Mu	m	Ω ω	Omega	ō

SOURCE: From Perry and Chilton.

TABLE D.5 Standard Symbols for Units

Unit	Symbol	Notes
Ampere	A	SI unit of electric current.
Ampere-hour	Ah	
Angstrom	Å	$1\text{Å} = 10^{-10}$ m.
Atmosphere, standard	atm	1 atm $= 101\ 325$ N/m^2 or Pa.
Atmosphere, technical	at	at $= \text{kg}_f/\text{cm}^2$.
Atomic mass unit (unified)	u	The (unified) atomic mass unit is defined as one-twelfth of the mass of an atom of the ^{12}C nuclide. Use of the old atomic mass unit (amu), defined by reference to oxygen, is depreciated.
Atto	a	SI prefix for 10^{-18}.
Bar	bar	1 bar $= 100\ 000$ N/m^2.
Barn	b	1 b $= 10^{-28}$ m^2.
Barrel	bbl	1 bbl $= 9702$ in^3 $= 0.15899$ m^3. This is the standard barrel used for petroleum, etc. A different standard barrel is used for fruits, vegetables, and dry commodities.
British thermal unit	Btu	
Calorie (International Table calorie)	cal$_{\text{IT}}$	1 cal$_{\text{IT}}$ $= 4.1868$ J. The 9th Conférence Générale des Poids et Mesures adopted the joule as the unit of heat. Use of the joule is preferred.
Calorie (thermochemical calorie)	cal	1 cal $= 4.1840$ J (see note for International Table calorie).
Centi	c	SI prefix for 10^{-2}.
Centimeter	cm	
Coulomb	C	SI unit of electric charge.
Cubic centimeter	cm^3	
Cycle	c	
Cycle per second	Hz, c/s	See hertz. The name "hertz" is internationally accepted for this unit; the symbol Hz is preferred to c/s.
Day	d	
Deci	d	SI prefix for 10^{-1}.
Decibel	dB	

Unit	Symbol	Notes
Degree (temperature):		
Degree Celsius	°C	Note that there is no space between the symbol
Degree Fahrenheit	°F	° and the letter. The use of the word *centigrade* for the Celsius temperature scale was abandoned by the Conférence Générale des Poids et Mesures in 1948.
Degree Kelvin		See kelvin.
Degree Rankine	°R	
Deka	da	SI prefix for 10.
Dyne	dyn	
Electronvolt	eV	
Erg	erg	
Farad	F	SI unit of capacitance.
Femto	f	SI prefix for 10^{-15}.
Gauss	G	The gauss is the electromagnetic cgs unit of magnetic flux density. Use of SI unit, the tesla, is preferred.
Giga	G	SI prefix for 10^9.
Gilbert	Gb	The gilbert is the electromagnetic cgs unit of magnetomotive force. Use of the SI unit, the ampere (or ampere turn), is preferred.
Gram	g	
Gram per cubic centimeter	g/cm³	
Hecto	h	SI prefix for 10^2.
Henry	H	SI unit of inductance.
Hertz	Hz	SI unit of frequency.
Hour	h	
Joule	J	SI unit of energy.
Joule per kelvin	J/K	SI unit of heat capacity and entropy.
Kelvin	K	In 1967 the CGPM gave the name "kelvin" to the SI unit of temperature which had formerly been called "degree Kelvin" and assigned it the symbol K (without the symbol °).
Kilo	k	SI prefix for 10^3.
Kilogram	kg	SI unit of mass.
Kilogram-force	kg$_f$	In some countries the name *kilopond* (kp) has been adopted for this unit.
Kilohm	kΩ	
Kilometer	km	
Kilometer per hour	km/h	
Kilovolt	kV	
Kilowatt	kW	
Kilowatthour	kWh	
Liter	L	$1 L = 10^{-3}$ m³.
Liter per second	L/s	
Lumen	lm	SI unit of luminous flux.
		SI unit of illuminance.
Maxwell	Mx	The maxwell is the electromagnetic cgs unit of magnetic flux. Use of the SI unit, the weber, is preferred.
Mega	M	SI prefix for 10^6.
Megohm	MΩ	
Meter	m	SI unit of length.

Unit	Symbol	Notes
Mho	mho	CGPM has adopted the name "siemens" (S) for this unit.
Micro	μ	SI prefix for 10^{-6}.
Microampere	μA	
Microgram	μg	
Micrometer	μm	
Micron	μm	See micrometer. The name "micron" was abrogated by the Conférence Générale des Poids et Mesures, 1967.
Microsecond	μs	
Microwatt	μW	
Milli	m	SI prefix for 10^{-3}.
Milliampere	mA	
Milligram	mg	
Milliliter	ml	
Millimeter	mm	
Conventional millimeter of mercury	mmHg	1 mmHg $= 133.322$ N/m^2.
Millimicron	nm	Use of the name "millimicron" for the nanometer is deprecated.
Millisecond	ms	
Millivolt	mV	
Milliwatt	mW	
Minute (time)	min	Time may also be designated by means of superscripts as in the following example: $9^h 46^m 30^s$.
Mole	mol	SI unit of amount of substance.
Nano	n	SI prefix for 10^{-9}.
Nanoampere	nA	
Nanometer	nm	
Nanosecond	ns	
Newton	N	SI unit of force.
Newton meter	N·m	
Newton per square meter	N/m^2	SI unit of pressure or stress; see pascal.
Newton second per square meter	N·s/m^2	SI unit of dynamic viscosity.
Oersted	Oe	The oersted is the electromagnetic cgs unit of magnetic field strength. Use of the SI unit, the ampere per meter, is preferred.
Ohm	Ω	SI unit of resistance.
Pascal	Pa	Pa $=$ N/m^2. SI unit of pressure or stress. This name accepted by the 14th Conférence Générale des Poids et Mesures.
Pico	p	SI prefix for 10^{-12}.
Picowatt	pW	
Revolution per second	r/s	
Second (time)	s	SI unit of time.
Siemens	S	$S = \Omega^{-1}$. SI unit of conductance. This name and symbol were adopted by the 14th Conférence Générale des Poids et Mesures. The name "mho" is also used for this unit in the United States.

TABLE D.5 Standard Symbols for Units *(Continued)*

Unit	Symbol	Notes
Square meter	m^2	
Tera	T	SI prefix for 10^{12}.
Tonne	t	1 t = 1000 kg.
(Unified) atomic mass unit	u	The (unified) atomic mass unit is defined as one-twelfth of the mass of an atom of the ^{12}C nuclide. Use of the old atomic mass unit (amu), defined by reference to oxygen, is deprecated.
Volt	V	SI unit of voltage.
Volt per meter	V/m	SI unit of electric field strength.
Voltampere	VA	IEC name and symbol for the SI unit of apparent power.
Watt	W	SI unit of power.
Watt per meter kelvin	W/(m·K)	SI unit of thermal conductivity.
Watthour	Wh	

SOURCE: From Fink and Beaty.

BIBLIOGRAPHY

Engineering Design Handbook—Metric Conversion Guide, Publication DARCOM P 706-470, Headquarters, U.S. Army Material Development and Readiness Command, Washington, July 1976.

D.G. Fink and W. Beaty (eds.): *Standard Handbook for Electrical Engineers,* 11th ed., McGraw-Hill, New York, 1978.

McGraw-Hill Metrication Manual, McGraw-Hill, New York, 1972.

R. H. Perry and C. H. Chilton (eds.): *Chemical Engineers' Handbook,* 5th Ed., McGraw-Hill, New York, 1973.

E

Conversion
Factors

TABLE E.1 Length Conversion Factors*

A. Length units decimally related to one meter

	Meters (m)	Kilometers (km)	Decimeters (dm)	Centimeters (cm)	Millimeters (mm)	Micrometers (μm)	Nanometers (nm)	Angströms (Å)
1 meter =	**1**	**0.001**	**10**	**100**	**1 000**	**1 000 000**	**10^9**	**10^{10}**
1 kilometer =	**1 000**	**1**	**10 000**	**100 000**	**1 000 000**	**10^9**	**10^{12}**	**10^{13}**
1 decimeter =	**0.1**	**0.000 1**	**1**	**10**	**100**	**100 000**	**10^8**	**10^9**
1 centimeter =	**0.01**	**0.000 01**	**0.1**	**1**	**10**	**10 000**	**10^7**	**10^8**
1 millimeter =	**0.001**	**10^{-6}**	**0.01**	**0.1**	**1**	**1 000**	**1 000 000**	**10^7**
1 micrometer (micron) =	**10^{-6}**	**10^{-9}**	**0.000 01**	**0.000 1**	**0.001**	**1**	**1 000**	**10 000**
1 nanometer =	**10^{-9}**	**10^{-12}**	**10^{-8}**	**10^{-7}**	**10^{-6}**	**0.001**	**1**	**10**
1 angström =	**10^{-10}**	**10^{-13}**	**10^{-9}**	**10^{-8}**	**10^{-7}**	**0.000 1**	**0.1**	**1**

B. Nonmetric length units less than one meter

	Meters (m)	Yards (yd)	Feet (ft)	Inches (in)	Mils (mil)	Microinches (μin)
1 meter =	**1**	1.093 613 30	3.280 839 89	39.370 078 7	3.937 007 87 × 10^4	3.937 007 87 × 10^7
1 yard =	**0.914 4**	**1**	**3**	**36**	**36 000**	**3.6 × 10^7**
1 foot =	**0.304 8**	**1/3** = 0.333 $\overline{3}$	**1**	**12**	**12 000**	**1.2 × 10^7**
1 inch =	**0.025 4**	**1/36** = 0.027 $\overline{7}$	**1/12** = 0.083 $\overline{3}$	**1**	**1 000**	**1 000 000**
1 mil =	**2.54 × 10^{-5}**	2.777 × 10^{-5}	8.333 × 10^{-5}	**0.001**	**1**	**1 000**
1 microinch =	**2.54 × 10^{-8}**	2.777 × 10^{-8}	8.333 × 10^{-8}	**10^{-6}**	**0.001**	**1**

C. Nonmetric length units greater than one meter (with equivalents in feet)

	Meters (m)	Rods (rd)	Statute miles (mi)	Nautical miles (nmi)	Astronomical units (AU)	Parsecs (pc)	Feet (ft)
1 meter =	**1**	0.198 838 78	6.213 711 92 × 10^{-4}	5.399 568 04 × 10^{-4}	6.684 491 98 × 10^{-12}	3.240 733 17 × 10^{-17}	3.280 839 89
1 rod =	**5.029 2**	**1**	**0.003 125**	2.715 550 76 × 10^{-3}	3.361 764 71 × 10^{-11}	1.629 829 53 × 10^{-16}	**16.5**
1 statute mile =	**1 609.344**	**320**	**1**	0.868 976 24	1.075 764 71 × 10^{-8}	5.215 454 50 × 10^{-14}	**5 280**
1 nautical mile =	**1 852**	368.249 423	1.150 779 45	**1**	1.237 967 91 × 10^{-8}	6.001 837 80 × 10^{-14}	6 076.115 48
1 astronomical unit† =	1.496 × 10^{11}	2.974 628 17 × 10^{10}	92 957 130.3	80 777 537.8	**1**	4.848 136 82 × 10^{-6}	4.908 136 48 × 10^{11}
1 parsec =	3.085 721 50 × 10^{16}	6.135 611 02 × 10^{15}	1.917 378 44 × 10^{13}	1.666 156 32 × 10^{13}	206 264.806	**1**	1.012 375 82 × 10^{17}
1 foot =	**0.304 8**	0.060 60$\overline{6}$	1.893 93$\overline{9}$ × 10^{-4}	1.645 788 33 × 10^{-4}	2.037 433 16 × 10^{-12}	9.877 754 72 × 10^{-18}	**1**

D. Other length units

1 cable = **720 feet** = **219.456 meters**
1 cable (U.K.) = **608 feet** = **185.318 4 meters**
1 chain (engineers') = **100 feet** = **30.48 meters**
1 chain (surveyors') = **66 feet** = **20.116 8 meters**
1 fathom = **6 feet** = **1.828 8 meters**
1 fermi = **1 femtometer** = **10^{-15} meter**
1 foot (U.S. Survey) = **0.304 800 6 meter**
1 furlong = **660 feet** = **201.168 meters**

1 hand = **4 inches** = **0.101 6 meter**
1 league (international nautical) = **3 nautical miles** = **5 556 meters**
1 league (statute) = **3 statute miles** = **4 828.032 meters**
1 league (U.K. nautical) = **5 559.552 meters**
1 light-year = 9.460 895 2 × 10^{15} meters (= distance traveled by light in vacuum in one sidereal year)
1 link (engineers') = **1 foot** = **0.304 8 meter**
1 link (surveyors') = **7.92 inches** = **0.201 168 meter**
1 micron = **1 micrometer** = **10^{-6} meter**

1 millimicron = **1 nanometer** = **10^{-9} meter**
1 myriameter = **10 000 meters**
1 nautical mile (U.K.) = **1 853.184 meters**
1 pale = **1 rod** = **5.029 2 meters**
1 perch (linear) = **1 rod** = **5.029 2 meters**
1 pica = 1/6 inch (approx.) = 4.217 518 × 10^{-3} meter
1 point = 1/72 inch (approx.) = 3.514 598 × 10^{-4} meter
1 span = **9 inches** = **0.228 6 meter**

*Exact conversions are shown in boldface type. Repeating decimals are underlined. The SI unit of length is the meter.
†As defined by the International Astronomical Union, 1964.
SOURCE: D. G. Fink and W. Beaty (eds.), *Standard Handbook for Electrical Engineers*, 11th ed., McGraw-Hill, New York, 1978.

TABLE E.2 Area Conversion Factors*

A. Area units decimally related to one square meter

	Square meters (m)²	Square kilometers (km)²	Hectares (square hectometers) (hn)	Square centimeters (cm)²	Square millimeters (mm)²	Square micrometers (μm)²	Barns (b)	Circular mils (cmil)
1 square meter =	1	10^{-6}	0.000 1	10 000	1 000 000	10^{12}	10^{28}	1.973 525 24 × 10^{9}
1 square kilometer =	1 000 000	1	100	10^{10}	10^{12}	10^{18}	10^{34}	5.111 406 91 × 10^{15}
1 hectare =	10 000	0.01	1	10^{8}	10^{10}	10^{16}	10^{32}	7.986 573 30 × 10^{12}
1 square centimeter =	0.000 1	10^{-10}	10^{-8}	1	100	10^{8}	10^{24}	4.991 608 31 × 10^{10}
1 square millimeter =	10^{-6}	10^{-12}	10^{-10}	0.01	1	10^{6}	10^{22}	1.650 118 45 × 10^{9}
1 square micrometer =	10^{-12}	10^{-18}	10^{-16}	10^{-8}	10^{-6}	1	10^{16}	1.833 464 95 × 10^{8}
1 barn =	10^{-28}	10^{-34}	10^{-32}	10^{-24}	10^{-22}	10^{-16}	1	1.273 239 55 × 10^{6}

B. Nonmetric area units (with square meter equivalents)

	Square meters (m)²	Square statute miles (mi)²	Acres (acre)	Square rods (rd)²	Square yards (yd)²	Square feet (ft)²	Square inches (in)²	Circular mils (cmil)
1 square meter =	1	3.861 021 59 × 10^{-7}	2.471 053 82 × 10^{-4}	3.953 686 10 × 10^{-2}	1.195 990 05	10.763 910 4	1 550.003 10	1.973 525 24 × 10^{9}
1 square statute mile =	2 589 988.1	1	640	102 400	3 097 600	27 878 400	4.014 489 60 × 10^{9}	5.111 406 91 × 10^{15}
1 acre =	4 046.856 41	1/640 = **0.001 562 5**	1	160	4 840	43 560	6 272 640	7.986 573 30 × 10^{12}
1 square rod =	25.292 852 6	9.765 625 × 10^{-7}	1/160 = **0.006 25**	1	30.25	272.25	39 204	4.991 608 31 × 10^{10}
1 square yard =	0.836 127 36	3.228 305 79 × 10^{-7}	2.066 115 70 × 10^{-4}	3.305 785 12 × 10^{-2}	1	9	1 296	1.650 118 45 × 10^{9}
1 square foot =	0.092 903 04	3.587 006 43 × 10^{-8}	2.295 684 11 × 10^{-5}	3.673 094 58 × 10^{-3}	1/9 = 0.111 $\underline{1}$1	1	144	1.833 464 95 × 10^{8}
1 square inch =	6.451 6 × 10^{-4}	2.490 976 69 × 10^{-10}	1.594 225 08 × 10^{-7}	2.550 760 13 × 10^{-5}	7.716 049 38 × 10^{-4}	1/144 = 0.006 944 $\underline{44}$	1	1.273 239 55 × 10^{6}
1 circular mil =	5.067 074 79 × 10^{-10}	1.956 408 51 × 10^{-16}	1.252 101 45 × 10^{-13}	2.003 362 32 × 10^{-11}	6.060 171 01 × 10^{-10}	5.454 153 91 × 10^{-9}	7.853 981 63 × 10^{-7}	1

Exact conversions are:
1 acre = **4 046.856 422 4** square meters
1 square mile = **2 589 988.110 336** square meters

C. Other area units

1 are = 100 square meters
1 centiare (centare) = 1 square meter
1 perch (area) = 1 square rod = 30.25 square yards = 25.292 852 6 square meters
1 rood = 40 square rods = 1 011.714 11 square meters
1 section = 1 square statute mile = 2 589 988.1 square meters
1 township = 36 square statute miles = 93 239 572 square meters

*Exact conversions are shown in boldface type. Repeating decimals are underlined. The SI unit of area is the square meter.
SOURCE: D. G. Fink and W. Beaty (eds.), *Standard Handbook for Electrical Engineers,* 11th ed., McGraw-Hill, New York, 1978.

TABLE E.3 Force Conversion Factors*

	Newtons (N)	Kips (kip)	Slugs-force (slug)	Kilograms-force (kg)	Avoirdupois pounds-force (lbf avdp)	Avoirdupois ounces-force (ozf avdp)	Poundals (pdl)	Dynes (dyn)
1 newton =	**1**	$2.248\ 089\ 43 \times 10^{-4}$	$6.987\ 275\ 24 \times 10^{-3}$	$0.101\ 971\ 62$	$0.224\ 808\ 94$	$3.596\ 943\ 09$	$7.233\ 014\ 2$	**100 000**
1 kip =	$444\ 8.221\ 62$	**1**	$31.080\ 949$	$453.592\ 370$	**1 000**	**16 000**	$32\ 174.05$	$444\ 822\ 162$
1 slug-force =	$143.117\ 305$	$0.032\ 174\ 05$	**1**	$14.593\ 903$	$32.174\ 05$	$514.784\ 80$	$1\ 035.169\ 5$	$14\ 311\ 730$
1 kilogram-force =	**9.806 650**	$2.204\ 622\ 62 \times 10^{-3}$	$6.852\ 176\ 3 \times 10^{-2}$	**1**	$2.204\ 622\ 62$	$35.273\ 961\ 9$	$70.931\ 638\ 4$	**980 665**
1 avdp pound-force =	$4.448\ 221\ 62$	**0.001**	$3.108\ 094\ 88 \times 10^{-2}$	$0.453\ 592\ 37$	**1**	**16**	$32.174\ 05$	$444\ 822.162$
1 avdp ounce-force =	$0.278\ 013\ 85$	$1/16\ 000 =$ **0.000 062 5**	$1.942\ 559\ 30 \times 10^{-3}$	$2.834\ 952\ 3 \times 10^{-2}$	$1/16 =$ **0.062 5**	**1**	$2.010\ 878\ 03$	$27\ 801.385$
1 poundal =	$0.138\ 254\ 95$	$3.108\ 094\ 9 \times 10^{-5}$	$9.660\ 253\ 9 \times 10^{-4}$	$0.140\ 980\ 81$	$0.031\ 080\ 95$	$0.497\ 295\ 18$	**1**	$13\ 825.495$
1 dyne =	**0.000 01**	$2.248\ 089\ 43 \times 10^{-9}$	$6.987\ 275\ 24 \times 10^{-8}$	$1.019\ 716\ 21 \times 10^{-6}$	$2.248\ 089\ 43 \times 10^{-6}$	$3.596\ 943\ 10 \times 10^{-5}$	$7.233\ 014\ 2 \times 10^{-5}$	**1**

The exact conversion is: 1 avdp pound-force = **4.448 221 615 260 5** newtons

*Exact conversions are shown in boldface type. The SI unit of force is the newton (N).
SOURCE: D. G. Fink and W. Beaty (eds.), *Standard Handbook for Electrical Engineers*, 11th ed., McGraw-Hill, New York, 1978.

TABLE E.4 Volume and Capacity Conversion Factors

A. Volume units decimally related to one cubic meter

	Cubic meters (steres) $(m)^3$	Cubic decimeters $(dm)^3$	Cubic centimeters $(cm)^3$	Liters (L)	Centiliters (cL)	Milliliters (mL)	Microliters (μL)
1 cubic meter =	1	1 000	1 000 000	1 000	100 000	1 000 000	10^9
1 cubic decimeter =	0.001	1	1 000	1	100	1 000	1 000 000
1 cubic centimeter =	0.000 001	0.001	1	0.001	0.1	1	1 000
1 liter =	0.001	1	1 000	1	100	1 000	1 000 000
1 centiliter =	0.000 01	0.01	10	0.01	1	10	10 000
1 milliliter =	0.000 001	0.001	1	0.001	0.1	1	1 000
1 microliter =	10^{-9}	0.000 001	0.001	0.000 001	0.000 1	0.001	1

B. Nonmetric volume units (with cubic meter and liter equivalents)

	Cubic meters (steres) $(m)^3$	Liters (L)	Cubic inches $(in)^3$	Cubic feet $(ft)^3$	Cubic yards $(yd)^3$	Barrels (U.S.) (bbl)	Acre-feet (acre-ft)	Cubic miles $(mi)^3$
1 cubic meter =	1	1 000	$6.102\ 374\ 41 \times 10^4$	35.314 666	1.307 950 62	6.289 810 97	$8.107\ 131\ 94 \times 10^{-4}$	$2.399\ 127\ 59 \times 10^{-10}$
1 liter =	0.001	1	61.023 744 1	0.035 314 66	$1.307\ 950\ 62 \times 10^{-3}$	$6.289\ 810\ 97 \times 10^{-3}$	$8.107\ 131\ 93 \times 10^{-7}$	$2.399\ 127\ 59 \times 10^{-13}$
1 cubic inch =	$1.638\ 706\ 4 \times 10^{-5}$	$1.638\ 706\ 4 \times 10^{-2}$	1	$1/1\ 728 = 5.787\ 037\ 03 \times 10^{-4}$	$1/46\ 656 = 2.143\ 347\ 05 \times 10^{-5}$	$1.030\ 715\ 32 \times 10^{-4}$	$1.328\ 520\ 90 \times 10^{-8}$	$3.931\ 465\ 73 \times 10^{-15}$
1 cubic foot =	$2.831\ 684\ 66 \times 10^{-2}$	28.316 846 6	1 728	1	$1/27 = 0.037\ 037$	0.178 107 61	$2.295\ 684\ 11 \times 10^{-5}$	$6.793\ 572\ 78 \times 10^{-12}$
1 cubic yard =	0.764 554 86	764.554 858	46 656	27	1	4.808 905 38	$6.198\ 347\ 11 \times 10^{-4}$	$1.834\ 264\ 65 \times 10^{-10}$
1 barrel (U.S.) =	0.158 987 29	158.987 294	9 702	5.614 583 33	0.207 947 53	1	$1.288\ 930\ 98 \times 10^{-4}$	$3.814\ 308\ 05 \times 10^{-11}$
1 acre-foot =	1 233.481 84	1 233 481.84	$7.527\ 168\ 00 \times 10^7$	43 560	1 613.333 33	7 758.367 34	1	$2.959\ 280\ 30 \times 10^{-7}$
1 cubic mile =	$4.168\ 181\ 83 \times 10^9$	$4.168\ 181\ 83 \times 10^{12}$	$2.543\ 580\ 61 \times 10^{14}$	$1.471\ 979\ 52 \times 10^{11}$	$5.451\ 776 \times 10^9$	$26.217\ 074\ 9 \times 10^9$	3 379 200	1

Exact conversion: 1 cubic foot = **28.316 846 592 liters**

C. United States liquid capacity measures (with liter equivalents)

	Liters (L)	Gallons (U.S. gal)	Quarts (U.S. qt)	Pints (U.S. pt)	Gills (U.S. gi)	Fluid ounces (U.S. floz)	Fluidrams (U.S. fldr)	Minims (U.S. minim)
1 liter =	1	0.264 172 05	1.056 688	2.113 376	8.453 506	33.814 023	270.512 18	16 230.73
1 gallon, U.S. =	3.785 411 8	1	4	8	32	128	1 024	61 440
1 quart, U.S. =	0.946 352 9	$1/4 = 0.25$	1	2	8	32	256	15 360
1 pint, U.S. =	0.473 176 5	$1/8 = 0.125$	$1/2 = 0.5$	1	4	16	128	7 680
1 gill, U.S. =	0.118 294 1	$1/32 = 0.031\ 25$	$1/8 = 0.125$	$1/4 = 0.25$	1	4	32	1 920
1 fluid ounce, U.S. =	$2.957\ 353 \times 10^{-2}$	$1/128 = 0.007\ 812\ 5$	$1/32 = 0.031\ 25$	$1/16 = 0.062\ 5$	$1/4 = 0.25$	1	8	480
1 fluidram, U.S. =	$3.696\ 691\ 2 \times 10^{-3}$	$1/102\ 4 = 9.765\ 625 \times 10^{-4}$	$1/256 = 3.906\ 25 \times 10^{-3}$	$1/128 = 0.007\ 812\ 5$	$1/32 = 0.031\ 25$	$1/8 = 0.125$	1	60
1 minim, U.S. =	$6.161\ 152 \times 10^{-5}$	$1/61\ 440 = 1.627\ 604\ 16 \times 10^{-5}$	$1/15\ 360 = 6.510\ 416\ 66 \times 10^{-5}$	$1/7680 = 1.302\ 083\ 33 \times 10^{-4}$	$1/1\ 920 = 5.208\ 333\ 3 \times 10^{-4}$	$1/480 = 2.083\ 333\ 3 \times 10^{-3}$	$1/60 = 0.016\ 666\ 6$	1

Exact conversion: 1 liquid quart, U.S. = **0.946 352 946 liters**

D. British Imperial liquid capacity measures (with liter equivalents)

	Liters (L)	Gallons (U.K. gal)	Quarts (U.K. qt)	Pints (U.K. pt)	Gills (U.K. gi)	Fluid ounces (U.K. floz)	Fluidrams (U.K. fldr)	Minims (U.K. minim)
1 liter =	1	0.219 969 2	0.879 876 6	1.759 753	7.039 018	35.195 06	281.560 5	16 893.63
1 gallon, U.K. =	4.546 092	1	4	8	32	160	1 280	76 800
1 quart, U.K. =	1.136 523	1/4 = 0.25	1	2	8	40	320	19 200
1 pint, U.K. =	0.568 261 5	1/8 = 0.125	1/2 = 0.5	1	4	20	160	9 600
1 gill, U.K. =	0.142 065 4	1/32 = 0.031 25	1/8 = 0.125	1/4 = 0.25	1	5	40	2 400
1 fluid ounce, U.K. =	$2.841\ 307 \times 10^{-2}$	1/160 = 0.006 25	1/40 = 0.025	1/20 = 0.05	1/5 = 0.2	1	8	480
1 fluidram, U.K. =	$3.551\ 634 \times 10^{-3}$	$1/1280 = 7.812\ 5 \times 10^{-4}$	1/320 = 0.003 125	1/160 = 0.006 25	1/40 = 0.025	1/8 = 0.125	1	60
1 minim, U.K. =	$5.919\ 391 \times 10^{-5}$	$1/76\ 800 = 1.302\ 083\ 33 \times 10^{-5}$	$1/19\ 200 = 5.208\ 333\ 33 \times 10^{-5}$	$1/9\ 600 = 1.041\ 666\ 66 \times 10^{-4}$	$1/2\ 400 = 4.166\ 666\ 66 \times 10^{-4}$	$1/480 = 2.083\ 333\ 33 \times 10^{-3}$	1/60 = 0.016 666 66	1

E. United States and British dry capacity measures (with liter equivalents)

	Liters (L)	U.S. dry measures Bushels (U.S. bu)	Pecks (U.S. peck)	Quarts (U.S. qt)	Pints (U.S. pt)	British dry measures Pecks (U.K. peck)	Bushels (U.K. bu)	Quarts (U.K. qt)	Pints (U.K. pt)
1 liter =	1	0.028 377 59	0.113 510 37	0.908 082 99	1.816 165 98	0.109 984 6	0.027 496 1	0.879 876 6	1.759 753 4
1 bushel, U.S. =	35.239 070	1	4	32	64	3.875 754 9	0.968 938 7	31.006 04	62.012 08
1 peck, U.S. =	8.809 767 5	1/4 = 0.25	1	8	16	0.968 938 7	0.242 234 7	7.751 509	15.503 02
1 quart, U.S. =	1.101 220 9	1/32 = 0.031 25	1/8 = 0.125	1	2	0.121 117 3	0.030 279 34	0.968 938 7	1.937 878
1 pint, U.S. =	0.550 610 5	1/64 = 0.015 625	1/16 = 0.062 5	1/2 = 0.5	1	0.060 558 67	0.015 139 67	0.484 469 3	0.968 938 7
1 bushel, U.K. =	36.368 73	1.032 057	4.128 228	33.025 82	66.051 65	4	1	32	64
1 peck, U.K. =	9.092 182	0.258 014 3	1.032 057	8.256 456	16.512 91	1	1/4 = 0.25	8	16
1 quart, U.K. =	1.136 523	0.032 251 78	0.129 007 1	1.032 057	2.064 114 2	1/8 = 0.125	1/32 = 0.031 25	1	2
1 pint, U.K. =	0.568 261 4	0.016 125 89	0.064 503 6	0.516 028 4	1.032 057	1/16 = 0.062 5	1/64 = 0.015 625	1/2 = 0.5	1

Exact conversion: 1 dry pint, U.S. = 33.600 312 5 cubic inches

F. Other volume and capacity units

1 gallon (Canadian, liquid) = $4.546\ 090 \times 10^{-3}$ cubic meter
1 perch (volume) = 24.75 cubic feet = 0.700 842 cubic meter
1 stere = 1 cubic meter
1 tablespoon = 0.5 fluid ounce, U.S. = $1.478\ 677 \times 10^{-5}$ cubic meter
1 teaspoon = 1/6 fluid ounce, U.S. = $4.928\ 922 \times 10^{-6}$ cubic meter
1 ton (register ton) = 100 cubic feet = 2.831 684 66 cubic meters

1 barrel, U.S. (used for petroleum, etc.) = **42** gallons = 0.158 987 296 cubic meter
1 barrel ("old barrel") = **31.5** gallons = 0.119 240 cubic meter
1 board foot = **144** cubic inches = $2.359\ 737 \times 10^{-3}$ cubic meter
1 cord = **128** cubic feet = 3.624 556 cubic meters
1 cord foot = **16** cubic feet = 0.453 069 5 cubic meter
1 cup = **8** fluid ounces, U.S. = $2.365\ 882 \times 10^{-4}$ cubic meter

*Exact conversions are shown in boldface type. Repeating decimals are underlined. The SI unit of volume is the cubic meter.

SOURCE: D. G. Fink and W. Beaty (eds.), Standard Handbook for Electrical Engineers, 11th ed., McGraw-Hill, New York, 1978.

TABLE E.5 Mass Conversion Factors*

	Kilograms (kg)	Tonnes (metric tons) (t)	Grams (g)
1 kilogram =	1	0.001	1 000
1 tonne =	1 000	1	1 000 000
1 gram =	0.001	0.000 001	1
1 decigram =	0.000 1	10^{-7}	0.1
1 centigram =	0.000 01	10^{-8}	0.01
1 milligram =	0.000 001	10^{-9}	0.001
1 microgram =	10^{-9}	10^{-12}	0.000 001

B. Nonmetric mass units less than one

	Grams (g)	Avoirdupois ounces-mass (oz_m, avdp)	Troy ounces-mass (oz_m, troy)
1 gram =	1	0.035 273 962	0.032 150 747
1 avdp ounce-mass =	28.349 523 1	1	0.911 458 33
1 troy ounce-mass =	31.103 176 8	1.097 142 86	1
1 avdp dram =	1.771 845 20	1/16 = 0.062 5	0.056 966 15
1 apothecary dram =	3.887 934 58	0.137 142 857	⅛ = 0.125
1 pennyweight =	1.555 173 83	0.054 863 162	1/20 = 0.05
1 grain =	0.064 798 91	1/437.5 = 2.285 714 29 × 10^{-3}	1/480 = 0.002 083 33
1 scruple =	1.295 978 20	4.571 428 58 × 10^{-2}	1/24 = 0.041 666 66

C. Nonmetric mass units of one pound-mass

	Kilograms (kg)	Long tons (long ton)	Short tons (short ton)
1 kilogram =	1	9.842 065 28 × 10^{-4}	1.102 311 31 × 10^{-3}
1 long ton =	1 016.046 9	1	1.12
1 short ton =	907.184 74	200/224 = 0.892 857 14	1
1 long hundredweight =	50.802 345 4	0.05	0.056
1 short hundredweight =	45.359 237	10/224 = 0.044 642 86	0.05
1 slug =	14.593 903	0.014 363 41	0.016 087 02
1 avdp pound-mass =	0.453 592 37	4.464 285 71 × 10^{-4}	0.000 5
1 troy pound-mass =	0.373 241 72	3.673 469 37 × 10^{-4}	4.114 285 70 × 10^{-4}

Exact conversions: 1 long ton = 1 016.046 908 8 kilograms
1 troy pound-mass = 0.373 241 721 6 kilogram

D. Other

1 assay ton = 29.166 667 grams
1 carat (metric) = 200 milligrams
1 carat (troy weight) = 31/6 grains = 205.196 55 milligrams
1 mynagram = 10 kilograms
1 quintal = 100 kilograms
1 stone = 14 pounds, avdp = 6.350 293 18 kilograms

*Exact conversions are shown in boldface type. Repeating decimals are underlined. The SI unit of mass is the kilogram.

SOURCE: D. G. Fink and W. Beaty (eds.), *Standard Handbook for Electrical Engineers,* 11th ed., McGraw-Hill, New York, 1978.

related to one kilogram

Decigrams (dg)	Centigrams (cg)	Milligrams (mg)	Micrograms (μg)
10 000	100 000	1 000 000	10^9
10^7	10^8	10^9	10^{12}
10	100	1 000	1 000 000
1	10	100	100 000
0.1	1	10	10 000
0.01	0.1	1	1 000
0.000 01	0.000 1	0.001	1

pound-mass (with gram equivalents)

Avoirdupois drams (dr avdp)	Apothecary drams (dr apoth)	Pennyweights (dwt)	Grains (grain)	Scruples (scruple)
0.564 383 39	0.257 205 97	0.643 014 93	15.432 358 4	0.771 617 92
16	7.291 666 66	18.227 166 7	437.5	21.875
17.554 285 7	**8**	**20**	480	24
1	0.455 729 17	1.139 322 92	27.343 75	1.367 187 5
2.194 285 70	**1**	**2.5**	60	3
0.877 714 28	**1/2.5 = 0.4**	**1**	24	1.2
3.657 142 85 $\times$ **10^{-2}**	**1/60 = 0.016 666 66**	**1/24 = 0.041 666 66**	1	0.05
0.731 428 57	1/3 = 0.333 333 33	5/6 = 0.833 333 33	20	1

and greater (with kilogram equivalents)

Long hundredweights (long cwt)	Short hundredweights (short cwt)	Slugs (slug)	Avoirdupois pounds-mass (lb$_\text{m}$, avdp)	Troy pounds-mass (lb$_\text{m}$, troy)
1.968 411 31 $\times$ 10^{-2}	2.204 622 62 $\times$ 10^{-2}	0.068 521 77	2.204 622 62	2.679 228 89
20	**22.4**	69.621 329	**2 240**	2 722.222 22
4 000/224 = 17.857 142 9	**20**	62.161 901	**2 000**	2 430.555 55
1	**1.12**	3.481 066 4	**112**	136.111 111
100/112 = 0.892 857 14	**1**	3.108 095 0	**100**	121.527 777
0.287 268 3	0.321 740 5	**1**	32.174 05	39.100 406
1/112 = 8.928 571 43 $\times$ 10^{-3}	**0.01**	3.108 095 0 $\times$ 10^{-2}	**1**	1.215 277 777
7.346 938 79 $\times$ 10^{-3}	8.228 571 45 $\times$ 10^{-3}	0.025 575 18	0.822 857 14	**1**

mass units

E-10

TABLE E.6 Pressure/Stress Conversion Factors

A. Pressure units decimally related to one pascal

	Pascals (Pa)	Bars (bar)	Decibars (dbar)	Millibars (mbar)	Dynes per square centimeter (dyn/cm²)
1 pascal =	1	0.000 01	0.000 1	0.01	10
1 bar =	100 000	1	10	1 000	1 000 000
1 decibar =	10 000	0.1	1	100	100 000
1 millibar =	100	0.001	0.01	1	1 000
1 dyne per square centimeter =	0.1	0.000 001	0.000 01	0.001	1

B. Pressure units decimally related to one kilogram-force per square meter (with pascal equivalents)

	Kilograms-force per square meter (kg/m²)	Kilograms-force per square centimeter (kg/cm²)	Kilograms-force per square millimeter (kg/mm²)	Grams-force per square centimeter (g/cm²)	Pascals (Pa)
1 kilogram-force per square meter =	1	0.000 1	0.000 001	0.1	9.806 65
1 kilogram-force per square centimeter =	10 000	1	0.01	1 000	98 066.5
1 kilogram-force per square millimeter =	1 000 000	100	1	100 000	9 806 650
1 gram-force per square centimeter =	10	0.001	0.000 01	1	98.066 5
1 pascal =	0.101 971 62	$1.019\ 716\ 2 \times 10^{-5}$	$1.019\ 716\ 2 \times 10^{-7}$	$1.019\ 716\ 2 \times 10^{-2}$	1

NOTE: 1 atmosphere (technical) = 1 kilogram-force per square centimeter = 98 066.5 pascals.

C. Pressure units expressed as heights of liquid (with pascal equivalents)

	Millimeters of mercury at 0°C (mmHg, 0°C)	Centimeters of mercury at 60°C (cmHg, 60°C)	Inches of mercury at 32°F (inHg, 32°F)	Inches of mercury at 60°F (inHg, 60°F)	Centimeters of water at 4°C (cmH₂O, 4°C)	Inches of water at 60°F (inH₂O, 60°F)	Feet of water at 39.2°F (ftH₂O, 39.2°F)	Pascals (Pa)
1 millimeter of mercury, 0°C =	1	0.100 282	0.039 370 1	0.039 481 3	1.359 548	0.535 775 6	0.044 604 6	133.322 4

1 centimeter of mercury, 60°C =	9.971 830	1	0.392 591 9	0.393 700 8	13.557 18	5.342 664	0.444 789 5	1 329.468
1 inch of mercury, 32°F =	**25.4**	2.547 175	1	1.002 824 8	34.532 52	13.608 70	1.132 957	3 386.389
1 inch of mercury, 60°C =	25.328 45	**2.54**	0.997 183 1	1	34.435 25	13.570 37	1.129 765	3 376.85
1 centimeter of water, 4°C =	0.735 539	0.073 762	0.028 958	0.029 040 0	1	0.394 083 8	0.032 808 4	98.063 8
1 inch of water, 60°F =	1.866 453	0.187 173	0.073 482	0.073 690 0	2.537 531	1	0.083 252 4	248.840
1 foot of water, 39.2°F =	22.419 2	2.248 254	0.882 646	0.885 139	30.479 98	12.011 67	1	2 988.98
1 pascal =	$7.500\ 615 \times 10^{-3}$	$7.521\ 806 \times 10^{-4}$	$2.952\ 998 \times 10^{-4}$	$2.961\ 34 \times 10^{-4}$	$1.019\ 74 \times 10^{-2}$	$4.018\ 65 \times 10^{-3}$	$3.345\ 62 \times 10^{-4}$	1

NOTE: 1 millimeter of mercury at 0°C = 133.322 4 pascals.

D. Nonmetric pressure units (with pascal equivalents)

	Atmospheres (atm)	Avoirdupois pounds-force per square inch (psi)	Avoirdupois pounds-force per square foot (lb/ft², avdp)	Poundals per square foot (pdl/ft²)	Pascals (Pa)
1 atmosphere =	1	14.695 95	2 116.217	68 087.24	**101 325**
1 avdp pound-force per square inch =	$6.804\ 60 \times 10^{-2}$	1	**144**	4 633.063	6 894.757
1 avdp pound-force per square foot =	$4.725\ 414 \times 10^{-4}$	$1/144 = 0.006\ 944$	1	32.174 05	47.880 26
1 poundal per square foot =	$1.468\ 704 \times 10^{-3}$	$2.158\ 399 \times 10^{-4}$	0.031 080 9	1	1.488 164
1 pascal =	$9.869\ 233 \times 10^{-6}$	$1.450\ 377 \times 10^{-4}$	0.020 885 4	0.671 968 9	1

NOTE: 1 normal atmosphere = 760 torr = **101 325** pascals.

*Exact conversions are shown in boldface type. Repeating decimals are underlined. The SI unit of pressure or stress is the Pascal (Pa).
SOURCE: D. G. Fink and W. Beaty (eds.), *Standard Handbook for Electrical Engineers*, 11th ed., McGraw-Hill, New York, 1978.

TABLE E.7 Energy/Work Conversion Factors

A. Energy/work units decimally related to one joule

	Joules (J)	Megajoules (MJ)	Kilojoules (kJ)	Millijoules (mJ)	Microjoules (μJ)	Ergs (erg)
1 joule =	1	0.000 001	0.001	1 000	1 000 000	10^7
1 megajoule =	1 000 000	1	1 000	10^9	10^{12}	10^{13}
1 kilojoule =	1 000	0.001	1	1 000 000	10^9	10^{10}
1 millijoule =	0.001	10^{-9}	10^{-6}	1	1 000	10 000
1 microjoule =	0.000 001	10^{-12}	10^{-9}	0.001	1	10
1 erg =	10^{-7}	10^{-13}	10^{-10}	0.000 1	0.1	1

NOTE: 1 watt-second = 1 joule.

B. Energy/work units less than ten joules (with joule equivalents)

	Joules (J)	Foot-poundals (ft·pdl)	Foot-pounds-force (ft·lbf)	Calories (International Table) (cal, IT)	Calories (thermochemical) (cal, thermo)	Electronvolts (eV)
1 joule =	1	23.730 36	0.737 562 1	0.238 845 9	0.239 005 7	$6.241\ 46 \times 10^{18}$
1 foot-poundal =	$4.214\ 011 \times 10^{-2}$	1	$3.108\ 095 \times 10^{-2}$	$1.006\ 499 \times 10^{-2}$	$1.007\ 173 \times 10^{-2}$	$2.630\ 16 \times 10^{17}$
1 foot-pound-force =	1.355 818	32.174 05	1	0.323 831 6	0.324 048 3	$8.462\ 28 \times 10^{18}$
1 calorie (Int. Tab.) =	4.186 8	99.354 27	3.088 025	1	1.000 669	$2.613\ 17 \times 10^{19}$
1 calorie (thermo) =	4.184	99.287 83	3.085 960	0.999 331 2	1	$2.611\ 43 \times 10^{19}$
1 electronvolt =	$1.602\ 19 \times 10^{-19}$	$3.802\ 05 \times 10^{-18}$	$1.181\ 71 \times 10^{-19}$	$3.826\ 77 \times 10^{-20}$	$3.829\ 33 \times 10^{-20}$	1

C. Energy/work units greater than ten joules (with joule equivalents)

	Joules (J)	British thermal units, International Table (Btu, IT)	British thermal units, thermochemical (Btu, thermo)	Kilowatthours (kWh)	Horsepower-hours, electrical (hp·h, elec)	Kilocalories, International Table (kcal, IT)	Kilocalories, thermochemical (kcal, thermo)
1 joule =	1	$9.478\ 170 \times 10^{-4}$	$9.484\ 516\ 5 \times 10^{-4}$	$1/3.6 \times 10^6 = 2.777 \times 10^{-7}$	$3.723\ 562 \times 10^{-7}$	$2.388\ 459 \times 10^{-4}$	$2.390\ 057\ 4 \times 10^{-4}$
1 British thermal unit, Int. Tab. =	1 055.056	1	1.000 669	$2.930\ 711 \times 10^{-4}$	$3.928\ 567 \times 10^{-4}$	0.251 995 8	0.252 164 4
1 British thermal unit, thermochemical =	1 054.35	0.999 331	1	$2.928\ 745 \times 10^{-4}$	$3.925\ 938 \times 10^{-4}$	0.251 827 2	0.251 995 7
1 kilowatthour =	3 600 000	3 412.141	3 414.426	1	$1/0.746 = 1.340\ 482\ 6$	859.845 2	860.420 7
1 horsepower-hour, electrical =	2 685 600	2 545.457	2 547.162	0.746	1	641.444 5	641.873 8
1 kilocalorie, Int. Tab. =	4 186.8	3.968 320	3.970 977	0.001 163	$1.558\ 981 \times 10^{-3}$	1	1.000 669
1 kilocalorie, thermochemical =	4 184	3.965 666	3.968 322	0.001 162 2	$1.557\ 938\ 6 \times 10^{-3}$	0.999 331	1

The exact conversion is 1 British thermal unit, International Table = 1 055.055 852 62 joules.

*Exact conversions are shown in boldface type. Repeating decimals are underlined. The SI unit of energy and work is the joule (J).
SOURCE: D. G. Fink and W. Beaty (eds.), *Standard Handbook for Electrical Engineers*, 11th ed., McGraw-Hill, New York, 1978.

TABLE E.8 Power Conversion Factors

A. Power units decimally related to one watt

	Watts (W)	Megawatts (MW)	Kilowatts (kW)	Milliwatts (mW)	Microwatts (μW)	Picowatts (pW)	Ergs per second (ergs/s)
1 watt =	1	0.000 001	0.001	1 000	1 000 000	10^9	10^7
1 megawatt =	1 000 000	1	1 000	10^9	10^{12}	10^{15}	10^{13}
1 kilowatt =	1 000	0.001	1	1 000 000	10^9	10^{12}	10^{10}
1 milliwatt =	0.001	10^{-9}	0.000 001	1	1 000	1 000 000	10 000
1 microwatt =	0.000 001	10^{-12}	10^{-9}	0.001	1	1 000	10
1 picowatt =	10^{-9}	10^{-15}	10^{-12}	0.000 001	0.001	1	0.01
1 erg per second =	10^{-7}	10^{-13}	10^{-10}	0.000 1	0.1	100	1

NOTE: 1 watt = 1 joule per second (J/s).

B. Nonmetric power units (with watt equivalents)

	British thermal units (International Table) per hour (Btu/hr, IT)	British thermal units (thermochemical) per minute (Btu/min, thermo)	Avoirdupois foot-pounds-force per second (ft·lbf/s avdp)	Kilocalories per minute (thermochemical) (kcal/min, thermo)	Kilocalories per second (International Table) (kcal/s, IT)	Horsepower (electrical) (hp, elec)	Horsepower (mechanical) (hp, mech)	Watts (W)
1 British thermal unit (Int. Tab.)-per hour =	1	0.016 677 8	0.216 158 1	$4.202\ 740\ 5 \times 10^{-3}$	$6.999\ 883\ 1 \times 10^{-5}$	$3.928\ 567\ 0 \times 10^{-4}$	$3.930\ 148\ 0 \times 10^{-4}$	0.293 071 1
1 British thermal unit (thermo) per minute =	59.959 853	1	12.960 810	0.251 995 7	$4.197\ 119\ 5 \times 10^{-3}$	0.023 555 6	0.023 565 1	17.572 50
1 foot-pound-force per second =	4.626 242 6	0.077 155 7	1	0.019 442 9	$3.238\ 315\ 7 \times 10^{-4}$	$1.817\ 450\ 4 \times 10^{-3}$	$1.818\ 181\ 8 \times 10^{-3}$	1.355 818
1 kilocalorie per minute (thermo) =	237.939 98	3.968 321 7	51.432 665	1	0.016 655 5	0.093 476 3	0.093 513 9	69.733 333
1 kilocalorie per second (Int. Tab.) =	14 285.953	238.258 64	3 088.025 1	60.040 153	1	5.612 332 4	5.614 591 1	**4 186.800**
1 horsepower (electrical) =	2 545.457 4	42.452 696	550.221 34	10.697 898	0.178 179 0	1	1.000 402 4	**746**
1 horsepower (mechanical) =	2 544.433 4	42.435 618	**550**	10.693 593	0.178 107 4	0.999 597 7	1	745.699 9
1 watt =	3.412 141 3	0.056 907 1	0.737 562 1	0.014 340 3	$2.388\ 459\ 0 \times 10^{-4}$	$1.340\ 482\ 6 \times 10^{-3}$	$1.341\ 022\ 0 \times 10^{-3}$	1

NOTE: The horsepower (mechanical) is defined as a power equal to **550** foot-pounds-force per second.
Other units of horsepower are:
1 horsepower (boiler) = 9 809.50 watts
1 horsepower (metric) = 735.499 watts
1 horsepower (water) = 746.043 watts
1 horsepower (U.K.) = 745.70 watts
1 ton (refrigeration) = 3 516.8 watts

*Exact conversions are shown in **boldface** type. Repeating decimals are underlined. The SI unit of power is the watt (W).
SOURCE: D. G. Fink and W. Beaty (eds.), *Standard Handbook for Electrical Engineers*, 11th ed., McGraw-Hill, New York, 1978.

TABLE E.9   Temperature Conversions*

Celsius (°C) °C = 5(°F − 32)/9	Fahrenheit (°F) °F = [9(C°)/5] + 32	Absolute (K) K = °C + 273.15
−273.15	−459.67	0
−200	−328	73.15
−180	−292	93.15
−160	−256	113.15
−140	−220	133.15
−120	−184	153.15
−100	−148	173.15
−80	−112	193.15
−60	−76	213.15
−40	−40	233.15
−20	−4	253.15
−17.77	0	255.372
0	32	273.15
5	41	278.15
10	50	283.15
15	59	288.15
20	68	293.15
25	77	298.15
30	86	303.15
35	95	308.15
40	104	313.15
45	113	318.15
50	122	323.15
55	131	328.15
60	140	333.15
65	149	338.15
70	158	343.15
75	167	348.15
80	176	353.15
85	185	358.15
90	194	363.15
95	203	368.15
100	212	373.15
105	221	378.15
110	230	383.15
115	239	378.15
120	248	393.15
140	284	413.15
160	320	433.15
180	356	453.15
200	392	473.15
250	482	523.15
300	572	573.15
350	662	623.15
400	752	673.15
450	842	723.15
500	932	773.15
1 000	1 832	1 273.15
5 000	9 032	5 273.15
10 000	18 032	10 273.15

*Conversions in boldface type are exact. Continuing decimals are underlined. Temperature in kelvins equals temperature in degrees Rankine divided by 1.8 [K = °R/1.8].

SOURCE: D. G. Fink and W. Beaty (eds.), *Standard Handbook for Electrical Engineers,* 11th ed., McGraw-Hill, New York, 1978.

Bibliography

Austin, L. G.: *Fuel Cells—a Review of Government-Sponsored Research, 1950–1964*, NASA SP-120, National Aeronautics and Space Administration, Washington, 1967.

Barak, M. (ed.): *Electrochemical Power Sources—Primary & Secondary Batteries*, Peter Peregrinus, Stevenage, U.K., 1980.

Bauer, H. H.: *Electrodics—Modern Ideas Concerning Electrode Reactions*, Halsted Press, Stuttgart, 1972.

Bechtel National, Inc.: *Handbook for Battery Energy Storage in Photovoltaic Power Sources*, Report no. SAND80-7022, Sandia National Laboratories, Albuquerque, N.M., 1980.

Bockris, J. O'M., and A. K. N. Reddy: *Modern Electro-chemistry*, vols. I and II, Plenum, New York, 1970.

Bode, H.: *Lead-Acid Batteries* (translated from German by R. J. Brodd and K. V. Kordesch), Wiley, New York, 1977.

Conway, B. E.: *Theory and Principles of Electrode Processes*, Ronald, New York, 1965.

Eveready Battery Engineering Data, Union Carbide Corporation, Battery Products Division, New York, 1976.

Falk, S. U., and A. J. Salkind: *Alkaline Batteries*, Wiley, New York, 1969.

Fleischer, A., and J. J. Lander (eds.): *Zinc-Silver Oxide Batteries*, Wiley, New York, 1971.

Gabano, J. P.: *Lithium Batteries*, Academic Press Ltd., London, 1983.

Garrett, A. B.: *Batteries of Today*, Research Press, Dayton, Ohio, 1957.

The Gould Battery Handbook, Gould, Inc., Mendota Heights, Minn., 1973.

Graham, R. W.: "Secondary Batteries—Recent Advances," Noyes Data Corp., Park Ridge, N.J., 1978.

———: "Rechargeable Batteries—Advances Since 1977," Noyes Data Corp., Park Ridge, N.J., 1980.

Gross, S. (ed.): "Battery Design and Optimization," The Electrochemical Society Symposium Proceedings Series, The Electrochemical Society, Pennington, N.J., 1978.

Hehner, N. E.: *Storage Battery Manufacturing Manual*, Independent Battery Manufacturers' Association, Inc. (IBMA), Largo, Fla., 1970.

Heise, G. W., and N. C. Cahoon (eds.): *The Primary Battery*, vols. I and II, Wiley, New York, 1971 and 1976.

Jasinski, R.: *High Energy Batteries*, Dekker, New York, 1963.

Kordesch, K. V. (ed.): *Batteries*, vol. I: *Manganese Dioxide*, vol. II: *Lead-Acid Batteries and Electric Vehicles*, Dekker, New York, 1974 and 1977.

Liebhafsky, H. A., and E. J. Cairns: *Fuel Cells and Fuel Batteries*, Wiley, New York, 1968.

Mantell, C. L.: *Batteries and Energy Systems*, 2d ed., McGraw-Hill, New York, 1983.

Nickel-Cadmium Battery Application Engineering Handbook, General Electric Company, Gainesville, Fla., 1975.

Sawyer, D. T., and J. L. Roberts: *Experimental Electrochemistry for Chemists*, Wiley-Interscience, New York, 1974.

Vinal, G. W.: *Primary Batteries,* Wiley, New York, 1950.
————: *Storage Batteries,* Wiley, New York, 1955.

Periodicals

Journal of the Electrochemical Society, Pennington, N.J.
Journal of Power Sources, Elsevier Sequoia, S. A., Lausanne, Switzerland.
NASA Goddard SFC Battery Workshop, 1968 to 1982, Goddard Space Flight Center, Greenbelt, Md.
Proceedings of the International Power Sources Symposium, 1962 to 1982, Academic, London.
Proceedings of the Power Sources Conference, 1946 to 1982, The Electrochemical Society, Pennington, N.J.
Progress in Batteries and Solar Cells, 1978 to 1981, JEC Press, Cleveland.
U.S. Department of Energy Battery and Electrochemical Contractors' Conference, 1977 to 1982, U.S. Department of Energy, Washington.

Other Reference Sources—Bibliographies

Batteries, Fuel Cells, and Related Electrochemistry: Books, Journals, Periodicals, and Other Information Sources, 1950–1979, U.S. Department of Energy, Report no. DOE/CS-0156, Washington, 1980.
Batteries—A DTIC Bibliography, Defense Technical Information Center, Report DTIC/BIB-79-OS (AD-A077600), Alexandria, Va. 1979.
Battery Literature Search—Bibliography and Abstracts (5 vols.), prepared by the Boeing Co. for NASA, Jet Propulsion Laboratory on JPL Contract 953984, W.O. 343-20, Boeing Document D180-18849-2, Seattle, July, 1976.
Electric Batteries, a Bibliography, Energy Research and Development Administration, Report no. TID-3361, Washington, 1977.
Fuel Cells, a Bibliography, Energy Research and Development Administration, Report no. TID-3359, Washington, 1977.

Standards

"American National Standard Specification for Dry Cells and Batteries," ANSI C18.1-1979, American National Standards Institute, May 1979.
International Electrotechnical Commission, IEC Standard Publication IEC 86, *Primary Batteries;* 86-1, *General;* 86-2, *Specification Sheets,* Bureau Centrale de la Commission Electrotechnique Internationale, Geneva, 1977.
NEDA Battery Cross-Reference Guide, National Electronic Distributors Association, Park Ridge, IL 60068, 1982.
U.S. Federal and Military Specifications, Military Specification MIL-B-18, *Batteries, Dry.*

Major Battery Manufacturers

Manufacturer	Major battery products*
ACR ELECTRONICS, INC. 3901 N. 29th Avenue Hollywood, FL 33022 (315) 921-6262	PP
ALTUS CORP. 1610 Crane Court San Jose, CA 95112 (408) 295-1300	PL, R
BATTERY ENGINEERING CO., INC. 1636 Hyde Park Avenue Hyde Park, MA 02136 (617) 361-7555	PL
BEREC GROUP, LTD. Group Technical Center St. Anns Road Tottenham, London N153TJ United Kingdom 01-800-1101	PC, PP, SP, SS, SI
ROBERT BOSCH GmbH Abteilung K 1/EWL 2 Postfach 30 02 40 7000 Stuttgart 30 West Germany	SS
BRENTRONICS, INC. 10 Brayton Court Commack, NY 11725 (516) 499-5155	PC, PP, SP

Manufacturer	Major battery products*
BRIGHT STAR INDUSTRIES 600 Getty Avenue Clifton, NJ 07015 (201) 772-3200	PC, PP
BROWN, BOVERI & CIE Neustadter Street 61 D 6800 Mannheim Germany	SA
BURGESS INC. Freeport, IL 61032	PC, PP
C & D BATTERIES DIV. Allied Corp. 3043 Walton Road Plymouth Meeting, PA 19462 (215) 828-9000	SS, SI, SA
CATALYST RESEARCH CORP. 1421 Clarkview Road Baltimore, MD 21209 (301) 296-7000	PL, R
CHLORIDE GROUP, LTD. 52 Grosvenor Gardens London, SW1 0AV United Kingdom	SP, SS, SI, SA
COMPAGNIE GENERALE d'ELECTRICITE Marcoussis, France	SA
CROMPTON PARKINSON LTD. PO Box 72 Crompton House Aldwych WC2B 4JH London, England	PC, PP, PL
DELCO-REMY DIV. General Motors Corp. 2401 Columbus Avenue Anderson, IN 46011 (317) 646-7656	SS, SI, SA
DOW CHEMICAL CO. 2800 Mitchell Drive Walnut Creek, CA 94598 (415) 944-2000	SA
DURACELL INC. Berkshire Industrial Park Bethel, CT 06801 (203) 796-4000 and South Broadway Tarrytown, NY 10591 (914) 591-7000	PC, PP, PL, SA, R

Manufacturer	Major battery products*
EAGLE-PICHER INDUSTRIES, INC. P.O. Box 47 Joplin, MO 64801 (417) 623-8000	PL, SI, SA, R
EIC CORP. 55 Chapel Street Newton, MA 02158 (617) 965-2710	PL, SA, FC
ELCA BATTERY CO. Electrochimica Corp. 2485 Charleston Road Mountain View, CA 94040 (415) 961-7400	PP, PL
ELECTROCHEM INDUSTRIES, INC. Division of Wilson Greatbach 9999 Wehrle Drive Clarence, NY 14031 (716) 759-2828	PL
ELPOWER 2117 S. Anne Street Santa Ana, CA 92704	SP
ENERGY RESEARCH CORP. 3 Great Pasture Road Danbury, CT 06810 (203) 792-1460	SI, SA, FC
ENGELHARD INDUSTRIES, INC. Menlo Park Edison, NJ 08817 (201) 321-5000	FC
EXIDE CORP. 5 Penn Center Philadelphia, PA 19103 (215) 972-8000	SP, SS, SI, SA
EXXON RESEARCH AND ENGINEERING CO. P.O. Box 45 Linden, NJ 07036 (201) 474-3900	SA
FORD AEROSPACE & COMMUNICATIONS CORP. Aeronutronic Div. Ford Road Newport Beach, CA 92663 (714) 759-5798	SA
FOURDEE, INC. Div. of Emerson Electric Co. 440 Plumosa Avenue Casselberry, FL 32707	SP

Manufacturer	Major battery products*
FREMONT BATTERY COMPANY 4711 Oak Harbor Road Fremont, OH 43420	PP
FUJI ELECTROCHEMICAL CO. LTD. Japan	PC, PP
FURUKAWA BATTERY CO. LTD. 246, 2-chome Hoshikawa, Hodogaya-ku Yokohama, Japan	SP, SS, SI, SA
GATES ENERGY PRODUCTS, INC. 1050 S. Broadway Denver, CO 80217 (303) 744-4806	SP
GENERAL BATTERY CORP. (Northwest Ind.) 645 Penn Street Reading, PA 19601	SP, SI
GENERAL ELECTRIC CO. Battery Business Dept. P.O. Box 861 Gainesville, FL 32602 (904) 462-4480	PL, SP
GENERAL ELECTRIC CO. R & D Center Schenectady, NY 12301 (518) 385-2211	SA, FC
GENERAL MOTORS Research Laboratories Warren, MI 48090 (313) 575-3154	SA
GLOBE BATTERY DIV. Johnson Controls, Inc. 5757 North Green Bay Avenue Milwaukee, WI 53201 (414) 228-2000	SP, SS, SI, SA
GOULD, INC. 40 Gould Center Rolling Meadows, IL 60008 (312) 640-4400	SS, SI, SA
GTE SYLVANIA, INC. Power Systems Operation 520 Winter Street Waltham, MA 02254 (617) 890-9200	PL

Manufacturer	Major battery products*
GULF & WESTERN Energy Development Associates 1100 W. Whitcomb Avenue Madison Heights, MI 48071 (313) 583-9434	SA
Accumulatoremfabriken WILHELM HAGEN, AG Thomastrasse 27/28 D-4770 Soest West Germany 0291-1021	PP, SS, SI
HITACHI-MAXELL, LTD. 1-1-88 Ushitora Ibaraki-shi Osaka, 567, Japan and 60 Oxford Drive Moonachie, NJ 07074 (201) 440-8020	PC, PP, PL
HONEYWELL, INC. Power Sources Center 104 Rock Road Horsham, PA 19044 (215) 674-3800	PL, R
JAPAN STORAGE BATTERY CO. LTD. Kisshoin, Minami-Ku Kyoto, Japan	PP, SP, SS, SI, SA
LECLANCHE SA 48, avenue de Grandson Yverson CH-1401 Switzerland 024 258121	PC, PP, SP
LOCKHEED MISSILES AND SPACE CO., INC. Research Laboratory 3251 Hanover Street Palo Alto, CA 94304 (415) 493-4411	SA, R
LUCAS BATTERIES LTD. Formans Road Sparkhill Birmingham B11 3DA England 021 777 3292	PC, PP, SP, SS, SI
MAGNAVOX CO. 1313 Production Road Fort Wayne, IN 46808 (219) 482-4411	R

Manufacturer	Major battery products*
MARATHON BATTERY CO. 8301 Imperial Drive P.O. Box 8233 Waco, TX 76710 (817) 776-0650	PP, SP
MATSUSHITA BATTERY INDUSTRIAL CO., LTD. 1 Matsushita-Cho Moriguchi Osaka 570, Japan 06-991-1141	PC, PP, PL, SP, SS, SI
McGRAW EDISON CO. Power Systems Div. P.O. Box 28 Bloomfield, NJ 07003 (201) 751-3700	PS, SI
MEDTRONIC, INC. Energy Technology 6700 Shingle Creek Parkway Brooklyn Center, MN 55430 (612) 574-4000	PL
MOTOROLA, INC. 8000 W. Sunrise Boulevard Fort Lauderdale, FL 33322 (305) 475-6100	SP
NIFE INC. George Washington Highway P.O. Box 100 Lincoln, RI 02865 (401) 333-1170	SP, SI
SAB NIFE AB S-57201 Oskarshamm Sweden 48-491-16000	SP, SI
OLDHAM BATTERIES LTD. Crown Point Denton, Manchester Lancashire, M34 3AT England 061-336-2341	SS
PANASONIC Div. of Matsushita Electric Corp. of America One Panasonic Way Secaucus, NJ 07094 (201) 348-7000	PC, PP, PL, SP, SS, SI
N.V. PHILIPS GLOEILAMPENFABRIEKEN Eindhoven The Netherlands 040-79-11-11	PC, PP, R

Manufacturer	Major battery products*
POLAROID CORP. Commercial Battery Div. 784 Memorial Drive Cambridge, MA 02139 (617) 577-4374	PS, PL
POWER CONVERSION, INC. 495 Boulevard Elmwood Park, NJ 07407 (201) 796-4800	PL
PRESTOLITE BATTERY DIVISION An Allied Co. 511 Hamilton Street Toledo, OH 43694 (419) 255-4463	SS
RAY-O-VAC CORP. 101 E. Washington Avenue Madison, WI 53703 (608) 252-7400	PC, PP, PL
ROCKWELL INTERNATIONAL Energy Systems Group 8900 Desoto Avenue Canoga Park, CA 91304 (301) 341-1000	SA
SAFT 156 avenue de Metz Romainville, 93230 France 843.93.61 and Rue Georges Leclanche 86009 Poitiers France 49-53-09-13	PC, PP, PL, SP, SS, SI
SAFT AMERICA, INC. 107 Beaver Court Cockeysville, MD 21030 (301) 667-0693 and 711 Industrial Boulevard Valdosta, GA 31601 (912) 247-2331	PC, PP, PL, SP, SS, SI
SANYO ELECTRIC CO. Moriguchi Osaka 573, Japan 06-991-1181 and 200 Riser Road Little Ferry, NJ 07643 (201) 641-2333	PL, SP, SI

Manufacturer	Major battery products*
Accumulatorenfabrik SONNENSCHEIN GmbH D-6470 Buedingen, Hessen Theirgarten, Germany 06 042 91	PL, SP, SI
SONY-EVEREADY, INC. Japan	PP
TADIRAN LTD. 3 Hashalom Road P.O. Box 648 Tel Aviv, 61000 Israel 26 72 72	PC, PP, PL, SP, SA
TOSHIBA 4-10 Minami Shinagawa 3-chome Shinagawa-ku Tokyo 104 Japan	PC, PP, PL
AB TUDOR S-440 41 Nol Sweden 46-303-40400	SI
ACCUMULATEURS TUDOR SA rue de Florival, 93 B-5981 Archennes Grez-Doiceau Belgium 84 11 C1	PC, SP, SS, SI
UNION CARBIDE CORP. Old Ridgebury Road Danbury, CT 06817	PC, PP, PS
UNITED TECHNOLOGIES CORP. Power Systems Div. P.O. Box 109 South Windsor, CT 06074 (203) 727-2200	FC
VARTA BATTERIES AG AM Leinevfer 51 3000 Hannover 21 West Germany 0511 79031 and 85 Executive Boulevard Elmsford, NY 10523 (914) 592-2500	PC, PP, PL, SP, SI, SA

Manufacturer	Major battery products*
WESTINGHOUSE ELECTRIC CORP. Advanced Energy Systems Div. P.O. Box 10864 Pittsburgh, PA 15236 (412) 892-5600 and	SA
WESTINGHOUSE R & D CENTER 1310 Beulah Road Pittsburgh, PA 15235	FC
WILSON GREATBATCH LTD. 10000 Wehrle Drive Clarence, NY 14031 (716) 759-6901	PL
LE PILES WONDER St. Ouen 93403 France 1257.11.50	PC, PP, SP, SI
YARDNEY ELECTRIC CORP. 82 Mechanic Street Pawcatuck, CT 06379 (203) 599-1100	PL, SI
YUASA BATTERY CO. LTD. 6-6 Josai-cho Takatsuki Osaka 569 Japan 0726 75 5501	PP, SP, SS, SI, SA

*The codes used to indicate major battery products are:
Primary:

Conventional (zinc-carbon)	PC
Premium	PP
Lithium anode systems	PL
Special types	PS

Secondary:

Portable types	SP
Automotive (SLI)	SS
Industrial	SI
Advanced systems	SA
Reserve batteries	R
Fuel cells	FC

OTHER SOURCES FOR LISTS OF MANUFACTURERS:

National Electronic Distributors Association, 1480 Renaissance Drive, Park Ridge, IL 60068.

SLIG Buyers' Guide, Independent Battery Manufacturers Association, Inc., 100 Larchwood Drive, Largo, FL 33540, (813) 586-1408.

The Storage Battery Manufacturing Industry, *Yearbook of the Battery Council International,* 111 East Wacker Drive, Chicago, IL 60601, (312) 644-6610.

INDEX

INDEX

Batteries and cell types are indexed alphabetically by anode.

About the Editor

David Linden, a consulting engineer, has been associated with battery research development and engineering for over 40 years, and is currently working with Duracell Inc. From 1942 to 1978 he was associated with the Power Sources Division, U.S. Army Electronic R&D Command, serving as director of the Division for the last nine of those years. Many of the new battery and power sources developed in recent years have resulted from work supported by that Command, including zinc/mercuric oxide batteries, sintered nickel-cadmium batteries, magnesium batteries, lithium batteries, fuel cells, and solar and other advanced energy conversion systems.

Mr. Linden has served as chairman of both the Electrochemical Working Group (Interagency Advanced Power Group), which was responsible for coordinating government-wide battery and fuel cell development, and the Power Sources Symposium. He served on a number of advisory groups, such as the NATO Electric Power Systems Commission. A member of the American Chemical Society and the Electrochemical Society, he has contributed numerous articles on batteries, electrochemical power sources, and battery applications to reference books and professional journals.